Birkhäuser

Mathematics and the Built Environment
Volume 8

Series Editors

Throughout history a rich and complex relationship has developed between mathematics and the various disciplines that design, analyse, construct and maintain the built environment.

This book series seeks to highlight the multifaceted connections between the disciplines of mathematics and architecture, through the publication of monographs that develop classical and contemporary mathematical themes – geometry, algebra, calculation, modelling. These themes may be expanded in architecture of any era, culture or style, from Ancient Greek and Rome, through the Renaissance and Baroque, to Modernism and computational and parametric design. Selected aspects of urban design, architectural conservation and engineering design that are relevant for architecture may also be included in the series.

Regardless of whether books in this series are focused on specific architectural or mathematical themes, the intention is to support detailed and rigorous explorations of the history, theory and design of the mathematical aspects of built environment.

Patricia Radelet-de Grave

Between Mechanics and Architecture

The Mathematical Search for Stability in Architecture

Birkhäuser

Patricia Radelet-de Grave
Irmp Institut De Recherche En
Mathématique Et En Physique
Université Catholique de Louvain
Ottignies-Louvain-la-Neuve, Belgium

ISSN 2512-157X ISSN 2512-1561 (electronic)
Mathematics and the Built Environment
ISBN 978-3-031-73529-5 ISBN 978-3-031-73530-1 (eBook)
https://doi.org/10.1007/978-3-031-73530-1

Mathematics Subject Classification: 01A45, 01Axx, 01A40, 01A99

The cover image is a graphical elaboration based on a study of curves by Charles-François Le Prud'homme D'hailly Nieuport, in his work "Essai analytique sur la méchanique des voûtes, présenté le 18 mai 1778", Mémoires de l'Académie impériale et royale des sciences et belles-lettres de Bruxelles, vol. II (1780).

This book is published under the imprint Birkhäuser, www.birkhauser-science.com by the registered company Springer Nature Switzerland AG
The registered company address is: Gewerbestrasse 11, 6330 Cham, Switzerland

Preface: What Do We Mean by *Between Mechanics and Architecture*?

Introduction

In Bressanone, at the ceremony to award the first Edoardo Benvenuto Prize,[1] I tried to point out the many different fields that influence architecture—artistic, mathematical, technological, sociological—in order to be able to focus on the interaction that Edoardo Benvenuto and I named *Between Mechanics and Architecture*. Here I wish to define more precisely what this title means, and what it includes.

First, I must point out an essential aspect of this study which, regretfully, does not appear in the title: history. In fact, the understanding of the actual relationship may be facilitated by the analysis of the historical relationships between the science of mechanics and the art of architecture, throughout the ages.

Interactions Between Mechanics and Architecture

The richness of the relationship between mechanics and architecture arises from the fact that it is an interaction, rather than an influence of one upon the other. Architecture did more than borrow knowledge from mechanics, it also stimulated mechanics. In his introduction to the first volume of the book series *Between Mechanics and Architecture*, Edoardo Benvenuto explains the stakes of this interaction and concludes:

> … Henceforth the alternative is no longer between science and art, but between two different epistemological positions: and it is not an alternative but a productive, although tricky, encounter, the history of which is very rich and probably unexplored as yet.[2]

---

[1] The ceremony took place in Bressanone, during the XVIII Congress *Scienza e Beni Culturali*, 4 July 2002. As of this writing, the Edoardo Benvenuto prize is now in its twentieth edition.

[2] Radelet-de Grave and Benvenuto, 1995, p. 18.

From the beginning of time, architecture influenced studies of mechanics. Evidence is given by the first mechanical law of the lever, defined by Aristotle and Archimedes, and by the first axiomatic theory of mechanics which assembles, through the lever law, all the simple machines—i.e., all the lifting devices—which were essential for the construction of houses and sacred buildings. These studies would lead in 1687 to statics, the name given by Varignon to the study of simple machines. With his *Eléments de statique*, published in 1803 and reprinted many times, this subject would be taught by Louis Poinsot to the students of the prestigious Ecole Polytechnique in Paris, marking a turning point in the interaction between mechanics and architecture.

Hydrostatics too, born in ancient times from the need to convey water, developed together with hydrodynamics, as an answer to engineers' and architects' needs: the first step towards the mechanics of continuous media.

But before that, other influences arose. At the end of Middle Ages, Galileo himself, who observed the construction of Gothic churches, drew inspiration from the many problems raised by the builders. Studies on clamped beams, catenaries, and beams on three supports were all initiated by him and further studied in the seventeenth and eighteenth centuries by the most important mechanical engineers: Leibniz; Jacob, Johann and Daniel Bernoulli; Leonhard Euler; and many others. All applied themselves to the development of differential calculus: an essential tool for solving the most complex problems. These mechanics, in turn, made possible the construction of new machines and the building of new architecture. The achievements of these researches would be collected by Lagrange, Poinsot, Poisson, Cauchy, and other famous professors of the Ecole Polytechnique, granted with much more sophisticated mathematics, who would contribute to the progress of mechanics. Mechanics would thereon develop in a more independent way, as a science in its own right, apparently liberated from applicative problems. Nevertheless, the Ecole Polytechnique is still an engineering school. The consequence would be the inversion of relationship. Architecture would no longer really influence mechanics, but mechanics—the better knowledge of materials, of their strength—will offer new possibilities to architects and engineers. And this interaction still exists nowadays.

It is the history of these mechanics, born from the demands of architecture, and developed together with architecture and the art of building, that Edoardo Benvenuto and I intended to study, in what was meant to be only one conference, whose proceedings, entitled Between Mechanics and Architecture, were published by Birkhaüser in 1995.

Prelude: The History of "Between Mechanics and Architecture"

The first lecture I ever gave on "Between Mechanics and Architecture" was held in Genova in 1992. It focused on Mascheroni's *Equilibrio delle volte*.

In 1993 we met as a group for the first time, with participants coming mainly from the Istituto di Costruzioni of the University di Genova but also from other parts of Italy, Germany, France, and Belgium. This took place in Zaragoza as part of the nineteenth International Congress on the History of Science, and we were trying to clarify this line of research situated at the crossroads “between mechanics and architecture”. I presented Jacob Bernoulli’s *De curvatura fornicis* concerning the introduction of the infinitely small in the calculation of vaults.

In 1994, my article on Coulomb’s essay regarding the application of the rules of maxima and minima to some static problems relating to architecture (Radelet-de Grave, 1994) was published in the journal *Sciences et Techniques en perspective*, directed by Jean Dhombres.[3]

Finally, in 1995 the book *Between mechanics and architecture*, edited by Edoardo Benvenuto and myself, was published by Birkhäuser (Fig. 1). It included the text of my lecture given in Zaragoza on the *de curvatura fornicis* by Jacob Bernoulli.[4]

We then met in Genova in June 1996 to study from the same point of view graphostatics and the work of Culmann. I gave a presentation on that in Liège in 1997, on the occasion of the twentieth International Congress on the History of Science, we continued our theme through the great Encyclopedias. A few days later in Louvain-la-Neuve on 28–29 July, by mutual agreement with our colleagues from Genova, we decided to commemorate the bicentenary of the birth of Adhémar Barré de Saint-Venant (Villers-en-Bière 1797—Saint-Ouen 1886). It seemed to us that this Polytechnician, master of hydrodynamics as well as the flow of fluids and elasticity (in short, of everything that will make up the mechanics of continuous media and to whom Todhunter and Pearson devote their history of elasticity) was one of the heroes, along with Cauchy and their followers, of the particularly fertile era of history that we were trying to reconstruct. I especially acknowledge the participants of those days, Massimo Corradi, Federico Foce, Antonio Becchi who from the first days helped us to clarify this idea.

This idea of “Between mechanics and architecture” did not result in just one conference, but five. The first one took place in Saragossa[5] and had the original title *Between Mechanics and Architecture*. Then we first went to Genoa for the meeting entitled *Culmann and Graphostatics*, next to Liège for *Between Mechanics and Architecture in Encyclopaedia*, then to Louvain-La-Neuve for *Saint-Venant*, and then to Pescara for *Perspective*. Until now the original idea has resulted in the publication of three books, of which we will speak later. Others are in preparation.

I extend particular thanks to those who participated in these first meetings and who therefore contributed to providing a precise framework for this research: Jean Dhombres, Paolo Freguglia, Karl-Eugen Kurrer and Bruna Gaino, as well

---

[3] See “A Study of Coulomb’s Application of the Rules of Maxima and Minima to Problems of Statics”, Chap. 10 in this present volume.

[4] See “The *De curvatura fornicis* of Jacob Bernoulli: The Introduction of the Infinitely Small into the Calculation of Arches”, Chap. 5 in this present volume.

[5] During the nineteenth international meeting on history of science. I wish to recall here the name of the Spanish organizers of this meeting: Elena Ausejo and her husband Mariano Hormigon who unfortunately left us too soon.

Fig. 1 The cover of the first publication of *Between Mechanics and Architecture*, 1995 publication

as those who quickly joined us: Bruno Belhoste, Maurizio Brocato, Konstantinos Chatzis, Antonella Mastrorilli, Chiara Melucci, Jean-François Thimus, M. Pittaluga, P. Ricciardi.

With time, our small group came to meet other scholars whose research shared an affinity with ours. Among others, I may name Jacques Heyman for his studies on the history of structural mechanics; Santiago Huerta and his team, founders and organizers of the series "International Congresses on Construction History", with the first congress held in Madrid, 20–24 January 2003; Pierre Smars, John Ochsendorf, and Tullia Iori, whose theses were awarded the Edoardo Benvenuto prize for the years 2002 and 2003, respectively; and Kim Williams, founder and director of the conference series "Nexus: Relationships Between Architecture and Mathematics" and the *Nexus Network Journal*, who is taking care of the publication of the present volume.

Sources, Influences, Previous Works

Of course, Benvenuto and I, and the people we invited to this first conference, did not start from nothing. Some important authors provided us with solid bases, even if their point of view was not always absolutely similar to ours:

- A. E. H. Love, *Treatise on the mathematical theory of elasticity*, Cambridge, 1892 (especially the "Historical Introduction", pp. 1–34).
- Isaac Todhunter and Karl Pearson, *History of the theory of elasticity and of the strength of materials*. 2 vols. Cambridge: Cambridge University Press, 1886 and 1893.
- Stephen P. Timoshenko, History of Strength of Materials, with a Brief Account of the History of Theory of Elasticity and Theory of Structures. New York: McGraw-Hill, 1953.
- Clifford Truesdell, *The rational mechanics of flexible or elastic bodies 1638–1788*. Zürich: Orell Füssli, 1960. (Introduction to vols. 10 and 11 of the *Leonhardo Euleri Opera Omnia*.)
- Benvenuto, Edoardo, *An Introduction to the History of Structural Mechanics*. Springer-Verlag: New York, 1991. This book is divided in two parts: I. *Statics and Resistance of Solids* and II. *Vaulted Structures and Elastic Systems* (Fig. 2).

The New Works

Our large group of participants produced many publications. I wish to point out, classify, and comment with a few words the foundational publications which meet our goal more closely, and constitute the foundations of the disciple "between mechanics and architecture".

Monographs

We will start with six monographs, five of which are mainly dedicated to vaults and domes.

- Heyman, Jacques. 1972. *Coulombs memoir on statics. An essay in the history of civil engineering*. Cambridge: Cambridge University Press. In this book, Heyman offers a fac simile, a translation, and a profound reflection on the famous text by Coulomb, *Essai sur une application des règles de Maximis et Minimis à quelques problèmes de statique relatifs à larchitecture.*
- Heyman, Jacques. 1995. *The stone skeleton*. Cambridge: Cambridge University Press. Spanish translation, CEHOPU, 1999. Here Heyman studies stone vaults and domes from still existing buildings and ancient treatises.

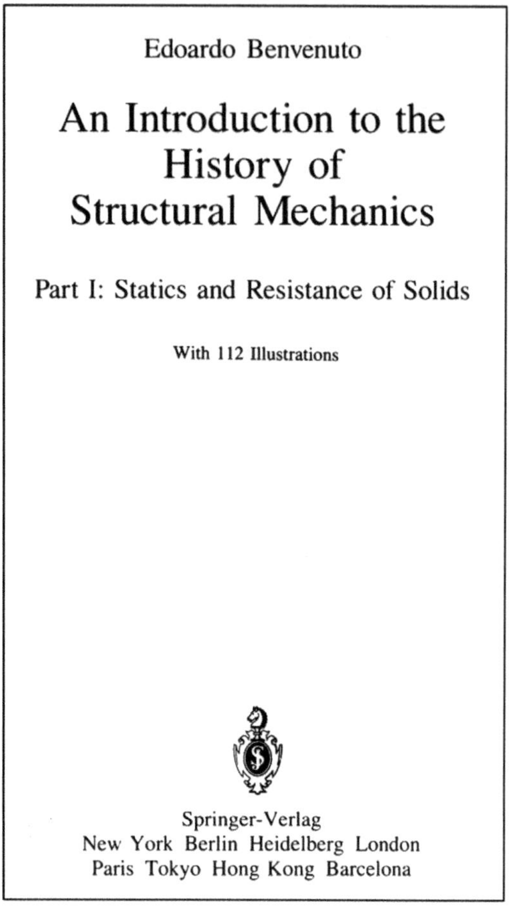

Edoardo Benvenuto

An Introduction to the History of Structural Mechanics

Part I: Statics and Resistance of Solids

With 112 Illustrations

Springer-Verlag
New York Berlin Heidelberg London
Paris Tokyo Hong Kong Barcelona

Fig. 2 Frontispiece of the book by Edoardo Benvenuto

- Di Pasquale, Salvatore. 1996. *Larte del costruire.* Tra conoscenza e scienza. Venice: Marsilio. The results of many years of studies on the dome of Santa Maria del Fiore in Florence and on masonry structures are collected in this book. It is the work of a building site man, a professor who likes to understand buildings from an historical approach.
- Smars, Pierre, *Etudes sur la stabilité des arcs et des voûtes, confrontation des méthodes de lanalyse limite aux voûtes gothiques du Brabant.* Thesis, Catholic University in Leuven. Smars makes a survey of all the Gothic vaults of Brabant, Belgium, studying each of them in detail as to their particularity and stability, and executes a meticulous analysis of their damage (cracks). This thesis was awarded the Edoardo Benvenuto prize in 2007.
- Becchi, Antonio and Federico Foce. 2002. *Degli archi e delle volte. Arte del costruire tra meccanica e stereotomia*, Venice: Marsilio. This book analyses the

relationship between architecture and stereotomy, a science whose development is tightly linked to the evolution of perspective.

- Kurrer, Karl-Eugen. 2002. *Geschichte der Baustatik*, Berlin: Ernst and Sohn. Kurrer gives an historical presentation from the standpoint of engineering, whereas Benvenuto, in his own book, follows mechanics principles and chooses the advent of iron to mark the division between his two volumes. Kurrer's book is rather oriented towards iron architecture. It includes very interesting short biographies of the major engineers and architects.

Books of Collected Papers

- Radelet-de Grave, P. and E. Benvenuto, eds. 1995. *Between Mechanics and Architecture*. Basel: Birkhäuser. In addition to the papers, of which we will speak below, the reader may find here a number of other modern and historical references, all pertinent to the subject. Through the bibliography, prepared by Antonio Becchi, Massimo Corradi, and Federico Foce, the reader can get a very precise definition of the subject.
- Graciani García, A., S. Huerta Fernandez, E. Rabasa Diaz, and M. A. Tabales Rodriguez, eds. 2000. *Actas del tercer congresso nacional de historia de la Construcción*, 2 vols, CEHOPU. The subject of this book is the construction site. It investigates mainly its social aspects.
- Huerta, S., ed. 2003. Proceedings of the First International Congress on Construction History, 3 vols. Madrid: Instituto Juan de Herrera. These three large volumes include many papers pertinent to our subject. We will return to them later on.
- Becchi, A., M. Corradi, F. Foce, and O. Pedemonte, *Towards a History of Construction*. Birkhäuser, Basel 2003. This book, which relates the many interests of Edoardo Benvenuto except for his theological concerns, covers a wider subject than the mere between mechanics and architecture, but is still related to it. The work is divided in five parts: Part I: Strength of Materials and Structural Mechanics, enquires into the problems of modern construction (resistance to earthquakes) regardless to history, unless when the object or monument which has to be preserved requires it; Part II: History of Applied Mechanics and Part III: Between Mechanics and Architecture both include studies that fall into the project "Between Mechanics and Architecture"; Part IV: Dedicated to Edoardo Benvenuto; and Part V: Architecture and History.
- Becchi, A., M. Corradi, F. Foce, and O. Pedemonte, eds. 2003. Essays on the history of mechanics. In Memory of Clifford Ambrose Truesdell and Edoardo Benvenuto. Basel: Birkhäuser. Quoting from the Preface:

> In the last 20 years, the works of these authors [Saint-Venant, Todhunter and Pearson, Duhem, Mach, Dugas, Timoshenko, Truesdell, Heyman, Szabô, Charlton, Benvenuto, and Di Pasquale] have led new generations of scholars to delve into themes of research that

> involve diverse disciplines—from mathematics to construction, from architecture to geometry, from the strength of materials to the mechanics of solids and structures—leading to the formulation of the research project "Between Mechanics and Architecture" (p. 7).

The papers in five collections which are related to the idea "Between Mechanics and Architecture" address a wide range of topics that helped define further research. These include:

- Simple machines and ancient building techniques: lifting machines and the way problems of lifting are solved in buildings whose context is either not at all or only slightly theoretically mechanized;
- Foundations: ancient problems of stability and earth mechanics;
- Thin vaults, stable under their own weight, and catenaries: Without a doubt this is the problem that has led to the most inquiry, probably because this is the very point where mechanics and architecture meet, along with mathematics, since the solutions require differential calculus;
- Other vaults, bridges;
- Domes;
- Principles of Mechanics;
- Structural analysis;
- Columns: A topic of which Edoardo Benvenuto was particularly fond, but that had not yet been inquired into at length;
- Timber frames;
- Elastica, membranes and their oscillations;
- Rose windows and ribs: The purpose is not so much to study their aesthetic aspect rather than their role in the stability of the structure;
- Iron structures: As Edoardo Benvenuto has shown, this topic is an important turning point in the history of "Between Mechanics and Architecture";
- Strength of materials, component equations: this topic was Truesdell's favourite. It is particularly wide, yet still too rarely studied, probably because of its difficulty;
- Mechanics of continuous media: in this category we include hydrostatics, hydrodynamics, elasticity and plasticity;
- Proportions and architecture: the call for geometry, and more precisely on proportion theory, goes back to Vitruvius. Aristotle and Archimedes apply this theory to mechanics while giving the law of the lever;
- Stereometry: perspective and descriptive geometry are among the mathematical developments stimulated by architecture;
- Graphic methods;
- Deformation method;
- Scientific biographies.

Conclusion

I hope I have been able to show that there are now several people aware of the essence of this research and willing to pursue it, but we still have to make some progress about the richest aspect that Edoardo Benvenuto transmitted to us: the analysis, by the means of history, of the relationships between a science and an art, or to use his own words in the introduction of Between Mechanics and Architecture, "between exact sciences and humanities".

I dedicate this present book to the memory of Edoardo Benvenuto with whom I began this research and who contributed so much to it, through his writings and through his questions during our meetings. He unfortunately left us on 27 November 1998, but his teaching remains present in our memories.

Translated from the French by Sylvie Duvernoy

Ottignies-Louvain-la-Neuve, Belgium Patricia Radelet-de Grave

Contents

1 Relations Between Mechanics and Architecture Through the History of the Parallelogram Law of Forces from Antiquity to 1800 1
1 Introduction 1
2 Antiquity: The Composition of Movements and First Law for Vaults 3
2.1 Third Century Before Christ, Pseudo Aristotle: Mechanical Problems 3
2.2 First Century Before Christ, Vitruvius's Ten Books *on Architecture* 5
3 The Composition of Forces in Statics and its Relation With the Catenary From a Purely Theorical Point of View 6
3.1 1586, Simon Stevin, *Het eerste bouck vande beghinselen der weegconst* 6
3.2 1604–1619, Isaac Beeckman's *Journal* (1588–1637) 12
3.3 1638, Galileo's *Discorsi e dimostrazioni matematiche intorno a due nuove scienze* 17
3.4 1636, Gilles Personne de Roberval, *Traité de mechanique des poids ...* 17
3.5 1634, Pierre Hérigone, *Cursus mathematicus* 19
3.6 1643, François Derand, *L'Architecture des voûtes:* An Architectural Point of View 20
3.7 1646, Christiaan Huygens's Letters to Marin Mersenne, and His Manuscript *De catena pendente* 22
3.8 1673, Ignace-Gaston Pardies, *La statique ou la science des forces mouvantes* 29
3.9 1676, Robert Hooke, *A Description of Helioscopes, and Some Other Instruments* 34

3.10 1687, Pierre Varignon, *Projet d'une nouvelle mecanique ou statique* ... 34

4 In Dynamics, Demonstration of the Composition of the Forces by Means of the Composition of the Movements; The Catenary; The Introduction of Differential Calculus ... 37

4.1 1687, Isaac Newton (1642–1727), *Philosophiae Naturalis Principia Mathematica* ... 37

4.2 Date Unknown, Johann Bernoulli, *Propositiones variae Mechanico-dynamicae* ... 39

4.3 1691, Johann Bernoulli, *Solutio Problematis Funicularii*, the *Lectiones mathematicae*, and the Discreet Ingress of Differential Calculus ... 41

4.4 1695, Philippe De la Hire, *Traité de mecanique* ... 44

4.5 1704, Jacob Bernoulli, *Problema de curvatura fornicis, cujus partes se mutuo propio pondere suffulciunt sine opere caementi* ... 47

4.6 1743, Jean-Baptiste Clairaut, Methodus generalis inveniendi catenarias ... 49

4.7 1714, Johann Bernoulli, *Essay d'une nouvelle theorie de la manœuvre des vaisseaux* ... 50

4.8 1717, James Stirling, *Methodus disponendi quotcunque Sphaeras in Fornicem* ... 52

4.9 1729, Bernard Forest de Bélidor, *La Science des ingenieurs dans la conduite des travaux de fortification et d'architecture civile* ... 54

4.10 1725, Pierre Varignon, *Nouvelle mecanique ou statique, dont le projet fut donne en 1687* ... 56

4.11 1726, Daniel Bernoulli, *Examen des Principes de la Mécanique* ... 61

4.12 1728; Daniel Bernoulli, *Methodus universalis determinandae curvaturae fili* ... 62

4.13 1728, Leonhard Euler, *Solutio problematis de invenienda curva* ... 64

4.14 1729, Claude Antoine Couplet, *De la poussee des voutes* ... 65

4.15 1748, Giovanni Poleni, *Memorie istoriche della gran cupola del Vaticano* ... 68

4.16 Other Works ... 70

References ... 72

2 Mathematics, Architecture and Mechanics in the School of François D'Aguilon and Grégoire De Saint-Vincent 75
1 Introduction 75
2 The Youth of D'Aguilon at the Time of the Construction of the Church of Douai 78
3 D'Aguilon's *Optics* 80
4 The Construction of St. Ignatius in Antwerp 81
5 Saint-Vincent's Thesis on Statics 85
6 The Construction of the College of Prague 86
7 The Affair of the College of Ghent 88
8 Conclusion 89
References 90

3 Is the Flat Vault a Lost Proposition in John Wallis's Work? 93
1 John Wallis and His Works 93
2 The Lever, Supported by Either One, Two or More Supports 97
3 Proposition X 99
4 Jean Truchet (1657–1729) 104
5 Conclusion 107
References 107

4 The Use of a Particular Form of the Parallelogram Law of Forces for the Building of Vaults (1650–1750) 109
1 Introduction 109
2 Analogy Between Vault and Catenary 110
3 Finding the Right Model for the Catenary and the Vault 113
3.1 For the Catenary 115
3.2 For the Vault 115
4 Fundamental Quantities 116
4.1 Mechanical Quantities: Tension and Pressure (Tuesdell 1968, p. 206) 116
4.2 Geometrical Quantities: The Radius of Curvature 117
5 Fundamental Principles 117
5.1 Purely Geometrical Principles 117
5.2 The Law of the Lever, or the Sum of the Moments of Momentum Must Be Zero 117
5.3 The Parallelogram Law of Forces or Sum of the Applied Forces Is Zero 122
5.4 The Necessity of Considering Both 127
5.5 Minimal Energy 129
6 Virtual Works 131
6.1 The Catenary 131
6.2 The Vault 131
References 132

5 The *De Curvatura Fornicis* of Jacob Bernoulli: The Introduction of the Infinitely Small into the Calculation of Arches ... 135
1 Introduction ... 135
2 Jacob Bernoulli, "Problem of the Curvature of Arches that Are Sustained by Their Own Weight Without Any Cement" (1744) ... 136
3 Claude Antoine Couplet, "On the Thrust of Arches" (1729) ... 139
4 Krafft, "On Funicular and Catenary Curves" (1755) ... 147
5 Abbé Charles Bossut, "Researches on the Equilibrium of Arches" (1774) ... 147
6 Nieuport's "Analytical Essay on the Mechanics of Arches" (1780) ... 150
7 Conclusion ... 154
Appendix: Chronological Table ... 155
References ... 156

6 The Catenary, or How Differential Calculus Appeared on the Public Square ... 157
1 Introduction ... 157
2 Differential Calculus on the Continent ... 158
2.1 Christiaan Huygens ... 159
2.2 Gottfried Wilhelm Leibniz ... 165
2.3 Ignace-Gaston Pardies and Johann Bernoulli ... 168
3 David Gregory and the Fluxions in England ... 177
3.1 Gregory's Memorandum and Newton's Conception of a Thin Arch ... 177
3.2 *Gregory's* Catenaria *of 1697 and 1698* ... 179
4 Conclusion ... 187
References ... 191

7 The Bernoullis and the Elastica: From Leibniz's *Nova Methodus* to Euler's *Methodus Inveniendi* ... 193
1 Introduction: The Various Difficulties Linked to the Problem of the Elastica ... 193
2 Presentation of the Various Curves Studied in This Chapter ... 194
3 First Period: From *Nova Methodus* by Leibniz in 1684 to Jacob Bernoulli's Elastica in December 1695 ... 197
3.1 Background ... 197
3.2 *Jacob Bernoulli's* Theorema Aureum *and Differential Calculus* ... 197
3.3 Various Problems Posed by Jacob and Johann Bernoulli ... 200

3.4 *The Public Treatment of the* Elastica 203
4 Second Period: From the Public Declaration of War in 1696 to the Last Publication of Johann Concerning the Isoperimetric Problems 1718 207
4.1 Background 207
4.2 The Problems Proposed by Johann Bernoulli 207
4.3 The Isoperimetrical Problems 209
4.4 The Constitutive Equations 212
4.5 About the Catenary 215
5 Third Period: From 1728, When Daniel Bernoulli and Leonhard Euler Reread the Work of Jacob Bernoulli, and the Birth of the *Methodus inveniendi* 218
5.1 Daniel Bernoulli 218
5.2 Leonhard Euler, E.8 219
5.3 Leonhard Euler, E.27 219
5.4 The Letters Exchanged by Euler and Daniel Bernoulli 219
5.5 Euler, E.65 220
6 Conclusion 220
References 220

8 The Theory of Vaults of Pierre Bouguer 223
1 Bouguer's Work on Vaults 223
2 History of the Calculation of Domes 225
3 Bouguer's Work 229
4 The Calculus of Vaults and the Principles of Mechanics 234
4.1 The Parallelogram Law of Forces 234
4.2 The Voussoir and the Wedge or Inclined Plane 236
4.3 The Lever 236
5 The Principle of Virtual Works and the Mechanisms of Rupture 236
6 The Principles of Mechanics According to Bouguer in His "Manoeuvring Vessels" 238
7 The Reality Test for Vault Theory 240
8 Theory and Practice in the *Traité du Navire*, the Role of Principles by Bouguer 240
References 245

9 The Entasis of Columns from Blondel to Lagrange: Mathematical Games in Search of the Greatest Stability 247
1 Introduction 247
2 Giacomo Barozzi da Vignola 247
3 "The Gift to All Architects" 251
4 Pieter van Musschenbroek and Leonhard Euler 257
5 Joseph-Louis de Lagrange 259
6 Conclusion 262
References 262

10 A Study of Coulomb's Application of the Rules of Maxima and Minima to Problems of Statics 265
1 Preamble 265
2 The Structure of the Coulomb's *Essai* 266
3 Commentary on the Essay 267
4 Preliminary Propositions I to III 269
5 Propositions IV (on Friction), and V (on Cohesion) 273
6 Proposition VII, Remarks on the Rupture of Bodies 274
7 Proposition VIII, Strength of Masonry Piers 281
8 Proposition IX, on Earth Pressures and Retaining Walls 283
9 Proposition XV, on the Surface of Greatest Thrust in Cohesive Fluids 286
10 Proposition XVI, on Arches 287
11 Proposition XVII, on Arches with Joints that Have Neither Friction nor Cohesion 287
12 Proposition XVIII, on the Stability of Arches, Allowing for Friction and Cohesion 290
13 Conclusion 294
Appendix 1: Three domains for the study of the stability of vaults 294
References 296

11 Jacques-Germain Soufflot and the Relationship Between Rules and Taste 299
1 Introduction 299
2 9 December 1739, Mémoire sur les proportions de l'architecture 300
3 12 April 1741, *Un parallèle des églises gothiques avec les églises moderne* 306
4 20 November 1775 and 18 January 1778, *De l'identité du goût et des règles dans l'exercice de l'architecture* 313
5 Conclusion 318
References 318

12 The Catenary as Inspiration 321
1 The Catenary Chosen for Aesthetical Reasons in Egyptian Antiquity 321
2 Vaults and Domes from Antiquity to the End of the Seventeenth Century 322
3 Mechanics Research on the Catenary from Simon Stevin to Robert Hooke 323
4 Back to Architectural Theory, as Elaborated in Important Treatises and Some Articles 327
5 More on Architecture 330
6 Conclusion 333
References 334

13 The Study of the Stability of Vaults by the Viscount of Nieuport ... 337
1 Introduction ... 337
2 The Equilibrium of Segments in General ... 338
3 Vaults, When the Voussoirs Meet at a Point ... 343
4 Vaults, When the Extensions of the Joints Do not Converge into a Single Point, the Voussoirs Being Infinitely Small ... 346
5 Previous Vaults in the Hypothesis of Voussoirs of Finite Size ... 348
6 The *Mémoire* on the Alleged Properties of Catenary Vaults ... 350
7 Conclusion ... 352
References ... 352

14 On Mascheroni's "New Researches on the Equilibrium of Vaults" ... 355
1 Preamble ... 355
2 Chronological Landmarks ... 355
3 The Catenary, the Elastica and Other Problems Posed by Jacob and Johann Bernoulli ... 357
4 Jacob Bernoulli's Treatment of the Elastica ... 359
5 Mathematics Appears in Architecture ... 362
6 Isoperimeters ... 364
7 The New Generation with Euler and Daniel Bernoulli ... 365
8 Jacob Bernoulli and the Problem of the Arch ... 366
9 Variations on the Catenary ... 370
10 Where One Finds the Vaults Again ... 376
11 Mascheroni's "New Researches on the Equilibrium of Vaults" ... 379
12 Conclusion, Part I: A Summary of Results ... 397
13 Conclusion, Part II: Constitutive Equations ... 398
References ... 399

15 The Supports Problem: A Dead-End Street in the History of Stability of Structures ... 403
1 1695: La Hire's Problem ... 403
2 1714: Johann Bernoulli and "Virtual Velocities" ... 404
3 1772: The Supports Problem ... 405
4 1773: Leonhard Euler's Influence ... 407
5 1788: The Non-influence of Lagrange's *Mécanique Analytique* ... 411
6 1811: Finally, with Its Second Edition, the Influence of Lagrange's "Analytical Mechanics" Extends ... 418
References ... 422

16 The Discovery of General Methods for Elastic Systems, 1852–1875 425
1 Prelude: Energy and Forms of Energies 425
2 1852–1875 the Discovery of General Methods for Elastic Systems 426
References 435

17 The Industrial Revolution and Its Consequences. Part I: Iron Frameworks and Cullmann's "Graphostatics" 437
1 Introduction 437
2 W. J. M. Rankine and August Ritter 438
3 Karl Culmann's "Graphostatics" 441
4 The Intentions, Aims, and Critiques in Culmann's Preface of 1865 441
5 A First Transition: August Ritter and Otto Mohr 444
6 The Preface of Culmann's Edition of 1875 447
7 Preface to the Edition of 1880 449
8 The Plans of the 1866 and 1875 Editions and Their Differences 453
9 The *Graphische Statik* of 1866 453
10 Culmann's *Ueber das parallelogramm und über die Zusammensentzung der Kräfte* 455
11 The Work of Alberto Castigliano 457
12 The French Translation of Culmann's Edition of 1875 459
13 Conclusion 464
References 465

18 The Industrial Revolution and Its Consequences. Part II: Elasticity Penetrates Construction 467
1 W. J. Macquorn Rankine 467
References 480

19 Arthur Vierendeel: The Man with Iron Nerves Who Invented the Truss Without Diagonals 483
1 Introduction 483
2 Vierendeel's Character and His First Works 485
2.1 Early Life and Education 485
2.2 The Royal Circus of Brussels (1876) 485
3 The First Trusses and the Invention of Vierendeel 487
3.1 What is a Vierendeel Truss? 488
3.2 The Discovery of the Truss 489
3.3 The Patent of Invention 490
3.4 The Experimental Truss of Tervuren Near Brussels (November 1897) 491
3.5 First Improvement Patent 492

3.6 The Second Improvement Patent for the Transition to Reinforced Concrete ... 495
3.7 The Theoretical Advantages ... 495
4 The Administration of Bridges and Roadways ... 498
5 The University of Louvain ... 501
5.1 Buckling Theory ... 501
5.2 Aesthetics in Architecture ... 504
6 The Evolution of the Vierendeel's Bridges and of His Criteria of Elegance ... 507
7 Bowstring Type Bridges ... 508
7.1 The Upper Flanges Are Formed of Two Parabolas Supposed to Be Parallel but Cannot Be ... 510
7.2 The Upper Flanges Are Formed of Two Parabolas Approximately Parallel and an Extension ... 511
7.3 The Upper Flanges Are Formed of Two Similar Parabolas ... 512
7.4 The Upper and Lower Flanges Are Made of Two Similar Parabolas and an Intermediate Parabola is Drawn ... 514
8 The Astonishing Posterity of Vierendeel's Truss ... 514
9 Conclusion ... 515
References ... 516

20 Pier Luigi Nervi's Idea of the *Verità delle Strutture* ... 519
1 Introduction ... 519
2 How Does Nervi Express His Ideas on Structures in His Writings? ... 520
2.1 Structure and Beauty ... 520
2.2 Structure and Economy ... 520
3 How Does Nervi Express Those Ideas in His Structures? ... 521
3.1 The Flow of Forces ... 521
3.2 Pier Luigi Nervi and History ... 523
3.3 Vaults ... 524
3.4 Domes ... 524
3.5 Catenaries ... 525
4 The Aula Paulo VI or Paul VI Audience Hall (1971) ... 527
5 Conclusion ... 527
References ... 529

21 Gustave Magnel (1889–1955): A Life in Structural Engineering 531
1 Beginnings 531
2 First Academic Experiences 532
3 The Laboratory for Reinforced Concrete: Teaching, Research and Experiments 534
4 Later Academic Career 538
5 The *SECO*: Building Sites Security and Insurance 539
6 The Prestressing Technology 542
7 The Magnel-Blaton Patent 547
8 The Magnel-Blaton System 548
9 Concrete Works to Which Magnel Made a Contribution 552
9.1 The Bridge in Rue du Miroir 553
9.2 Underpinning During Restoration Works 557
9.3 Other Works 558
10 Conclusion 562
References 562

Index 565

Chapter 1
Relations Between Mechanics and Architecture Through the History of the Parallelogram Law of Forces from Antiquity to 1800

1 Introduction

In an article published in the journal *Palladio* in 1988, Edoardo Benvenuto recalls *"the entrance of history in the structural disciplines"*. On the basis of a beautiful text by Schopenhauer, he reminds the three main themes of structural mechanics:

> The struggle between weight and rigidity constitutes ... properly the only aesthetic theme of art in architecture; to bring out such a contrast in the most varied and most evident way: this is his office. Office to which it fulfils, closing the direct way of their free expansion to those indestructible forces, curbing them by diverting them; thus it prolongs the struggle, and makes the indefatigable effort of the two enemy forces visible in a thousand varied forms. Abandoning to its natural tendency, the whole building would form a compact mass, pressing to the maximum on the ground, on which the weight relentlessly lays ...; rigidity instead ... opposes to such an effort with energetic resistance. The architecture prevents the immediate manifestation of the natural inclination (of gravity), only granting it a manifestation mediated by a winding path. For example: the scaffolding can only rest on the ground by means of columns; the vault must stand by itself, and the pillars are the only means by which it can satisfy its tendency towards the earth; etc. But precisely by virtue of these twisted and forced ways, by virtue of these obstacles, the forces immanent to those rough boulders of stone have the way to manifest themselves in the most evident and most varied form ... So the beauty of a building will consist in the obvious final adaptation of man to arbitrary (the work, in such a case, would belong to practical and applied architecture); but referring to the static of the whole, in order to which each element must, by location, size and shape, be in such a necessary relationship, that if you remove any part from anywhere, the whole building will fall into disrepair.[1]

[1] Benvenuto (1988, p. 7): *La lotta fra il peso e la rigidità costituisce ... propriamente l'unico tema estetico dell'arte in architettura; far risaltare tele contrasto nel modo più vario e più evidente: questo è il suo ufficio. Ufficio a cui adempie, chiudendo a quelle indistruttibili forze la* via *diretta della loro libera espansione, frenandole col deviarle; cosi prolunga la lotta, e rende visibile sotto mille forme svariate lo sforzo infaticabile delle due forze nemiche. Abbandonando alla sua tendenza naturale, tutto l'edificio verrebbe a formare una massa compatta, premente al massimo sul suolo, su cui lo spinge inesorabile il peso... la rigidità invece ... oppone a tale sforzo un'energica resistenza. L'architettura impedisce, della naturale inclinazione (della gravità), la manifestazione immediata,*

P. Radelet-de Grave, *Between Mechanics and Architecture*, Mathematics and the Built Environment 8, https://doi.org/10.1007/978-3-031-73530-1_1

The three themes are:

(1) The struggle between the natural tendency and the resistance opposed by rigidity: the resistance of the solids started by Galileo. Schopenhauer says: The "struggle" between the "natural tendency" that would reduce the building into a pile of stones and the "energetic resistance" opposed by "rigidity".
(2) The principles of composition and decomposition of forces: The "tortuous ways" by which architecture offers a "mediated manifestation" of gravity
(3) Structural mechanics in all its complexity: The "immanent finality" that directly "refers to the static of the whole" giving coherence and teleonomic harmony to every single part.

The role played by the parallelogram law of forces is thus central to the structural disciplines and it is not surprising that Edoardo Benvenuto devoted several articles or book chapters to it. It is also interesting to note the growing importance that he attaches to it. In a preprint of 1977–78, we read *Se le leggi della statica e della meccanica siano di verità necessaria o contingente* (Benvenuto 1978) but this has been abandoned, I fear.

A few years later, in *La scienza delle costruzioni e il suo sviluppo storico* (Benvenuto 1981), he limits himself to a few lines on Simon Stevin (1548–1620), and to a few remarks on Gilles Personne de Roberval (1602–1675) and Pierre Varignon (1654–1722) in order to devote, in the chapter dedicated to graphostatics, half a page to the demonstration of the parallelogram law by Daniel Bernoulli (1700–1772).

In 1985, three years before the article published in *Palladio*, he published in *Meccanica*, an article entitled *The parallelogram of forces* (Benvenuto 1985) and in *An introduction to the History of Structural Mechanics* (Benvenuto 1991) the same title is given to the whole of Chap. 4 of the first volume.

I would like to complete the highlighting of the importance of this law in structural mechanics, by recalling that, more generally, the parallelogram law is the law of vectorial addition, that is, the law of composition mostly used in the structural disciplines.

I propose to review here the history of the parallelogram law by emphasizing the important role of the more specific study of the parallelogram of forces mostly carried out in studies related to stability and in particular to the stability of construction.

This will not be a linear story because it springs from pure mathematics with the discovery of the generality of the parallelogram law for vectors to the practical knowledge of stability, passing through the obscure, for historians, region of the

---

non concedendole che una manifestazione mediata per una via *tortuosa. Per esempio: l'impalcatura non può che poggiare al suolo che per mezzo di colonne; la volta deve reggersi da sé, e i pilatri son l'unico mezzo per cui può soddisfare la sua tendenza verso la terra; ecc. Ma appunto in virtù di queste vie contorte e forzate, in virtù di questi ostacoli, le forze immanenti a quei rudi massi di pietra hanno il modo di manifestarsi nella forma più evidente e più varia … Quindi la bellezza di un edifizio consisterà nell'evidente adattazione finale dall'uomo ad arbitrio (l'opera, in tal caso, apparterebbe all'architettura pratica e applicata); ma riferentesi alla statica dell'insieme, in ordine al quale ogni elemento deve, per collocazione, per grandezza e per la forma, essere in una relazione così necessaria, che al togliere une qualsiasi parte da qualsiasi luogo, tutto l'edificio cada in rovina.*

cathedral builders. The historian must find a way through those various points of view and also imagine the cross roads and links between them. That is not an easy task. I decided to write it purely chronologically rejecting the more tempting way of writing it per field.

2 Antiquity: The Composition of Movements and First Law for Vaults

2.1 *Third Century Before Christ, Pseudo Aristotle: Mechanical Problems*

The story goes back to Antiquity with the study of simple machines or lifting gear by the Pseudo-Aristotle. A study inspired also by the science of constructions since his *Quaestiones mechaniche* treats lifting gears as the lever, the balance, the pulley, the screw and the corner.

The Pseudo-Aristotle gives there the law of vectors' composition or parallelogram law in the case of two movements of which he underlines the important fact that these must be proportional. More, for us today, the vector composition or parallelogram law is also the law that governs the composition of forces, essential for the stability of constructions.

> 1. … Now whenever a body is moved in two directions in a fixed ratio it necessarily travels in a straight line, which is the diagonal of the figure which the lines arranged in this ratio describe.
>
> Let the ratio according to which the body moves be represented by the ratio of AB to $A\Gamma$. Let $A\Gamma$ move towards B while AB be moved towards the position $H\Gamma$; now let A travel to Δ, and let AB travel a distance determined by the point E. Then if the ratio of the movement is that of AB to $A\Gamma$, then the line $A\Delta$ must bear the same ratio to AE. Then the small parallelogram has the same proportions as the larger, so that its diagonal is the same, and the body will move to Z. It can be shown that it will behave in the same way at whatever point its movement be interrupted; it will always be on the diagonal.[2] (Fig. 1).

Let us note here that it has never been considered necessary to demonstrate this law of composition of movements or displacements. The description given by Pseudo Aristoteles was considered as a demonstration.

Pseudo Aristoteles makes also an important remark concerning the fact that magnitudes like displacements, velocities and for us forces are vectors also means that they not only have a magnitude but also a direction.

He makes a particularly insightful remark that will preoccupy many authors for a long time. If we compose two movements or two vectors, the resulting motion or vector may be very different if the angle between the components is different. The resulting motion or vector depends also of their mutual direction. This is essential

[2] Aristotle (1955, pp. 337–339).

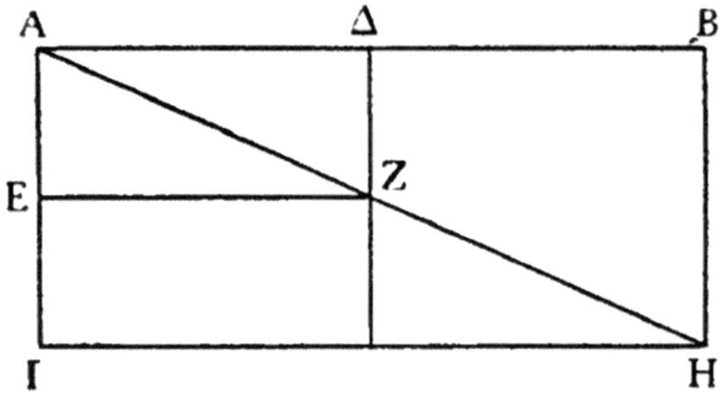

Fig. 1 Pseudo-Aristotle considers first a motion from A to B then from B to Γ (Γ looks like a I on the picture. (Aristotle 1955, p. 339). Public domain

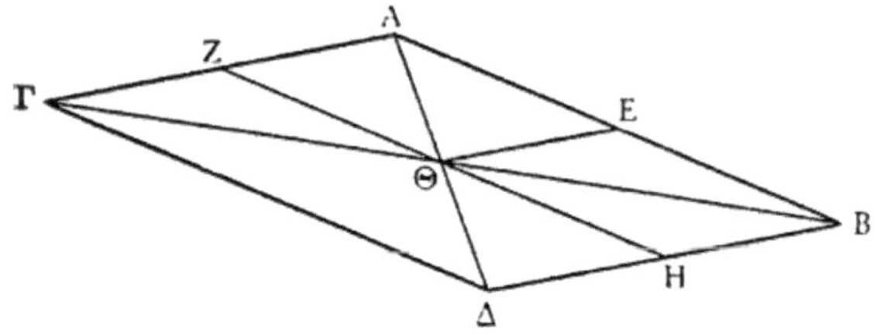

Fig. 2 Pseudo-Aristotle considers points traveling in a rhombus (Aristotle 1955, p. 383). Public domain

for the science of construction but it won't be fully understood before the eighteenth century.

> 23. Why is it that in a rhombus, when the extreme points travel in two movements, they do not each travel along an equal straight line, but one travels much farther than the other? It is only another way of asking the same question to inquire why the travelling point passes through a distance less than the side? For the diagonal is the less distance and the side the greater; the one travels with one motion and the other with two. (Fig. 2)
>
> Let A travel towards B, and B towards A with the same velocity along the line AB; again let AB travel along AΓ parallel to ΓΔ with the same velocity as these. The point A must be carried along the diagonal AΔ and B along BΓ and each must arrive at the end at the same time, and AB moves along the side AΓ. For let the point A be carried along AE, and AB along to AZ, so as to make ZH parallel to AB, and a line drawn from E to complete the parallelogram. The parallelogram thus formed is similar to the whole. Then AZ is equal to AE, so that the point A is borne along the side AE. Then AB would travel along AZ and will therefore be on the diagonal at Θ. And it must always travel along the diagonal. At the same time the side AB will travel along the side AΓ, and the point A will travel along the diagonal AΔ. Similarly it can be proved that the point B is borne along the diagonal BΓ; for BE is equal to BH. When, then, the parallelogram is completed by a line drawn from H, the enclosed parallelogram is similar to the whole. The point B will be on the diagonal at the intersection of the diagonals, and the side will travel along the side at the same time as the point B will travel along the diagonal BΓ. Then the point B will travel many times more than AB, and the side will travel along the lesser side, though carried at the same velocity, and the side in one journey has travelled further than A. The more acute-angled the rhombus is the less the diagonal AΔ becomes and the greater BΓ, but the side is less than BΓ. For it is odd, as has been said, that the point travelling along two components should sometimes move more slowly than that travelling along one, and that when both points are given an equal velocity one should travel a greater distance than the other.

> But the reason is, that when a point moves from an obtuse angle, the two paths are more or less opposite, I mean the path which the point travels and that in which it is impelled along the side; when on the other hand the point moves from the acute angle it is almost being borne in the same direction. For the angle made by the sides assists to move the point along the diagonal; and in proportion as the one makes the angle more acute and the other more obtuse, so the former travels more slowly and the latter more quickly. For they are more in opposite directions because the angle is more obtuse; but in the other case they approximate more nearly to the same direction because the lines are closer together. For the point B in both its movements is travelling nearly in the same direction; for the one movement is assisted by the other, and the more acute the angle the more this becomes true. But with A the opposite is the case for the point itself is travelling towards B, while the side tends to divert it to Δ. The more obtuse the angle, the more opposed to each other do the two movements become; for the lines approach more nearly to the straight. If they were entirely straight, they would be entirely opposite. But the side travelling in one direction is checked by nothing. Naturally therefore it traverses the greater distance.[3]

This important text shows how crucial is the role played by the direction of particular entities which in mathematics we have call "vectors" since the term was coined by William Rowan Hamilton in 1843.

2.2 *First Century Before Christ, Vitruvius's Ten Books* on Architecture

François Blondel commented on Vitruvius with these words:

> Chapter II. Doctrine of arches or arches according to Vitruvius: Buildings built with piers; that is to say, as I believe, those which are of arches or arcades whose joints all correspond to the same point. It is necessary, he says, to make the outer piers, which he calls angular, larger than the space, that is to say, wider than the bay of the arch; so that they are strong enough to support the thrust and prevent the large top load from causing the segments or transoms to move out of their places.[4]

Here we can see the first ideas about stability of vaults:

(1) The lines between the arch stones (voussoirs) should meet in a single point and
(2) the piers should be as large as the arch.

We'll see that not until 1643 will a more mathematical but still oversized law for the measure of the width of pillars be given by Derand.

[3] Aristotle, Mechanical Problems, p. 381–387.

[4] Blondel (1675, p. 316): Doctrine des arcs ou arcades suivant Vitruve: Aux bâtiments qui sont faits par Piles; c'est-à-dire comme je crois, à ceux qui sont à Arcs ou Arcades, dont les joints repondent tous à un même point; Il faut, dit-il, faire les dernières Piles, qu'il appelle angulaires, plus grosses que l'espace, c'est-à-dire plus larges que la baye de l'Arc; afin qu'elles soient assez fortes pour an soutenir la poussée et empêcher que la grande charge de dessus ne fasse sortir les voussoirs ou les impostes hors de leurs places.

3 The Composition of Forces in Statics and its Relation With the Catenary From a Purely Theorical Point of View

3.1 1586, Simon Stevin, Het eerste bouck vande beghinselen der weegconst

The next to go further in studying the composition of forces is Simon Stevin (Fig. 3). With help of that law of composition he will study the behaviour of a hanging chain or of the catenary.

Stevin's influence on the development of construction science is undeniable because he has been as interested in the purely theoretical developments of statics as in the study of fortifications.

In his *First Book of the Elements of the Art of Weighing* (Stevin 1586), after a long analysis of the balance of vertically supported sets of weights, Stevin begins the study of weights supported by oblique forces with a particularly original demonstration of the weight balanced on two differently inclined planes:

THEOREM XI. PROPOSITION XIX.

Given a triangle, whose plane is at right angles to the horizon, with its base parallel 'thereto, while on each of the other sides there shall be a rolling sphere, of equal weight to one another:

Fig. 3 The cover of Simon Stevin's *First Book of the Elements of the Art of Weighing* (Stevin 1586). Public domain

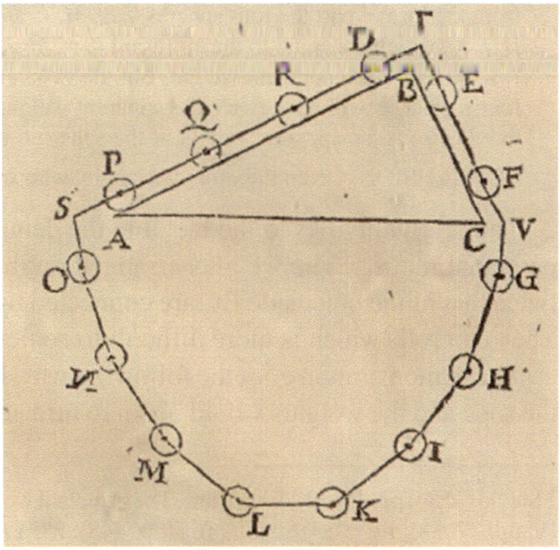

Fig. 4 Illustration for Theorem XI, Proposition XIX (Stevin 1634, p. 448). Public domain

as the right side of the triangle is to the left side, so is the apparent weight of the sphere on the left side to the apparent weight of the sphere on the right side.

SUPPOSITION. Let *ABC* be a triangle, whose plane shall be at right angles to the horizon, and the base *AC* parallel to the horizon, and on the side *AB*, which shall be double of the side *BC*, let there lie a sphere D, and on the side *BC* a sphere *E*, of equal weight and equal size to the sphere D.

WHAT IS REQUIRED TO PROVE. We have to prove that as the side *AB* is to *BC*, so is the apparent weight of the sphere *E* to the apparent weight of the sphere D.

PRELIMINARY. Let us make about the triangle *ABC* a wreath of fourteen spheres, of equal size and equal weight, and equidistant from one another, as *E, F, G, H, I, K, L, M, N, 0, P, Q, R, D,* all of them strung on a line passing through their centres, in such a way that they can revolve; about those centres; let there also fit two spheres on> the side *BC* and four on *BA,* i.e. as line to line, so spheres to spheres. Let there also be three fixed points at *S, T, V,* over' which the line or the string of the spheres can slide, in such a way that the two parts of the string above the triangle shall be parallel to the sides *AB, BC,* so that if the wreath is' pulled down on 'one side or the other, the spheres roll on the lines *AB, BC.* (Fig. 4)

PROOF. If the apparent weight of the four spheres *D, R, Q, P* were not equal to the apparent weight of the two spheres *E, F,* either one or the other will be the heavier. Let us suppose (if this were possible) this to be the one of the four spheres *D, R, 0, P.* But the four spheres *0, N, M, L* are of equal weight to the four spheres *G, H, I, K.* The side therefore of the eight spheres *D, R, Q, P, 0, N, M, L* is heavier in appearance than the side of the six spheres *E, F, G, H, I, K.* But because that which is heavier always preponderates over that which is lighter, the eight spheres will roll downwards and the other six will rise. Let this be so, and let *D* have fallen where is now, then *E, F, G, H* will be where *P, Q, R, D* are now, and *I, K* where *E, F* are now. But this being so, the wreath of spheres will have the same appearance as before, and on this account the eight spheres on the left side will again have greater apparent weight than the six spheres on the right side, in consequence of which the eight spheres will again roll down and the other six will rise. This descent on the one and ascent on the other side will continue for ever, because the cause is always the same, and the spheres will automatically perform a perpetual motion, which is absurd. The part of the wreath *D, R, Q, P, 0, N, M, L* therefore is of equal apparent weight to the part *E, F, G, H, I, K.* But if from such equal weights there are subtracted equal weights, the remainders will have equal weight. Let us therefore subtract from the former part the four spheres *0, N, M, L,* and from the latter part the four spheres *G, H, I, K* (which are equal to the aforesaid *0, N, M, L);* then the remainders *D, R, Q, P* and *E, F* will be of equal apparent weight. But the two latter being of equal apparent weight to the four former, *E* will have twice the apparent weight of *D.* As therefore the line *BA* is to the line *BC,* so is the apparent weight of the sphere *E* to the apparent weight of the sphere *D.*

CONCLUSION. Given therefore a triangle, whose plane, etc.[5]

The originality lies in the fact that the demonstration is based on the rejection of perpetual motion. The weights arranged on the two inclined planes AB of a length twice that of the other side BC are connected by a rope and are equal and equidistant. They can roll (which is more difficult to achieve, but it is a thought experiment).

The demonstration goes the following way. If the weights were not in equilibrium, the rope and the weights would begin to turn around the inclined planes, either in the

[5] Stevin (1586, pp. 40–42); Eng. trans. The principal works, vol. I, pp. 175–179; also in Stevin (1605, IV, pp. 37–38); Fr. trans. Stevin (1634, p. 448): *XI. VERTOOCH. XIX. VOORSTELL. WESENDE een driehouc wiens plat rechthouckich op den sichteinder is met sijn grondt daer afeuewidich, ende op elck der ander sijden een rollende cloot lnet malcanderen euewichtich: Ghelijck des driehoucx rechter sijde tot de slincker, also t'staltwicht des op de slincker sijde, tottet staltwicht des cloots op de rechter sijde.*
T'GHEGHEVEN. Laet ABC een driehouck wesen diens plat sy rechthouckich op den sichteinder, ende den grondt AC evewydich vanden sichteinder, ende op de sijde AB, die dobbel sy an BC, ligghe een door D, ende op de sijde BC een cloot E, evewichtich ende evegroot met den cloot D.
T'BEGHEERDE. Wy moeten bewysen dat ghelijck de sijde AB, tot BC, alsoo t'staltwicht des cloots E, tottet staltwicht des cloots D.
T'BEREYTSEL. Laet ons maecken rondom den driehouck ABC eene crans van veerthien clooten, evegroot, evewichtich, ende evewijt van malcanderen, als E, F, G, H, I, K, L, M, N, O, P, Q, R, D, algesnoert an een lini, streckende door haer middelpunten, also dat sy op die middelpunten drayen mueghen; Datter oock twee clooten passen op de sijde BC, ende vier op BA, dat is gelijck lini tot lini, also clooten tot clooten; laet oock an S, T, V, drie vast punten staen, over welcke de lini ofte t'snoer der clooten flieren mach, also dat de twee deelen des snoers die boven den driehouck staen, evewydich sijn vande sijden AB, BC; Inder voughen dat alsmen den crans an d'een ofte d'ander sijde neertrect, soo rollen de clooten op de linien AB, BC.
T'BEWYS. Soo t'stalwicht der vier knoten clooten D, R, Q, P, niet even en waer met het staltwicht der twee clooten E, F, t'een of t'ander sal swaerder sijn, latet wesen (soot mueghelick waer) der vier D, R, Q, P; Maer de vier clooten O, N, M, L, sijn evewichtig met d vier clooten G, H, I, K, de sijden dans der acht clooten D, R, Q, P, O, N, M, L, is swaerder na de gestalt dan de sijde der ses clooten E, F, G, H, I, K: maar want het swaerste altijd je licheste overweeght, de acht clooten sullen neerwaert rollent, ende d'ander ses rijsen: Later soo wesen, ende D sy ghevallen daer nu O is, ende E, F, G, H, sullen sijn daer nu P, Q, R, D, ende I, K, daer nu E, F, sijn. Maer dit soo wesende, den crand der clooten sal alsulcken ghestalt hebben als sy te vooren dede, ende sullen om de selue redenende acht clooten ter flincker sijde wederom staltwichtigher sijn dan de ses clooten ter rechter, waer duer de acht clooten wederom neer sullen rollen, ende d'ander ses sijsen, welcke valling ter eender, ende rijsing ter ander, om dat de reden altijdt de selue is, altijdt ghedueren sal, ende de clooten sullen uyt haer seluen een eeuwich roersel maken, t'welck valsch is. Het deel dan des crans D, R, Q, P, O, N, M, L, is evestaltwichrich met het deel E, F, G, H, I, K: Maer van sulcke evewichtighe getrocken evewichtighe, de resten sijn evewichtich, laet ons dan van dat deel trekkende vier clooten O, N, M, L, ende van dit de vier clooten G, H, I, K, (welke even sijn ande voornoemde O, N, M, L,) de resten D, R, Q, P, ende E, F, sullen evestaltwichtich sijn, Maer wesende dese twee evestaltwichtich met die vier, E sal tweemael staltwaerder sijn als D. Ghelijck dan de lini BA, tot de lini BC, also t'staltwicht des cloots E, tottet staltwicht des cloots D.
T'BESLUYT. Wesende dan een driehouck wiens plat, etc.

left hand side either in the right hand side. Both are impossible because when E would have turned and would have taken the place of F, all weights would have changed their place to the place of the next weight and the situation would be exactly the same as before and they would go on turning in a perpetual motion. The same would happen if they turned in the other direction. Therefore, Stevin says, this movement would not end, and that is absurd.

They cannot turn either to the left or to the right, so they remain still; they are in equilibrium. Since the weights ONMLKIHG are symmetrical, they are in equilibrium and so are the four weights on AB with the two weights on BC.

Stevin has demonstrated that they are in equilibrium if there are twice as many weights on AB as on BC. So that we have the relation $\frac{AB}{BC} = \frac{4}{2}$, four being the number of weights on AB and two the number of weights on BC. We know that AB is twice as long as BC, thus a weight on BC is twice as powerful that the same weight on AB.

In the second book, Simon Stevin focuses on determining the centres of gravity of different objects, which is also an essential element for construction.

In his addition to the statics, Stevin gives a first decomposition of a force into its two components by means of the parallelogram law:

> 3rd COROLLARY [of the 27th proposition of the lst book of the Art of Weighing]
>
> But in order to set forth the properties of weights hanging on cords, let AB be a prism, whose centre be C and which be hanging in the two fixed points D, E, with two lines CD, CE coming from the centre of gravity C; these lines CD and CE are centre lines of gravity of the prism by the 5th definition 4). Therefore, if Hl be drawn between DC and CF, parallel to CE, by the 13th definition Cl is vertical lifting line, CH oblique lifting line, with which we say that as Cl is to CH, so is the former's vertical lifting weight to the latter's oblique lifting weight. But the vertical lifting weight of Cl is equal to the weight of the whole prism. Therefore, as Cl is to CH, so is the weight of the whole prism to the weight acting on D. And in the same way the weight acting on E is also found, provided there be drawn from I to CE the line IK, parallel to DC; we can then say: as the vertical lifting line Cl is to the oblique lifting line CK, so is the weight of the whole prism to the weight acting on E.
>
> But CK is always equal to Hl, therefore it is not necessary to draw this latter line IK, but we have all the requisite known terms in the three sides of the triangle HIC, so that we can say as follows:
>
> As Cl is to CH, so is the weight of the prism to the weight acting on D. Further as Cl is to IH, so is the weight of the prism to the weight acting on E. Again, as CH is to Hl, so is the weight acting on D to the weight acting on E.[6] (Fig. 5)

---

[6] Stevin (1586, pp. 181–182); Eng. trans. The principal works, vol. 1, pp. 541–543; also in Stevin (1605, pp. 181–182); Fr. trans. Stevin (1634, p. 505): *3 VERVOLGH. Maer om nu te commen tot verclaring vande ghedaenten der gewichten an tauwen hanghende, foo laet AB een pylaer sijn, diens middelpunt C, en hanghende ande twee vastpunten D, E, met twee linien CD, CE, commende uyt het swaerheyts middelpunt C; de selve CD n CE sijn swaerheyts middelliinen des pylaers deur des bepaling: Daerom tusschen DC. en C F, ghetrocken HI evewijdeghe met CE, foo is deur de 13 bepaling CI rechtheflijn, CH scheefheflijn, waer me wy segghen, dat ghelijck Cl tot CH, also diens rechthewicht, tot desens scheefhefwicht: Maer t'rechthefwicht van Cl is even antghewicht des heelen pylaers: Daerom ghelijck Cl tot CH, alsoo t'ghewicht des heelen pylaers, tottet ghewicht op D ancommende: Ende inder selver voughen vintmen oock t'gewicht op E ancommende, midts te trecken van I tot in CE, de lini IK; evewijdeghe met DC, en meughen dan segghen, ghelijck rechtheflijn CI, tot scheefheflijn CK, also t'gewicht des heelen pylaers, tottet ghewicht op E ancommende. Maer*

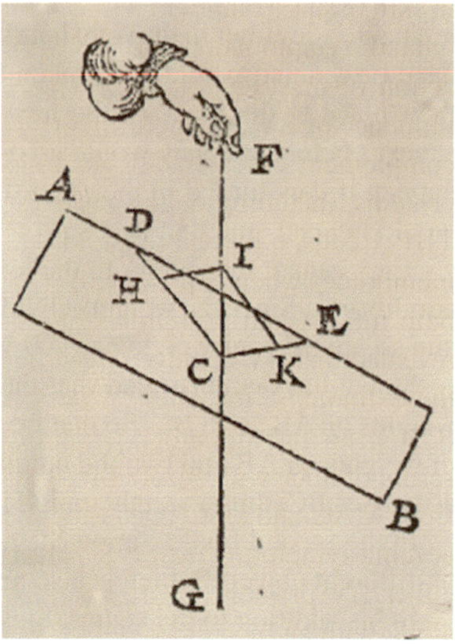

Fig. 5 Illustration for the 3rd Corollary (Stevin 1634, p. 505). Public domain

Stevin draws HI between DC, CF, parallel to CE, so CI will be the right elevation, CH the oblique; that as CI is to CH, so is this direct elevation to the oblique one: but the direct elevation of CI is equal to the all weight of the column: Thus, as CI is to CH, so the weight of the whole column to the weight that acts on D. In the same way we find the weight which acts on E, leading from I to CE the line lK, parallel to DC; and thus, as the direct elevation CI is to the oblique elevation CK, so the weight of the whole column to the weight that acts on E.

Finally, he finds the following proportion: $\frac{CH}{HI} = \frac{P_D}{P_E}$.

The weights acting in D and E are in the same ratio as the lengths CH and HI.

4th COROLLARY

But in order that we may make our proposed explanation of the property of weights hanging on cords even clearer, let the prism AB be pulled down in such a way that it be now in the place shown below; then by the 3rd postulate it does not cause on that from which it is hanging any different weight from that it first did, when hanging higher. Therefore, the foregoing proportion of the 3rd corollary still holds in the figure of the 4th corollary.[7] (Fig. 6)

CK valt altijt even an HI, daerom en ist niet noodich te treken dese laetste lini IK, maer hebben alle noodighe bekende palen inde drie sijden des driehoucx HIC, met welcke wy meughen aldus segghen: Ghelijck CI tot CH, alsoo t'ghewicht des pylaers, tottet ghewicht op D ancommende. Voort ghelijck CI tot IH, alsoo t'ghewicht des pylaers, tottet ghewicht op E ancommende. Weerom ghelijck CH tot Hl, alsoo t'ghewicht op D ancommende, tottet ghewicht op E ancommende.

[7] Stevin (1586, p. 186); Eng. trans. The principal works, vol. 1, p. 543; also in Stevin (1605, p. 182); Fr. trans. Stevin (1634, p. 505): *4 VERVOLGH. Maer op dat wy ons voorghenomen verclaring der ghedaente van ghewich ten an tauwen hanghende noch naerdet commen so laet den pylaer AB neerwaert ghetrocken worden, alsoo datsè nu sy ter plaets als hier onder; en deur de 3 begeerrete, soo en veroirsaeckse an t'ghene daerse an hangt, gheen ander ghewicht danse eerst en dede hoogher hanghende: Daerom de voorgaende everedenheyt des 3 vervolghs is noch ïnde form des vervolghs.*

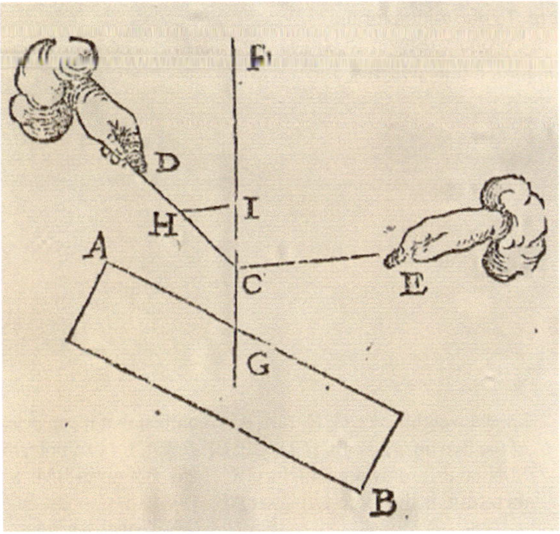

Fig. 6 Illustration for the 4th Corollary (Stevin 1634, p. 505). Public domain

It is possible to lower the column without changing anything and thus maintaining the same proportion.

5th COROLLARY

Now let us hang instead of the prism AB in the 4th corollary another body of equal gravity, but of any form and specific gravity, viz. AB in this 5th corollary. Then it is also manifest that as Cl is to CH, so is the weight AB to the weight acting on D. Further, as Cl is to IH, so is the weight AB to the weight acting on E. Again, as CH is to Hl, so is the weight acting on D to the weight acting on E.

From this it is manifest that if from a line DCE as cord there were hanging a known weight AB, and if the angles FCD; FCE were also known, it can be said how much weight each part DC, CE has to carry.[8] (Fig. 7).

Stevin finally replaces the beam with an equal weight and thus gives a formulation equivalent to the parallelogram law of forces. This Corollary V of Stevin will be taken up by Roberval and many others after him.

6th COROLLARY

But if there were thus hanging on a line two or more weights, as in the following figure the line ABCDEF, whose extreme fixed points are A and F, on which line there are hanging four

---

[8] Stevin (1586, p. 182); Eng. trans. The principal works, vol. 1, p. 535; also in Stevin (1605, p. 182); Fr. trans. Stevin (1634, p. 505): *5 VERVOLGH. Laet ons nu in plaets des pylaers AB int 4 vervolgh, hanghen een ander lichaem der selfde swaerte, maer van form en stoffwaerheyt soot valt, als AB in di 5s vervolgh. Ende is openbaer dat ghelijck Cl tot CH, also t'ghewicbt AB, tottet ghewicht op D ancommende. Voort gelijck CI tot IH, also t'gewicht AB, tottet gewicht op E ancommende. Weerom ghelijck CH tot HI, alsoo t'ghewicht op D ancommende, tottet gewicht op E ancommende. Hieruyt is openbaer, dat sooder aen een lini DCE als coorde, hinghe een bekent ghewicht AB, en de houcken FCD, FCE, oock bekent sijnde, dat men can segghen hoe veel gewelt elck deel DC, CE te draghen heeft.*

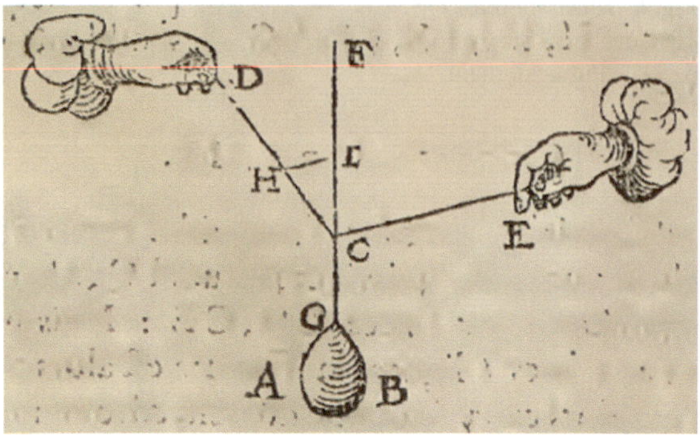

Fig. 7 Illustration for the 5th Corollary (Stevin 1634, p. 505). Public domain

> known weights, viz. G, H, I, K, it is manifest that it can be said how much force acts on each of the five lines AB, BC, CD, DE, EF. For if, for example, the line GB be produced to L, and M N be then drawn parallel to BC: I say BN gives BM, what the weight G? What follows therefrom is the force acting on AB.
>
> Again BN gives MN, what the weight G? What follows therefrom is the force acting on BC.
>
> Let HC again be produced to O, and BP drawn parallel to CD. I say: CP gives CB, what the weight H? What follows therefrom is the force acting on BC. From which it is evident that the same will then have to be found that was previously found of BC. And so on with all the others. This and several other things his PRINCELY GRACE tested in practice and found to be entirely in agreement with the theory.[9] (Fig. 8).

Stevin thus generalizes his theory of the parallelogram of forces to the equilibrium of a chain on which are suspended a series of weights and thus opens the door to the discovery of the catenary curve that will play an important role in understanding the stability of vaults and bridges. We'll follow that story.

3.2 *1604–1619, Isaac Beeckman's* Journal *(1588–1637)*

According to the editors of Isaac Beeckman's *Journal*, Beeckman became interested in the catenary following a question asked by Descartes. In the notes on Descartes

[9] Stevin (1586, p. 182); Eng. trans. The principal works, vol. 1, p. 535–537; also in Stevin (1605, p. 182–183); Fr. trans. Stevin (1634, p. 505): *6 VERVOLGH. Maer by aldien an een lini alsoo hinghen twee of meer ghewichten, als inde Volghende form de lini ABCDEF, diens uyterste vastpunten sijn A en F, an welke lini hanghen vier bekende ghewichte, als G, H, I, K: TIS openbaer datmen can segghen hoe veel gewelt datter comt an elck der vijf linien AB, BC, CD, DE, EF: Want treckende, by voorbeelt gheseyt, de lini GB voorwaert, als tot L, daer na MN evewijdeghe met BC: Ick segh BN gheeft B M, wat t'gewicht G? T'ghene daer uyt volght is voor t'ghewelt op AB ancommende. Weerom BN gheeft MN, wat T'ghewicht G ? T'ghene daeruyt volght is voor t'ghewelt op B C ancommende. Laet andermael HC voorwaert ghetrocken worden tot O, en BP evewijdeghe met CD: Ick segh CP gheeft CB, wat t'ghewicht H? T'gene daeruyt comt is voor t'ghewelt op BC ancommende. Waeruyt blijckt datmen alsdan t'selve sal moeten vinden, datmen te vooren van BC vant: En soo voort met al d'ander. Hier af en van meer ander heeft sijn VORTELICKE GHENADE dadelicke proef ghedaen, en bevonden die gantschelick t'overcommen mette spiegheling.*

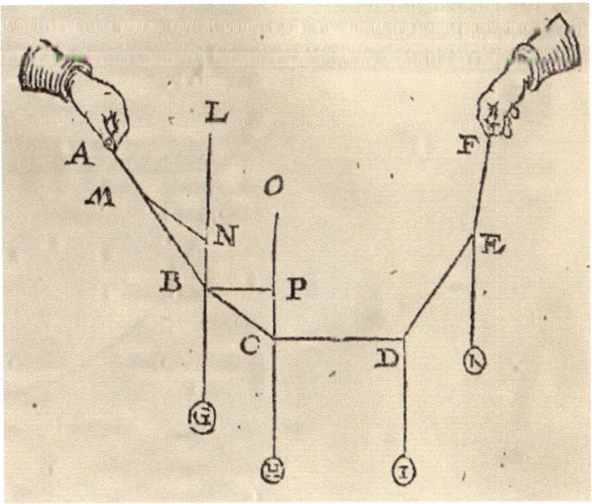

Fig. 8 Illustration for the 6th Corollary (Stevin 1634, p. 505). Public domain

Fig. 9 The catenary illustrated in Beeckman (1939, p. 362). Public domain

written by Beeckman at the end of 1618, we find the following remark together with a picture of the catenary: "[Descartes] asked Isaacus of Middelburg [Beeckman] if the chain acb, attached at points a and b, describes a part of the conic section".[10] (Fig. 9).

In his *Journal* entries from April 1614 to January 1615, Beeckman gives, following Stevin, a new explanation of the form of the catenary:

> A chain *abcdecba* is attached to a beam and allowed to hang loosely and freely. It is of constant thickness and uniform weight. So, the weight which is in d is also partially on c, so that it descends lower and the weight in c is also partially on d, so that d is slightly heavier; it follows that cb is stiffer than cd, and the weight in b increases the weight in c and the weight in c causes the point b to descend, it follows that ab is steeper than bc, and therefore when a rope is firmly tied at the ends a and d, and therefore d becomes lighter and if we put supports only at d and a, the higher they are, the more they will be stretched.[11] (Fig. 10)

---

[10] Beeckman (1939, I, p. 362): *Petijt a me Isaacus Middelburgensis an funis acb affixus clavis a, b, sectionis conicae partem describat.*

[11] Beeckman (1939, p. 43); Beeckman (1644, p. 24): *Hangt aen eenen blacken koorde a. b. c. d. c. b. a. En laet se los en vry hanghen, dat se oock sy van eender dickte ende ghelijcker swaerte al-om;*

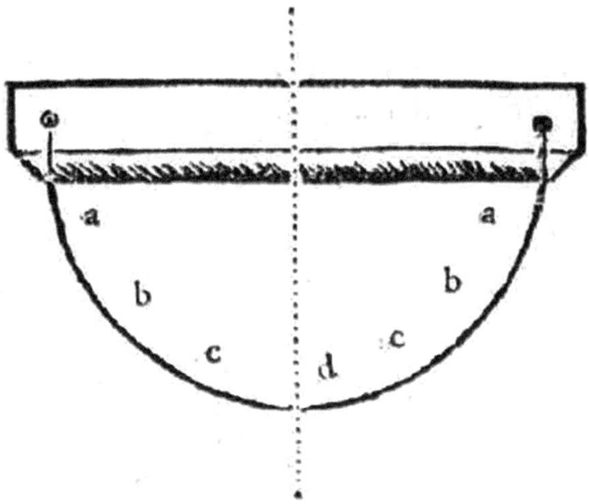

Fig. 10 The hanging rope illustrated in Beeckman (1619, p. 24). Public domain

A bit later at an unprecise date, he notes in his Journal: What line is described by a rope suspended at two points?[12]

> Make the rope DCHHCD whose HH and GC are equal and suspend on it at H and C equally heavy weights and draw the line AB through the middle of HH, and extend CH and DC to AB.
>
> I say that *the sunbeams falling parallel to AB, will be reflected on the line CH.*
>
> Let AB hang vertically towards the centre of the earth. Hanging in this way ECK will be parallel to AB and to the rays of the sun. This line ECK falling on DCA will make an angle ECD larger than the angle ACB, thus the ray EC is reflected on CH, because it makes in the reflection the same angle as when it falls on the same line.[13] (Fig. 11)

He continues:

> I will demonstrate that the DCE angle is larger than ACB. Draw from E the line ED parallel to CH, make HG equal to CE, and also draw GF parallel to CH. Now, the weight in CE is

---

dewijle dan de swaerte die aen d is ten dele komt oock op c. Alsoo dat c nederwaerder buycht, en de swaerte die aen c, is komt te, deele op d also dat sy d wat op-waerder heft, so volcht daer uyt dat c b steylder is dan cd soo oock de swaerte aen b doet c op-waerts gaen, en de swarte aen c doet b neder-waerts gaen, waer uyt volcht dat ab steylder is dan bc als dan een touvve vast ghemaeckt worrt aen a en d alsoo dat d leegher hanght dan eenich deel tusschen da en dat men de buysen soo leght, soo liggen se hoe hoogher hoe steylder.

[12] Beeckman (1939, p. 357)*: chorda ex duobus tabulati locis dependens, quam lineam describat quaesitum.*

[13] Beeckman (1939, p. 357)*: Maeckt een koorde DCHHCD ende laet HH en GC even groot zyn ende hanght aen H en C even swaer gewichten, ende treckt de linie AB doort midden van HH, ende treckt CH ende DC voorwaerts tot aen AB.*
Ick segge dat de sonnestrale evenwydich vallende met AB, wederomsteuten sal boven de linie CH. Laet AB recht nederwaerts hangen naer het middelpunt van het aerdtryck, soo sal ECK daert aenhanght, evenwydich syn met AB ende verstrecken voor de sonnestralen. Ist dan dat de linie ECK vallende op DCA den hoeck ECD grooter maeckt dan den hoeck ACB is, soo steut de strale EC boven CH, want sy maeckt int steuten sulck eenen hoeck als int vallen op deselve linie.

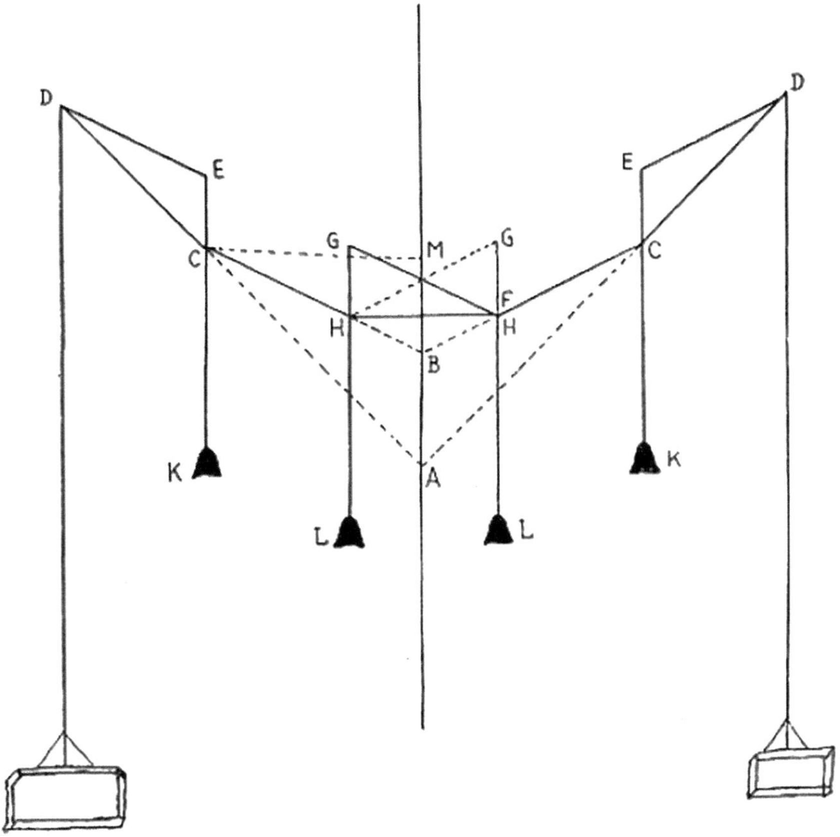

Fig. 11 The hanging rope illustrated in Beeckman (1939, p. 356). Public domain

equal to the weight in GH, and so [length] CE is equal to [length] GH; reciprocally: as the weight placed in HG is equal to that placed in FG, thus [the length] HG is equal to [length] FG. But the weight in CE is as heavy as the one in HG; therefore, [length] GF is equal to [length] GH. CE is also in the same ratio to GF. So, CE and GH and GF are the same length. But since GF is always opposite the right angle in the triangle GHF, GF is always longer than GH, and for the same reason DE is always longer than CE. But the angle DCE is as large as the angle CAB and BCA is equal to EDC and angle ECD is larger than the angle CDE. Therefore, angle ECD is bigger than angle ACB.

Thus, the ray EC, reflecting, will make with CA a greater angle than ACB, and for this reason will be again reflected in M, on CH, which will be such that CM and AM are equal to each other.[14]

[14] Beeckman (1939, p. 357): Dat den hoeck DCE grooter is dan ACB bewys ick aldus. Treckt van E de lynie ED evenwydigh met CH ende maeckt HG soo groot als CE ende treckt GF oock evenwydigh met CH. Nu gelyck hem heeft de swaerte, die op CE kompt, tot die op GH compt, soo heeft hem CE tot GH; ende wederom: gelyck de swaerte op HG komende, tot die op FG komende, soo heeft hem HG tot FG. Maer de swaerte, die op CE kompt, is soo groot als de swaerte die op HG kompt; soo is dan GF soo groot als GH. Deselve reden heeft oock CE tot GF. Soo syn dan CE

Beeckman deduces but does not demonstrate that all the rays of sun converge finally at a single point. This is shown in the next figure (Fig. 12). This property being a characteristic of the parabola, he deduces that the curve adopted by the rope—the catenary—is a parabola.

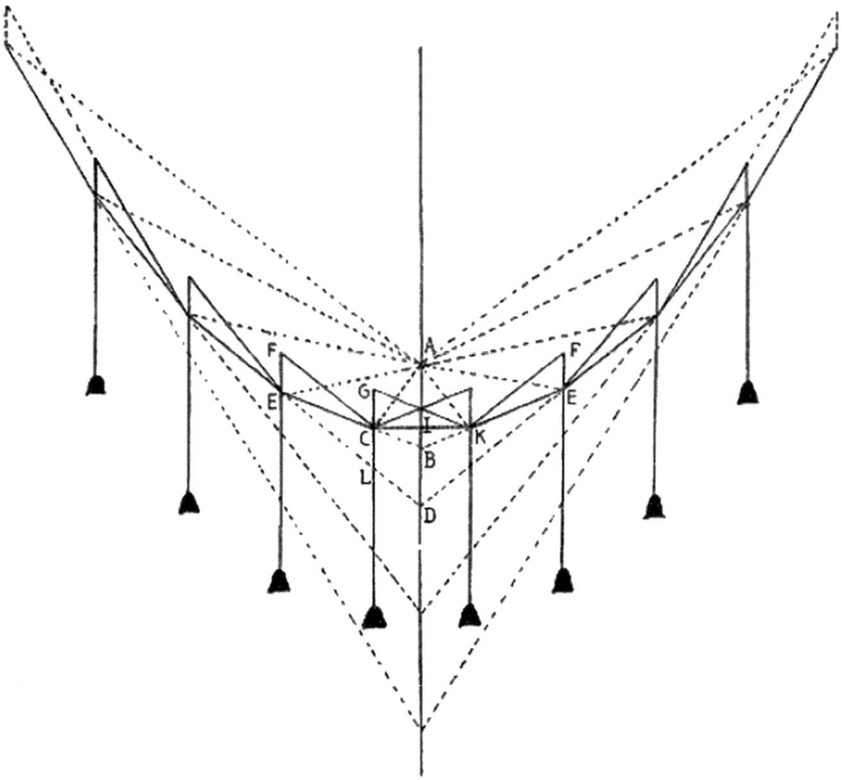

Fig. 12 The hanging rope and rays of the sun, illustrated in Beeckman (1939, p. 358). Public domain

ende GH ende GF even groot. Maer dewyle GF in den dryhoeck GHF altyt over den rechthoeck staet, so is GF altyt grooter dan GH, ende daerom oock DE altyt grooter dan CE. Maer den hoeck DCE is soo groot als den hoeck CAB ende BCA gelyck EDC ende den hoeck ECD grooter dan den hoeck CDE; soo is dan oock den hoeck ECD grooter dan den hoeck ACB. Soo sal dan de strale EC, steutende, eenen grooteren hoeck maecken met CA dan ACB, sal derhalven boven CH wederom steuten in M, twelck is als ghy CM ende AM even groot maeckt te zyn.

3.3 *1638, Galileo's* Discorsi e dimostrazioni matematiche intorno a due nuove scienze

Unfortunately for Beeckman, gravity doesn't behave the same way as light, nor does the suspended rope form a parabola. Galileo also thought the form of a hanging rope was a parabola, as he explains in the *Discorsi*, on the fourth day:

> Besides I must tell you something which will both surprise and please you, namely, that a cord stretched more or less tightly assumes a curve which closely approximates the parabola. This similarity is clearly seen if you draw a parabolic curve on a vertical plane and then invert it so that the apex will lie at the bottom and the base remain horizontal; for, on hanging a chain below the base, you will observe that, on slackening the chain more or less, it bends and fits itself to the parabola; and the coincidence is more exact in proportion as the parabola is drawn with less curvature or, so to speak, more stretched; so that using parabolas described with elevations less than 45° the chain fits the parabola almost perfectly.[15]

The common error of Beeckman and Galileo shows how difficult it was for them to admit new curves different from the conics.

3.4 *1636, Gilles Personne de Roberval,* Traité de mechanique des poids …

In his *Treatise of mechanics of weights* (Fig. 13) Roberval starts from the study of a weight placed on an inclined plane and held by a rope that does not act parallel to the plane (Proposition II), and wonders what part of the weight is supported by this rope. In Proposition III, he frees himself from the inclined plane supporting the weight by means of two ropes. He wonders then what part of the weight is supported by each rope.

In the second scholia, he first recalls the result of the first scholia:

> In the preceding scholia we showed that in every case in which the two ropes that sustain the weight, eventually extended, meet with the line CF perpendicular to the line of direction AF, the one on one side and the other on the other side of the point F, the weight and the two powers are homologous to the three sides of a triangle. But in this second scholia we will demonstrate that, in general, in whatever arrangement the weight and the two powers

[15] Galilei (1638, p. 284); Eng. trans. p. 290: *Ma più voglio dirvi, recandovi insieme maraviglia e diletto, che la corda così tesa, e poco o molto tirata, si piega in linee, le quali assai si avvicinano alle paraboliche: e la similitudine è tanta, che se voi segnerete in una superficie piana ed eretta all'orizonte una linea parabolica, e tenendola inversa, cioè col vertice in giù e con la base parallela all'orizonte, facendo pendere una catenella sostenuta nelle estremità della base della segnata parabola, vedrete, allentando più o meno la detta catenuzza, incurvarsi e adattarsi alla medesima parabola, e tale adattamento tanto più esser preciso, quanto la segnata parabola sarà men curva, cioè più distesa; sì che nelle parabole descritte con elevazioni sotto a i gr. 45, la catenella camina quasi ad unguem sopra la parabola.*

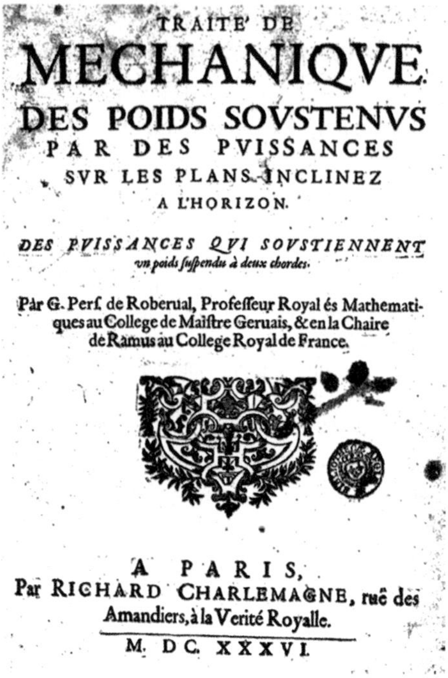
TRAITÉ DE
MECHANIQVE.
DES POIDS SOVSTENVS
PAR DES PVISSANCES
SVR LES PLANS INCLINEZ
A L'HORIZON.
DES PVISSANCES QVI SOVSTIENNENT
vn poids ſuſpendu à deux chordes.
Par G. Perſ. de Roberual, Profeſſeur Royal és Mathematiques au College de Maiſtre Geruais, & en la Chaire de Ramus au College Royal de France.
A PARIS,
Par RICHARD CHARLEMAGNE, ruë des Amandiers, à la Verité Royalle.
M. DC. XXXVI.

Fig. 13 The cover of Roberval's treatise on the mechanics of weights (1636). Public domain

are that sustain it on two ropes, as long as the two ropes are not parallel to each other in a straight line, the weight and the two powers are always homologous to the three sides of a triangle.[16] (Fig. 14).

The "line of direction" is a technical term given by Roberval to the direction in which the body would fall if freed from all support. When the body hangs on ropes, it is obviously the vertical passing through its centre of gravity. It is a parallel to the inclined plane if the body rolls on it.

The statement of that scholia II is important because it is equivalent to the parallelogram law of forces and from our particular point of view, it is even more important because, as we will see, it will generally be preferred to the parallelogram law in the domain of architecture.

Nevertheless, at the end of his second scholia, Roberval states the solution of this problem in a simpler way, getting closer to the current formulation of the parallelogram of the forces.

---

[16] Roberval (1636, pp. 27–28): *Par le Scholie precedent nous avons fait voir qu'en tous les cas ausquels les deux chordes qui soutiennent le poids, estant prolongees, s'il en est besoin, concourent avec la ligne CF perpendiculaire à la ligne de direction AF, l'une d'une part, et l'autre de l'autre du point F; le poids et les deux puissances estoient homologues aux trois costez d'un triangle. Mais en ce second Scholie nous demonstrerons en general qu'en quelque disposition que soient le poids et les deux puissances qui le soutiennent sur deux cordes, pourveu que les chordes ne soient pas entre elles en ligne droite, le poids et les deux puissances sont toujours homologues aux trois cotez d'un triangle.*

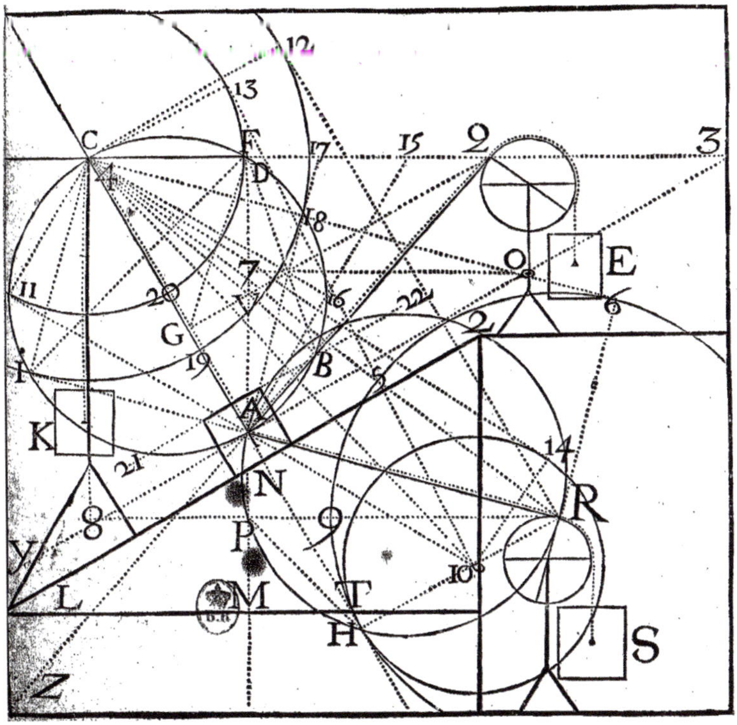

Fig. 14 Roberval's figure for the second scholia. Unfortunately, Roberval has grouped all the figures necessary to his text on one picture (Roberval 1636, p. 14). Public domain

> That if from any point taken in the line of direction of weight, a line is drawn, parallel to one of the strings, until the other string, the triangle formed by this parallel, by the line of direction, and by the string, will be similar to the aforesaid triangle, and consequently its sides will be homologous to the weight and to the two powers; what a Geometer will prove easily, with several other properties that we leave.[17]

3.5 1634, Pierre Hérigone, Cursus mathematicus

At the end of his third volume on mathematics (Fig. 15), after expounding on the construction of the tables of sines and logarithms and how to use them, the measurement of right triangles, practical geometry, fortifications, Pierre Hérigone finally gives the parallelogram figure we are used to, whose sides are proportional to the

[17] Roberval (1636, p. 28): *Que si de quelque point pris en la ligne de direction du poids, on mene une ligne parallèle à l'une des chordes iusques à l'autre chorde, le triangle formé de cette parallèle, de la ligne de direction, & de la chorde, sera semblable au triangle susdit, et par conséquent ses côtez seront homologues au poids & aux deux puissances; ce qu'un géomètre prouvera facilement, avec plusieurs autres propriétés que nous laissons.*

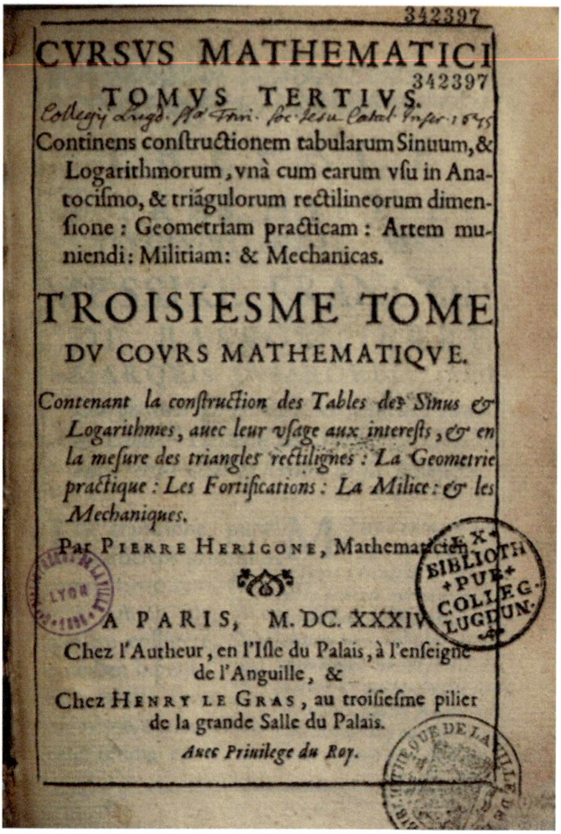

CVRSVS MATHEMATICI
TOMVS TERTIVS.

Continens conſtructionem tabularum Sinuum, & Logarithmorum, vnà cum earum vſu in Anatociſmo, & triāgulorum rectilineorum dimenſione: Geometriam practicam: Artem muniendi: Militiam: & Mechanicas.

TROISIESME TOME
DV COVRS MATHEMATIQVE.

Contenant la conſtruction des Tables des Sinus & Logarithmes, auec leur vſage aux intereſts, & en la meſure des triangles rectilignes: La Geometrie practique: Les Fortifications: La Milice: & les Mechaniques.

Par Pierre Herigone, Mathematicien.

A PARIS, M.DC. XXXIV.
Chez l'Autheur, en l'Iſle du Palais, à l'enſeigne de l'Anguille, &
Chez Henry le Gras, au troiſieſme pilier de la grande Salle du Palais.
Auec Priuilege du Roy.

Fig. 15 Cover of the third volume of Hèrigone's *Cursus mathematicus*, 1634. Public domain

weights attached to the pulley and whose diagonal is proportional to the central weight. This is the exact transcription of the last remark by Roberval we reproduced above:

> Proposition XII. Given a weight supported by two ropes, find out how much each rope carries. The weight A is supported by the weights C and D, one asks the weights C and D.[18] (Fig. 16)

3.6 *1643, François Derand,* L'Architecture des voûtes: *An Architectural Point of View*

As this example from Derand's treatise on "the architecture of vaults, or the art of lines and cutting of stones", we can see that either Derand has no idea of the parallelogram

[18] Hérigone (1634, III, p. 313): *Propos. XII. Dato pondere duobus funibus suspenso, invenire quantum ponderis singuli funes ferant.*

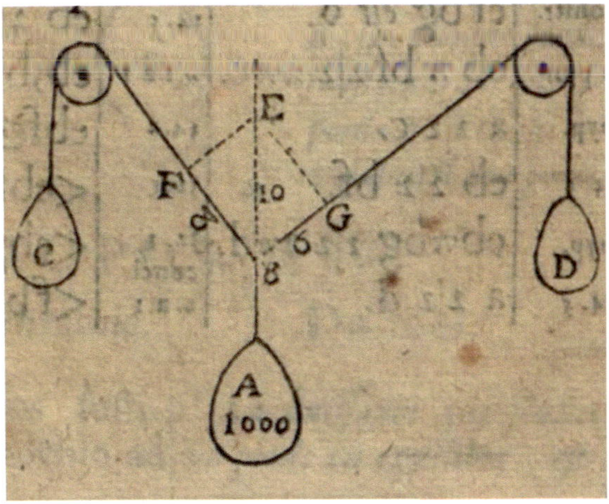

Fig. 16 Hérigone's figure for Proposition XII (Hérigone 1634, III, p. 313). Public domain

law of forces or he doesn't want to share this knowledge. His inspiration goes back to Vitruvius and his law, a rather famous one at the time, is only justified by practice. But we must remind that secret had been and still is important in architecture at those times:

> Chapter VI: On the line of the thrust of the vaults, and where we see what thicknesses the walls and flying arches which support them must have.
>
> Given the semicircular vault ABCD (marked in the middle with a P) divide it into three equally at points B, C, then by a third like CD draw the straight line CDF, and using the same point D for centre, and opening the compass of the extent of the rope CD, made below, and outside the work of said hanger, the arc EF. And through point F, where the said arc intersects the line CF, draw the plumb FG for the outside of the wall, which must support the vault ACD: and therefore, the thickness of the said wall can be included between the lines EH, and FG and thus it will be sufficient to resist the thrust of the vault, as practice and experience show.
>
> That if the vaults are lowered, like IKA in figure Q: using the same method, the wall will be thicker; this is necessary; because the pressure of these vaults is greater than that of the arched vaults, as in the precedent; and much more than that of the vaults which are raised or in Gothic arches, as it is possible to notice with the two lines LMI and NOP in the figures marked with the letters N, and O. However, it is not always necessary that the aforesaid thicknesses found in practice, which we have just argued, are kept in all the extent of the walls which carry the vaults: thus it will suffice to maintain them at the place of the principal arches, where they form advances, which are named vulgarly, salient bodies or flying buttress.[19] (Fig. 17)

[19] Derand (1643, p. 10): *Chapitre VI: Du Trait de la Poussée des voûtes, et où l'on voit quelles épaisseurs doivent avoir les Murs et Arcs boutants qui les portent. Soit donnée la voûte en plein ceintre ABCD (marquée en son milieu d'un P) divisez la en trois également és points B, C, puis par un tiers comme CD tirez la ligne droite CDF, et vous servant du mesme point D pour centre, et ouvrant le compas de l'estenduë de la corde CD, faites au-dessous, et au dehors œuvre dudit cintre, l'arc EF. Et par le point F, où le dit arc coupe la ligne CF, tirez l'aplomb FG pour le dehors de la muraille, qui doit porter la voûte ACD: et partant l'épaisseur de ladite muraille, pourra estre comprise entre les lignes EH, et FG et ainsi elle sera suffisante pour résister à la poussée de la voûte, comme la pratique et l'expérience le font voir.*

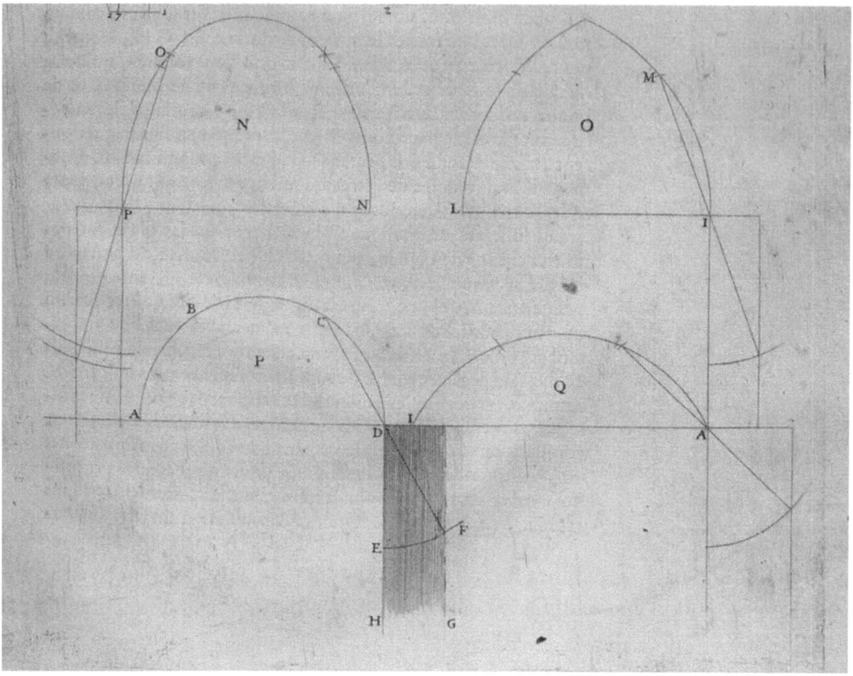

Fig. 17 Derand's illustration of the lines of thrust in vaults and the thicknesses they must have (Derand 1643, p. 11). Public domain

3.7 1646, Christiaan Huygens's Letters to Marin Mersenne, and His Manuscript De catena pendente

The first discovery Christiaan Huygens made in mechanics concerns the hanging chain and thus the reversed arch, a discovery made when he was only 17 years old. We have four testimonies of that discovery: a manuscript and three letters to Mersenne. The manuscript *De catena pendente* published in vol. 11 of his *Oeuvres complètes* is more complete and coherent but the three letters contain elements that are implied in it.

In a letter to Mersenne written around 28 October 1646, Huygens announces his discovery:

Que si les voûtes sont surbaissées, comme IKA en la figure Q: usant de la même industrie, la muraille se trouvera plus épaisse; cela estant necessaire; puis que la poussée de ces voûtes est plus grande, que celle des voûtes qui ont leur plein ceintre, comme est la precedente; et beaucoup d'avantage que celle des voûtes, qui sont surhaussées, ou en tiers-points, comme il se peut remarquer és deux traits LMI et NOP compris sous les lettres N, et O/ Or il n'est pas tousiours necessaire que les susdites épaisseurs trouvées dans la pratique, que nous venons d'alleguer, se gardent en toute l'estenduë des murs qui portent les coûtes: ainsi il suffira de les conserver à l'endroit des arcs principaux, où elles formeront des avances, lesquelles se nomment vulgairement, corps saillants ou arcsboutans.

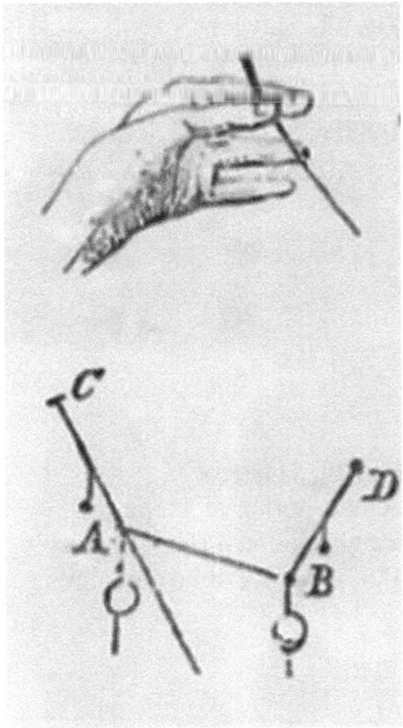

Fig. 18 Huygens's demonstration of a rope with weights, points 1 and 2 (Huygens 1888, p. 34). Public domain

> I'll send you, in another letter the demonstration that a suspended rope or chain does not make a parabola and what should be the pressure on a mathematical cord or a cord without gravity to do one; of which I also have the demonstration.[20]

In his letter to Mersenne written in November 1646, Huygens begins his exposition of the hanging chain with a supposition:

> 1. First of all, I suppose that all ropes to which some weight hangs freely, tend to the centre of the earth and are parallel to each other.[21] (Fig. 18)

This means that, due to gravitation, the strings suspended at A and B in Fig. 18 and carrying the weights are parallel to each other and vertical.

> 2. Secondly, that two or more weights, A and B, attached to the string CABD, fixed in C and D, can only stay at rest in a unique situation.[22]
>
> 3. Third, that if some weights hang on a string ADF, according to the situation naturally required, and that one fixes in their situation any two points B, E of them, that it will not change the situation of the points C and D which are between them.[23] (Fig. 19)

[20] Huygens (1888, no. 14, p. 28): *Ie fineray icij de peur de ne vous detenir pas trop longtemps, et vous envoyeray par une autre lettre la demonstration de ce qu' une corde ou chaine pendue ne faict point une parabole, et quelle doit estre la pression sur une corde matematique ou sans gravité pour en faire une; d'ont j'ay aussi trouve la demonstration, il n'y a pas longtemps.*

[21] Huygens (1888, no. 20, p. 34): *1. Je suppose donc premierement que toutes les cordes dont quelque gravité depend librement, tendentes au centre de la terre, sont paralleles l'un à l'autre.*

[22] Huygens (1888, no. 20, p. 34): *2. Secondement que deux ou plusieurs gravitez A et B attachez à la corde CABD qui est tenue en C et D, ne peuvent demeurer en repos que d'une seule façon.*

[23] Huygens (1888, no. 20, p. 35): *3. Troisiemement, que si d'une corde ADF dependent quelques gravitez, selon leur situation naturellement requise, et qu'on arreste quelques deux poincts B, E, dans leur situation, que cela ne changera point celle des poincts C et D, qui sont entre deux.*

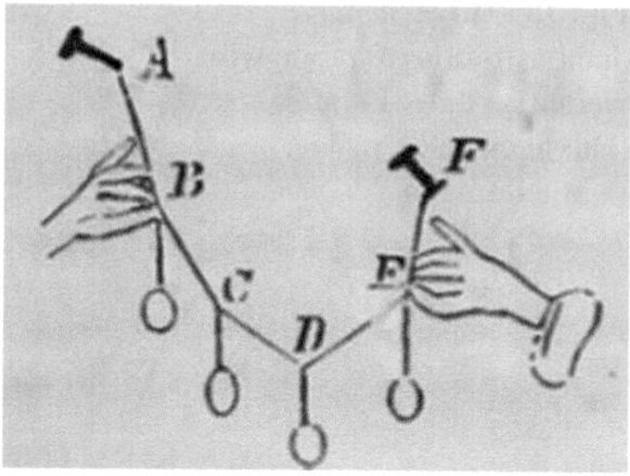

Fig. 19 Huygens's demonstration of a rope with weights, point 3 (Huygens 1888, p. 35). Public domain

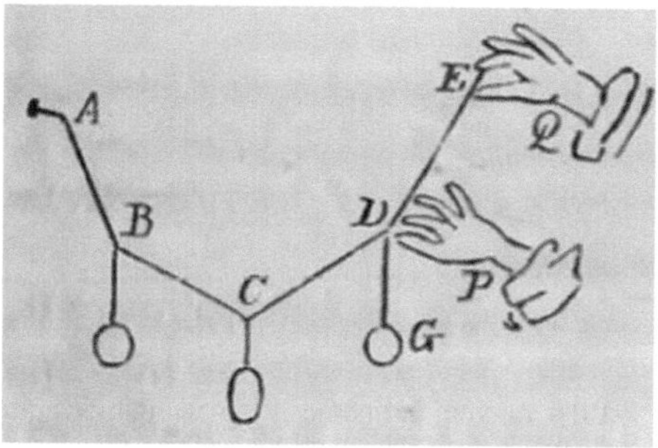

Fig. 20 Huygens's demonstration of a rope with weights, point 4 (Huygens 1888, p. 35). Public domain

As the weights hang vertically, and thus pull the rope ABCDE in parallel directions there is only one solution to the problem. This property, which makes it possible to change the points of suspension A and F to B and F for example, will be important for the further development of the study of the catenary.

> 4. Let some weights be suspended at the rope ABCDE in B, C, and D, and that those in B and C hang according to the situation naturally required when the rope is held at point D where the weight G is attached; I suppose it is possible that the hand Q holds the end E so that the point D remains at the same place, after the hand P has left it.[24] (Fig. 20)

The problem isn't changed by changing the place where hand Q holds the rope if it maintains it in the same direction. In his letter no. 22 to Mersenne, Huygens will imagine that hand Q could maintain it in any direction as the dots indicate in Fig. 21.

In his manuscript, *De catena pendente,* Huygens changes his first supposition to a theorem:

> First Theorem: If a weight is suspended on two cords, those prolonged will intersect at the vertical to the centre of gravity of this weight.[25]

---

[24] Huygens (1888, no. 20, p. 35): *4. Soyent suspendues de la corde ABCDE quelques gravitez comme en B, C, et D, et que celles en B et C pendent selon leur situation naturellement requise quand la corde est tenue au poinct D ou est attachée la gravité G; je suppose qu'il est possible que la main Q tiene en quelque façon le bout E, que le poinct D, demeure en sa mesme place, apres que la main P l'aura quitté.*

[25] Huygens (1908, p. 37): *Theorema 1^mum^ Si pondus suspendatur ex duobus funibus, hi producti secabunt sese in pendula ponderis gravitatis diametro.*

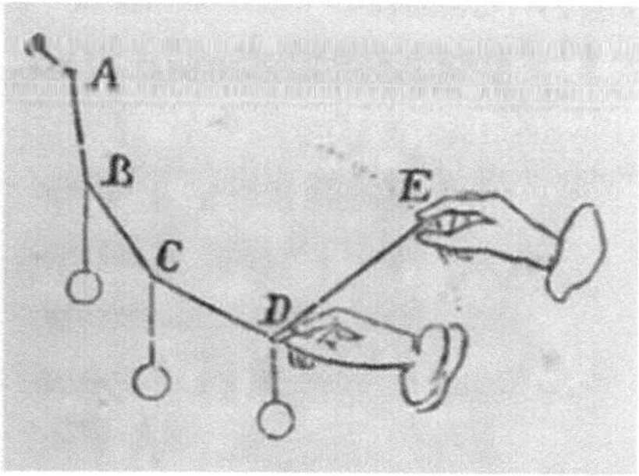

Fig. 21 Huygens's demonstration of a rope with weights, changing the location of the hand (Huygens 1888, p. 41). Public domain

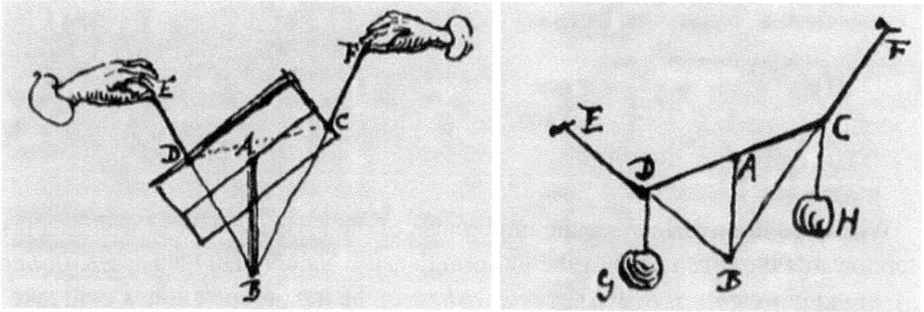

Fig. 22 Huygen's illustrations from *De catena pendente*. **left**, the image corresponding to Huygens' demonstration (Huygens 1908, p. 37); **right**, the one corresponding to Stevin's demonstration (Huygens 1908, p. 39). Public domain

But his demonstration (Fig. 22, left) is not convincing so that he gives also an equivalent to his theorem and rightly recalls that it was carefully demonstrated by Stevin (Fig. 22, right), as we have seen. He also has another reason to give that demonstration; he will use this theorem in the second part of his article.

In his letter to Marsenne, no. 21, Huygens goes further with propositions 5 to 12. His aim is now to study the form of the curve that would be formed by the rope if it were, we would say today, of uniform density. His first difficulty is to describe that idea of uniform density. He does this in propositions 5 and 6.

Proposition 5:

> If there are as many weights S, R, P, Q as one wishes suspended to a rope ABCD, I say that MD and BC prolongated will intersect in L, vertically under the centre of gravity of the weights P and Q, AB and DC in K, vertically under the centre of gravity of R and P, and so on. Because if one chooses two points A and D (leaving two other points B and C between them), in the situation in which they are, it won't change the situation of the points B and C; but the points A and D having been chosen, the intersection of the extended lines AB and DC must be vertically under the centre of gravity of R and P. That is the sign that it was there also before. And so will it be proved for the others.[26] (Fig. 23).

[26] Huygens (1888, p. 35): *Si il y a tant de gravitez qu'on veut comme S, R, P, Q pendues à une corde ABCD, je dis que MD et BC continuées s'entre-couppent en L au diametre pendule des gravitez P*

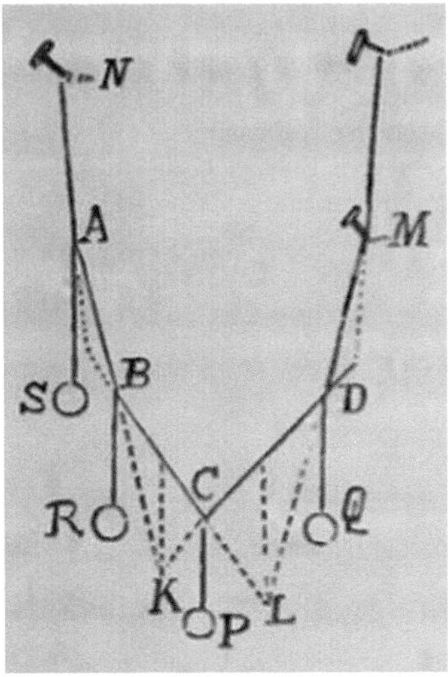

Fig. 23 Huygens's demonstration of a rope with weights, Proposition 5 (Huygens 1888, p. 35). Public domain

Where we immediately see the importance of the *First theorem* as well as a first step towards the illustration of the uniform density. Namely Huygens says: *If there are as many weights S, R, P, Q as one wishes.* In the next proposition he will take a second step calling "heavy commas" (*virgulae ponderantes*) the segments AB, BC, and so on, between the weights.

Proposition 6

> The same method is still correct if AB, BC, CD etc. "heavy commas" all of the same weight, whatever two of them as CD, FE, extended intersect vertically under the centre of gravity of the segment (the heavy comma) that was left between them as DE. Thus, if all the weights are equal, DR must be equal to RE, EP equal to PF and so on.[27] (Fig. 24).

The "heavy commas" having the meaning of an infinitesimal weight dx giving the idea of a uniform density.

In his text Proposition 8 Huygens tries to show that the curve formed by the "heavy commas", our catenary is not a parabola, and gives an interesting figure showing the difference (Fig. 25). In order to get a parabola, he shows in Proposition 12 that we must hang equal weights at equal horizontal distances. As we will see below, Euler will call that the hanging bridge.

---

et Q. AB et DC en K au diametre pendule des gravitez R et P et ainsi du reste. Car si on arreste quelque deux poincts A et D (en laissant deux autres entre deux comme B et C), en la situation ou ils sont, cela ne changera poinct celle des poincts B et C; mais les poincts A et D estants arrestez, l'intersection des continuées AB et DC doit estre au diametre pendule des gravitez R et P. C'est donc signe qu'aussi auparavant elle ij a esté: Et ainsi le prouvera ton des autres.

[27] Huygens (1908, pp. 37–44): *Eadem methodo probatur si AB, BC, CD* etc. *sint virgulae ponderantes aequalis ponderis, quasque duas productas ut CD, FE, sese intersecare in pendula gravitatis diametro ejus quae intermedia relicta est ut DE. Ergo si omnes aequalis sunt ponderis, DR debet aequalis esse RE, EP aequalis PF* etc.

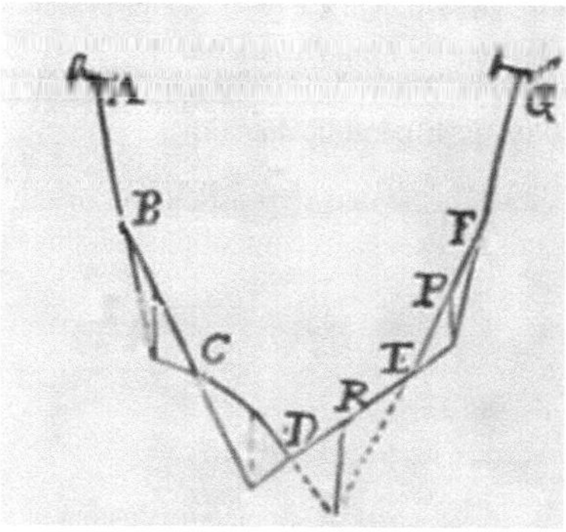

Fig. 24 Huygens's demonstration of a rope with weights, Proposition 6 (Huygens 1888, p. 35). Public domain

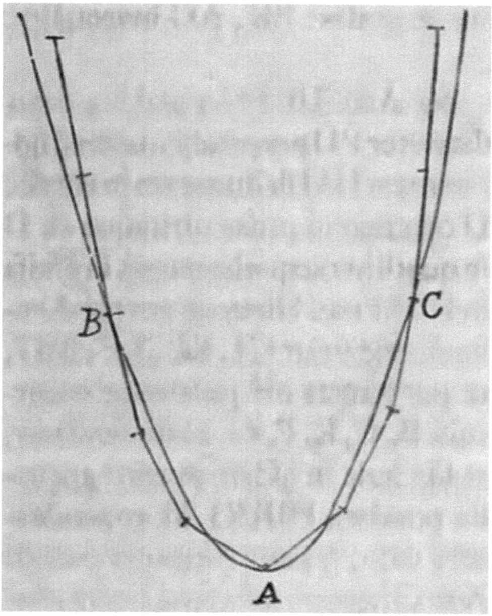

Fig. 25 Huygens's demonstration of the parabola and the catenary, Proposition 8 (Huygens 1888, p. 37). Public domain

With help of the *theorema primum* of the manuscript *De catena pendente*, Huygens demonstrates the following proposition 2:

Proposition 2. Now let there be proposed a wire ADG along which equal weights are fixed at equal distances [of the wire]. It is therefore obvious that its situation must depend on the fact that between any two nodes, an interval extended to the left intersects the vertical passing through the centre of gravity of the two weights. Thus, BC and ED intersect at the point N, which is in the vertical from the centre of gravity of the two weights C and D. Let us say that BC and ED do not intersect in N: thus, if we fix the points B and E in the points at which they hang now, this does not change the position of points C and D; but these fixed, the weights fixed at C and D can be suspended differently only so that the intersection of prolonged BC and ED are on the vertical from the centre of gravity [H], as it appears and as it was suspended before.

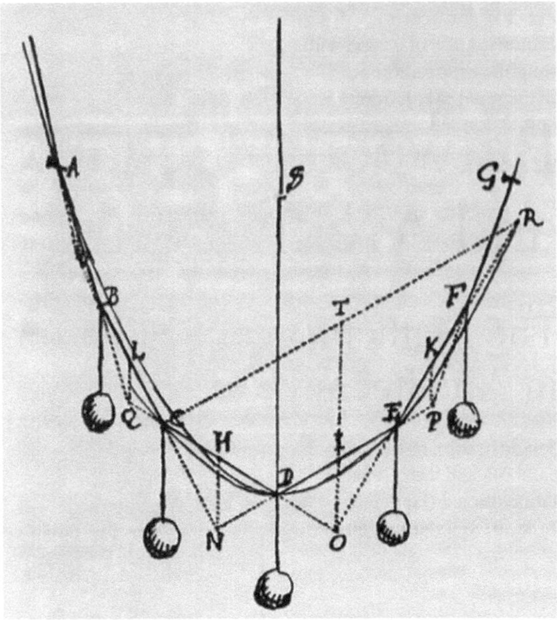

Fig. 26 Huygen's illustrations from *De catena pendente*, Prop. 2 (Huygens 1908, p. 40). Public domain

> In the same way, we can prove that AB and DC meet in the vertical passing through the centre of gravity of the weights suspended in B and C, and so on with the other ones going on to infinite. It was thus easy, if we know three points C, D and F of that catena, to determine the other ones as B, F, A, G. One has just to divide DE in two in I and to draw IO parallel to SD, and starting from O, where CD extended, cuts that IO, trace OF passing through E and take EF equal to DE, and we obtain the point F. In the same way one can obtain the rest of the curve.[28] (Fig. 26).

Huygens concludes that if one knows three points of the curve one can determine it completely, and his Proposition 3 notes that, "The line that passes through those points is not very different from a parabola, but is not a parabola".[29]

---

[28] Huygens (1908, p. 40): *Propositus nunc sit funis ADG in quo aequalibus intervallis ligata sunt aequalia pondera.*
Manifestum itaque est eo situ ea pendere debere ut bina quaeque internodia, relicto uno intermedio, producta, sese intersecent in pendula duorum ponderum, gravitatis diametro: Sic BC, ED sese intersecant in puncto N, quod est in pendul. gr. diametro ponderum suspensorum ex C et D. dicantur enim BC et ED invicem non secare in N: Si ergo firmentur puncta B et E in eisdem punctis ubi nunc pendent, hoc quidem situm punctorum C et D nihil immutabit; sed firmatis iis, pondera ex C et D aliter non possunt suspendi quam ut intersectio productarum BC, ED sit in pend. grav. diametro, ergo apparet et ante sic suspensa fuisse.
Eodem modo demonstrari potest, AB et DC se intersecare in pend. gr. diametro ponderum ex B et C pendentium, et ita de quotlibet aliis in infinitum ascendentibus. Facile hinc erit, si nota habeamus tria puncta C, D et E catenae hujus, caetera quoque invenire ut B, F, A, G. Oportet enim DE bifariam dividere in I et agere IO parallelam ipsi SD, et ex O, ubi producta CD illam IO secat, per E ducere OF et ponere EF aequalem DE, et habebimus punctum F. Simili modo et reliqua reperiri possunt.

[29] Huygens (1908, p. 41): *Prop. 3. Linea quae per haec puncta ducitur valde parum distare videtur a linea parabolica, quam tamen non refert.*

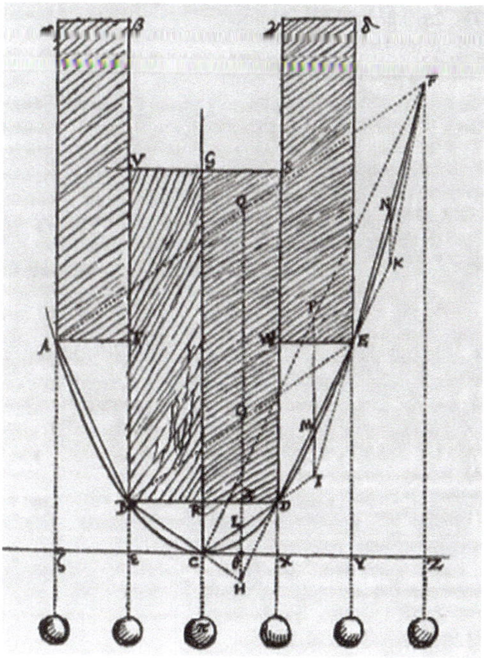

Fig. 27 Huygen's illustrations from *De catena pendente*, Prop. 4 (Huygens 1908, p. 42). Public domain

Huygens doesn't reproduce the figure of Proposition 8 of *De catena pendente* in his letter no. 21 to Mersenne, but he does demonstrate that point F of the catenary is not on the parabola. He also shows as one can see on the figure that in order to get a parabola, one should hang the weights at equal horizontal distances and not at equal distances of the curve (Fig. 27).

3.8 *1673, Ignace-Gaston Pardies,* La statique ou la science des forces mouvantes

Pardies uses the parallelogram law of forces to study complicated equilibriums of beams. But that law is still only a useful recipe.

> LXX. The situation of bodies suspended by two ropes.
>
> When a body on, whose centre of gravity is *e*, is suspended by two cords *oA*, *na*, these cords incline so that, being continued, they would cross in the line of direction *eB*. For if in the first figure we lengthen the cord *oA* to *B*, and stop there, it is obvious that the body would remain in the same situation, since the directions would not change at all. Likewise, the rope also lengthened to *B* and stopped there, would maintain the body in the same situation as before. So instead of attaching the strings to the two points *a* and *A*; if we attached them to the single point *B*, the body would remain suspended as before, and consequently the centre *e* would be perpendicularly under *B*.

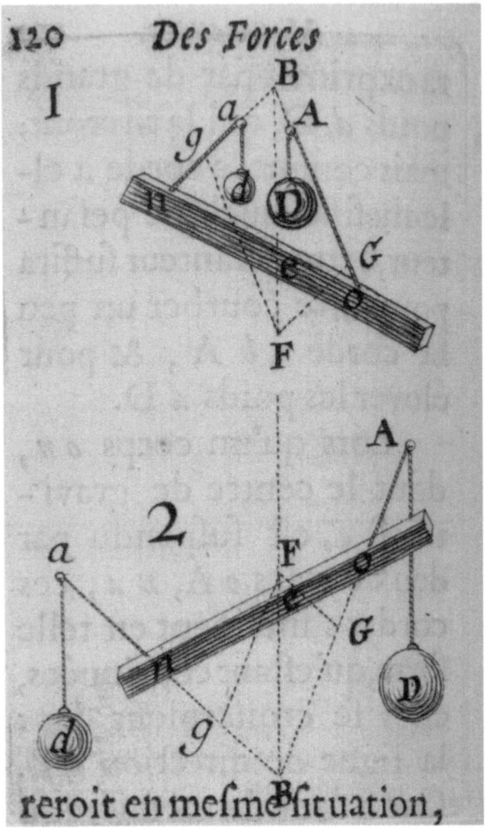

Fig. 28 Pardies' Figs. 1 and 2 (Pardies 1673, p. 120). Public domain

But in the second figure, it should be imagined that the strings extended to the common point *B*, stiffen like bars to be able to maintain the body on; for this body thus pressed on *oB*, *nB* would remain, as when it is held by the strings *Ao*, *an*; so the centre *e* would be in the perpendicular passing through point *B*.[30] (Fig. 28)

---

[30] Pardies (1673, pp. 119–122): *LXX. Situation des corps suspendus par deux cordes. Lors qu'un corps* on, *dont le centre de gravité est* e, *est suspendu par deux cordes* oA, na, *ces cordes s'inclinent en telle sorte qu'estant continuées, elles se croiseroient dans la ligne de direction* eB. *Car si dans la première figure on allongeoit la corde* oA *jusqu'en* B, *et qu'on l'arrestast là, il est manifeste que le corps demeureroit en même situation, puisque les directions ne changeroient nullement. De même la corde* na *allongée aussi jusqu'en* B *et arrestée-là, soûtien droit le corps en même situation qu'auparavant. Ainsi au lieu d'attacher les cordes aux deux points* a *et* A*; si on les attachoit au point unique* B, *le corps demeureroit suspendu comme auparavant, et par conséquent le centre de* e *seroit perpendiculairement sous* B.
Mais dans la 2. Fig. il faudroit imaginer que les cordes prolongées jusqu'au point commun B, *se roidissent comme des barres pour pouvoir soûtenir le corps on; car ce corps ainsi appuyé sur* oB, nB *demeureroit, comme lors qu'il est soûtenu par les cordes* Ao, an*; ainsi le centre* e *se trouveroit perpendiculairement sur le point* B.

Next, he shows some properties of the catenary, which we have already encountered in Huygens:

> LXXIII, Property of the tangents of this curve.
>
> This weight *abA* thus suspended from the ends attached in *aA*, would remain in the same situation if we pulled the tangents *ae*, *AE*, and we suspended it by the points *e*, *E*. (We must imagine that these tangents have zero gravity.) Because the cord *abA* would remain in the same situation, when we imagine that the part *aC* is stiffened, and that the only part *CbA* is flexible; although it is assumed that this part *aC* thus stiffened can turn around a, to rise towards *aA*, or to lower towards *aB*. That if instead of this curved and stiffened part *aDC*, we put a straight rod *aC*, all the rest *CbA* will still remain in the same situation, provided nevertheless that we imagine that all the force, of which the part *aDC* pulled down point *C* , or picked up at the same point *C*, by a weight suspended by *C*, which pulls down, as did all the curved part *aDC*: because it does not matter at all whether the rod which stood by the ends *a* and *C* is straight or curved, or of some other nature, provided that the force of its end *C* is always the same, as we suppose that it is here. So, also by extending this rod in a straight line towards *e*, and fixing it in *c*, everything else will remain as it was before; and finally, if this rod becomes flexible like a net, nothing will move.[31] (Fig. 29)

Pardies shows, as many had done before him that shifting along the curve the point where the rope is fixed doesn't change the form of the curve. Later one will say that the tension in the rope remains the same.

He also signals, in LXXIV, using the example of the previous figure and the one that follows one that the tangents cross each other on the vertical issuing from the centre of gravity, or as he puts it, they cross on the line of direction of that centre of gravity.

> LXXV. Ordinary chains and ropes do not bend into parabolas. Some have thought that the ropes and chains attached at both ends curve in a parabolic line.
>
> But this is not true in chains or in ropes which cannot be easily extended. Because if a chain composed of small, very delicate rings were in the figure *abA*, by drawing the tangents at *b*, namely *bD*, and at *a*, namely *aD*, these two tangents would intersect at *D* in the line of direction *DC*, of the chain *aCb* (by the previous proposition). Because we can imagine that the chain is now fixed at *a* and *b*: and then this part *aCb* would remain in the same situation as it was, being freely attached only to the ends *a* and *A*. Thus, the centre of gravity of the chain *ab* would be in *C*. Now if the figure *aCb* were parabolic, the line *DCE* would divide

---

[31] Pardies (1673, p. 125–129): *LXXIII, Propriété des tangentes de cette courbure. Ce poids* abA *ainsi suspendu des bouts attachez en* aA, *demeureroit en même situation, si l'on tiroit les tangentes* ae, AE, *et qu'on le suspendist par les points* e, E. *(Il faut imaginer que ces tangentes n'ont nulle pesanteur;) Car la corde* abA *demeureroit en même situation, quand on imagineroit que la partie* aC *est roidie, et que la seule partie* CbA *est flexible; quoy que l'on suppose que cette partie* aC *ainsi roidie puisse se tourner autour d'a, pour se hausser vers* aA, *ou pour s'abaisser vers* aB. *Que si au lieu de cette partie courbe et roidie* aDC, *on met une verge droite* aC, *tout le reste* CbA *demeurera encore en même situation, pourveû néanmoins qu'on imagine que toute la force, dont la partie aDC tiroit en bas le point C, soit ramassée au même point* C, *par un poids suspendu par* C, *qui tire en bas, comme faisoit toute la partie courbe* aDC*: car il n'importe de rien que la verge qui soûtient par les bouts* a *et* C *soit droite ou courbe, ou de quelque autre nature, pouveû que l'effort de son extrémité* C *soit toûjours le même, comme nous supposons qu'il l'est icy. Donc aussi en prolongeant cette verge en droite ligne vers* e, *et l'attachant en* c, *tout le reste demeurera comme il estoit aupravant; et enfin, si cette verge vient à se rendre flexible comme un filet, rien ne bougera.*

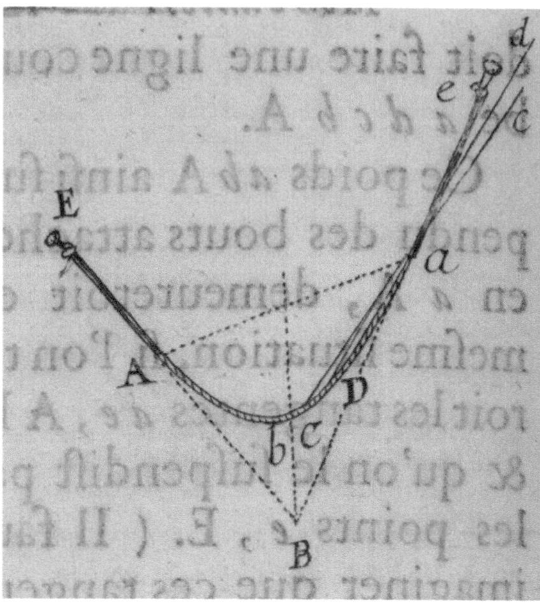

Fig. 29 Pardies' properties of the tangents of a curve (Pardies 1673, p. 126). Public domain

equally *aF* in two, but the part of the parabola *aC* would be greater than *Cb*; and it is very easy to demonstrate that the centre of gravity of the parabola *ab* cannot be at *C*.[32] (Fig. 30).

Finally, he uses those properties to show differences between the catenary and the parabola.

LXXVI. In which case a wire would bend in a parabola.

But if we conceived a wire without gravity, on which would press an infinity of equally heavy lines *EC* and *ec*, parallel and equally distant from each other; then the net *aCbA* would be perfectly parabolic, because the centres of gravity of all these heavy lines would be in the line *FbB*, that is, in the middle of *aA*. In this situation the tangents *aB*, *AB* would intersect in the line *FbB*. Likewise, the centre of the lines which are between *a* and *F* is on the line *EC*, that is, in the middle between *a* and *F*. Thus, the tangents *bD* and *aD* will have to cross on

---

[32] Pardies (1673, p. 130–132): *LXXV. Les chaînes et les cordes ordinaires ne se courbent pas en parabole. Quelques-uns ont pensé que les cordes et les chaînes attachées par les deux bouts se courboient en ligne parabolique.*
Mais cela n'est pas vray dans les chaînes ny dans les cordes qui ne se peuvent pas allonger aisément. Car si une chaîne composée de petits anneaux fort délicats estoit dans la figure abA, *en tirant les tangentes par* b, *sçavoir* bD, *et par* a, *sçavoir* aD, *ces deux tangentes se couperoient en* D *dans la ligne de direction* DC, *de la chaîne* aCb *(par la précédente proposition). Car on peut imaginer que la chaîne est maintenant arrestée en* a *et en* b*: et alors cette partie* aCb *demeureroit dans la même situation qu'elle estoit estant attachée librement aux seuls bouts* a *et* A. *Ainsi le centre de gravité de la chaîne ab seroit en* C. *Or si la figure* aCb *estoit parabolique, la ligne* DCE *diviseroit* aF *en deux également, mais la partie de la parabole* aC *seroit plus grande que* Cb*; et il est fort aisé de démontrer que le centre de gravité de la parabole* ab *ne peut pas estre en* C.

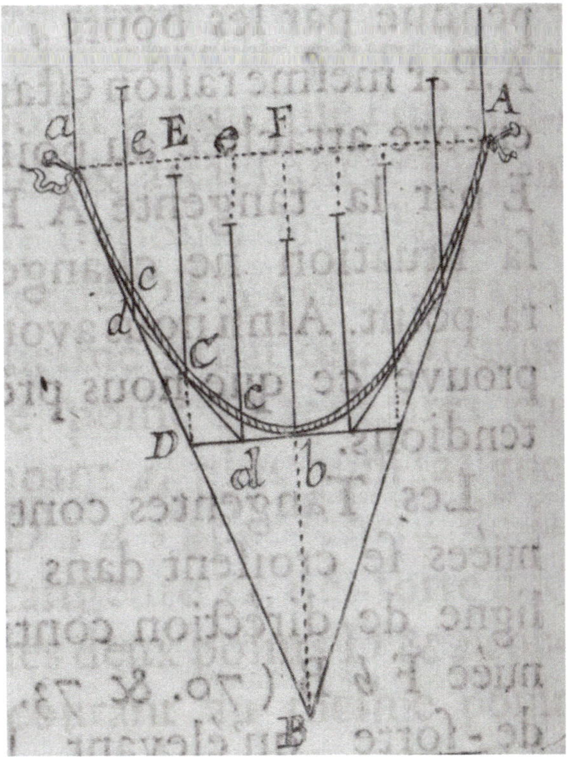

Fig. 30 Pardies' properties of the tangents of a curve (Pardies 1673, p. 130). Public domain

this line *ECD*. Similarly, the tangents *bd*, *Cd* will cross on the line *ecd*, that is, in the middle between *E* and *F*, etc. Now this is a property of the parabola and the geometers know that there is no other line where it is like that.[33]

Pardies will further clarify situations where, for example, an elastic string will bend into a parabola. He then even considers giving a rope the shape of an ellipse or a hyperbola.

[33] Pardies (1673, p. 132–134): LXXVI. *En quel cas un filet se courberoit en parabole. Mais si nous concevions un filet sans pesanteur, sur lequel une infinité de lignes également pesantes* EC, ec, *paralleles, et également distantes les unes des autres; alors le filet* aCbA *seroit parfaitement parabolique: car les centres de gravité de toutes ces lignes pesantes seroit dans la ligne* FbB, *c'est-à-dire, au milieu de* aA. *Ainsi les tangentes* aB, AB *se couperoient en cette ligne* FbB. *De même le centre des lignes qui sont entre* a *et* F, *est dans la ligne* EC, *c'est-à-dire, au milieu entre* a *et* F. *Ainsi les tangentes* bD, aD *se devront croiser dans cette ligne* ECD. *De même les tangentes* bd, Cd *se croiseront dans la ligne* ecd, *c'est-à-dire, au milieu entre* E *et* F, etc. *Or c'est là une propriété de la parabole et les Geometres sçavent qu'il n'est point d'autre ligne où cela se rencontre.*

3.9 1676, Robert Hooke, A Description of Helioscopes, and Some Other Instruments

We are now turning more directly to architecture. As is well known, Hooke collaborated with Christopher Wren on the construction of St Paul's Cathedral, and in particular on the middle dome. He hints at the catenary in a rebus:

> 2. *The true Mathematical and Mechanical form of all manner of Arches for Building, with the true butment necessary to each of them.* A Problem which no *Architectonick* Writer hath ever yet attempted, much less performed, Abcccddeeeeefggiiiiiiiillmmmmnnnnnooprrssstttttttuuuuuuuux.[34]

The solution given by Hooke is *Ut pendet continuum flexile, sic stabit contiguum rigidum inversum* (As hangs the flexible line, so but inverted will stand the rigid arch).

Thus, even if Hooke and Wren were unable to derive the equation of the catenary, they could experimentally draw its shape following the form of the hanging flexible cable.

3.10 1687, Pierre Varignon, Projet d'une nouvelle mecanique ou statique

Although Varignon's "project for a new mechanics, or statics" (Fig. 31) opens with a dynamic axiom, it will be entirely devoted to static problems:

> Axiom: The spaces that one and the same body, or equal bodies in equal times, are to each other as the forces that move them; and reciprocally when the spaces are to each other as the forces, they make the body or equal bodies go through them in equal times.[35]

He explains this in the lemmas:

> Lemme III: If the point A (Fig. 7) without gravity is pushed in the same time and uniformly by two powers E and F along the lines AC and AB, that make between them some angle CAB whatever, and that the force with which the power E acts, be to the one with which acts the power F, as AC to AB. This point A will follow the diagonal AD of the parallelogram BC drawn on those two lines.[36]

---

[34] Hooke (1679, p. 31).

[35] Varignon (1687, p. 1): *Axiome: Les espaces que parcourt un même corps, ou des corps égaux dans des tems égaux, sont entr'eux comme les forces qui les meuvent; et réciproquement lorsque les Espaces sont entr'eux comme ces forces, elles les font parcourir au même corps, ou à des corps égaux en temps égaux.*

[36] Varignon (1687, p. 6): *Lemme III: Si le point A* (Fig. 7*) sans pesanteur est poussé en même tems, et uniformément par deux puissances E et F suivant les lignes AC et AB, qui fassent entre-elles quelque angle CAB que ce soit, et que la force dont agit la puissance E, soit à celle dont agit la puissance F, comme AC à AB. Ce point A suivra la Diagonale AD du parallelogramme BC fait sous ces deux lignes.*

Fig. 31 Cover of Pierre Varignon's first edition of *Projet d'une nouvelle mécanique* (1687)

> Lemma IV: Presently point A (Fig. 8) is the centre of direction of the body EF, and that body is pushed at the same time and uniformly by two powers applied at E and F, along lines EC and FB which pass through the point A, with forces which are between them like AC and AB: that we conclude the parallelogram BC, and that we look for a moment this body as if it had no gravity. Whatever angle BAC these lines make to them, this body thus pushed will follow the diagonal AD.[37] (Fig. 32)

Varignon then gives a different form to the parallelogram law; we shall call it the "the triangle law of forces":

> Lemma V: The three sides of a rectilinear triangle (Fig. 9), whatever it is, are between them, like the sines of the angles to which they are opposed.[38] (Fig. 33).

Next, Varignon sets forth the "fundamental proposition":

> Fundamental proposition for weights suspended by any number of ropes; and for all the possible angles that they can make between them.
>
> The weight K supported with cords PB and CR by the powers P and R, and in equilibrium with them, is always to each of them as the sine of the angle PAR that their cords make between each other, to each of the sines of the angles RAK and PAK which each of these

---

[37] Varignon (1687, pp. 7–8): *Lemme IV: Présentement que le point A* (Fig. 8) *soit le centre de direction du corps EF, et que ce corps soit poussé en même-tems et uniformément par deux puissances appliquées en E et en F, suivant les lignes EC et FB, qui passent par le point A, avec des forces qui soient entr'elles comme AC et AB: que l'on achève le parallelogramme BC, et que l'on regarde pour un moment ce corps comme s'il n'avoit aucune pesanteur. Quelque angle BAC que ces lignes fassent ent'elles, ce corps ainsi poussé suivra la diagonale AD.*

[38] Varignon (1687, p. 9): *Lemma V: Les trois côtez d'un triangle rectiligne* (Fig. 9), *quel qu'il soit, sont entr'eux, comme les sinus des angles ausquels ils sont opposez.*

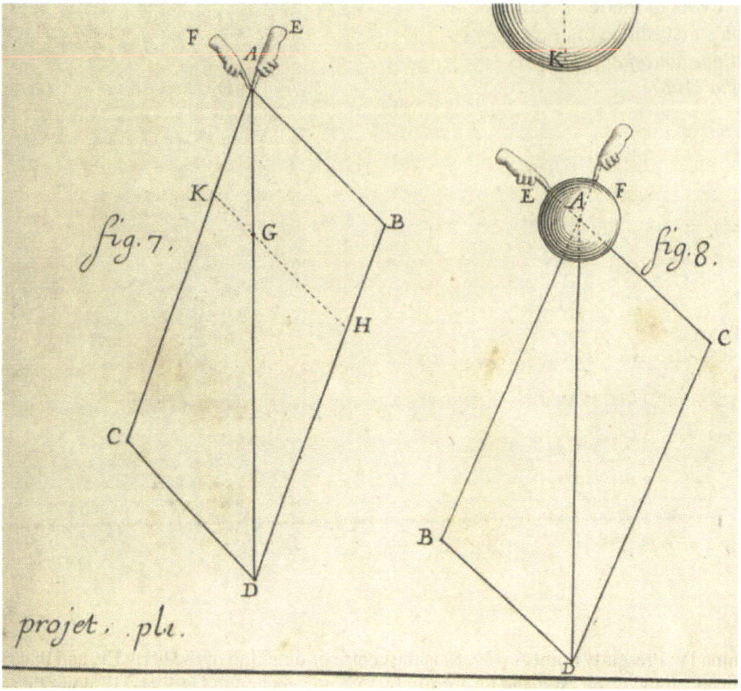

Fig. 32 Varignon's Figs. 7 and 8 for Lemmas III and IV (Varignon 1687, Pl. 1). Public domain

Fig. 33 Varignon's Fig. 9 for Lemma V (Varignon 1687, Pl. 1). Public domain

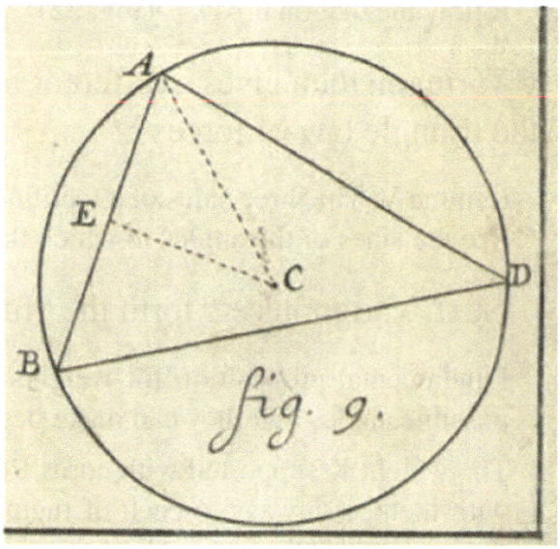

cords, taken reciprocally, make with the line of direction AK of this weight, For example, it is to power P as the sine of RAP is to the sine of RAK; and to the power R, as the sine of RAP is to the sine of PAK.[39]

[39] Varignon (1687, p. 11): *Proposition fondamentale des poids suspendus avec des cordes en quelque nombre qu'elles soient; et pour tous les angles possibles qu'elles peuvent faire entre-elles.*
Le poids K soutenu avec des cordes PB et CR par les puissances P et R, et en équilibre avec elles, est toujours à chacune d'elles, comme le sinus de l'angle PAR que leurs cordes font entre-elles,

We could write this as:

$$\frac{K}{P} = \frac{\sin PAR}{\sin RAK} \quad \text{and} \quad \frac{K}{R} = \frac{\sin PAR}{\sin PAK}$$

In order to illustrate this fundamental proposition, which is in fact the parallelogram law of forces, Varignon gives a series of drawings for seven particular cases (Fig. 34). Johann Bernoulli will mention this fundamental proposition in his *Essay d'une nouvelle théorie de la manœuvre des vaisseaux*, as we will see.

We will see that this lemma will often be used in architecture to build vaults.

4 In Dynamics, Demonstration of the Composition of the Forces by Means of the Composition of the Movements; The Catenary; The Introduction of Differential Calculus

4.1 *1687, Isaac Newton (1642–1727),* Philosophiae Naturalis Principia Mathematica

At this time, differential as well as integral calculus begin to develop under the impulse of Isaac Newton and of Gottfried Wilhelm Leibniz, and with Newton's *Principia,* the parallelogram law of forces becomes a fundamental law of all mechanics. However, we should not forget that Newton compounds movements, as we will see in his *Principia*:

> COROLLARY I.
>
> A body pushed by two forces conjoined will describe the diagonal of a parallelogram, in the same time that it would describe the sides, by those forces apart.
>
> If a body in a given time, by the force M impressed apart in the place A, should with an uniform motion be carried from A to B; and by the force N impressed apart in the same place, should be carried from A to C; complete the parallelogram ABCD, and, by both forces acting together, it will in the same time be carried in the diagonal from A to D. For since the force N acts in the direction of the line AC, parallel to BD, this force (by the second law) will not at all **alter the velocity generated by the other force M,** by which the body is carried towards the line BD. The body therefore will arrive at the line BD in the same time, whether the force N be impressed or not; and therefore at the end of that time it will be found somewhere in the line BD. By the same argument, at the end of the same time it AY will be found somewhere in the line CD. Therefore it will be found in the point D, where both lines meet. But it will move in a right line from A to D, by Law I.[40] (Fig. 35).

à chacun des sinus des angles RAK et PAK que font avec les lignes de direction AK de ce poids, chacune de ces cordes réciproquement prises. Par exemple, il est à la puissance P, comme le sinus de RAP au sinus de RAK; et à la puissance R, comme le même sinus de RAP au Sinus de PAK.

[40] Newton (1687, p. 13) (Eng. Trans. 1846, p. 84): *Corol. I. Corpus viribus conjunctis diagonalem parallelogrammi eodem tempore describere, quo latera separatis.*

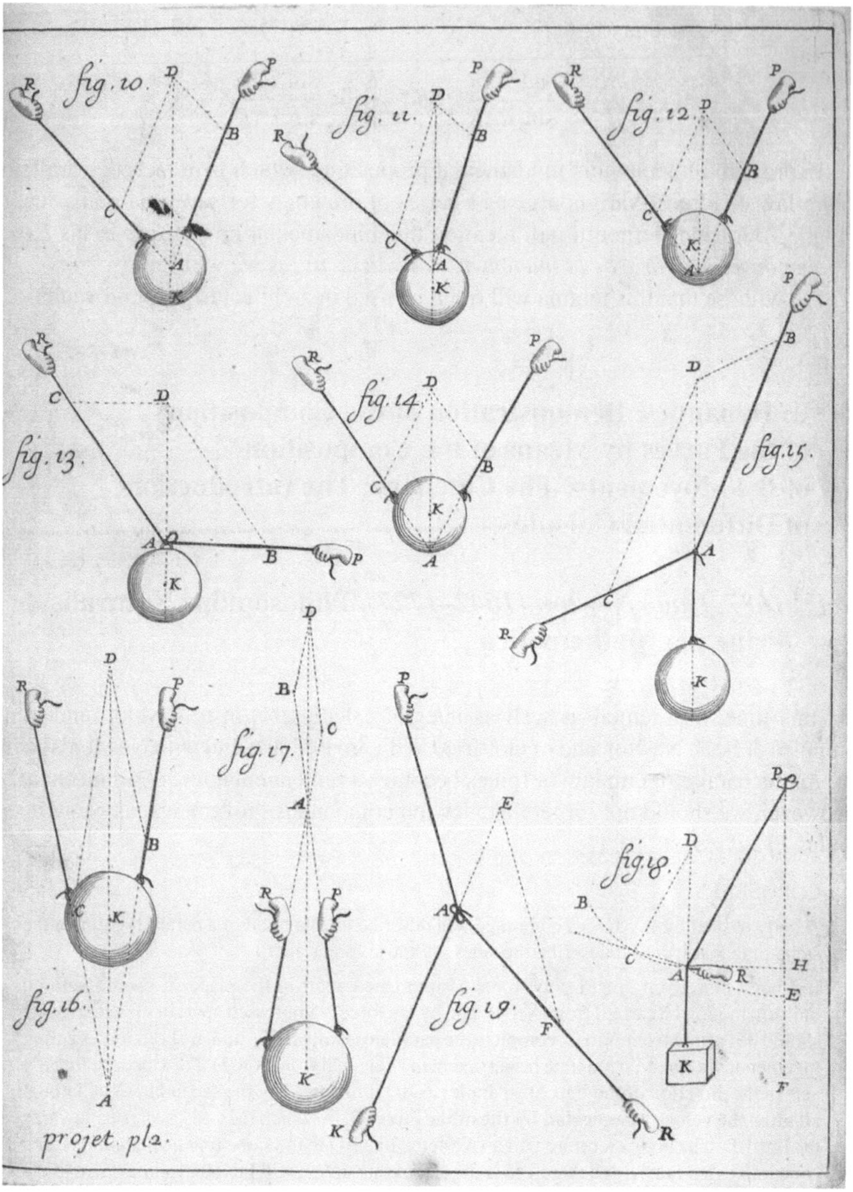

Fig. 34 Varignon'sFigs. 10, 11, 12, 13, 14, 15, 16 and 17 illustrations his "fundamental proposition" (Varignon 1687, Pl. 2). Public domain

Fig. 35 Newton's figure for his Corollary I (Newton 1687, p. 13). Public domain

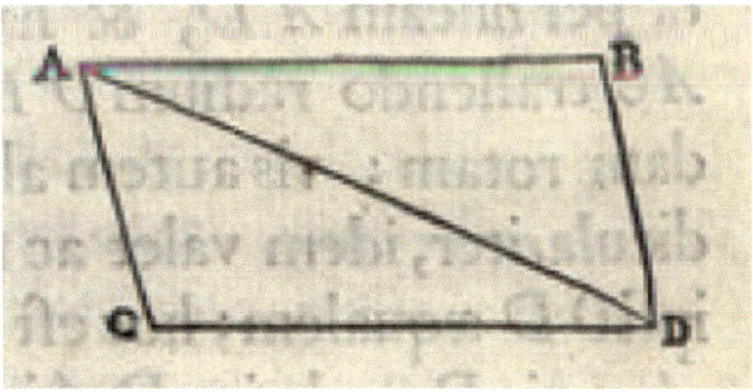

4.2 *Date Unknown, Johann Bernoulli,* Propositiones variae Mechanico-dynamicae

As we can see in Newton's Corollary V, he compounds the velocities generated by forces. Johann Bernoulli will note this, writing in Scholia V of his *Propositiones variae Mechanico-dynamicae*: "They are in error, those who confuse the composition of forces with that of movements".[41]

Johann considers, in his "various dynamic-mechanical propositions, the composition and decomposition of forces in this way:

> II. On the composition and resolution of forces
>
> Two forces, or powers, acting at the same point P; whose directions and intensities are given by the lines PA and PB, are said to be *compounded*, if once removed, we can replace them with a third force C, which acting alone in P produces the same strength or effort along the same direction PC, which the two forces A and B generated simultaneously. Similarly, the power C, expressed by the line PC, can be said to be *resolved* or *decomposed* into two PA and PB, which once the first removed, produce the same effect at the point P in the direction PC, which produced only the power PC.[42] (Fig. 36).
>
> III. Let now be three powers A, B and D, represented by the lines PA, PB, PD, acting at the point P able to move in all directions and which is in equilibrium under the simultaneous action of these three forces; We can by the previous definition consider that one of these

---

Si corpus dato tempore, vi sola M, ferretur ab A ad B, et vi sola N, ab A ad C, compleatur parallelogrammum ABDC, et vi utraque feretur id eodem tempore ab A ad D. Nam quoniam vis N agit secundum lineam AC ipsi BD parallelam, haec vis nihil mutabit velocitatem accedendi ad lineam illam BD a vi altera genitam. Accedet igitur corpus eodem tempore ad lineam BD sive vis N imprimatur, sive non, atque adeo in fine illius temporis reperietur alicubi in linea illa BD. Eodem argumento in fine temporis ejusdem reperietur alicubi in linea CD, et idcirco in utriusque lineæ concursu D reperiri necesse est.

41 Joh. Bernoulli (1742a, p. 256): *Peccant, qui confundunt compositione virium cum compositione motuum.*

42 Joh. Bernoulli (1742a, p. 253–254): *Duae vires, vel potentiæ, in commune punctum P agentes; quarum directiones et intensitates designentur per rectas_PA et PB, dicuntur* componi, *si, iis remotis, substítuitur vis tertia C, quæ sola eodem nisu vel conatu agit in P, secundum directionem PC, quo antea egerant simul duæ A et B. Atque vicissim, potentia C, expreffa per rectam PC,* resolvi *vel* decomponi *dicitur in duas PA et PB, quæ, remota priori, eundem nisum exerunt in punctum P, in directione PC, quam fecerat sola potentia PC.*

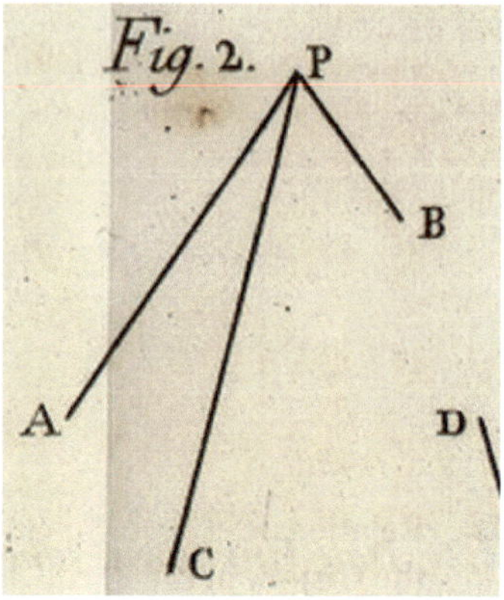

Fig. 36 Johann Bernoulli's composition of forces (1742a, Tab. LXXXV). Public domain

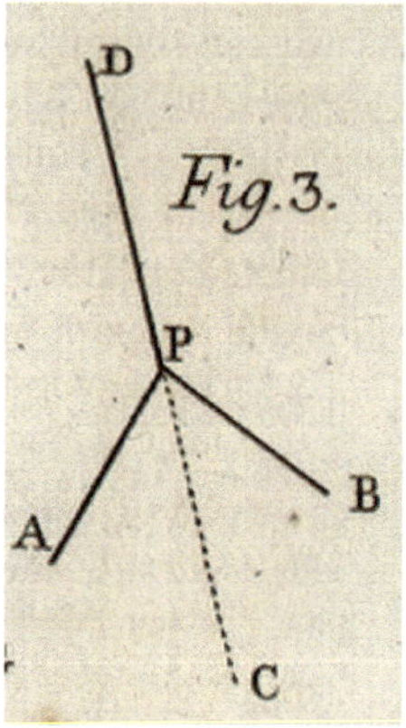

Fig. 37 Johann Bernoulli's equilibrium of forces (1742a, Tab. LXXXV). Public domain

> three forces is equal and of opposite direction to that which is composed of the other two. Let us consider in fact any two of them, for example that PA and PB are withdrawn and replaced by their equivalent PC, which must in any case resist alone the third power PD, and both must maintain the point P in equilibrium; moreover, they are opposite in direction and equal to each other.[43] (Fig. 37)

In all those figures, PC the diagonal is the resultant of PA and PB and they all are vectors, as we would say today. They have a length and a direction. Thus, Bernoulli has two things to demonstrate:

[43] Joh. Bernoulli (1742a, p. 253–254): *Sint jam tres potentiæ A, B et D, repræsentatæ per rectas PA, PB, PD, agentes in punctum P mobile in omnes plagas, quod in æquilibrio servetur a tribus istis potentiis simul agentibus; liquet, ex præcedenti Definitione, unamquamque ex istis tribus potentiis æqualem esse et directe oppositam ei, quæ ex reliquis duabus componitur. Finge enim duas quascunque, ex. gr. PA et PB removeri, earumque loco substitui æquipollentem PC, quæ utique sola resistet potentia tertiæ PD, et ambæ punctum P in æquilibrio retinere pergent; adeoque in direoctum erunt positæ, et altera alteri æqualis erit.*

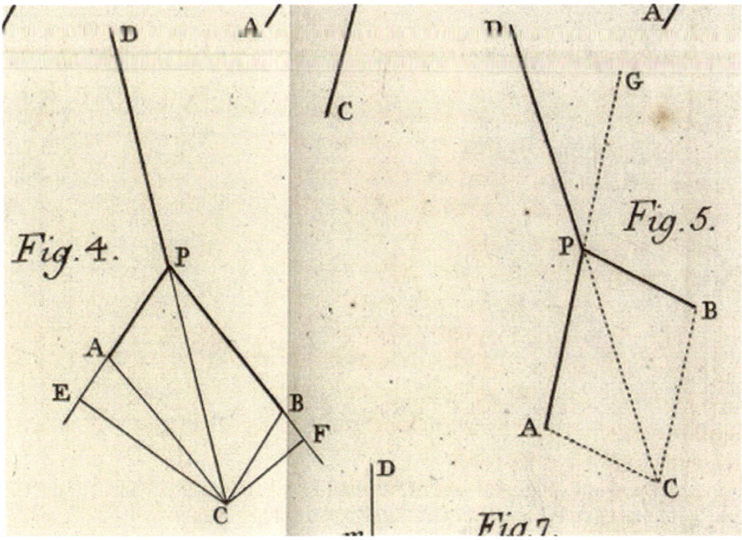

Fig. 38 Johann Bernoulli's decomposition of forces (1742a, Tab. LXXXV). Public domain

> 1. The power PC, compounded, in the sense specified above of the powers PA and PB, has the following position, as PA is to PB so is the sine of the angle BPC to the sine of the angle APC; or like the sine of the angle BPD to the sine of the angle APD. (Fig. 38, left)
> 2. I complete the parallelogram, the sides of which are PA, PB, and be the diagonal of this parallelogram PC itself; which represents in position and magnitude the power composed.[44] (Fig. 38, right)

This is the first tentative to demonstrate the parallelogram law of forces. Johann's son Daniel will go further in this direction.

4.3 *1691, Johann Bernoulli,* Solutio Problematis Funicularii, *the* Lectiones mathematicae, *and the Discreet Ingress of Differential Calculus*

In his *Solutio Problematis Funicularii* (1691) Johann Bernoulli only gives his results without demonstrations, but in his lectures to the Marquis de l'Hôpital, on the contrary, he gives the demonstrations. He first describes carefully the physical situation (Hypothesis 1) and gives a first property (Hypothesis 2), the tensions at the

[44] Joh. Bernoulli (1742a, p. 254): *Duo demonstranda jam sunt, 1°. Potentiam PC* (Fig. 4*), compositam, in* sensu *praedicto, ex potentiis PA et PB, eum habere situm, ut PA it ad PB sicuti sinus anguli BPC ad sinum anguli APC; seu ut sinus anguli BPD ad sinum anguli APD. 2°. Completo parallelogrammo, cujus latera sint PA, PB* (Fig. 5)*, fore ejus parallelogrammi diagonalem ipsissimam PC; quae et situm et quantitatem potentiæ compositæ designabit.*

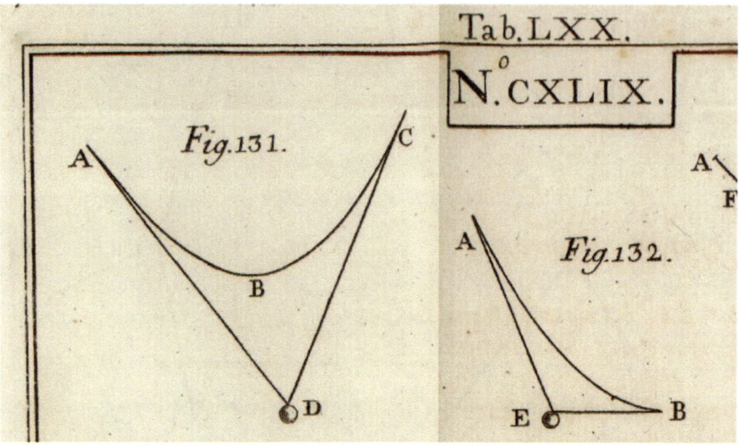

Fig. 39 Johann Bernoulli's Figs. 131 and 132 illustrating his Hypotheses 1 and 2 of the lectures to Marquis de l'Hôpital (Joh. Bernoulli 1742b, Tab. LXX). Public domain

fixing points are the same for a catenary or for a stone of the same weight attached to the same fixed points:

> Hypothesis 1: The thread, the rope, the chain or whatever that represents the curve is assumed to be perfectly flexible in each of its points and inextensible. That is, it does not undergo expansion due to its weight.
>
> Hypothesis 2: If the catenary curve ABC (Fig. 131 and 132) is suspended in any two of its points A and C, the forces required in A and C will be the same as those required to support a weight D equal to the weight of the ABC chain and suspended at competition point of two threads without weight AD, CD tangent to the curve ABC at points A and C.[45] (Fig. 39)

He then gives the well-known property that no matter what point the cord in held in, the form of the curve will not change. What is strange is the fact that he gives it as a hypothesis:

> Hypothesis 3: If the chain attached at its two ends A and C (Fig. 133) is fixed at one of its points, F, so that the portion AF can be removed, the curve that the rest of the chain represents, FBC, will not be changed; that is, the other points will retain the same situation they had before the fixation.[46] (Fig. 40)

He continues:

---

[45] Joh. Bernoulli (1742b, p. 492): *1° Filum, Funis, Catena, vel quicquid curvam repraesentat, supponitur in omnibus suis punctis flexile, & inextensibile, id est, quod ob gravitatem suam extensionem non patitur.*
2° *Si in duobus punctis A et C quibuscunque curva Catenaria ABC sustinetur, potentiae, quae in punctis A et C requiruntur, erunt eaedem que requiruntur ad sustinendum pondus D aequale ponderi catenae ABC, et in concursu D duorum filorum nullius gravitatis AD, AC, curvam ABC in punctis A et C tangentium appensum.*

[46] Joh. Bernoulli (1742b, p. 492): *Si catena duobus terminis A et C alligata in alio quovis puncto F figatur, ita ut pars AF auferri possit, curva, quam residua catena FBC repraesentat, non mutabitur; id est, reliqua puncta in eodem, quem ante fixationem habuerunt, situ manebunt.*

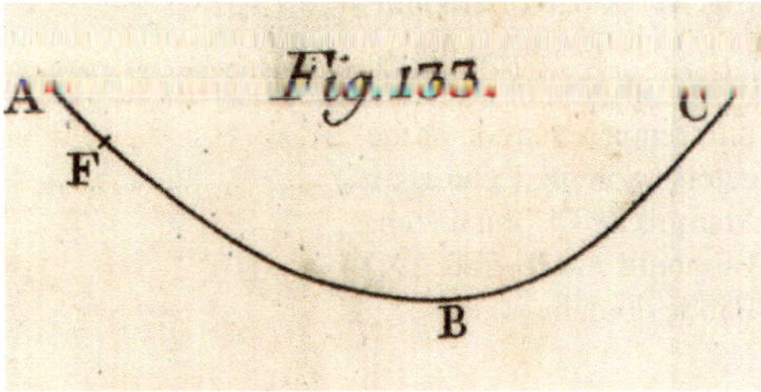

Fig. 40 Johann Bernoulli's Fig. 133 illustrating his Hypothesis 3 of the lectures to Marquis de l'Hôpital (Joh. Bernoulli 1742b, Tab. LXX). Public domain

> Hypothesis 4: ... After prolonging or shortening the chain BFA as much as we want, that is, wherever we take the fixing point in F, the force at the lowest point B will not be increased or decreased but will always remain the same and equal.[47]

Now comes the interesting demonstration that gives, for the first time, the differential equation of the catenary.

> Let BAa be the sought-after curve, whose lowest point is B; the axis, or the vertical line passing through B is BG; the tangent to the lowest point BE, which will be horizontal and let AE be the tangent at any point A. Let AG be the ordinate and LE parallel to the axis, let BG = x, GA = y, Gg = dx, Ha = dy; the weight of the rope or, since this is homogeneous, the length of the curve element BA = s. As, therefore, the power required in B is always the same and constant [by hypothesis 4] that we lengthen or shorten the chain AB, that we put this force or the straight line that represents it C = a.
>
> Now imagine that the weight of the rope AB is concentrated and suspended at the meeting point E of the tangent threads AE, BE, the power required [by hypothesis 2] at point B to support the weight E is the same as that you had to exercise before to support the BA rope. Now, [hypothesis 5] the weight E is to the force in B like the sinus of the angle AEB, or of its complement to two rights EAL, to the sinus of the angle AEL, that is to say like EL to AL.

Therefore, wherever the suspension point A of the curve is taken (the curve, in fact by the hypothesis 3, remains the same), the weight of the rope AB is to the force in B as EL is to AL, that is to say, $\frac{s}{a} = \frac{EL}{AL} = \frac{AH}{Ha} = \frac{dx}{dy}$, and inversely, $\frac{dy}{dx} = \frac{a}{s}$.[48] (Fig. 41).

---

[47] Joh. Bernoulli (1742b, pp. 492–493): *Hinc quantumcunque catena BFA, vel prolongetur, vel decurtetur, hoc est, ubicunque punctom ficationis F sumatur, potentiae in infimo puncto B neque augetur, neque diminuetur, sed semper manebit eadem et aequalis.*

[48] Joh. Bernoulli (1742b, pp. 493): *Sit BAa curva quaesita, cujus infimum punctum B; axis, vel linea verticalis, transiens per B, BG; tangens in infimo puncto BE, quae erit horizontalis; et in quocunque alio puncto A tangens AE. Ductis applicata AG et axi parallela EL, sit BG* = x, *GA* = y, *Gg* = dx, *Ha* = dy; *pondus catenae, vel quia aequaliter crassa, longitudo curvae BA = s. Quoniam itaque in puncto B semper aequalis et constans potentia requiritur [per hypoth. 4] sive catena BA continetur, sive decurtetur; si tilla potentia, vel linea recta ipsam exprimens C = a. Intelligatur nunc pondus catenae AB concentratum & appensum esse in concursu E filorum tangentium AE, BE, requititur [per hypoth. 2] in puncto B, Eadem potentia ad sustinendum pondus E, quae antea requirebatur ad sustinendam catenam BA. Verum [per hypoth. 5] pondus E est ad potentiam in B, ut sinus anguli AEL, id est, ut EL ad AL: Proinde ubicunque in curva punctum fixationis A sumatur [curva enim, per hypoth. 3, semper manet eadem] pondus catenæ AB est ad potentiam in B, ut EL ad AL, id est:* s: a = *EL: AL* = *AH:H*a = *dx:dy, & inverse* dy:dx = a:s.

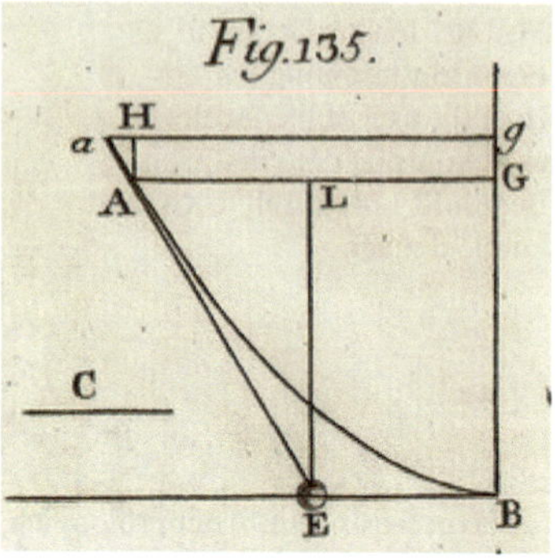

Fig. 41 Johann Bernoulli's Fig. 135 illustrating his demonstration of the differential equation of the catenary, in the lectures to Marquis de l'Hôpital (Joh. Bernoulli 1742b, Tab. LXX). Public domain

Without warning, Johann Bernoulli passes from the calculation of a physical proposition on the catenary, the sines being represented by segments, to the use of differential calculus, by applying as usual the similarity of the triangles aHA and AGB.

In this text, differential calculus enters into a problem that will be useful later in architecture, mainly in the construction of vaults. But it will be necessary to wait until 1770 before this problem is treated mathematically by builders.[49]

4.4 1695, Philippe De la Hire, Traité de mecanique

The demonstration of the parallelogram law of forces given in proposition XXI of Philippe De La Hire's "treatise on mechanics where we explain everything that is necessary in the practice of the Arts and the properties of heavy bodies which have a greater use in physics" is very similar to the one given by Johann Bernoulli in his "various propositions concerning mechanics and dynamics" (Joh. Bernoulli 1742a), written at un unknown date and certainly unknown to La Hire. We read:

> Prop. XXI: We must find three powers A, X, B which, by pulling a point K in three given directions CK, DK, EK, are in equilibrium with each other.
>
> Having taken some point E into one of the directions, as EK, and from this point having drawn the lines EF, EG parallel to the two other directions, and which meet them, extended if necessary, in F and G: I say that the three powers we are looking for, will be to each other as the three lines EF, EG or FK which is equal to it, and EK taken in the same order, and that will be parallel or are in the same direction as the powers to which they correspond. … thus, the three powers A, X, B will be between them like the sines of the three angles EKF, FEK, EFK taken in the same order, that is, the power A must respond to the sine of the angle EKF, the power X at the sine of the angle FEK, et finally the power B at the sine of the angle EFK.
>
> But we know that in every triangle the sine of the angles are between them like the sides opposite to these same angles; that is why there will be the same reason between the three powers A, X, B, as between the three sides EF, FK, EK of the triangle EFK taken in the

[49] For an in-depth study, see Patricia Radelet-de Grave, "The Panthéon's stability already questioned by Pierre Patte in 1770" (2015).

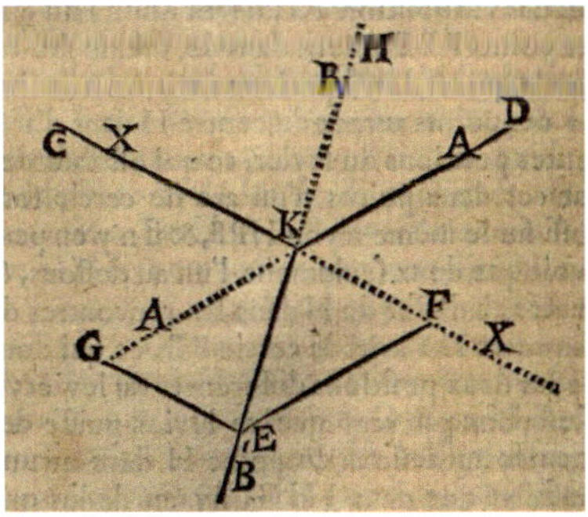

Fig. 42 La Hire's illustration of his Proposition XXI (La Hire 1695, p. 70). Public domain

same order; that is, the power A responds to the EF side, the X power to the FK side, and the B power to the EK side. These sides being taken on the very directions of the powers or on their parallels, which is what was to be demonstrated.[50] (Fig. 42)

La Hire immediately transforms this result into an equivalent one that is better suited to the construction of vaults:

Prop. XXII: The same things as in the previous one being posed, one can find in another way the ratio of the powers.

Taking the point K as centre and a radius of whatever length we want, we describe the circle CDE which cuts the three directions at points C, D, E; and through these points C, D, E having drawn the tangents to the circle or the perpendiculars to the directions, GCF, FDH, HEG which form the triangle FGH; I say that the powers AXB must be between each other as the sides of this triangle, taking for a power the side which is perpendicular to the direction of this power, that is to say that these three powers A, X, B will be to each other as the three sides FH, FG, GH taken in the same order.[51] (Fig. 43)

---

[50] La Hire (1695, pp. 70–72): *Prop. XXI: Il faut trouver trois puissances A, X, B qui, tirant un point K par trois directions données CK, DK, EK, soient en équilibre entr'elles.*
Aïant pris quelque point E dans l'une des directions, comme EK, et de ce point aïant mené les lignes EF, EG parallèles aux deux autres directions, et qui les rencontrent, étant prolongées, s'il est nécessaire, en F et G: Je dis que les trois puissances cherchées seront entr'elles comme les trois lignes EF, EG ou FK son égale, et EK qui sont prises dans le même ordre, et qui sont parallèles, ou font partie des directions des puissance ausquelles elles répondent ... donc les trois puissances AXB seront entr'elles comme les sinus des trois angles EKF, FEK, EFK pris dans le même ordre, c'est-à-dire que la puissance A doit répondre au sinus de l'angle EKF, la puissance X au sinus de l'angle FEK, et enfin la puissance B au sinus de l'angle EFK.
Mais on sçait que dans tout triangle les sinus des angles sont entr'eux comme les côtés opposés à ces mêmes angles; c'est pourquoi il y aura même raison entre les trois puissances AXB, qu'entre les trois côtés EF, FK, EK du triangle AFK pris dans le même ordre; c'est-à-dire que la puissance A répond au côté EF, la puissance X au côté FK, et enfin la puissance B au EK. Ces côtés étant pris sur les directions mêmes des puissances ou sur leurs paralleles. Ce qu'il falloit démontrer.

[51] La Hire (1695, Proposition XXII, p. 49–50): *Les mêmes choses que dans la précédente étant posées, on peut trouver d'une autre manière le rapport des puissances.*
Si du point K pour centre et pour raïon telle grandeur qu'on voudra, on décrit le cercle CDE qui coupe les trois directions aux points CDE; et par ces points CDE aïant mené des touchantes au cercle, ou des perpendiculaires aux directions, à sçavoir GCF, FDH, HEG qui forment le triangle FGH; je dis que les puissances AXB doivent être entr'elles comme les côtés de ce triangle, en

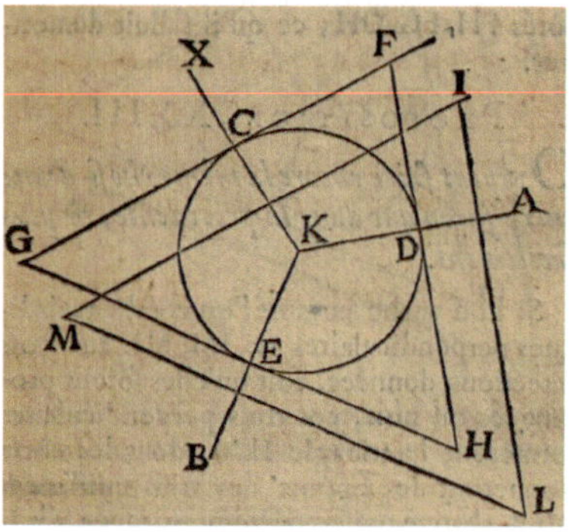

Fig. 43 La Hire's illustration of his Proposition XXII (La Hire 1695, p. 73). Compare this to Varignon's triangle law of forces in our Fig. 33. Public domain

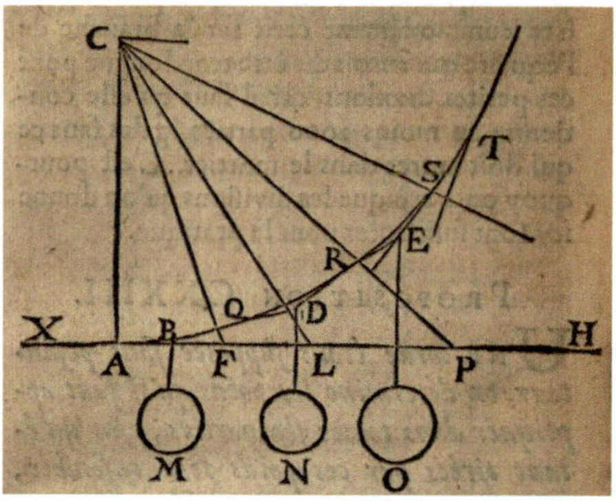

Fig. 44 La Hire's illustration of his Proposition CXXIII (La Hire 1695, p. 456). Public domain

We will now see that this form of the parallelogram law is actually selected for the culmination of the work of La Hire, namely, propositions CXXIII and CXXV dealing with the construction of vaults:

> Prop. CXXIII: A rope being assumed without gravity, we determine the weights to be applied to it in all its parts, so that, being drawn by all these weights together it takes the curved shape that we wish.[52] (Fig. 44)

In proposition CXXV La Hire demonstrates that the most stable arch is a reversed catenary:

---

prenant pour la puissance le côté qui est perpendiculaire à la direction de cette puissance, c'est-à-dire que ces trois puissances AXB seront entr'elles comme les trois côtés FH, FG, GH pris dans ce même ordre.

[52] La Hire (1695, p. 455): *Une corde étant supposée sans pesanteur, on détermine les poids qu'il faut lui appliquer dans toutes sas parties, afin qu'étant tirées par ces poids tous ensemble, elle prenne une figure courbe telle qu'on voudra.*

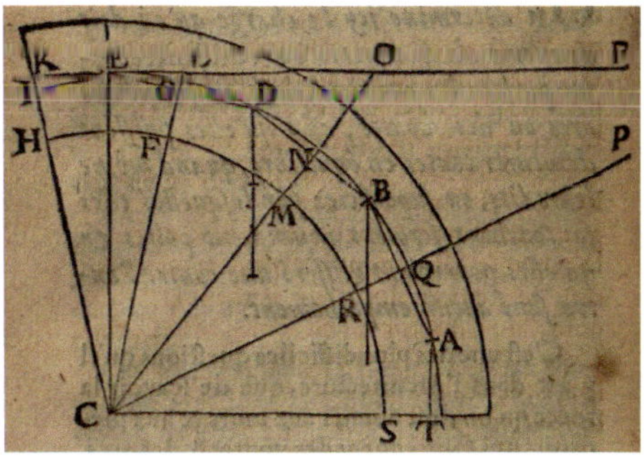

Fig. 45 La Hire's illustration of his Proposition CXXV (La Hire 1695, p. 456). Public domain

> Prop. CXXV: We ask here the load that must be given to each stone or archstone, as the workers speak, of which arches or vaults are formed, so that they can remain all in equilibrium, even if their beds, or surfaces by which they touch each other would be infinitely polished and that they could slide against each other without any impediment.[53] (Fig. 45)

We will see that Bernard Forest de Bélidor, the author of *La Science des ingénieurs dans la conduite des travaux de fortification et d'architecture civile* (Bélidor 1729) uses that property in order to find the best form of a vault, and that later Euler will prove mathematically that the best shape for a vault is the catenary or hanging chain.

4.5 1704, Jacob Bernoulli, Problema de curvatura fornicis, cujus partes se mutuo propio pondere suffulciunt sine opere caementi

Using differential calculus,[54] Jacob calculates the equilibrium of an infinitely thin vault built without cement:

> Let ABC be an arch constructed with cut stones DKLE of the figure of an oblong rectangle parallelepiped having KL slightly larger than DE, its width DK sufficiently narrow or infinitely small, and name EN = dx, ND = dy, DE [the weight of the particle DL] = ds, and the weight of[55] BE = s. (Fig. 46)

He writes equilibrium to translation and for rotation for a segment DKLE of an arch and deduces the two following equations:

---

[53] La Hire (1695, p. 465): *On détermine icy la charge qu'on doit donner à chaque pierre ou voussoir, comme parlent les ouvriers, dont on forme des arcs ou des voutes, afin qu'elles puissent demeurer toutes en équilibre, quand même leurs lits, ou superficies par lesquelles elles se touchent seroient infiniment polies et qu'elles pouroient glisser l'une contre l'autre sans aucun empêchement.*

[54] For a detailed study of this work of Jacob's, see "The *De curvatura fornicis* of Jacob Bernoulli: The Introduction of Infinitely Small into the Calculus of Arches", Chap. 5 in this present volume.

[55] Jac. Bernoulli (1744, p. 1119): *Sit Fornix ABC constructus ex lapidibus cæsis DKLE figuræ fere parallelogrammæ rectangulæ oblongæ, nisi quod KL tantillo major sit quam DE, latitudinis DK valde exiguæ vel infinite parvæ; et vocetur EN* = dx, *ND* = dy, *DE [pondus particulæ DL]* = ds, *adeoque pondus ipsius BE* = *s.*

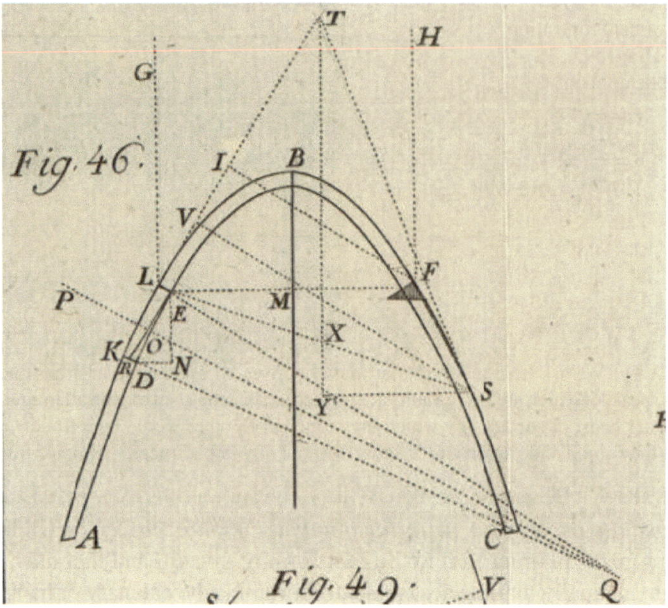

Fig. 46 Jacob Bernoulli's Fig. 46 (1744: Table XLVIII). Public domain

$$\frac{ds}{s} = \frac{d^2x}{dx} \quad \text{and} \quad \frac{ds}{2s} = \frac{d^2x}{dx}.$$

Despite an error in the last equation, he finds the equilibrium equation of this vault thanks to the equation of balance to translation and the expression of the radius of curvature that he himself had discovered between 1689 and 1691:

$$dy = \frac{adx}{\sqrt{x^2 - a^2}},$$

which characterises the catenary curve as found by David Gregory, who had, in fact, found this equation by means of the calculation of the fluxions which reigned in England since its invention by Newton (Fig. 47).

Gregory posed the question as follows: "Determine the relationship between axial fluxion and ordered fluxion in the catenary curve".[56]

This equation for the catenary, also found by Jacob's brother Johann, as we have seen, being equivalent to the one Jacob himself just found for the arch, it became obvious to him that the most stable arch had the shape of an inverted catenary.

[56] Gregory (1697, p. 637): *Prop. I. Problema. Invenire relationem inter fluxionem axeos et fluxionem ordinatae in Curva Catenaria.*

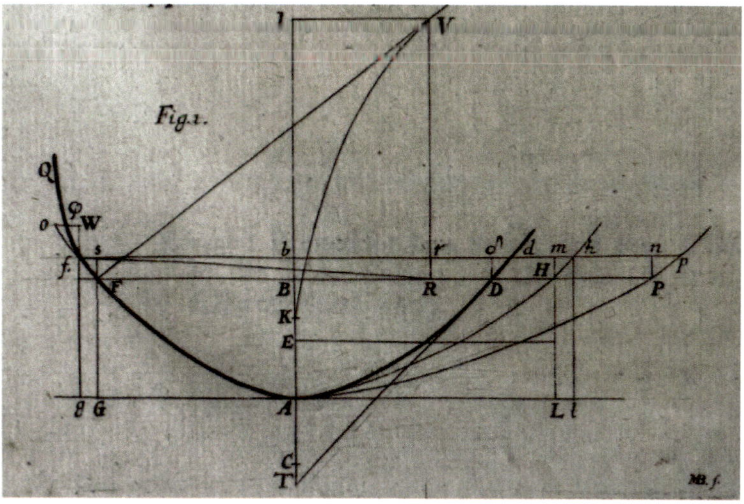

Fig. 47 David Gregory's figure of the catenary (Gregory 1697, Fig. 1). Public domain

Fig. 48 Clairaut's figure of the parallelogram law (Clairaut 1734, Tab. IV). Public domain

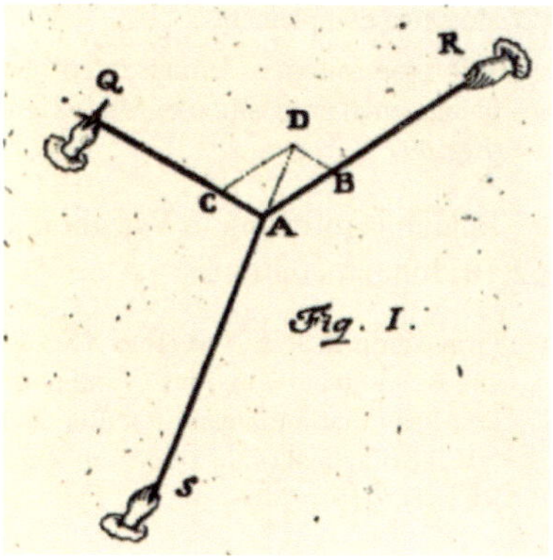

4.6 *1743, Jean-Baptiste Clairaut, Methodus generalis inveniendi catenarias*

In this text Clairaut gives a new form to the law of the parallelogram. This time it is an algebraic one.

> If three powers Q, R, S are in equilibrium, and their directions are QA, RA, AS: I say that these three powers will be between them like the sines of the angles SAR, QAS, QAR.[57] (Fig. 48)

Today we would write this in the following way:

---

[57] Clairaut (1734, p. 270): *Si tres potentiæ Q, R, S sint in equilibrio, et directio virium QA, RA, AS: Dico has tres potentia inter se fore, uti sinus angulorum SAK, QAS, QAR.*

$$\frac{\sin SAR}{\sin QAS} = \frac{QA}{AR}; \frac{\sin QAS}{\sin QAR} = \frac{AR}{AS}; \frac{\sin QAR}{\sin SAR} = \frac{AS}{QA}.$$

4.7 *1714, Johann Bernoulli,* Essay d'une nouvelle theorie de la manœuvre des vaisseaux

With this text, we see the parallelogram law entering a new practical domain: navigation.

> Chapter XV. On the axis of the equilibrium of the pressures of the wind on a curved sail determined by a theorem which is demonstrated by some propositions of statics. Theorem. The line CD, which also bisects the angle ACB, will be the average direction and the axis of equilibrium of the pressures.
>
> To demonstrate this theorem, I will use some propositions deduced from common statics.
>
> Proposition I. If three forces A, B, F (Fig. XIX) together pull a point C, and they are in mutual equilibrium, I say that if the direction of one FC is extended in D. The three forces A, B, and F will respectively be as the sines of these three angles DCB, DCA and ACB or of its two complements.
>
> This proposition is demonstrated in almost all mechanics books. See among others the fundamental proposition of Mr. Varignon on p. 11 of his *Projet d'une nouvelle mécanique*.[58] (Fig. 49)

Johann is referring to Varignon's "fundamental proposition", which we quoted in §2.10. Johann continues:

> Proposition II. If ACDB (Fig. XX.) is a wire or rope attached or supported at both ends A and B, while at two points C and D it is bent or stretched by two powers or forces CR and DS; I say that the mean direction of these two powers or their axis of equilibrium will be XV, the diagonal of the trapezoid which is formed by the extension of the lines AC, BD and RC, SD.[59] (Fig. 50)

---

[58] Joh. Bernoulli (1714, pp. 122–123): *Chapitre XV, De l'axe de l'équilibre des impressions du Vent sur une Voile courbe déterminé par un Théorème que l'on demontre par quelques propositions de Statique. Theoreme. La ligne CD, qui coupe l'angle ACB en deux également, sera la direction moyenne et l'axe de l'équilibre des impressions.*
Pour démontrer ce théorème, je me servirai de quelques propositions déduites de la statique commune.
Proposition I. Si trois forces A, B, F, (Fig. XIX) tirent ensemble un point C, et qu'elles soient en équilibre mutuellement: Je dis que si la direction de l'une FC est prolongée en D; Les trois forces A, B, et F seront respectivement comme les sinus de ces trois angles DCB, DCA et ACB ou de son complément à deux droits.
Cette proposition se trouve démontrée dans presque tous les livres de méchanique: Voyez entre autres la proposition fondamentale de Mr. Varignon dans son projet d'une Nouvelles Méchanique p. 11.

[59] Joh. Bernoulli (1714, p. 123): *Proposition II. Si ACDB (Fig. XX.) est un fil ou une corde attachée ou soutenue aux deux extrémités A et B, pendant qu'aux deux points C et D; elle est bandée ou tendue par deux puissances ou forces CR et DS; Je dis que la direction moyenne de ces deux puissances*

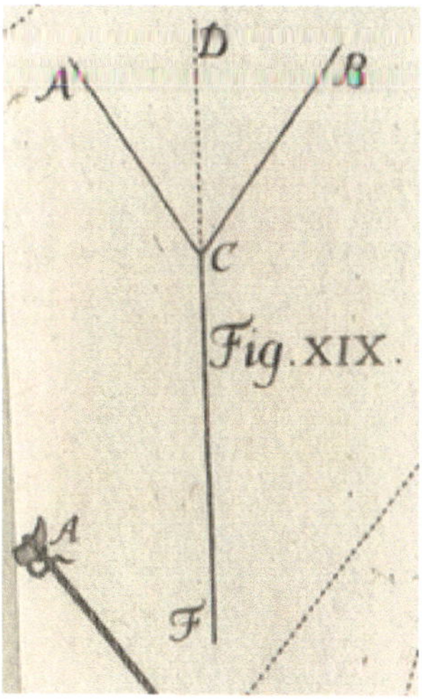

Fig. 49 Johann Bernoulli's figure XIX of the parallelogram law (Joh. Bernoulli 1714, Tab. 7). Public domain

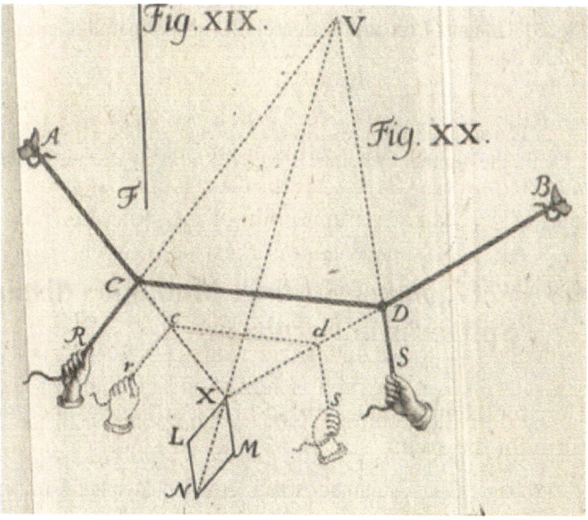

Fig. 50 Johann Bernoulli's figure XX of the parallelogram law (Joh. Bernoulli 1714, Tab. 7). Public domain

Proposition III. The wire or rope ACDEFB (Fig. XXI) attached or supported at the ends A and B and any several powers CR, DS, ET, FP etc. applying traction the wire, so that it takes the form of a polygon ACDEF.

I say 1) The mean direction VX or the axis of equilibrium of these forces or powers will pass through the crossing X of the extreme portions of the chord AC, BF extended, and through the centre of gravity O of the points L, M, N, K, after having drawn the lines XL, XM, Xn, XK, parallel and equal respectively to CR, DS, ET, FP.

---

ou leur axe d'équilibre sera XV diagonale du trapeze, qui se forme par la prolongation des lignes AC, BD et RC, SD.

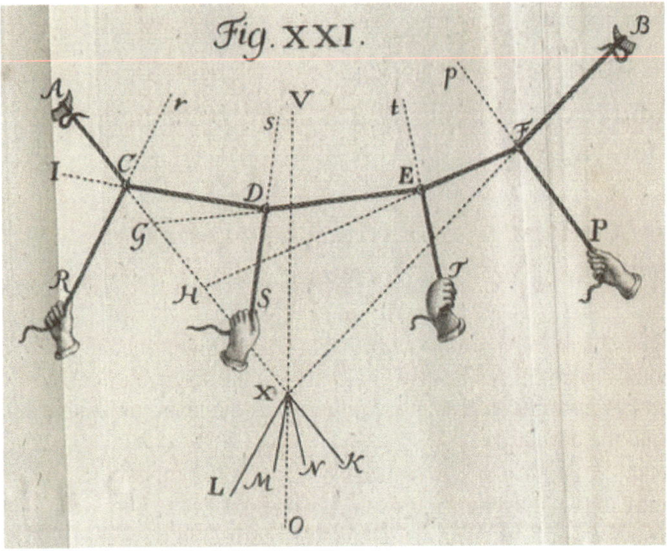

Fig. 51 Johann Bernoulli's figure XXI of the parallelogram law (Joh. Bernoulli 1714, Tab. 8). Public domain

> 2) The amount of force or average power will be expressed by the fourth proportional of the unit, the number of points L, M, N, K and the distance from their centre of gravity to point X.[60] (Fig. 51)

4.8 *1717, James Stirling,* Methodus disponendi quotcunque Sphaeras in Fornicem

Here for the first time, James Stirling arranges spheres in order to find the most stable form for the arch:

> Problem. Given a number whatever of equal spheres to be disposed into a vault in such a manner that they sustain each other's gravity.
>
> The centre of the highest sphere being A, the centres of the contiguous ones being B, b. Draw DC crossing AB in C and parallel to Ab, and let the radii AD, BG be perpendicular to the horizon. Join D and G and in it, extended, let GF = AC. Draw BF and in it let BE

[60] Joh. Bernoulli (1714, pp. 127–128): *Proposition III. Le fil ou la corde ACDEFB (Fig. XXI.) attachée ou soutenue aux extrémintés A & B étant tendue par plusieurs puissances CR, DS, ET, FP* etc. *quelconques, ensorte qu'elle prenne la forme d'un polygone ACDEF. Je dis que 1°. La direction moyenne VX ou l'axe d'équilibre de ces forces ou puissances passera par le concours X des portions extrêmes de la corde AC, BF prolongées, et par le centre de gravité O des points L, M, N, K, aprés avoir tiré les lignes XL, XM, Xn, XK, parallèles et égales à leurs respectives CR, DS, ET, FP.*
2°. La quantité de la force ou puissance moyenne sera exprimée par la quatrième proportionnelle de l'unité, du nombre des points L, M, N, K et de la distance de leur centre de gravité au point X.

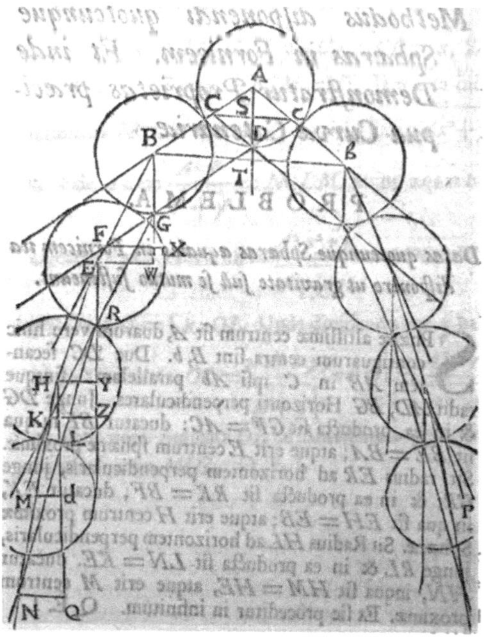

Fig. 52 Stirling's arrangement of spheres to form a stable arch (Stirling 1717, p. 12). Public domain

= BA, and let E be the centre of the next sphere. Let the radius ER be perpendicular to the horizon, join G and R, and in it, prolonged, let RK = BF; trace EK in which EH = EB; and let H be the centrum of the next sphere. Let the radius HL be perpendicular to the horizon, join R and L and in it be prolonged let LN = KE. Trace HN in which HM = HE, and let M be the next centre. And proceed like this infinitely. Q.E.I.[61] (Fig. 52)

Proof. All those spheres are submitted to three forces, **and we know from mechanics that three forces in equilibrium are in the same ratio to each other as three lines respectively parallel to the directions of the forces and ending at their mutual ends**. The sphere AD tends to draw the weight from A to D, and the neighbouring spheres act in the lines AB, Ab. More if the radius AD shows the weight of the absolute sphere, AC shows the force with which the sphere AD presses the sphere BG from B to E. Similarly, BF shows the force with which the sphere ER presses the sphere HL, and EK the force with which the sphere HL presses the next sphere. And so on in infinitum. The force BF is decomposed in the forces BG, GF: or BG, AC: and equivalent to the same. If BG is really the force of gravity of the spheres and AC the force with which the sphere AD presses the sphere BG: and those two forces compose all the weight to sustain the sphere ER, namely each sphere must sustain all the weights that sustain the next superior sphere one with its absolute weight. And the force

[61] Stirling (1717, p. 11): *Problema. Datas quotcunque Sphaeras æquales in Fornicem ita disponere ut gravitate suâ se mutuo sustineant.*
Sphaeræ altissimæ centrum sit A, duarum vero huic contiguarum centra sint B, b. Duc DC secantem AB in C ipsi Ab parallelam: sintque radii AD, BG Horizonti perpendiculars. Junge DG et in ea producta sit GF = AC: ducatur BF in qua sit BE = BA; atque erit E centrum sphæræ proximæ. Sit radius ER ad horizontem perpendicularis, junge GR, et in ea producta sit RK = BF; ducatur EK, in qua sit EH = EB; atque erit H centrum proximæ Sphæræ. Sit Radius HL ad horizontem perpendicularis, junge RL et in ea producta sit LN = KE. Ducatur HN, inqua sit HM = HE, atque erit M centrum proximæ. Et sic proceditur in infinitum. Q. E. I.

> EK equilibrates the forces ER, RK or ER, BF that is the weight of the sphere ER and of all the weights that it supports. And so on in infinitum, you see that in this way that everywhere the position of the spheres can be compared, in order that any of them can sustain all the weight the immediate superior to it was sustaining, each with its own absolute weight: and I add that those spheres are in equilibrium and that their weights sustain each other.[62]

Stirling's proof uses the law of the triangle of forces, as we can see from the first lines of the proof in the sentence, which we have set in bold, and, for example, from the triangle BFG. Some years later, in 1748, Giovanni Poleni will give a similar figure based on the parallelogram, as we shall see in §3.14 below.

4.9 *1729, Bernard Forest de Bélidor,* La Science des ingenieurs dans la conduite des travaux de fortification et d'architecture civile

Bélidor criticizes Blondel's use of the law given by Derand in 1643 to calculate the thickness of the columns supporting a vault, which we discussed in §2.6. Bélidor gives a figure in which we recognize Varignon's and de la Hire's triangle law of forces (Fig. 53, right) and next to it the parallelogram law (Fig. 53). Bélidor identifies both laws.

Bélidor gives also another important answer to the following question:

> Propositon 5, Problem: Find what curve should be given to an arch, so that all segments, being equal in weight, are in equilibrium.

[62] Stirling (1717, p. [12]): *Demonstratio. Omnes hae Sphaeræ triplici potentia urgentur: et constat ex Mechanica, quod tres potentiæ in aequilibrio consistentes eam ad invicem rationem habent, quam tres rectæ potentiarum directionibus respective parallelæ et ad ipsarum intersectiones mutuas terminatæ. Sphaera AD urgetur gravitate ab A versus D tendente, et actione sphærarum contiguarum in rectis AB, Ab. Adeoque si radius AD exponat gravitatem Sphaeræ absolutam, AC exponet vim Sphæra AD urget Sphæram BG à B versus E. Item BF exponet vim qua Sphæra ER urget sphæram HL, EK vim qua Sphæra HL urget Sphæram sequentem. Et sic in infinitum. Vis BF resolvitur in vires BG, GF: id est BG, AC: et iisdem æquipollet. Est vero BG vis gravitatis globi et AC vis qua Sphæra AD urget Sphæram BG: et hæ duæ vires componunt omne pondus sustinendum à sphæra ER; scilicet unaquæque Sphæra sustinere debet omne pondus quod sustinet proxime superior Sphæra una cum gravitate absoluta. Et vis EK æquipollet viribus ER, RK vel ER, BF id est vi gravitatis Sphæræ ER et omni pondere quam eadem sustinet. Atque sic pergendo in infinitum, videbis ubique Sphæratum situ mita esse comparatum, ut earum quælibet sustinere potest omne pondus quod sustinet Sphæra immediate superior, una cum ipsius pravitate absoluta: adeoque hæ Sphæræ erunt ipsius gravitate absoluta: adeoque hæ Sphæræ erunt in equilibrio, et vi gravitatis sese sustinebunt.*

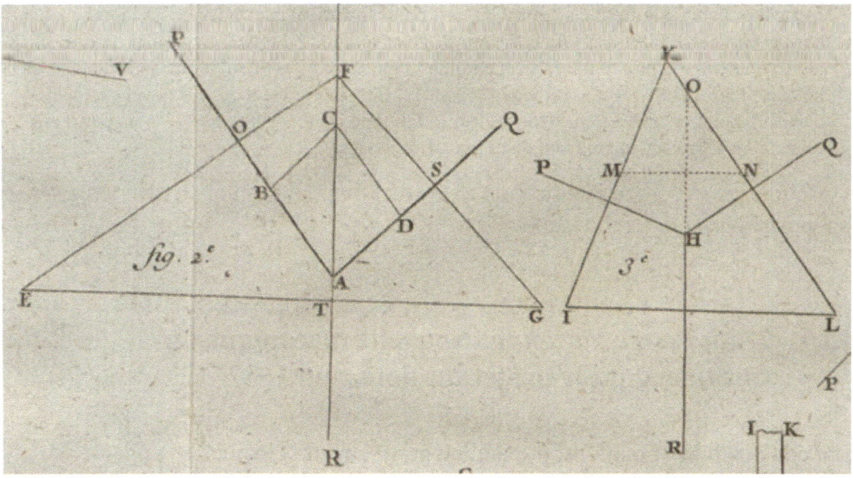

Fig. 53 Bélidor's for the triangle law and the parallelogram law (Bélidor 1729, Bk. II, Pl. IV). Public domain

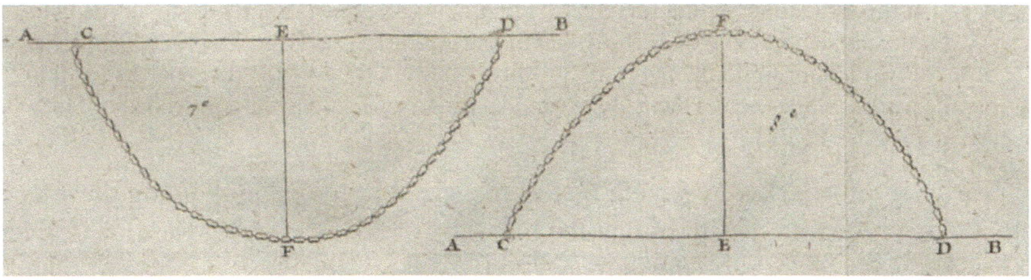

Fig. 54 Bélidor's representation of the hanging chain and its inversion (Bélidor 1729, Bk. II, Pl. VI). Public domain

40. If we draw on a vertical plane a line AB (Fig. 7) parallel to the horizon, and that at two points C and D, taken in this line, we attach the ends of a chain composed of small links, and let them have the freedom to take the situation that suits them, whose form represents that which should be given to a vault, so that all segments are in balance, although equal in weight.[63] (Fig. 54)

We know that we are not disturbing anything in the balance of powers by only changing their direction to the opposite direction. Thus, assuming that all the links are united in such a way that they are not able to disturb the curvilinear figure that they compose all together, if we

[63] Bélidor (1729, II, p. 43): *Proposition cinquième, Problème: Trouver quelle est la courbe qu'il conviendrait de donner à une voûte, pour que tous les voussoirs, étant égaux en pesanteur, soient en équilibre.*
40. Si l'on tire sur un plan vertical une ligne AB (Fig. 7) *parallele à l'horizon, et qu'à deux points C et D, pris dans cette ligne, on attache les extrêmités d'une chaîne composée de petits couplets, et qu'on leur laisse la liberté de prendre la situation qui leur convient, dont la figure représente celle qu'il faudroit donner à une voûte, pour que tous les voussoirs soient en équilibre, quoiqu'égaux en pesanteur.*

> rotate the chain CFD around the line CD, as around an axis, to take the opposite situation, but always vertical, CFD, all the links, keeping between them the same situation as before, will tend towards the centre of the earth according to the same lines of direction, and whether we increase their gravity or not, provided this increase is the same, they will always remain in balance, and they will not make any effort to fall than if they were not maintained by some cause which prevents them from disturbing themselves.[64]

Bélidor gives a good justification to the fact that the form of an inverted catenary is the best possible for the vault.

4.10 *1725, Pierre Varignon,* Nouvelle mecanique ou statique, dont le projet fut donne en 1687

The composition of motions, or parallelogram law of velocities.

> Lemme I, In order to prepare imagination for compounded movements, let us conceive the point A (Fig. 1) without weight moving uniformly towards B along the straight line AB, while this line is moving uniformly towards CD along AC, remaining always parallel to itself, that is, always making the same angle with the fixed line AC. Of these two motions begun at the same time, let the speed of the first be to the speed of the second as AB is to AC. Of the parallelogram ABCD, I say that by the concurrence of the two producing forces of these two motions into the moveable A, this point will traverse the diagonal AD of this parallelogram, while each of them alone would have travelled along the corresponding sides AB and AC.[65] (Fig. 55)

Varignon proves this by saying that as the speed of the point A going towards B along AB is equal to the speed that the line AB [or KH] has going towards CD, the point A always remains on the diagonal AD.

> Lemma II. If point A (Fig. 1) without gravity is simultaneously and uniformly pushed by two forces or powers E and F, both acting on it, along the lines AC, AB, which make between

---

[64] Bélidor (1729, II, p. 43): *On sçait qu'on ne dérange rien dans l'équilibre des puissances en changeant seulement leur direction en sens contraire. Ainsi, dans la supposition que tous les couplets sont unis à ne pouvoir se déranger de la figure curviligne qu'ils composent tous ensemble, si l'on fait tourner la chaîne CFD sur la ligne CD, comme sur un axe, pour prendre sa situation opposée, mais toujours verticale, CFD, tous les couplets gardant entr'eux la même situation qu'ils avoient auparavant, tendront au centre de la terre selon les même lignes de direction, et soit qu'on augmente leur pesanteur ou non, pourvu que cette augmentation soit la même, ils se maintiendront toujours en équilibre, et ils ne feront pas plus d'effort pour tomber que s'ils n'étoient point entretenus par quelques cause qui les empêche de se déranger.*

[65] Varignon (1725, I, p. 13): *Lemme I. Pour préparer l'imagination aux mouvemens composez, concevons le point A sans pesanteur uniformement mû vers B le long de la droite AB, pendant que cette ligne se meut uniformément vers CD le long de AC, en demeurant toujours parallele à elle-même, c'est à dire, faisant l'angle toûjours le même angle quelconque avec cette ligne immobile AC: de ces deux mouvemens commencez en même tems, soit la vîtesse du premier à la vîtesse du second, comme les côtez AB,AC, du parallelogramme ABCD, je dis que par le concours des deux forces productrices de ces deux mouvemens dans le mobile A, ce point parcourra la diagonale AD de ce parallelogramme, pendant le temps que chacune d'elles lui en auroit fait parcourir seule chacun des côtez AB, AC, correspondans.*

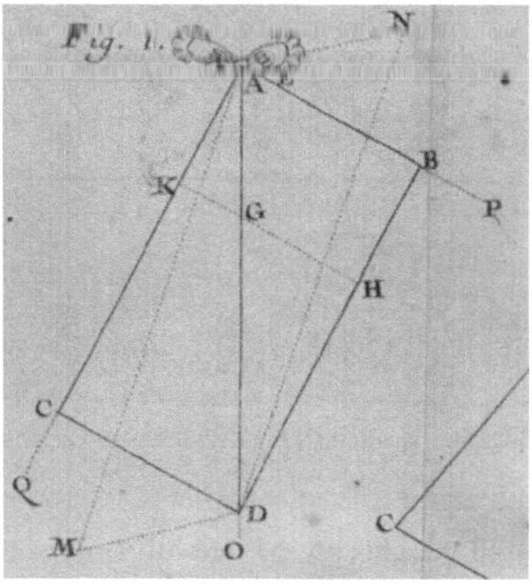

Fig. 55 Varignon's parallelogram of motions (Varignon 1725, Pl. I). Public domain

> them any angle CAB, and that the force or power E, acting along AC, is to the force or power F, acting along AB, as AC is to AB. This point A, by means of these two forces E, F, without the aid of any moving line, will traverse the diagonal AD of the parallelogram ABCD in the same time as they would have made it traverse separately the sides AC, AB, which they are assumed to be proportional.[66]

His demonstration is very close to the previous one and, like Newton's demonstration, does not compose forces but velocities.

Although in the rest of his book, Varignon often speaks of funicular polygons, he does not mean the question of the chain or of the catenary but rather of wires or ropes without weight on which different sets of forces act. Varignon is then looking for the resultant of these forces. Even in the cases of his Figs. 89 and 98 (Fig. 56), he is not looking for the catenary.

However, by considering in this way a very large number of particular configurations of sets of forces acting on a weightless rope, Varignon arrives at the following result:

> Theorem X: I. Any two powers K, L (Figs. 90, 91) directed at will, and applied at any two points C, D, of a loose and perfectly flexible cord ACDB, fixed by the two ends to two nails or hooks A, B, and remaining in equilibrium between them, as in Th. 8. 9. From any point S are drawn SE, SF, SG, parallel to the three sides AC, CD, DB, of the polygon ACDB which these powers cause to be made by this rope; and from a point F also taken at will on SF, let FE, FG, be drawn parallel to the directions CK, DL, of the powers K, L, until these two lines meet SE, SG, in E, G. That done, I say that in this case of equilibrium the powers K, L, will be between them like EF, FG, that is, K . L:: EF. FG.

---

66 Varignon (1725, I, pp. 14–15): *Lemme II, Si le point A sans pesanteur est poussé en même tems et uniformément par deux forces ou puissances E et F, toutes employées sur lui, suivant des lignes AC, AB, qui fassent entr'elles quelqu'angle CAB que ce soit, et que la force ou puissance E suivant AC, soit à la force ou puissance F suivant AB, comme AC est à AB. Ce point A par le concours de ces deux forces E, F, sans le secours d'aucune ligne mobile, parcourra la diagonale AD du parallelogramme ABCD dans le même tems qu'elles lui en auroient fait parcourir separement les cotez AC, AB, qu'on leur suppose proportionnels.*

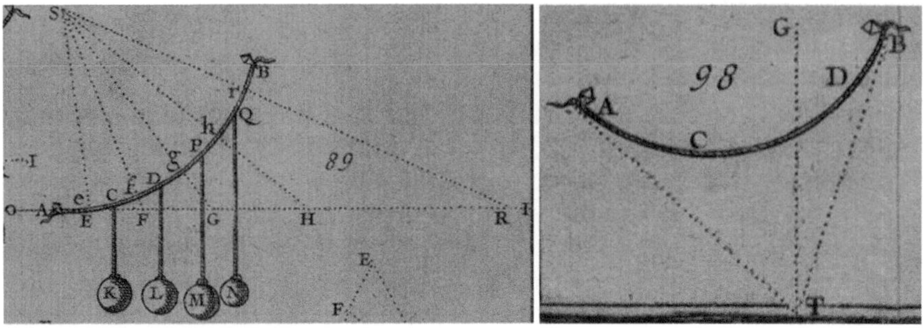

Fig. 56 Varignon's Figs. 89 and 98 (Varignon 1725, Pl. 11). Public domain

> II. Reciprocally the cord ACBD (Figs. 90, 91) being given of position, that is to say, the polygon which it forms being given, if from a point S taken at will, we make SE, SF, SG, parallel to the three sides AC, CD, DB, of this polygon; and that from a point F taken at will on SF, we lead any two straight lines FE, FG, which meet SE, SG in E, G: two powers K, L, which would be between them like these two line FE, FG, and which would have their directions CK, DL, parallel to these same lines, each to each, will retain the cord ACDB in this given position, remaining in equilibrium therebetween.[67] (Fig. 57)

This result will become fundamental a little less than 150 years later when Culmann will invent graphic statics.[68]

He follows this with Corollary I:

> Corollary I: Let there be as many powers K, L, M, N, etc. (Figs. 92 and 93) as we want to apply along the directions CK, DL, PM, QN, as we choose, and as many arbitrary points C, D, P, Q, etc. of the loose and perfectly flexible rope A, C, B, which are in equilibrium between them. From any point S let there be drawn SE, SF, SH, SR, etc. parallel to the sides AC, CD, DP, PQ, QB, etc., each to each of the polygon ACDPQB as these powers K, L, M, N, etc. give form (Th. 1, Cor. II.) to this string ACB. Let also EF, FG, GH, HR, etc. be parallel to the directions CK, DL, PM, QN, etc. of these same powers, which meet SE, SF, SG, SH, SR, etc. in E, F, G, H, R, etc.

[67] Varignon (1725, I, pp. 190–191): *Théorème X: I. Deux puissances quelconques K, L, dirigées à volonté, et appliquées en deux points quelconques C, D, d'une corde lâche et parfaitement flexible ACDB, attachée par les deux bouts à deux clous ou crochets A, B, demeurant encore en équilibre entr'elles, comme dans les Th. 8. 9. D'un point quelconque S soient faites SE, SF, SG, paralleles aux trois côtez AC, CD, DB, du polygone ACDB que ces puissances font faire à cette corde; et d'un point F pris aussi à volonté sur SF, soient menées FE, FG, parallèles aux directions CK, DL, des puissances K, L, jusqu'à ce que ces deux lignes rencontrent SE, SG, en E, G. Cela fait, je dis qu'en ce cas d'équilibre les puissances K, L, seront entr'elles comme EF, FG, c'est-à-dire, K. L.:: EF. FG. II. Réciproquement la corde ACBD étant donnée de position, c'est-à-dire, le polygone qu'elle forme étant donné, si d'un point S pris à volonté, on fait SE, SF, SG, parallele aux trois côtez AC, CD, DB, de ce polygone; et que d'un point F pris à volonté sur SF, on mene deux droites quelconques FE, FG, qui rencontrent SE, SG en E, G: deux puissances K, L, qui seroient entr'elles comme ces deux ligne FE, FG, et qui auroient leurs directions CK, DL, paralleles à ces mêmes lignes, chacune à chacune, retiendront la corde ACDB dans cette position donnée, y demeurant en équilibre entr'elles.*

[68] See "The Industrial Revolution and its Consequences. Part I: Iron frameworks and Cullmann's "Graphostatics" (1850–1900)", Chap. 17 in this present volume.

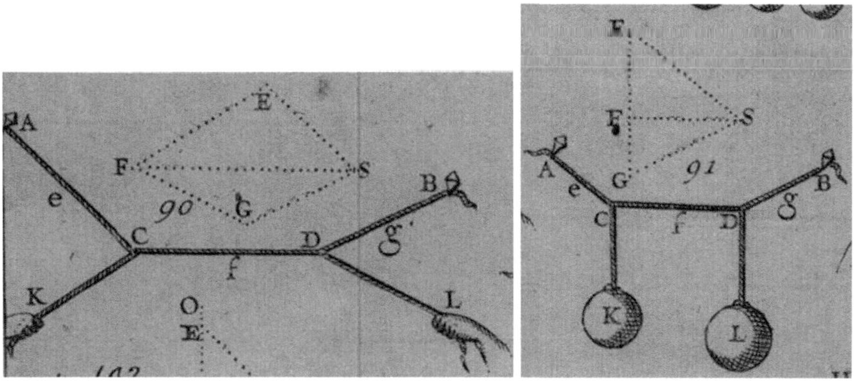

Fig. 57 Varignon's Figs. 90 and 91, the polygon of forces in the two simplest cases (Varignon 1725, Pl. 11). Public domain

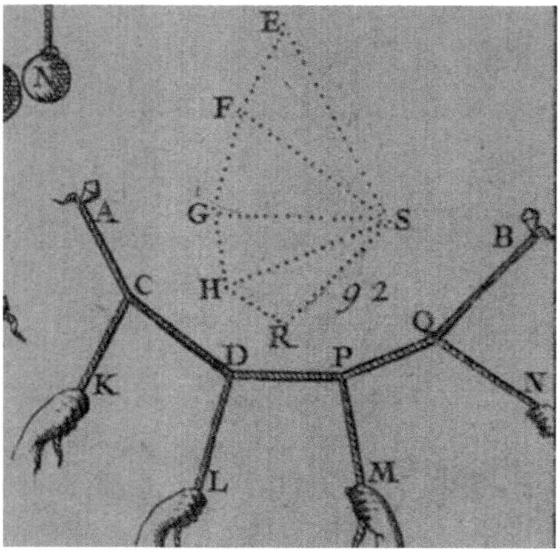

Fig. 58 Varignon's Fig. 92, the polygon of forces in a general case (Varignon 1725, Pl. 11). Public domain

That done, it follows from part. I. of this Th. 10 that in case of equilibrium between all these powers K, L, M, N, etc. they will be to each other as the lines EF, FG, GH, HR, etc. parallel to their directions.[69] (Fig. 58)

---

[69] Varignon (1725, I, pp. 191–192): *Corollaire I: Soient présentement tant de puissance K, L, M, N,* etc. (Figs. 92 et 93) *qu'on voudra appliquées suivant telles directions CK, DL, PM, QN, qu'on voudra aussi, autant de points quelconque C, D, P, Q,* etc. *de la corde lâche et parfaitement flexible A, C, B, et en équilibre entr'elles. D'un point quelconque S soient menées SE, SF, SH, SR,* etc. *parallèles aux côtez AC, CD, DP, PQ, QB,* etc. *chacune à chacun, du polygone ACDPQB que ces puissances K, L, M, N,* etc. *font former (Th. 1, Cor. II.) à cette corde ACB. Soient aussi EF, FG, GH, HR,* etc. *parallèles aux directions CK, DL, PM, QN,* etc. *de ces mêmes puissances, et qui rencontrent SE, SF, SG, SH, SR,* etc. *en E, F, G, H, R,* etc.
Cela fait, il suit de la part. I. du présent Th. 10 qu'en ce cas d'équilibre entre toutes ces puissances K, L, M, N, etc. *elles seront entr'elles comme les lignes EF, FG, GH, HR,* etc. *parallèles à leurs directions.*

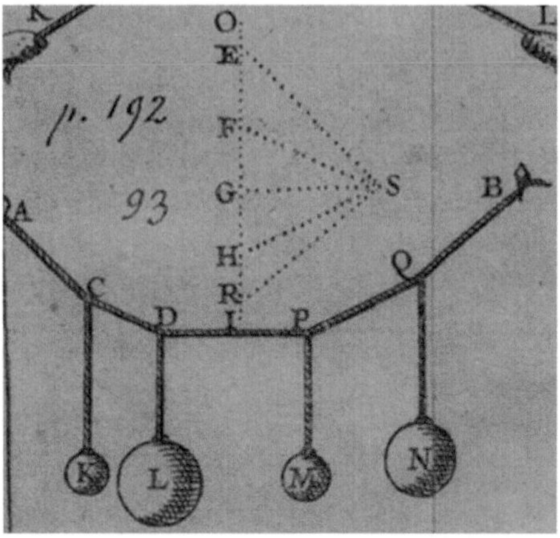

Fig. 59 Varignon's Fig. 93, the polygon of forces in a general case (Varignon 1725, Pl. 11). Public domain

Varignon deduces therefore that K. L:: EF. FG; L. M:: FG. GH; M. N:: GH. HR, etc. He continues with Corollary II:

> Corollary II: All these lines EF, FG, GH, HR, etc. here being parallel to the directions CK, DL, PM, QN, etc. of all powers K, L, M, N, etc. each to each, it is obvious that if all these directions are parallel to each other, all their parallels EF, FG, GH, HR, etc. will then only form a single straight line OI, of which they will be as many parts. So (corol. I) the powers K, L, M, N, etc. assumed to be in equilibrium between them, will also therefore be like the corresponding parts EF, FG, GH, HR, etc. of this straight line OI parallel to all and to each of their directions. From which we can see in Fig. 93 as weights K, L, M, N, etc. from the right OI parallel to their directions.[70] (Fig. 59)

Corollary III gives the reciprocal of Corollaries I and II and Corollary V generalizes to a polygon of an infinite number of infinitely small sides, in other words to a continuous curve. In Corollary VI, Varignon replaces, in the case of the continuous curve of Corollary V, the weights suspended to the rope by heavy line elements of the rope.

A bit further on, Varignon points out that, as we have seen in §3.7, Johann Bernoulli found this same direction in his *Manœuvre des vaisseaux*:

---

[70] Varignon (1725, I, pp. 192–193): *Corollaire II: Toutes ces lignes EF, FG, GH, HR,* etc. *étant ici (constr.) parallèles aux directions CK, DL, PM, QN,* etc. *de toutes les puissances K, L, M, N,* etc. *chacune à chacune; il est manifeste que si toutes ces directions sont parallèles entr'elles, toutes leurs parallèles EF, FG, GH, HR,* etc. *ne feront alors ensemble qu'une seule et même ligne droite OI, de laquelle elles seront autant de parties. Donc (corol. I) les puissances K, L, M, N,* etc. *supposées en équilibre entr'elles, seront aussi pour lors entr'elles comme les parties correspondantes EF, FG, GH, HR,* etc. *de cette ligne droite OI parallèle à toutes et à chacune de leurs directions. D'où l'on voit dans la Fig. 93 que des poids K, L, M, N,* etc. *de la droite OI parallèle à leurs directions.* (Vol. 1, p. 192–193.).

> Mr. Bernoulli found the previous common direction GT in another way in his *Essay de la Manoeuvre des vaisseaux*, chap. 15, prop. 3, where it is called *Mean Direction*.[71]

4.11 *1726, Daniel Bernoulli,* Examen des Principes de la Mécanique

Daniel Bernoulli's "Examination of the principles of mechanics and geometrical demonstrations of the composition and resolution of forces" is crucial for the history of mechanics and given its importance and its fundamental character it will have an influence on the scientific world in general as well as on the domain of construction.

> I. The part of Mechanics which deals with the balance of forces can be entirely deduced from the sole composition and resolution of forces; Pierre Varignon has sufficiently demonstrated this. If then we add to this principle the other, that the increases in velocities are proportional to the elements of time according to the forces or pressures soliciting the bodies to movement, we will obtain a second part of the mechanics devoted to the movement of bodies. Thence also can be deduced the rules of the communication of motions, and many others, which have since been eminently confirmed by infinite experiments; therefore, if I judge rightly, the Galilean principle is founded on experience rather than on reason and, therefore, it must be classified among the truths generally called *contingent truths* as opposed to *necessary truths*, because nature could have made it such that the increases in speed in moving bodies were proportional to the elements of time according to some function of pressures. So that, setting time = t, pressure = p and velocity = v, we would no longer have $dv = p\ dt$, but, for example, $dv = p^2\ dt$ or $dv = p^3\ dt$, etc., each of which equations governs different laws of motion.[72]

I underlined the fact that all important demonstrations of the parallelogram law of forces were reduced to demonstrations of the parallelogram of motions or of velocities. Daniel Bernoulli finds that this is not permitted, because, as he says, there

---

[71] Varignon (1725, I, p. 200): *M. Bernoulli a trouvé la précédent direction commune GT d'une autre manière dans son Essay de la Manoeuvre des Vaisseaux,* Chap. 15, *prop. 3, où l'appelle* Direction Moyenne.

[72] Dan. Bernoulli (1726, pp. 126–127): *I. Llam Mechaicae partem quae versatur circa aequilibrium potentiarum, totam ex sola compositione & resolutione virium duduci posse, abunde monstravit Petrus Varignon; huic dein principio si addimus alterum, quod incrementa velocitatum proportionalia sint elementis temporum ductis in vires seu pressiones corpora ad motum solicitante, obtinemus secundam mechanicae partem circa motum corporum occupatam. Posteriori principio Galileus primum, ni sallor, usus est pro determinando motu corporum uniformiter acceleratorum. Possunt quoque exinde deduci regulae communicationis motuum & alia quam plurima, quae dein infinitis experimentis egregie suerunt confirmata; experientia ergo potius, si recte iudico, fundatum est principium Galileanum, quam ratione, atque proin annumerandum illis veritatibus, quae dicuntur vulgo* contingenter verae, *quibus opponuntur* necessario verae. *Potuisset enim natura essicere, ut incrementa velocitatum in corporibus motis proportionalia essent elemntis temporum ductis in functionem quamcunque pressionum, ita ut dicto tempore,* t, *pressione* p *& velocitate* v *non amplius esst* du = pdt, *sed verbi gratia* du = ppdt *vel* du = p^3dt *&c, quarum aequationum singulae alias atque leges motuum subministran.*

can exist different relations between time, pressure, and velocity. For that reason, Daniel Bernoulli's aim is to give a rigorous demonstration of the parallelogram law of forces:

> II. The question presents itself quite differently with the principle of the composition of forces from which the first part of mechanics, which is usually called statics, as I said, depends. Because for that one I found an entirely geometric proof by the force of which I finally recognized that the theorems of statics are not less necessarily true, than those of geometry. Until now, the composition of forces has not been demonstrated by anyone, except from the composition of movements. However, although this mode of demonstration has been accepted by eminent masters, such as Newton, Varignon and others, it has hardly any geometric rigour and does not confer to the proposition anything more than a *contingent truth* as I will show hereafter.[73]

4.12 *1728; Daniel Bernoulli,* Methodus universalis determinandae curvaturae fili

Lemma 1. Let the wire ABCD (Fig. 1) be fixed at points A and D and divided into three equal parts AB, BC, CD. Suppose, applied in B and C, the forces BK and CI whose directions are along the bisector of the angles ABC and BCD and the forces BH and CL whose directions are parallel to each other. The lines and the perpendiculars of the figure are extended and the forces BK, CI, BH and CL are designated by A, B, C and D, let the wire ABCD fixed in points A and D be divided in three equal parts AB, BC, CD. Thus I say that $\frac{sin.ang.ABR}{sin.ang.ABm}A + \frac{sin.ang.ABM}{sin.ang.ABm}C = \frac{sin.ang.BCd}{sin.ang.BCn}B + \frac{sin.ang.DCX}{sin.ang.BCn}D$.. Or supposing that Ao + BC = CD for the *sinus totus*,[74] we get $\frac{AO}{Am}A + \frac{AM}{Am}C = \frac{Ba}{Bn}B + \frac{DX}{Bn}D$. The demonstration falls under mechanics and can easily be derived from the composition of forces, with help of which one easily determines the force that acts in some point of BC, assuming that the forces pulling in the directions of B and G must be equal to each other.[75] (Fig. 60).

[73] Dan. Bernoulli (1726, pp. 127): *II. Aliter vero seres babet cum principio compositionis virium, cuit priorem mechanicae partem, vulgo statiscam nominant, inniti dixi; illius enim demonstrationem inveni plane geometricam, cuius vi demum cognovi, teeoremata statica non minus necessario vera esse, quam sun geometrica. Hactunus a nemine demonstrata suit compositio viuium, quam ex compositione motus, quad demonstrandi genus tametsi receptum a viris summis, veluti Newtono, Varignonio, aliisque: minime tamen rigore geometrico munitum est, propositionemque non aliter quam* contingenter v ram *reddit, uti inserius demonstrabo.*

[74] Nowadays the value of the radius of the trigonometrical circle and thus the value of the *sinus totus* is taken equal to 1.

[75] Dan. Bernoulli (1728, pp. 63–64): *Lemma 1. Sit filum ABCD affixum punctis A et D divisumque in tres partes aequales AB, BC, CD. Intelligantur in punctis B et C, applicatae potentiae BK et CI quarum directiones bisecent potentiae BK et CI quarum directiones bisecent aequaliter angulos ABC et BCD: ut et potentiae BH et CL, quarum directiones sunt inter se parallelae, erit productis lineis demissisque perpendicularibus quas figura ostendit atque designatis potentiis BK, CI, BH et CL per A, B, C et D erit inquam* $\frac{sin.ang.ABR}{sin.ang.ABm}A + \frac{sin.ang.ABM}{sin.ang.ABm}C = \frac{sin.ang.BCd}{sin.ang.BCn}B + \frac{sin.ang.DCX}{sin.ang.BCn}D$ *seu sumpta Ao + BC = CD pro sinu toto, erit* $\frac{AO}{Am}A + \frac{AM}{Am}C = \frac{Ba}{Bn}B + \frac{DX}{Bn}D$. *Demonstratio constat ex*

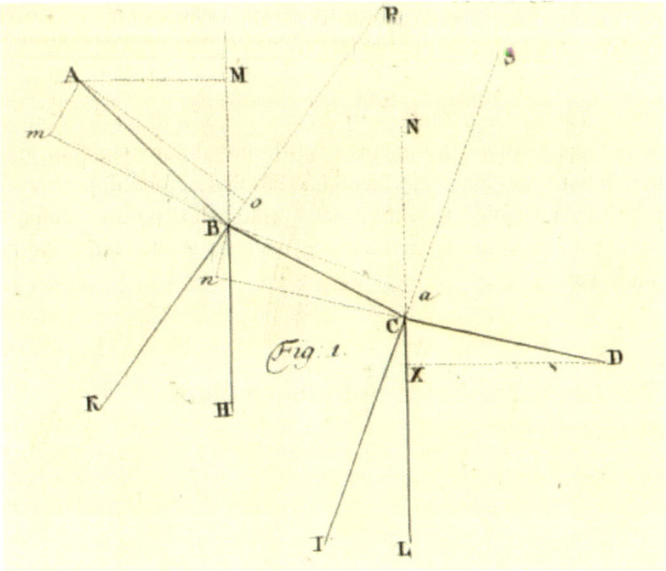

Fig. 60 Daniel Bernoulli's Fig. 1 (Dan. Bernoulli 1728, Tab. III). Public domain

Lemma 2. If the angles Abm and OCn are infinitely small neglecting what has to be neglected, let AO = AB, Am = 2 Bo; BQ = BC = AB and Bn = 2 Ca, if also the radius of the circle passing through the three points A, B, C is posed = R and let S be the radius of the circle passing through three other points B, C, D, then $2BO = \frac{AB^2}{R}$ and $Ca = \frac{BC^2}{S} = \frac{AB^2}{S}$, so that substituting these values in the equation of the preceding lemma, it will transform into the following one $\frac{R.A}{AB} + \frac{AM.R.C}{AB^2} = \frac{S.B}{AB} + \frac{DX.S.D}{AB^2}$.[76]

4. Problem. To find the curvature of the wire, of which each point is pulled by any two given forces, the one acting perpendicularly to the curve and the other one parallel to one given line.[77]

Daniel Bernoulli finds first a general equation, $-ARds - CRdx = ds \int Cdy$, and then gives two corollaries in which he studies two particular cases:

---

mechanicis atque facile deducitur ex compositione potentiarum, cuius ope determinantur potentiae punctum aliquod in BC assumtum versus B et G trahentes, quae inter se debent esse aequales.

[76] Dan. Bernoulli (1728, p. 64): 3. *Lemma 2. Si anguli Abm et OC*n *sint infinite parvi neglectis negligendis, erit AO* = *AB,* Am = 2 Bo*; BQ* = *BC* = *AB et* Bn = 2 Ca*, si praeterea radius circuli per tria puncta A, B, C transeuntis ponatur* = *R et radius circuli per tria alia puncta B, C, D transeuntis* = *S, erit* $2BO = \frac{AB^2}{R}$ *et* $Ca = \frac{BC^2}{S} = \frac{AB^2}{S}$, *unde substitutis hisce valoribus mutabitur aequatio praecedentis lemmatis in hanc* $\frac{R.A}{AB} + \frac{AM.R.C}{AB^2} = \frac{S.B}{AB} + \frac{DX.S.D}{AB^2}$.

[77] Dan. Bernoulli (1728, p. 64): *Problema. Invenire curvaturam fili, cuius singula puncta datis duabus quibuscunque potentii, altera ad curvam normali, altera ad datam lineam ubique parallela trahuntur.*

Corollary 1. If C = 0, if all forces are perpendicular to the curve, then AR = constant, which occurs for the elastica, the velaria, the lintearia and for an infinity of other curves.[78]

Corollary 2. If A = 0, thus if all forces acting on the wire are parallel to each other so that $-CRdx = ds \int Cdy$. ... Which occurs in the case of the catenary of unequal thickness.[79]

In this way Daniel solves, by means of differential equations, all the problems that his father Johann and his uncle Jacob had studied before him.

It is difficult to determine at what precise time calculus, differential as well as integral, entered the domain of architecture, because at the beginning it was kept hidden by architects.

4.13 *1728, Leonhard Euler,* Solutio problematis de invenienda curva

In his, "Solution of the problem to find the curve which forms an elastic blade when it is stressed at all points by any forces", Euler starts from a most general problem:

The General Problem. Let BMA be any elastic blade and with displacements from forces applied at any fixed points whatever, moreover with end in B fixed, and at A two weights E and F are hung on pulling on A, of which E is along the vertical and F the horizontal. It is required to determine the curve AMB that said lamina has been turned into by being bent.

Solution: The horizontal line AC is taken as the axis in which the intercepts $AP = x$ are taken. With $AP = x$, $PM = y$, $AM = s$; s being a constant element of the curve; the radius of curvature in $M = r$, the angle between two elements in M vary as $\frac{1}{r}$; the elastic strength in $M = v$ and the strength producing this angle varies as $\frac{v}{r}$.[80] (Fig. 61).

He finds first a general equation: $\frac{\Delta v}{r} = Ex + Fy + \int Pdx + \int Qdy$, in which the sum of the moments of the vertical forces $= \int Pdx$, and the sum of the moments of the horizontal forces $= \int Qdy$.

---

[78] Dan. Bernoulli (1728, p. 65): *Corollarium I. Si C = 0, id est, si solae poentiae ad curuam normaliter insistentes adsint, sit AR = quantitati constanti, id quod indicat esse in huiusmodi curuis semper et ubique radium osculi reciproce proportionalem potentiae: estque haec proprietas pro esastica, velaria, lintearia aliisque infinitis.*

[79] Dan. Bernoulli (1728, p. 65): Coroll. 2. *Si vero A = 0, id est, si solae potentiae inter se parallellae filum trhant, inserviet haec aequatio* $- CR\,dx = ds \int C\,dy$ *... inseruitque aequatio pro determinanda curvatura catenarum utcunque inaequaliter crassarum.*

[80] Euler (1728, p: 73): *Problema Generale. Sit lamina BMA utcunque elastica et in singulis punctis a potentiis quibusuis follicitata, fixa autem ea sit in B, atque in A duo pondera appensa habens E et F, quorum E secundum verticalem AE, et alterum F secundum horizontalem AC trahit. Oportet determinare curvam AMB iuxta quam lamina hoc modo sollicitata flectitur.*
Solutio. Sumatur horizonalis AC pro axe, in quo capiantur abscissae AP = x. Sit ei orthogonalis PM = y et AM = s, cuius elementa constantia accipiantura. Ponatur radius curvedinis in M = r, erit angulus, quem duo elementa in M constituunt, reciproce vt r.

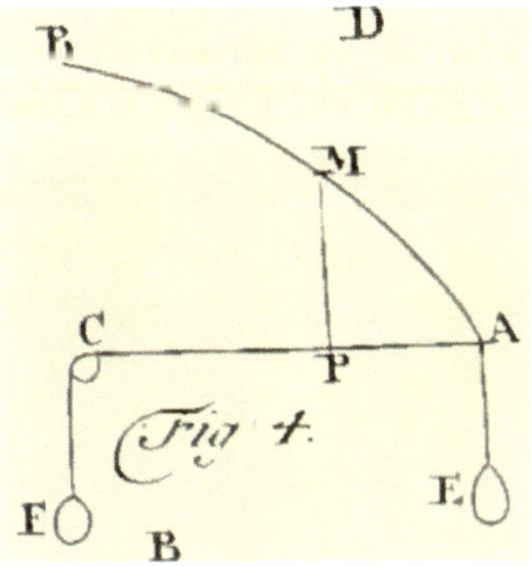

Fig. 61 Euler's Fig. 4 for his general problem (Euler 1728, Tab. III). Public domain

Next Euler considers only vertical strengths. He therefore takes $\int Qdy = 0$ and finds catenaries of all kinds. He considers two particular cases:

(1) $dP = a\,ds$, thus equal masses hanging at equal distances taken along the cord;
(2) $dP = a\,dx$, thus equal masses hanging at equal distances taken along the x axis.

In the first case he finds the usual catenary and in the second he finds the parabola. This is a well-known result as we have seen, but this time Euler demonstrates it in full detail with the help of differential equations. Euler also calls case 2 the problem of the hanging bridge.

4.14 *1729, Claude Antoine Couplet,* De la poussee des voutes

Couplet starts his text by showing the equivalence of the parallelogram law of forces with what we have called the "triangle law of forces".

> Lemma: If the force x breaks down into two forces y and z, these three forces will be together like the sides of a triangle formed by the perpendiculars drawn on the directions of the three forces.
>
> This is demonstrated in all Mechanics.[81]
>
> …
>
> Problem I. Determine the relationship that there must be between the gravity of the voussoirs which form any arch and what the thrust of those voussoirs is so that they balance each other without the help of the meshing of their parts.[82]

---

[81] Couplet (1729, p. 80): Lemme: *Si la force* x *se décompose en deux forces* y *and* z, *ces trois forces seront entre elles comme les côtés d'un triangle formé par les perpendiculaires menées sur les directions des trois forces.*
Ceci est démontré dans toutes les Méchaniques.

[82] Couplet (1729, p. 81): *Problème I. Déterminer le rapport qu'il doit y avoir dans les pesanteurs des Voussoirs qui forment une voûte quelconque, et quelle est la poussée des voussoirs, afin qu'ils fassent équilibre entre eux sans le secours de l'engrènement de leurs parties.*

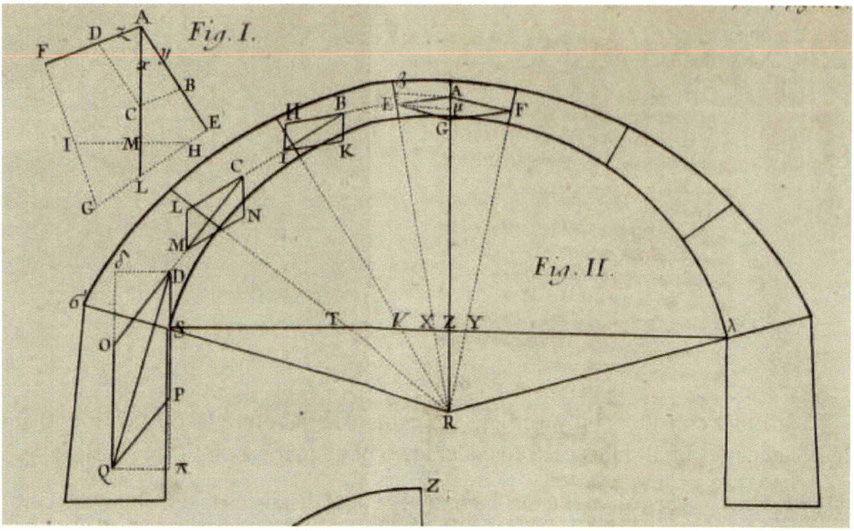

Fig. 62 Couplet's figures I and II from "De la poussée des voûtes" (1729: Pl. 6). Public domain

For the first time Couplet shows here the study of the equilibrium of the arch with help of the parallelogram law of forces (Fig. 62).

> Therefore, by expressing the gravity of the Key A by the vertical diagonal AG, the sides AE, AF of the parallelogram will express the efforts that this Key A makes perpendicularly on the neighboring segments. Now let the direction AE of the effort that the Key A makes on its neighboring segment B be extended to H. And from point B, where the line meets the vertical BK, take BH = AE, which is the effort that voussoir B had received from Key A, and the HIKB parallelogram is completed, for this voussoir B will press its neighbour C with a force BI composed of gravity BK, and the force BH it received from Key A.[83]

It then suffices to proceed in an analogous manner by extending BI to find CL and the following parallelogram. And so on to the pillar.

In problem IV, Couplet treats the problem of the velaria that he presents as an inverted vault or a way of finding the right curvature of the most stable vault.

Here we shall not follow the rather short story of the problem of the velaria. It was posed to Johann Bernoulli by his brother Jacob when Johann was traveling to Paris. Johann found the solution immediately and showed that it was the same curve as the catenary.[84]

---

[83] Couplet (1729, p. 82): *Pour lors en exprimant la pesanteur de la Clef A par la diagonale verticale AG, les côtés AE, AF, du parallelogramme exprimeront les efforts que cette Clef A fait perpendiculairement sur les Voussoirs voisins.*
Maintenant soit prolongée la direction AE de l'effort que la Clef A fait sur son Voussoir voisin B jusqu'en H. Et du point B, où la ligne rencontre la verticale BK, soit pris BH = AE, qui est l'effort que le Voussoir B a reçû de la Clef A, et soit achevé le parallelogramme HIKB, pour lors ce Voussoir B pressera son voisin C avec une force BI composée de la pesanteur BK, et de l'effort BH qu'il a reçû de la Clef A.

[84] Joh. Bernoulli (1692).

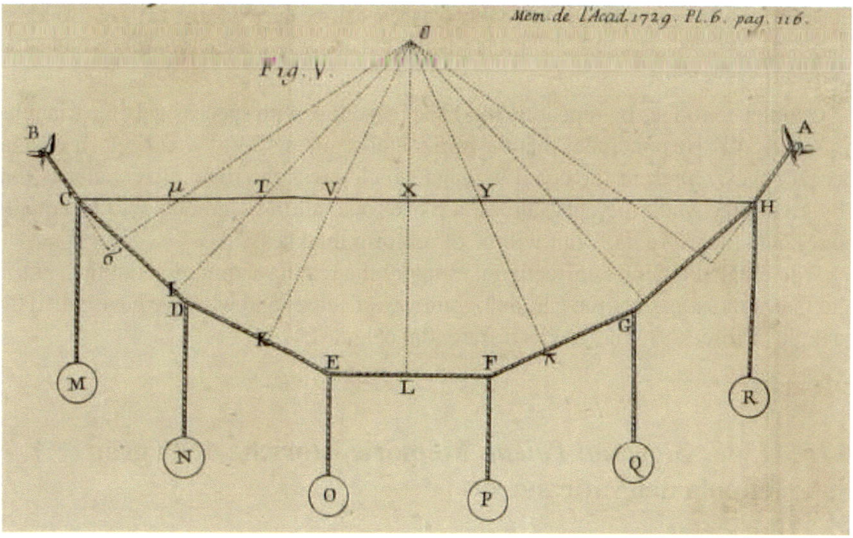

Fig. 63 Couplet's Fig. V from "De la poussée des voûtes" (1729: Pl. 6). Public domain

Problem IV. Determine the uniform curvature of a vault, such that it is kept in equilibrium, and of which we consider the Segments as polished, that is, without bond.[85] (Fig. 63)

Solution. Take a rope charged with equal weights which will be attached to it all along at equal distances, these small weights will represent the effective segments of the Vault that we ask.

Apply the two ends of this cord on the two ends of the lower surface, that is, on the inner ends of the chevets, from where the vault takes birth, and let go of this rope until it touches the lower surface of the key, which I suppose overturned, or until it gives the desired vault. This curve is the one to use, as we will demonstrate below. We can conclude that if we want a uniform thickness for the arch, we must give towards the middle of the arch the curvature of a loose cord, and make the lower and upper surfaces parallel to this loose cord.[86]

Demonstration. Since the loose cord is balanced in all its parts, the weight *N* will be in balance with the forces from which the *CD* and *DE* sides of the Polygon are pulled.

[85] Couplet (1729, p. 82): *Probleme IV. Déterminer la courbure uniforme d'une Voûte, telle qu'elle se maintienne en équilibre, et dont nous considérons les Voussoirs comme polis, c'est-à-dire, sans liaison.*

[86] Couplet (1729, p. 95): *Solution. Prenés une Corde garnie de poids égaux qui y seront attachés tout du long à égale distances, ces petits poids représenteront les Voussoirs effectifs de la Voûte que nous demandons. Appliqués les deux bouts de cette Corde sur les deux extrémités de l'intrados, c'est-à-dire, sur les extrémités intérieures des Chevets, d'où la Voûte prend sa naissance, et lâchés cette corde jusqu'à ce qu'elle touche l'intrados de la Clef, que je suppose renversée, ou jusqu'à ce qu'elle donne une Voussure à volonté. Cette Courbe est celle qu'il convient d'employer, comme nous en donnerons la Démonstration ci-après. L'on en peut conclure que si l'on veut une épaisseur uniforme à la Voûte, il faut donner vers le milieu de la Voûte la courbure d'une corde lâche, et faire l'intrados et l'extrados parallele à cette corde lâche.*

> These three powers will therefore be between them like the three sides of a Triangle which would be perpendicular on their directions, that is, like the sides of the triangle *STV*, according to the Lemma above.[87]

Couplet continues by repeating the same reasoning with successively the triangles *SVX* and *SXY*. He concludes that the horizontal segments *TV*, *VX*, *XY*, etc. are equal because they represent the equal weights which stress the rope. He concludes that "by giving the Vault the curvature of a loose rope, all the segments will be equally heavy, and therefore the vault will be of uniform thickness".[88]

We note that Couplet correctly takes up all the results acquired about the *velaria* and the vault using only the parallelogram law of forces and what we have called the triangle of forces. But he avoids differential calculus.

4.15 1748, Giovanni Poleni, Memorie istoriche della gran cupola del Vaticano

In Poleni's "Historical memoire of the great come of the Vatican", in Chap. VIII (1748, cols. 30–42), after having given a list of authors who had contributed to the study of the form of vaults and of their stability (Derand, Blondel, Fontana, La Hire, Parent, and others), Poleni reminds us of the parallelogram law of forces, with help of figure X. And about vaults, he reminds us the *spheres of James Stirling*. But contrarily to Stirling, Poleni uses the parallelogram law instead of the triangle one and his picture is much clearer and readable, for us at least (Fig. 64).

His figure XI shows also a phenomenon we have not yet heard of. The upper surfaces HR and mg of the two piers that sustain the vault IAp are not horizontal, but slightly inclined. Poleni explains:

> 68. The same will be used with respect to sphere I [the last one on the left]. It will be assumed that the diagonal EK is prolonged to *u*, and that in it the I*u* equal to EK is taken, and that the parallelogram *u*ILN is completed. It will be found that sphere I will not be able to move, if it touches another immobile sphere, and touches it at a point *h*, which is in the central line IN (I believe necessary to add also this case) will be perpendicular to the plane HR of pier HRMX; on which the sphere I (last of this part of the Arc) remains set, which touches it slowly in the point h. The same things are to be understood with respect to the spheres of the other side (terminating on the plane *mg* of the pier opposite) so as to form the arch, or vault, of equal spheres, which with their weight reciprocally sustain themselves, as was proposed

[87] Couplet (1729, p. 96): *Démonstration. Puisque la corde lâche est en équilibre dans toutes ses parties, le poids* N *sera en équilibre avec les forces dont les côtés* CD *et* DE *du Polygone sont tirés. Ces trois puissances seront donc entre elles comme les trois côtés d'un Triangle qui seroient perpendiculaires sur leurs directions, c'est-à-dire, comme les côtés du Triangle* STV*, suivant le Lemme ci-devant.*
Mais ces trois côtés du Triangle STV sont perpendiculaires sur les directions CD, DN, DE, *de ces trois puissances. Donc* ST *représentera la force avec laquelle le côté* CD *est tiré;* TV *représentera la pesanteur du poids* N *et* SV *la force avec laquelle le côté* DE *est tiré.*

[88] Couplet (1729, p. 97): *...donc en donnant à la Voûte la courbure d'une corde lâche, tous les Voussoirs seront également pesants, et par conséquent la voûte sera d'une épaisseur uniforme.*

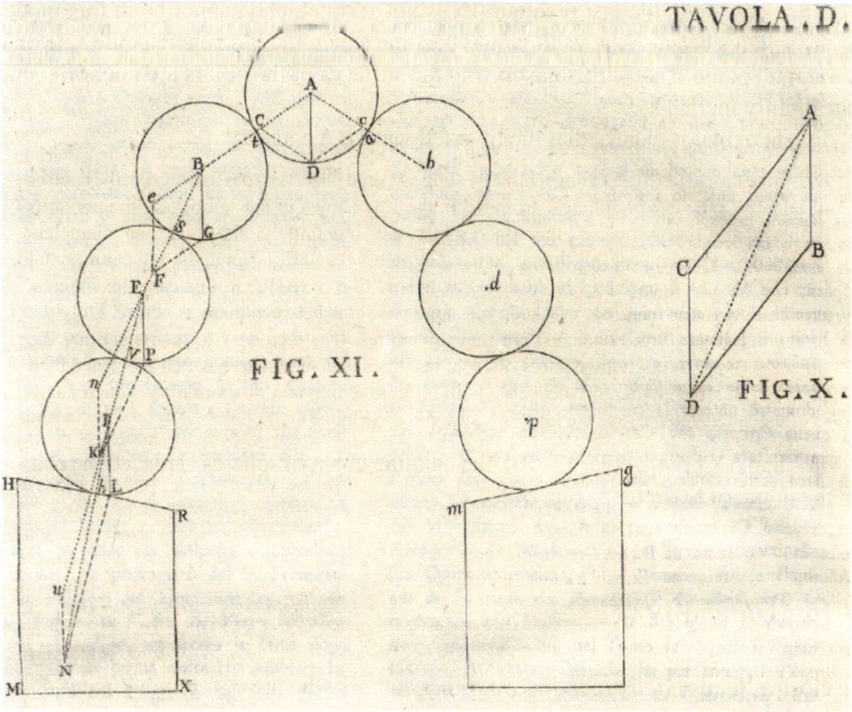

Fig. 64 The parallelogram law and James Stirling's spheres illustrated in the study of vaults by Poleni (Poleni 1748, Tav. D). Public domain

> in the problem. In which it is clearly seen that the size of the spheres does not enter at all in its solution; whether they are large, or indefinitely small, those spheres, the solution of the Problem itself just set forth remains firm and certain.[89]

The text is clear and easy to understand. Poleni adds an important point. He knows its importance and the fact that it is new, as he writes: "I believe necessary to add

---

[89] Poleni (1748, col. 38: 68). *Lo stesso si adopererà per rispetto alla sphera I. Si supporrà, che sia prolungata ad u la diagonale EK, e che in essa sia presa la I*u *uguale alla EK, e che sia compito il Parallelogrammo* u*ILN. Si troverà, che la Sfera I non potrà moversi, se toccherà un'altra Sfera immobile, e la toccherà in un punto* h, *che sia nella diagonale linea IN (che credo necessario l'aggiungere anche questo caso) sarà perpendicolare al piano HR del Pilastro HRMX; su cui resti impostata la Sfera I (ultima di questa parte dell'Arco) che tocchi esso piano nel punto* h. *Le medesime cose si debbono intendere per rispetto alle sfere dell'altra parte (terminanti sul piano* mg *dell'opposto Pilastro) onde si formi intieramente l'Arco, o Volta delle Sfere uguali, che con la loro gravità si sostentino scambievolmente, come proposto fu nel problema. Nel quale chiaramente si vede, che la grandezza delle Sfere per niente entra nella di lui soluzione; onde sieno grandi, o sieno indefinitamente picciole desse Sfere, sempre resta ferma e certa quella già esposta soluzione del Problema medesimo.*

also this case": As the last resultant IN is not vertical, and cannot be in this model of the spheres because the vault would become a pier, the last sphere could role on a horizontal plane. That is the reason why the surfaces of the pillars are not horizontal but slightly inclined.

Poleni explains clearly that the inverted vault is a catenary:

> 69. A fine thread (as Mr. Stirling learnedly supposed and observed) passes through the centres of quantity [gravity] of spheres, arranged in the way shown in the solution of the previous problem: and the number of the spheres increases indefinitely, and thus their diameters diminish. Let the ends of this thread be fixed in points B and C, taken in the same horizontal line DE; and let the spheres be turned downwards like BGAHC: in this case the spheres, though now freely hanging, will keep the same positions between them as before. Only the directions of the forces changes, not their quantities. It is then easy to recognize that the thread, passing through the centres of each, becomes that same curve which is formed by a perfectly flexible chain, and incapable both of lengthening and of shortening, which curve is often called catenary, as we have already said.[90] (Fig. 65)

In Chap. IX (1748, cols. 42–50), the hanging chain of spheres is placed in relation to the dome of St. Peter's (Fig. 66).

4.16 Other Works

In closing, let us briefly mention two other authors who are covered in greater detail in another chapter.[91]

In *De curvis funiculariis et catenaries* (1755), Georg Wolfgang Krafft puts all previous results about catenaries, vaults and some others together with help of the differential equation we have seen in Euler. He takes the element of the curve $= ds$ and the osculating radius $= r$, and thus the force $= \frac{ds}{r}$.

Krafft's text is more interesting for the references it contains than for its results.

The two works by Abbé Charles Bossut *Recherches sur l'équilibre des voûtes* (Bossut 1774) and *Nouvelles recherches sur l'équilibre des voûtes* (Bossut 1776) concerns various forms of vaults, such as barrel vaults and domed vaults:

> Having taken on the direction of the force F, the part XE to represent it, I decompose it into two other forces Xu, Xt, perpendicular respectively to the two joints mM, nN of the segment

[90] Poleni (1748, cols. 38–39: 69). *Un sottil filo (come dottamente il Signor Stirling suppose, ed avverti) passi per i centri di quantità di Sfere, disposte nel modo dimostrato della soluzione del precedente Problema: ed accrescasi indefinitamente il numero delle Sfere, e così diminuiscansi i diametri delle medesime. Abbia esso filo le estremità sue fermate ne' punti B, C costituiti in una medesima orizzontale linea DE; e siano le Sfere rivoltate all'in giù come BGAHC: in tal caso le medesime, benchè liberamente pendenti, conserveranno tra loro le posizioni stesso come prima. Conciossiachè si mutano solamente le direzioni delle potenze, non le grandezze. Indi si conoscerà facilmente, che il filo, passante per i centri di tutte, diviene quella stessa Curva la quale formata esser può da una catenella perfettamente flessibile, ed incapace tanto d'allungamento, quanto d'accorciamento, la qual Curva suole chiamarsi Catenaria, come già detto abbiamo.*

[91] See "The "De curvatura fornicis" of Jacob Bernoulli: The Introduction of the Infinitely Small into the Calculation of Arches", Chap. 5 in this present volume.

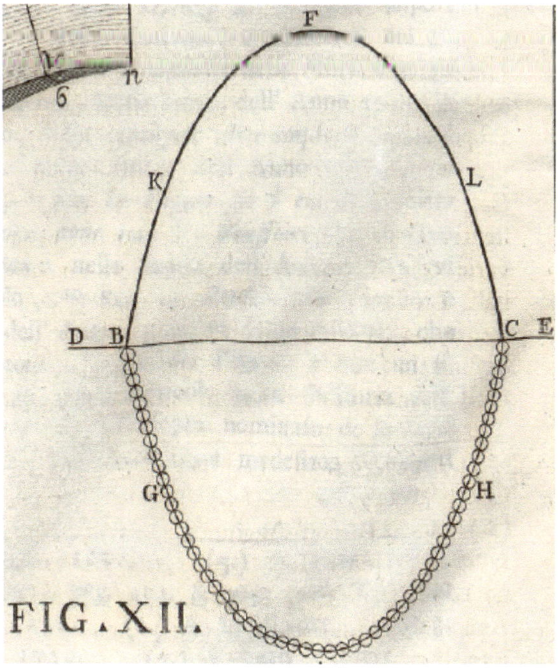

Fig. 65 The hanging chain of spheres illustrated in the study of vaults by Poleni (Poleni 1748, Tav. D). Public domain

X. Let X′ be the point where the direction of the force Xt meets that F′X′ of force F′. I take on F′X′, the part X′E′ to represent the force F′, and I break it down into two other forces X′q and X′l, perpendicular respective to the two joints nN, pP of segment X′. Having said this, it is clear that the two segments X, X′ will balance if the two forces Xt and X′q, directly opposite, by which they act against each other, are equal.[92]

[92] Bossut (1774, p. 537): *Ayant pris sur la direction de la force F, la partie XE pour la représenter, je la décompose en deux autres forces Xu, Xt, perpendiculaires respectivement aux deux joints mM, nN du voussoir X. Soit X′ le point où la direction de la force Xt rencontre celle F'X' de la force F'. Je prends sur F'X', la partie X'E' pour représenter la force F′, et je la décompose en deux autres forces X′q, X′l, perpendiculaires respectivement aux deux joints nN, pP du voussoir X′. Cela posé il clair que les deux voussoirs X, X′ se feront équilibre si les deux forces Xt, X′q, directement opposées, par lesquelles ils agissent l'un contre l'autre, sont égales.*

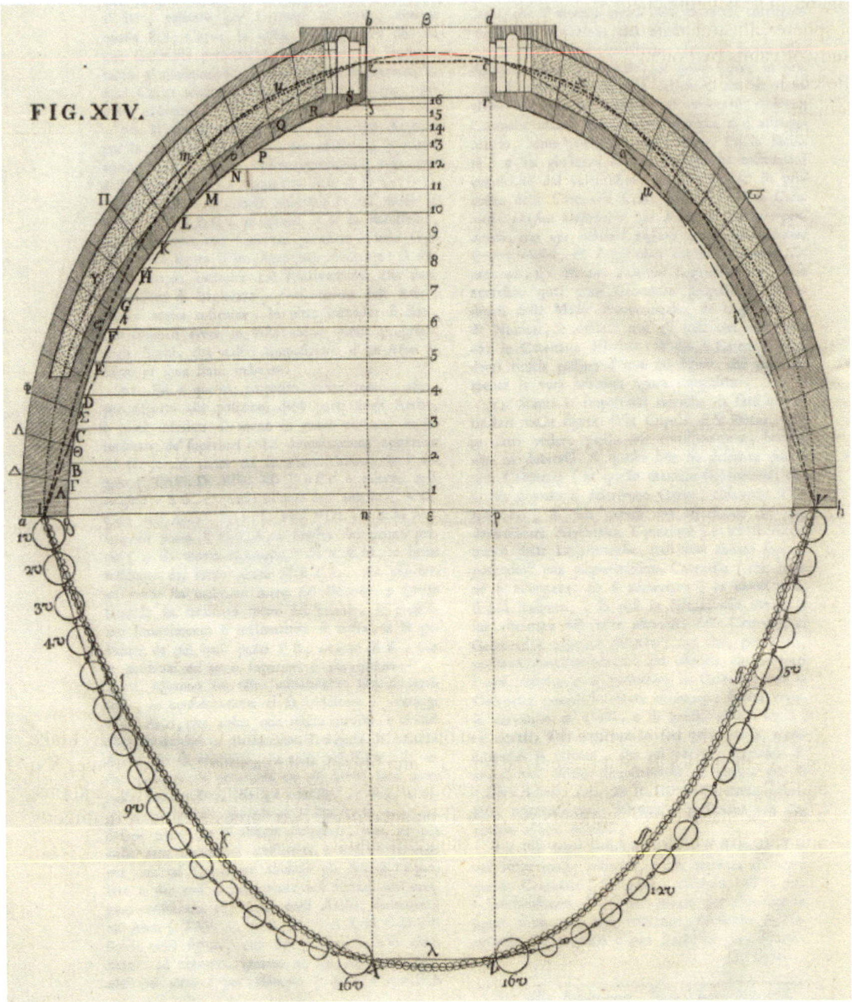

Fig. 66 The hanging chain of spheres related to the dome of St. Peter's (Poleni 1748, Tav. E). Public domain

References

Aristotle. 1955. Quaestiones Mecanicae. In *Aristotle, minor works, mechanical problems.* W. S. Hett, trans. Cambridge, MA: Harvard University Press.

Beeckman, Isaac. 1644. *D. Isaaci Beeckmanni … Mathematico-physicarum meditationum meditationum, quaestionum, solutionum centuria.* Trajecti ad Rhenum [Utrecht]: apud Petrum Danielis Sloot.

Beeckman, Isaac. 1939. *Journal tenu par Isaac Beeckman, 1604–1619*, vol. 1. C. de Waard, ed. The Hague: Martinus Nijhoff.

Bélidor, Bernard Forest de. 1729. *La Science des ingénieurs dans la conduite des travaux de fortification et d'architecture civile*. Paris: Jombert

Benvenuto, E. 1978. *Se le leggi della statica e della meccanica siano di una verità necessaria o contingente?* Pre-print: Facoltà di Architettura, Università di Genova.

Benvenuto, E. 1985. The parallelogram of forces. *Meccanica* 20 (2): 99–109.

Benvenuto, Edoardo. 1988. L'ingresso della storia nelle discipline strutturali. *Palladio* (Nuova Serie), 1, 7–14.

Benvenuto, Edoardo. 1991. *An introduction to the history of structural mechanics*, vol. 2, New York: Springer.

Bernoulli, Johann. 1742a. Op. CLXXVII, Propositiones variae Mechanico-dynamicae. In: *Opera omnia, tam antea sparsim edita quam hactenus inedita (cura G. Cramer)....*, vol. 4, 253–386.

Bernoulli, Johann. 1691. Solutio Problematis Funicularii. *Acta Eruditorum*, Junii 1691, 274–276.

Bernoulli, Johann. 1692. Solution du probleme de la courbure que fait une voile enflée par le vant. *Journal des Sçavans*, 28 avril, 189.

Bernoulli, Johann. 1714. *Essay d'une nouvelle théorie de la manoeuvre des vaisseaux, avec quelques lettres sur le même sujet.* Basel: J. G. König.

Bernoulli, Daniel. 1726. Examen des Principes de la Mécanique et démonstrations géométriques de la composition et de la résolution des forces. *Commentarii Academiae scientiarum imperialis Petropolitanae* 1, 1726–1728, 126–142.

Bernoulli, Daniel 1728. St. 15, Methodus universalis. Determinandae curvaturae fili a potentiis quamcunque legem inter se observantibus extensi. *Commentarii Academiae scientiarum imperialis Petropolitanae* 3, 1729–1731, 62–69.

Bernoulli, Johann. 1742b, Lectio Tregisima Sexta; Lectio Tregisima Septima (Lectures XXXVI and XXXVII to Guillaume de l'Hôpital), *Opera omnia, tam antea sparsim edita quam hactenus inedita (cura G. Cramer)....*, vol. 3, 491–497. Lausanne and Geneva: Bousquet.

Bernoulli, Jacob. 1744. Problema de Curvatura fornicis, cujus partes se mutuo proprio pondere suffulciunt sine opere caementi (5/12/1704), *Varia Postuma Jacob Bernoulli*, *Opera*, vol. II, XXIX, 119–1123.

Blondel, François. 1675. *Cours d'architecture enseigné dans l'Académie royale d'architecture.* Paris: P. Auboin et F. Clouzier.

Bossut, Charles (Abbé). 1774. Recherches sur l'équilibre des voûtes. *Mémoires de l'Académie de Paris,* année 1774 (1778), 534–566.

Bossut, Charles (Abbé). 1776. Nouvelles recherches sur l'équilibre des voûtes en dôme, *Histoire de l'Académie royale des sciences ... avec les mémoires de mathématique & de physique*, année 1776, Paris 1779, 587–596.

Clairaut, Jean-Baptiste. 1743. Methodus generalis inveniendi catenaries. *Miscellanea Berolinensia* VII, 270–272.

Couplet, Claude Antoine. 1729. De la poussée des voûtes, *Histoire de l'Académie Royale des sciences, avec les mémoires de mathématique et de physique* (1731), 79–117.

Derand, François. 1643. *L'Architecture des voûtes ou l'Art des traits et coupe des pierres.* Paris: Cramoisy.

Euler, Leonhard. 1728. Solutio problematis de invenienda curva, quam format lamina utcunque elastica in singulis punctis a potentiis quibuscunque sollicitata. *Commentarii Academiae scientiarum imperialis Petropolitanae* 3, 1729–1731, 70–84.

Galilei, Galileo. 1638. *Discorsi e dimostrazioni matematiche, intorno a due nuove scienze attenti alla mecanica e i movimenti locali.* Leiden: Elsevier. Eng. trans. Stillman Drake, *Two New Sciences, Including Centres of Gravity and Force of Percussion.* Madison: University of Wisconsin Press, 1974.

Simon Stevin, 1605. *Wisconstige Gedachtenissen*, vol. 4. Leiden, Bouwensz, , Vierde [Derde in the Cortbegryp at the begining of the book.] Stuck der Wisconstige Gedachtenissen van de Weeghconst, Eerste bouck der Weegconst, 37–38.

Radelet-de Grave, Patricia. 2015. The Panthéon's stability already questioned by Pierre Patte in 1770. In *Masonry Structures: Between Mechanics and Architecture,* eds. Danila Aita, Orietta

Pedemonte, Kim Williams, 127–185. Basel: Birkhäuser. ISBN: 978-3-319-13002-6, 978-3-319-13003-3 (ebook).

Gregory, David. 1697. Catenaria, *Philosophical Transactions of the Royal Society of London*, 19, 231, 637–652.

Hèrigone, Pierre. 1634. *Cursus mathematicus, nova, brevi et clara methodo demonstratus... Cours mathématique, demonstré d'une nouvelle, briefve, et claire methode..., tome III, contenant la construction des Tables des Sinus et logarithmes, avec leur usage aux interest, et en la mesure des triangles rectilignes: la géométrie pratique: Les fortifications: la Milice: et les Mechaniques.* Paris: chez l'autheur, chez Henry le Gras.

La Hire, Philippe de. 1695. *Traité de mecanique où l'on explique tout ce qui est necessaire dans la pratique des Arts et les propriétés des corps pesants lesquelles ont un plus grand usage dans la physique.* Paris: Imprimerie royale.

Hooke, Robert. 1676. *A Description of Helioscopes, and Some Other Instruments.* London: John Martyn.

Huygens, Christiaan. 1888. *Oeuvres complètes de Christiaan Huygens. Correspondance [de Christiaan Huygens], 1666–1669/publ. par la Société hollandaise des sciences*, vol. 1. The Hague: M. Nijhoff.

Huygens, Christiaan. 1908. *Oeuvres complètes de Christiaan Huygens. / publ. par la Société hollandaise des sciences, Travaux mathematiques 1645–1651,* vol. 11. The Hague: M. Nijhoff.

Krafft, Georg Wolfgang. 1755. De curvis funiculariis et catenariis, vel illis, quae corporibus flexibilibus inducitur, cum potentiis quibusvis solicitantur. *Novi Commentarii academiae scientiarum Petropolitanae* 5 (1754–1755), 145–163.

Newton, Isaac. 1687. *Philosophiae Naturalis Principia Mathematica.* London: Jussu Societatis Regiæ ac Typis Josephi Streater. Eng. trans. *Newton's Principia. The Mathematical Principals of Natural Philosophy*, trans. Andrew Motte, New York: Danile Adlee, 1846.

Pardies, Ignace Gaston. 1673. *La statique ou la science des forces mouvantes.* Paris: Mabre-Cramoisy.

Poleni, Giovanni. 1748. *Memorie istoriche della gran cupola del tempio vaticano, e de'danni di essa, e de'ristoramenti loro, divise in libri cinque.* Padova: Stamperia del seminario.

Roberval, Gilles Personne de. 1636. *Traité de méchanique des poids sous-tenus par des puissances sur les plans inclinez à l'horizon, des puissances qui soutiennent un poids suspendu à deux cordes.* Paris: Charlemagne.

Stevin, Simon. 1586. *Het eerste bouck vande beghinselen der weegconst, Beschreven door Simon Steuvin...* Eng. trans. in *The Principal Works of Simon Stevin*, vol. I, ed. Ernst Crone et al. Amsterdam: C. V. Swets & Zeitlinger, 1955.

Stevin, Simon. 1634. *Les œuvres mathématiques de Simon Stevin augmentées par Albert Girard.* Leiden: Elsevier.

Stirling, James. 1717. Methodus disponendi quotcunque Sphaeras in Fornicem. In *Lineae tertii ordinis Neutoninae.* Oxford: Whistler.

Varignon, Pierre. 1687. *Projet d'une nouvelle mécanique ou statique.* Paris: Veuve Martin, Jean Boudot, Estienne Martin.

Varignon, Pierre. 1725. *Nouvelle mécanique ou statique, dont le projet fut donné en 1687, ouvrage posthume de M. Varignon.* Paris: Claude Jombert.

Chapter 2
Mathematics, Architecture and Mechanics in the School of François D'Aguilon and Grégoire De Saint-Vincent

1 Introduction

The aim of this article is not to follow a precise scientific theme, but instead, as the example of Tullio Viola invites us to do, it is to establish the connections and points of contact between different kinds of historical works. The principal theme is that of points of contact between works of art historians,[1] historians of the Jesuits,[2] and historians of science.[3]

My story begins shortly after the founding of the order of the Jesuits in 1538 and its recognition on the part of Pope Paul III in 1540. Between 1542 and 1585 the Jesuits installed themselves in numerous cities in the Flemish-Belgium and French-Belgium provinces and began to construct churches, colleges and *maisons professes* (Table 1).

Until 1 January 1580, the year of his death, Everard Mercurian, the Superior General of the Jesuits, imposed the use of *traças communes* (common plans) furnished by Rome.[4] His successor, Claudio Aquaviva, modified the system: the designs were no longer established by Rome, but had to be sent there for approval, and an exemplar of them had to be conserved in the Eternal City to guarantee that they would not be modified:

> When a design for construction of one of our houses is sent to us for approval, it is in any case transmitted in two copies to Rome so that one is conserved there and the other, after having been approved and, if necessary, corrected, is sent back to the province.[5]

On 14 July 1668, Giovanni Paolo Oliva, the successor to Aquaviva, continued in the same tone:

---

[1] Plantenga (1926), Braun (1907); de Maisière (1943).

[2] Baldini (1992), Garcia-Villoslada (1954).

[3] Ziggelaar (1983).

[4] Vallery-Radot (1960, p. 8).

[5] Vallery-Radot (1960, p. 12).

P. Radelet-de Grave, *Between Mechanics and Architecture*, Mathematics and the Built Environment 8, https://doi.org/10.1007/978-3-031-73530-1_2

Table 1 The dates of installation of the Jesuits in the various cities of these provinces, and the dates of the construction of the buildings mentioned in the present chapter

Cities	Dates of establishment	Dates of construction
Brussels	28 July 1542	
Leuven	13 August 1542	
Tournai	1554	1650–1666
Recognition of the Society of Jesuits in the Low Countries	1556	1609–1614
Kamerijk (Cambrésis)	1560	
Antwerp	1566	1583–1591
Douai		
St. Omer	1567	
Liège	1569	
Bruges	1570	
Maastricht	1574	
Lille	1578	
Valenciennes	1581	
Mons	1583	1608–1617
Courtrai	1583	
Ghent	1585	1660–1662

> With regard to the designs of our churches and establishments or of some significant building, in order that no modification can be made, the design, examined first by experts where you are, will be sent to me in two copies: one of these will be conserved here in the archives, and the other, returned with the approval and, as need arises, with the necessary corrections, will be conserved by you with the greatest diligence. Further, it is not permitted for anyone to deviate from it, a circumstance that has, however, already happened, as has been written to me more than once. For the rest, the construction of our establishments must be adequate for our religious uses; must be simple, sober, functional, and not show in any of its parts the desire to dazzle in either materials or style. It shall be a motive for edification and not realised for pomp, nor to be admired.[6]

Since each Jesuit college was obliged to conserve a copy of the design, there still exist some collections of such designs, approved and not, and then sent back. One of the most important of these is found in the *Cabinet des estampes* in Paris. There are others in the *Archives du Royaume* in Brussels and in a very prestigious collection, an "Exhibition of paintings or collection of various drawings and icons that the library of the professed house of the Society of Jesus of Antwerp has given to Pierre Dolmans, provincial officer for the Flemish-Belgian Province of this same society" (the *Promptuarium*).[7] This collection is composed of three very large volumes, two

---

[6] Vallery-Radot (1960, p. 15).

[7] Dolmans (1747. I would like to thank the reverend father Butaye for having drawn my attention to it. Fr. Butaye is the conservator of the Jesuit's library in Leuven and he allowed me to photograph the *Promptuarium*.

Table 2 The revisers of designs for Jesuit churches and establishments, 1558–1665

Giovanni Tristano	1558–1586
Giovanni De Rosis	1586–1609
Christophe Grienberger	1602–1605; 1624–1625; 1628–1633
Odon van Maelcote	1605–1610
Vincentius Filliucce	1610–1611
Orazio Grassi	1616–1624; 1626–1628
Nicolaus Zucchi	1635–1626
Ioannes Giattini	1633–1635; 1647–1649
Fabrizio Brizi	1635–1637
Gabriel Beati	1638–1639; 1646–1647; 1660–1661
Athanase Kircher	1639–1640; 1644–1646
Ioannes Antonio Martini	1640–1642
Loretus Lauri	1649–1650
Paolo Casati	1650–1651; 1653–1655
Conradus Confalonieri	1651–1653
Ignatius Tellini	1655-
Iosephus Ferroni	1657–1660
Gilles-François de Gottignies	1661–1665

of which are conserved by the Jesuits in Leuven, and the third in Antwerp in the church of St. Charles Borromeo, which contains the designs.

On the other hand, there was a Roman reviser entrusted with the responsibility of checking the designs for every building. The first two of these, Tristano and De Rosis, were architects, but starting in the seventeenth century this responsibility fell to the *Mathematicus*, that is, the professor of mathematics[8] in the college (Table 2).

Those who created the designs—the architects—only rarely signed their work. Jean Vallery-Radot curated the inventory of the signatures of the Paris collection. Found there are the names Orazio Grassi, François d'Aguilon, Adam Adamand Kochanski, Christophe Grienberger, Pierre Huyssens, all of whom we shall refer to in what follows. Also found there is Vincent Léotaud, author of a "Magnetology in which one exposes the new philosophy of magnets"[9]; he signed, in 1640, the design for the college in Embrun.[10] Also found there is François Derand, author of the well-known treatise on vaults (1643) which contains the famous rule for determining the thickness to give to a pier in order that it can support a given vault.

Having set the scene, it is now necessary to define our theme and concentrate on François d'Aguilon and Grégoire Saint-Vincent in order to show what part the architect devoted to science and what part the scientist devoted to architecture. In

[8] Garcia-Villoslada (1954, p. 335).

[9] Léotaud (1668.

[10] Vallery-Radot (1960, p. 31).

short, the aim is to overcome the barrier between history of art and history of science and to portray these figures in all their profundity.

The Jesuit father François d'Aguilon (1566–1617) is known to art historians for his contribution to the realisation of the designs for the church of St. Charles Borromeo in Antwerp. Although there are some who attribute its design to Rubens and Huyssens (1577–1637), d'Aguilon certainly played a role in this work. His "Six Books on Optics"[11] is also known to art historians since the etchings that adorn it are the work of Rubens.[12] Instead, unfortunately, few historians of science have turned their attention to the contents of this work.

For his part, Grégoire de Saint-Vincent (1584–1667) is known for the work *Problema Austriacum plus ultra quadratura circuli.*[13] This work may provoke a smile because of the attempt to square the circle, but it led to admiration among the readers of Leibniz, who acknowledged the influence of Saint-Vincent in his creation of differential calculus.

2 The Youth of D'Aguilon at the Time of the Construction of the Church of Douai

François d'Aguilon[14] was born in Brussels on 4 January 1566. He was the son of Pedro d'Aguilon,[15] secretary to Philip II, king of Spain. For three years he studied Latin and humanities at the Collège de Clermont in Paris, and later pursued his studies for two years at the university of Douai, precisely at the time when the first stone of the church in that city was laid (1583).

On 15 September 1586 d'Aguilon entered the novitiate of Tournai, where he met Henri Hoeimaker (1559–1626), who had entered the Jesuits as a builder and who would construct the Jesuit college and church of the city. After a stay in Courtrai from 15 March to 14 May 1587, d'Aguilon returned to Douai, where he met Odon van Maelcote (1572–1615). Together they attended the course in mathematics (1587–1588), which at that time also included astronomy and optics, probably with a young Laurent Delepré. D'Aguilon, who taught the "sphere of the world"[16] by Sacrobosco and, like van Maelcote, constructed instruments for astronomy, spheres, astrolabes and more. In 1600 van Maelcote was invited to Rome to present the astrolabe that

[11] D'Aguillon, *Opticorum Libri Sex* (1613).

[12] Ziggelaar (1983) provides an interesting analysis of it that is quite useful; see also Ziggelaar (2011).

[13] The title often cited, *P. Gregorii A S^{to} Vincentio Opus geometricum quadraturae circuli et sectionum coni decem libris comprehensum*, is incorrect. Jean Dhombres has written extensively about Saint-Vincent and the analysis of his work. See, for example, Dhombres (1993) and (1995), where other references may be found.

[14] Aguilon is a small city in Spain, 40 km south of Saragossa.

[15] De Maeyer (1933).

[16] Johannes de Sacrobosco, 1230 ca. The book is known by two titles: *Tractatus de Sphaera* and *De sphaera mundi*. It counts more than 200 editions.

he had constructed to Clavius and in 1602, he returned there to study mathematics in the Roman College with Gaspar Alferio. In 1603, he taught mathematics with Grienberger; among his students was Orazio Grassi, the famous opponent of Galileo, but also the author of the design of Sant'Ignazio, the church of the Roman College.

For his part, d'Aguilon, after having taken his vows on 15 September 1588, first taught Latin grammar (1589), and then physics (1591–1592), with a commentary on Aristotle's physics.

In the meantime, for the whole of this period, the construction of the church of Douai went forward. It was consecrated on 15 September 1591. The church is in the Baroque style, and the designs for it had been sent from Rome.

After a voyage to Spain (1592) and a year of theology in Salamanca (1593), d'Aguilon returned to Belgium, traveling through Ghent, where he was ordained subdiaconate in 1596; through Ypres, where he was deacon; through Tournai, where he returned to the novitiate on 6 March 1596 and took his orders twelve days later; and to Douai, where he taught philosophy. In 1598, he left Douai for Antwerp, where he became confessor of the Spanish and Italians.

The youth of Grégoire de Saint-Vincent at the time of the elaboration of the designs of the church of Mons and the church of the construction of the novitiate in Tournai

Grégoire de Saint-Vincent was born in Bruges on 8 September 1584, a year after the laying of the first stone of the church of Douai, where he would undertake his early studies (1601–1605). He entered the novitiate in Rome on 21 October 1605, and there, in 1607, he studied under Clavius.

In that epoch, in 1606, Carlo Scribani (1561–1629), d'Aguilon's predecessor as rector of the College of Antwerp, attempted to achieve, with d'Aguilon's help, the opening of a school of mathematics. To this end he circulated a petition among the merchants of the city that highlighted the positive consequences that such a school would have on the reputation of the city and its commercial activities.[17]

After that, d'Aguilon left Antwerp for Mons, where preparations were underway for the construction of a new church. The earliest designs, by the hand of Johannes du Blocq (1583–1606), date to 1607. The design drawings were sent to Rome to father Aquaviva in 1608, and the first stone was laid on 1 March 1608. The architects of this Gothic church, which today no longer exists, were d'Aguilon and Henri Hoeimaker, his companion in the novitiate of Tournai. Du Blocq did not produce a design. These three were Jesuits, but while Du Blocq had entered the Society as *faber lignarius* (joiner) and Hoeimaker as a builder, d'Aguilon was not an artisan. Construction was finished in 1617.

In the meantime, in 1608 d'Aguilon signed several designs for the novitiate of Tournai (Fig. 1). The construction of the church, also Gothic, began in 1609 and its inauguration took place in 1612. It is credited to Du Blocq, but designs for it signed by d'Aguilon have been found, and those of Du Blocq are clearly derived from them.

[17] Ziggelaar (1983, p. 47).

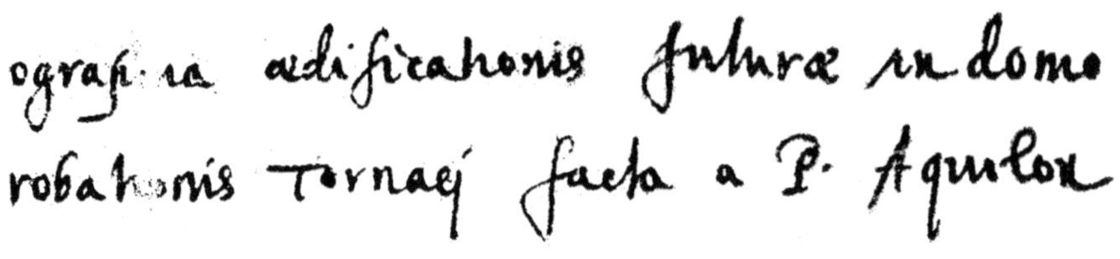

Fig. 1 D'Aguilon's signature on a design for the *domo probationis* of Tournai, never built. Public domain

In May 1611, Galileo famously presented his *Sidereus nuncius* to the Roman College. Van Maelcote chaired this session, in which Grégoire de Saint-Vincent also participated. Later Saint-Vincent twice recalled the event with enthusiasm. In a letter dated 23 July 1611, he described the day to Jooris Van der Straeten,[18] and near the end of his life, in a letter dated 4 October 1659, he described it to Huygens:

> I very much enjoyed examining your book, which I leafed through very quickly. I recall some similar kinds of phases that we observed for our pleasure, if not for hundreds of months, at least for numerous whole nights, some forty years ago, first in Belgium, then in Venice, and finally in Rome, by means of a telescope that had been given by the teacher Scholiers to one of his teachers at Antwerp, Father Odon Maelcote, who later became professor of mathematics. I would find it difficult to believe that someone had discovered this star before us, who were called the disciples of Clavius. We compared the telescope of Galileo with ours (which were by no means inferior) and, in the presence of Galileo, in front of the scholars gathered there, we explained to the Roman College all these new phenomena. We then demonstrated with all evidence that Venus travels around the Sun, a fact that did not fail to create a murmur among the philosophers.[19]

3 D'Aguilon's *Optics*

D'Aguilon's *Optics* of 1613 shows, from the very first page, the ties with Peter Paul Rubens. Both participated in the construction of the church of St. Ignatius, later St. Charles Borromeo, in Antwerp, but their relationship was not limited to this.

In his *Optics,* d'Aguilon furnishes a theory of colour (Fig. 2) that Rubens applied in two paintings: the *Annunciation* now in the Kunsthistorisches Museum in Vienna, and *Juno and Argus* in the Wallraf-Richartz-Museum in Cologne.[20]

On the other hand, both the painter and the architect were concerned with optical illusions and the effects of perspective: Rubens, to create the illusion of depth on a flat surface; d'Aguilon to accentuate, for example, the effect of depth in a church. We see things in perspective thanks to our binocular vision, as Rubens makes clear in an illustration in d'Aguilon's treatise (Fig. 3).

[18] Letter from Grégoire de St. Vincent to Jooris Van der Straeten, 23 July 1611, published in Bosmans (1902).

[19] Letter 673 from Grégoire de Saint-Vincent to Christiaan Huygens, published in Huygens (1889, II, p. 489–491).

[20] Ziggelaar (1983, tavola III, Figs. 1 and 2. 22, pp. 76–77).

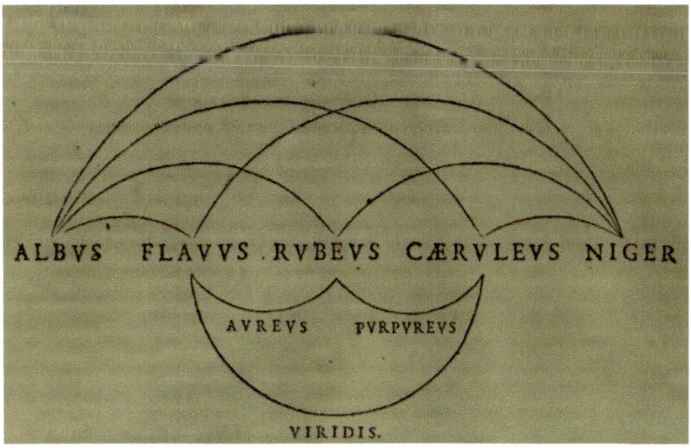

Fig. 2 Figure showing d'Aguilon's theory of colour (1613, p. 40). Public domain

By playing on the point of intersection of two straight lines that are supposed to be parallel, it is possible to give an illusion of greater depth. This idea makes it possible to depict depth in two dimensions, as well as to accentuate depth in three dimensions.

The *Promptuarium* of Pierre Dolmans (1697–1751), mentioned earlier, contains numerous examples of drawings of pavements, some of which follow ideas of Rubens for the church of Leuven (Fig. 4).

Rubens was also the author of a design for the pavement of the church of St. Ignatius (St. Charles Boromeo). This is the only element that escaped the fire of 1718 that destroyed the nave of the church, including Ruben's paintings on the ceiling.

In his *Optics*, d'Aguilon posed the inverse problem: what curve must be drawn in order for the lines to appear parallel to the observer. To do this d'Aguilon defines the "horopter", the line that passes through the point that is fixed, that is, that passes through the intersection of the visual rays coming from each of the two eyes and parallel to the line that joins the center of these. He then defines the plane of the horopter. He solves the problem in a special case, that is, when one of the lines is perpendicular to the horopter. In this case the other "straight line" must be a segment of a hyperbole in order to obtain the desired illusion (Fig. 5).

4 The Construction of St. Ignatius in Antwerp

The first design for St. Ignatius in Antwerp dates back to 1613.[21] At that time, Scribani was rector; d'Aguilon succeeded him on 22 February 1614. Starting in 1613, Pierre Huyssens (1577–1597) was proposed as *Architectus* to the *maison professe* of this city.

[21] Brigode (1934.

Fig. 3 The first page of Book III in d'Aguilon's *Optics* with an illustration by Rubens. Public domain

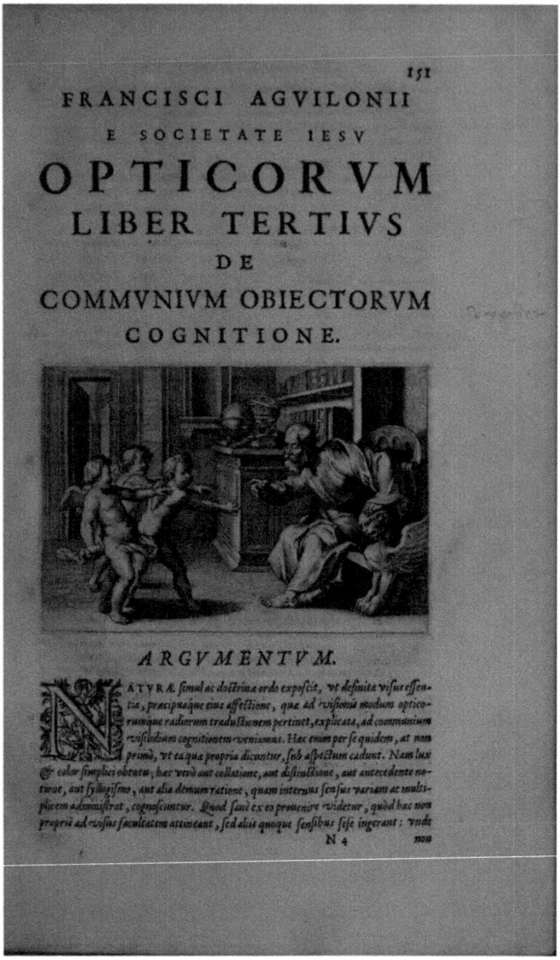

151

FRANCISCI AGVILONII

E SOCIETATE IESV

OPTICORVM

LIBER TERTIVS

DE

COMMVNIVM OBIECTORVM

COGNITIONE.

ARGVMENTVM.

NATVRÆ ſimul ac doctrinæ ordo expoſcit, vt definita viſus eſſentia, præcipuaque eius affectione, quæ ad viſionis modum opticorumque radiorum traductionem pertinet, explicata, ad communium viſibilium cognitionem veniamus. Hæc enim per ſe quidem, at non primò, vt ea quæ propria dicuntur, ſub aſpectum cadunt. Nam lux & color ſimplici obtutu; hæc verò aut collatione, aut diſtinctione, aut antecedente notione, aut ſyllogiſmo, aut alia demum ratione, quam internus ſenſus variam ac multiplicem adminiſtrat, cognoſcuntur. Quod ſanè ex eo prouenire videtur, quòd hæc non propriè ad viſus facultatem attineant, ſed aliis quoque ſenſibus ſeſe ingerant: vnde

N 4 non

After having taught mathematics in Louvain and later, in 1613, Greek in Brussels, and having been, in 1614, supervisor of the College of Bois le Duc, in 1615 Saint-Vincent attended the third year of probation in Courtrai, at the very moment in which the College obtained the authorization to build the church of St. Ignatius. He arrived in Antwerp in August, a few months after the first stone was laid on 15 April 1615.

During the period of construction of the church, d'Aguilon, who was one of the architects, was teaching mathematics. After his death in 1617, Saint-Vincent succeeded him as professor of mathematics, a post he held from 1618 until 1620, and perhaps also as architect. In those years, he defended two theses: the *Theses de Cometis* in 1619, and the *Theses Mechanicae* in 1620. Among the students in Antwerp between 1619 and 1621 was another architect: Guillaume Hesius (1601–1690).

The church of St. Ignatius, constructed in a local style onto which were grafted Baroque forms and ornaments, was due to Huyssens, d'Aguilon and Rubens. The

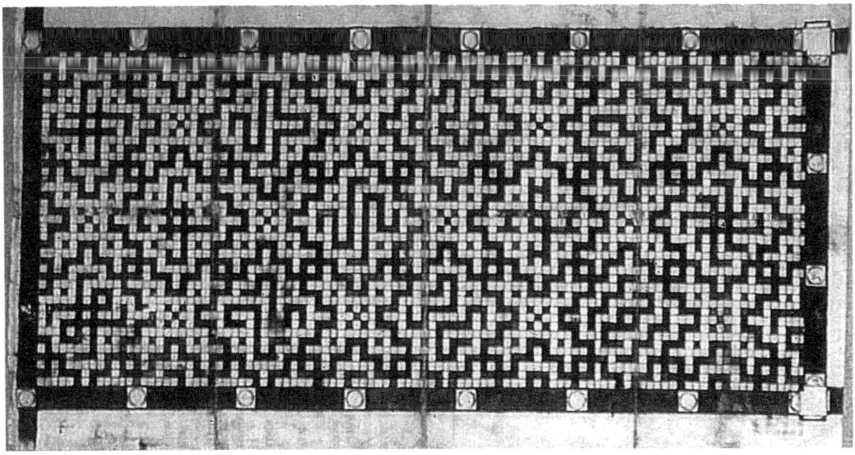

Fig. 4 A pavement design from the *Promptuarium*

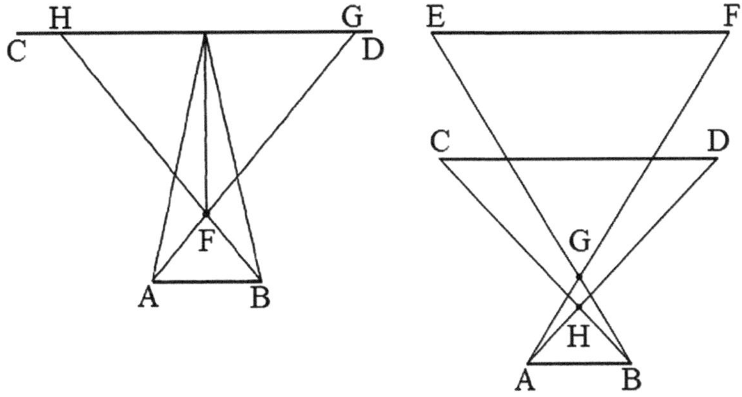

Fig. 5 Illustration of the theory of the horopter, after Ziggelaar (1983). Author

contribution that d'Aguilon made to the execution of this work is often forgotten in favour of Rubens, or only Huyssens. Plantenga quotes the test of the *Acta Sanctorum*:

> The Society of Jesus possesses in Antwerp near the *maison professe*, as we call it, a magnificent temple of antique style, constructed in the marble of Liguria; the forms of this temple were designed according to the rules of Vitruvius (contrary to what is usually done in this country, where Gothic construction usually enjoys a preference) and its foundations were laid in MDCXV by François d'Aguilon, the last rector of the College who remained until that time.[22]

[22] Plantenga (1926, p. 104).

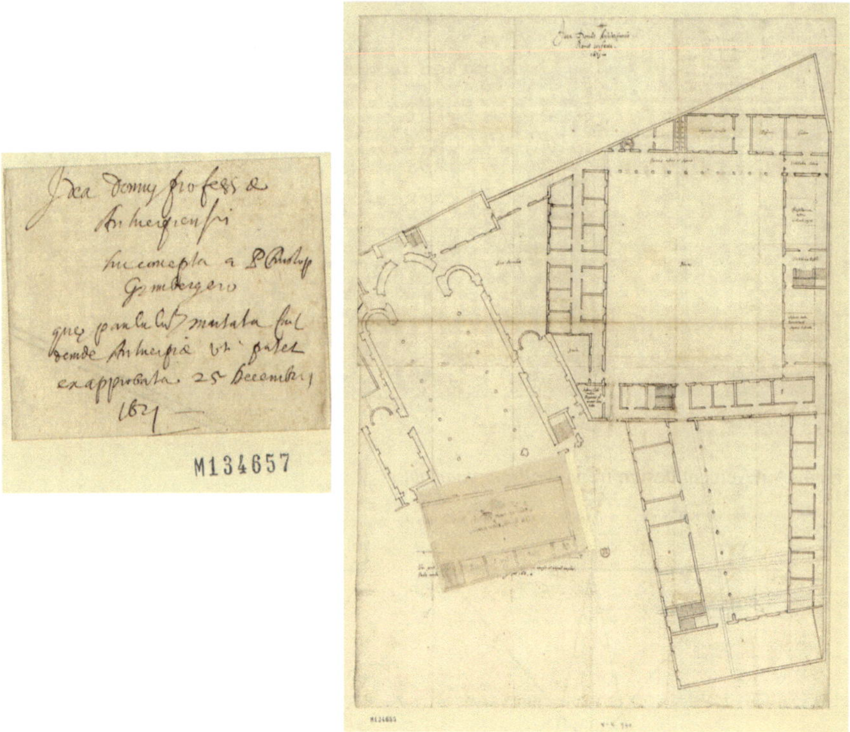

Fig. 6 Designs for the *Maison professe* of Antwerp, signed by Grienberger (Gallica)

Here we again find, as in the case of Tournai, a man of the trade, Huyssens, whose father was a master builder and who became a master builder himself, and a theoretician, d'Aguilon.

One noteworthy event took place in 1621, during the construction of the *maison professe* of Antwerp (Fig. 6). Grienberger, the *Mathematicus* of the Roman College, sent a construction drawing of his own, for this building,[23] thus entering into conflict with d'Aguilon and his student, Saint-Vincent.

Grienberger is the author of other designs in Aurillac and Charleville but he is above all known for having been a professor of mathematics in Austria and in the Roman College. We will meet him again in Prague.

[23] Vallery-Radot (1960, p. 290).

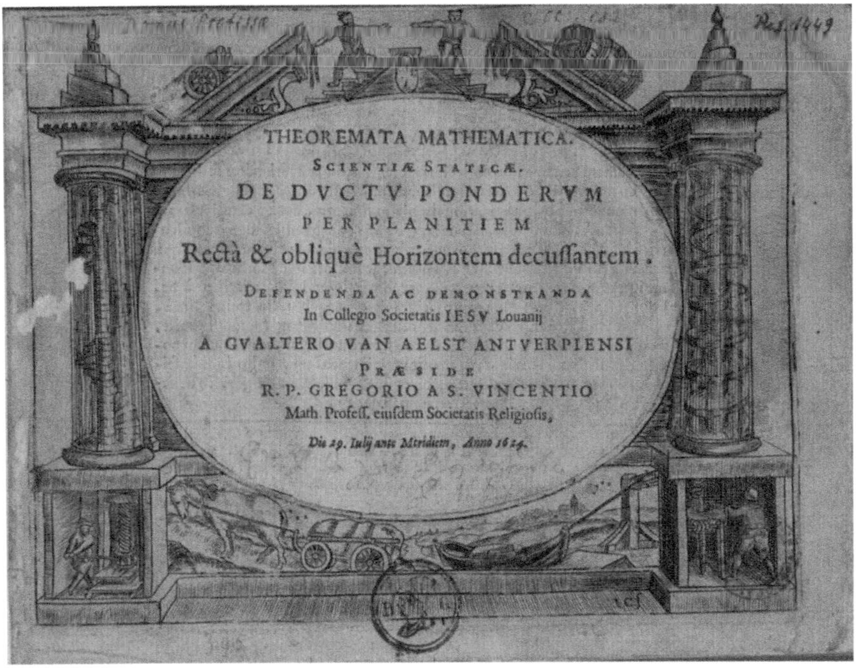
THEOREMATA MATHEMATICA.
SCIENTIÆ STATICÆ.
DE DVCTV PONDERVM
PER PLANITIEM
Rectà & obliquè Horizontem decussantem.
DEFENDENDA AC DEMONSTRANDA
In Collegio Societatis IESV Louanij
A GVALTERO VAN AELST ANTVERPIENSI
PRÆSIDE
R. P. GREGORIO A S. VINCENTIO
Math. Profess. eiusdem Societatis Religiosis,
Die 29. Iulij ante Meridiem, Anno 1624.

Fig. 7 Title page of the booklet containing the thesis defended by Van Aelst in 1624. Public domain

5 Saint-Vincent's Thesis on Statics

From 1621 to 1624 Saint-Vincent was professor of mathematics in Louvain, where he wrote several theses on mathematics. Then, in 1624, he wrote one on statics, "Theorems of the mathematical science of statics,[24] which was discussed in the morning by one student, Guillaume Van Aelst (Fig. 7), and in the afternoon by another, Jan Ciermans, focusing on the study of the inclined plane.

The interest in the inclined plane appears in the work of Adamus Adamandus Kochanski[25] (1631–1700), a mathematician in the court of Poland who signed the design for the College of Florence.[26] It is easy to image that this interest might be in some way connected to the fine examples of structures in wood that are calculated in the *Promptuarium*. The text of this thesis is certainly due to Saint-Vincent, the instructor of the two students, since both versions discussed are the same. It opens with a law of equilibrium that states that the center of gravity of the object and its support must be aligned with the center of the world. It also contains important

---

[24] Saint-Vincent (1624, *Theoremata mathematica scientiae staticae*. Except for the title pages that give the name of the student, both booklets are identical.

[25] Kochanski (1685.

[26] Vallery-Radot (1960, p. 29).

Fig. 8 Illustration from the thesis defended by Van Aelst (Saint-Vincent 1624: facing p. 14). The same figure appears in the text of the thesis defended by Ciermans. Public domain

observations about the energy acquired as the result of the fall from a certain height, which does not depend on either the length of the path travelled or the height fallen.

Further, the energy acquired as a result of the fall can be restored by making the object go back up to the height from which it departed, as is shown in the numerous children's games illustrated in Figs. 8 and 9.

During the same period Saint-Vincent was working on his most important work, that circulated under the false title *Opus geometricum* ... In 1625 he was invited to Rome to discuss it with Grienberger. Was it perhaps there that he perfected his knowledge of architecture? Between the end of 1627 and the beginning of 1628, he left Rome for Prague.

6 The Construction of the College of Prague

In Prague, Saint-Vincent taught mathematics at the College Saint Clément from 1628 to 1631. The letters of the Superior General Mutius Vitelleschi provide us with some information about his activities. On 6 April 1630 he wrote from Rome to the provincial father Christophoro Grenzino:

> also to the councilors of the College, and expressly by father Grégoire de Saint-Vincent, author of the approved design.[35]

Two days later, on 9 April 1661, Oliva informed Saint-Vincent of the essence of the contents of this letter.[36] On 26 November, he communicated directly to Hesius[37] this time, the order to obey. Oliva's defense of Saint-Vincent at this point is certainly due in part to his reputation as a teacher, as he himself had occasion to write to father Fransiscus de Cleyn in Rome, on 9 August 1662:

> That which we owe to mathematics Your Reverence knows well. This science has procured great authority for the Society and copious fruits both outside Europe and within. I congratulate the Flemish-Belgian province on the still living father Grégoire de Saint-Vincent, veteran mathematician, creator of very profound results in the mathematical sciences and still ready to acquire merit.[38]

8 Conclusion

In this article, I tried to show the important role played by Grégoire de Saint Vincent in the abandonment of the common plan (*traças communes*) imposed by Rome for the construction of new Jesuit churches.

After having observed the role of architect assumed by numerous Jesuit scientists in the seventeenth century, how can one refrain from comparing the image from the treatise on statics by Grégoire de Saint-Vincent (Fig. 7) and the dome of Sant'Ivo in Rome (Fig. 10), begun eighteen years later by Borromini?

---

[35] *Epistolae Generalium ad Flandro-Belgicam 1643–1664*, fol. 825v: *Renuntiatum est nobis P. Hesium, rectorem Gandavensis collegii, velle aliquid immutare in idea Gymnasii sui collegii, hic a nobis approbata, R^a V^a non patiatur id fieri; nolumus enim quidquam mutari, nobis inconsultis et sine nostro assensu, adeoque si videbitur aliquid immutandum, prius nobis explicentur causae, sed non a solo rectore, verum etiam a R^a V^a et auditis consultoribus collegii ac nominatim Patre Gregorio a Sancto Vincentio qui ideam approbatam formavit.* See also Henri Bosmans Archives, vol. 5, fol. 3. and vol. 7, fol. 279. With regard to the College of Ghent, Bosmans drew these notes from the Archives d'Exaten, Fland. Belg. Epist. gen. 1643–64. It would be equally interesting in this regard to analyse the correspondence of Hesius, conserved in the Archives générales du Royaume, Inv. I 40/01, Archives jésuitiques, Province Flandro-Belge, No. 704, 708, 712.

[36] See *Epistolae Generalium ad Flandro-Belgicam 1643–1664*, fol. 825: *Displicet nobis quod in fabrica vestri gymnasii, Pater Rector non sequitur ideam hic communicatam et approbatam; nec patiemur quidquam in illa mutari, nobis inconsultis et sine nostro assensu jamque de ea re opportune … mandata dedimus Patri Vice-Provenciali.* See also Henri Bosmans Archives, vol. 5, fol. 3 and fol. 130.

[37] Henri Bosmans Archives, vol. 7, fol. 280 v.

[38] Henri Bosmans Archives, vol. 7, fol. 253: *Quid Mathesi debeamus satis novit R.V. Magnam Societati authoritatem, ingentes fructus peperit illa scientia tum extra tum intra Europam. Gratulor Flandro Belgicae Provinciae de superstite adhuc P. Gregorio a S. Vincentio Mathematico veterano, bene de Mathesi profundiora commerito, et adhuc mereri parato.*

Fig. 10 Drawing, ca. 1655, of the cupola of Sant'Ivo, attributed to Domenico Barrière (Kunstbibliothek Hz. 1025); compare to the illustration shown in Fig. 7. Public domain

References

Baldini, Ugo. 1992. *Legem impone subactis. Studi su filosofia e scienza dei gesuiti in Italia, 1540–1632*. Rome: Bulzoni.

Bosmans, Henri. 1902. Documents inédits sur Grégoire de Saint-Vincent. *Annales de la société scientifique de Bruxelles* 27, 2e partie (1902): 21–64.

Braun, J. 1907. *Die belgischen Jesuitenkirchen: Ein Beitrag zur Geschichte des Kampfes zwishen Gotik und Renaissance*. Freiburg im Breisgau: Herdersche Verlagshandlung.

Brouwers, Louis. 1980. *De Jezuiten te Gent 1585–1773*. Ghent: Sint Barbara College.

D'Aguilon, François. 1613. *Francisi Aguilonii e societate Iesu Opticorum libri sex philosophis iuxta ac Mathematicis utiles*. Antwerp: Plantin Moretus.

Derand, François. 1643. *Architecture des voûtes*. Paris, Sébastien Cramoisy.

Dhombres, J. 1993. Is one proof enough? Travels with a baroque Mathematician. *Educational Studies in Mathematics* 26 (1993): 401–419.

Dhombres, Jean. 1995. L'innovation comme produit captif de la tradition: entre Apollonius et Descartes, une théorie des courbes chez Grégoire de Saint-Vincent. In : *Geometria, flussioni e differenziali*, eds. Clara Silvia Roero and Marco Panza, 13–83. Naples: La città del sole.

Dolmans, Pierre. 1747, *Promptuarium pictorum seu collection variorum delineationum et iconum quam bibliothecae domus professorum societatis Jesu Antwerpiae dedit R. P. Dolmans per Provinciam Flandro-Belgicam ejusdem societatis praepositus provincialis.* Jesuit library in Leuven and KBR Brussels.

Garcia-Villoslada, R. 1954. *Storia del Collegio Romano dal suo inizio (1551) alla soppressione della Compagnia di Gesú (1773).* Rome: Aedes Universitatis Gregorianae.

Gilissen, J. 1938. Le Père Guillaume Hesius, Architecte du XVIIe siècle. *Annales De La Société Royale D'archéologie De Bruxelles* 42:216–255.

Huygens, Christiaan. 1889. *Oeuvres complètes de Christiaan Huygens publiées par la Société hollandaise des Sciences.* The Hague.

Kochanski, Adam Adamand. 1685. Consideratio Speciminis Libri de Momentis Gravium, Authore J.F.V. Lucensi. *Acta Eruditorum*, 262–265.

Léotaud, Vincent. 1668. *Magnetologia in qua exponitur nova magneticis philosophia,* Lugduni, Anisson.

De Maeyer, C. 1933. Le Père François Aguilon, architecte jésuite du XVIIe siècle. *Bulletin de la Société royale d'Archéologie de Bruxelles* 9, December 1933, 113–122.

De Maisière, Thibaut. 1943. *L'Architecture religieuse en Belgique à l'époque de Rubens.* Brussels: Editions du cercle d'art.

Plantenga, J. H. 1926. *L'Architecture religieuse dans l'ancien duché de Brabant depuis le régne des archiducs jusqu'au gouvernement autrichien (1598–1713).* The Hague: M. Nijhoff.

Saint-Vincent, Grégoire de. 1624. *Theoremata mathematica. Scientiae staticae. De ductu ponderum per planitiem rectà & obliquè horizontem decussantem. Defendenda ac demonstranda in Collegio Societatis Iesu Louanij a Gualtero van Aelst Antuerpiensi praeside R.P. Gregorio a S. Vincentio ...* Leuven, Hasteni.

Simon, B. 1934. Projet de l'église des Jésuites à Anvers. *Bulletin De L'institut Historique Belge De Rome* 14:157–174.

Vallery-Radot, J. 1960. *Le recueil de plans d'édifices de la Compagnie de Jésus conservés à la BN Paris.* Paris: Bibliothèque Nationale.

Ziggelaar, A. 2012. François De Aguilón's *Opticorum Libri Sex*—First Description of the Horopter, the Area of Single Vision, in 1613. *Strabismus* 20 (1): 166–167.

Ziggelaar, August. 1983. *François de Aguilon S. J.* (1567–1617), *Scientist and Architect.* Rome: Bibliotheca Instituti historici.

Chapter 3
Is the Flat Vault a Lost Proposition in John Wallis's Work?

1 John Wallis and His Works

Without attempting a biography of John Wallis (1616–1703) (Fig. 1), of whom we have an autobiography (Scriba 1970), we would like to draw attention to some elements of his life and work. After having considered mathematics as a *divertissement* and focusing his attention on theology, chance led him, around 1642 to become interested in cryptography. In 1647, he discovered Oughtred's *Clavis Mathematicae* (*Key to mathematics*), which inspired his interest in the subject. It was between these two events that, in 1644, as he himself recounts in the introduction to Proposition X, he was set himself to solving the problem of the flat vault. In the meantime, he published a grammar of the English language (Wallis 1653), but then he let himself be carried away by the study of mathematics, especially the sections of the angle, conic sections, infinite developments with the *Arithmetica infinitorum* (Wallis 1656), then developments in series and mechanics. This was followed by the publication of a treatise on logic (Wallis 1687). His mathematical work is collected in his three-volume *Opera Mathematica* (Fig. 2). This esteemed scholar was part of the group of men who formed the Royal Society and for fifty years, until his death, he was professor of mathematics, holding Henry Savile's chair at Oxford.

In addition to the eclecticism of his interests, which oscillated from theology, through cryptography and logic, to the purest mathematics of his time, Wallis's purely mathematical *Opera Mathematica* published in three volumes in 1695 clearly shows us the scholar of the seventeenth century. Indeed, the third volume is entirely dedicated to re-editions of ancient texts with commentaries, such as Ptolemy's *Harmonics* with Porphyry's commentary, Archimedes' *The Sand Reckoner* and *Measurement of a Circle*, and Aristarchus's *On the magnitudes and Distances of the Sun and Moon*. On the other hand, the *Arithmetica infinitorum*, the *De Algebra*, the *Sectionibus Angularibus Tractatus*, and the *sectionibus conicis*, which contain *inter alia* his research on series, making him a forerunner of differential and integral calculus.

P. Radelet-de Grave, *Between Mechanics and Architecture*, Mathematics and the Built Environment 8, https://doi.org/10.1007/978-3-031-73530-1_3

Fig. 1 Portrait of John Wallis by D. Loggan, engraved in 1668 and published in *Opera Mathematica*. Public domain

The final part of volume I of the *Opera Mathematica* is dedicated to a re-edition of the *De Motu*, which Wallis had published in three parts in 1669, 1670, and 1671 (Fig. 3).

The first part is dedicated to the three following themes:

- movement in general (*De Motu, Generalia*);
- falling weights (*De Gravium Descensu*), and the motion on inclined planes (*Motuum Declivitate*);
- the balance (*De Libra*).

The second part is entirely dedicated to the center of gravity:

- the center of gravity (*De Centro Gravitatis*);
- its calculation (*De Calculo Centri Gravitatis*).

The third part is more varied, taking up the pseudo-Aristotelian problems again, but adding new questions directly inspired by Galileo:

- the lever (*De Vecte*) supported by either one, two or more supports;
- axles and wheels (*In Axe cum Peritrochio, (& Potentiis cognatis…)*) (Fig. 4);
- the pulley (*De Trochlea, seu Polyspasto*);
- Archimedes' screw (*De Cochlea*);
- compound movement, accelerated movement and movement of a projectile (*De motibus compositis, Acceleratis, Retardatis, et Projectorum*);
- percussion (*De Percussione*);
- the wedge (*De Cuneo*);
- *the elastic force* (*De Elatere, & Resilitione seu Reflexione*);
- hydrostatics (*De Hydrostaticis*);
- Various questions of mechanics (*Epilogus, ex Miscellaneis*).

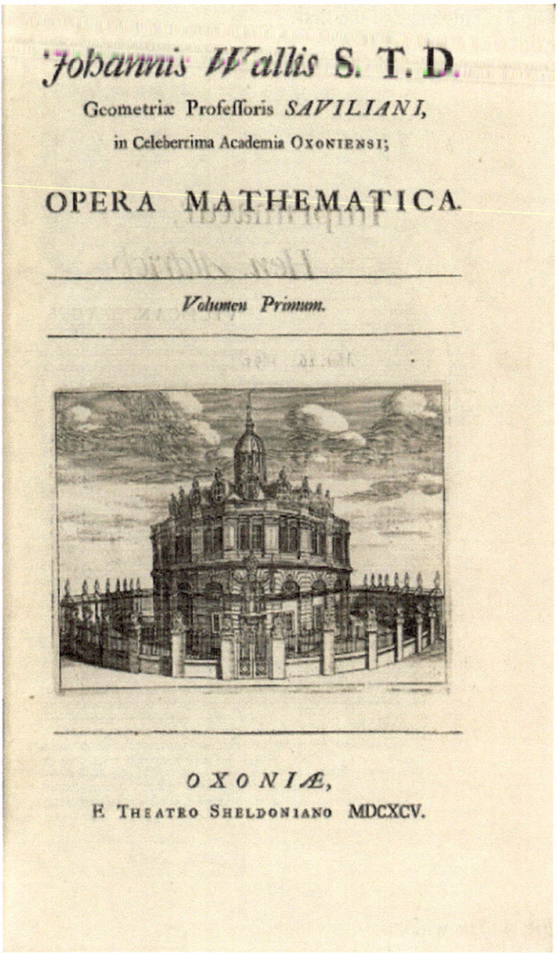
Johannis Wallis S. T. D.
Geometriæ Professoris *SAVILIANI*,
in Celeberrima Academia OXONIENSI;
OPERA MATHEMATICA.
Volumen Primum.
OXONIÆ,
E THEATRO SHELDONIANO MDCXCV.

Fig. 2 Title page of Wallis's *Opera Mathematica* (1695), published in three volumes. Public domain

If at first sight this list of contents appears to fall within the lineage of the commentators of Pseudo-Aristotle with the study of simple machines, an Archimedean influence is clear with the calculation of centers of gravity, which immediately brings to mind the premises of integral calculus. Two Galilean themes are added, those of accelerated and decelerated movements; these—and this is significant—occupy the place of the chapter which should be entitled "on the inclined plane", and collisions, which this time presume the influence of Descartes.

Let us now concentrate on the first section of the third part.

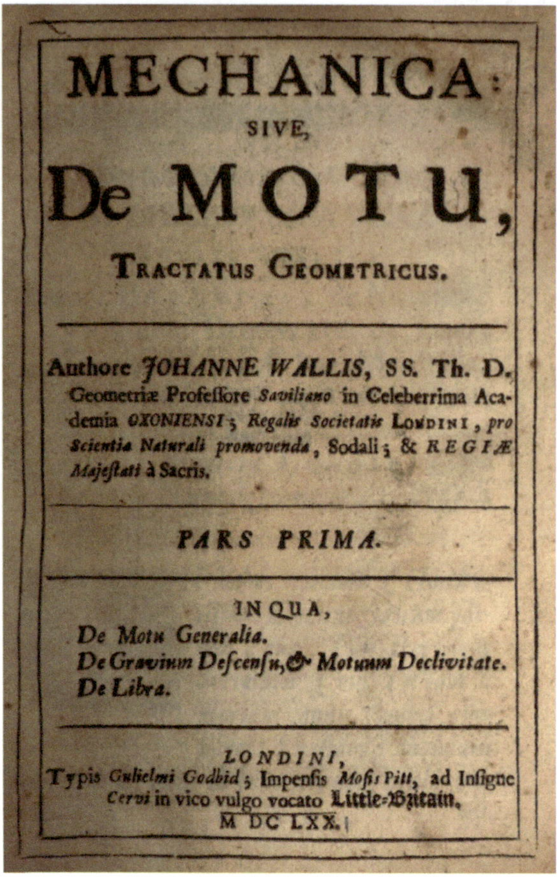

MECHANICA:
SIVE,
De MOTU,
TRACTATUS GEOMETRICUS.

Authore *JOHANNE WALLIS*, SS. Th. D.
Geometriæ Professore *Saviliano* in Celeberrima Academia *OXONIENSI*; *Regalis Societatis* LONDINI, *pro Scientia Naturali promovenda*, Sodali; & *REGIÆ Majestati* à Sacris.

PARS PRIMA.

IN QUA,
De Motu Generalia.
De Gravium Descensu, & Motuum Declivitate.
De Libra.

LONDINI,
Typis *Gulielmi Godbid*; Impensis *Mosis Pitt*, ad Insigne *Cervi* in vico vulgo vocato Little-Britain.
M DC LXX.

Fig. 3 Title page of the first edition of *De Motu*, 1670. Public domain

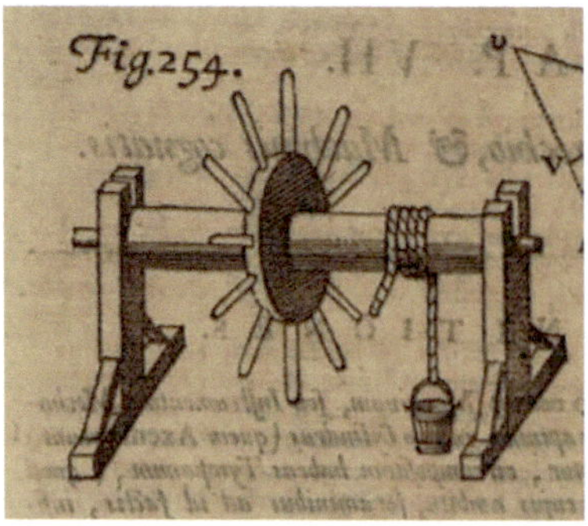

Fig. 4 The windlass depicted in Ch. 7, *De Axe in Peritrochio* (Wallis 1695, p. 966). Public domain

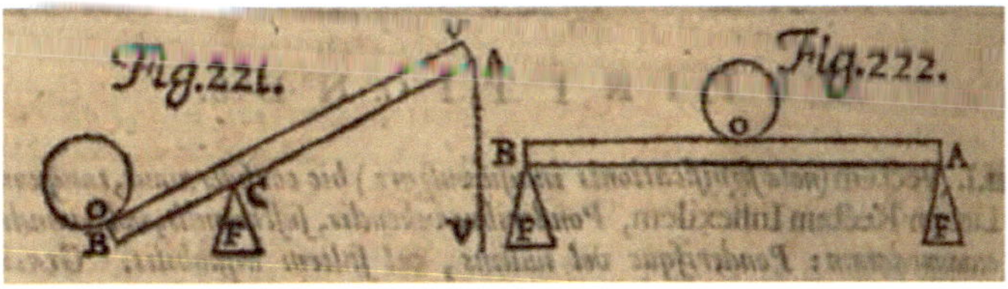

Fig. 5 Wallis's Figure 221 shows the classic lever, and Figure 222, a beam loaded with the ball O, which rests on two supports (Wallis 1695, p. 942). Public domain

2 The Lever, Supported by Either One, Two or More Supports

The title of this first section, which includes our proposal X on flat vaults, immediately draws the attention of the historian of mechanics. The lever, supported by either one, two or more supports. What might a lever supported in two points be, and especially, what can its utility be?

The part of the *De Motu* dedicated to the balance opens with definitions, the second of which states that the balance arm or lever is an oblong beam. Having defined it in this way, Wallis can then also consider the case of a beam on two or more supports. But here we must emphasize the originality of Wallis's approach, because he is the only one to make the transition to architecture in the context of the study of simple machines. Neither the beam on two supports nor the flat vault are, to our knowledge, found in any other author of this period dealing with simple machines. Vitruvius undertakes the analysis of *quaestiones mechanicae* in the context of architecture but does not attempt to generalize the lever to the beam on two supports. The author who comes closest to this is Galileo, and we may consider his influence, but the problem of the breaking of a beam that haunts Galileo is different from that of its stability which motivates our author. Proposition X on flat vaults, which is also the last proposal of the lever section, is not a mere exercise but a culmination. Its particularly long and explicit title—"To build a flat floor by means of beams connected to each other and much shorter than the side of the surface of this floor…: and to estimate by calculation, the load that falls on each of its joints"[1]—shows clearly that this is a question of architecture. Let us see how Wallis prepares this study in the chapter on the lever and the beam on two or more supports.

After the three definitions, Wallis considers the two problems in parallel (Fig. 5).

Proposition 1 (p. 942) goes back to the lever having only one support to the balance, which closes the subject, since the balance has already been treated. There remains the beam, *vectis—Tignum, palanga, seu Trabs oblonga*, as he defines it in Prop. V (p. 946), of which he distinguishes four different cases (Fig. 6).

[1] Wallis, *Opera Mathematica* (1695, p. 953): *Contignationem planam ex Tignis multo Brevioribus quam sit Areae Latitudo invicem conjunctis construere: Et computo aestimare, Quantum cuique Juncturae Onus incombat.*

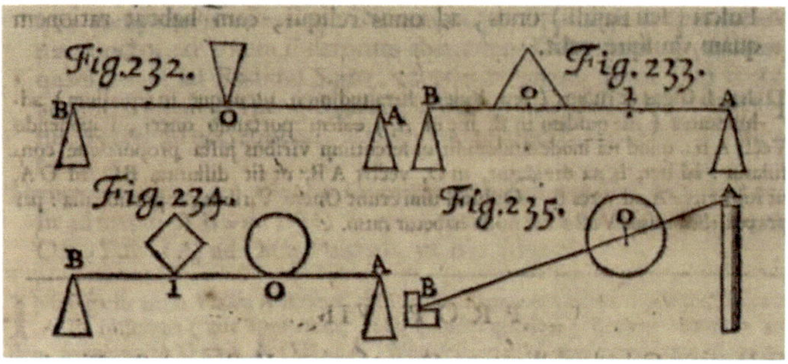

Fig. 6 Figures 232 to 235 describe the various problems posed by the loaded beam (Wallis 1695, p. 947)

- case 1 (his Fig. 232): The unloaded but homogeneous beam weighs equally on each of the supports. The weight of the beam can be reduced to a central load which also weighs equally on each support. In both cases, each support supports half the load.
- case 2 (his Fig. 233): If instead the load is not centered, the support A bears a load equivalent to the force that would have to be exerted on A on a lever having only the support B and loaded with the said weight. That is, the load supported by the support is the load applied as OB to OA. This follows from the theory of the balance or of the lever.
- case 3 (his Fig. 234): If there are several loads, it suffices to consider the load collected at the center of gravity of these and to apply the preceding rule.
- case 4 (his Fig. 235): The reader will continue the exercise, for this case does not apply in the case of the flat vault.

Proposition VII (p. 948) considers breaking and indicates, in a scholium, that in architecture one must estimate the resistance of the beams and choose them so that they resist. Here Wallis shows that architecture is the goal of the chapter.

Proposition IX (p. 951) considers the case of more than two supports and begins with the case of the equipartition of the load on three supports in the case of three beams consolidated in O and whose bearing points form an equilateral triangle. To go further, the author needs his first proposition on the centers of gravity (Fig. 7):

- if a weight (or a set of consolidated solids) is supported at a single point, such as in its center of motion, it can at least move from any side around that point but not otherwise;
- if it is supported at two points, or in several aligned points, of its axis of motion, it will be able to move on both sides around this axis but not otherwise;
- if it is supported by three (or more) non-aligned points, it will not be able to move on any side.[2]

[2] Wallis, *De Centro Gravitatis*, Proposition I (*De Motu*, Partie II, 1670, p. 112; *Opera Mathematica*, 1695, p. 646):

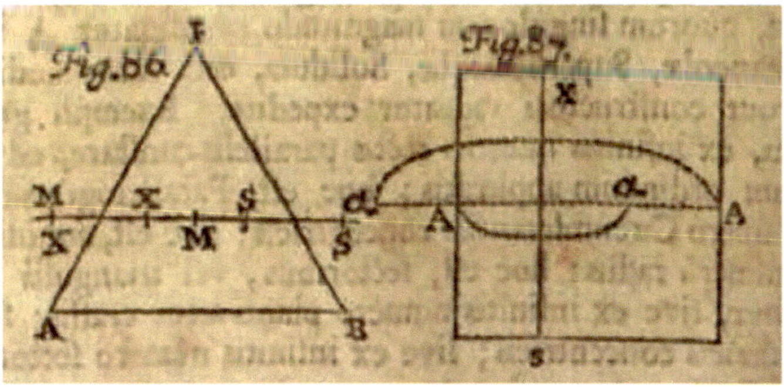

Fig. 7 Wallis's Figs. 86 and 86 illustrating Proposition I of *De Centro Gravitatis* (Wallis 1695, p. 646). Public domain

Let us now return to proposition IX and to the three beams consolidated in O and whose support points form an equilateral triangle (Fig. 8, left). Wallis extends the beam AO to its intersection with BC which links the support points of the other two beams. Thus, it is no longer, from what we have just seen, to prevent the lever AOD loaded at O from turning around the axis BC. The load at A will therefore be to the total load in O as OD to AD. Likewise for the loads in B and C, *mutatis mutandis.* This reasoning remains valid regardless of the shape of the triangle ABC. Wallis complicates the problem by adding points of support (Fig. 8, right).

3 Proposition X

Finally, we come to Proposition X: "Construct a flat floor consisting of beams much shorter than the width of its surface and connected together: And estimate by calculation, the load that falls on each joint".

As an introduction to this proposal, Wallis points out that the problem can be solved in various ways, and that he believes he gave the best way at the Queen's College in Cambridge, of which he was a member for a year, until his marriage in 1645. He says he built such a wooden beam vault during the evening committees at Oxford in 1652 and finds that the theory can be reduced to practice. He explained the same thing again the following year, still at Oxford, under the same circumstances. The subject pleased, was imitated, and was reported to King Charles II who, after

---

Si Puncto unico, ut Centro Motus, sustineatur Grave (vel ex pluribus conjunctis gravibus aggregatum:) poterit nihilo minus in quavis partes, circa illud rotando moveri ; non alias.

Si duo punctis, pluribusve in eadem recta, ut motus Axe sustineatur: poterit, circa illam rectam, in utramvis partem rotando moveri; non alias.

Si tribus (pluribusve) non in eadem recta punctis sustineatur: in nullas partes movebitur.

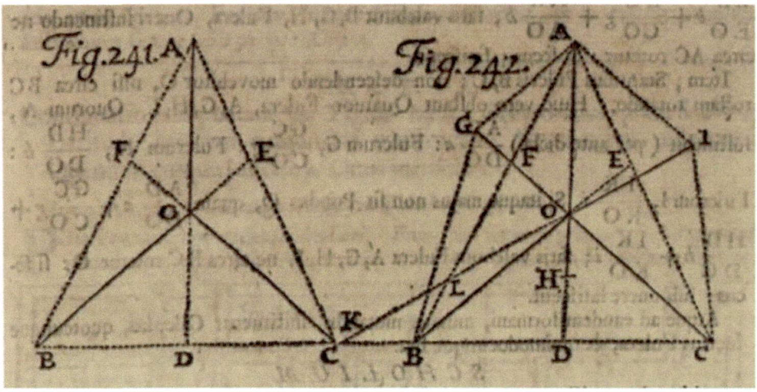

Fig. 8 Wallis's Fig. 241 and 242 relating to the problem of three beams consolidated in O and whose supports form an equilateral triangle and generalization in Fig. 242 to 5 points of suspension (Wallis 1695, p. 951). Public domain

his return to the throne, on 18 October 1660 had the text translated into English and judged it worthy of appearing among his precious objects and treasures (his *keimêlia*, writes Wallis in Greek). Wallis describes the construction of this vault and calculates the load at each important point of the structure.

The flat and square vault is constructed using two types of wooden pieces (Fig. 9). The length of the longest beam is one-fourth of the side of this square and the length of the shorter one is two-thirds of the longest. The beams are of homogeneous material, so their weights are in the same ratio as their lengths. The notches that we see in the image are always carved to half the thickness of the beam, in order that the beams fit smoothly without unevenness. The plan of the vault is shown in Fig. 10.

The surface thus created can be supported on walls arranged in a square shape, or on columns. The structure begins on one wall and continues gradually to the wall opposite. It is also possible to prepare more complex elements, such as squares, that are then fixed to each other. Wallis points out that the square-based grid that he shows in his Fig. 246 could be replaced by the simplest grid, namely the triangular grid of his Fig. 247 (Fig. 11).

Then he explains that one cannot build in this way to infinity, not for mathematical reasons but because of the limited resistance of materials. He leaves this kind of problem to the architects' expertise.

These preliminaries completed, Wallis addresses the calculation of the arrangement of the vault. He sets the weight of a large beam as T, and therefore the weight

Fig. 9 Wallis's Figs. 244 and 245 showing the shapes of the two types of wooden pieces that make up the vault (Wallis 1695, p. 954). Public domain

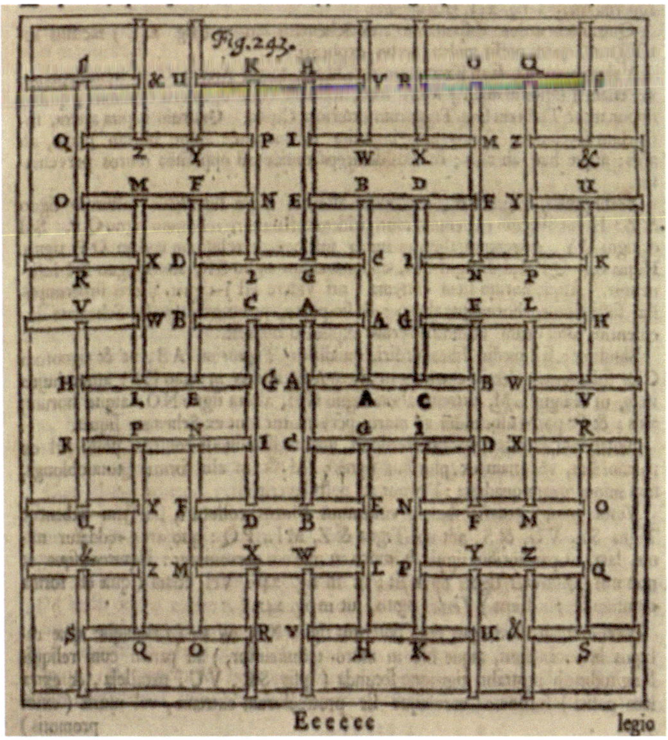

Fig. 10 Wallis's Fig. 243 showing the plan of the vault (Wallis 1695, p. 953)

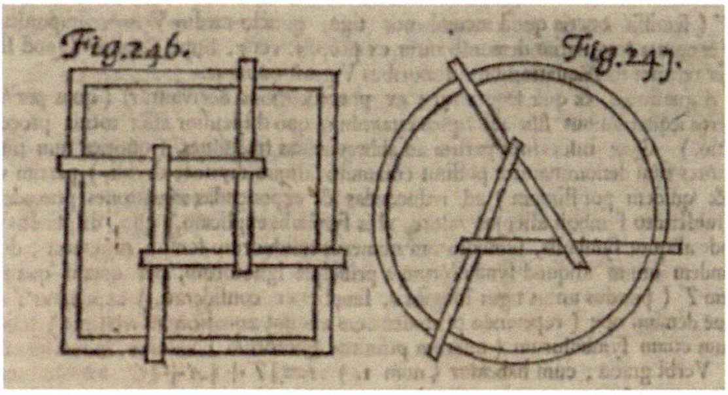

Fig. 11 Wallis's Figs. 246 and 247 showing two elemental grids, one square supported by a square, and the other triangular supported by a circle (Wallis 1695, p. 955). Public domain

of a small beam is 2T/3. Because of the symmetry of the whole, in each case there are four points that are subject to the same load because they have the same position within the whole. He gives the same name to the points that support the same load. All that remains then is to apply the theory that has just been developed. Let's explain one or two equations in detail.

Let us consider one of the four beams AB of the central square AAAA, and calculate the load in B (Fig. 12). We know that A and B each support half of T, the weight of the beam. But each one also supports a part of the load placed on A (another point A of the central square) and in C.

By the above, the load in B is to the load in C, as the distance from C to A is to the distance from B to A. Thus B/C = 2/3. The load in B is to the load in A as the distance from A to A is to the distance from B to A, that is to say B/A = 1/3.

Thus, the load in B is the following: B = T/2 + 2C/3 + A/3. That corresponds to Wallis's first equation.

We can still calculate the load in W, that is to say Wallis's equation 21. This time it is a small beam whose weight is 2T/3.

The small beam with one end located at W and the other placed on the wall is loaded at V (Fig. 13). W supports half the weight of the beam and part of the load at V. The load in W is at the load in V as the distance from W to V is to the distance from W to the wall: W/V = ½. The total load in W will thus be: W = T/3 + V/2. Wallis thus obtains a system of 25 linear equations with 25 unknowns, which he will then solve in order to determine the loads with respect to the weight T in 25 points (Figs. 14 and 15).

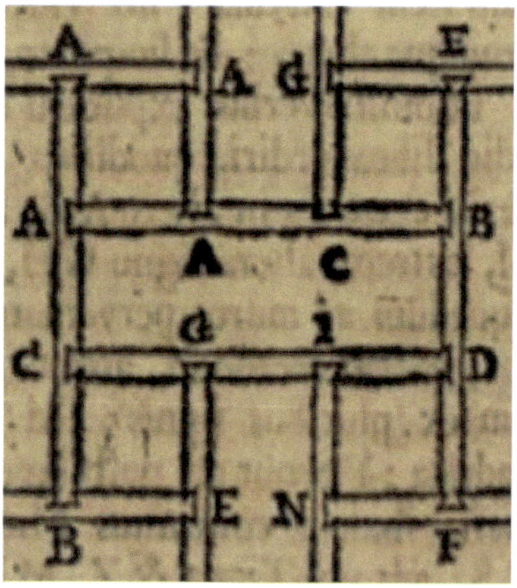

Fig. 12 Detail of the plan shown in Fig. 10, where we see the beam AB whose load in B is evaluated

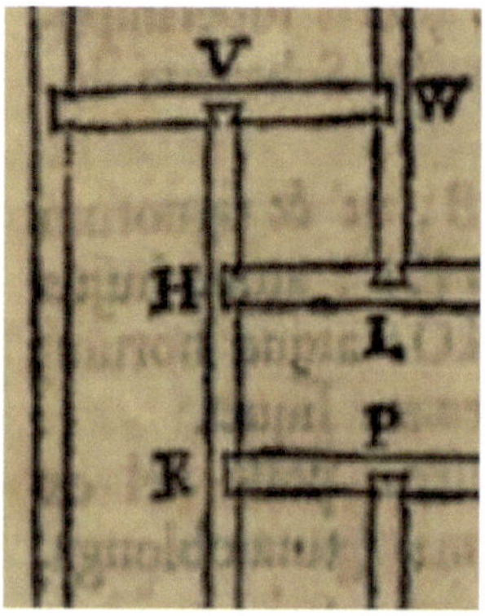

Fig. 13 Detail of the plan shown in Fig. 10, where we see the beam with one end at W and the other placed on the wall, which has not received a name. It is loaded in V

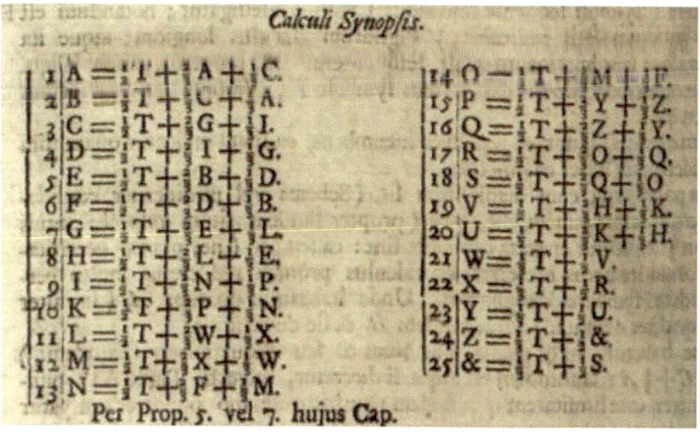

Fig. 14 The 25 equations established by Wallis (Wallis 1695, p. 956). Public domain

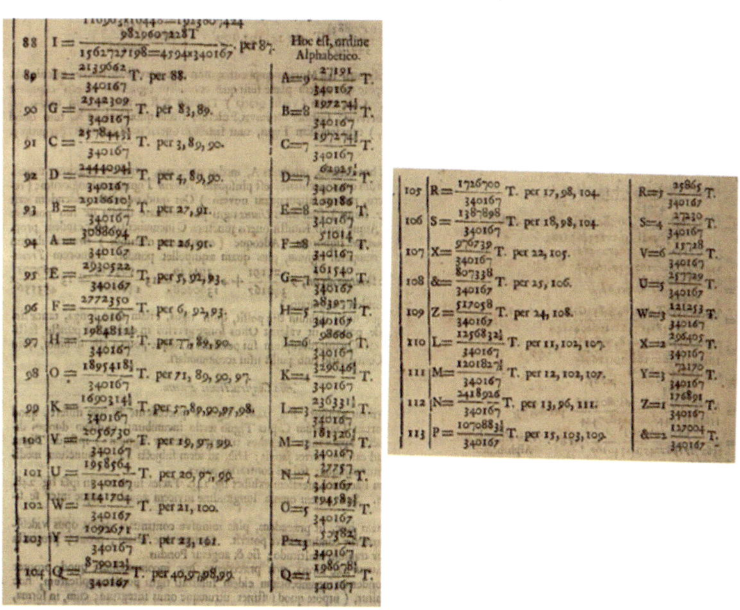

Fig. 15 The solutions he finds for these equations (Wallis 1695, pp. 959–960). Public domain

Wallis describes a second plan for a flat vault, which this time uses only one kind of beam and for which he also calculates the loads (Fig. 16).

The author ends by giving four other sketches of plans for flat arches, but he does not calculate them, and is content with comparisons with the previous ones (Fig. 17).

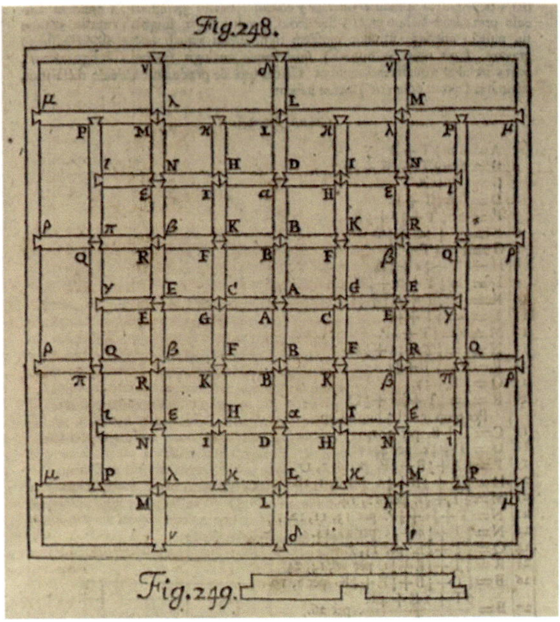

Fig. 16 Wallis's second plan of a flat vault and the shape the elementary beam (Wallis 1695, p. 961). Public domain

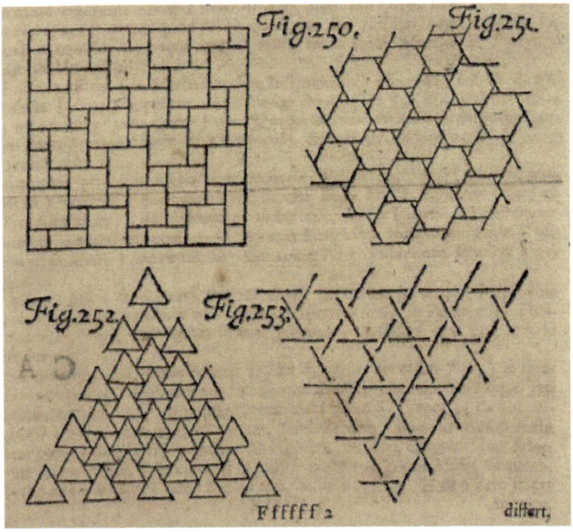

Fig. 17 Four propositions of plans for flat vaults (Wallis 1695, p. 963). Public domain

4 Jean Truchet (1657–1729)

The patterns shown in Fig. 17 show us clearly that a problem of tiling hides behind the construction of these flat vaults. We find proof of this in the collection of machines of the Academy of Sciences published in 1735, containing a short article devoted to two flat vaults invented by Jean Truchet (known as Père Sébastian, the name he adopted when he entered the Carmelites), noted as no. 31 of 1699.

The first vault (Fig. 18) is composed of a relatively simple piece, while the one forming the second vault (Fig. 19) is very sophisticated. The plates show each time and from various points of view the details of the piece that is reproduced, the apparatus of the piece and the general result. The text that accompanies it was written by the publisher of the journal and indicates only that these vaults are supported on four walls, while often, as in the case of the arch of Wallis, they are supported only on two walls. Wallis is not mentioned; it is difficult to know if Truchet read the Oxford professor's work.

Truchet published only a few works. He was a man of technical skills, *un grand Mécanicien* (a great mechanic) says the praise in the "Eloge de P. Sébastien Truchet, Carme" (Anon 1731) published by the Académie, and was known mainly as a watchmaker. He did, however, leave a perfectly mathematical dissertation entitled "Mémoire sur les combinaisons" (Truchet 1704). This paper is one of the first systematic studies on pavements, as can be seen on the plates reproduced in Fig. 20. It is a paving whose grid is as simple as possible: a square divided in two halves according to its diagonal, of which one half is black while the other is white.

Truchet's Table 1 shows us at a glance his systematic way of studying all possible tilings, while Table 4 shows some of the designs he produced from the results.

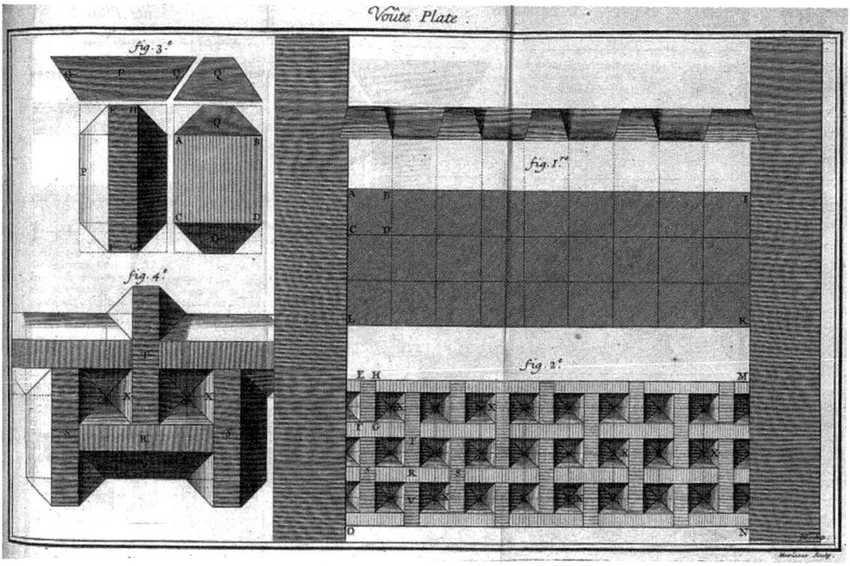

Fig. 18 Truchet's first plate where he describes his first flat vault (Truchet 1735, Pl. 1). Public domain

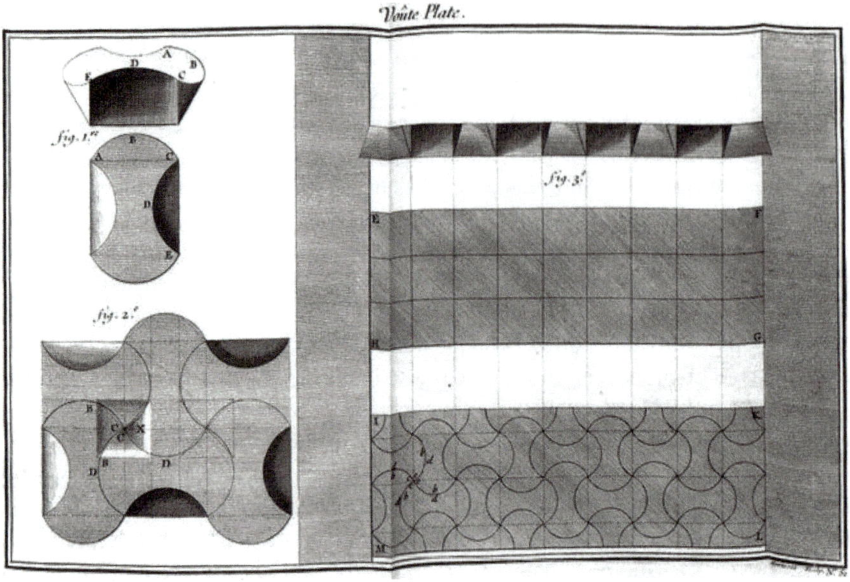

Fig. 19 Truchet's second plate where he describes another flat vault whose basic element is of very sophisticated form

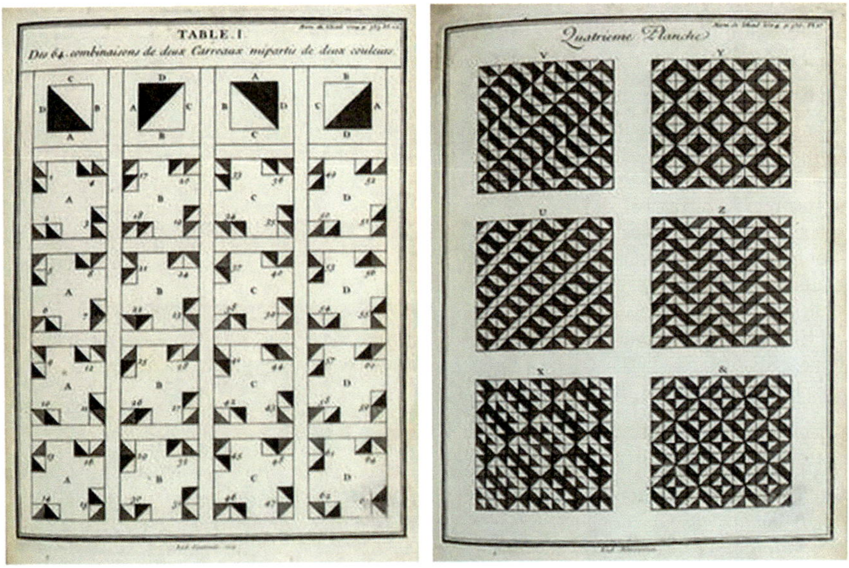

Fig. 20 Tables 1 and 4 of Sebastian Truchet's article on paving patterns (Truchet 1704, Tab. I and Pl. 4). Public domain

5 Conclusion

Proposition X on the flat vault is not lost in the work of Wallis. He has organized the entire study of the lever, one of the traditional simple machines, with the aim of introducing, in addition to the simple lever, the beam on two or more supports. After reducing this to the lever, an instrument that he finally reduces to the balance, he is able to calculate the loads at each important point of the models of flat vault that he proposes.

The last figures of Proposition X (Fig. 17) allows us to better understand the source of Wallis's interest. We realize thanks to these, and thanks to the work of Father Sébastien Truchet, that the problem of the flat vault conceals a problem of tiling and therefore of combinatorics. We had implied this by talking about grids. The regularities or symmetries that these problems raise *ipso facto* have repercussions on permutation properties at the level of the writing and the solving of the equations. This is where the interest of our cryptographer, a master in handling series, lies.

References

Anon. 1731. Eloge de P. Sébastien Truchet, Carme. *Histoire se L'Academie Royale Des Sciences. Année MDCCXXIX. Avec les Memoires de Mathematique & de Physique, pour la même Année*, 93–101. Paris: Imprimerie Royale.

Scriba, Christoph J. 1970. The autobiography of John Wallis, F.R.S., *Notes and Records of the Royal Society of London* 25, pp. 17–46.

Truchet, Jean (Père Sébastien). 1704. Mémoire sur les combinaisons. *Histoire de l'Académie royale des Sciences pour MDCCIV avec les mémoires de mathématiques et de physique pour la même année*, 363–373. Paris: Martin, Coignard, Guerin.

Truchet, Jean (Père Sébastien). 1735. Voûte plate inventée par le Père Sébastien de l'Académie royale des sciences. *Machines et inventions approuvées par l'Académie royale des sciences depuis son établissement... avec leur description, Tome Premier depuis 1666 à 1701*, 163–164. Paris: Boudet.

Wallis, John. 1653. *Grammatica Linguae Anglicanae*. Lichfield: Thomas Robinson.

Wallis, John. 1656. *Arithmetica infinitorum, sive Nova Methodus Inquiriendi in Curvilineorum Quadraturam, aliaq; difficilora Matheseos Problemata*. Lichfield: Thomas Robinson.

Wallis, John. 1687. *Istitutio logicae ad communes usus acommodata*. Oxford: E Theatro Sheldoniano, prostant apud Amos Curteyne.

Wallis, John. 1695. *Mechanica: sive De Motu Tractatus Geometricus* in Wallis, *Opera Mathematica*, Vol. I (Oxford: Sheldon, 1695), 571–1063 (1st ed. London: G. Godbid and M. Pitt, 1670).

Chapter 4
The Use of a Particular Form of the Parallelogram Law of Forces for the Building of Vaults (1650–1750)

1 Introduction

When I told Clifford Truesdell that I was working on the history of the parallelogram law of forces, he asked me if I knew Edoardo Benvenuto, who had written an article on that argument. That is the reason why, some years later, I met Edoardo Benvenuto. Therefore, at the conference in Genoa 30 November–1 December 2001 in honour of Benvenuto and Truesdell, as a record of that first meeting, I wanted to speak of the parallelogram law of forces in the context of the study of the stability of vaults, But as we all know, one of the principal, if not the only, interests in the history of mechanics of Clifford Truesdell as well as of Edoardo Benvenuto, was to understand how people discovered fundamental notions, laws and principles. Reading those texts on vaults once more, I realised that the history of the study of stability of a vault under its own weight during the period 1650–1750 was important for the discovery of other fundamental notions, so I decided to enlarge the aim of the present paper in order to account for it. I'll try to answer the question of what the study of that problem brought to the knowledge of fundamental mechanical notions.

As mentioned, the story will take place during a period from 1650 to 1750. I'll tell that story from my point of view using original texts but also referring to the work of Truesdell and Benvenuto. I'll use principally the two books that defined our colloquium and two other books, one by Truesdell, *Essays in the history of Mechanics* (1968), and one by Benvenuto, *La scienza delle costruzioni ed il suo sviluppo storico* (1981).

But I should explain first the particular problem of which I intend to study the development, because the problem of the stability of an arch has many components. The one I want to concentrate upon is which is the best form to give to an arch in order for it to be stable under its own weight. But the correct definition of the problem is, as always, an essential part of the difficulties scientists will have to overcome.

P. Radelet-de Grave, *Between Mechanics and Architecture*, Mathematics and the Built Environment 8, https://doi.org/10.1007/978-3-031-73530-1_4

2 Analogy Between Vault and Catenary

To understand the history of the stability of vaults and, more precisely, of the best form to give to an arch, it is essential to know that in studying that argument, we are dealing with the history of techniques and that empirical knowledge existed since antiquity, in all civilisations. We can still see it in some important monuments. On the other hand, theorisation of that subject arose rather late. In fact, the first who tried to theorise the problem was Philippe de la Hire in his *Traité de Mécanique* (1651), that is, eleven years after Leibniz invented differential calculus,

It is also important to know that the history of the theorisation of the form of the most stable vault is linked to the history of the catenary, the form assumed by a chain hanging freely, and that that history is linked to the birth of the calculus. Omitting Galileo's question about the catenary and his attempt to solve it by the composition of movements, which lead to the erroneous answer, the parabola, we turn directly to Jacob Bernoulli, who posed that question in May 1690, first to Leibniz, then to the scientific community, in order to develop Leibniz's differential calculus.

The stories of the catenary and of the vault were first linked empirically. We can find some ancient vaults that are inverted catenaries. Then by an affirmation of Hooke, who in 1675, at the end of an article famous for giving for Hooke's law for the spring, added "a *decimate* of the *centesme* of the Inventions" he intended to publish. The second item in the list gives an anagram, which when deciphered reads: *ut pendet continuum flexile, sic stabit contiguum rigidum inversum* (as it hangs in a continuous flexible form, so it will stand contiguously rigid when inverted). (Hooke 1675). Regarding this, Truesdell remarks,

> While none of the available papers of Hooke reveals how he reached this conclusion, there is no reason to doubt that he had sufficient mastery of statics to show that an arch of infinitely small stones in order to exert purely tangential thrust should be formed like an inverted catenary subject to inverted loads. Thus the problem of the catenary and the arch are reduced to one, but neither is solved. (Truesdell 1960: 57).

Leibniz answered Johann Bernoulli's question about the form of the catenary in this way:

> The problem of the funicular or catenary, presents a double interest, first the one to enlarge the art of inventing, in other words Analysis, until now unable to affront correctly such questions, secondly the one to make the technique of building progress (Leibniz 1691: 243)

Reading this, it seems that the affinity between vault and catenary was well known at that moment, so we wonder why Benvenuto—after having told us that La Hire links the two problems in his *Traité* but, as Johann Bernoulli wrote to Leibniz in October 1698, "sees something but what he has seen he himself does not understand"[1]—tells us that "the first to establish the affinity between arches and catenaries was another scholar, an Englishman, David Gregory" (Benvenuto 1991, II, p. 327).

[1] Leibniz 1745, vol. 1 letter 82, p. 412: *La Hirius in suo Tractatus Mechanico, Propp. 123, 124 & 125, affinitatem inter fornices & catenarias subolvisse videtur; nec tamen rem satis assequi potuit, nostro Calculo destitutus. Vidit aliquid; quid autem viderit, ipse non intelligit.*

It is important to distinguish, here as well as in what follows, different problems. There is a mathematical problem, that of solving the problem of the catenary or of the vault using differential calculus. And there are mechanical problems:

(1) What particular model of vault would stand best if it has the form of a catenary? This is a modelisation problem. The vault has to be made of infinitely thin blocks perfectly rigid and without friction between them.

Infinitely thin, I say. But is it possible then to split the mathematical from the mechanical problems? We'll try, as at the worst it could bring us is to learn something about the relations of mathematics with mechanics.

(2) What are the mechanical similarities and differences between the vault and the catenary?

There are different answers to that question, depending on the fundamental principle you use: (a) the similarity is that the stresses are tangential in both problems. The difference is that those stresses are tensions in the catenary and pressures in the vault; (b) if one uses a variational principle to solve the problem, the similarity is that the equation to solve is the same, while the difference is that the catenary corresponds to the minimum and the vault to the maximum solution.

This said, let us go back to the historical problem. For Leibniz, the reason why La Hire could not understand that affinity is that he was *lacking our calculus.* But couldn't he see the mechanical affinity just as Hooke could have seen it? La Hire only says, "This proposition is only a converse of the previous one that was about a hanging rope." I would say that La Hire saw a mechanical affinity but could not formulate it because, as we will see in the next section, he could not completely formulate the idealised model of either the catenary or the vault. But I don't agree with Leibniz that the reason why La Hire couldn't express the affinity is due to his lack of knowledge of the calculus. The analogy is expressed by calculus, but its mechanics can be understood without it, even if infinitesimals are implied.

La Hire saw that the important point is the direction of the pressure, He says, "As the courses of the stones or voussoirs are supposed to be infinitely polished, one does only have to considerate them for the direction of the weight of the voussoirs".[2] In fact, it was first realised that if the pressure doesn't follow the curve of the vault, the stones would spring out of it, just as the parts of the chain in the catenary would do if the tensions didn't follow the curve. To follow the curve they have to be tangential. That way of thinking will become clearer reading Gregory.

The fact that Johann Bernoulli does not speak of the vault in his lectures to de l'Hôpital (Bernoulli 1741) and so neither does de l'Hôpital in his *analyse da infiniment petits*, also shows that the affinity is a mechanical one and doesn't bring anything to the calculus.

Let's turn now to Gregory's text on the catenary written in 1695, after he had read the three answers of Johann Bernoulli, Leibniz, and Christiaan Huygens to

---

[2] De la Hire 1695, p. 467: *Puisque les lits des pierres ou voussoirs sont supposés infiniment polis, on les doit seulement considérer pour la direction de la pesanteur des voussoirs.*

Jacob Bernoulli's question of May 1690. As Truesdell says, "For 1690, these three solutions, in the order received [Johann Bernoulli, Leibniz, Huygens], exhibit the mathematics of the future, the present, and the past" (Truesdell 1960, p. 66). In fact, Johann Bernoulli and Leibniz used differential calculus but only published the results, and Huygens didn't use it. To know how Bernoulli and Leibniz used the calculus we have to go to letters by Leibniz and to Bernoulli's lectures to l'Hôpital (Bernoulli 1741), texts that Gregory couldn't have read. Gregory's aim is to show that he could demonstrate the same results with the help of Newton's calculus of fluxions.

> Since the problem of the shape of the chain (that is, of the line of the chain, heavy towards the far center, and bent by its own weight while it hangs at the two motionless ends) is especially noble among the philosophers of this age, and from the famous men Huygens, Leibniz, and Bernoulli, most of the figures proper to these have been woven, and published in *Actis Eruditorum Lipsiae* (but without demonstration): He wished to interweave the demonstrations of all these, with the aid of the Method of Newton, familiar to the geometer to-day, to determine fluxes and currents in relation to data, and vice versa.[3]

The sixth corollary of proposition 11 tells us that,

> the catenary in a vertical plane, but inverted, makes it possible to find but doesn't govern the figure of the thinnest arch or vault. That is to say, that very small, rigid and polished spheres, arranged to form an inversed catenary, constitute an arch of which no part is being pushed inside or outside by the other parts; but leaves those infinitesimal points fixed, supported by virtue of its form.[4]

This idea is nicely illustrated by Poleni (Figs. 1 and 2).

The affinity of the vault and the catenary is common knowledge for Bernard Forest de Bélidor, in his *Science des ingenieurs* (1729), as one can see from the way he advises to design arches (Fig. 3):

> If we want to build a natural vault whose width and height are given, we must, on a vertical surface, draw a line CD equal to the width of the vault, drop from the middle of this line a perpendicular EF equal to the height that we want to give it, then attach the end of a chain to point C, and carry the other end towards D, so that by increasing or decreasing the chain, its own weight makes it pass through point F when it will be stopped at places C and D. After that we can, with a pencil which we will draw all along the chain (without wavering), trace a curve, and above it establishes the figure of the false hanger of the vault, the section of the voussoirs and the rest.[5]

---

[3] Gregory 1697, p. 637: *Cum problema de figura Catenae (id est lineae fexilis, versus centrum longinquum gravis, & pondere suo dum duobus extremis immotis dependet incurvatae) sit inter hujus aevi Philosophos imprimis nobile, ac à Celeberrimis Viris Hugenio, Leibnitio & Bernoullio, plurimae figurae stius proprictates fuerint detextae, & in Actis Eruditorum Lipsiae (at sine demonstratione) editae: Libuit harum omnium demonstrationes pertexere, ope Methodi Neutoniae Geometris hodie familiaris, fluxiones è fluentium relatione data determinandi & vicissim.)* (my translation).

[4] Gregory 1697, p. 640; my translation.

[5] Bélidor 1729, pp. 149–150: *Si l'on veut construire une voûte naturelle dont la largeur et la hauteur soient données, il faut sur une surface verticale tracer une ligne CD égale à Îa largeur de la voûte, abaisser du milieu de cette ligne une perpendiculaire EF égale à la hauteur qu'on veut lui donner, ensuite attacher l'extrémité d'une chaine au point C, et porter l'autre extrémité vers D, de maniéré*

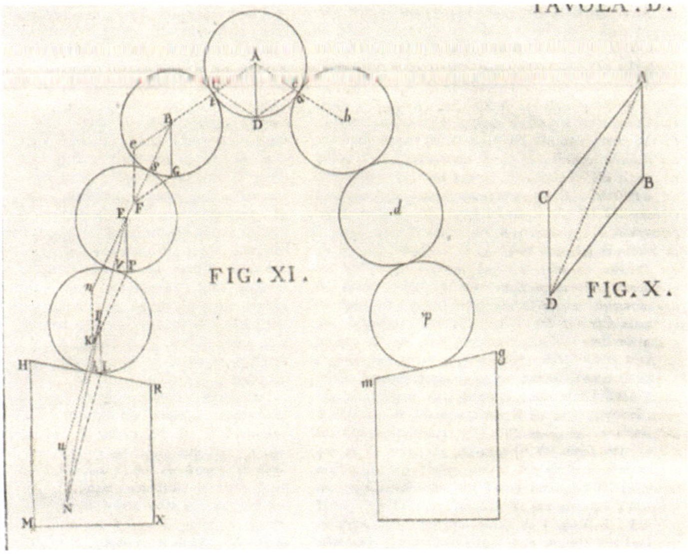

Fig. 1 The idea of a stable arch composed of spheres in Poleni's *Memorie istoriche* (1748, Tav. D). Public domain

To understand the equilibrium of the catenary and of the vault, Hooke probably, and Gregory certainly, are considering the equilibrium of the forces. They use the fundamental principle:

$$\sum \vec{F} = 0$$

or the parallelogram law. But we know that there is another way of solving the problem, using the principle of virtual works. It is more difficult to tell when people understood the similarity between the vault and the catenary in term of energy.

Lorenzo Mascheroni, in *Nuove ricerche sull'equilibrio delle volte* (1785) insists on that analogy, as one can see in his figures, but he doesn't explain it.

3 Finding the Right Model for the Catenary and the Vault

If one loads a rope or chain with various weights, one can obtain different forms.

In *La statique ou la science des forces mouvantes* (1673) Pardies showed that one would also obtain a parabola if, instead of a chain of uniform weight having *ds*, and

qu'en augmentant ou diminuant la chaine, son propre poids la fasse passer par le point F lorsqu'elle sera arrétée aux endroits C et D. Après cela on pourra, avec un crayon que l'on conduira tout du long de la chaine (sans pourtant vaciller), tracer une courbe, et li-dessus établir la figure du faux cintre de la voute, la coupe des voussoirs et le reste.

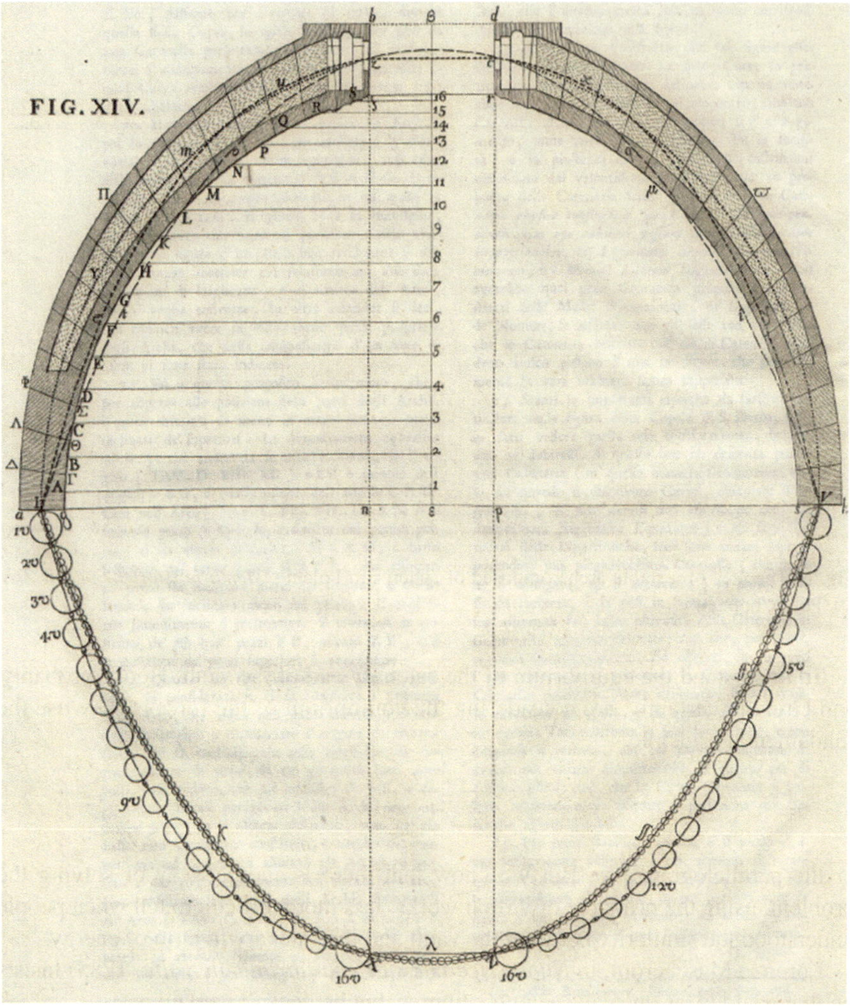

Fig. 2 The catenary in the dome of St. Peter's according to Poleni's *Memorie istoriche* (1748, Tav. E). Public domain

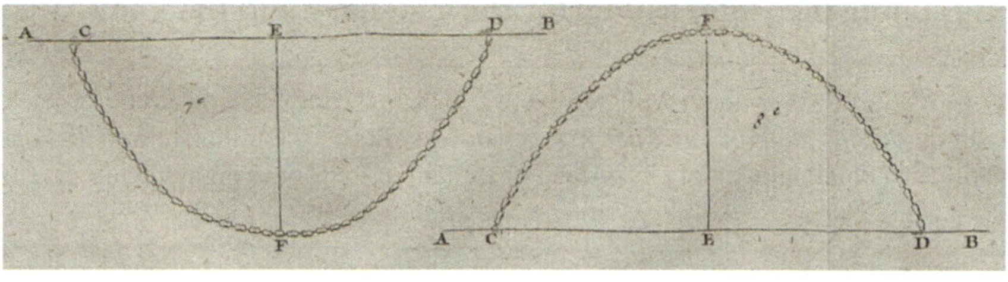

Fig. 3 Bélidor's representation of the hanging chain and its inversion to form a vault (Bélidor 1729, Bk. II, Pl. VI). Public domain

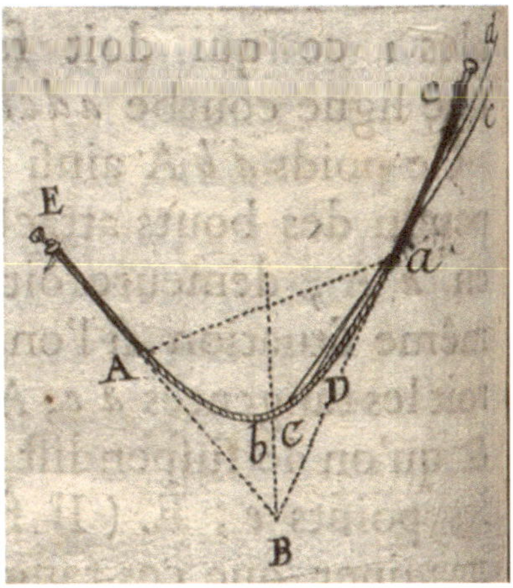

Fig. 4 Pardies's figure of the catenary (1673, p. 126). Public domain

element of the curve, of constant mass, we suspend to the chain weight such that *dx*, an element of the horizontal axis, is of constant weight, such as in the suspension bridge. The same is true for the vault. One can make it circular and stable if one correctly chooses the weight of the different voussoirs. In fact, one has to describe precisely the model of the catenary or of the vault, that is, one has to select particular constitutive equations (Fig. 4).

3.1 For the Catenary

The hanging rope will take the form of a catenary, the hyperbolic cosine, if one & S a chain of uniform mass, perfectly flexible and of constant length. Pardies doesn't state that explicitly, nor does Johann Bernoulli in his public answer to Jacob's question but he does it in his lectures to de l'Hôpital: "The wire, the rope, the chain or whatever represents the curve is supposed to be perfectly flexible in all its points and inextensible" (Johann Bernoulli 1741; 4921).

3.2 For the Vault

To get the same form, the vault ought to be infinitely thin, the voussoirs perfectly rigid and polished so that they slide one upon the other without friction.

The model gives a mathematical description of the constitution of the voussoirs: they are infinitely thin, the distance between two points is constant and we neglect friction.

La Hire considers the vault as circular and the voussoirs as equal and reduced to their centre of gravity. Usually. architects such as Philippe de La Hire, Claude Antoine

Couplet and Bélidor consider the catenary to be the curve linking the centre of gravity of a vault having a certain thickness; mathematicians such as Jacob Bernoulli and David Gregory consider the vault as being infinitely thin.

4 Fundamental Quantities

Both problems helped people to understand better some fundamental quantities for mechanics and for geometry from the most difficult ones.

4.1 Mechanical Quantities: Tension and Pressure (Tuesdell 1968, p. 206)

Pardies's study of the catenary describes the tension with ate and shows its vectorial character, even if he denominates it differently:

> Force of their traction: These suspended bodies being inclined, will pull the ropes which support them differently; & the force of traction is measured as in article 67 by taking in the line of direction[6] at point F, & drawing the parallels FG, Fg. Because the force of the weight on being expressed by the line FB, the line BG will express the force by which the rope oA is pulled, & the line Bg that of the rope na. Which can also be expressed by the two weights D & d which would be at body no, like the lines BG, Bg at line BF.[7]

Pardies explains clearly that the tensions in the chain must be tangential. This property was very important for Johann Bernoulli's solution of the catenary.

The affinity between the two problems showed that tension and pressure were two manifestations of a more general notion: the stress. Jacob Hermann generalized both notions in his *Phoronomia*:

> The tension or compression of a thread or body at any of its points or at an element of the curve is that force of the thread or body which resists that power or force growing from all the applied powers and tending by pulling the thread in opposite directions to tear it apart. This tension exactly equals or is equipollent to that tearing force resulting from all the powers applied to the body (Hermann 1716).

---

[6] The "line of direction" was a technical term introduced by Tartaglia (1554: 84) to account for the vectorial character of the weight. It was generalized to other forces by Roberval in 1636.

[7] Pardies 1673, p. 177: *Force de leur traction: Ces corps suspendus étant inclinez, tireront diversement les cordes qui les soûtiennent; & la force de la traction se mesure comme dans l'article 67 en prenant dans la ligne de direction au point F, & tirant les paralléles FG, Fg. Car la force du poids on étant exprimée par la ligne FB, la ligne BG exprimera la force dont la corde oA est tirée, & la ligne Bg celle de la corde na. Ce qui se peut exprimer encore par les deux poids D & d qui seroient au corps no, comme les lignes BG, Bg à la ligne BF.*

4.2 Geometrical Quantities: The Radius of Curvature

This important mathematical tool was known long before the period we are considering and Christiaan Huygens had already showed how useful it could be for mechanics. Jacob Bernoulli had discovered its differential expression working on the logarithmic spiral but published that expression with his study of the elastica, underlining in this way its importance for mechanics. As we will see later, that importance is shown once more by the completely original use he makes of the ray of curvature in the study of the vault (Radelet 1999).

5 Fundamental Principles

5.1 Purely Geometrical Principles

As Benvenuto (1991, pp. 313–315) shows clearly, before La Hire and his followers, vaults and even the greatest domes were built following purely rules such as the one given by Derand, still quoted by Bélidor in 1729 (Fig. 5).

Blondel's treatise on the *Résolution des quatre principaux problémes d'architecture* (1673) is also purely geometrical, and so is a rather difficult paper by La Hire, *Remarques sur la forme de quelques arcs dont on se sert dans l'Architecture* (1702).

5.2 The Law of the Lever, or the Sum of the Moments of Momentum Must Be Zero

Truesdell (1968, pp. 239–271), and more recently Giorgio Maltese (Maltese 2001), showed how difficult the way was towards the general law of moment of momentum. But a special case of that law was the first mathematical law of the whole of mechanics. It is called the law of the lever, and up until the period we are considering it was the only mathematical law in mechanics. It is perhaps for this reason that the catenary couldn't be treated mathematically before the invention of calculus, which allows the use of the property that the tensions are tangential.

Nevertheless, I don't think anybody studied the problem of the catenary with the help of the lever. On the contrary, La Hire's first attempt to mat the equilibrium of the vault is based on the law of the lever. In fact, we must be more precise. La Hire's mechanics rest upon this one mechanical principle: "the lever that can be considered

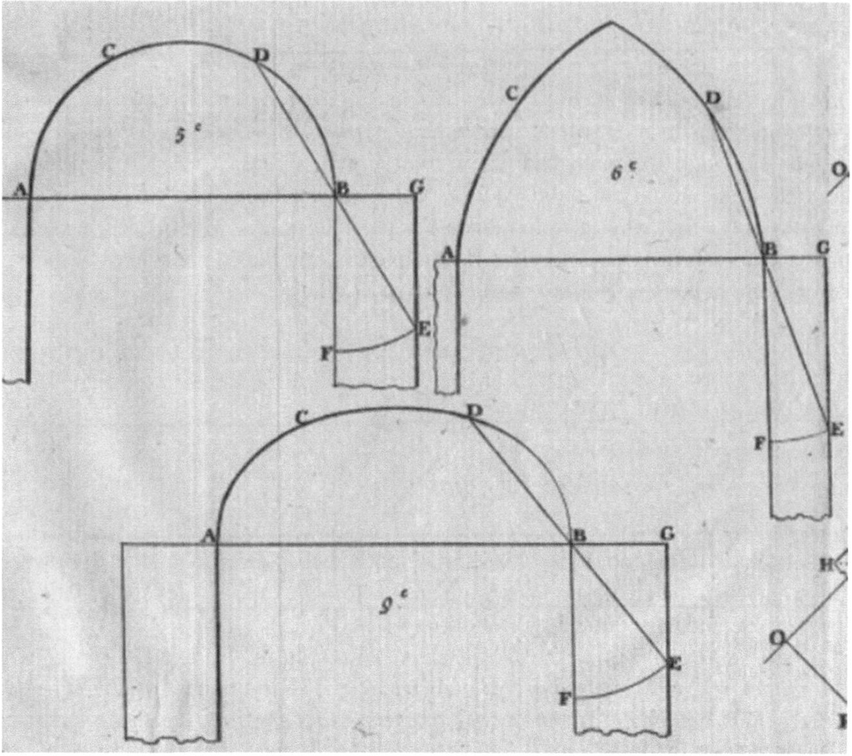

Fig. 5 Belidor's illustrations of the thickness of counterforts (Bélidor 1729, Bk. II, Pl. 4). Public domain

as the fundamental proposition of all Mechanics, since the other parts can be easily reduced to it”.[8]

From the law of the lever, La Hire derives a partial form of the parallelogram law of forces. His proposition XXI says, “We must find three forces AXB, which pulling a point K by three given directions CK, DK, EK, are in equilibrium among themselves”.[9] (Fig. 6).

To do so, he draws the diagrams shown in Fig. 7. One might think he uses the parallelogram law, but instead he uses the proposition XIX, which says,

> In an angular lever CHD, force A will be to force X as the sine of the angle DKH made by for the directions of the two forces X and B, or its the supplement, which is the same thing, to the sine of the angle CKD, or its supplement DKN, which is the same thing, made in the

[8] De la Hire 1695: *Preface: le levier qu'on peut regarder comme la proposition fondamentale de toute la Mécanique, puisque les autres parties s'y peuvent réduire facilement.*

[9] De la Hire 1695: Prop. XXI, p. 70: *Il faut trouver trois puissances AXB, qui tirant un point K par trois directions données CK, DK, EK, soient en équilibre entre elles.*

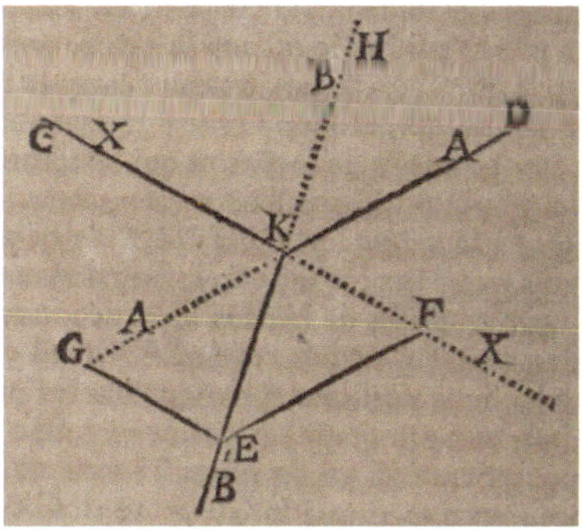

Fig. 6 La Hire's partial parallelogram law (1695, p. 70). Public domain

direction of the two forces A & B; & force B will be to force X, as the sine of the angle CKH is to the sine of the angle CKD or its supplement CKP, or CKV, which is the same sine.[10]

Then, La Hire gives another form to the theorem equating the proportion of the forces to that of the sides of a triangle having its sides perpendicular to those forces (Fig. 8).

That last theorem is the one he uses to solve the problem of the vault and to deduce that the weight of the different voussoirs must be proportional respectively to KL, LO, OP on a horizontal line tangent to the top E of the vault (Fig. 9).

Antoine Parent also studied vaults and catenary but, as Truesdell (1960: 80, 109–110) explains, his works are difficult to find. In fact, one finds in *Histoire de l'Academie de Paris* a short remark without mathematics saying that Parent had found the curvature of the external part of a vault, the internal one being circular and the weights of the voussoirs different from each other and given by La Hire's law (Parent 1704). One should check with the *Essai* of Parent (1713).

Jacob Bernolli also applies the law of the lever twice to solve that problem.[11] First, he uses it to compute the tangential pressure exerted by the top of the vault BEF on the voussoir KDEL, rather than using the parallelogram law. This is rather similar to what La Hire did. He imagines that the top BEF of the vault is acting on the lever EF. Thus, the weight of BEF, $2s$, is acting on M, the middle of the lever, which sits under the centre of gravity E of BEF. This is equivalent to half the force acting at twice the distance, thus s is acting in E. But s does not act vertically; it acts along LI. So LI: s = FE: FI. But as EFI and EDN are similar we have LI: s = DE: DN = ds: dx. Thus, he finds that the pressure along LI is $\frac{sds}{dx}$. Another force acts on the voussoir; to find that one, Jacob imagines a string maintaining the weight ds of the voussoir KDEL

[10] De la Hire 1695, pp. 63–64: *dans un levier angulaire CHD, la puissance A sera à la puissance X comme le sinus de l'angle DKH fait par les directions des deux puissances X & B, ou de son supplément qui est le même, au sinus de l'angle CKD, ou de son supplément DKN qui est le même, fait de la direction des deux puissances A & B; & la puissance B sera à la puissance X, comme le sinus de l'angle CKH au sinus de l'angle CKD ou de son supplément CKP, ou CKV qui est le même sinus.*

[11] See "The De curvatura fornicis of Jacob Bernoulli: The Introduction of the Infinitely Small into the Calculation of Arches", Chap. 5 in this present volume.

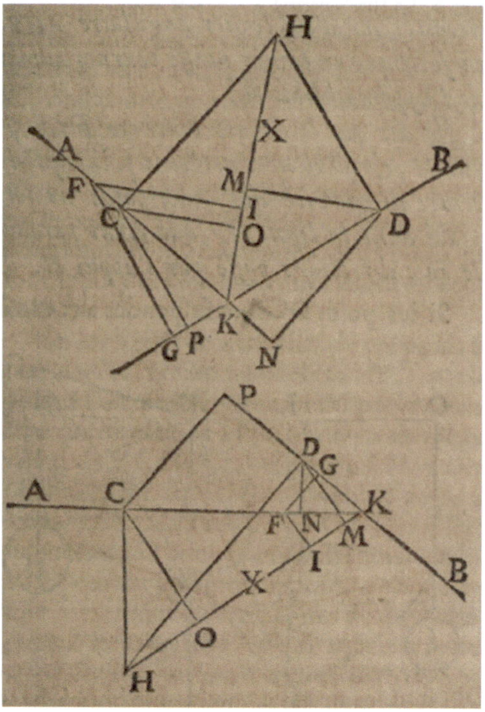

Fig. 7 La Hire's figures for his Prop. XIX (1695, p. 65). Public domain

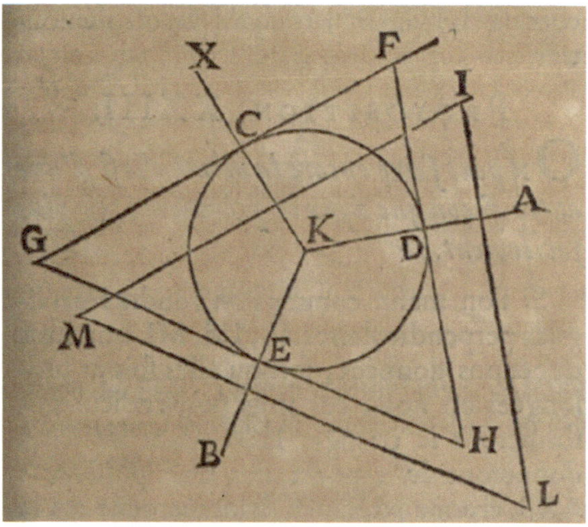

Fig. 8 La Hire's triangle of forces (1695, p. 73). Public domain

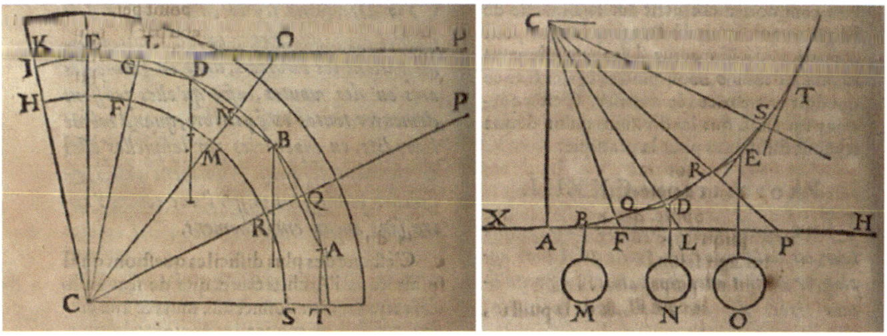

Fig. 9 La Hire's approach to the problem of the vault (1695, **left**, p. 466; **right**, p. 456). Public domain

along the perpendicular to the vault OP. The tension in that string is to the weight *ds* of the voussoir as DN is to DE or as *dy* to *ds*. Therefore, the tension is *dy* (Fig. 10).

But then he uses it in a completely different way, He imagines that the equilibrium of the vault could be broken if the voussoir KDEL rotates around the point D fixed

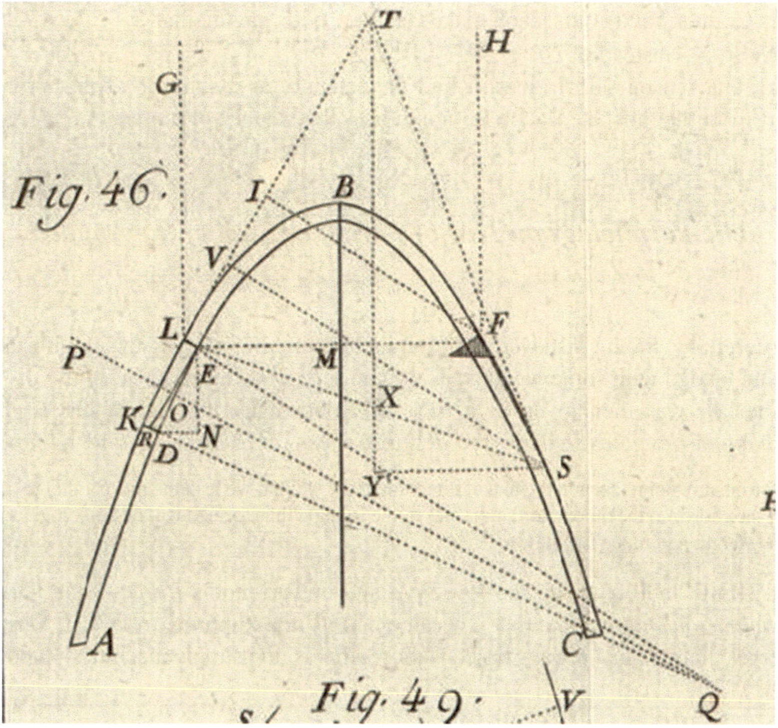

Fig. 10 Jacob Bernoulli's Fig. 46 (1744a: Table XLVIII). Public domain

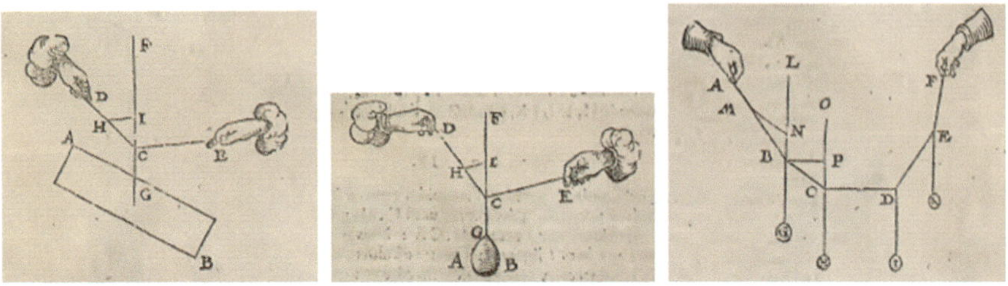

Fig. 11 Stevin's drawings showing the discovery of the parallelogram law and the study of the catenary in his *Byjwough der Weeghconst* (Stevin 1608, pp. 182–183; Stevin 1634, p. 505). Public domain

by friction. To avoid that, he has to maintain in equilibrium the lever DE on which two forces are acting, $\frac{sds}{dx}$ and dy. The first one acts on E along IL or RD, and the other one on O, the middle of the lever along PO. If one applies the law of the lever, one gets

$$\frac{sds}{dx} \cdot DR = dy \cdot OD.$$

As Cramer points out, Jacob makes an error here, saying that.
$DR = \frac{DE^2}{DQ}$ instead of $DR = \frac{DE^2}{2DQ}$.
For that reason, Jacob doesn't find the catenary as a solution of that problem. Apart from that mistake, Jacob Bernoulli uses the balance of momentum correctly.

5.3 The Parallelogram Law of Forces or Sum of the Applied Forces Is Zero

The catenary. Stevin discovered the parallelogram law studying the equilibrium of a string loaded with some weights, as we can see in his figures (Fig. 11).

The law was then given as a trick by Roberval at the end of his *Traité de Méchanique des poids soutenus par des puissances sur les plans inclinés à l'horizon*:

> If, from any point taken in the line of direction of the weight, we lead the line parallel to one of the chords to the other chord, the sides of the triangle thus formed will be homologous to the weight and to the two powers.[12]

It is only in 1687, with the *Principia* and in Varignon's *Projet d'une nouvelle mécanique ou statique* that that law came into its real importance. In fact, Varignon observes that one had always based statics—that is, the simple machines—upon the

[12] Roberval, 1636, p. 28: *Si, de quelque point pris en la ligne de direction du poids, on mène la ligne parallèle à l'une des cordes jusqu'à l'autre corde, les côtés du triangle ainsi formé seront homologues au poids et aux deux puissances.*

law of the lever. He proposes to base it on the parallelogram law. We know that in fact one needs both laws to do statics.

After Varignon's publication, one sees Jacob Bernoulli struggling in his *Meditationes*, trying to understand it and confronting the results it gave with the one obtained with the law of the lever (Radelet 1989). But we saw in his treatment of the vault that even at the end of his life, he still preferred the law of the lever, even to find the component of a force.

In another of his last *Meditationes*, Jacob tries to generalise the problem of the catenary studying the equilibrium of a rope under the action of various forces acting in different directions:

> The string ACDEFGB is suspended by its ends A & B from the infinite potentials C, D, E, F, G, and is extended along any directions acting HC, HD, IE, KF, LG. The question is the curvature of the string, its direction in the middle of the LP, and the force with which it is pushed along the LP.[13]

This argument will be taken up again by Jacob Hermann in his *Phoronomia* (Hermann 1716: bk I, sec. I, chap. III) and in 1728 by both his nephew, Daniel Bernoulli (1728), and by Euler (1728). But there things are quite different. Two years before, Daniel had given a demonstration of the parallelogram of forces and he will use it to find the components of the different forces along a first direction orthogonal to the curve and another one always parallel (vertical) to a fixed direction. Therefore, he easily finds the particular cases of the catenary if only the components parallel to each other are different from zero or the *lintearia* if only the components perpendicular to the curve are different from zero (Fig. 12).

After having read Daniel's article, Euler, in the same volume of the *Commentarii Academiae Petropolitanae*, gives a program for the solution of the problem of finding the curve assumed by an arbitrarily elastic band loaded by arbitrary forces at its several points. But it will take him fifty years to fulfil it completely. Truesdell says that Daniel's article contains nothing new:

> That he was ignorant of his uncle's unpublished work is to be expected, since Jacob Bernoulli's paper were kept from Johann Bernoulli and his circle; that Daniel Bernoulli should not know the general solution in the book by his senior colleague, Hermann, is surprising, especially since Euler refers to it (Truesdell 1960: 146–147).

But in my opinion, Daniel added something to what Hermann had done. He chose two particular directions on which he tacks the components of the various forces, so he doesn't have the difficulty Hermann had. Truesdell explains, "the difficulty may lie in the failure to realize that his polar co-ordinate diagram must be drawn over again at each point" (Truesdell 1960: 86). So Daniel's choice of two particular directions, a choice he made in order to find the well-known particular cases easily, will open the way to generalisation for Euler. He'll change one thing: he will replace

---

[13] Jacob Bernoulli, 1744b: *Filum ACDEFGB extremitatibus suis A & B suspensum ab infinitiis potentiis C, D, E, F, G, juxta directiones quasvis HC, HD, IE, KF, LG agentibus extenditur. Quaeritur fili curvatura, cijus directio media LP, & vis, qua secundum LP impellitur.*

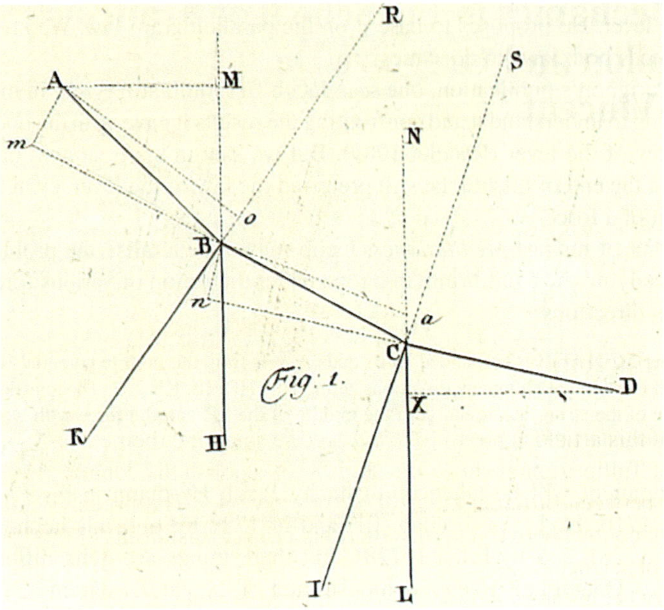

Fig. 12 Daniel Bernoulli's "universal method" for finding the curve of a string (1728, Pl. III). Public domain

the components parallel to a fixed direction by tangents to the curve, obtaining an orthogonal system of reference.

The vault. We saw that La Hire demonstrated his particular form of the parallelogram law from the law of the lever. Couplet, on the contrary, starts with the parallelogram law. But it was sufficiently new to be mentioned in the introduction to Couplet's article published in the *Histoire de l'Académie royale des sciences* (Fig. 13):

> The thrust [of the key] on this voussoir can only be a perpendicular drawn from the centre of gravity of the key on the surface of the voussoir. This line is at the same time the diagonal of a parallelogram, the two sides of which would be the vertical tendency of the key to fall, & a horizontal effort to push the voussoir in or out.[14]

But Couplet prefers La Hire's form of the parallelogram law, as seen in his Figure I (upper right corner in Fig. 13):

> If force x is decomposed into two forces y & z, these three forces will be like the sides of a triangle formed by the perpendiculars to the directions of these three forces.[15]

[14] Couplet 1729, p. 76: *Son impulsion [de la clef de voite] sur ce voussoir ne peut étre qu'une perpendiculaire tirée du centre de gravité de la Clé sur la surface du voussoir. Cette ligne est en même temps la diagonale d'un parallélogramme dont les deux cotés seroient la tendance verticale de la Clé pour tomber, & un effort horizontal pour pousser le voussoir ou l'écarter.*

[15] Couplet 1729, p. 80: *Si la force x se décompose en deux forces y & z, ces trois forces seront entre elles comme les côtés d'un Triangle formé par les perpendiculaires menées sur les directions de ces trois forces.*

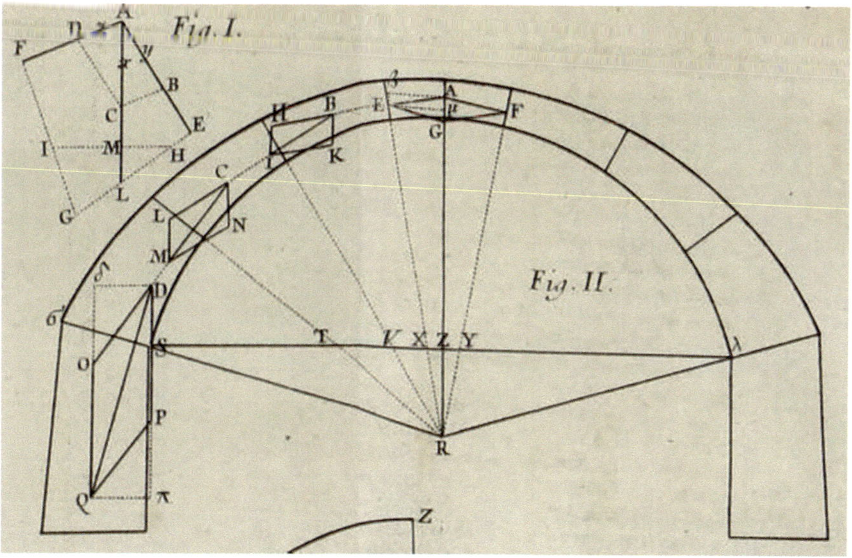

Fig. 13 Couplet's Figs. I and II showing lines of force in a vault (1729, Pl. IV). Public domain

The reason is that it leads to a very easy graphical law to find the weight of the different voussoirs of a stable vault:

> When all the joints of the vault are directed towards the same point R, as is said to be the case in the vault whose lower surface is circular, if any horizontal line S is drawn, the weights of the voussoirs A, B, C and D will be expressed by the parts XY, VX, TV and ST of this horizontal line enclosed between the extended joints of these voussoirs.[16]

In Fig. 14, we can see that Couplet uses La Hire's graphical method and that he is trying to consider infinitely small voussoirs, in order to obtain a continuous curve. He uses differential calculus in his work but couldn't find the equation of the curve of the extrados. To Couplet's question, "Determine the uniform curvature of a vault, such that it remains in equilibrium, & whose voussoirs we consider to be polished, i.e. without bonding",[17] he answers, "Take a rope furnished with equal weights…". One explanation could be that he doesn't seem to have the notion of radius of curvature and so uses always circles.

The same year, Bélidor, in his *Science des ingénieurs* (1729) dedicates an entire book to the mechanics of vaults. After a survey of ancient geometrical methods, the

[16] Couplet 1729, p. 86: *Lorsque tous les joints de la Voûte sont dirigés vers un même point R, comme l'on dit qu'ils doivent étre dans la Voûte, dont l'intrados est circulaire, si l'on tire une ligne horizontale quelconque S, les pesanteurs des Voussoirs A, B, C, D, seront exprimées par les parties XY, VX, TV, ST, de cette ligne horizontale renfermées entre les joints prolongés de ces Voussoirs.*

[17] Couplet 1729, p. 951: *Déterminer la courbure uniforme d'une voûte, telle qu'elle se maintienne en équilibre, & dont nous considérons les voussoirs comme polis, c'est-d-dire, sans liaison.*

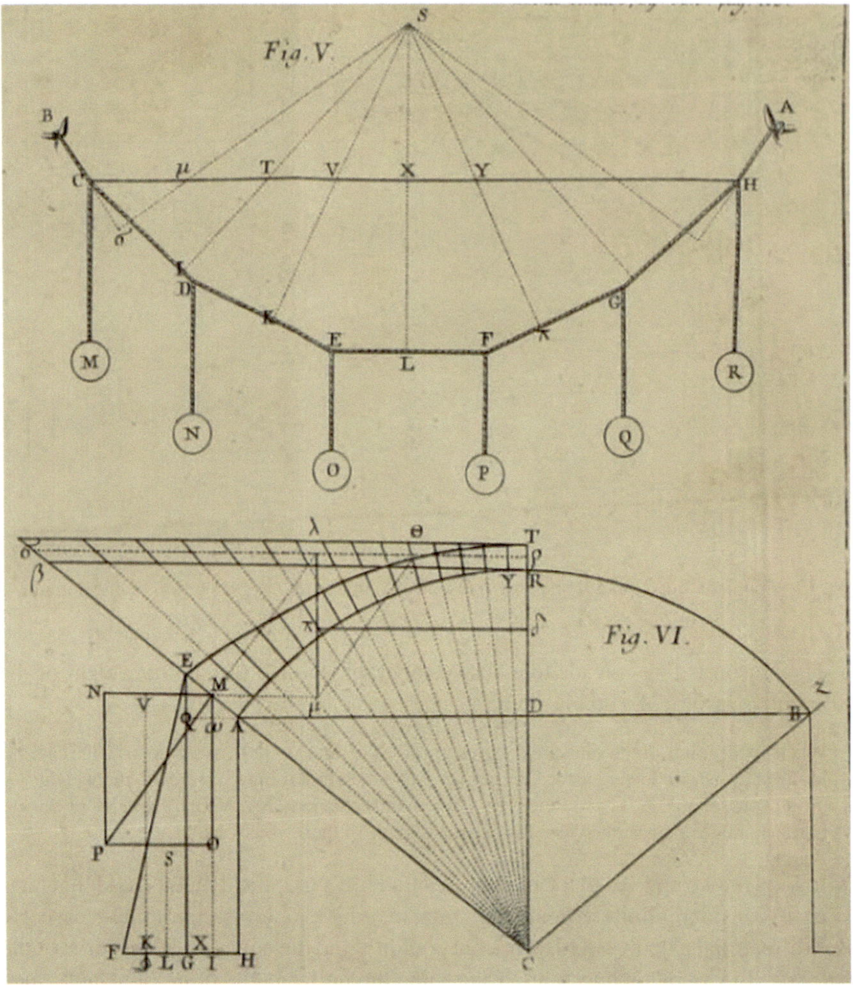

Fig. 14 Couplet's Figs. V and VI for determining the uniform curvature of a vault (1729, Pl. VI). Public domain

book starts with principles taken from the mechanics. The first one is the parallelogram law and La Hire's variant (Fig. 15). And as we have already said, to find the best curve for equal voussoirs he takes the chain.

Pierre Bouguer and Abbé Charles Bossut are familiar with both the parallelogram law and the radius of curvature, as one immediately sees in their illustrations (Figs. 16 and 17).[18]

[18] For more on Bouguer and Bossut, see "The Theory of Vaults of Pierre Bouguer", Chap. 8 in this present volume.

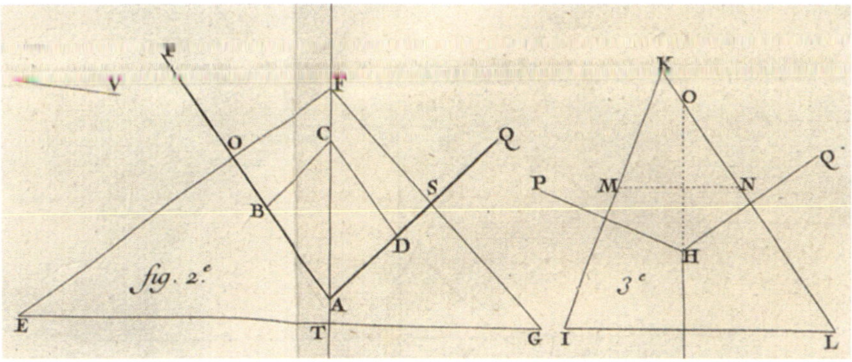

Fig. 15 Bélidor's illustration of the parallelogram law of forces and La Hire's triangle of forces (1729, Bk. II, Pl. 4). Public domain

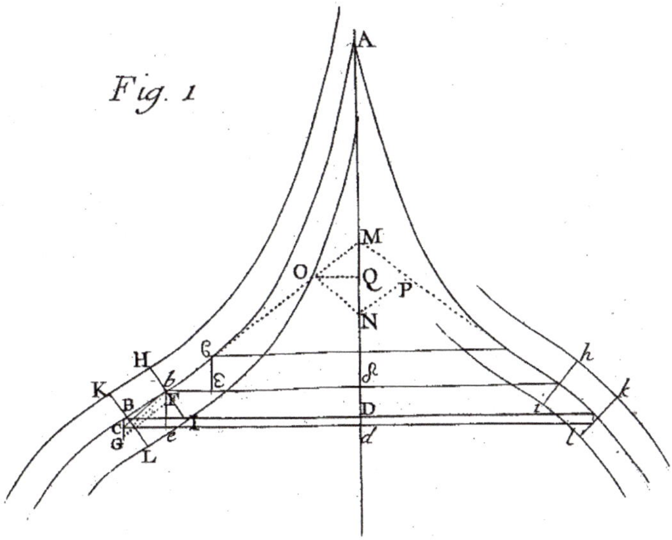

Fig. 16 Bouguer's illustration of forces on the voussoirs (Bouguer 1736, Fig. 1). Public domain

5.4 *The Necessity of Considering Both*

Truesdell often insists on the fact that one must consider both $\sum \overrightarrow{F} = 0$ and $\sum \overrightarrow{M} = 0$, and from that point of view, I think that he doesn't insist enough on Jacob's way of reasoning for the most stable vault. We have already seen that he used the balance of momentum but let's go back to the beginning of his work where he says:

> I suppose that if the part EBF of the vault were removed, the stone DL, which has no cement adhering to the lower part AK, can only fall under its own weight; which it can only do by

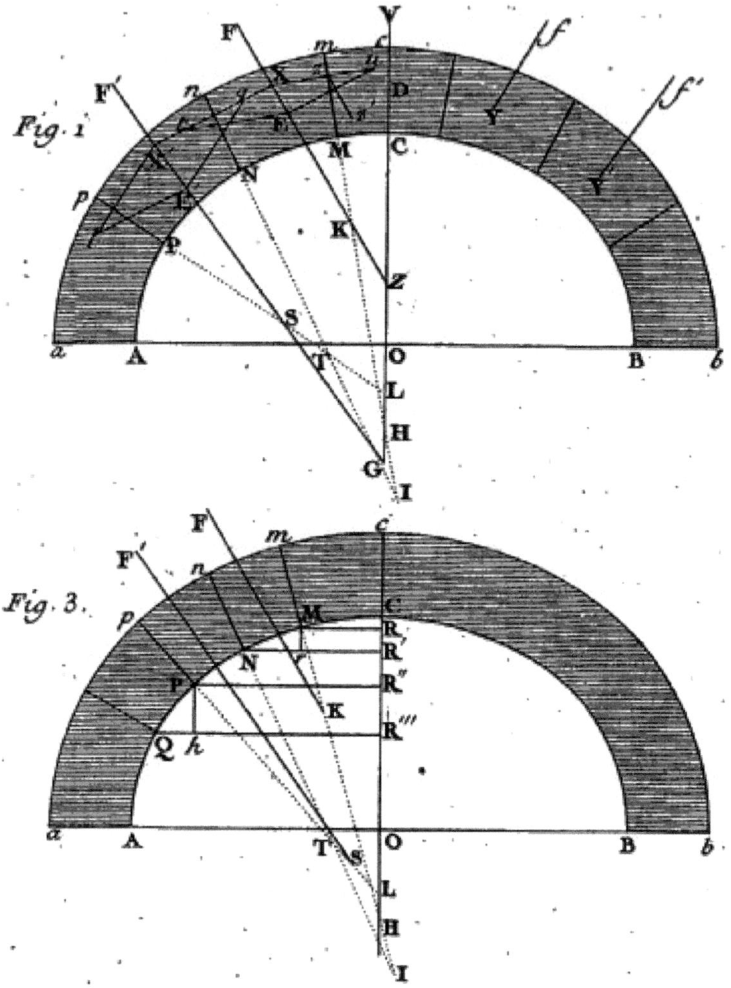

Fig. 17 Bossut's parallelogram law and the radius of curvature Figs. 1 and 3 (1778, Pl. 1). Public domain

> sliding along the plane KD, if this plane is perfectly smooth, or by turning around the point D, ... if there is friction with DK.[19]

In other words, the voussoir can only have two movements: one of translation, sliding on the surface KD, or one of rotation around D. To prevent sliding Jacob uses the balance of forces, a law that the engineers call "equilibrium to translation",

[19] Jacob Bernoulli 1744c: *Je suppose que si l'on retirait la partie EBF de la voûte, la pierre DL n'adhérant à la partie inférieure AK par aucun ciment, ne peut tomber que sous l'effet de son propre poids; ce qu'elle ne peut faire qu'en glissant sur le plan KD, si ce plan est parfaitement lisse, ou en tournant autour du point D, … si il y a friction avec DK.*

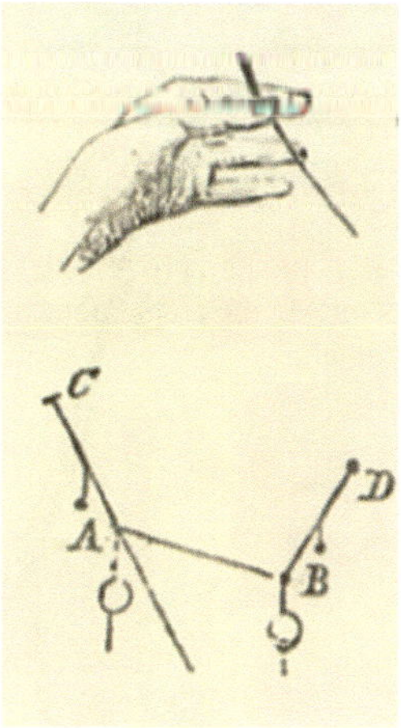

Fig. 18 Huygens's demonstration of a rope with weights, points 1 and 2 (Huygens 1888, p. 34). Public domain

and to prevent it from rotating, he uses balance of momentum, "equilibrium to rotation", the engineers say. (It would be interesting to find out who introduced those two expressions, equilibrium to translation and to equilibrium to rotation, into the language of engineers but I think it's rather recent.)

5.5 *Minimal Energy*

When the young Huygens started to study the problem of the catenary in 1646, following a suggestion made by Mersenne, he immediately gave an axiomatic form to his study. The first axiom he used, after Torricelli, is an axiom of minimal energy. Torricelli wrote in 1644, "We will assume that two gravities, linked together, cannot move by themselves unless their common centre of gravity descends".[20]

Huygens changes that dynamical principle into a statistical principle: the centre of gravity occupies the lowest possible place. With the help of this axiom, he demonstrates that the centre of gravity of two weights lies on a vertical passing through the intersection of the prolongation of the adjacent sides, and with the help of that property traces the curve point by point (Figs. 18 and 19).

Johann Bernoulli once used this hypothesis in his lectures to de l'Hôpital for demonstrating the last property 13:

> If we imagine an infinite number of curves traced on EF, equal to the funicular EBF, which we extend into a straight line, and if we place in ordinate at each of the points of each extended [curve] straight lines equal to their respective distances from the line EF, among all the areas which are thus formed, the one generated by the funicular will be the largest.

[20] Torricelli 1644, bk. 1, prop. I, p. 99: *Nous poserons en principe que deux graves, liés ensemble, ne peuvent se mouvoir d'eux mêmes, à moins que leur commun centre de gravité ne descende.*

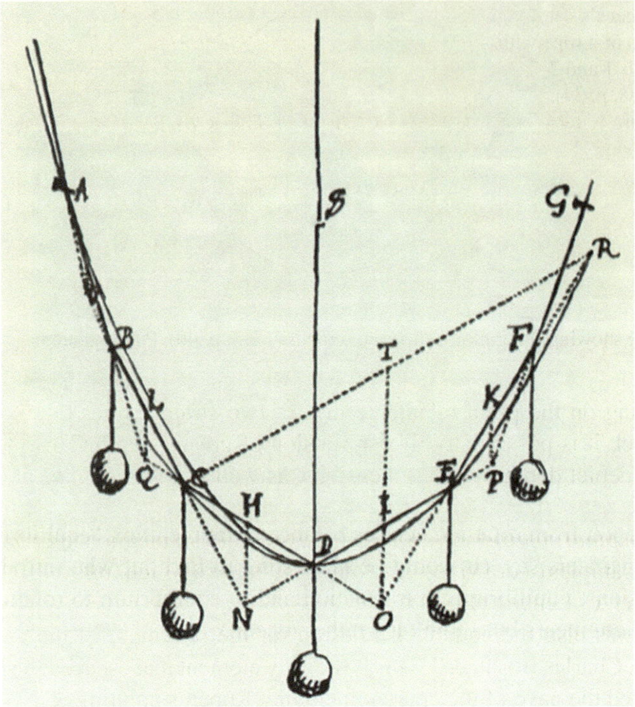

Fig. 19 Huygen's illustration from *De catena pendente*, Prop. 2 (Huygens 1908, p. 40). Public domain

> This axiom shows that the centre of gravity occupies the lowest possible position.[21]

However, he doesn't give the demonstration.

Jacob is much more interested in this axiom and in all other extremal properties he encounters. Already in his first text on the elastica, he enunciates without demonstrating them the following properties:

> of all the isoperimeters resting on *Af* the elastica is the one whose centre of gravity is furthest from *Af*, while the catenary is the one whose centre of gravity is furthest from the curve itself.[22]

---

[21] Johann Bernoulli 1741, p. 497: *Si l'on se représente une infinité de courbes tracées sur EF, égales à la funiculaire EBF, que l'on étend en une droite, et que l'on met en ordonnée en chacun des points de chaque [courbe] étendue des droites égales à leurs distances respectives de la ligne EF, parmi toutes les aires qui sont ainsi formées, celle qui est engendrée par la funiculaire sera la plus grande. Ce que l'on montre au moyen de cet axiome que le centre de gravité occupe l'emplacement le plus bas possible.*

[22] Jacob Bernoulli 1694, p. 599: *parmi toutes les isopérimétres qui s'appuient sur* Af *l'élastica est celle dont le centre de gravité est le plus éloigné de* Af *alors que la caténaire est celle dont le centre de gravité est le plus éloigné de la courbe elle-méme.*

In his Meditationes CCXXIX, *Inter omnes Figuras Isoperimetras vel acqualium arearum invenire illam, quae habeat maximum vel minimum aliquod Methodus Nova*, written in 1697, one finds various interesting examples:

(1) Among all isoperimeters, find the one of which the centre of gravity occupies the lowest place. He finds the catenary.
(2) Among all isoperimeters, find the one which contains the largest surface. He finds the circle.
(3) Among all isoperimeters, find the one of which the centre of gravity of the contained surface occupies the lowest place. He doesn't find the elastica as he thought, but the catenary,
(4) Among all curves containing the same surface, find the one of which the centre of gravity of the surface occupies the lowest place. He finds an impossibility.

Truesdell told us of the influence of those ideas on Euler and how, with the help of Daniel Bernoulli, he followed that way to solve the problem of the elastic ribbon in his *additamentum* to his *Methodus inveniendi lineas curvas maximi minimive proprietate gaudentes, sive solutio problematis isoperimetrici latissimo* sensu *accepti* at exactly the same time as Maupertuis published his first article on his principle of least action (Radelet 1998).

6 Virtual Works

6.1 The Catenary

I have shown elsewhere in some detail how Huygens, progressively and principally studying the catenary, begins to use the principle of virtual works.

6.2 The Vault

As regards the vault, let return once more to Jacob Bernoulli's text. Truesdell says that for preventing the voussoir from sliding, Jacob uses virtual works. He first translates Bernoulli:

> The [infinitesimal] stone KL in figure 29 "... is to be regarded as a wedge trying to force itself into the triangle DQE. As it come from KL into the position DE, that is, while it traverses the space KD, it pushes back the force pressing along IL by a distance KL-DE (Truesdell 1960, p. 83).

The idea of comparing the voussoir to a wedge was used the same year by Parent but Bernoulli couldn't have known that. Truesdell continues:

> Then the virtual work done by the normal force —F, pointing inward equals that done against the compression –T. That is, $-F \cdot KD = -T \cdot (KL - DE)$ (Truesdell 1960, p. 83).

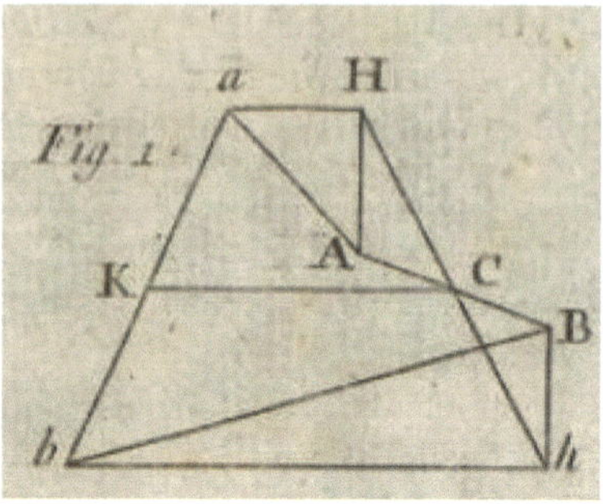

Fig. 20 Mascheroni's theorem of weights on an inclined plane (1785, Fig. 1). Public domain

This is, with other notations, exactly the equation written by Bernoulli by the nature of the wedge.

The champion of virtual works within the frame of the study of stability of vaults arrive much later: in 1785, Lorenzo Mascheroni wrote his work on the equilibrium of the vault. The work is based on the following Theorem 1 (Fig. 20):

> If weight *B* is placed in *b* onto an plane inclined on the horizon by an infinitesimal space *Bb*, and at the same time by way of a machine, the weight *A* is placed on the inclined plane *Aa*, and if one traces the horizontals *AH and bh* and the perpendiculars AH and Bh, the force of weight *A* will be to the force of weight *B* in the compound ratio of weight *A* to weight *B*, and of the perpendicular distance *AH* to the perpendicular distance *Bh*.[23]

But even here only work done by gravity and vertical paths is taken into consideration, just as Jordanus de Nemore had already done for the inclined plane. And his idea comes directly from the Aristotelian conception of the gravity.

References

Bélidor, Bernard. Forest de. 1729. *Les Science des ingénieurs dans la conduite des travaux de fortification et d'architecture civile ... par M. Belidor*. Paris.

Benvenuto, Edoardo. 1981. *La scienza delle costruzioni e il suo sviluppo storico.* 2 vols. Florence: Sansoni. Engl. Trans. 1991. *An Introduction to the History of Structural Mechanics. Part I: Statics and Resistance of Solids. Part II: Vaulted Structures and Elastic Systems.* New York: Springer.

Bernoulli, Daniel. 1728. Methodus universalis determinandae curvaturae fili. *Academiae Imperialis Scientiarum Petropolitanae* 3, 62–69.

Bernoulli, Jacob. 1744a. Op. LVIII, Curvatura Laminae Elasticae. Ejus Identitas cum Curvatura Lintei a pondere inclusi fluidi expansi. Radii Circulorum Osculantium in terminis simplicissimis exhibiti, una cum novis uibusdam Theorematis huc pertinentibus, &c. In *Opera Omnia*, 576–600. Geneva.

[23] Mascheroni 1785, pp. 2–3: *Theorema 1. Se il peso* B *si porti in b sopra un piano inclinato all'orizzonte per lo spazio infinitesimo* Bb, *e nello stesso tempo per costruzione di macchina il peso* A *si porti in a sopra un altro piano inclinato* Aa, *e se si tirino le orizzontali* aH, bh, *e le perpendicolari* AH, Bh; *sarà la forza del peso A alla forza del peso B in ragione composta del peso A al peso B, e del viaggio perpendicolare* AH *al Viaggio perpendicolare* Bh.

Bernoulli, Jacob. 1744b. Varia Posthuma No. XI. In *Opera Omnia*, 1036–1048. Geneva.

Bernoulli, Jacob. 1744c. Varia Posthuma No. XXIX. In *Opera Omnia*, 1119–1123. Geneva.

Bernoulli, Johann. 1741. Op. CXLIX—Leçons à de l'Hôpital no. 36–45. In *Opera Omnia*, IV, 491–516. Lausanne and Genève.

Blondel, François. 1673. Résolution des quatre principaux problèmes d'architecture. In *Mémoire de l'Acad. Roy. Des Sciences* (1729), 355–530. Paris.

Bossut, Charles (Abbé). 1777. Recherches sur l'équilibre des voûtes. *Histoire de l'Académie royale des sciences, avec les mémoires de mathématique et de physique* (1778), 534–566.

Bouguer, Pierre. 1736. Sur les lignes courbes qui sont propres à former les voûtes en dôme. *Mémoires de l'Académie Royale des Sciences, année 1734,* 149–164.

Couplet, Claude Antoine. 1729. De la poussée des voutes. In *Memoire de l'Acad. Roy. Des Sciences* (1731), 79–117. Paris.

Duhem, Pierre. 1906. *Les origins de la statique*, 2 vols. Paris.

Euler, Leonhard. 1728. E 8, Solutio problematic de invenienda curva.... In *Academiae Imperialis Scientiarum Petropolitanae*, 3, 64–77.

Gregory, David. 1697. Catenaria. In *Philosophical Transactions*, 19, 637–652, 1698. Repr. In *Acta Eruditorum* Leipzig, 1698, 305–321.

Hermann, Jacob. 1716. *Phoronomia.* Amsterdam.

Hooke, Robert. 1675. *A Description of Helioscopes, and Some Other Instruments.* London.

Huygens, Christiaan. 1888. *Oeuvres complètes de Christiaan Huygens. Correspondance [de Christiaan Huygens], 1666–1669 / publ. par la Société hollandaise des sciences*, 1. The Hague: M. Nijhoff.

Huygens, Christiaan. 1908. *Oeuvres complètes de Christiaan Huygens. / publ. par la Société hollandaise des sciences, Travaux mathematiques 1645–1651,* 11. The Hague: M. Nijhoff.

La Hire, Philippe de. 1695. Traité de Mechanique in Opera diverses .... In *Histoire de l'Académie royale des sciences ... avec les mémoires de mathématique & de physique*, IX (1730), 1–333. Paris.

La Hire, Philippe de. 1702. Remarques sur la forme de quelques arcs don't on se set dans l'Architecture. In *Mémoire de l'Acad. Roy. des Sciences* (1720), 100–103. Paris.

Leibniz, Gottfried Wilhelm von. 1691. De solutionibus problematic catenarii vel funicularis in actis junii A. 1691, aliisque a Dn. J.B. propositis. In *Acta Eruditorum*, 435–439. Leipzig.

Leibniz, Gottfried Wilhelm von. 1745. *Virorum celeberrimorum, G. G. Leibnitii et J. Bernoullii Commercium Philosophicum et mathematicum.* Lausanne and Geneva.

Maltese, Giulio. 2001. *Da "F=ma" alle leggi cardinali del moto: sviluppo della tradizione newtoniana nella meccanica del '700.* Milan: Hoepli.

Mascheroni, Lorenzo. 1785. *Nuove ricerche sull'equilibrio delle volte.* Bergamo.

Pardies, Ignace-Gaston. 1673. *La statique ou la science des forces mouvantes.* Paris.

Parent, Antoine. 1704. Sur la figure de l'extrados d'une voute circulaire, don't tous les voissoirs sont en equilibre entre eux. In *Histoire de l'Acad. Roy. des Sciences* (1712), 93–96. Paris.

Parent, Antoine. 1713. *Essais et Recherches de Mathématique et de Physique.* Paris.

Radelet, Patricia. 2000. Déplacement, vitesses et travaux virtuels. Paper presented at the symposium "Varier pour mieux trouver", 4–9 September 2000, Patzcuaro.

Radelet, Patricia. 1999. La mesure de la courbure et la pratique du calcul différentiel du second ordre. *Proceedings of the workshop, "Le pensée numerique"*, 7–10 September 1999, Peyresq.

Radelet, Patricia. 1989. La composition des mouvements, des vitesses et des forces. In *Proceedings of the Symposium "Der Ausbau des Calculus durch Leibnez und die Brüder Bernoulli"*, 15–17 June 1987, Basel, *Studia Leibnitiana* 17, 25–47.

Roberval, Gille Personne de. 1636. Traité de mechanique des poids soustenus par des puissances sur les plans inclinez à l'horizon. In *L'Harmonie universelle*, ed. M. Mersenne, 1–36.

Stevin, Simon. 1608. De Byvough Der Weeghconst. In *The Principal Works of Simon Stevin,* 1, ed. E. J. Dijksterhuis et al., 528–549. Amsterdam: E.V. Swets & Zeitlinger, 1955.

Stevin, Simon. 1634. *Les œuvres mathématiques de Simon Stevin augmentées par Albert Girard.* Leiden: Elsevier.

Torricelli, Evangelista. 1644. *Opera Geometrica, De motu gravium naturaliter descendentium et projectorum libri duo*. Florence.
Truesdell, Clifford A. 1960. *The Rational Mechanics of Flexible or Elastic Bodies 1638–1788*. Introduction to *Leonhardi Euleri Opera Omnia*, second series, 11 (2). Ziirich: Orell Füssli.
Truesdell, Clifford A. 1968. *Essays in the History of Mechanics*. Berlin and New York: Springer.

Chapter 5
The *De Curvatura Fornicis* of Jacob Bernoulli: The Introduction of the Infinitely Small into the Calculation of Arches

1 Introduction

For various reasons, it seemed interesting to me to analyse the first applications of infinitesimal calculus to the evaluation of the stability of arches, especially those stable under their own weight. This problem, central in the history of architecture, was one of the first to be treated by means of mechanics and more precisely by statics. Further, the laws of statics were the first to be treated mathematically by Archimedes and Euclid, for example. On the other hand, the problem of the stability of an arch under its own weight is linked to the problem of the catenary or of rope that is perfectly free and flexible suspended between two points. To determine the equation of that curve was one of the first problems even the very first, that Jacob Bernoulli, encouraged by Leibniz, posed to the scientific community, in order to develop the new calculus, only sketched by Leibniz in his *Nova methodus*.[1]

When we say that arches are stable under their own weight, we are saying that they are composed of perfectly rigid voussoirs separated by perfectly smooth joints, thus eliminating all problems of elasticity and friction. Those problems are also central in the history of architecture, but friction is intrinsically too complicated to constitute the first step of our research. And elasticity, equally complicated, is involved with the history of differential calculus itself and does not allow us to dissociate the various ingredients. It was by developing the calculus that Jacob Bernoulli discovered the equation of the *elastica*, which furnished, as its name indicates, the starting point of elasticity. We cannot, as that example shows, isolate two periods, one for the scientific development and the other for application or the solution to a technical problem. To be precise, the second period came after the introduction of steel and concrete into construction, that is, at a time where differential calculus was by then in use. But the aim of my article is to analyse an interaction between science and technique.

[1] See chronological table in the Appendix at the end of this chapter.

P. Radelet-de Grave, *Between Mechanics and Architecture*, Mathematics and the Built Environment 8, https://doi.org/10.1007/978-3-031-73530-1_5

In the case of arches, the problem was solved with help of differential calculus made possible by the idealisation of a perfectly flexible and inextensible rope, an analogy that had been already shown byHooke (1676) and Pardies (1673). We shall analyse the work of Jacob Bernoulli, who applied the calculus directly to the arch and clearly showed the idealisation that allows the derivation. After that we will show other researchers, such as Georg Wolfgang Krafft, Claude Antoine Couplet, Charles-François Le Prud'homme D'hailly, Vicomte de Nieuport and others, who attempted to get rid of that strict idealisation in order to be more useful in practice. Nevertheless, our study will allow certain careful parallelisms with the study of the *elastica.* We will come back to that point in the conclusions.

We will conduct our study by carefully analysing four texts:

- The *De curvatura fornicis* of Jacob Bernoulli (1704) announced in the title of this chapter. It was written in his *Meditationes*, Jacob's scientific diary, on 5 December 1704. It is the last but one of his *Meditationes* and one of the very few that is dated. It was published in 1744 with remarks by Gabriel Cramer in Jacob's *Opera.* It was through that publication that the text would come to be known;
- The *De la poussée des voûtes* by Claude Antoine Couplet (1729);
- The *Recherches sur l'équilibre des voûtes* published by Bossut (1774), which had been read at the Academy in July 1770;
- The *Essai analytique sur la méchanique des voûtes* by Nieuport (1780), a Belgian engineer. Unlike the first three texts, this one is not well known.[2]

In addition to these main texts, we shall also quote from a commentary of Jacob Bernoulli's work due to Krafft and from a commentary of the commentary proposed by Nieuport.

2 Jacob Bernoulli, "Problem of the Curvature of Arches that Are Sustained by Their Own Weight Without Any Cement" (1744)

Let us sum up Jacob's text; its clarity allows us to be brief. He starts with the conditions of the experiment:

> Let an arch ABC be constructed with cut stones DKLE in the form of a rectangular oblong parallelepiped having KL slightly larger than DE, its width DK sufficiently narrow or infinitely small; and let us call EN $= dx$, ND $= dy$, DE [the weight of the particle DL] $= ds$, and the weight of BE $= s$.[3](Fig. 1)

[2] I would like to thank B. Van Tiggelen-Urbain who drew my attention to this text.

[3] Bernoulli (1744, p. 1119): *Sit Fornix ABC constructus ex lapidibus cæsis DKLE figuræ fere parallelogrammæ rectangulæ oblongæ, nisi quod KL tantillo major sit quam DE, latitudinis DK valde exiguæ vel infinite parvæ; et vocetur EN* = dx, *ND* = dy, *DE [pondus particulæ DL]* = ds, *adeoque pondus ipsius BE* = s. This and all other translations are mine.

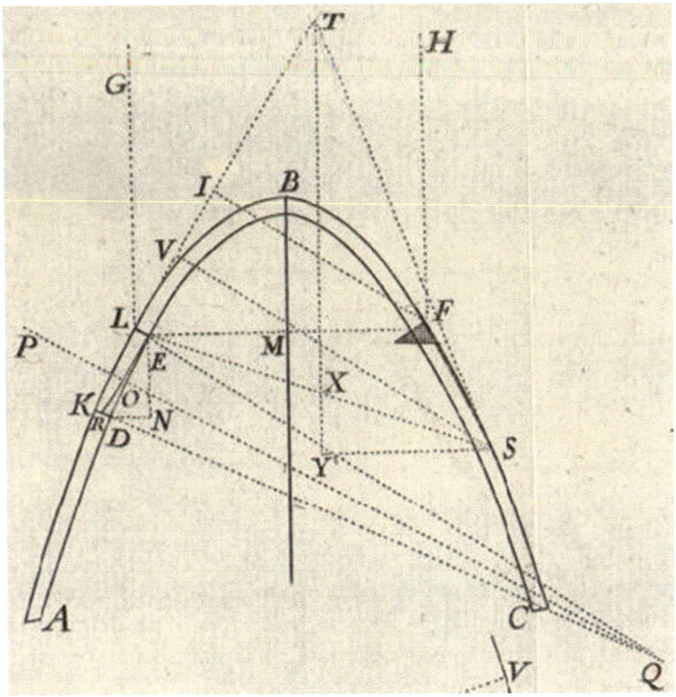

Fig. 1 Jacob Bernoulli's Fig. 46 (1744: Table XLVIII). Public domain

In Fig. 1 we see an infinitely thin arch made of infinitely small voussoirs, which corresponds to a catenary having only one dimension. Those voussoirs are not held by any cement so they can only fall because of their own weight:

> ... stone DL, not adhering to the lower part AK by any cement, can only fall under the effect of its own weight; which it can only do either by sliding on the plane KD, if this plane is perfectly smooth, or by turning around point D....[4]

Jacob writes a first equation describing the equilibrium at translation or equilibrium of forces using—and this is important—the osculating radius or radius of curvature. He establishes the expression of this radius in differential terms very early on[5] but published it only in 1694 in his first text on the *elastica* (1694). He was very proud of this result and baptized it the *theorema aureum*,[6] probably because that expression allowed him to translate the results of the old geometry, which often

---

[4] Bernoulli (1704: 1119): *...lapis DL nullo cæmento cum parte inferiore AK cohærens, proprio pondere cadere necessum haberet; idque vel labendo super plano KD, si hoc planum sit perfecte lubricum; vel rotando circa punctum D.*

[5] Differential calculus as well as the formula giving the radius of curvature $\frac{dxds}{ddy}$ or $\frac{dyds}{ddx}$, appear together in Bernoulli (1691).

[6] For more on Jacob's *theorem aureum*, see On Mascheroni's "New Researches on the Equilibrium of Vaults", Chap. 14 in this present volume.

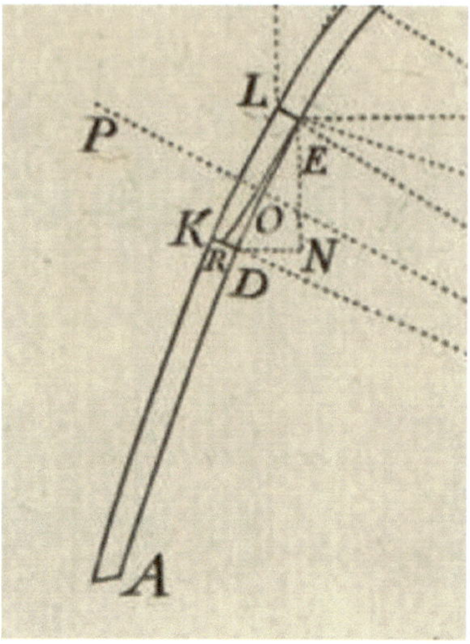

Fig. 2 Detail of Jacob Bernoulli's Fig. 46 (1744: Table XLVIII). Public domain

made use of the evolute (*développée*), into the terms of the new differential geometry. Thus, this theorem allowed him to check his results. We'll see that in the study that we are presenting here the radius of curvature plays an essential role, and this is no coincidence (Fig. 2).

Jacob wrote a second relation that describes the equilibrium at rotation or equilibrium of moments, but he made an error in his calculus. Jacob thinks that DR, perpendicular to the rope DE, is collinear to the joint KD, aligned with the ray of curvature DQ. This mistake, which doesn't concern differential calculus, will be corrected by Gabriel Cramer in his edition of Jacob's works but it will still be commented on by Krafft in 1755 and by Charles Augustine Coulomb in 1773.

The consequence of this error is that he correctly obtains the equation of the catenary by integrating the relation found for the translation equilibrium but that he doesn't find it again for the rotational equilibrium. Differential calculus confirms the result announced by Hooke, Pardies and Philippe de La Hire, who affirmed, by comparing both problems from the point of view of statics, that the curve assumed by a loose chain in equilibrium under the action of its weight alone must be the same as that of the arch stable under its own weight. Jacob does not content himself comparing problems; he now compares the equations he derives.

The problem is now solved in an "almost" perfect way. But—and this justifies what follows—that solution is, as usual, the solution of an idealized case and that idealization is not really helpful in practice. Infinitely thin voussoirs do not exist in reality. Krafft will notice it and will add—although here he is wrong—that it would have been easy for Jacob to go ahead and to consider an arch of a finite thickness. But he makes an error in his generalization. Nieuport will underline the distance between idealisation and practical cases and will continue to think about it.

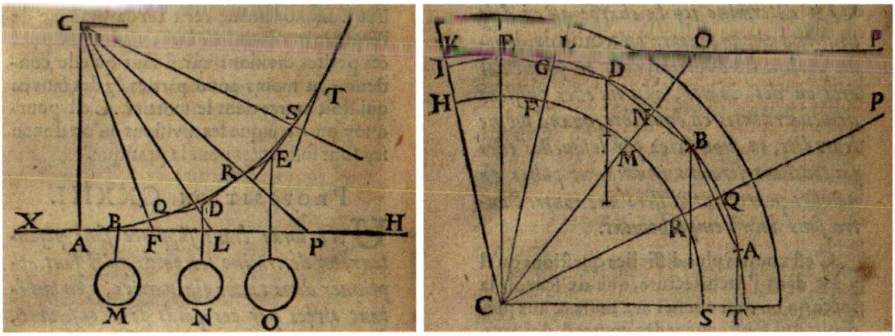

Fig. 3 Philippe de La Hire’s figures from *Traité de Mechanique*, pp. 457 (left) and 466 (right). Public domain

The first reaction going in that direction will come from Couplet. He purposely ignores—because certain elements show that he knows them—the work of Jacob and differential calculus. He goes back to another work, that of La Hire:

> One supposes here that the directions of the weights are parallel to each other, that is the reason why if the power X that is pulling point B along AB is in equilibrium with power M suspended in B, with a power Q pulling in the direction of the tangent BD, by proposition twenty-three the triangle CAF whose sides are perpendicular to the directions of the powers and of the weight, will give the relationship between those three powers taking the weight as a power. In that way X will be represented by CA, the power Q by CF, and the power or weight M by AF.[7]

La Hire immediately adapts this proposition to arches, as shown in our Fig. 3 which juxtaposes his figures: “This proposition is nothing but the converse of the previous one: because if the arch is circular …”.[8]

3 Claude Antoine Couplet, “On the Thrust of Arches” (1729)

Contrary to Jacob Bernoulli, who founded statics on the law of the lever, Couplet founds it on the law of the parallelogram of forces. He particularises that law to his own study thanks to a relation of similarity of triangles whose sides are perpendicular:

---

[7] La Hire (1695): 456–457: *On suppose icy que les directions des poids sont paralleles entr’elles, c’est pourquoy si la puissance X qui tire le point B selon AB demeure en équilibre avec le poids M suspendu en B, avec une puissance Q qui tire selon la touchante BD, par la vingt-troisième proposition le triangle CAF dont les côtés sont perpendiculaires aux directions des puissances et du poids, donnera le rapport de ces trois puissances en prenant le poids pour une puissance. Ainsi la puissance X sera représentée par CA, la puissance Q par CF, and la puissance ou le poids M par AF.*

[8] La Hire (1695: 466): *Cette proposition n’est qu’une converse de la précédente: car que la voute soit circulaire ….*

Fig. 4 Couplet's Fig. I from "De la poussée des voûtes" (1729: Pl. 4). Public domain

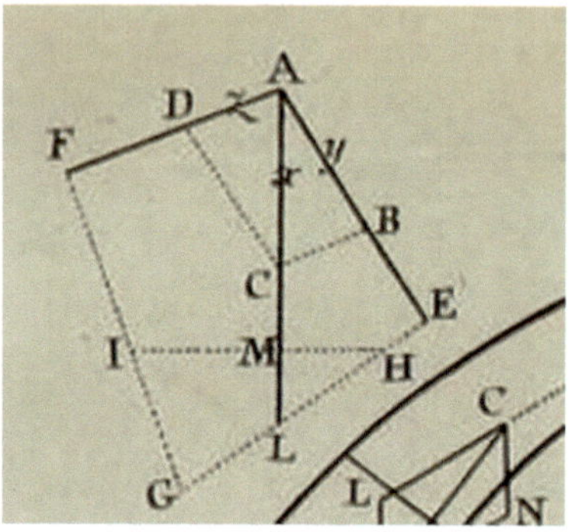

LEMMA: If the force x is decomposed into two forces y and z, those three forces will be to each other as the sides of a triangle formed by the perpendiculars drawn to the directions of the three forces. This is demonstrated in almost all of Mechanics.[9]

In the particular case of arches, the idea of passing from the triangle (or parallelogram) of forces to the triangle whose sides are perpendicular to the directions of the forces is due to La Hire. This idea has the advantage of making use of the radius of the circle of the arch. The fact is important because the use of the radius of this circle will open the way to the use of the osculating circle or the radius of curvature in the work of Bossut and others, and this idea opens the door to the use of differential calculus. Even if Couplet himself does not go so as far as to introduce differential calculus into that problem, there is no doubt that he consciously furnishes various keys that will allow his successors to use it.

In Problem I, Couplet considers the weight of the voussoirs as proportional to their surface.[10] Thanks to the parallelogram law that he gives as a lemma, he will:

determine the ratio that should be between the weights of the voussoirs that form whatever arch and what pressure the voussoirs should exert in order to be in equilibrium without the help of mesh of their parts (Fig. 4).[11]

Having determined that ratio, Couplet gives a theorem that will allow him to introduce, in Problem II, the thickness of the voussoirs (Fig. 5). This was the first step to take:

---

[9] Couplet (1729: 80): *Lemme: Si la force x se décompose en deux forces y et z, ces trois forces seront entre elles comme les côtés d'un Triangle formé par les perpendiculaires menées sur les directions de ces trois forces. Ceci est démontré dans presque toutes les Méchaniques.*

[10] A precision of terminology is helpful in order to clarify the text. There will never be any question of the third dimension of an arch, its depth. The segments are therefore reduced to surfaces and their weight will always be proportional to this surface. Therefore, I call the thickness of a voussoir the dimension which goes in the direction of the height of the arch, and width its other side, which goes in the direction of the fragmentation of the arch.

[11] Couplet (1729: 81): *Déterminer le rapport qu'il doit y avoir dans les pesanteurs des Voussoirs qui forment une Voûte quelconque, et quelle est la poussée des Voussoirs, afin qu'ils fassent équilibre entre eux sans le secours de l' engrènement de leurs parties.*

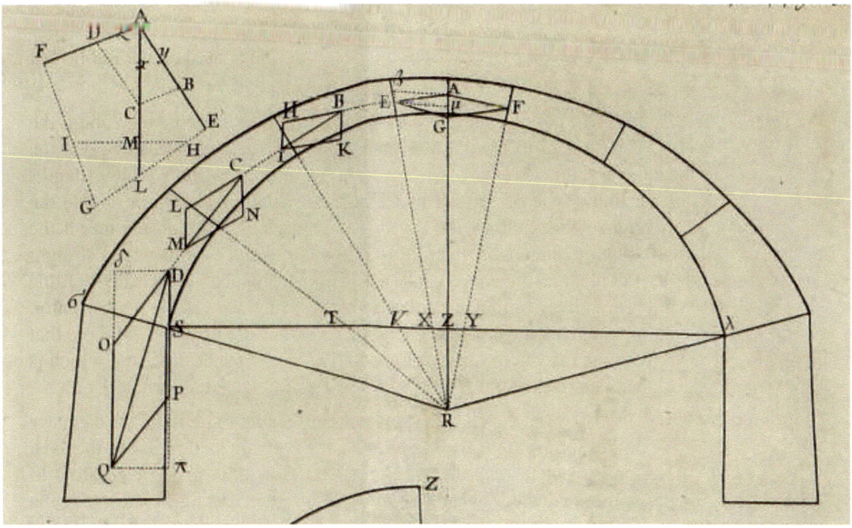

Fig. 5 Couplet's Fig. II from "De la poussée des voûtes" (1729: Pl. 4). Public domain

> Theorem I. When all the joints of the arch are directed towards the same point R, as one says they must be in an arch whose intrados is circular, if one traces any horizontal line SI, the weights of the voussoirs A, B, C, D, will be expressed by the parts XY, VX, TV, SI of that horizontal line enclosed between the prolonged joints of those voussoirs.
>
> Their pressures will be expressed by RX, RV, RT, RS, that are the prolongations of their inferior joints intercepted between the crossing point R and the horizontal line SI, if all voussoirs are in equilibrium with each other.[12]

A particular theorem, once more due to La Hire, will help him study the case of an arch with circular intrados whose joints are perpendicular to the intrados circle. This is the only case he will treat, even if he gives himself the illusion of treating a more general case. La Hire's theorem reduces the report of the weights of the voussoirs to the report of the lengths of segments intercepted by the rays or the directions of the joints on whatever horizontal line. He is still dependent on the existence of a unique centre R, thus of a circle.[13]

Thanks to these elements, he will study the thickness to be given to the voussoirs of such an arch, when the width of the voussoirs is everywhere the same. That way, he arrives at Problem II: "Determine the length of the voussoirs that by their own

---

[12] Couplet (1729: 84): *Theorem I: Lorsque tous les joints de la Voûte sont dirigés vers un même point R, comme l'on dit qu'ils doivent être dans la Voûte, dont l'intrados est circulaire, sil' on tire une ligne horizontale quelconque S l, les pesanteurs des Voussoirs A, B, C, D, seront exprimées par les parties XY, VX, TV, SI, de cette ligne horizontale renfermées entre les joints prolongés de ces Voussoirs. Leurs poussées seront exprimées par RX, RV, RT, RS, qui sont les prolongements de leurs joints inférieurs interceptés & compris entre le point de concours R & la ligne horizontale S l, si tous ces Voussoirs font équilibre entre eux.*

[13] One may ask if there exists a generalization of that theorem?

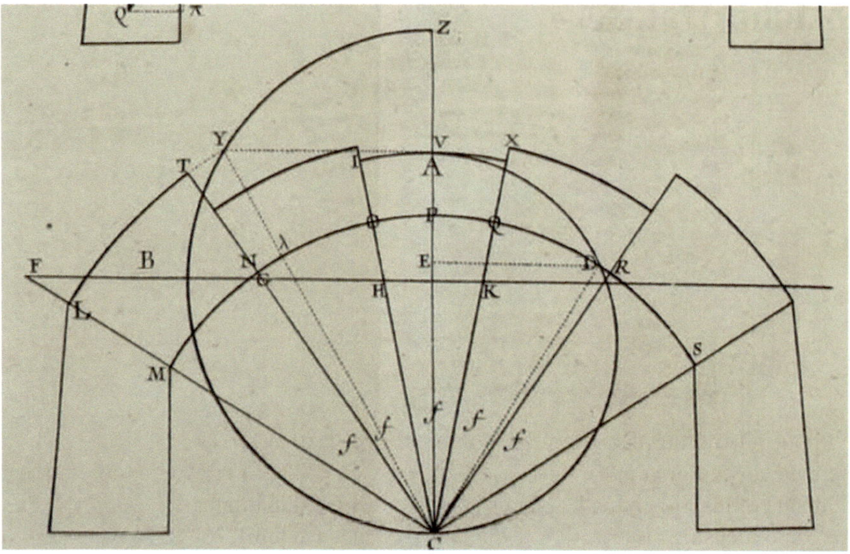

Fig. 6 Couplet's Fig. III from "De la poussée des voûtes" (1729: Pl. 4). Public domain

weight maintain a circular arch in equilibrium, without considering the mesh of their parts".[14]

Couplet immediately transforms this theorem placing the circular arch in the centre of the arch (Fig. 6).

Next he arrives at Problem III: "Find the length of the voussoirs of an arch, such that an arc of circle be equidistant to the intrados and the extrados of each voussoir" (Fig. 7).[15]

He punctuates the solution of this problem in the following way:

> As the demonstration requires that each voussoir has its intrados and extrados on arcs of concentric circles, both having their common centre in Z. In these conditions the intrados and the extrados will not be perfect curves, unless the thickness of each voussoir is infinitely thin.[16]

This remark, made by one who knows differential calculus but decides to forget it and to start over from the beginning in the particular case that interests him, is quite remarkable. The figure he proposes is also eloquent (see Fig. 7).

---

[14] Couplet (1729: 87): *Problème II: Déterminer la longueur des Voussoirs qui par leur propre poids soutiennent en équilibre dans une Voûte circulaire, sans y considérer l'engrènement des parties ?*

[15] Couplet (1729: 90): *Problème III, Trouver les longueurs des Voussoirs d'une Voûte, telle qu'un Arc de Cercle soit également distant de l'intrados et de l'extrados de chaque Voussoir.*

[16] Couplet (1729: 91): *Comme la démonstration exige que chaque voussoir ait son intrados* et *son extrados en Arcs de cercles concentriques, ayant tous les deux leur centre commun en Z, pour lors l'intrados et l'extrados ne seront point des courbes parfaites, à moins que l'épaisseur de chaque voussoir ne soit infiniment mince.*

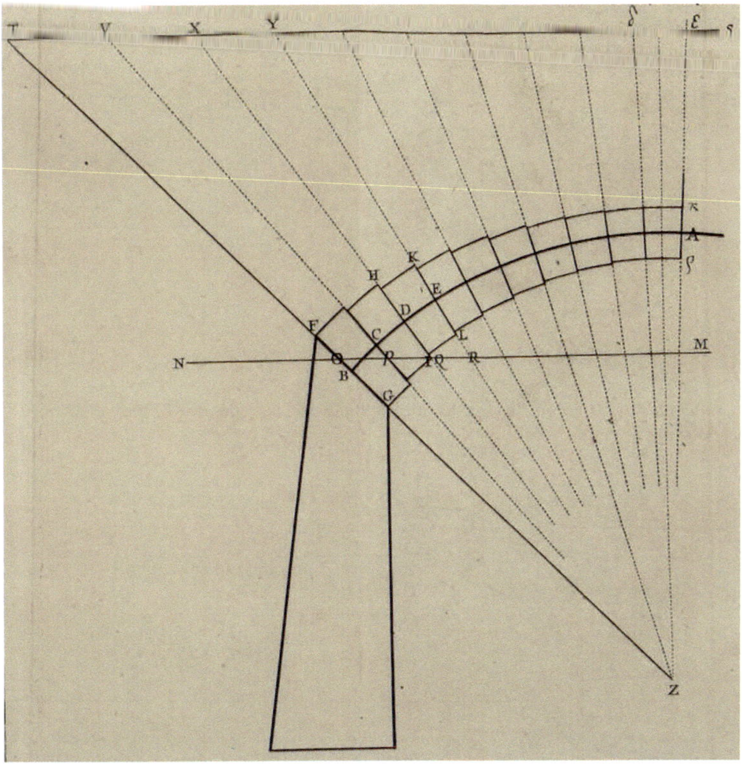

Fig. 7 Couplet's Fig. IV from "De la poussée des voûtes" (1729: Pl. V). Public domain

If one applies the same reasoning to Problem II, Fig. III, one finds at the limit Nicomedes' conchoid as Coulomb shows in his famous "Essai sur une application des règles de *Maximis* et *Minimis*" (1773) (Fig. 8).

The calculation of this curve is the only place that Coulomb leaves to differential calculus. His work is founded upon Fermat's theory of *maximis* and *minimis*, which, as he knows, is much less elaborate but sufficient, in his opinion, for artisans. It allows him to calculate the extreme values that insure stability. We will not comment here on the originality of his calculus.[17]

Theorem II, with which Couplet pursues his study, shows the same type of approach:

> We shall get half the weight of the arch if from above and below the key we trace two horizontal lines, one tangent to the intrados and the other tangent to the extrados, both extended out to the intersection of the last lowest joint of the prolonged voussoir; and half the sum of these two tangents being multiplied by the height of the key, which is the first

[17] On Coulomb and his "Essai", see Chap. 13 in this present volume.

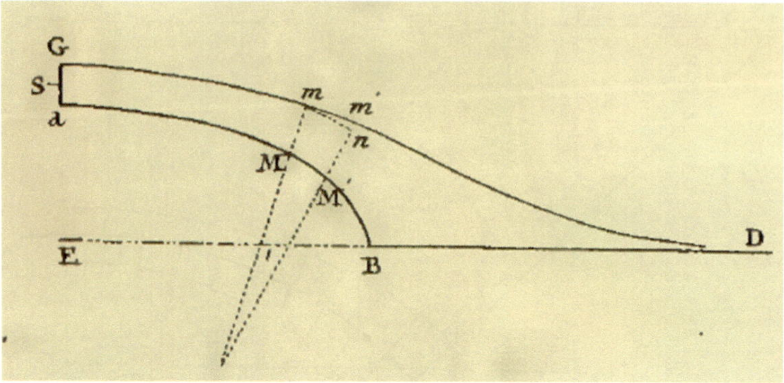

Fig. 8 Coulomb's Fig. 11 from "Essai sur une application des règles de *Maximis* et *Minimis*" (1773: Tab. XVI). Public domain

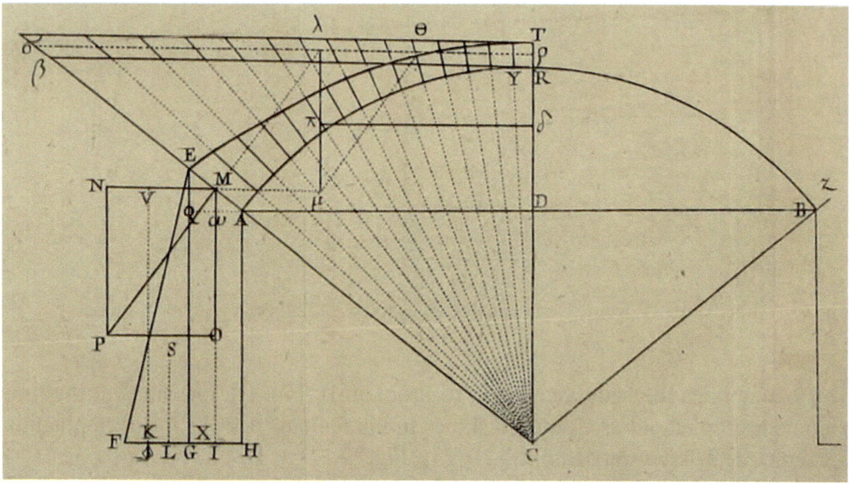

Fig. 9 Couplet's Fig. VI from "De la poussée des voûtes" (1729: Pl. 6). Public domain

> voussoir, will give a surface equal to the cut of the arch, and so will give the strength of the arch and consequently its weight.[18] (Fig. 9)

This theorem shows that no matter how one traces the arch into the beam of rays, if the elements are of equal surface, the ratio of the weights will be the same as that of a flat arch built between the tangents to the intrados and to the extrados. His Corollary

[18] Couplet (1729: 92): *L'on aura la pesanteur de la moitié de la Voute, si du dessous et du dessus de la Clef l'on tire deux lignes horizontales, c'est à-dire, une tangente à l'intrados, et une tangente à l'extrados, toutes deux menées jusqu'à la rencontre du dernier joint inférieur du Voussoir prolongé; et la moitié de la somme de ces deux tangentes étant multipliée par la hauteur de la Clef qui est le premier Voussoir donnera une surface égale à la coupe de la Voûte, et donnera par conséquent la solidité de la Voute, et partant sa pesanteur.*

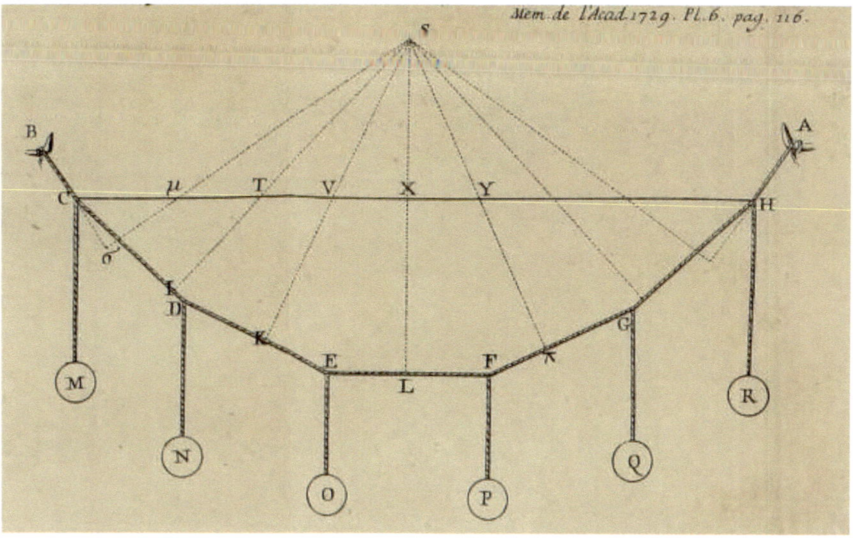

Fig. 10 Couplet's Fig. V from "De la poussée des voûtes" (1729: Pl. 6). Public domain

II may perhaps better express what the theorem contains. Couplet shows there that the surface and thus the weight of the half-arch is equal to the area between the two tangents, one to the intrados and the other to the extrados. From that he deduces that "if the intrados is any curved line, the extrados will also be a curve that will depend on the curvature of the intrados.[19] This dependence merits attention.

Couplet's Problem IV is: "Determine the uniform curvature of an arch such that it is maintained in equilibrium, and of which we consider the voussoirs as polished, that is to say, without connection" (Fig. 10).[20]

He obviously knows that the answer is the catenary, or the loose rope as he calls it, but his explanation shows that he hasn't fully understood it. He in fact considers the perpendiculars to that curve as intersecting each other in one point. He does not see the role played by the radius of curvature in the generalization that he considers, and which would allow him to overcome to hard restriction.

His Problem VI is: "Find the centre of gravity of the half-arch whose intrados is a circle" (Fig. 11).[21]

He will find it by summation, but the calculation of centres of gravity had already given rise to integrations long before the calculus of Leibniz. He then goes on to find

[19] Couplet (1729: 94): *que si l'intrados est une ligne courbe quelconque, l'extrados sera aussi une courbe qui dépendra de la courbure de l'intrados.*

[20] Couplet (1729: 95): *Déterminer la courbure uniforme d'une Voûte, telle qu'elle se maintienne en équilibre, et dont nous considérons les Voussoirs comme polis, c'est-à-dire, sans liaison.*

[21] Couplet (1729: 99): *Trouver le centre de gravité de la demi-Voûte* ANDE, *dont l'intrados est circulaire.*

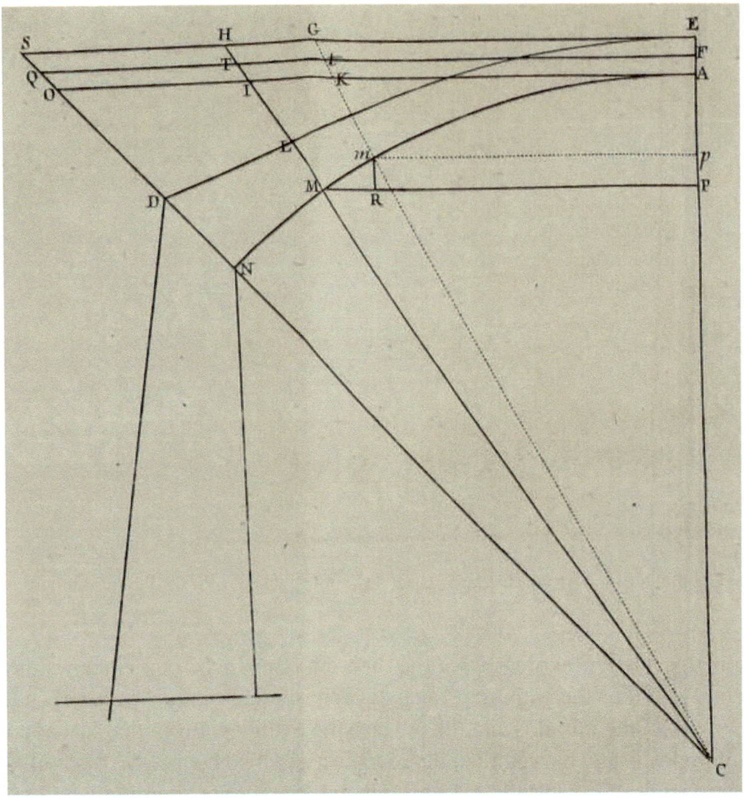

Fig. 11 Couplet's Fig. VII from "De la poussée des voûtes" (1729: Pl. 7). Public domain

the centre of gravity in order to find the force that the arch exerts on its right foot. This part of his work is beyond the scope of our article.

To summarize, we observe in Couplet a clear decline in the way differential calculus is used with respect to Jacob Bernoulli. However, it is evident that such preoccupations gave rise to his reflection and that he does not arrive at the application of this calculus to the real problem of voussoirs that have a certain thickness.

His thinking leads him to underline the independence of the curves of intrados and extrados, and he points out that the catenary, which characterises the greatest stability of an arch infinitely thin under its own weight, must, if the arch is not infinitely thin, transform itself in a curve passing through all centres of gravity of the different voussoirs.

4 Krafft, "On Funicular and Catenary Curves" (1755)

At the end of his text dedicated to funicular and catenary curves, Krafft, who knows the property of stability of that curve in the case of the arch, pursues the same goal as Couplet,[22] that is, to give thickness to an arch. For that reason, he considers voussoirs that are infinitely narrow but have some thickness and are included between two radiuses of curvature infinitely close. Underlining that last point, which was already known to Jacob Bernoulli, he allows the recourse to differential calculus. Unfortunately, he does not see the contradiction that exists between the hypothesis made by Bernoulli on the weight proportional to the element of line ds and the fact of giving a weight to the arch.

This contradiction will be highlighted by Nieuport, in his "Memoire on the alleged property of catenary arches" (Nieuport 1783). After having textually reproduced the extract of Krafft and the hypotheses of Jacob Bernoulli, Nieuport writes:

> Here is what caused the paralogism in question. In fact, it is evident that in all curves other than the circle, the radius of the evolute varies in each moment while the arc of the intrados and the thickness of the arch are constants; it is necessary that the one of the extrados varies, and consequently those voussoirs could not be proportional to ds.[23]

He then takes over Krafft's calculus, replacing the line element by a volume element. He concludes that even if the equations obtained are almost the same it is not possible to identify them. The equation of Nieuport may be:

> reduced to $Vddy = -dy\,dV$, or $Vddy + dy\,dV = 0$, whose integral is $Vdy = adx$, which differs from the result of Mr. Krafft by the fact that the voussoir V is found in the place of s, which is the arc of the intrados. The paralogism thus comes only, as we have said, from the supposition that the one is proportional to the other.[24]

5 Abbé Charles Bossut, "Researches on the Equilibrium of Arches" (1774)

It is interesting to observe that the organisation and the structure of Bossut's text is similar to that of Couplet, and to that, posterior, of Nieuport. Like Couplet, he starts from the description of the equilibrium of the voussoirs with help of the parallelogram law of forces, which he writes in function of the sines and that induces a relation

[22] We have already seen that Couplet could not abstract himself from the circle.

[23] Nieuport (1783: 22): *...voilà ce qui a occasionné le paralogisme en question. En effet il est évident que dans toute autre courbe que le cercle, le rayon de Développée variant à chaque instant, tandis que l'arc d'intrados et l'épaisseur de la voûte sont constants; il faut nécessairement que celui d'extrados varie, et conséquemment ces voussoirs ne sauroient être proportionnels à* ds.

[24] Nieuport (1783: 23): *qui se réduit à* Vddy $= -$ dydV, *ou* Vddy + dydV $= 0$, *dont l'intégrale est* Vdy = adx, *qui ne diffère du résultat de Mr. Krafft qu'en ce que le voussoir* V *s'y trouve à la place de* s *qui est l'arc d'intrados. Le paralogisme vient donc uniquement, comme nous l'avons dit, de la supposition que l'un soit proportionnel à l'autre.*

between the sides of another triangle whose sides are perpendicular to those of the first one, exactly as in the works of La Hire and Couplet (Fig. 12).

Bossut first calculates the equilibrium of three voussoirs lying next to each other, without any help from differential calculus, and which he particularises to the case of Couplet: given a circular intrados and the joints perpendicular to it, determine the size, in this case the thickness of the voussoirs. He finds a stepped curve (Fig. 13), which, made uniform, would give the curve of Nicomedes, dear to Coulomb and already mentioned by us.

Once this is done—and it is here that he begins to be original—he writes the same relation again between the forces applied to two contiguous voussoirs, this

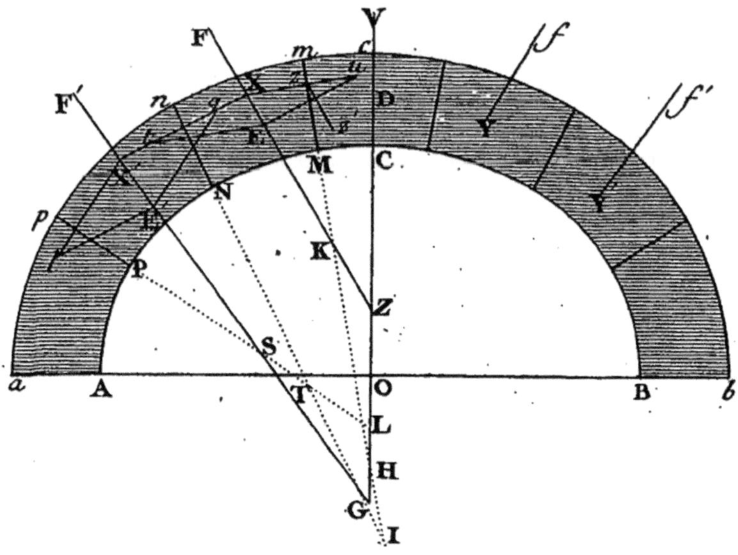

Fig. 12 Bossut's from "Recherches sur l'équilibre des voûtes" (1774: Pl. V). Public domain

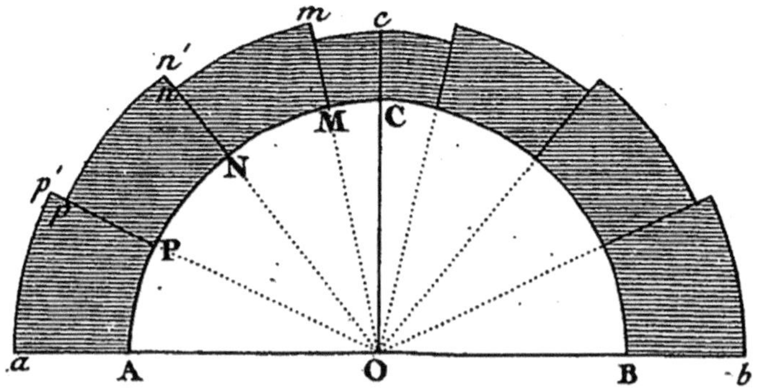

Fig. 13 Bossut's Fig. 2 from "Recherches sur l'équilibre des voûtes" (1774: Pl. V). Public domain

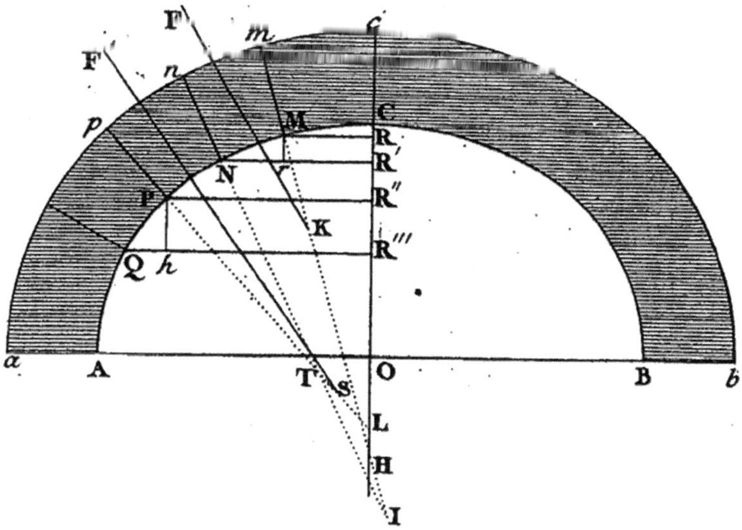

Fig. 14 Bossut's from "Recherches sur l'équilibre des voûtes" (1774: Pl. V). Public domain

time considering those voussoirs as being infinitely thin and situated between two radii of osculation or radii of curvature and infinitely close. Thus, he rewrites this relation in function of the line of the curve and of the radius of curvature (Fig. 14):

> Let us come back to the general hypothesis of the first figure and suppose the number of voussoirs to be infinite. Then the arcs *MN*, *NP*, etc. are infinitely small, and the angles *NIM*, *PTN*, etc. are those formed by consecutive radii of osculating circles.[25]

He then explains a point that will be matured by Nieuport:

> …there are two principal questions to examine. One consists in finding the shape of the arch when one knows the law of forces pressing the voussoirs; the other one, on the contrary, in finding the law of the forces that must press the voussoirs when one knows the shape of the arch. One sees that the second question is the inverse of the first. The equation will serve us to solve both problems.[26]

He will next play with differential calculus and imagine various cases of particular fields of forces: "vertical" as in the case of gravitation, in which case he finds the catenary again; "central", as in the case of arches pressed by water, where he finds the elastica; and finally, by "combinations" of these two types, thanks to the composition or parallelogram of forces.

---

[25] Bossut (1778: 539): *Revenons à l'hypothèse générale de la figure première; et supposons que le nombre des voussoirs soit infini. Alors les arcs* MN, NP, etc. *sont infiniment petits; et les angles* NIM, PTN, etc. *sont ceux que forment entr' eux les rayons osculateurs consécutifs.*

[26] Bossut (1778: 541): *…il y a deux questions principales à examiner: L'une consiste à trouver la figure de la voûte lorsqu'on connoît la loi des forces qui pressent les voussoirs; l'autre, au contraire, à trouver la loi des forces qui doivent presser les voussoirs lorsqu'on connoît la figure de la voûte. On voit que la seconde question est l'inverse de la première. L'équation* (C) *va nous servir à les résoudre l'une & l'autre.*

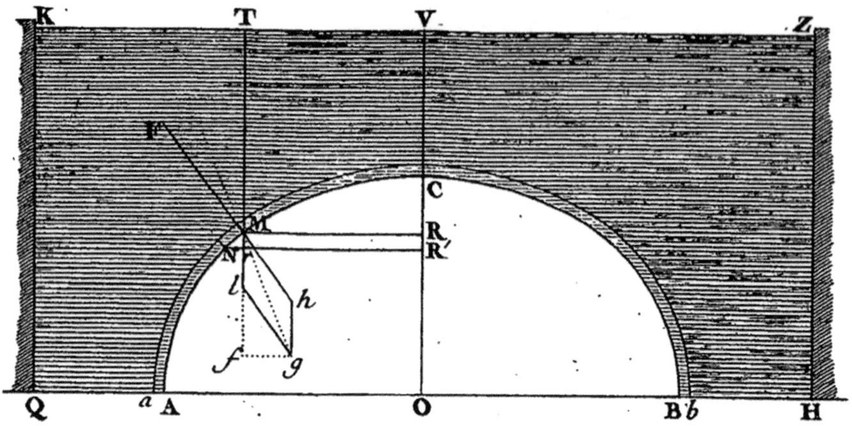

Fig. 15 Bossut's from "Recherches sur l'équilibre des voûtes" (1774: Pl. V). Public domain

When considering infinitely thin voussoirs, Bossut is not explicitly clear concerning the thickness of the arches. He never alludes to their thickness and if one looks at his Fig. 4 (Fig. 15), one may imagine that they have no thickness or that the curve which he is looking for and which he calls the intrados is the one going through the centres of gravity of the various voussoirs. As he considers whatever forces applied to this curve, those forces may take the weight of the voussoirs into account.

Let us sum up. While La Hire and Couplet introduce the radius of a circle in either the intrados or the extrados, thanks to the theorem of La Hire, Bossut, in contrast, studies the local equilibrium of three voussoirs and introduces the radius of curvature. The curves that underpin his arches are no longer exclusively circular.

6 Nieuport's "Analytical Essay on the Mechanics of Arches" (1780)

Nieuport, as his name indicates, was a Belgian, though he born in Paris on 13 January 1746. He died in Brussels on 20 August 1827. He was originally from Ghent and spoke the dialect of this city fluently. He was an engineer and started to do mathematics only at the age of 40, but this did not prevent him from being the first to develop differential calculus in Belgium.

Even if the essay we are examining was published at almost the same time as the one by Bossut, we know that it had been read already seven years earlier, and if we believe the account of Adolphe Quetelet,[27] Nieuport and Bossut knew each other. We may therefore think that Bossut had an influence on Nieuport. But it is curious that

[27] See Quetelet and Jacques (1864, pp. 333–334).

while Bossut, in order to use the radius of the osculating circle, set about writing a local relation of equilibrium, Nieuport wonders about the value of the results obtained by differential calculus, precisely because they are inevitably local:

> Before leaving this topic, it is time to make some reflections, otherwise one could link some false ideas to certain expressions in this chapter. First, what does equilibrium between infinitely small voussoirs mean? If it were possible to get such voussoirs and to form an arch with them, would it really be in equilibrium? Surely not. Only an infinitely small part of that arch would have that property. This will not surprise the reader who knows that millions of such voussoirs have only an infinitely small size. Thus, those millions of voussoirs will be in equilibrium, but never a finite portion of the arch. Indeed, let us suppose that an infinity of those small voussoirs forms the finite portion, etc.: the equilibrium would cease to take place because the sum, etc. of small voussoirs could not be in equilibrium without both parts E and H of it also being in equilibrium. And we have seen that instead of the proportion AB: BC that exists between the parts E and H, we should have AB: BV. Thus etc.

But would one perhaps ask, why do we find by calculus portions that are finite and entire curves that have the same property? This is why. Those curves do not have the property of equilibrium; they have the one expressed by $PT^2 - PV^2$; and that formula only coincides with the property of equilibrium for infinitely small portions of the curve. It is only in that sense that we must understand that chapter; and the only exact statement of the problem is to search for the curves that have the property of equilibrium for every infinitely small part of their circumference subdivided in voussoirs even smaller but of the same order and not for a finite sum of its parts. One may judge from that if it is reasonable to wish to appropriate the catenary and other similar curves for the practice of arches.[28]

I can only explain Nieuport's hesitations by the fact that he does not see the role of the edge conditions.

---

[28] Nieuport (1780: 87): *Avant de quitter cette matière, il est à propos de faire quelques réflexions.faute de quoi on pourroit attacher des idées peu exactes à certaines expressions qui se trouvent dans ce chapitre. D'abord que signifie cet équilibre entre des voussoirs infiniment petits? s'il étoit possible d'avoir de pareils voussoirs, et d'en former une voûte, seroit-elle vraiment en équilibre? non sans doute. Il n'y auroit réellement qu'une partie infiniment petite de cette voûte qui auroit cette propriété. Ceci n'étonnera par le lecteur qui sait que des millions de pareils voussoirs ne sont encore qu'une grandeur infiniment petite. Ainsi, ces millions de voussoirs seront en équilibre, mais jamais une portion finie de la voûte. En effet, supposons qu'une infinité de ces petits voussoirs formassent la portion finie* etc., *dès-lors l'équilibre cesseroit d'avoir lieu; car la somme* etc. *de petits voussoirs ne sauroit être en équilibre, sans que les deux parties E et H n'y soient aussi. Or, nous avons vu qu'alors, au lieu de la proportion AB: BC qui règne entre les parties E, H, il faudroit celle AB: BV. Donc* etc.

Mais dira-t-on peut-être, pourquoi trouvons nous par le calcul des portions finies, des courbes entières qui ont la même propriété? le voici. Ces courbes n'ont pas la propriété de l'équilibre proprement dite; elles ont celle qui est exprimée par la formule $PT^2 - PV^2$; *& cette formule ne coïncide avec la propriété de l'équilibre que des portions infiniment petites de la courbe. Voilà le seul sens dans lequel il faut entendre ce chapitre; & l'énoncé exact du problème revient à chercher les courbes qui ont la propriété de l'équilibre, pour chaque partie infiniment petite de leur circonférence subdivisée en voussoirs encore plus petits, mais de mime ordre, et non pas pour la somme finie de ces parties. On peut juger delà s'il est raisonnable de vouloir approprier la chainette & autres courbes semblables à la pratique des voûtes.*

From the point of view of their structure, the texts of Bossut and Nieuport have elements in common, but as Nieuport's contents of his article shows,[29] he has jumped a step further: he has found a classification that exploits differential calculus:

Table of Contents

Chap. I	On the equilibrium of voussoirs in general
Chap. II	On the arches whose voussoirs meet at a point[30]
Chap. III	On the arches whose voussoirs do not meet at a point, those voussoirs being infinitely small
Chap. IV	On the preceding arches in the hypothesis of the voussoirs with a finite dimension
Chap. V	The choice of the curves of the points of intersection
Chap. VI	The choice of the curve of the intrados
Chap. VII	On the thrust of arches and everything related to it
Section I	On the way of estimate this thrust
Section II	On the length of the voussoirs
Section III	The thrust of the arches and all that is related to it
Chap. VIII	On ramped arches

Like both of his predecessors, Couplet and Bossut, Nieuport starts with the parallelogram law of forces, which he adapts using the radius of curvature thanks to La Hire's theorem. After that, he considers that in this problem there are generally three important curves: the intrados, the extrados and the curve formed by the points of intersection of the joints of the voussoirs. If one considers those joints as perpendicular to the intrados – which is generally requested for aesthetic reasons – this curve, which links the centres of curvature, is the evolute of the intrados. But Nieuport considers more general cases (Fig. 16).

The structure of his text immediately results from the consideration of those three curves. First are the cases where one of those curves, namely the curve linking the intersections of the joints, is reduced to one point. Those cases are the only ones Couplet could consider without using the radius of curvature. As Nieuport says, in this case it is not indispensable to consider the voussoirs as infinitely small. In this case, he studies, as did his predecessors, the curve of extrados when the intrados is a circle and finds Nicomedes' conchoid. As Coulomb did before him, he gives its equation (Fig. 17).

In contrast, if the joints do not meet at a point, one must use the radius of curvature and consider the voussoirs as infinitely small. One does not speak here of an infinitely small thickness, as Jacob Bernoulli did, but of an infinitely small width separating two consecutive radii of curvature. Nieuport adds to that a remark that we had seen appear at the horizon of Couplet and asserted by Bossut, namely that those three curves are linked together and that if two of them are known, the relations of equilibrium make

[29] Nieuport (1780, p. 142).

[30] As de Nieuport's text shows clearly, it is the joints separating the voussoirs which, extended, meet at one point.

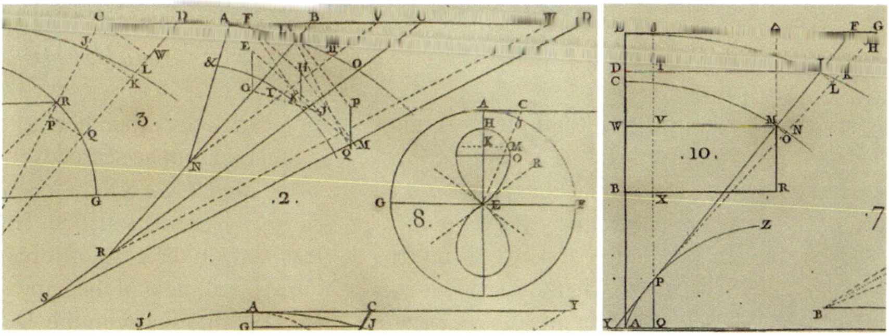

Fig. 16 Figs. 2 (left) and 10 (right) from Nieuport's "Essai analytique sur la méchanique des voûtes" (1780: Pl. 1). Public domain

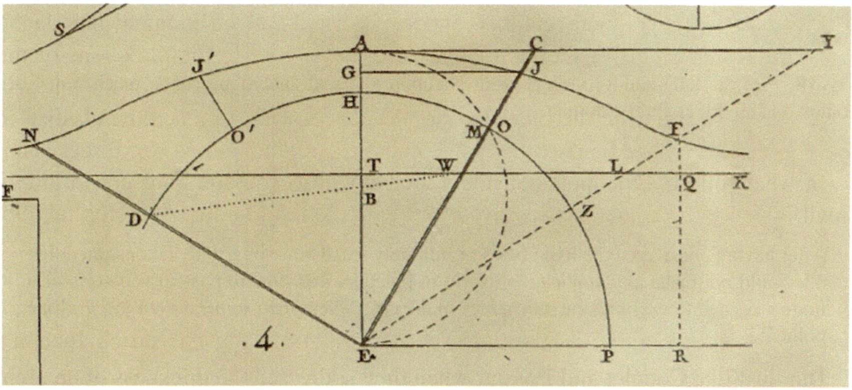

Fig. 17 Fig. 4 from Nieuport's "Essai analytique sur la méchanique des voûtes" (1780: Pl. 1). Public domain

it possible to determine the third one. Nieuport is on the road to a long-term analytic study of which we will only show some graphical results (Fig. 18).

This study recalls that of Euler (1744) in the case of the elastic lamina, when he generalised the common elastica that had been found by Jacob Bernoulli, just as here Nieuport generalises the common catenary, although this remark has to be limited in various senses, because the problem treated is not as broadly. This remark intends to emphasise the appeal made in a different physical context to the same type of purely mathematical analysis.

Nieuport also takes the opportunity to show that the catenary does not correspond to the intrados curve of the arch stable under its own weight when this has a certain thickness. In order to find the catenary, it is necessary to search for the curve that passes through the centres of gravity of the voussoirs. We have already mentioned another article that Nieuport will publish on that argument (1783) and in which he comments the work of Krafft.

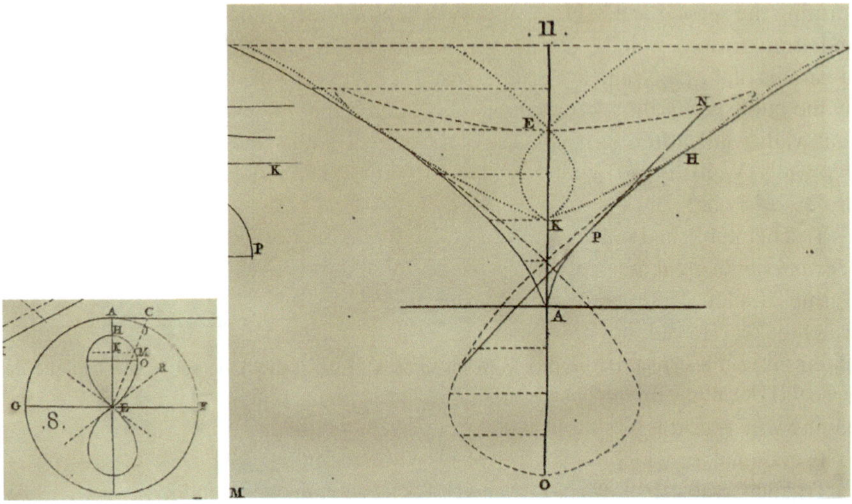

Fig. 18 Figs. 8 (**left**) and 11 (*right*) from Nieuport's "Essai analytique sur la méchanique des voûtes" (1780: Pl. I). Public domain

In what follows, Nieuport considers arches whose voussoirs are not infinitely small:

> After having considered the hypothesis of infinitely small voussoirs, and demonstrated that one could not make any application useful in practice, it is time to pass to voussoirs of a finite size, and to examine the mechanics in the case where their joints do not meet in one point.[31]

But, just like Couplet and Bossut, when they addressed the thickness of an arch, Nieuport can only treat some particular cases without using infinitely small voussoirs and thus leaving behind the mathematical tool that he had so brilliantly used before. The engineer has taken up the reins again and he will try to adapt the mathematical tools.

7 Conclusion

It seems to me that this analysis of the introduction of calculus into a domain that, at first sight, appears to have lagged behind tells us much concerning the relationships between pure science and technique or applications. Here are some examples.

We have seen how the earliest authors relate to a problem already developed mathematically with help of the catenary. We observe the same phenomenon with

[31] Nieuport (1780: 88): *Après avoir considéré l'hypothese des voussoirs infiniment petits, & démontré qu'on ne peut en faire aucune application utile dans la pratique, il est temps de passer à ceux d'une grandeur finie, et d'en examiner la méchanique, dans le cas où leurs joints ne concourent pas en un point.*

Daniel Bernoulli and Euler: when they started their joint study of magnetism, they went back to descriptions where they were able to apply theories they had already developed for hydrodynamics or elasticity.

Jacob Bernoulli proposed an idealisation of the arch that made it possible mathematically to find the catenary again. This is the infinitely thin arch, composed of infinitely narrow voussoirs. This idealisation will be loosened in order better to adapt to real problems. In order to do that, the authors started by renouncing differential calculus and then progressively went back to it, in order to arrive at a classification derived from it that made many solutions possible.

It seemed interesting to underline this way an analytic construction and show the prescience that the first two authors had of it to a certain degree. Three curves intervene in the problem: the intrados, the extrados and the curve of intersection or orientation of the joints. If two of those curves are known, the third is determined by a condition of equilibrium. Couplet had seen that when the curve of intersection of the joints is reduced to a point, the curve of the extrados is determined once the curve of the intrados is given.

The fact that Couplet had a differential geometrical view of the problem while he did not use the mathematical tool of differential calculus also seemed to us worthy of attention. In contrast, with Nieuport, the mathematical tool is acquired but the real representation poses a problem, to the point of considering it a mathematical game as devoid of practical utility.

Appendix: Chronological Table

1684 Leibniz , *Nova Methodus*

1686 Leibniz poses the problem of the isochrone

1690 Jacob Bernoulli poses the problem of the catenary

1691 Jacob Bernoulli poses the problem of the velaria

1692 Leibniz, Johann Bernoulli and Huygens solve the problem of the catenary

1692 Jacob poses the problem of the *elastica*

1694 Jacob publishes his formula of the radius of curvature in his first work on the *elastica*

1695 La Hire publishes his *Traité de mécanique*

1696 Johann poses the problem of the brachistochrone

1704 Jacob published the *de curvatura fornicis*

1744 publication of the *Opera* by Jacob Bernoulli and of the *Methodus inveniendi* by Euler.

References

Bernoulli, Jacob. 1691. De Spiralis Parabolicae (Parabolae helicoidis) dimensione (Med. CLXI) (T.P. 1689–T.A. 1/1691). *Die Werke von Jakob Bernoulli: Bd. 5: Differentialgeometrie*, 167ff. Basel: Birkhäuser.

Bernoulli, Jacob. 1694. Curvatura Laminae Elasticae. Ejus Identitas cum Curvatura Lintei a pondere inclusi fluidi expansi (Op. LVIII), *Acta Eruditorum*, Junii 1694, 262–276; *Opera*, 576–600

Bernoulli, Jacob. 1744. Problema de Curvatura fornicis, cujus partes se mutuo proprio pondere suffulciunt sine opere caementi (5/12/1704), *Varia Postuma Jacob Bernoulli, Opera*, vol. II, XXIX, 1119–1123.

Bossut, Charles (Abbé). 1774. Recherches sur l'équilibre des voûtes. *Mémoires de l'Académie de Paris,* année 1774 (1778), 534–566.

Couplet, Claude Antoine. 1729. De la poussée des voûtes, *Histoire de l'Académie Royale des sciences, avec les mémoires de mathématique et de physique* (1731), 79–117.

Coulomb, Charles Augustin. 1773. Essai sur une application des règles de Maximis et Minimis à quelques problèmes de statique, relatifs à l'architecture, *Mémoires de Mathématique et de physique, présentés par divers savans*, vol. 7, année 1773 (1776), 343–382.

Euler, Leonhard. 1744. *Methodus inveniendi lineas curvas maximi minlmive proprietate gaudentes, Additamentum I de curvis elasticis.* Lausanne et Genève, Marcum-Michaelem Bousquet.

Hooke, Robert. 1676. *A description of helioscopes, and some other instruments*, London: John Martyn.

Krafft, Georg Wolfgang. 1755. De curvis funiculariis et catenariis, vel illis, quae corporibus flexilibus inducitur, cum a potentiis quibusvis solicitantur, *Novi Commentarii acaderniae scientiarum Petropolitanae*, vol. 5 (1754–55), 145–163.

La Hire, Philippe de. 1695. *Traité de mécanique, où l'on explique tout ce qui est necessaire dans la pratique des Arts.* Paris: Imprimerie Royale.

Leibniz, Gottfried Wilhelm. 1684. Nova methodus pro maximis et minimis. *Acta Eruditorum*, October 1684, 467–471.

Nieuport, Charles-François Le Prud'homme D'hailly, Vicomte de. 1780. Essai analytique sur la méchanique des voûtes, présenté le 18 mai 1778, *Mémoires de l'Académie impériale et royale des sciences et belles-lettres de Bruxelles*, vol. II (1780), 43–136.

Nieuport, Charles-François Le Prud'homme D'hailly, Vicomte de. 1783. Mémoire sur la propriété prétendue des voûtes en chaînettes, présenté le 6 novembre 1780, *Mémoires de l'Académie impériale et royale des sciences et belles-lettres de Bruxelles* vol. IV (1783), 18–26.

Pardies, Ignace-Gaston. 1673. *La statique ou la science des forces mouvantes.* Paris: Mabre-Cramoisy.

Quetelet, Lambert Adolphe Jacques. 1864. *Histoire des sciences mathématiques et physiques chez les Belges.* Brussels: M. Hayez.

Chapter 6
The Catenary, or How Differential Calculus Appeared on the Public Square

1 Introduction

Before even mentioning the problem, we will examine its interest by giving some important dates concerning the discovery of calculus. Leibniz started his long journey to the discovery of differential calculus in his manuscript III 6, *De progressionibus et de arithmetica infinitorum*, written between autumn 1672 and December 1672. In his manuscript III 16, *Methodus tangentium inversa* of April–May 1673, he began integral calculus, and it would be only after 11 years that he published the *Nova methodus pro maximis et minimis* (Leibniz 1684), in which he founds differential calculus; two years later he founds integral calculus in *De geometria recondita et Analysi Indivisibilium atque infinitorum* (Leibniz 1686), which he presents as *Addenda* to the *Nova methodus.*

Our chosen example is situated between 1691 and 1695, that is seven years after the release, in the *Acta Eruditorum,* which he had founded, of the *Nova methodus pro maximis et minimis.* But such a long time was surely necessary in order for the idea to be understood and for someone to plan to develop it. For this purpose, Leibniz had given some examples in his article. Later, in the *Nouvelles de la république des lettres* of September (Leibniz 1687), he proposed a problem to Catelan:

> To find the isochrone curve along which a heavy body may fall uniformly, that means in equal times to approach equally the horizon and to be drawn downwards without being accelerated, at a constant velocity.[1]

Having read Leibniz's *Nova methodus* in 1687, Jacob Bernoulli had realized that, combined with Hooke's law, calculus enabled specialists to treat continuous various phenomena such as flexibility of strings and elasticity of surfaces. But the inventor of the catenary problem as well as the one who also had solved it was Jacob's younger

[1] Leibniz(1687, p. 956): *Trouver une ligne de descente, dans laquelle le corps pesant descende uniformément et approche également de l'horison en temps égaux*. Leibniz will give his solution in (Leibniz 1689, p. 196).

P. Radelet-de Grave, *Between Mechanics and Architecture*, Mathematics and the Built Environment 8, https://doi.org/10.1007/978-3-031-73530-1_6

brother and one time pupil, Johann Bernoulli. At that moment the brothers were still getting along; their famous quarrel was to start later.

When asking the question of the catenary, in May 1690, Jacob Bernoulli pursues the same goal as Leibniz, namely, to develop differential calculus by applying it to a new problem and in this way showing its power.

Jacob received three answers to his new problem: the first by one of the greatest geometers of his time, Christiaan Huygens; another of Leibniz himself; and the last one from his own brother, his junior by 12 years, Johann Bernoulli.

One might have expected a fourth answer from Isaac Newton, but either he didn't accept the challenge or he didn't know about it. There doesn't seem to be any mention of the challenge in his correspondence before the publication of the three answers in the issue of June 1691 of the *Acta Eruditorum.*

2 Differential Calculus on the Continent

After a short introduction by Leibniz, those three solutions will be published together in the *Acta Eruditorum* of June 1691 in their order of arrival: Johann Bernoulli, Gottfried Wilhelm Leibniz, Christiaan Huygens. Truesdell (1960, p. 66) said that in 1690 it represented the mathematics of the future (Johann Bernoulli), of the present (Leibniz) and of the past (Huygens). It seemed interesting to me, in order to understand the beginnings of differential calculus, to make a precise comparison of those three texts and along the way to deepen Truesdell's remark. We shall limit our analysis, in the framework of the first part of this chapter, to the constructions of the catenary and not to its mathematical properties.

This gives us a well-defined and precise *corpus*, which I shall try to explain with help of other texts because it has a defect that will be underlined by David Gregory in his article on the catenary (1697), which we shall analyze in the second part of this article: those three articles do not give any demonstration. The new calculus must be kept secret. In order to reach our goal, we shall have to understand the various ways of doing mathematics to study that particular problem of the catenary; we must therefore look for those demonstrations in other, more explicit texts.

With regard to Johann Bernoulli, things are simple since this subject is part of the lectures that he gave to the Marquis de l'Hôpital between winter 1691 that the second part of 1692. There he repeats the text of the *Acta Eruditorum*, adding the demonstrations. With regard to Huygens, things are much more complicated because one finds elements of answers to Jacob Bernoulli's question scattered among many handwritten notes. Nevertheless, an explanation of his construction, together with many properties, figures in a letter to Basnage de Beauval written in February 1693[2] but only using what he knew at the time of the publication and not taking into account, as Huygens himself says, elements concerning the link to the quadrature of the hyperbola and to the logarithm, which he learned later from Leibniz. Concerning

[2] Huygens to Basnage de Beauval, [February] 1693 (Huygens 1905, letter no. 2793, pp. 407–417).

Leibniz himself, we shall content ourselves with the original text of the *Acta* because the explanation that he gives about it many years later[3] does not explain the essential link with the logarithmic curve.

My aim in this first part is to show how three high-level mathematicians solve the same problem at the same time but in different ways. In the second part we will see other mathematicians from another country, namely England, reacting in a different mode. The interest resides in the facts that this occurs exactly at the moment differential calculus enters the scene, and that both inventors – Leibniz of differential calculus, and Newton of the calculus of fluxions—are concerned, together with their followers, the brothers Bernoulli on the one side and David Gregory on the other. In between we have Huygens, accustomed to the infinitely small but who will never assimilate calculus. And on the English side, we have another type of follower, namely engineers, interested in the catenary.

The story begins in the *Acta Eruditorum* of May 1690, where at the end of an article dedicated to the problem of the trajectory of a body falling with uniform velocity, Jacob Bernoulli asks the following question:

> To find the curve assumed by a loose string hung freely from two fixed points. I also assume this string be flexible and of uniform cross-section.[4]

2.1 Christiaan Huygens

After having sent to Leibniz two figures characterizing his solution,[5] Huygens will finally send the text he destines for the publication. The handwritten explanations concerning the first figure[6] and more precisely concerning the construction of the catenary drafted by Huygens will help us to better understand the letter he wrote to Basnage. This starts with a presentation of the problem and of the context in which it was proposed:

> But as the subject led me to the consideration of the catenary, which gave rise to one of the nice geometrical researches of this time, I had like to add here a rather singular manner that I found in order to arrive at the construction of that curve, which is the most difficult part of what one had proposed to search about it.
>
> Among what I have contributed to be inserted into the *Acta* of Leipsig, together with the nice and learned discoveries of gentlemen Leibnitz and Bernoulli, I said that I had reduced the construction, or the invention of the points of that curve, to the quadrature of a curve which equation is

---

[3] See the letter from Leibniz to Huygens, 14 September 1694 (Huygens 1905, letter no. 2876, pp. 675–683).

[4] Bernoulli (1690, p. 219): *Invenire, quam curvam referat funis laxus et inter duo puncta fixa libere suspensus. Sumo autem, funem esse lineam in omnibus suis partibus facillime flexilem.*

[5] Huygens to Leibniz, 9 October 1690 (Huygens 1901, letter no. 2623, pp. 496–499) and Huygens to Leibniz, 26 march 1691 (Huygens 1905, letter no. 2667, pp. 55–58).

[6] Huygens to Leibniz, 9 October 1690 (Huygens 1901, *appendice* no. 2625 to letter no. 2623, pp. 500–501).

$$a^4 = aaxx + yyxx;$$

and that I had recognized that that quadrature depended of the knowledge of the sum of the secants of the arcs of the circle that decrease by minima[7]; sum that had been reduced since a long time to the quadrature of the Hyperbola by Jac. Gregorius in his Geometrical exercises, wherefrom he deduces the measure of the loxodromic line; a fact that I did not remember at the time.

Misters Leibniz and Bernoulli, as far as I can judge, arrived at that construction using a curve that Bernoulli represents in the first figure (see our Fig. 1), which he gives in order to solve that problem; because Mr Leibniz wrote to me that he also had met it. And I find that it is the same as the one I reported before and whose equation is $a^4\infty = aaxx + yyxx$; having its quadrature depending, as I said, of that of the hyperbola, though I could not, until now, imagine how calculation led them to that curve. But I go to my construction, which, without considering other curved line, allows to find the points of the catenary using the dimension of the parabolic line.[8]

After that Huygens gets to the heart of the matter and explains his construction:

The first foundation of all researches concerning that line is (Fig. 2) that if there is a chain composed of equal weights, attached to a thread as *BCDEF*, it always occurs that from three interstices following each other, the two extreme lines, as *CD*, *PE*, when prolonged, meet in the line *IH* perpendicular to the horizon, that cuts the central interstice *ED* in two equal parts. Considering now a chain composed of weights attached to it at equal distances, that one must imagine as infinitely little, and placed so that the lowest interstice *BC* is parallel to the horizon: if on each of the other interstices, taken as the hypotenuse, one conceives triangles rectangles *CDK*, *DEL* etc. from which one side be horizontal; one will find that starting from the lowest one, the angles *DCK*, *EDL*, *FEM* etc. are such that their tangents grow also as the numbers 1, 2, 3, 4, etc., which is easy to demonstrate by said principle, though perhaps one should not have noticed it without the algebraic calculation.[9]

---

[7] Huygens means that it decreases by infinitesimal quantities, we would say by *dx*.

[8] Huygens (1905, letter no. 2793, pp. 413–414): *Mais puis que le sujet m'a mené à la considération de la Chainette, qui a donné occasion à une des jolies recherches Geometriques de ce tems, je veux ajoûter icy un manière assez singuliere que j'ay trouvée, pour parvenir à la construction de cette courbe; qui est ce qu'il y avoir de plus difficile dans ce qu'on s'est proposé d'en chercher. Parmi ce que j'ay contribué pour etre inseré dans les Acta de Leipsich, avec les belles & savantes decouvertes de Messiuers Leibnitx & Bernoully, j'ay dit que j'avois reduit la construction, ou l'invention des points de cette Ligne, à la quadrature d'une courbe dont l'Equation est* $a^4 = aaxx + yyxx$; *& que j'avois reconnu que cette quadrature dependoit de la connoissance de la somme des Secantes des arcs de cercle qui croissent également per minima; laquelle somme avoit été reduite il y a long-tems à la quadrature de l'Hyperbole par Jac. Gregorius dans ses Exercitations Geometriques, où il en deduit la mesure de la Ligne Loxodromique; de quoy je ne me resouvenoir pas alors.*
Mrs. Leibnits et Bernoully, à ce que je puis juger, sont parvenus à cette construction par le moyen de la courbe que ce dernier représente dans la premiere Figure, qu'il donne pour resoudre ce Problême; car Mr Leibnitz m'a écrit qu'il l'avait rencontrée aussi. Et je trouve que c'est la même que celle que j'ay raportée cy-devant dont l'Equation est $a^4\infty = xxyy - aayy$; *ayant sa quadrature dependante, comme j'ay dit, de celle de l'Hyperbole: quoyque je n'aye encore pu m'imaginer, comment le calcul les a conduits à cette ligne. Mais je passe à ma construction, qui sans considerer d'autre ligne courbe, fait trouver les points de la chaînette par la dimension de la ligne Parabolique.*

[9] Huygens (1905, letter no. 2793, p. 414–415): *Le premier fondement de toute recherche à l'égard de cette ligne est* (Fig. 2) *que s'il y a une chaîne composée de poids égaux, attachez à un fil comme sont BCDEF, il arrive toûjours, que de trois interstices qui se suivent, les deux lignes extrêmes, comme CD, PE, étant continuées se rencontrent dans la ligne IH perpendiculaire, à l'horison, qui coupe*

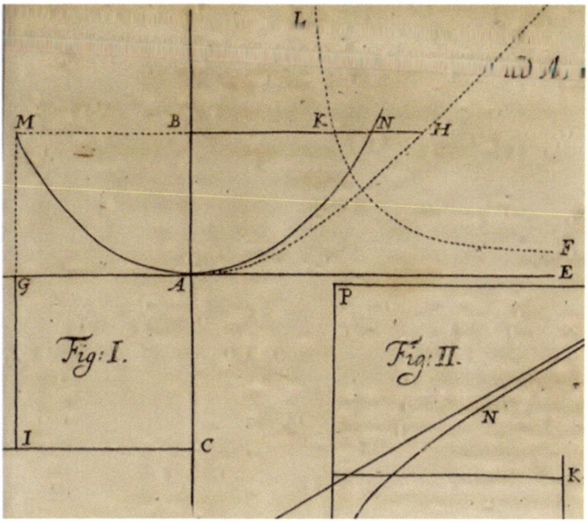

Fig. 1 Bernoulli's Fig. I (Bernoulli 1691a, 6, Tab. VI, facing page p. 274). Public domain

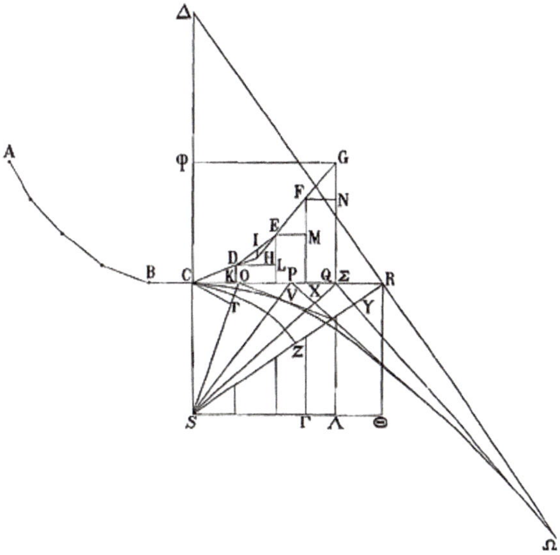

Fig. 2 Huygen's figure from the letter to Blasnage (Huygens 1905, p. 414). Public domain

l'interstice du milieu ED en deux parties égales. Considerant maintenant une chaine ainsi composée de poids nouëz à égales distances, qu'on doit imaginer infiniment petites, et disposée en sorte que l'interstice le plus bas BC soit parallele à l'horison: si sur chacun des autres interstices, comme hypotenuse, on conçoit des triangles rectangles CDK, DEL etc. *desquels un côté soit horisontal; on trouvera que depuis le plus bas les angles DCK, EDL;. FEM* etc. *sont tels que leurs Tangentes croissent également comme les nombres 1, 2, 3, 4,* etc. *ce qui est aisé à demonstrer par le dit principe, quoy que peut-être on ne s'en serait pas avisé sans le calcul d'Algebre.*

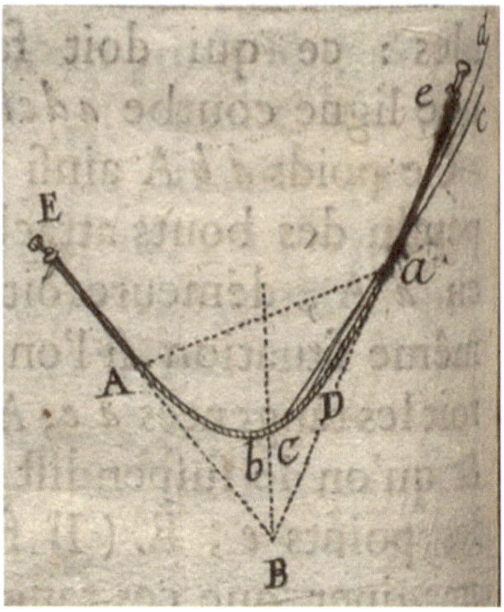

Fig. 3 Pardies's figure of the catenary (1673, p. 126). Public domain

In Fig. 2 where the curve *ABCDIEFG* represents the catenary, Huygens considers three consecutive gaps or elements of the curve. He takes *CD, DE* and *EF*. He extends the two extreme lines *CD* and *EF* which intersect in *H*, at the vertical of the center of gravity or at the middle of the intermediate segment *DE*.

This property, which will be used by Johann Bernoulli and by Gregory, had been highlighted by Pardies (Fig. 3).

It says that two tangents in any two points *A* and *a* of the catenary always meet at the vertical *BC* of the center of gravity *C* of the portion of the curve that those two points delimitate. Huygens envisions the case in which the curve is formed by homogeneous infinitely small rectilinear elements, the center of gravity necessarily is in the middle of the segment. Huygens considers the little triangles *CDK, DEL,* etc. in Fig. 2, and asserts, without explaining his demonstration, that the tangents of the angles *DCK, EDL, FEM*, etc. increase as the numbers 1, 2, 3, 4, etc. The demonstration figures in the manuscript notes explaining the code sent by Huygens to Leibniz:

> Foundation of all we found about the catenary.
>
> Equal weights are attached to a thread without weight, in three intervals between nodes whatever but successive, the forces are inclined with respect to the horizontal plane in such a way the tangents of those inclinations increase each time of the same quantity?
>
> Let equal weights *A, B, C, D* (Fig. 4) be attached to a chain to make three continuous and upwards tending internodes *AB, BC, CD* and extend the two outer internodes until they meet in *P*. Through *P* one traces the perpendicular to the horizon *NPK* until it meets the horizontals *BK* and *CN*. Thus the internode *AB* is inclined from the horizon by the angle *PBK*, and the internode *BC* is inclined by angle *FBK*, and the internode *CD* is inclined by the angle *PCN*. I say thus that the tangent of the angle *PBK* is as much larger than the tangent of the angle *FBK* as the tangent of the angle *FBK* is larger than the tangent of the angle *PCN*.
>
> Thus the intersection of the intersection *P* of the straight lines *AB* and *DC*, is on the perpendicular to the horizontal line that passes through the center of gravity of the weight *BC*; as those weights are equal it is necessary that the perpendicular *KPN* cuts the segment *BC* in

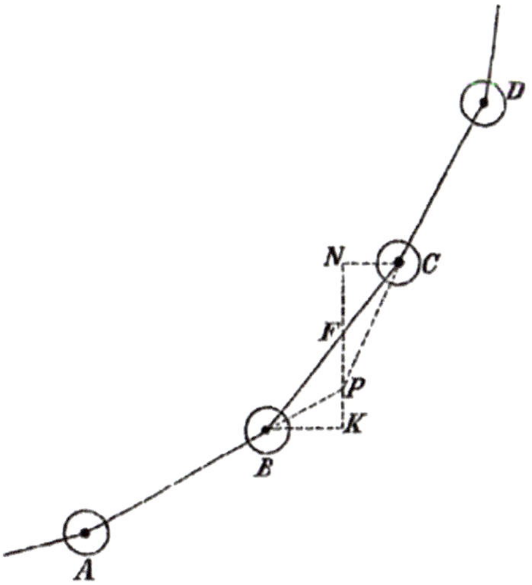

Fig. 4 Huygen's Fig. 1 (Huygens 1901, p. 502). Public domain

its middle *F*, thus *BK* and *CN* are equal. That is the reason why whatever the radius chosen, it appears that *PK*, the tangent of the angle *PBK* is surpassed by the same straight line *PF*, by the tangent of the angle *FBK*, as the tangent of this last *FBK* or the one of *FCN* that is equal to it, is surpassed by the tangent of the angle *PCN*. Those angles are, as we already said, the inclinations of the internodes *AB*, *BC*, *CD* with the horizontal plane. That proves the proposition.

If a chain or a thread suspended to which equal weights are attached, has its lowest internode parallel to the horizon, then the angles between the others internodes and the horizontal plane are such that their tangents increase as the ratio of the numbers starting with unity 1, 2, 3, 4, 5, etc.[10]

---

[10] Huygens (1901, letter no. 2625, pp. 502–503): *Fundamentum omnium eorum quae de curvae catenae reperimus.*
Fili gravitate carentis, et aequalia pondera innexa habentis, tria qualibet internodia continua ac sursum tendentia ita ad planum horizontale inclinantur, ut tangentes angulorum hujus inclinationis crescant aequali excessu. Sint catenae pondera aequalia innexa habentis A, B, C, D, internodia tria sursum tendentia AB, BC, CD, producanturque internodia extrema et conveniant in P. Et per P ducatur ad horizontem perpend. NPK eique occurrant horizontales BK, CN. Est ergo internodii AB inclinatio ad horizontem angulus PBK, et internodii BC inclinatio angulus FBK, et internodii CD inclinatio angulus PCN. Dico itaque tangentem anguli PBK tantum superari a tangente anguli FBK quantum haec superatur à tangente anguli PCN.
Quia enim concursus rectarum AB, DC, hoc est punctum P incidit in perpendicularem horizonti quae per centr. gr. ponderum BC transit; sunt autem pondera haec aequalia, necesse est perpendicularem KPN secare internodium BC aequaliter in F, unde et BK, CN aequales erunt. Quarum utrovis pro radio adsumta, apparet PK tangentem anguli PBK eadem recta PF superari a tangente anguli FBK, qua tangens hujus anguli, sive illi aequalis, FCN superatur a tangente anguli PCN. Sunt autem hi anguli ut jam diximus inclinationes intemodiorum AB, BC, CD ad planum horizontis. Ergo constat propositum.
Catenae seu fili suspensi aequalia pondera innexa habentis, si infimum internodium horizonti parallelum fuerit erunt deinceps anguli reliquorum internodiorum cum horizontali plano tales, ut eorum tangentes crescunt secundum rationem numerorum ab unitate incipientium 1, 2, 3, 4, 5, etc.

Huygens wants to compare the tangents of the inclination angles, that means the ratios $\frac{PK}{BK}, \frac{FK}{BK}, \frac{NF}{NC}, \frac{NP}{NC}$ where the denominators are all equal since $NC = BK$. It is thus sufficient to compare *PK*, *FK*, *NF* and *NP*. He compares the inclination of the first and the second segment:

$$FK = PK + FP.$$

The inclination of the second segment is equal to the one of the first, to which one adds *FP*. He compares then the inclination of the third segment to the second.

$$FK = PK + FP.$$

The inclination of the second segment is equal to that of the first, to which one adds *FP*.

Let us refer again to Fig. 2 and continue the reading of the text of the letter to Basnage:

> If one imagines than the equal parts of the chain *CDEFG*, extended on the horizontal line *COPQR*, and that from the first division *O* one draws *OS*, which meets the perpendicular *CS* so that the angle *COS* is equal to *CDO*, and that one traces the other straight lines *SP*, *SQ*, *SR*: the triangles *SCO*, SCP, *SCQ*, *SCR* will necessarily be similar to each of the *COD*, *DLE*, *EMF*, *FNG*, since *SCO* is similar to *COD* by construction, and that the tangents of the others *SCP*, *SCQ*, etc. increase equally.[11]

In order to rectify it, Huygens extends his curve on a horizontal passing through the apex. On this manner, he determines the points *O*, *P*, *Q*, *R* distant from one another by a rectilinear element of the curve. From *O* he traces a straight line *OS* in order to have the angle *COS* equal to the angle *CDO*. He finds thus the point *S* through which he traces the straight lines *SP*, *SQ*, *SR* and forming the triangles *SCO*, *SCP*, *SCQ*, *SCR* similar to the triangles *CKD*, *DLE*, *EMF*, *FNG*. The text goes on:

> If one traces *CT*, *OV*, *PX*, etc. perpendicular on *SO*, *SP*, *SQ*, etc., it is evident that the triangles *CTO*, *OVP*, *PXQ*, etc. will be equal and similar to the triangles *COD*, *DLE*, *EMF*, etc., taking the same order. From this one concludes that if the gaps *CD*, *DE*, etc. are infinitely small, and the parts *CO*, *OP*, etc. too that means if *CG* is the catenary, and *CR* is equal to its length; than the sum of *TO*, *VP*, *XQ*, etc. will be equal to the sum of the perpendiculars *KD*, *LE*, *MF*, etc., that is to say to the straight line $G\Sigma$, or to the axis φC, (because the gap *BC* counts for nothing) and that the sum of the *CT*, *OV*, *PX*, etc. will be equal to the sum of *CK*, *DL*, *EM*, etc. that is to say, to the applied $G\varphi$.[12]

---

[11] Huygens (1905, letter no. 2793, p. 415): *Que si on s'imagine en suite les parties égales de la chaine CDEFG, étenduës sur la droite horisontale en COPQR, et que de la premiere division O on mene OS, qui concoure avec la perpendiculaire CS, en sorte que l'angle COS soit égal à CDO, et qu'on tire les autres droites SP, SQ, SR: les triangles SCO, SCP, SCQ, SCR, seront necessairement semblables à chacun des COD, DLE, EMF, FNG, puis que SCO est semblable à COD par la construction, et que les autres SCP, SCQ,* etc. *ont leur tangentes qui croissent également.*

[12] Huygens (1905, Letter no. 2793, p. 415): *Si de plus on mene CT, OV, PX,* etc. *perpendiculaires sur SO, SP, SQ,* etc. *il est évident que les triangles CTO, OVP, PXQ* etc. *seront égaux et semblables aux triangles COD, DLE, EMF* etc. *en prenant les mêmes en ordre. D'où l'on conclud, que si les interstices CD, DE* etc. *sont infiniment petits, et de même les parties CO, OP,* etc. *c'est-à-dire si*

2.2 Gottfried Wilhelm Leibniz

Leibniz's construction is rather different because he has immediately seen the link with the curve that he calls logarithmic (and that has the same form as our exponential, considering the chosen axis). Unfortunately, Leibniz will not explain how he got that result. But this fact is not astonishing because the logarithm was at the center of his preoccupations. The fact that the demonstration of the constancy of the sub-tangent closes the *Nova methodus:*

> As an appendix, I had like to add the solution of the problem which M. De Beaune proposed to Descartes, which this last tried to solve in tome 3 of his letters, but in vain: "Find the curve *WW* such that tracing the tangent *WC* to the axis, *XC* be always equal to the same constant segment *a*". Let *XW* or *w*, be thus with respect to *XC or* a as *dw* to *dx*; thus, if one takes *dx* (which one may take as one wishes) constant, that is, invariable, or equal to *b*, in other words if the *x*, or *AX*, increase uniformly, we shall get: $w = \frac{a}{b}dw$, and the ordinates *w* will be proportional to their increase or differentials *dw* or also if the *x* are numbers, the *x* will be their logarithms; the curve *WW* is thus the logarithmic curve.[13]

Here is the text of Leibniz's answer to the question asked by Jacob Bernoulli and published in the *Acta Eruditorum*:

> Here is a geometrical construction of the curve, without help of any thread nor any catenary, and without presupposing any quadrature, a construction that one must, in my opinion judge as the more perfect one may obtain for transcendents and the most consistent with Analysis. Let two segments whatever having between them a certain report determined and invariable, represented here by D and K, as soon as one knows the report of those two segments all the rest flows out by simply applying ordinary geometry.[14]

Leibniz starts by giving, not a segment, which would mean a unit, but two and by the way a relation, one of the two lengths. One sees those two segments on the left side of Fig. 5. After that he fixes his reference thanks to a first horizontal axis, *ON*, which

---

CG est la courbe de chaine, et CR égale à sa longueur; alors la somme des TO, VP, XQ &c. sera égale à la somme des perpendiculaires KD, LE, MF etc. *c'est-à-dire à la droite G*Σ, *ou à l'axe* ϕC, *(car l'interstice BC est alors compté pour rien) et que la somme des CT, OV, PX,* etc. *sera égale à la somme des CK, DL, EM* etc. *c'est-à-dire à l'appliquée G*ϕ.

[13] Leibniz (1684, p. 473): *Appendicis loco placet adjicere solutionem Problematis, quod Cartesius a Beaunio sibi propositum, Tom. 3. Epist. Tentavit, sed non solvit. Lineam invenire WW talis naturae, ut ducta ad axem tangente WC, sit XC semper aequalis eidem rectae constanti, a. Jam XW seu w ad XC semper aequalis eidem rectae constanti, a. Jam XW seu w ad XC seu a, ut dw ad dx: Ergo si dx (quae assumi potest pro arbitrio) assumatur constans sive semper eadem nempe b, seu si ipsae x sive AX crescant uniformiter, fiet W aequ. a/b dw, et erunt ipsae W ordinatae, ipsis dw, suis incrementis sive differentiis, proportionales, hoc est si x sint progressio arithmeticae, erunt w progressionis Geometricae, seu si w sint numeri, x erunt logarithmi: linea ergo WW logarithmica est..*

[14] Leibniz (1691a, p. 278): *Linea sic construitur Geometrice, sine auxilio fili aut catenae, et sine suppositione quadraturarum, eo constructionis genere, quo pro Transcendentibus nullum perfectius et magis Analysi consen taneum mea sententia haberi potest. Sint duae quaecunque lineae rectae, determinatam quandam et invariabilem inter se habentes rationem, eam scilicet quam D et K* (Fig. 5) *hic expositae, qua ratione semel cognita caetera omnia per Geometriam ordinariam procedunt.*

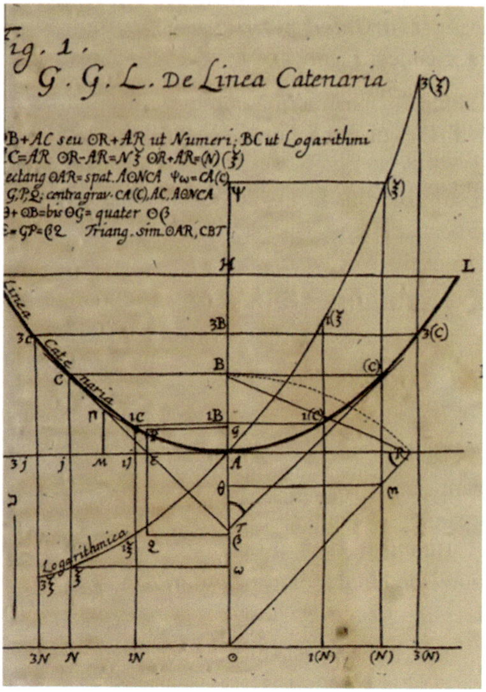

Fig. 5 Leibniz's Fig. 1 (1691a, foldout facing p. 278). Public domain

he graduates from either side of O in order to have $O_1(N) = {}_1(N)_3(N)$[15] (Fig. 5) and symmetrically $O_1N = {}_1N_3N$, (N) and N being in whatever point between ${}_1(N)_3(N)$ and between ${}_1N_3N$. He then builds a square, taking on an axis orthogonal to the previous one, $OA = O_3(N)$. Next, he must determine the abscissa corresponding to $O_3(N)$. He poses it as having the same report to OA as D to K. He fixes at the same time the unit that equals the side of the square. Since

$$\frac{e^1}{e^0} = e = \frac{D}{K}$$

the relation that he gives corresponds thus to the base of the exponential. He constructs the logarithm point by point, by appealing to the principal property of the curve: whatever an abscissa is mean proportional between the abscissas at an equal distance. That corresponds to the definition in use at the time: the logarithm is the curve that, at ordinates in arithmetical progression, makes abscissa correspond in geometrical progression:

> Let ON be an undefined line parallel to the horizon, OA a segment perpendicular and equal to O_3N, and on ${}_3N$, a vertical segment ${}_3N_3\xi$, having with OA the relation of D to K. Let us seek the mean proportional ${}_1N_1\xi$ between OA and ${}_3N_3\xi$, as well as between ${}_1N_1\xi$ and ${}_3N_3\xi$, as

[15] We remind the reader that Leibniz writes the indices before the letter instead of after it so that the equation $O1(N) = {}_1(N)_3(N)$ means $O(N)_1 = (N)_1\,(N)_3$.

> well as between $_1N_1\xi$ and OA we also ask the mean proportional, and so on we ask the mean proportional, and finding third proportional, and so describing the line $\xi\xi A(\xi)(\xi)$ of such a nature that, for example, the intervals $_3N_1N$, $_1NO$, O, $(N)\ _1(N)$, $_3(N)$, etc. are supposed equal, the ordinates $_3N_3\xi$, $_1N_1\xi$, OA, $_1(N)\ _1(\xi)$, $_3(N)\ _3(\xi)$, are in a continued geometrical progression, which I use to call the logarithmic line. Having assumed ON, $O(N)$ to be equal, upon N or (N), let us raise NC or $(N)(C)$ equal to half the sum of those $N\xi$, $(N)(\xi)$, and C or (C) will be the point on the catenary $FCA(C)L$, of which we can now determine geometrically as many points as we want.[16]

Once the logarithmic curve is traced, he traces the catenary. His construction corresponds exactly to the hyperbolic cosine, the arithmetical mean of an exponential and of its inverse:

> On the contrary, if one constructs a catenary physically, suspending a thread or a chain, one can in that way show whatever mean proportional as we wish and find the logarithm of given numbers or find numbers given the logarithm. Let us for example look for the logarithm of the number $O\omega$, or, what amounts to the same thing, the logarithm of the relation between OA and $O\omega$, the one of OA (which I choose as unity and that I'll also call parameter) being posed equal 0, one must take the third proportional $O\varphi$ of $O\omega$ and OA, then choose as abscissa the semi-sum OB of $O\omega$ and $O\varphi$, the corresponding ordinate BC or ON on the catenary will be the searched logarithm corresponding to the given number. Reciprocally, if a logarithm ON is given, one must take the double of the vertical segment NC lowered from the catenary and cut it in two segments whose mean proportional be equal to OA, which is given (it is the unity) (which is very easy); both segments will be the numbers we were looking for, the one greater, the other one less than 1, corresponding to the proposed logarithm. Other method: after having obtained NC as I said, also let be OR (the point R taken on the horizontal AR in such a way that OR be equal to OB or to NC), the sum and the difference of the segments OR and AR will the two numbers, the one greater than, the other less than to 1, corresponding to the given logarithm. The difference between OR and AR are in their turn half the sum and half the difference of $(N)(\xi)$ and $N\xi$.[17]

---

[16] Leibniz (1691, p. 278): *Sit recta indefinita ON horizonti parallella, eique perpendicularis OA, aequalis ipfi O*$_3$*N, et super* $_3$*N verticalis* $_3N_3\xi$*, quae sit ad OA, ut N ad J. Inter OA et* $_3N_3\xi$ *quaeratur media proportionalis* $_1N_1\xi$*; et inter* $_1N_1\xi$ *et* $_3N_3\xi$*, itemque inter* $_1N_1\xi$ *et OA quaeratur rursus media proportionalis, et ita porro quaerendo medias, et inventis tertias proportionales, deficribatur continueturque linea* $\xi\xi A(\xi)(\xi)$ *quae erit talis naturae, ut ipsis intervallis verbi gr.* $_3N_1N$*,* $_1NO$*, O, (N)* $_1$*(N),* $_3$*(N),* etc. *sumtis aequalibus, sint ordinatae* $_3N_3\xi$*,* $_1N_1\xi$*, OA,* $_1(N)\ _1(\xi)$*,* $_3(N)\ _3(\xi)$*, in continua progressione Geometrica, qualem lineam Logarithmicam appellare soleo. Jam sumtis ON, O(N) aequalibus, super N vel (N) erigatur NC vel (N)(C) aequalis dimidiae summae ipsarum N*ξ*, (N)(*ξ*), et C vel (C) erit punctum lineae catenaria FCA(C)L, cujus ita puncta quotcunque assignari Geometrice possunt.*

[17] Leibniz (1691, pp. 278–279): *Contra si linea catenaria physice construatur ope fili vel catenae pendentis, ejus ope exhiberi possunt quotcunque mediae proportionales, et Logarithmi inveniri datorum numerorum vel numeri datorum Logarithmorum. Sic si quaeratur Logarithmus numeri O*ω*, posito ipfius OA, (tanquam Unitatis, quam et parametrum vocabo) Logarithmum esse nihilo aequalem. Seu quod eodem redit, si quaeratur Logarithmus rationis inter OA et O*ω*, sumatur ipsis O*ω *et OA tertia proportionalis O*φ*, et ipsarum O*ω *et O*φ *summae dimidiae OB, tanquam abscissae, respondens Lineae Catenariae ordinata BC vel ON, erit Logarithmus quaesitus Numeri dati. Contra, dato Logarithmo ON inde ductae ad Curvam Catenariam verticalis NC duplam, oportet secare in duas partes tales, ut media proportionalis inter segmenta sit aequalis datae (Unitati) OA, (quod facillimum est) et duo segmenta erunt respondentes dato Logarithmo Numeri quaesiti, unus major, alter minor unitate. Aliter: inventa, ut dictum est, NC seu OR (sumto ita puncto R in horizontali AR, ut habeamus OR aequalem OB vel NC) erunt summa ac differentia rectarum OR et AR duo*

This last part also shows that Leibniz's interest focalizes on the logarithmic curve and that the catenary that one can trace "experimentally" following the curve given by a small chain hanging freely is for him only a tool that makes it possible to trace the logarithmic curve impossible to trace physically or experimentally.

2.3 Ignace-Gaston Pardies and Johann Bernoulli

Let us turn now to Johann Bernoulli's construction as it appears in the *Lectiones* (Bernoulli 1691–1692), a text published in Johann's *Opera Omnia* based on notes he had written in Paris in order to give lectures about differential and integral calculus to the Marquis Guillaume de l'Hôpital. Johann introduces there the subject saying that he exposes here the solution of this problem *together with the calculations that were not annexed in the Actas*, that is, in the publication of his text on the catenary published in the *Acta Eruditorum* (Bernoulli 1691b). He then gives the principal properties of the curve that had already been exhibited by Pardies (1673), in form of five hypotheses:

> 1. The thread, the cord, the chain or whatever represents the curve is supposed to be perfectly flexible in each of its points and non-extendable. That means that it does not extend because of its weight.[18]
>
> 2. If one suspends the catenary *ABC* (see our Fig. 6) in two points whatever *A* and *C* the forces requested in *A* and *C* will be the same as those necessary to sustain a weight *D* equal to the weight of the chain *ABC* and suspended at the meeting point of the two threads without weight *AD* and *CD* tangent to the curve *ABC* in the points *A* and *C*. The reason is evident; since the weight of the chain *ABC* exerts its virtue in *A* and *C* along a direction, namely along the tangents *AD* and *CD* and the traction of the same weight or of an equal weight *D* in *A* and in *C* occurs along the lines *AD* and *CD*. That is the reason why the forces requested must be the same in both cases. It follows that the requested force acting at the lowest point *B* will be obtained by searching for the force the weight *E* (see our Fig. 7) exerts in the same point, sustained by two threads without weight, one of them being tangent in *B* to the curve and thus horizontal, and the other being tangent in *A*.[19]

---

respondentes Logarithmo dato Numeri, unus major, alter minor Unitate. Nam differentia ipsarum OR et AR est Nξ, et summa earum est (N)(ξ); uti vicissim OR est semisumma, et AR semi differentia ipsarum (N)(ξ) et Nξ.

[18] Bernoulli (1691–1692, p. 492): 1°. *Filum, Funis, Catena, vel quicquid curvam repraesentat, supponitur in omnibus suis punctis flexile, et inextensibile, id est, quod ob gravitatem suam extensionem non patitur.*

[19] Bernoulli (1691–1692, p. 492): 2°. *Si in duobus punctis A et C quibuscunque curva Catenaria ABC sustinetur; potentiae, quae in punctis A et C requiruntur, erunt eadem quae requiruntur ad sustinendum pondus D aequale ponderi catenae ABC, et in concursu D duorum filorum nulius gravitatis AD, CD, curvam ABC in punctis A et C tangentium appensum. Ratio hujus est evidens; nam pondus catenae ABC exerit virtutem suam in A et C secundum directionem, id est, secundum tangentes AD, CD, et ejusdem vel aequalis ponderis D tractio in A et C est etiam secundum rectas AD et CD: oportet itaque ut potentiae requisitae in punctis A et C sint in utroque casu eaedem. Hinc potentia quae requiritur ininfimo puncto B habebitur, si quaeratur potentia quam pondus E in eodem puncto exerit, sustentatum a duobus filis nullius gravitatis, quorum unum tangit curvam in B, et proinde est horizontale, alterum vero tangit in punto A.*

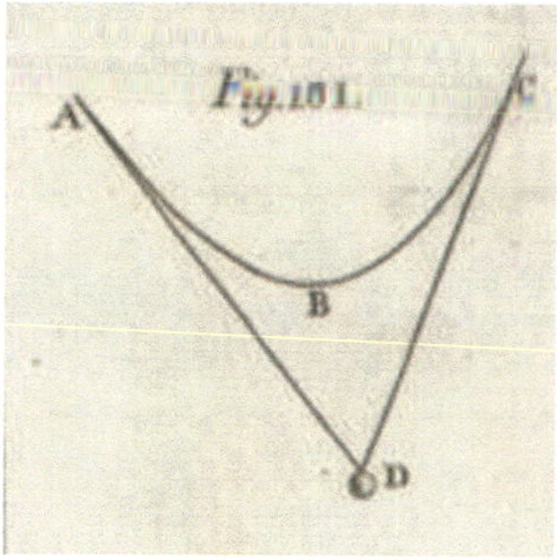

Fig. 6 Johann Bernoulli's Fig. 131 (1691–1692, Tab. LXX). Public domain

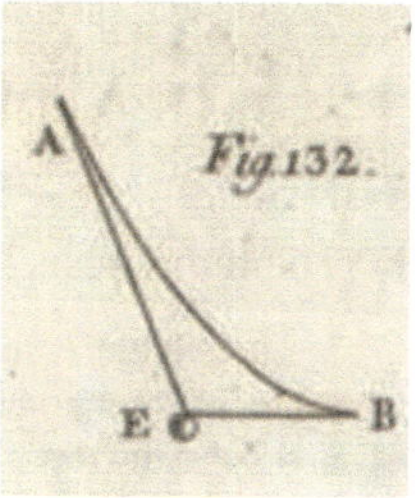

Fig. 7 Joh. Bernoulli's Fig. 132 (1691–1692, Tab. LXX). Public domain

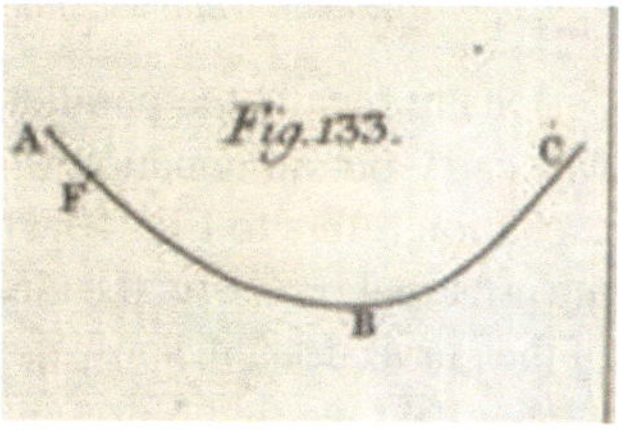

Fig. 8 Joh. Bernoulli's Fig. 133 (1691–1692, Tab. LXX)

Johann explains that the tensions are tangent to the curve, from what follows, as Pardies had explained it, that the tangents cross in line with the center of gravity.

> 3. If the chain attached at its extremities *A* and *C* (see our Fig. 8) is fixed in whatever point *F*, so that part *AF* could be removed, the curve left over remaining the same and the remaining part of the chain, *FBC*, will not be modified; that is to say that the other points will keep the same situation they had before fixing in *F*. This does not need any demonstration; reason convinces us and daily experiment shows it.[20]

This property comes also from the fact that the tensions are tangent to the curve.

> 4. In the same conditions as before, the same force, that is to say the preceding force is requested before and after the fixation in every particular point of the curve, or, which amounts to the same, the force by which a point is sustained before fixation is the same that

[20] Bernoulli (1691–1692, p. 492): *3°. Si catena duobus terminis A et C alligata in alio quovis puncto F figatur, ita ut pars AF auferri possit; curva, quam resitua catena FBC rapraesentat, non mutabitur; id est, reliqua puncta in eodem, quem ante fixationem habuerunt, situ manebunt.*

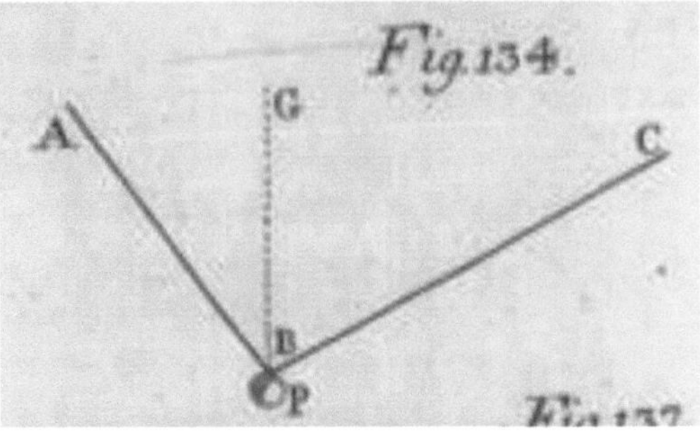

Fig. 9 Joh. Bernoulli's Fig. 134 (1691–1692, Tab. LXX). Public domain

the one that sustains it after fixation. This is but a corollary of what precedes. Then, if one extends of shortens the chain *BFA* as much as one wonts, that is to say wherever one takes the point of fixation in *F*, the force in the lowest point *B* will not be augmented nor diminished but will stay always the same and equal.[21]

The hypothesis 4 signals that tensions are the same in every point of the curve.

5. The weight *P* suspended to two threads *AB, CB* (see our Fig. 9) placed in a way that it exerts its forces on the points *A* and *C* in the following proportion so that the force requested in *A* be to the force requested in *C* as inversely (when one traces the vertical *BG*) the sine of the angle *CBG* to the sine of the angle *ABG* and the weight *P* to one of the two forces [let us say] in *C* as the sine of the total angle *ABC* to the sine of the opposed weight *ABG. This is demonstrated in whatever statics.*[22]

Johann recalls the parallelogram law of forces. Armed with these properties, he will carry out an equation of which there is no trace in the *Acta Eruditorum.* This execution refers to Fig. 10 where we recognize the Fig. 7 that announces the use of the principal property: the tangents are crossing at the vertical of the center of gravity of the curve. Johann starts defining the elements of the curve as well as the constant tension that we find represented by the segment C in Fig. 10:

Once this is presupposed, one finds the common catenary in the following way:

Let *BAa* be the line sought (see our Fig. 10), whose lowest point is *B*; the axis, or the vertical line going through *B*: *BG*; the tangent in the lowest point BE that will be horizontal and let *AE* be the tangent at whatever point *A*. One traces *AG* the ordinate and *LE* parallel to the axis, let $BG = x$, $GA = Y$, $Gg = dx$, $Ha = dy$; the weight of the thread or, as it is homogenous, the length of the element of the curve $BA = s$. As, thus the force required in *B* is always

[21] Bernoulli (1691–1692, pp. 492–493): 4° *iisdem positis quae prius, ante et post fixationem in singulis curvæ punctis eadem, id est, pristina potentia requiritur, vel, quod eodem recidit, qua vi unum punctum ante fixationem trahitur, eadem vi trahetur post fixationem. Hoc nihil aliud est, quam Corollaríum prioris. Hínc quantumcunque catena BFA, vel prolongetur, vel decurtetur, hoc est ubicunque punctum fixationis F sumatur, potentía in infimo puncto B' neque augebitur, neque diminuetur, sed semper manebit eadem et aequalís.*

[22] Bernoulli (1691–1692, p. 493): 5° *Pondus P sustentatum a duobus filis AB, CB quomodocunque positis, hac proportione vires suas exerit in puncta A et C, ut potentia requisita in A sit ad potentiam requisitam in C, ut viceversa [ducta verticali B G] sinus anguli CBG ad sinum anguli ABG, et pondus P ad potentíam alterutram in C, ut sinus anguli totius ABC ad sinum anguli oppositi ABG. Hoc in quavís Statica demonstratur.*

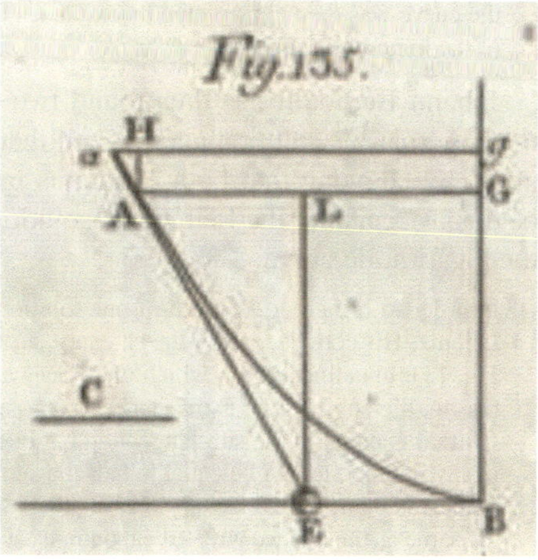

Fig. 10 Joh. Bernoulli's Fig. 135 (1691–1692, Tab. LXX). Public domain

the same and constant (by hypotheses 4) that one extends or shorten the chain *AB*, that one poses that force or the strait line that represents it $C = a$.[23]

To express the problem in an equation is child's play:

Let us imagine now that the weight of the rope *AB* be concentrated and suspended at point *E* where the tangent threads *AE, BE* meet, the force requested (by hypothesis 2) at point *B* in order to sustain the weight *E* is the same as the one that one had to exercise before to sustain the thread *BA*. Now, (by hypothesis 5) the weight *E* is to the force *B* as the sine of angle *AEB*, or of its complement to two right angles *EAL* to the sine of angle *AEL*, that is, as *EL* is to *AL*, thus wherever one take the point *A* of suspension of the curve (the curve, in fact by hypothesis 3, remains always the same) the weight of the thread *AB is to the force in* B as *EL* to *AL*, thus

$$\frac{s}{a} = \frac{EL}{AL} = \frac{AH}{Ha} = \frac{dx}{dy}$$

and inversely

$$\frac{dy}{dx} = \frac{a}{s}.$$

From one clearly sees that the catenary curve *BA* is the one of which we gave the construction and the nature before in the inverse method of the tangents (Lecture XII) where we draw back the proportion $dy : dx = a : s$ to the equality $dy = adx : \sqrt{(2ax + xx)}$ and of which

23 Bernoulli (1691–1692, p. 493): *Sit BAM curva quaesita, cujus infimum punctum B; axis, vel linea verticalís, transiens per B, BG; tangens in infimo puncto BE; quae erit horizontalis; et in quocunque alio puncto A tangens AE. Ductis applicata AG et axi parallela sit* $BG = x$, $GA = y$, $Gg = dx$, $Ha = dy$; *pondus catenae, vel quia aequaliter crassa, longítudo curvae* $BA = s$. *Quoniam itaque in puncto B semper aequalis et constans potentia requiritur [per hypoth. 4] sive catena BA continuetur, sive decurtetur; sit illa potentia, vel linea recta ipsam exprimens* $C = a$.

the curve was then constructed by extension of the parabolic curve and also by quadrature of the hyperbolic space.[24]

Johann Bernoulli has thus found two equations for the catenary. He then tries to show that his solution agrees with that of Leibniz. This he will do—and this is new—by showing that both equations may each be drawn back to the other. He begins by recalling the Leibnizian solution, without taking over the construction of the logarithmic curve:

> In order to bring out the truth of our solution we will look to see if it agrees with that of Leibniz. His construction of the catenary curve is the following. Let there be *NCP* (see our Fig. 11), undefined, upon which one traces a logarithmic curve *OMBQ*, of which for that reason, the sub-tangent is the same, or constant, in each point. One considers the ordinate *CB* that is equal to the sub-tangent and after having taken on both sides *CP* and *CD* equal in any manner, let *DA* be equal to half the sum of the ordinate *DM* and *PQ*; he says that the point *A* is than on the catenary *BA*. That is the reason why, in order to check if this curve is the same as the one we have given, on must see if the nature of the curve *BA* is expressed by the same differential equation.[25]

He recalls that the sub-tangent of the logarithmic is constant, fixes *CB*, which corresponds to Leibniz's *AO* (linked, as we have seen, to *e*), with help of the sub-tangent. The ordinate from the catenary as well as from the logarithmic is x, while y is the abscissa of the catenary and z the one of the logarithmic.

> To go on, let *CB*, which is the sub-tangent, $= a$, $BG = x$, $GA = CD = y$, $DM = z$, $Gg = dx$, $Dd = Ha = dy$, we'll have according to the nature of the logarithmic curve: $zdy = adz$ and, thus $dz = \frac{zdy}{a}$.[26]

Starting from the property of mean proportional of the logarithmic curve and from the expression given by Leibniz for the abscissa of a point of the catenary, Johann shows that one can go back to its equation:

---

[24] Bernoulli (1691–1692, pp. 493–494): *Intelligatur nunc pondus catenæ AB concentratum et appensum esse concursu E í filorum tangentium AE, BE, requiritur [per hypoth-2] in puncto B, eadem potentia ad sustinendum ponduus E, quae antea requirebatur ad sustinendam catenam BAL. Verum [per hypoth. 5] pondus E est ad potentiam in B, ut sinus anguli AEB, vel complementi ad duo rectos EAL, ad sinum anguli AEL, id est, ut EL ad AL: proínde ubícunque in curva punctum fixationis A sumatur [curva enim, per hypoth.3, semper manet eadem] pondus catenae AB est ad potentiam in B, ut EL ad AL, id est,* $s : a = EL : AL = AH : Ha = dx : dy$*, et inverse* $dy : dx = a : s$*. Ex quo patet quod curva Catenaria BA fit illa ipsa, cujus constructionem et naturam supra dedimus in Methodo tangentium inversa [Lect. XII*] ubi prius proportionem hanc* $dy : dx = a : s$ *redegimus ad hanc æqualitatem* $dy = adx : \surd(2ax + xx)$*, et dein constructa fuit curva, per extentionem curvae parabolicae, ut et per quadraturam spatii hyperbolici.*

[25] Bernoulli (1691–1692, p. 494): *Ut veritas nostrae solutionis eo magis elucestat, examinabimus illam, an cum solutione Dni. LEIBNITII conveniat. Cujus constructio curvae Catenariae est talis: Sit NCP recta indefinita horizontalis, superque ea describatur curva Logarithmica OMBQ, cujus ideo subtangens ubique est aequalis seu constans; eligatur applicata CB, quae subtangenti est aequalis, et sumptis hinc inde quomodocunque aequalibus CD, CP, fiat D A = dimidiae summae applicatarum DM, PQ; dicit puctum A esse in curva Catenaria BA. Ad disquirendum itaque, an haec curva sic eadem cum nostra quam dedimus; vídendum est, num natura curvae BA per eandem aequationem differentialem exprimatur.*

[26] Bernoulli (1691–1692, p. 494): *Sit proinde CB, vel subtangens* $= a, BG = x, GA = CD = y, DM = z, Gg = dx, Dd = Ha = dy$*, erit,per naturam Logarithmícae:* $zdy = adz$ *proinddz* $= \frac{zdy}{a}$.

Fig. 11 Bernoulli's Fig. 136 (1691–1692, Tab. LXX). Public domain

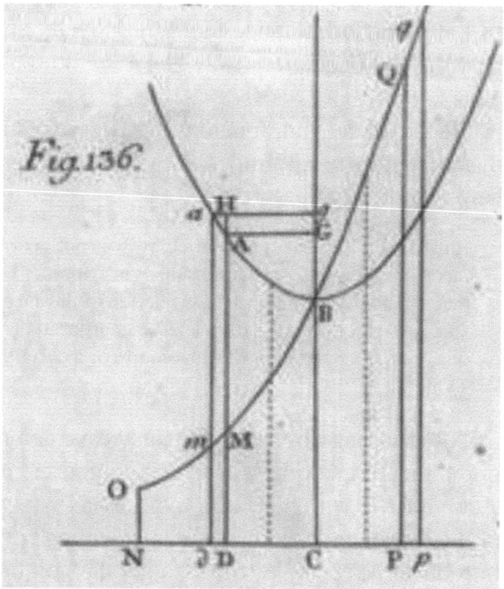

Since by construction, $CD = CP$. One gets $DM/CB = CB/PQ$ and therefore $PQ = \frac{aa}{z}$ and $\frac{1}{2}DM + \frac{1}{2}PQ$, that is, by construction, $DA = \frac{(aa+zz)}{2z} = CB + BG = a + x$ and so $zz = 2az + 2xz - aa$, an equation that gives, if one solves it, $z = a + x + \sqrt{(2ax + xx)}$ and, thus, $dz = dx + \frac{(a+x)dx)}{\sqrt{(2ax+xx)}}$. If one replaces z by its value in the previous equation, one gets or

$$\frac{adx\sqrt{2x+xx} + aadx + axdx}{\sqrt{2x+xx}} = ady + scy + dy\sqrt{(2ax+xx)}$$

and if one divides both members by $a + x + \sqrt{2x+xx}$, one obtains

$$\frac{adx}{\sqrt{2x+xx}} = dy$$

an equation that, because it is the same as the one we had found, allows us to deduce also that the curve *BA* is our catenary and, thus that, although the Leibnizian construction is completely different from the one we gave here, it nevertheless does not give a different curve.[27]

---

[27] Bernoulli (1691–1692, pp. 494–4955): *Sit proinde CB, vel subtangens = a, BG = x, GA = CD = y, DM = z, Gg = dx, Dd = HA = dy; erit, per naturam Logaríthmícae, zdy = dz, proínde dz = zdy: a. Quoniam per constructionem CD = CP; erit DM: CB = CB: PQ, ideoque PQ = zaa: z, et ½ DM + ½ PQ, hoc est, per constructionem, D A = (aa + zz): 2z = CB + BG = a + x; ergo zz = 2az + 2xz - aa; quae aequatio si resolvatur dat z = a + x + √(2ax + xx), proinde dz = dx + (a + x) dx: √(2ax + xx), Substituto valore ipsius z in priore aequatione dz = zdy: a, proveniet dx + (a + x) dx: √(2ax + xx) = (ady + xdy + dy √(2 ax + xx)): a, vel (adx√(2ax + xx) + aa dx + ax dx): √(2ax + xx) = ady + xdy + dy √(2ax + xx), et diviso utroque per a + x + √(2ax + xx) habetur a dx: √(2ax + xx) = dy, quae aequation quia eadem est cum illa quam nos invenímus,*

Let us underline once more the novelty and the importance of this last remark: very different equations may generate the same curve, if one can trace each back to other.

On the other hand, Johann will never explain the two constructions of the catenary that he prefers to call funicular, which he proposes in the *Acta* and that we reproduce here:

> But I found that our funicular curve was not geometrical but was part of the curves called mechanics, because its nature cannot be expressed by a determined algebraic expression and that it can only be obtained by a relation of the curve to the straight line or from a curvilinear space to a rectilinear. It follows that in order to describe it one supposes the rectification of one or another curve, or the quadrature of a curvilinear space as the following constructions show.[28]

After having observed that this curve cannot be obtained without an integration or rectification or quadrature, Johann fixes a reference frame *CB, DE* and constructs a curve *LKF* with help of a hyperbola, using the mean proportional, the abscissas of the three curves being the same, and calculates next the area under the curve. He introduces a rectangle *ICAG,* to get the point *M* of the catenary (Fig. 12).

> Let us trace the normals *CB* and *DE* that cut each other at point A; the center *C,* placed anywhere on the axis *CB*, and equilateral hyperbola *AH* with vertex A. Let us construct the curve *LKF*, so that CA is always the mean proportional between *BH* and BK. Let the rectangle *CG* have the same surface as *EABKF*. Extending *IG* and *HB*, their meeting point will be on the funicular *MAN.*[29]
>
> The equilateral hyperbola *BG* was traced, as before, on the axis *BA* (Fig. 13). Trace on the same axis the parabola *BH*, whose *Latus rectum* is equal to four times the *latus recti* or *transversi* of the hyperbola. Let us extend regularly the applicate *HA* until *E*, so that the line *GE* be equal to the parabolic line *BH*. I say that point *E* is on the funicular curve *EBF.*[30]

The second construction uses the hyperbola already used and a parabola whose parameter is linked to the one of the hyperbolas and having the same abscissa as both the hyperbola and the catenary. The ordinate of point *E* of the catenary is given thanks to the length of the parabola.

---

sequitur quoque Curvam BA esse nostram Catenariam, et proinde constructionem Leibnitianam, utut diversissimam ab illa quam supra dedimus, non tamen dı́versam generare lineam.

[28] Bernoulli (1691b, p. 274): *Reperi autem Curvam nostrum Funicularíam non esse geometricam, sed ex earum censu, quae mechanicae dicuntur; utpote cujus natura determinata aequatione algebraica exprimi nequit, nec nisi per relationem curvae ad rectam, vel spatii curvilinei ad reectilineum, habetur; sicut, ad illam describendam, alterius curvae rectificatio, vel spatii curvilinei quadratura supponatur, ut ex sequentibus Construectionibus liquet.*

[29] Bernoulli (1691b, p. 274): *Ductis normalibus CB, DE sese secantibus in A, centroque C ubivis sumpto in axe CB, & vertice A descripta Hyperbola aequilatera AH, construatur curva LKF, quae talis sit, ut ubique CA sit media proportionalis inter BH & BK; fiat rectangulum CG aequale spatio EABKF, erit productis IG, HB punctum consursus M in Curva Funicularia MAN.*

[30] Bernoulli (1691b, pp. 274–275): *Constr. II. Descripta, ut prius, ad axem BA Hyperbola aequilatera BG; construatur ad eundem axem Parabola BH, cujus latus rectum aequetur quadruplo lateris recti vel transversi Hyperbolae; ordinatimque applicata HA producatur ad E, ita ut recta GE sit aequalis lineae Parabolicae BH; dico punctum E esse in Curva Funicularia EBF.*

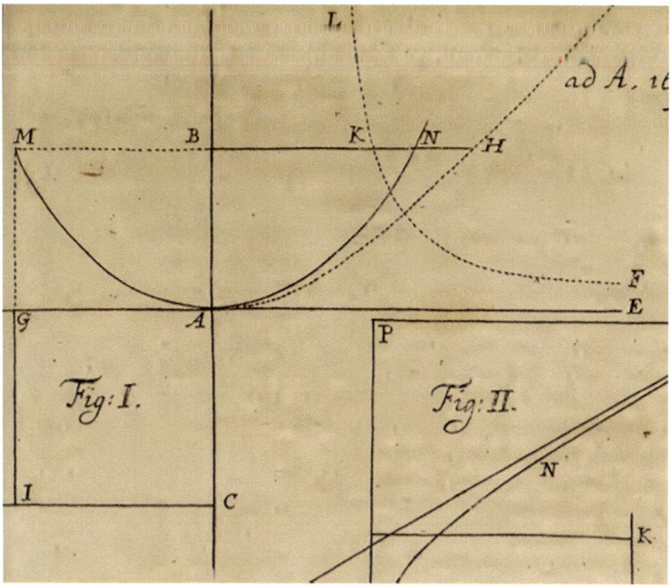

Fig. 12 Bernoulli's Fig. 1 (1691, Tab. VI, facing page p. 274). Public domain

Fig. 13 Bernoulli's Fig. II (1691, Tab. VI, facing page p. 274). Public domain

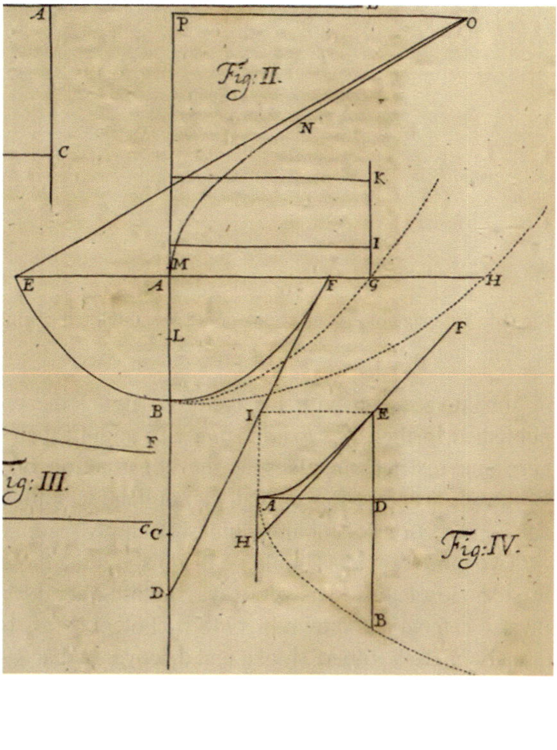

MEMORIE ISTORICHE
DELLA
GRAN CVPOLA
DEL
TEMPIO VATICANO,
E DE' DANNI DI ESSA, E DE' RISTORAMENTI LORO,
DIVISE IN LIBRI CINQVE.
ALLA SANTITA' DI NOSTRO SIGNORE
PAPA
BENEDETTO XIV.

IN PADOVA. CIↃIↃCCXLVIII.
Nella Stamperia del Seminario.
CON LICENZA DE' SVPERIORI.

Fig. 14 Titlepage of Poleni's book (1748) containing the principal properties of the catenary

Johann goes on with his lectures, giving the properties of the curve already published in the *Acta Eruditorum* and adding their demonstrations. We will stop here and restrict ourselves to the construction of the catenary. The reading of Gregory's article stripped of all camouflage will allow us to better understand those last constructions of Johann which we simply paraphrased without explaining them.

Two more remarks should be made about the three answers to Jacob's challenge. First, none of those answers says anything about the stability of the thin arch or inverse catenary as drawn in 1748 by Poleni (Figs. 14 and 15), despite the fact that already in 1675 Robert Hooke had discovered that "as hangs the flexible line, so but

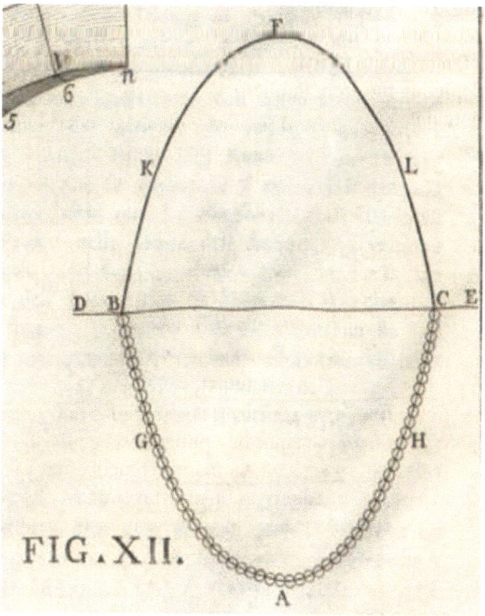

Fig. 15 Poleni's description of the identity of the catenary with the stability of the thin arch (1748, Tav. D). Public domain

inverted will stand the rigid arch".[31] Secondly, all authors are close to Leibniz; this is noteworthy even if prior to the priority debate between Leibnizians and Newtonians after 1700.

3 David Gregory and the Fluxions in England

3.1 Gregory's Memorandum and Newton's Conception of a Thin Arch

Newton's reaction was not immediate, but he didn't overlook Hooke's property. It is only in May 1694 that we find a drawing and a remark on this problem, in a Memorandum due to both Gregory and Newton (1694). The remark and the drawing as Newton's editors reproduce them reduce to the following: "Newton's demonstration that an upright catenary curve is the shape of the strongest arch".[32]

On the basis of Fig. 16, it is rather difficult to make any hypothesis about Newton's proof. We can only observe little circles on the left, which remind us of Poleni's famous figure about a vault (Fig. 17).

[31] Hooke (1676, p. 31), in the form of an anagram: *Ut pendet continuum flexile, sic stabit contiguum rigidum inversum.* Eng. trans. in Truesdell (1960, p. 75).

[32] Gregoryand Newton (1694, p. 344): *Newtoni Demonstratio Catenariam erectam esse arcum fortissimum.*

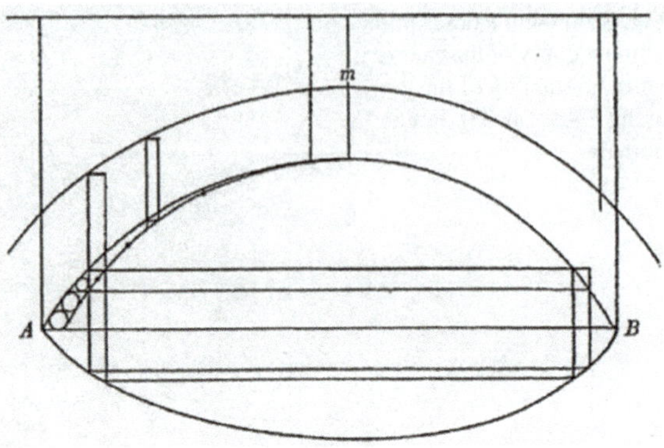

Fig. 16 Figure 3 from Newton and Gregory's Memorandum (1694) depicting the arch. Public domain

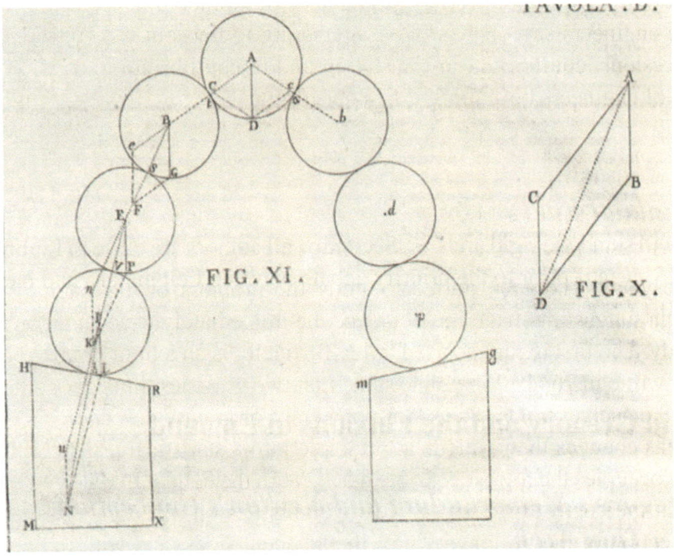

Fig. 17 The arch and the parallelogram law of forces in Poleni's *Memorie istoriche* (1748, Tav. D). Public domain

But Poleni's text would not be published until 1748. Nevertheless, there is another text by James Stirling, which is directly influenced by Newton and was published in 1717. We'll have a quick look at these two texts but let us first analyze Gregory's work on the catenary.

3.2 Gregory's *Catenaria* of 1697 and 1698

There are several proofs that the *Acta Eruditorum* were read in England but certainly with some delay. Further, they were not easily available, in the same way as the *Philosophical Transactions* were difficult to get on the continent.

Gregory's article for example, as well as Leibniz's answer to it, was published in both the *Philosophical transactions* (1697) and in the *Acta Eruditorum* (1698) as well. Nevertheless, it seems that Newton didn't know about Jacob's challenge, on 27 August 1691 when David Gregory writes to Newton:

> Yesterday I got letters from those Amsterdamers of whom I spoke to you … they sent me too three solutions of the problem concerning the figure of a loose hanging cord by Bernoulli, Leibniz and Huygens all in the *Acta lipsiae* for June 1691.[33]

It is no wonder then that some years later, Gregory published his text in the *Philosophical Transactions* for 1695–1697, but surely written just after the publication of the previous ones. Gregory is a defender of Newton's calculus of fluxions. In his *Historia et Origo Calculi differentialis* Leibniz writes about Gregory's text:

> The new inventions due to differential calculus were not known to the disciples of Newton, who could not produce discoveries of any importance, nor avoid paralogisms until they learned Leibniz's calculus, as one finds it in David Gregory's research on the problem of the catenary.[34]

Gregory's text has the advantage of giving demonstrations, contrary to the texts published in the *Acta Eruditorum* by Bernoulli, Leibniz and Huygens. His demonstrations are done with Newton's method of fluxions:

> As the problem of the form of the catenary … has been discussed by the greatest philosophers of our century as Bernoulli, Leibniz et Huygens who found various properties of that form and that they published them (without demonstration) in the *Acta Eruditorum*. I wanted to carry out the demonstrations of those properties by means of the Newtonian method, today familiar to the geometers, of the relations between fluxions and fluents.[35]

---

[33] Gregory to Newton, 27 August 1691, Cambridge University Library, Department of Manuscripts and University Archives, MS Add.3980.

[34] Leibniz (1846, p. 3): *Unde etiam nova per calculum differentialem inventa discipulos Newtoni latuerunt, nec aliquid ipsi alicujus momenti proferre nec etiam paralogismos evitare potuerant, donec calculum Leibnitianum didicerunt, ut in Davide Gregorio Catenariam affectante compertum est.*

[35] Gregory (1697, p. 637): *Cum problema de figura Catenae (id est lineae flexilis, versus centrum longinquum gravis, et pondere suo a duobus extremis immotis dependet incurvatae) sit inter hujus aevi Philosophos imprimis nobile, ac a Celeberrimis Viris Hugenio, Leibnitio et Bernoullio, plurimae figurae istius proprietates fuerint detectae, et in Actis Eruditorum Lipsiae (at sine demonstratione) editae: Libuit harum omnium demonstrationes pertexere, ope methodi Neutonianae Geometris hodie familiaris, fluxiones e fluentium relatione data determinandi et vicissim; et alias insignes Curvae hujus proprietates nunc primum detectas adjicere, tibique Reverende Decane, harum rerum Judici idoneo mitere.*

Gregory unfortunately made a mistake in his first and principal proposition, as Leibniz explains in his *Animadversiones*: "It is sufficient to observe attentively how the author demonstrated the first and principal Proposition, foundation of all others.[36]

Nevertheless, Gregory's result is correct: a differential equation for the catenary which had already been published by Johann. Surprisingly, while he had been working in collaboration with Newton, Gregory's mistake is due to a faulty knowledge of both tension and the parallelogram law of forces. Leibniz underlines this in his *Animadversiones* (Leibniz 1699), a sharp critique of Gregory's work. Gregory will answer in the *Philosophical Transactions* of (Gregory 1699) and the text was reproduced in the *Acta Eruditorum* (Gregory 1700). Jacob Bernoulli also mentions Gregory's mistake in his paper about infinite series, *Positionum de seriebus infinitis* (Bernoulli 1698), where more politely than Leibniz, he summarises the facts:

> David Gregory's analysis of the catenary curve, recently published in the Leipzig *Acta* for July, neatly shows how it is possible for us to be misled through an unobvious and false though plausible argument to a true conclusion.[37]

Let us turn to Gregory's text. He also uses the properties of the tensions, which are constant and tangent to the curve (Fig. 18):

> To find the relation between the fluxion of the axis and of the ordinate for the catenary curve
>
> Let the catenary *FAD* hanging from its extremities *F* and *D*, and from its inferior point (or vertex of the curve) *A* one erects the perpendicular *AB* and one traces the *BD*. Find the relation between *Bb* or $D\delta$ and $d\delta$; supposing *b* be the point infinitely near to *B* that *bd* be parallel to *BD* and $D\delta$ to *BA*.[38]

After that Gregory explains the parallelogram law of forces, which he applies immediately:

> I pose by the lemma of mechanics just mentioned that $D\delta$, or the fluxion of the axis *AB*, expresses the force along the direction which *Dd* has to move in order for it to become parallel to the vertical direction; force that has to compose itself with the preceding conatus of the heavy line that could prevent it from doing so; and is thus (all things being equal) proportional to the line *DA*. Thus, the fluxion of the ordinate is to δD, the fluxion of the abscissa, as a constant line *a* to the curve *DA*. q.e.f.[39]

---

[36] Leibniz (1699, p. 87–88): *Suffecerit autem attente confiderare, quomodo demonstraverit Dn. Autor propositionem primam qui reliquae superstruuntur.*

[37] Bernoulli (1698), n.p. (last page): *Davidis Gregorii Analysis Curvae Catenariae, nupere Act. Lips. Iulio vel Augusto inserta, opportune ostendit, fieri utique posse, ut quandoque per inevidens et falsam, plausibile licet, ratiocinium ad veram conclusionem perducatur.*

[38] Gregory (1697, pp. 637–638): *Prop. I. Problem, Invenire relationem inter fluxionem axeos et fluxionem ordinatae in Curva Catenaria. Sit Catena FAD ab extremitatibus F et D dependens, cujus puntum imum (seu Curvae vertex) A, axis AB ad horizontem erectus; eique applicate BD horizonti parallela. Invenienda est relatio inter Bb seu* $D\delta$ *et* $d\delta$*; posito b puncto ipsi B proximo; et bd ad BD, item* $D\delta$ *ad BA parallela.*

[39] Gregory, 1697, p. 638: *Porro ex supra citato Lemmate Mechanico,* $D\delta$ *sive fluxio axeos AB exponet vim secundum directionem ipsius dD exerendam, quae priori conatui gravis dD ad componendam se in situm verticalem aequipolleat, eumque impedire possit. Haec vero vis oritur a linea gravi DA secundum directionem dD trahente, estque proinde (caeteris manentibus) linee DA proportionalis. Est igitur* δd *fluxio ordinatae ad* δD *fluxionem abscissae, sicut recta a ad DA curvam. q.e.f.*

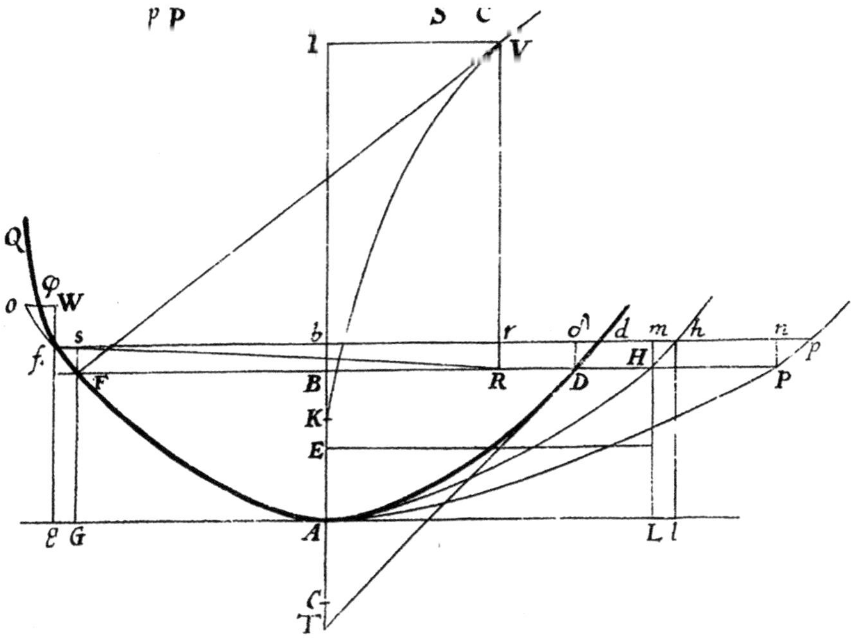

Fig. 18 Gregory's Fig. 1 for Proposition I, Problem (1697, p. 637). Public domain

His reasoning is the following: as the tension is tangent, its components are the fluxions and as the tensions are constant the fluxions are also constant. The equation of the catenary

$$\frac{\dot{x}}{\dot{y}} = \frac{a}{s}$$

follows immediately. This equation is Johann Bernoulli's $\frac{dx}{dy} = \frac{a}{s}$.

Corollary: If the straight line *DT* is tangent to the catenary and meets *AB* extended in *T*, one has[40]

$$\frac{DB}{BT} = \frac{d\delta}{\delta D} = \frac{a}{DA}.$$

In the corollary he puts in the equation the property of the tangent, while Prop. 2 explains the second construction published by Jacob Bernoulli in the *Acta Eruditorum* (1690):

Prop. 2 Theorem

[40] Gregory (1697, p. 638): *Corollarium. Si recta DT tangat Catenariam, et axis BA producto occurrat in T, erit DB.BT:: (dδ.δD::) a.DA Curvam.*

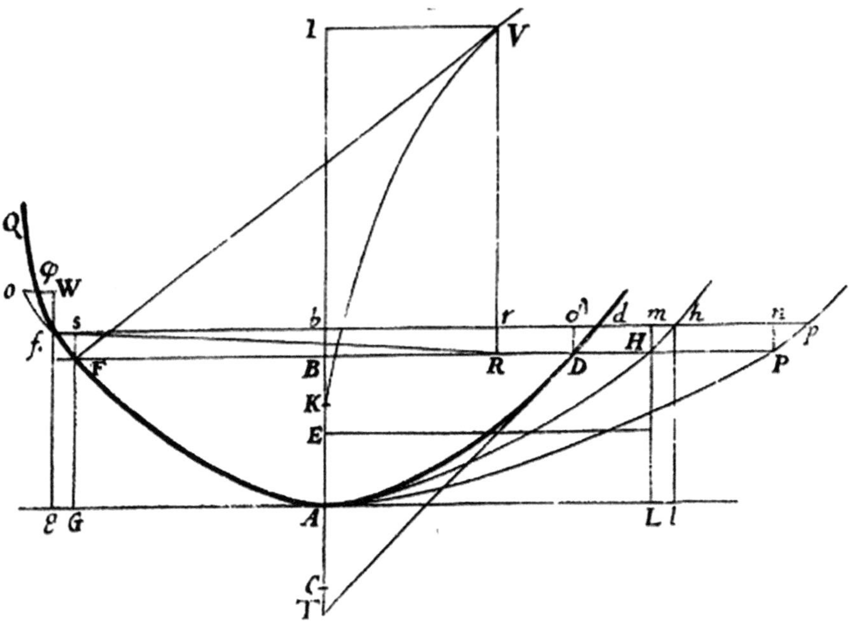

Fig. 19 David Gregory's explanation of Prop. 2 (Gregory 1697, Fig. 1). Public domain

> Taking the perpendicular *AB* as axis, *A* as the summit, construct the equilateral hyperbola *AH* whose semi-axis AC is equal to *a*; and with the same axis and vertex, a parabola *AP* whose parameter equals three times the axis of the hyperbola, and extend each time the ordinate *HB* of the hyperbola until *HF* equals the curve *AP*: I say that the curve *FAD* in which the points *F* and *D* fall (if one takes *BD* and *BF* equal) is the catenary.[41] (Fig. 19)

Gregory specifies the link between the semi-axis of the hyperbola *a*, and the parameter of the parabola whose value is *8a* and demonstrates that in this way he obtains a point of the catenary. In order to do this, he shows that, from that construction, he can deduce the equation of the catenary already found:

> One calls *AB*, *x;* and thus $Bb = \dot{x}$, and $BH = \sqrt{2ax + x^2}$. Thus by the method of fluxions, the fluxion of this *BH* (that is to say *mh*)
>
> $$= \frac{a\dot{x} + x\dot{x}}{\sqrt{2ax + x^2}}.$$
>
> As the parameter of the parabola is 8*a* we shall get $BP = \sqrt{8ax}$. Thus *np* (that is the fluxion of *BP*)

[41] Gregory (1697, p. 638): *Si ad perpendiculum AB tanquam axem vertice A, describatur hyperbola aequilatera AH, cujus semi-axis AC aequalis a; et ad eundem axem et verticem, parabola AP cujus parameter aequalis quadruplo axi hyperbolae, et producatur semper hyperbolae ardinata HB, donec HF aequalis Curvae AP: Dico Curvam FAD in quo puncta F et D versantur (positis BD, BF aequalibus) esse Catenariam.*

$$= \frac{2a\dot{x}}{\sqrt{2ax}}.$$

That is why the fluxion of the curve $AP\left(= Pp = \sqrt{np^2 + Pn^2}\right)$

$$= \sqrt{\frac{4a^2\dot{x}^2}{2ax} + \dot{x}^2} = \sqrt{\frac{2a\dot{x}^2 + x\dot{x}^2}{x}}$$

(which, if one multiplies the numerator and the denominator by $\sqrt{2ax + x^2}$)

$$= \frac{2a\dot{x} + x\dot{x}}{\sqrt{2ax + x^2}}.$$

And as HF is always equal to AP, the fluxion of HF is a line, it means that

$$mh + sf = \frac{2a\dot{x} + x\dot{x}}{\sqrt{2ax + x^2}}.$$

But then we have found $mh = \frac{a\dot{x}+x\dot{x}}{\sqrt{2ax+x^2}}$. Then sf, the fluxion of BF referred to the axis of the catenary is

$$\frac{a\dot{x}}{\sqrt{2ax + x^2}}.$$

And therefore the fluxion of the curve AF (or $Ff = \sqrt{sf^2 + Fs^2} = \sqrt{\frac{a^2x^2}{2ax+x^2} + \dot{x}^2}$)

$$= \frac{a\dot{x} + x\dot{x}}{\sqrt{2ax + x^2}},$$

Of which, by the given method, the fluent is $\sqrt{2ax + x^2}$. And thus $AF = \sqrt{2ax + x^2}$. And it follows that the fluxion of the ordinate BF or $\frac{a\dot{x}}{\sqrt{2ax+x^2}}$ is to $\dot{x}$, the fluxion of the abscissa AB as the given a to the curve AF, and this is the property of the catenary shown above. Thus, the point of the catenary is correctly determined by the preceding construction. Q.E.D.[42]

---

[42] Gregory (1697, p. 639): *Vocetur AB x, erit Bb* $= \dot{x}$, *etBH* $= \sqrt{2ax + x^2}$. *Unde ex methodo fluxionum, fluxio ipsius BH (sive mh)* $= \frac{a\dot{x}+x\dot{x}}{\sqrt{2ax+x^2}}$. *Rursus quia parabola AP parameter* $= 8a$, *erit BP* $=$ *BP* $= \sqrt{8ax}$. *Unde np (hoc est fluxio ipfius BP) aequalis* $\frac{2a\dot{x}}{\sqrt{2ax}}$. *Quare fluxio Curvae* $AP\left(= Pp = \sqrt{np^2 + Pn^2}\right) = \sqrt{\frac{4a^2\dot{x}^2}{2ax} + \dot{x}^2} = \sqrt{\frac{2a\dot{x}^2+x\dot{x}^2}{x}}$ *quae, ducendo tam numeratorem quam denominatorem in,* $=)\frac{2a\dot{x}+x\dot{x}}{\sqrt{2ax+x^2}}$. *Et cum HF sit ubique* $=$ *AP, erit fluxio HF rectae, hoc estmh* $+$ $sf = \frac{2a\dot{x}+x\dot{x}}{\sqrt{2ax+x^2}}$. *Sed hactenus inventa estmh* $= \frac{a\dot{x}+x\dot{x}}{\sqrt{2ax+x^2}}$. *Unde sf sive fluxio ipsius BF ordinatae ad axem Catenariae, est aequalis* $\frac{a\dot{x}}{\sqrt{2ax+x^2}}$. *Et igitur Curvae AF (sive ipsa Ff* $= \sqrt{sf^2 + Fs^2} =$ $\sqrt{\frac{a^2x^2}{2ax+x^2} + \dot{x}^2}$ *cujus fluens modo ostensa est* $\sqrt{2ax + x^2}$*). Et igiturAF* $= \sqrt{2ax + x^2}$. *Patetque fluxionem ordinatae BF sive* $\frac{a\dot{x}}{\sqrt{2ax+x^2}}$ *esse ad* $\dot{x}$ *fluxionem abscissae AB sicut data a ad Curvam AF, quae est superius inventa Catenariae proprietas. Igitur Catenariae puncta recte determinantur per praecedentem constructionem. q. e. d.*

Corollary 6 of the second theorem carefully explains the physics involved in the "inverted arch of a catenary":

> In a vertical plane, but in an inverted situation, the chain will preserve its figure without falling, and therefore will constitute a very thin arch, or fornix; that is, infinitely small rigid and polished spheres disposed in an inverted arch of a catenary curve will form an arch, no part of which will be thrust outwards or inwards by other parts, but, the lowest part remaining firm, it will support itself by means of its shape. For since the situation of the points of the catenary is the same, and the inclination of the parts to the horizon, whether in the situation FAD, or in an inverted situation, so that the curve may be in a plane which is perpendicular to the horizon, it is plain it must keep its figure unchanged in one situation as in the other. And, on the contrary, none but the catenary is the figure of a true legitimate arch, or fornix. And when an arch of any other figure is supported, it is because in its thickness some catenary is included. Neither would it be sustained if it were very thin and composed of slippery parts. From Corol. 5 it may be collected, by what force an arch, or buttress, presses a wall outwardly, to which it is applied; for this is the same with that part of the force sustaining the chain, which draws according to a horizontal direction. For the force, which in the chain draws inwards, in an arch equal to the chain drives outwards. All other circumstances, concerning the strength of walls to which arches are applied, may be geometrically determined from this theory, which are the chief things in the construction of edifices.[43]

Gregory elaborates Newton's model of an arch composed of infinitely small rigid and polished spheres explaining that it cannot break: "For the force, which in the chain draws inwards, in an arch equal to the chain drives outwards".[44]This is a sentence that could have influenced Jacob Bernoulli's study of the thin arch. Jacob's arch is made of thin blocks without cement, so that it can only break in two ways:

> ...the stone DL ... can only fall under the action of its weight, which it can do either sliding on the plane KD, if this plane is perfectly polished, or turning around the point D...[45]

This gives him two conditions for the equilibrium to determine the equation of the vault; one would have been enough[46] and as Jacob's second relation does not

---

[43] Gregory (1697, pp. 640–641): *6. Catena in plano verticali, sed situ inverso, figuram sevat nec decidit, adeoque arcum seu fornicem facit tenuissimum: Hoc est spherae minimae rigidae et lubricae in inversa Curva Catenaria dispositae, arcum constituent cujus nulla pars ab aliis extrorsum vel introrsum propellitur; sed manentibus infimis punctis immotis, virtute suae figurae sustinetur. Cum enim punctorum Curvae Catenariae situs, partiumque inclination ad Horizontem eadem sit, sive* in situ *FAD, sive* in situ *inverso, dummodo Curva sit in plano ad Horizontem recto, patet illam aeque servare figuram immutatam in uno situ ac in altero. Et e converso solae Catenariae sunt fornice sive arcus legitimi: Et cujuscunque alterius figurae Arcus ideo sustinetur, quod in illius rassitie quaedam Catenaria inclusa sit: Neque, si tenuissimus esset, partesque haberet lubricas sustineretur. Ex praecedente Corol. 5. Colligitur quali vi arcus, murosquibus insistit extra propellit; nempe haec eadem est cum parte vis Catenam sustinentis, quae secundum directionem Horizontalem trahit. Quae enim in Catena introrsum trahit vis, in arcu Catenae aequali, extrorsum propellit. Alia omnia de murorum quibus fornices imponuntur firmitate requisite, ex hac Theoria Geometrice determinantur, quae in aedificiorum extrctione praecipua sunt.*

[44] Gregory (1697, p. 641): *Quae, enim in Catena introrsum trahit vis, in arcu Catenae aequali, extrorsum propellit.*

[45] Bernoulli (1704, p. 1119): ... *lapis DL ... proprio pondere cadere necessum haberet; idque vel labendo super plano KD, si hoc planum sit perfecte lubricum.*

[46] Jacob Bernoulli also makes a mistake here and finds only once the catenary, in the case of sliding. Both conditions should give the catenary equation. For more on this, see "The *De curvatura fornicis*

give the catenary, it is false. Sadly enough, what Jacob had written on 5 December 1704, a few months before his death, would remain unpublished until 1744 when his *Opera* were published. But let's go back to Gregory's article.

Corollary 7 (Gregory 1697, p. 641) of the same proposition shows, from physical arguments, the identity of the catenary with the *velaria*, a section of a veil inflated by wind. In fact, forces act on a *velaria* in the same way as forces which shape a catenary. Thus, the induced shape must be the same.

Finally, Gregory finds the same equation as did Johann Bernoulli, starting from the Leibnizian construction. Here also, he gives, as did Johann, the link between the logarithm and the hyperbola, since the sub-tangent of the logarithm is *a,* semi-axis of the hyperbola as defined not only in Corol. 2 Prop. 2 but also before in proposition 2:

> Proposition 7 Theorem
>
> If in the logarithmic curve *LAG* of which the sub-tangent *HS* is equal to the line *a,* defined in Corol. 2 Prop. 2, one supposes that point *A* whose distance to the asymptote *HP* namely *AC*, is equal to the subtangent *HS*, and from the points *H* and *P* whatever on the asymptote, equidistant from *C* one erects *HL, PG* ordinates of the logarithmic, whose semi-sum is posed equal to *HD* or *PF*, then *D* and *F* are on the catenary corresponding to the line *AC*.
>
> Let $AB = x$, and *CB* or *DH* the semi-sum of the ordinates *HL, PG* be $x + a$. The semi-difference is called y. Thus $HL = a + x + y$, and $PG = a + x - y$. And thus, by the nature of the logarithmic, *CA* is the mean proportional between them, or
>
> $$a^2 + 2ax + x^2 - y^2 = a^2.$$
>
> Thus $y = \sqrt{2ax + x^2}$.
>
> Further, $HL = a + x + \sqrt{2ax + x^2}$ and $PG = a + x - \sqrt{2ax + x^2}$. That is the reason why fluxion of *HL* or *lm* is
>
> $$\frac{ax + x\dot{x} + x\sqrt{2ax + x^2}}{\sqrt{2ax + x^2}}.$$
>
> And by equality of the angles of triangles *lmL, LHS* one has
>
> $$\frac{LH}{HS} = \frac{lm}{mL}.$$
>
> Thus mL or dδ fluxion of $BD = \frac{a\dot{x}}{\sqrt{2ax+x^2}}$.
>
> Thus the curve *AD* generated by the logarithms as enounced before, has for nature, if one calls the axis x, and its fluxion $\dot{x}$ the fluxion of the ordinates *BD* is
>
> $$\frac{a\dot{x}}{\sqrt{2ax + x^2}}$$

of Jacob Bernoulli: The Introduction of the Infinitely Small into the Calculation of Arches", Chap. 5 in this present volume.

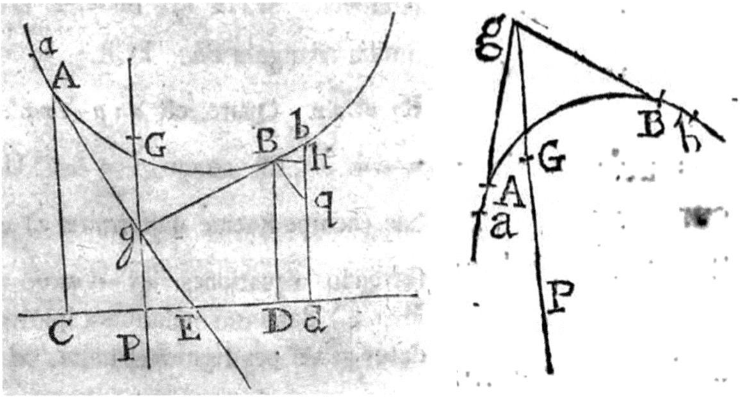

Fig. 20 Symmetrical figures of Brook Taylor's work showing the identity of the catenary and the vault made of spheres. **Left** Brook Taylor's Fig. 7 (1717, Prop. XVIII, Prob. XIII, p. 75) Public domain; **right** Taylor's Fig. 8 (1717, Prop. XIX, Prob. XIV, p. 81). Public domain

> But this is precisely the property of the catenary that we have demonstrated at proposition 1. Thus the curve *FAD* described above is that catenary. Q.E.D.[47]

Writing in 1715, Brook Taylor reasoned in a similar way in his *Methodus incrementorum directa et inversa* (Method of Increments direct and inverse) (Taylor 1717). Taylor intends to explain Newton's calculus of fluxions and one of his examples is the catenary curve. Then immediately follows the drawing of an arch supported by its own weight. For the catenary curve, he had explained the property already used by Pardies: two tangents to the curve cross each other on the vertical passing through the center of gravity of the arch of the curve between those two points (Fig. 20).

Then he considers a vault and draws a symmetrical figure saying (Fig. 20, right):

> Let *AB* be a certain part of an arch, and *a* and *b* points of it, close to *A* and *B*, and let the tangents *Ag* and *Bg* be drawn. Then if the center of gravity of the portion *AB* is at *G*, through this point is drawn the perpendicular to the horizontal *GP*, which cuts the point of

[47] Gregory (1697, pp. 647–648): *Si in Curva Logarthmica LAG cujus data subtangens HS aequalis rectae a, Corol. 2. Prop. 2. hujus difinitae, sumatur punctum A cujus distantia ab HP asymptoto, nempe AC, aequalis sit subtangenti HS, & ex puncto C aequaliter distantibus, erigantur HL, PG ordinatae ad Logarithmicam, quarum semisummae ponatur aequalis HD vel PF, erunt D & F ad Catenariam rectae AC correspondentem.*
Vocetur ABx, adeoque CB vel DH semisumma ordinatarum HL, PG erit $a + x$*; semidifferentia earundem vocetur y. Unde* $HL = a + x + y$*, &* $PG = a + x - y$*. Cumque ex natura Logarithmicae, CA sit inter has media proportionalis, erit* $a^2 + 2ax + x^2 - y^2 = a^2$*. Unde* $y = \sqrt{2ax + x^2}$*. Adeoque* $HL = a + x + \sqrt{2ax + x^2}$ *&* $PG = a + x - \sqrt{2ax + x^2}$*. Quare fluxio ipsius HL, sive ipsa lm est* $\frac{ax + x\dot{x} + x\sqrt{2ax+x^2}}{\sqrt{2ax+x^2}}$.
Et ab aequiangula triangula lmL, LHS, est LH.HS::lm.mL, unde mL sive dδ fluxio ipsius BD $= \frac{a\dot{x}}{\sqrt{2ax+x^2}}$*. Hoc est Curva AD ex Logarithmica supradicto modo genita, ejus est naturae, ut si axis Vocetur X, ejusque fluxio* $\dot{x}$*, flucio ordinatae BD sit* $\frac{a\dot{x}}{\sqrt{2ax+x^2}}$*. Sed haec ipsa est proprietas Catenariae ad quam a pertinet, Prop. 1, jujus demonstrata. Ergo Curva FAD superius descripta est haec ipsa Catenaria. q.e.d.*

> concurrence of the tangents *g*, since the weight of the arch *AB* is supported by the short lines *aA* and *bB*. As these forces may be interpreted in the same way as in the preceding proposition, it is established that the figure of the arch is the same as that of the rope yet inverted in position.[48]

James Stirling, also belonging to Newton's environment, will pick up the arch problem again using Newton's spheres model (Fig. 21). His text is a chapter coming at the end of the book entitled *Newton's Lineae tertii ordinis Neutonianae,* as an illustration of Newton's classification of curves of degree three.[49] This context underlines the geometrical importance of the study of the catenary curve. He uses the fact that three forces are in equilibrium if they are in the same proportion as three lines having their direction parallel to those of the forces. Each sphere is submitted to its weight and the reactions of the two neighbouring spheres. Stirling almost obtains the same differential equation as Johann Bernoulli (Fig. 22).

> Method to dispose some spheres in order to form an oven and thus demonstrating the principal properties of the catenary curve.
>
> Problem: Given some equal spheres, to form an arch such that their weights support each other.
>
> …
>
> Thus, the properties of the *catenaria* is: As the increment of the ordinate is to the increment of the abscissa, so is the radius of curvature at the vertex to the length of the curve between the vertex and considered ordinate.[50]

4 Conclusion

Truesdell's remark concerning the novelty or outdatedness of those texts is rather vague and sharp at the same time. It has the advantage of arousing interest and being confirmed by an internalist analysis such as the one we just provided. But that analysis brings us much more, since it allows us to see what their ancient, contemporary or modern character is due to, and observe along the way typical elements of mathematics before differential calculus as well as what they will become

---

[48] Taylor (1717, p. 81): *Sit fornicis portio quaedam AB, et sint a et b puncta ipsis A et B proxima, et ducantur tangentes Ag, Bg. Tum si portionis AB centrum gravitatis sit G, per id ducta horizonti perpendicularis GP transibit per tengentium concursum g, (per Lem. 5 et 6.) quoniam sustinetur pondus arcus AB per lineolas aA, bB. Viribus itaque eodem modo interpretatis, atq; in Propositione praecedenti, constat figuram hujusmodi fornicis, eadem esse atq; funiculi,* in situ *tamen inverso.*

[49] The book's full title is *Lineae tertii ordinis Neutonianae, sive, Illustratio tractatus D. Neutoni De enumeratione linearum tertii ordinis: cui subjungitur, solutio trium problematu.* One of those problems is: *Methodus disponendi quotcunque Phaeras in Fornicem. Et inde Demonstratur Proprietas praeci pua Curvae Catenariae*, pp. 11–14, appearing after Newton's text *Lineae tertii ordinis.*

[50] Sterling (1717, pp. 11–14): *Methodus disponendi quotcunque sphaeras in fornicem et inde Demonstratur proprietas praecipua Curvae Catenariae. Problema: Datas quotcunque Sphaeras aequales in Fornicem ita disponere ut gravitate suae e mutuo sustineant. … Unde Catenae proprietas haec est; Ut incrementum Ordinatae ad incrementum Abscissae, ita duplus Radius curvaturae in vertice ad longitudinem curvae inter verticem et ordinatam illam interceptae. Q.E.D.*

LINEÆ
TERTII ORDINIS
NEUTONIANÆ,
SIVE
Illuſtratio Tractatus D. *Neutoni*
De Enumeratione *Linearum*
Tertii Ordinis.
Cui Subjungitur,
Solutio Trium Problematum.

Authore JACOBO STIRLING, è Coll. Ball. *Oxon.*

OXONIÆ,
E THEATRO SHELDONIANO, Impenſis *Edvardi Whiſtler*
Bibliopolæ *Oxonienſis*, MDCCXVII.

Fig. 21 Title page of Stirling's publication of Newton's work on the lines of the third order (Stirling 1717), where Stirling added the solution of problems concerning the catenary. Public domain

once this calculus is really assimilated. It also teaches us something about the period in between, in the period of transition. How can we make those elements clearer? They concern only the mathematics, because the mechanics that is used is the same for our three authors: it is that of Pardies (1673).

Huygens only uses elementary geometry, theory of proportions and a definition of the tangent as the ratio of two sides of a triangle. It is thus not a functional definition of the tangent. His reasoning is only applicable to the catenary. The way both Johann Bernoulli and David Gregory put the problem in equation form is adaptable to many other problems. The one applied by Johann Bernoulli coincides with what one would do nowadays. But Truesdell is right in underlining the transient character of the Leibnizian construction, whether or not it was due to his desire to keep the calculus

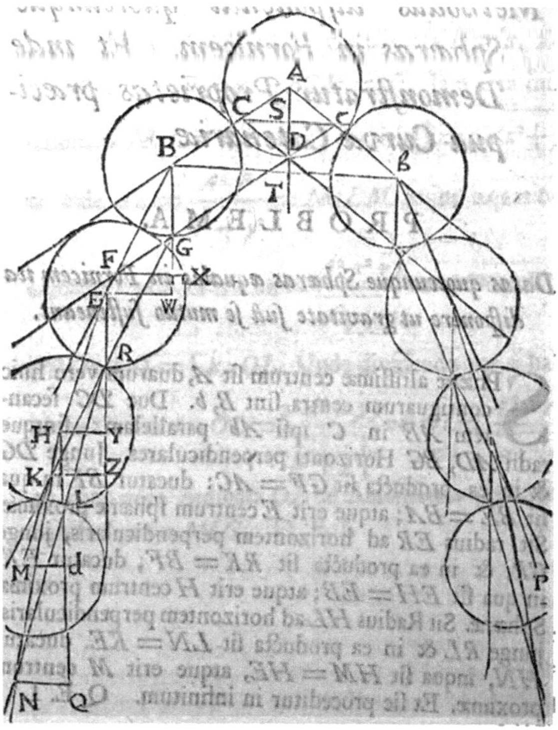

Fig. 22 Stirling's vault composed of spheres (1717, p. 12). Public domain

secret. But we must not forget that he gives the hyperbolic cosine that we still use to describe the catenary.

Our analysis leads us to question the importance accorded by Leibniz, and at the time of Leibniz, to the logarithm, the curve which, thanks to the problem of de Beaune, marks a decisive turning point in the birth of calculus. In order to better understand the Leibnizian point of view, let us quote a text in which he compares the answers of Johann Bernoulli and Huygens to his own. He recalls the numerous relations between the catenary, the logarithmic curve, the hyperbola, the parabola but also the loxodrome, etc., underlining along the way the importance of the logarithmic curve and shedding light on the role played by all those curves in various demonstrations. This last aspect has remained a bit in the dark in our analysis; it appears more clearly in an analysis of the properties of the curve than in one restricted to the constructions. Leibniz highlights the mathematical culture of his time, which reports on what today appear as "jumps" difficult to understand in the demonstrations that preceded them. In fact, those authors juggle properties of those curves that we can certainly find again but that today are no longer part of our mathematical toolkit, the one present to our minds:

> I am delighted to notice, in reading them, the concordance of the three solutions of the problem inaugurated by Galileo and renovated by M. Bernoulli; it is a pledge of its exactness that will convince those who do not enter into the details of questions of that kind. Even if

one does not take the time to compare them point after point, their concordance about the essential is obvious. The three of us have established the law of the tangents, as well as the rectification of the catenary. I had shown formerly, in the *Acta* of June 1686, p. 489 (thanks to the introduction of a new type of contact that I called osculation), how to evaluate the curvature of a curve either with help of the ray of its osculating circle, or among all the tangent circles, by finding the one that is the closest and consequently forms with a straight line the same angle of contact as the curve itself. The famous Huygens (observing that the centers of those circles are always situated on the curves of which he was the first inventor, curves whose development gives the initial curve again) had thus the idea to apply here that theory, and to search for the ray of curvature, or osculating circle, of the catenary, which is once more its developed curve; this one is also part of the solution of the Bernoulli.

The solution of Huygens also gives the distance between the center of gravity of the catenary and the axis, the one of the Bernoulli, as well as mine, gives not only its distance to the axis, but also to the base, and to whatever other straight line. They thus make it possible to determine where the point is situated, as well as the quadrature of the figure drawn by the catenary. And to this, I added in my solution, the center of gravity of that figure, that is to say, its surface. Mr. Huygens gives the construction of the figure supposing the quadrature of the following one:

$$xxyy = a4 - aayy,$$

while Mr Jean Bernoulli and myself have traced it back to the quadrature of the hyperbola; this last one uses quite rightly the rectification of the parabolic curve, while I finally trace all back to the logarithms. I obtain in this manner the type of expression but also the most perfect way of constructing the transcendents. Because it is only necessary then to know or to suppose a unique constant relation in order to be able to describe, using only the usual geometry, without involving neither quadrature nor rectification, an infinity of exact points. One will perhaps take pleasure in noting through my construction this singular and elegant concordance of the catenary and logarithms. By the way, Mr. Huygens, (giving us hope of a considerable simplification using the table of sines) observed that the problem could also be traced back to the sums of secants increasing uniformly by minimal increments. I had made the same remark, and as it came to my mind that it is also starting from the same that one determines the rhombic or loxodromic curve used by the navigators, a curve that I remembered having established many years ago using logarithms, I exhumed my old drafts and finally published all that in last April, on p. 181 of the *Acta eruditorum*.[51]

[51] Leibniz (1691b, pp. 435–437): *Valde delectatus sum lectis tribus Problematis a Galilaeo propositi, a D. Bernoullio renovati, solutionibus inter se consentientibus, quod indicium est veritatis, apud eos valiturum, qui talia accurate non examinant. Etsi autem omnia conferre non vacaverit, in summa tamen rei manifesta est concordia. Legem tangentium, et extensionem curvae catenariae in rectam invenimus omnes, et cum curvedinis mensura olim in Actis Junii A, 1686, p. 489, (introducto novo contactus genere, quem osculum appellare placuit) explicuerim per radium circuli curvam osculantis, seu ex omnibus circulis tangentibus maxime ad curvam accedentis, eundemque adeo quem ipsa curva ad rectam facientis angulum contactus, placuit celeberrimo Hugenio (animadvertenti centra horum circulorum semper incidere in lineas a se primum inventas, quarum evolutione describuntur datae) speculationem huc applicare, et investigare radium curvitatis vel circulum osculatorem curvae catenariae, sive ejus curvam evolutione generantem, quam et dedit solutio Bernoulliana. In Hugeniana autem, distantia quoque habetur centri gravitatis catenariae ab axe. In Bernoulliana et mea, ejusdem distantia tam ab axe quam basi aut alia recta adeoque puncti determinatio, item quadratura figurae catenariae. Quibus ego in mea centrum gravitatis etiam hujus figurae seu areae adjeci. Constructio lineae Dn. Hugenius exhibet ex supposita quadratura curvae, qualis est* $xxyy = a4 - aayy$, *Dn. Joh. Bernoullius et ego reduximus ad quadraturam hyperbolae; illo perbene adhibente etiam extensionem curvae parabolicae in rectam, me denique*

One should not underestimate Leibniz's research because if the importance of the logarithmic will be relativized by the fact that other functions will come besides it, nevertheless, it is the juxtaposition that Leibniz starts to do with parabola, hyperbola, etc. and the logarithmic that will permit the notion of function to emerge.

References

Bernoulli, Jacob. 1690. Analysis Problematis, de inventione lineae descensus a corpore gravi percurrendae uniformiter (Op. XXXIX). In *Acta Eruditorum* (maji), 217–219.

Bernoulli, Jacob. 1691a. Additamentum ad Problema Funicularium exhibita a Johanne Bernoulli (Op. XLVII). In *Acta Eruditorum* (Juin), 274–276; Gerhard, *Mathematische Schriften*, vol. V, 252–254.

Bernoulli, Johann. 1691b. Solutio problematis funicularii (Op. IV). *Acta Eruditorum* (juin 1691), 274–276. Repr. in Gerhardt, *Mathematische Schriften*, V, 248–250; *Opera Omnia* I, 48–51.

Bernoulli, Jacob. 1698. *Positionum de Seriebus infinitis earumque usu in quadraturis spatiorum et rectificationibus curvarum, Pars quarta* (Op. XC). Basel: Conradi à Mechel.

Bernoulli, Jacob. 1704. Problema curvatura fornicis, cujus partes se mutuo proprio pondere suffulciunt sine opere caementi (VP XXIX). In ed. by *Opera* II, Jacob Bernoulli, 1119–1123.

Bernoulli, Johann. 1691–1692. Lectiones Mathematicae, de Methodo Integralium, aliisque, conscriptae in usum Ill. Marchionis Hospitalii (Op. CXLIX), Cum Auctor Parisiis ageret Annis 1691 et 1692. In *Opera Omnia*, vol. III, 385–558.

Gregory, David, and Newton Isaac. 1694. Memoranda by David Gregory and Newton, May 1694. *The Correspondence of Sir Isaac Newton* III (448): 344–348.

Gregory, David. 1697. Catenaria, ad Reverendum virum D. Henricum Aldrich, S.T.P. *Philosophical Transactions* 19: 637–652.

Gregory, David. 1698. Catenaria, ad Reverendum virum D. Henricum Aldrich, S.T.P. *Acta Eruditorum*, 305–321.

Gregory, David. 1699. Responsio ad animadversionem ad DAVIDIS GREGORII catenariam. *Philosophical Transactions* 21 (258): 419–426.

Gregory, David. 1700. Responsio ad animadversionem ad DAVIDIS GREGORII catenariam. *Acta Eruditorum* (Juli), 301–306.

Hooke, Robert. 1676. *A Description of Helioscopes and Some Other Instruments*. London: T.R. for John Martyn.

Huygens, Christiaan. 1901. *Œuvres complètes de Christiaan Huygens publiées par la Société Hollandaise des Sciences*, vol. IX: Correspondance 1685–1690. La Haye: Martinus Nijhoff.

Huygens, Christiaan. 1905. *Œuvres complètes de Christiaan Huygens publiées par la Société Hollandaise des Sciences*, vol. X: Correspondance 1691–1695. La Haye: Martinus Nijhoff.

---

rem omnem reducente ad logarithmos eaque ratione obtinente perfectissimum in Transcendentibus exprimendi pariter et construendi genus, Sic enim unica tantum semel supposita vel habita ratione constante, de reliquo infinita puncta vera exhiberi possunt per communem Geometriam sine interventu ulteriore quadraturarum, aut extensionum in rectas, Lineae Catenariae mirum et elegantem cum Logarithmis consensum, ex mea constructione animadvertere fortasse non injucundum videbitur. Caeterum a Dn Hugenio (egregii ex Tab. Sinuum compendii nobis spem faciente) observatum est, rem etiam reduci ad summam secantium arcuum, per minima aequaliter crescentium. Idem et a me notatum fuerat, et cum in mentem venisset, ab iisdem pendere et linea rhombicae seu loxodromicae determinationem in usum nautarum, quam jam multis abhinc annis etiam ex logarithmis definiisse recordabar, excussi veteres schedas et in proxime praecedente Aprili p. 181 Actorum Erud. Hujus anni rem tandem publice exhibui.

Leibniz, Gottfried Wilhelm. 1684. Nova methodus pro maximis et minimis, itemque tangentibus, quae nec fractas nec irrationales quantitates moratur et singulare pro illis calculi genus. *Acta Eruditorum* (octobre), 467–473. Repr. in Gerhardt, *Mathematische Schriften*, vol. V, 220–226.

Leibniz, Gottfried Wilhelm. 1686. De geometria recondite et Analysi Indivisibilium atque infinitorum, Addenda his quae dicta sunt a. 1684, Maji, 233; Octob, 264; Decemb, 586. *Acta Eruditorum* (Junii), 292–300.

Leibniz, Gottfried Wilhelm. 1687, Réponse de M. Leibniz à la remarque de M. l'abbé Catelan contenue dans les Nouvelles. ... Mois de Juin 1687 où il prétend soutenir une Loi de la Nature avancée par Descartes. *Nouvelles de la république des lettres de septembre 1687*, 952–956. Repr. Gerhardt, *Leibnizens Philosophische Schriften*, III, 49–51.

Leibniz, Gottfried Wilhelm. 1689. De linea isochrona in qua grave sine acceleratione descendit et de controversia cum Dn. Abbate D.C. *Acta Eruditorum* (Avril), 195–198. Repr. Gerhardt, *Leibnizens mathematische Schriften*, vol. V, 234–237.

Leibniz, Gottfried Wilhelm. 1691a. De linea in quam flexile se pondere proprio curvat, ejusque usu insigni ad inveniendas quotcunque medias proportionales et logarithmes. *Acta eruditorum* (June), 277–281; Gerhardt, *Leibnizens Mathematische Schriften*, vol. V, 243–247.

Leibniz, Gottfried Wilhelm. 1691b. De solutionibus problematis catenarii vel funicularis in actis junii an. 1691, aliisque a Dn. Jac. Bernoullio propositis. *Acta Eruditorum* (September), 435–439; Gehrard, *Mathematische Schriften*, vol. V, 255–258.

Leibniz, Gottfried Wilhelm. [unsigned]. 1699. Animadversio ad Davidis Gregorii schediasma de catenaria, quod habetur in Actis Eruditorum an. 1698. *Acta Eruditorum* (Februar), 87–89; Gerhardt, *Mathematische Schriften*, vol. V, 336–339.

Leibniz, Gottfried Wilhelm. 1846. *Historia et Origo Calculi differentialis* (1713–1714). Repr. Gerhardt, ed. Hannover: Hahnschen Hofbuchhandlung.

Pardies, Ignace-Gaston. 1673. *La statique ou la science des forces mouvantes*. Paris: Mabre-Cramoisy.

Poleni, Giovani. 1748. *Memorie istoriche della gran cupola del tempio Vaticano: e de' danni di essa, e de' ristoramenti loro*. Padova: Stampa del Seminario.

Stirling, James. 1717. *Lineae tertii ordinis Neutonianae, sive, Illustratio tractatus D. Neutoni De enumeratione linearum tertii ordinis: cui subjungitur, solutio trium problematum. Cui Subjungitur Solutio Trium Problematum*: Oxford: Theatrum Sheldonianum and Edward Whistler.

Taylor, Brook. 1717. *Methodus incrementorum directa et inversa*, 2nd ed. London: Gugliemi Innys.

Truesdell, Clifford Ambrose. 1960. *The Rational Mechanics of Elastic Bodies, 1638–1788. Introduction to Leonardi Euleri Opera Omnia*, series secunda, vols. X et XI. Zurich: Orell Füssli.

Chapter 7
The Bernoullis and the Elastica: From Leibniz's *Nova Methodus* to Euler's *Methodus Inveniendi*

1 Introduction: The Various Difficulties Linked to the Problem of the Elastica

We can easily understand Jacob Bernoulli's interest for the elastica just enouncing the number of difficulties linked to its study. The problem is physically complicated, it is mathematically complicated and for us it is historically complicated.

(A) That problem is **physically** complicated because the preferred place to observe the elastica namely the embedded beam, gives rise to various physical phenomena that have to be distinguished from each other. One ought to distinguish the form of the curvature adopted by the beam under the action of certain forces from the different types of deformation of various materials as well as the various circumstances of rupture of the beam. It is in fact easier to break a beam by acting perpendicularly to its length then by pressing a column, a beam, a blade along its length, etc.

By tackling this problem, we understand, as Truesdell has always emphasized, that it is necessary to distinguish universal laws of physics from constitutive laws that vary from one material to another. Even without going into plastic deformations, we shall see that Jacob Bernoulli will inexorably be drawn to this type of distinction.

The study of these different phenomena and of the curvature of the beam in particular calls for various notions that should be noted in full and distinguished with precision. In order to best represent that difficulty, I cannot do better than citing Benvenuto's funny presentation: notice that mass of the beam plays an important role, but its color is already a difficult thing.

Part of these notions are of the most complex and the less intuitive of classical physics. Let us limit ourselves mentioning the principal one, namely tension, of which Benvenuto said that it is the only one that one never can measure directly.

(B) This problem is **mathematically** complicated partly because it is physically complicated, but principally because it is difficult to idealize. This difficulty of

P. Radelet-de Grave, *Between Mechanics and Architecture*, Mathematics and the Built Environment 8, https://doi.org/10.1007/978-3-031-73530-1_7

idealization is not only due to a lack of imagination but also to their mathematical reflection. I think, for example, of the edge conditions, to which Huygens, Jacob and Euler will be sensitive.

To transcribe into mathematical language the constitutive laws that take into account the physical properties of the different matters is also complex.

The differential equations resulting from this study are not simple especially with the knowledge that was at disposal at the time. Even if they are beautiful, the elliptic integrals are not of the simplest.

(C) That problem is **historically** complex, because it is situated at the birth of differential and integral calculus. The auteurs are thus applying a new language and they are making their first steps on a complicated mathematical problem.

They surround their calculations with a certain mystery which makes the reading of their texts difficult. To all that has to be added the context of the quarrel between the brothers Bernoulli. They both refuse to publish the justifications of their affirmations and just leave us as only resources the *Lectiones* that Johann gave to Guillaume de l'Hôpital, Jacob's *Meditationes*, and a few letters.

2 Presentation of the Various Curves Studied in This Chapter

The aim of what follows is to help the reader to find his way through that world of curves studied by the first mathematicians developing the calculus. The curves are given in alphabetical order.

Brachistochrone: curve of fastest descent under the action of gravity.

Given in a vertical plane two points A and B determine the way AMB, along which under the action of his gravity, it would, starting from A, arrive in the least time at a different point B.

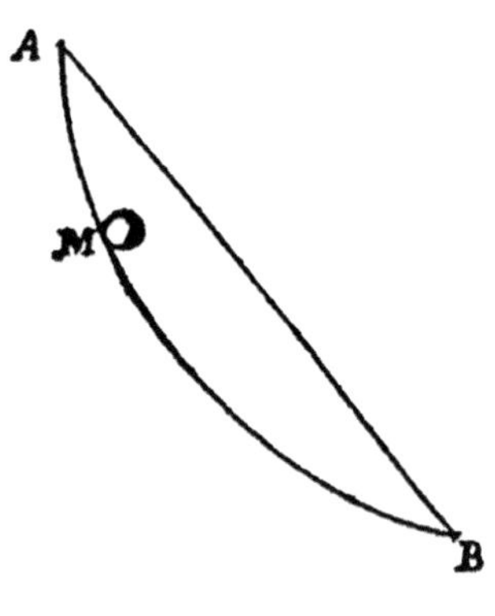

Fig. 1 The brachistochrone (Bernoulli 1696, Tab. V, Fig. 5). Public domain

Catenary: The catenary is the form of a perfectly flexible thread under the action of gravity. When turned upside down, it is the most stable arch (Fig. 2).

Elastica is the curve taken by a beam supported at one end under the action of a weight placed at its free end (Fig. 3).

Isochrone is the curve formed when a body, under the action of gravity, runs through equal vertical spaces in equal times (Fig. 4).

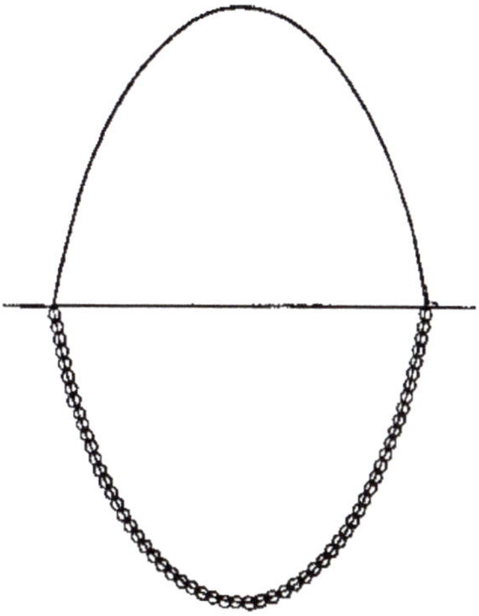

Fig. 2 The catenary. Author

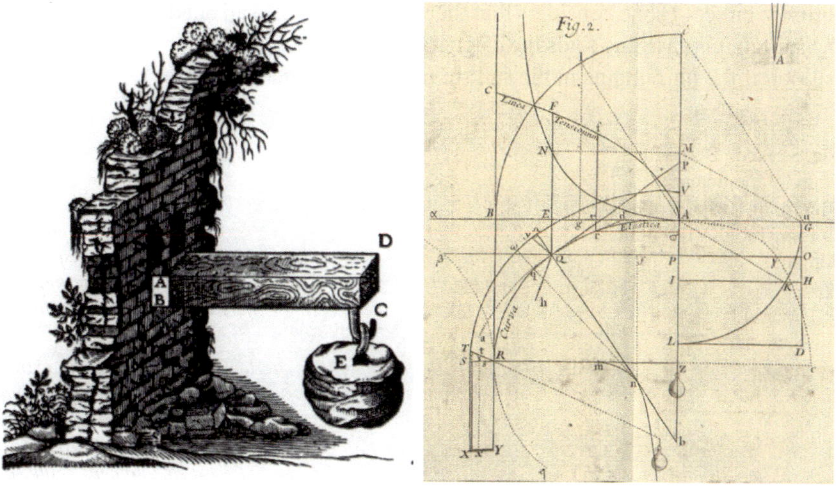

Fig. 3 **Left** Galileo investigates the rupture of a beam nei *Discorsi* (1637, p. 114); **right** and Jacob Bernoulli looks for the equation of the curve when bent by a weight, in Op. LVIII (1694; *Opera*, 1744, Tab. XXIII). Public domain

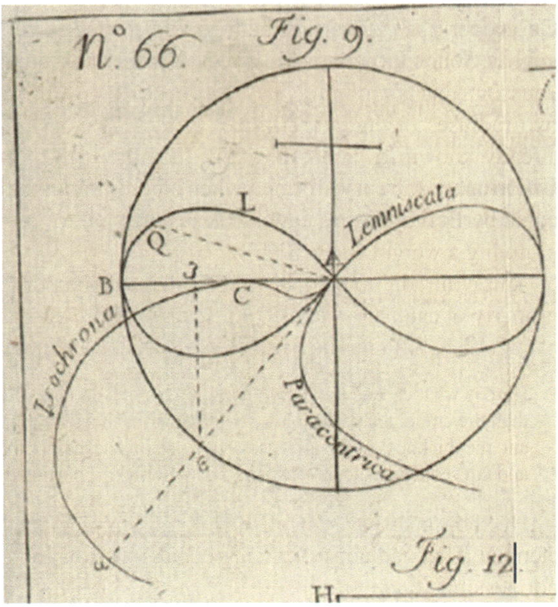

Fig. 4 The isochrone of Jacob Bernoulli (Op. LXVI, *Opera*, 1744, Tab. XXVIII). Public domain

Lintearia is the curve formed by a linen cloth full of liquid. It is identical with the elastica as Jacob Bernoulli, shows in his famous *Curvatura Laminae Elasticae* (Bernoulli 1694a, b).

Velaria is a sail pushed by the wind (Fig. 6). The idealization made by Jacob is as follows: either (1) one considers that the wind inflates the sail but does not stay inside the sail; in this case one finds the catenary; or (2) one considers the sail as an inflated balloon if the air remains in the sail; in this case one gets a sphere.

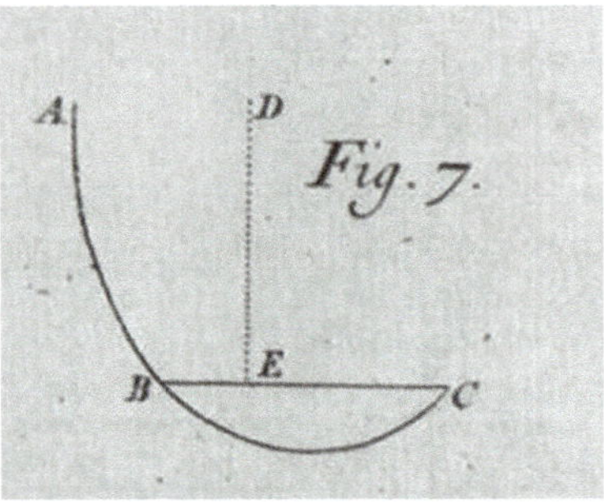

Fig. 5 Jacob Bernoulli's figure of the lintearia (Bernoulli 1694a, b (*Opera*, 1744, I, Tab. XVI). Public domain

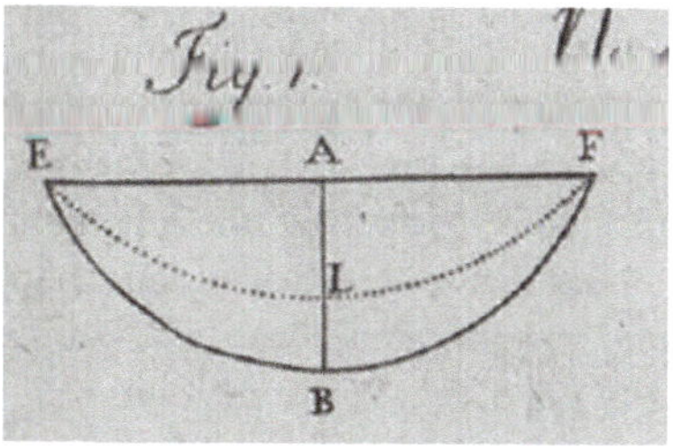

Fig. 6 The velaria (Bernoulli 1691a, b (*Opera*, 1744, I, Tab. XVIII). Public domain

3 First Period: From *Nova Methodus* by Leibniz in 1684 to Jacob Bernoulli's Elastica in December 1695

3.1 Background

Leibniz' *Nova methodus* of 1684 gexplains the fundaments of differential Calculus in a short text of five pages. The brothers Jacob and Johann Bernoulli read Leibniz article two years later and study it together. Jacob begins on 15 December 1687 a correspondence with Leibniz. At the end of 1691 Johann Bernoulli arrives in Paris where he will give lessons on the calculus to the Marquis de l'Hôpital. At the end of 1692 he comes back to Basel. At the beginning, 1693 until half of 1695 they are working together but only Jacob publishes. Johann presents works and thesis at the university.

3.2 Jacob Bernoulli's *Theorema Aureum* and Differential Calculus

The term *theorema aureum* selected by Jacob to qualify his theorem on the **radius of curvature** shows his proudness but also the importance he rightly gave to it. Nevertheless, he had found it and used it before baptizing it that way. This term attracted and still attracts the attention. Of course, it is but « golded» under the pen of Jacob's younger brother Johann. That theorem is an element of the profound quarrel that opposed both brothers.

But let us place ourselves in the context of the mystery surrounding the research which followed the publication of the *Nova Methodus* by Leibniz, in order to see some interesting facts about the radius of curvature and the use of this calculation.

Differential calculus in its Leibnizian form appears in Jacob Bernoulli's Med. CLXI,[1] *De Spiralis Parabolicae (Parabolae helicoidis) dimensione,* written between

[1] Jacob Bernoulli, *Meditationes, Annotationes, Animadversiones Theologicae et Philosophicae, à me J. B. concinnatae et collectae ab anno 1677* is the title given by Jacob to the journal, in which he wrote the copies or preparations of his articles. He has kept it from his first studies on theology until his last days. All of Jacob's works mentioned in this chapter are published in the *Opera* (1744)

1689 and 1691. One can often observe a striking difference, concerning the use of differential calculus, between the *Meditationes* of Jacob and his published articles as well as between the *Lectiones* that Johann gave to the Marquis de l'Hôpital and his published texts at the same period.

But that ascertainment must be refined because in Op. XXXIX, *Jacobi Bernoulli analysis problematis antehac propositi, De inventione Lineae descensus a corpore gravi percurrendae uniformiter, sic ut temporibus aequalibus aequales altitudinens emetiatur: & alterius cujusdam problematis propositio* of 1690, Jacob studies the vertical fall in equal times that had been proposed by Huygens, using the calculus. This means there was not a great distance in time between these two appearances, public and private.

On the contrary, if one considers the problem of the catenary, which he poses in that same text, saying that it is similar to the preceding one, one notes that it will be treated by Johann, at the same period, publicly without using the calculus, and using it in the lessons given to de l'Hôpital. The same occurs a bit later with Jacob. He treats the *Velaria,* first using the calculus privately in the *Meditationes,* while publishing a purely geometrical version. One observes the same shift for a long time, for instance with Jacob, in his Op. LXXVIII, *Demonstratio Synthetica Problematis de Infinitis Cycloidibus absque adminiculo infinite parvorum; item Constructio aliorum huic affinium* of 1698, on the brachistochrone.

With regard to the radius of curvature, one finds in Med. CXIX, *In hyperbola ABC, cujus Asymptotae DE, DF ducere perpendicularem GH per datum in illa punctum C, inque hac perpendiculari reperire punctum H, super quo descriptus circulus radio HC, secet Hyperbolam,* written between June 1686 and June 1688, different constructions of the tangent circle to a hyperbola. One of them is done in the framework of the ancient theory of Maxima and Minima due to Fermat. Another solution is unique to Jacob.

The formula for radius of curvature in the form $-\frac{dxds}{ddy}$ or $\frac{-dyds}{ddx}$ appears at the same time as differential calculus in Med. CLXI, *De spiralis parabolicae Dimensione.* We notice, and Jacob confirms it in his Op. XLII, "*Specimen alterum Calculi differentialis in dimetienda Spirali Logarithmica, Loxodromiis Nautarum, & Areis Triangulorum Sphaericorum...* of 1691, that he found the radius of curvature while he was carrying out his research about spirals. This theorem, which he will nevertheless associate with the elastica, since it is in his first text on the elastica that he will publish it for the first time, is particularly important for different reasons. It is one of the first, if not the first, to be expressed in a differential form. This fact that would not be that important, if it did not allow him to translate all the theorems established in the ancient (or vulgar, as he will qualify it) geometry into differential geometry. In fact, most of the theorems made use of the evolute of the curve, that is, the envelope of the radii of curvature. Giving the expression of the radius of curvature, which we gave above and which depends only on the elements of the curve itself, Jacob is able to translate into that language all the theorems in which the evolute is involved.

---

and are available at: https://www.e-rara.ch/zut/content/structure/1039822. Jacob's *Meditationes* are available on an interactive platform: https://beol.dasch.swiss/.

Despite this, in his Op. XLI, *Specimen Calculi differentialis in dimensione Parabolae helicoidis, ubi de flexuris curvarum in genere, earundem evolutionibus, aliisque* of 1691, in which Jacob studies the radius of curvature and its value at the points of *flexus contrariis* (points of inflexion), that theorem is not found in that form. That problem will trouble the Marquis de l'Hôpital after the publication of the theorem of the radius of curvature. On the other hand, this text calls on the differential calculus.

And so does Jacob's Op. XLVII, *Additamentum ad Solutionem Curvae Causticae fratris Jo. Bernoulli, una cum Meditatione de Natura Evolutarum, & variis osculationum generibus* of 1692, where he returns to the point of inflexion and describes the evolute as the locus of the centers of curvature of the curve. Let us recall that it is in this text that he shows that the catacaustic of the logarithmic spiral with respect to the radius coming out of its pole reduces back to the same spiral. It is for this reason that in his Op. XLIX, *Lineae Cycloidales, Evolutae, Ant-Evolutae, Causticae, Anti-Causticae, Peri-Causticae. Earum usus & simplex relatio ad se invicem. Spira mirabilis. Aliaque* of 1692 he asks that his tomb stone be adorned with a logarithmic spiral and the words EODEM MUTATA RESURGO. One finds in this text a theorem on the osculating radius that makes it possible to connect the caustics to the evolutes. It is stated in terms of the theory of proportions.

In his lessons to the Marquis de l'Hôpital, Johann speaks also of the osculating circle and gives it a good place with respect to setting into equation of the different problems that will concern us. He also gives it the same differential form as Jacob, but without using the lineaer element, neither here nor later. This linear element appears in Jacob's in Med. CLXXXI, *Invenire Evolutam Logarithmicae*, written between June 1691 and May 1692. He thus has the expression. But—and this restraint is important—he calculates it independently for different curves (parabola, hyperbola, etc.) and while he correctly observes the permanence of the expression, it does not give it the generality which Jacob gives it.

Jacob gives his differential expression of the radius of curvature the name *theorema aureum* for the first time in Med. CXCII, *Demonstratio Theorematis a Fratre animadversi (vid. Lit. ejus de 15 Martii 1692) Caustica Cycloidis vulgaris nata ex reflexis radiis parallelorum axi itidem est vulgaris cyclois, cuius basis est prioris dimidia.*

In the texts that he publishes later, Op. L, *Additio ad Schedam de Lineis Cycloidalibus* of 1692, and Op. LVI, *Curvae Dia-Causticae, earum relatio ad Evolutas* of 1693, Jacob will first show the importance of the radius of curvature in the old language of geometry. In order to do that, he perpetually refers back to his theorem of Op. XLIX, *Lineae Cycloidales, Evolutae*… of 1692, which links the caustic and the evolute thanks to the radius of curvature. This is the end of a genre, but it also shows us the link to a great number of problems, such as the elastica and the velaria which we shall also meet below.

While in his *Meditationes* he prepares the translation into differential language, as we can see it in Med. CCII, *Formulae quaedam pro constructionis Aequationibus differentialibus absque quadraturis curvarum* and in Med. CCIIa, *vid. supra pag. 221. ad signum* …. and, in his famous text on the elastica, Op. LVIII (Bernoulli 1692), he finally decides to publish his *theorem aureum* then to use it to translate from one style to another. He does it in Op. LX, *Constructio Curvae Accessus & Recessus aequabilis, ope rectificationis Curvae cujusdam Algebraicae* of 1694, where we can read

> There are three principal manners to construct mechanical or transcendental curves. The first that is not practical, is done using the quadrature of curvilinear spaces. But it is better to use the rectification of algebraical curves; one can in fact rectify those curves more rapidly and exactly with help of a fil or of a catenary that one applies along the curve, than squaring the space.[2]

This is the reason why his theorem is so important is because it makes it possible to translate instantly all the theorems that had been acquired by the ancient method.

As Johann shows in his Op. XIX, *Constructio facilis curvae recessus aequabilis a puncto dato, per rectificationem curvae algebraïcae* of 1694, in which he announces that he will not follow his brother's method consisting of rectifying the mechanical curves. And that he will not need the radius of the osculating circle. Johann did not measure the importance of the *theorema aureum*, or the reason why it is qualified as golden. In contrast, Gottfried Wilhelm Leibniz in his *Constructio propria Problematis de Curva Isochrona Paracentrica* of 1694, underlines the use one can make of the radius of the osculating circle on the first page of that text, where he comments on the results of Jacob Bernoulli before giving his own solution of the isochrone.

3.3 Various Problems Posed by Jacob and Johann Bernoulli

Two problems had been mentioned by Leibniz himself in the *Nova Methodus*: the problem of Fermat that gives the Snell law when applied to a ray of light passing from one medium to another; and the problem posed by De Beaune to Descartes. Later, Leibniz will also pose the problem of the isochrone. Then Jacob will take over by posing the problem of the catenary in his Op. XXXIX, already mentioned:

> To find the curve adopted by a thread hanging freely suspended between to two points. I also assume that the thread is a line perfectly flexible in every of its parts.[3]

Johann answers rapidly in Op. IV, *Solutio Problematis Funicularii* of 1691, with a geometrical construction that apparently does not use the calculus. But we know by the lessons to l'Hôpital that he had such a solution at the time. He then studies the string that is inhomogeneous or non-uniformly loaded. Finally, he mentions the

---

[2] Bernoulli (1694a, b), Op. LX (*Opera*, 1744, p. 608): *Triplex praecipua modus habetur construendi curvas mechanicas, sive transcendentes. Primus, sed a praxis parum idoneus, fit per curvaturas spatiorum curvilineorum. Melior est, qui instituitur per rectifications curvarum algebricarum; accuratius enim et expeditius rectificari possunt curvae, ope fili vel catenulae ipsis circumplicatae, quam quadrari spatia.*

[3] Jacob Bernoulli, *Invenire quam curvam referat funis laxus inter duo puncta fixa libere suspensus. Sumo autem, funem esse lineam in omnibus suis partibus facillime flexilem.*

inverse problem: given the curve that one wants to obtain, find the repartition of the loads. Jacob goes further in Op. XLII, *Specimen alterum Calculi differentialis in dimetienda Spirali Logarithmica, Loxodromiis Nautarum, & Areis Triangulorum Sphaericorum: una cum Additamento quodam ad Problema Funicularium, aliisque* of 1691. He studies the string of non-uniform mass, and of uniform mass but extensible. He then poses the problem of the elastica and of the velaria starting from the catenary (the funicular). Now we arrive to the problems we are interested in. The lintearia is occasionally mentioned, it never was the object of a published question.

Johann gives a geometrical solution of the velaria in Op. VII, *Solution du Probléme de la Courbure que fait une voile enflée par le vent*[4] in 1692. He joins to his solution another one which he qualifies as mechanical and in which he applies the law of the lever. During that time Jacob questions himself about the direction of the forces acting on the *velaria* in his *Meditationes* CLXIV, *De aestimandis liquorum pressionibus in folle figurae cujuscunque*, written between 1689 and 1692. Then he searches for a solution of the same problem, using differential calculus Med. CLXV, *De figura veli vento inflati*, in which is found the radius of curvature in the form already mentioned. However, he only publishes a series of geometrical results in Op. XLVIII, *De Curvatura Veli*, in 1692. The radius of curvature does not appear in that text. Jacob only publishes his geometrical results in a series of affirmations, exactly as Johann did it in Op. VII or in Op. IV for the catenary. Nevertheless, for his brother, he gives the characterization of the velaria by means of the Leibnizian calculus, in the following way: "Taking equal small portions of the curve, the cubes of the first ordinate differences are proportional to the second differences of the abscissas",[5] that is, $(dx)^3 = ad^2y$.

At the end of this text, one finds an interesting remark: "For as a linen sheet swelled by the wind is to a funicular, so the same stretched out by the weight of the liquid pressing down is to the elastic curve".[6] It seems that Jacob makes here a difference based on the direction of the force, or of what he calls pressure. This distinction is also promising in practical terms, and we shall see the fruits later on. Unfortunately, it is a little tainted in Jacob's work by the fact that he was not very sure of his idealization of the sail, as shown by Med. CLXXXVIII, *De linea directionis, juxta quam velum impellitur, de impulsus vi & de veli robore*, written between January and June 1692.

I think that Jacob already sees things in this way, even he does not formulate it exactly like that. One can see that in a series of *Meditationes* in which he studies the geometrical properties due to the fact that the forces are perpendicular to the lintearia. It is in Med. CCXXVIII, *Curvae Elasticae puncta infinita & quantumvis proxima possunt geometrice inveniri, & quidem solis circulis & lineis rectis,* written between September 1694 and December 1695. There are errors in that *Meditatio* but he starts

---

[4] Johann Bernoulli, Op. VII, *Solution du Probléme de la Courbure que fait une voile enflée par le vent,* 1692.

[5] Jac. Bernoulli, Op. XLVIII, *Opera*, 1744, p. 482: *[Sumptis aequalibus Curvae portiunculis, Cubi ex primis differentiis ordinatarum sunt proportionales secundis differentiis abscissarum].*

[6] Jac. Bernoulli, Op. XLVIII, *Opera*, 1744, p. 490: *Ut enim linteum vento tumidum funiculariae, sic idem ab incumbentis liquoris pondere expansum* flexi Elateris *curvaturam induit.*

from an important, true point: all forces acting on the lintearia are perpendicular to the curve. This is not true in the case of the elastica. For that reason, he searches for one more property for the elastica. After various errors, in Med. CCXXXI, *Lineam mediarum directionum invenire*, written in the same period, he seems satisfied. His aim is to find the mean direction (we would say the resultant) in the case of forces perpendicular to the curve. We must see Jacob's research in the context of the knowledge of the parallelogram law of forces.[7] That Jacob was not familiar with this law is clear in these *Meditationes* as well as from his discussions with Leibniz. One sees it also in other *Meditationes,* where he systematically translates the results of a decomposition of forces into the language of Varignon (law of sines) or into that of Leibniz on the resulting motion of the centre of gravity.

We shall return to such remarks, but let us recall for the moment that Jacob classifies the problems of the various curves with respect to the directions of the acting forces, as shown by the short sentence at the end of Op. XLVIII on the lintearia on the one hand, and on the other by the fact that he tries to place the elastica in that classification. And on the other hand, that the parallelogram law of forces is not acquired.

Let us go back to the lessons given by Johann to the Marquis de l'Hôpital, Op. CXLIX, *Lectiones Mathematicae, de Methodo Integralium, aliisque, conscriptae in susm Ill. Marchionis Hospitalii, cum Auctor Parisiis ageret Annis 1691 & 1692.* The text of those *Lectures* will only appear in Johann's *Opera Omnia* in 1742, and while they will exert no influence, they are nevertheless useful for measuring Johann's knowledge of the calculus in 1691–92, a knowledge that he tries to conceal.

We have already mentioned what he said about the radius of the osculating circle. He next treats the various problems that we just considered, beginning chronologically, I would say, with the vertical fall during equal times, then the isochrone, the homogeneous, inhomogeneous, and extensible catenary, the catenary under the pressure of a fluid, the elastica and the lintearia. All those solutions use differential calculus.

Let us recall that he idealizes very precisely those different problems. It is thus reasonable to think that if he carefully avoids the elastica, it is because he has difficulties idealizing and thus mathematicising it. The only page dedicated to it is part of the *Anekdota* and is thus not dated, Op. CLXXIV, *Solutio Problematis curvaturae laminae elasticae a pondere appenso curvatae.* He prefers to insist on the lintearia, whose idealisation is easier, and he will always say of the equation of the lintearia that is it the equation that his brother also attributes to his elastica. Sadly, this simplification makes it necessary for him to group all those problems in order to generalize with the hope of finding the most general elastica. In my opinion, that means avoiding the motivation that will push Jacob to pose the problem of the isoperimeters.

[7] See "The History of the Parallelogram Law of Forces from Antiquity to Charles Bossut", Chap. 1 in this present volume.

3.4 The Public Treatment of the *Elastica*

Jacob Bernoulli's text on the elastica is one of his most important publications, if not the most important and that for various reasons.[8] I shall restrict myself to those related to architecture.

Jacob Bernoulli Op. LVIII, *Curvatura Laminae Elasticae. Ejus Identitas cum Curvatura Lintei a pondere inclusi fluidi expansi:*

> The problem here is more difficult than that of the funicular [catenary], and not without reason. For, to say nothing of other things, it should be noted that there are two keys at hand in the investigation of the catenary that pave the way to two different equations: one which expresses the nature of the curve by the relation of itself to its coordinates; the other by the relation of the developing wire to them. But only the latter key affords access to the investigation of the nature of the elastic curve. For from this it clearly follows that it may well be possible for anyone to overcome the difficulties of the greater problem, and yet not immediately escape the other; destitute of the other key, which is the said relation of the evolving wire or the osculating radius of the circle (not somehow, as done in the *Acta Eruditorum* of January 1691...) but in very simple and purely differential terms. These things had already been made known to us at the time when we were busy speculating on the string, and my brother soon shared them with some of the researchers; moreover, while the immense usefulness of the discoveries in analyzing the Velaria, and this, which we have before me, the elastic curvature, and other more sublime ones, daily more and more revealed itself to me, and made it so that I could no longer conceal from the public the *theorema aureum*...[9]

He finds:

(a) if ds is constant, for the radius of the osculatory circle, $z = \frac{dxds}{ddy}$, or $\frac{dyds}{ddx}$, because $\frac{ho}{bc} = \frac{ho}{hc} + \frac{hc}{bc}$.

(b) if dx is constant, the radius of the osculating circle is $z = \frac{ds^3}{dxddy}$ and the equivalent expression if one takes dy as constant (Fig. 7).

After that, he adds something that even his brother does not know: a theorem for curves whose *applicatae* are not parallel but converge in one point, a as for a spiral, x is the length taken on the curve, and y the length from the origin a to the point.

---

[8] There is a rather long purely mathematical preparation to that treatment in Jacob's *Meditationes*.

[9] Bernoulli (1694a, b), Op. LVIII (*Opera*, 1744, pp. 577–578): *Problema hocce plus difficultatis habere* Funiculario, *nec sine ratione. Ut enim alia taceam, notandum, duas praesto esse claves in* Catenarii *investigatione, quae ad duas differentes aequationes viam sternunt, quarum una naturam curvae per relationem psiusmet ad coordinatas ejus, altera per relationem fili evolventis ad easdem exprimit; cum ad curvae Elasticae naturam indagangam posterior tantum clave aditus pateat. Hinc enim manifesto sequitur, fieri bene posse, ut quispiam rioris Problematis difficultates superet, nec tamen protinus alterius quoque victor evadat; destitutus nimirum clavium altera, quae dictam relationem fili evolventis seu radii circuli osculatoris [non quomodocunque, ut factum in Actis Lipsiensibus Mensis Januarii 1691 ** pag. 22,] sed in terminis simplicissimis et pure differentialibus exhibeat. Haec nobis jam eo tempore, quo speculation funis vacabamus, innotuerat, eamque mox nonnullis inter peregrinandum communicavit Frater; praeterea dum immense inventi utilitas in resolvenda* Velaria, *et hac, quam prae minibus habemus,* Elaterum Curvatura, *aliisque sublimioribus, quotidie magis magisque sese mihi explicuit, effecitque, ut non possim publico aureum Theorema diutius invidere. ...*

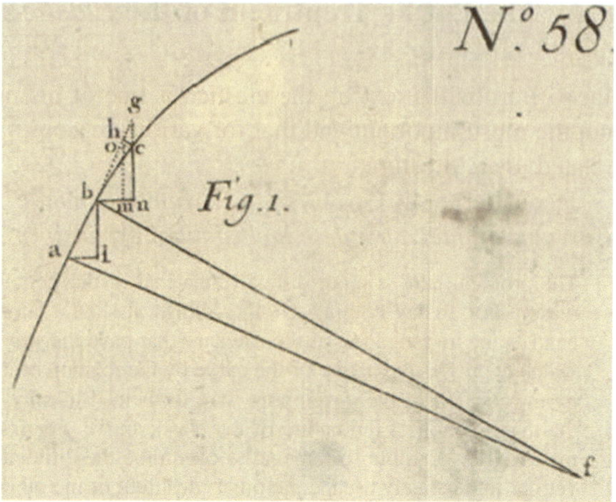

Fig. 7 Jacob's Fig. 1, in Bernoulli (1694a, b) (*Opera*, 1744, Tab. XXIII). Public domain

After that, the article continues in three parts: I. *Generaliter*; II. *Specialius*; III. *Specialissime.*

The solution *Generaliter* is exactly in the purely geometrical style Johann uses for the catenary or the velaria. The presentation of that construction contains two parts: the construction itself; and the description of the physical problem that it characterizes, a vertically fixed blade to which a weight has been applied such that it bends until this weight is vertical.

The two keys he is speaking of are on one hand the law of the equilibrium of the lever, a general law of mechanics, and on the other hand, the law of tension, a constitutive law. Those elements will be clearer in Op. LXVI (Bernoulli 1695). He takes a law of tension and the radius of curvature in a passage which involves the evolute. We shall see in Daniel Bernoulli and Euler how the radius of curvature intervenes in our current vision of the problem.

Sixteen geometrical properties come next. Here are some examples[10]:

1) If one fixes the beam in another point, all other things remaining the same, the curve remains the same. This is of course an important note for architecture.

2) If on the contrary, one extends the curve symmetrically and fixes it further on that curve, it stays in equilibrium in the same conditions.

5) If one applies an oblique force instead of the weight, one obtains a lightly different curve.

7) Given a law of tension determine the elastic curve. In abstract geometry it is nothing else that given AFC, find AQR where the radii of the osculating circles are inversely proportional to the respective applied of the other. The inverse is equally simple: An elastic curve being given together with the radius of the osculatory circles, determine the law of tension Nature observes in order to produce it.

[10] Jac. Bernoulli, Op. LVIII, 1694 (*Opera*, 1744, pp. 582–583).

In the second part, *Specialius*, he takes a “paraboloid” (he means a conic curve) as the curve of tension. He thus imposes a law normally given by the nature, thus a constitutive equation between the solicitation and the material. He gives the geometrical results of his study, as before, and finishes with particular cases:

1° $m = 0$, the answer of the material is always the same. The conic becomes a line parallel to the axis AB and the elastica becomes a circle.
2° $m = -\frac{1}{2}$, the conic is a hyperbola then the curve becomes rectifiable but this case where the tension decreases when the acting force increases does not correspond neither to the laws of nature nor to the experiment.
3° the case $m = 1$, i.e., the linear law, the parabola becomes a straight line AC.[11]

This is the classic case that he studies in the third, *Specialissime* part. The linear law was used by Leibniz and himself until "he arrives at the general construction of the problem" we understand until he discovers his theorem. But he points out that nature does not act in this linear way, certainly not always. The different materials react differently as experience shows us and as Leibniz has pointed out. He gives the equation[12]:

$$dy = \frac{xxdx}{\sqrt{a^4 + x^4}}.$$

He then gives once more the geometrical properties and the identity of the lintearia and the elastica with a justification. He also remarks that the tension in the tarpaulin forming a lintearia is the same in every point.

Let us underline some other properties. Among all the isoperimeters which rely on Af, the elastica is the one whose center of gravity is furthest from Af while the catenary is the one whose center of gravity is furthest from the curve itself. This is a property to which we will come back. He finishes with a series of problems linked to the elastica: to find the curve of the blade under its own weight; to find its form if it is not of equal thickness or of constant width, as well as the inverse problem; find the force which must be applied to a curved blade in order to straighten it; how to obtain this or that curvature.

Christiaan Huygens reacts to Bernoulli's work in the same issue of the *Acta Eruditorum*, and makes two important remarks. He points out, on the one hand, that it is unlikely that only the external face of the beam is put in tension but rather that there will be tension on a part then compression on the internal face. This is the birth of the important notion of neutral fiber. It remains to be understood the progressive compression on the one side of that fiber and the progressive extension on the other side. He also shows a good series of elastic curves which he found by observation and asks to find them by the theory of Bernoulli (Fig. 8).

Jacob will thus go ahead with that study and publish his second important work on the elastica, Op. LXVI, *Explicationes, Annotationes, et Additiones ad ea, quae*

[11] Bernoulli (1694a, b), Op. LVIII (*Opera*, 1744, pp. 588–589).
[12] Bernoulli (1694), Op. LVIII (*Opera*, 1744, p. 592).

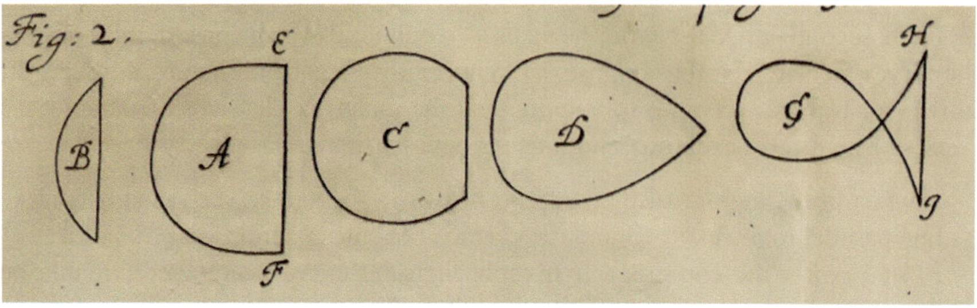

Fig. 8 Examples of elastic curves shown by Huygens for Jacob's elastica (Huygens 1694, Tab. X). Public domain

in Actis sup. anni de Curva Elastica, Isochrona Paracentrica, & Velaria (Bernoulli 1695).

He puts one fiber in equation thanks to the principle that the forces that tends the fiber SY are proportional to QP the arm of the lever for weight P that means inversely proportional to the radiuses of the osculator circles. This is a hypothesis that Huygens and Leibniz as well admit and that Jacob already used in Med. CCV. This time he gives the equation found thanks to the *Theorema Aureum.*

He starts from the physical hypothesis $\frac{1}{2}aa = xz$, which means

$$\frac{1}{2}aa = \frac{xdxds}{ddy} \text{ or } aaddy = 2xdxds.$$

He integrates and takes its square $a^4dy^2 = x^4ds^2$.

Then subtracting x^4dys^2 to the right and to the left, one obtains:

$$\left(a^4 - x^4\right)dy^2 = x^4ds^2 - x^4dy^2 = x^4dx^2.$$

He thus starts his demonstration again in order to take into account Huygens' remark concerning the extension and the contraction of the fibers. To do this, he considers the case where the compressed thickness is equal to the extended thickness, thus putting what will be called the neutral fiber in the middle and bringing the problem back to the case of Op. LVIII (Bernoulli 1694a, b).

He answers then Huygens' question concerning the various forms of the elastica, and for that reason considers the edges conditions. He gives himself the forces acting perpendicularly to the tangents at the extremities of the curve. We shall see more about that in Euler's text studied below. He finds in that a more general equation that coincides with the one of the lintearia.

The end of the text is dedicated to various problems, such as the velaria, which is identical with the catenary if the liquid traverses the tarpaulin, and thus it depends on the orientation of the forces of pressure.

4 Second Period: From the Public Declaration of War in 1696 to the Last Publication of Johann Concerning the Isoperimetric Problems 1718

4.1 Background

A promise of a contract between Johann and de l'Hôpital stipulated that in return for a substantial lifetime stipend, the scientific content of all Johann's present and future scientific discoveries would belong to de Hôpital. In any case, this was how de Hôpital understood it, though Johann only discovers this gradually. The contract will not start immediately, but would wait until an important inheritance for de Hôpital arrived. Johann writes concerning the contract: "He even honored me with a pension of 400 pounds, which I enjoyed until his death".[13]

This contract, which Jacob did not discover until around 1694, is the real cause of the quarrel between the two brothers because, given their collaboration, the contract implicated that Jacob's findings would also belong to de Hôpital. The article published by Johann in the *Acta Eruditorum* of 1695, Op. LXVI, *Iacobi Bernoulli explicationes, annotationes et additiones ad ea quae in actis superiorum annorum de curva elastica*, makes their dispute public.

4.2 The Problems Proposed by Johann Bernoulli

Many historians think that the publication by Johann Bernoulli of his article Op. XXX, *Problema novum ad cujus solutionem Mathematici invitantur*, was the origin of the quarrel between the Bernoulli brothers but we have seen it initiated long before and for different reasons.

This article, which was published on the *Acta Eruditorum*, was also published as a poster that was plastered around the city of Groningen where Johann was lecturing at the time (Fig. 9). Johann poses the problem of the brachistochrone, challenging the scientific community as well as his brother. Even if this is not the origin of the quarrel, this publication at least marks a turning point in it, and will give rise to many publications, not all interesting, from both brothers. Some are purely polemical and contain more allusions and anagrams than scientific literature.[14]

The first problem stated by Johann concerns the brachistochrone, or line of quickest descent. It was published in the *Acta Eruditorum* of June 1696. Johann's problem was:

---

[13] Quoted in Spiess (1955, p. 156): *Ja sogar mich mit einer jährlichen pension von 400 livres beehret, welche ich schier bis zu seinem Tod genossen.*

[14] The interested reader will find them all in *Die Streitschriften von Jacob und Johann Bernoulli. Variationsrechnung* (Goldstine 1991). The texts are in their original language, the Introduction is in English.

Acutiſſimis qui toto Orbe florent Mathematicis

S. P. D.

JOHANNES BERNOULLI, Math. P. P.

CUm compertum habeamus, vix quicquam eſſe quod magis excitet generoſa ingenia ad moliendum quod conducit augendis ſcientiis, quàm difficilium pariter & utilium quæſtionum propoſitionem, quarum enodatione tanquam ſingulari ſi qua aliâ via ad nominis claritatem perveniant ſibique apud poſteritatem æterna extruant monumenta: Sic me nihil gratius Orbi Mathematico facturum ſperavi quam ſi imitando exemplum tantorum Virorum Merſenni, Paſcalii, Fermatii, præſertim recentis illius Anonymi Ænigmatiſtæ Florentini aliorumque qui idem ante me fecerunt, præſtantiſſimis hujus ævi Analyſtis proponerem aliquod problema, quo quaſi Lapide Lydio ſuas methodos examinare, vires intendere & ſi quid invenirent nobiſcum communicare poſſent, ut quiſque ſuas exinde promeritas laudes à nobis publicè id profitentibus conſequeretur.

Factum autem illud eſt ante ſemeſtre in Actis Lipſ. m. Jun. pag. 269. Ubi tale problema propoſui cujus utilitatem cum jucunditate conjunctam videbunt omnes qui cum ſucceſſu ei ſe applicabunt. Sex menſium ſpatium à prima publicationis die Geometris conceſſum eſt, intra quod ſi nulla ſolutio prodiret in lucem, me meam exhibiturum promiſi: Sed ecce elapſus eſt terminus & nihil ſolutionis comparuit; niſi quod Celeb. Leibnitius de profundiore Geometriâ præclarè meritus me per literas certiorem fecerit, ſe jam feliciter diſſolviſſe nodum pulcherrimi hujus uti vocabat & inauditi antea problematis, inſimulque humaniter rogavit, ut præſtitutum limitem ad proximum paſcha extendi paterer, quo interea apud Gallos Italoſque idem illud publicari poſſet nulluſque adeo ſupereſſet locus ulli de anguſtiâ termini querelæ; Quam honeſtam petitionem non ſolum indulſi, ſed ipſe hanc prorogationem promulgare decrevi, viſurus num qui ſint qui nobilem hanc & arduam quæſtionem aggreſſuri, poſt longum temporis intervallum tandem Enodationis compotes fierent. Illorum interim in gratiam ad quorum manus Acta Lipſienſia non perveniunt, propoſitionem hîc repeto.

Problema Mechanico-Geometricum de Linea Celerrimi deſcenſûs.

Determinare lineam curvam data duo puncta in diverſis ab horizonte diſtantiis & non in eadem rectâ verticali poſita connectentem, ſuper qua mobile propriâ gravitate decurrens & à ſuperiori puncto moveri incipiens citiſſime deſcendat ad punctum inferius.

Senſus problematis hic eſt, ex infinitis lineis quæ duo illa data puncta conjungunt, vel ab uno ad alterum duci poſſunt eligatur illa, juxta quam ſi incurvetur lamina tubi canaliſve formam habens, ut ipſi impoſitus globulus & liberè dimiſſus iter ſuum ab uno puncto ad alterum emetiatur tempore breviſſimo.

Ut vero omnem ambiguitatis anſam præcaveamus, ſcire B. L. volumus, nos hîc admittere Galilæi hypotheſin de cujus veritate ſepoſitâ reſiſtentiâ jam nemo eſt ſaniorum Geometrarum qui ambigat, *Velocitates ſcilicet acquiſitas gravium cadentium eſſe in ſubduplicata ratione altitudinum emenſarum*, quanquam aliàs noſtra ſolvendi methodus univerſaliter ad quamvis aliam hypotheſin ſeſe extendat.

Cum itaque nihil obſcuritatis ſuperſit, obnixè rogamus omnes & ſingulos hujus ævi Geometras, accingant ſe promtè, tentent, diſcutiant quicquid in extremo ſuarum methodorum receſſu abſconditum tenent; Rapiat qui poteſt præmium quod Solutori paravimus, non quidem auri non argenti ſummam quo abjecta tantum & mercenaria conducuntur ingenia, à quibus ùt nihil laudabile ſic nihil quod ſcientiis fructuoſum expectamus, ſed cùm virtus ſibi ipſi ſit merces pulcherrima, atque gloria immenſum habeat calcar, offerimus præmium quale convenit ingenui ſanguinis Viro, conſertum ex honore, laude & plauſu, quibus magni noſtri Apollinis perſpicacitatem publicè & privatim, ſcriptis & dictis coronabimus, condecorabimus & celebrabimus.

Quod ſi verò feſtum paſchatis præterierit nemine deprehenſo qui quæſitum noſtrum ſolverit, nos quæ ipſi invenimus publico non invidebimus; Incomparabilis enim Leibnitius ſolutiones tum ſuam tum noſtram ipſi jam pridem commiſſam protinus ut ſpero in lucem emittet, quas ſi Geometræ ex penitiori quodam fonte petitas perſpexerint, nulli dubitamus quin anguſtos vulgaris Geometriæ limites agnoſcant, noſtraque proin inventa tanto pluris faciant, quanto pauciores eximiam noſtram quæſtionem ſoluturi extiterint etiam inter illos ipſos qui per ſingulares quas tantopere commendant methodos, interioris Geometriæ latibula non ſolum intimè penetrâſſe, ſed etiam ejus pomœria Theorematis ſuis aureis, nemini ut putabant cognitis, ab aliis tamen jam longè priùs editis mirum in modum extendiſſe gloriantur.

Problema alterum purè Geometricum, quod priori ſubnectimus & ſtrenæ loco Eruditis proponimus.

Ab Euclidis tempore vel Tyronibus notum eſt; Ductam utcunque à puncto dato rectam lineam, à circuli peripheriâ ita ſecari ut rectangulum duorum ſegmentorum inter punctum datum & utramque peripheriæ partem interceptorum ſit eidem conſtanti perpetuo æquale. Primus ego oſtendi in eod. Actor Jun. pag. 265. hanc proprietatem infinitis aliis curvis convenire, illamque adeo circulo non eſſe eſſentialem: Arrepta hinc occaſione, propoſui Geometris determinandam curvam vel curvas, in quibus non rectangulum ſed ſolidum ſub uno & quadrato alterius ſegmentorum æquetur ſemper eidem; ſed à nemine hactenus ſolvendi modus prodiit; exhibebimus eum quandocunque deſiderabitur: Quoniam autem non niſi per curvas tranſcendentes quæſito ſatisfacimus, en aliud cujus ſolutio per merè algebricas in noſtra eſt poteſtate.

Quæritur Curva, ejus proprietatis, ut duo illa ſegmenta ad quamcunque potentiam datam elevata & ſimul ſumta faciant ubique unam eandemque ſummam.

Caſum ſimpliciſſimum exiſtente ſc. numero potentiæ 1, ibidem in actis pag. 266. jam ſolutum dedimus, generalem verò ſolutionem quam etiamnum premimus, Analyſtis eruendam relinquimus.

Dabam Groningæ ipſis Cal. Jan. 1697.

GRONINGÆ,

Typis Catharinæ Zandt, Provinciæ & Academiæ Typographæ 1697.

Fig. 9 The poster publicizing the challenge posed by Johann Bernoulli. Public domain

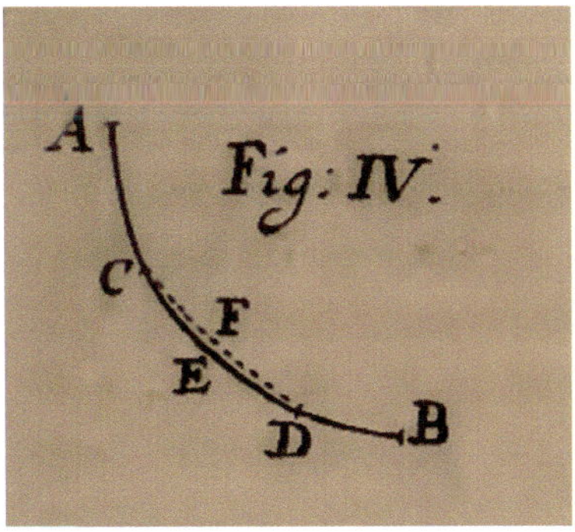

Fig. 10 Jacob Bernoulli's Fig. IV of his answer to his brother's challenge (Jac. Bernoulli, Op. LXXV, 1697, *Opera* 1744, I, Tab. IV). Public domain

> Given two points A and B (see figure) in a vertical plane, determine the way AMB, along which the gravity will drive the mobile M starting in A in order to reach the point B in the shortest time.[15]

Immediately after its publication one finds many attempts by Jacob in his *Meditationes* and in May 1697, in his Op. LXXV, *Solutio Problematum Fraternorum, una cum Propositione reciproca aliorum*, he gives a first solution. In this work Jacob introduces the calculus of variation, as we can see on the figure IV of his article (Fig. 10).

Jacob starts from the following: if the curve ACEDB is the one we are looking for, thus the one in which the gravity drives the mobile M from A to B in the shortest time. Then, the time taken along CFD will be longer than along CED (Fig. 11).

He uses then the figure V where he considers that segments CG and GD and writes that $tCG + tGD$ is a minimum.

In his Fig. VI, the last one (our Fig. 11, right), he shows the cycloid ACP, the solution of the problem, whose generating circle is RQVP.

The second person to solve that problem after Johann was Leibniz, followed by Jacob and Newton.

The brachistochrone, is of course the solution of a particular problem, but the general idea of Fig. IV will give rise to the *calculus of variation.*

4.3 *The Isoperimetrical Problems*

Jacob goes further generalizing his problems. In Op. LXXV, he starts studying isoperimetric problems. In fact, as Euler will show, all the questions Jacob had asked until that time may be grouped together as isoperimetric problems. All the problems posed by Johann concerns the calculus of variations, but they cannot be grouped under a same theme.

---

[15] Bernoulli (1696, p. 269): *Datis in plano verticali duobus punctis A et B (vid. Fig. 5) assignare Mobili M viam AMB, per quam gravitate sua descendens et moveri incipiens a puncto A, brevissimo tempore perveniat ad alterum punctum B.*

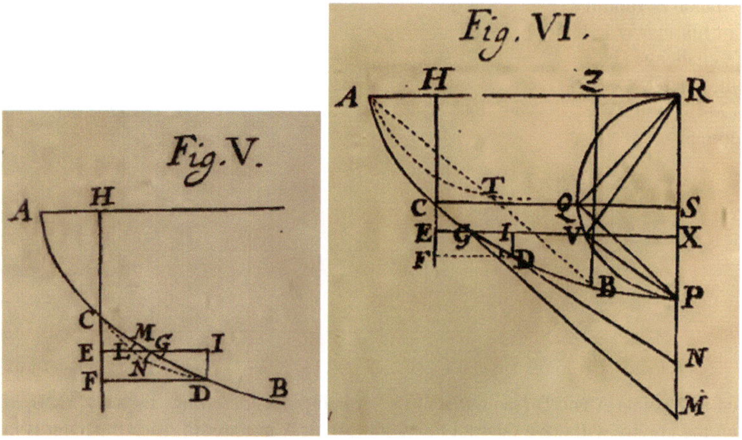

Fig. 11 Jacob Bernoulli's figures V **left** and VI **right** of his answer to his brother's challenge (Bernoulli 1697, Tab. IV). Public domain

Jacob prepares his question in Med. CCXXXIX, *Inter omnes Figuras Isoperimetras vel aequalium arearum invenire illam, quae habeat maximum vel minimum aliquod. Methodus Nova*, written between February and May 1697. In this text, one finds the first differential treatment of the problem of isoperimetric figures, followed by several examples.

- Among the isoperimeters, find the one with the lowest centre of gravity: he finds the catenary.
- Among the isoperimeters, find the one that contains the largest surface: he finds the circle.
- Among the isoperimeters, find the one whose centre of gravity of the surface is the lowest: he notes that this does not give the elastica, as he thought, but he finds the catenary.
- Among the curves containing the same surface, find the one with the lowest centre of gravity of the surface: he arrives at an impossibility.

In his Op. LXXV, *Solutio Problematum Fraternorum, una cum Propositione reciproca aliorum*, Jacob solves the problem of the brachistochrone and proposes new problems. First, among an infinity of cycloids or circles, parabolas coming out of the same point, find the one that drives a point to a given vertical in the least amount of time.[16] Second, find the curve of a ray of light in a medium of variable density, i.e., Johann's analogous solution for brachistochrone.

From there, he moves on to other speculations. He introduces the isoperimetric problem, and gives the example of the catenary of which his brother has shown that the centre of gravity of both the curve and the surface of the curve on a line must be

[16] This problem is solved in Meds. CCXL and CCXLI.

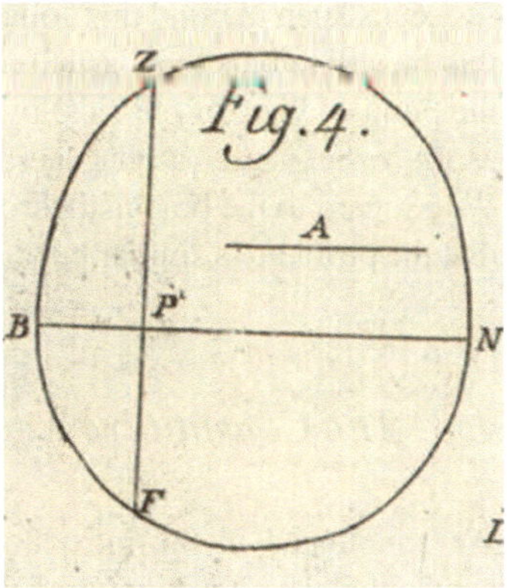

Fig. 12 Bernoulli's figure for the problem of curve BN (Bernoulli 1702; *Opera*, 1744, Tab. XXXIII). Public domain

furthest from this line, but he asks for a solution using the method of maxima and minima. Further, he poses a more precise problem:

> The question is that, of all the isoperimeters constructed on the common base BN, BFN does not itself include the maximum area, but causes another curve BZN to be the maximum, by applying which PZ is said to be in any ratio multiplied or submultiplied the straight line PF, or the arc BF; that is, what is the proportion proportional to: the given A and the straight line, PF, or the curve BF?[17] (Fig. 12).

Johann will soon react with his Op. XXXVIII, *Lettre sur le Probléme des Isopérimètres* of 1697, in which he says he has solved the problem of the brachistochrone in two ways, directly and indirectly (that is, using an analogy with a ray of light.). He gives that last solution and talks of generalizing the problem of the isoperimeters, which he will do later that same year in his Op. XL, *Lettre sur le Probléme des Isopérimètres, qui contient la Solution de ce Probléme*, in which, without any explanation, he gives the solutions of some cases in the form of an integral. It is interesting to note that those cases are the semi-circle and the lintearia, curvatures that his brother attributes also to his elastica. He obviously does not want to enter into this problem.

He generalizes then the problem considering any "functions" whatsoever of the ordinates.

Jacob will not publish anymore on that argument and Johann will only publish two texts on it. One is his Op. LXXV, *Solution & Démonstration du Probléme des Isopérimètres, De la courbure d'un linge chargé de liqueur* was published only in 1706 but it had been deposited in 1701. The other is his Op. CIII, *Remarques sur les Solutions données jusqu'ici du Problème des Isopérimètres, avec une nouvelle méthode de le resoudre, &c.*

[17] Jac. Bernoulli, Op. LXXV, *Opera*, 1744, p. 775: *Quaeritur ex omnibus Isoperimetris super communi basi BN cosstitutis illa BFN quae non ipsa quidem maximum comprehendat spatium, sed faciat, ut aliud curva BZN comprehensium sit maximum, cujus applicate PZ ponitur esse in ratione quavis multiplicate vel submultiplicata rectae PF, vel arcus BF; hoc est, quae sit quotacunque proportionalis ad datam A & rectam, PF, curvamve BF?*

Let us keep in mind that Johann is first of all interested in the brachistochrone and that he still avoids the elastica replacing its study (and we can understand him from the point of view of physics) by that of the lintearia, which gives the same equation as the *vulgus* case (a bit generalised) without the constitutive laws of the elastica. With regard to the brachistochrone, just as for the geodesics, we shall see with Euler that the problem is similar but does not belong to the coherent set proposed by Jacob.

4.4 The Constitutive Equations

We announced in the introduction that the problem of the elastica immediately confronted the enquiring mind with constitutive equations. We also saw that Leibniz as well as Jacob realized that various materials reacted in different ways and thus that the law of tension that they generally supposed as linear had to be adapted to each material. Jacob tries to do it in Op. LVIII (Bernoulli 1694a, b) by introducing a law of tension. I would even go as far as to say that if Jacob goes on generalising the problem using the isoperimeters, it is in the hope of determining the law of tension given by nature, measuring the curvature.

This impression is confirmed in the last part of his life after the publication of Op. LXVI, *Explicationes, Annotationes, et Additiones ad ea, quae in Actis sup. Anni de Curva Elastica, Isochrona Paracentrica, et Velatia.*

Jacob will not change a lot his theory of the elastica. We shall see the unique exception soon, but he will study the elastic behavior of material objects. This is, of course, important for architecture and will completely change it, though not immediately.

Jacob had already started this study in his Med. CXCVIII (*Lintei parallelogrammi ad extensi figuram reperire*) but it is in a series of four *Meditationes* that will lead to Op. CII that he will attack those problems once more. This will lead to a better study of the elastica. He realized, as he explains it in his Med. CCL, *Curvatura Laminae Elasticae cum supponitur non tantum superficiem externam extendi [sed] & internam contrahi*, written between May 1697 and May 1698. Thus, he considered erroneously that the beam was uniformly extended in half of its thickness and uniformly compressed in the other half keeping the neutral fiber in the middle. In fact, the elongation due to the tension augments from zero to a certain positive value when one moves away from the neutral fiber let say in the upper part of the beam and from zero to a negative value in the other part of the beam.

He takes this fact into account in this *Meditationes* introducing an integration and a dependance on the distance from the neutral fiber. Jacob the continues this reflection in the Med. CCLI, *Quaestio Num elasterum, sublata subito vi tendente, eodem tempore in omnibus suis partibus in rectitudinem se restituat; an vero in aliis partibus citius, in aliis tardius?*

Jacob is thus interested in the way the elastic beam returns to—if he exactly returns to—its initial position. This is an enormous problem that is different for each material. As Edoardo Benvenuto explains in *An Introduction to the History of*

Structural Mechanics,[18] those equations were solely submitted to experiment until Jacob began to theorize them and rational mechanics tried to axiomatize that part of mechanics.

In this sense, Jacob Bernoulli furnished two important elements:

(1) An element I already mentioned, namely the description of a perfectly rigid body in his Op. XLV, *Demonstratio centri oscillationis ex natura vectis*, considering once more a lever fixed in the body.
(2) Here the description of the beam perfectly elastic that returns to its place in function of the displacement it has undergo.

It concerns two particularly important idealizations in such a theory of constitutive equations. He will confront them in Med. CCLXXXVI, *Linea datae rigidae ab infinitis potentiis 2dum quasvis directiones impulsae tractaeve determinare directionem mediam axem aequilibrii & vim impulsus*, written on the 12 of December 1704. In that text, Jacob calculates in that particular hypothesis the time necessary to the line to recover its place.

Jacob goes ahead with that study in Med. CCLXXII, *Quamquam non maneat eadem quantitas motus absoluta in universo, tamen semper eadem motus quantitas ad easdem partes sive corpora ponantur Elastica sive elatere destituta*, written between 1702 and 1703, and in Med. CCLXXIII, *Inventio Centri Tensionis,* of the same period, in which he tries to determine the centre of tension, that is, the point of the section through which passes the neutral fibre that is neither elongated nor shortened. He is thus looking for the position of the neutral fibre, which—let us repeat—depends on the material.

Jacob publishes his last one article on 12 March 1705: Op. CII, *Veritable hypothèse de la résistance des solides…*, published in the *Mémoires de Paris*. He will die on 16 August 1705. In this text he explains clearly what he noted in his *Meditationes* CCL, CCLI, CCLXXI, CCLXXIII. Jacob has enlarged his field of research to, what we call today, constitutive laws.

In order to do that he goes back to Galileo's remark concerning his observation concerning the difference of resistance to rupture if one pulls a beam in the sense of its length or if one folds it perpendicularly to it:

> …whatever the nature of this resistance which solids offer to large tractive forces there can at least be no doubt of its existence; and though this resistance is very great in the case of a direct pull, it is found, as a rule, to be less in the case of bending forces.[19]

[18] Benvenuto (1991, p. 274–277).

[19] Galilei (1637, p. 108): *Posta qualunque ella sia la resistenza de i corpi solidi all'essere spezzati per una violenta attrazzione, basta che indubitabilmente ella in loro si trova: la quale ben che grandissima contro alla forza di chi per diritto gli tira, minore per lo più si osserva nel violentargli per traverso.* Eng. trans. Galilei (1914, p. 109).

After having mentioned the works of Leibniz,[20] Mariotte,[21] and Varignon,[22] Jacob continues the story and criticizes what he said eleven years earlier in Med. CCL. One must take each fiber contained into the thickness of the blade into account. Further, he gives laws of elongation:

> Lemma I. Fibers of the same matter and of the same width, or thickness, extended of pressed by the same force, extend or compress proportionally to their length.[23]
>
> Lemma II. Homogene fibers of the same length but of different width or thickness, extend or compress equally for forces proportional to their widths.[24]
>
> Lemma III. Homogene fibers of the same length and same thickness, but differently charged, do not extend nor compress proportionally to those weight; but the extension or compression caused by the greatest weight, is to the extension or compression caused by the smallest, in less reason than that weight is to this one.[25] (Fig. 13)

In a Scholia he says: "This is also what the experiment confirms",[26] and in the Corollary he comments:

> Besides, it is probable that this curve [the line of tension and compression] is different with regard to different bodies, because of the different structure of their fibers.[27]

In lemma IV, he writes:

> The same force which causes a beam or pole ABCD to bend from AB to GF, extending one part of its fibers by the amount of triangle BSF, and compressing the other by the amount of triangle ASG, would be able to extend the assembly of all the fibers, on the support A, by the quantity of triangle ABF; or to compress this assembly, on the support B or F, by the quantity of the triangle BAG or FAG.[28]

---

[20] Leibniz (1684).

[21] Mariotte (1700, Part. V, Disc. 2.

[22] Varignon (1702.

[23] Bernoulli (1702, p. 178) (*Opera*, 1744, p. 979): *Lemme 1. Des fibres de même matière et de même largeur, ou épaisseur, tirées ou pressées par la même force, s'étendent ou se compriment proportionnellement à leurs longueurs.*

[24] Bernoulli (1702, p. 179) (*Opera*, 1744, p. 980): *Lemme 2. Des fibres homogènes et de même longueur, mais de différentes largeurs ou épaisseurs, s'étendent ou se compriment également par des forces proportionnelles à leurs largeurs.* Mém. Paris.

[25] Bernoulli (1702, p. 180) (*Opera*, 1744, p. 981): *Lemme 3. Des fibres homogènes de même longueur et de même largeur, mais chargées de différens poids, ne s'étendent, ni ne se compriment pas proportionnellement à ces poids; mais l'extension ou la compression causée par le plus grand poids, est à l'extension ou à la compression causée par le plus petit, en moindre raison que ce poids-là n'est à celui-ci.*

[26] Bernoulli (1702, p. 180) (*Opera*, 1744, p. 981): *C'est aussi ce que l'expérience confirme.*

[27] Bernoulli (1702, p. 181) (*Opera*, 1744, p. 982): *Au reste, il est probable que cette courbe [il s'agit de la ligne de tension et de compression] est différente à l'égard de différents corps, à cause de la différente structure de leurs fibres.*

[28] Bernoulli (1702, p. 181) (*Opera*, 1744, p. 982): *La même force qui fait plier une poutre ou perche ABCD de AB en GF, en étendant une partie de ses fibres de la quantité du triangle BSF, et en comprimant l'autre de la quantité du triangle ASG, seroit capable d'étendre l'assemblage de toutes les fibres, sur l'appui A, de la quantité du triangle ABF; ou bien de comprimer cet assemblage, sur l'apui B ou F, de la quantité du triangle BAG ou FAG.*

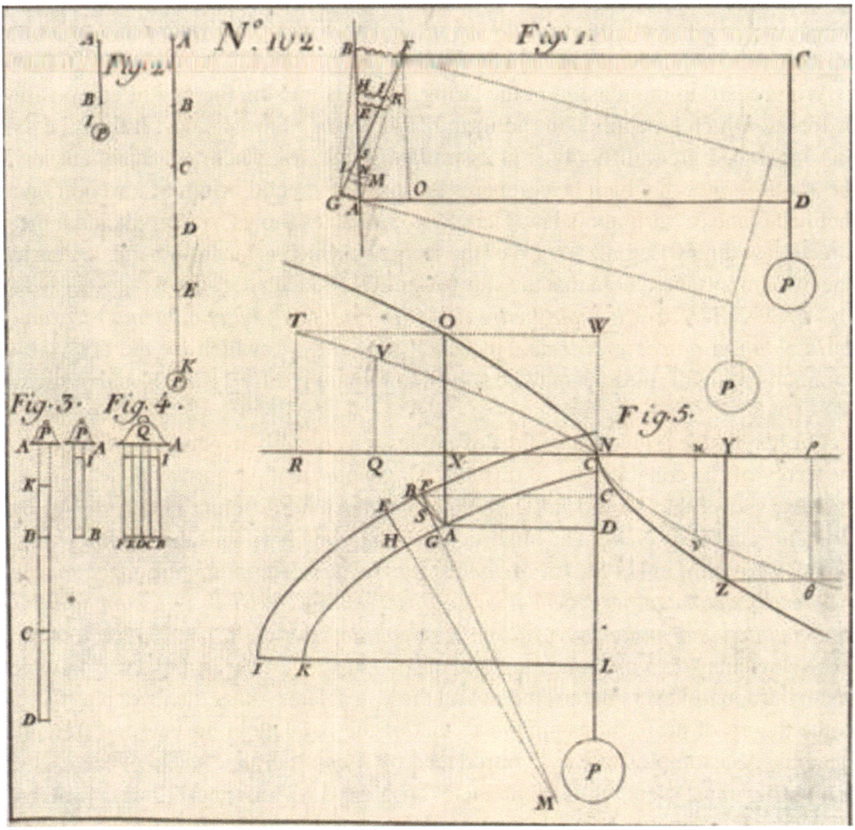

Fig. 13 Figures for Jacob's Op. CII (Bernoulli 1702). Public domain

4.5 About the Catenary

I would like to close this period with a text, written by Jacob shortly before but that did not result in any of his publications, in which he gives another way of grouping all problems that he had proposed. I have already said that it was possible to group them thanks to the isoperimeters. It is also possible using the catenary, as Jacob explains in one of his *Varia Posthuma,* published in his *Opera.*[29] I conclude this part with it because it is the starting point from which Daniel Bernoulli and Leonhard Euler will start their work on the topic. Later they will group all those problems under the commune idea of isoperimeters.

In this text, Jacob studies the equilibrium of a string under the action of an infinity of forces acting in any directions whatsoever. He thus completely generalizes the catenary. After that he studies the particular cases and finds the lintearia when the

[29] Jac. Bernoulli, Med. CCXLV (*Opera*, 1744, *Varia posthuma* XI, p. 1036–1048).

pressures are perpendicular to the string, and then the other particular cases, the elastica, the circle, the velaria and so forth.

With regard to the tension in the string, he finds it using the law of composition of forces, which he applies in the form of Varignon's law of sines. It must be said that Jacob had great difficulties in assimilating that law. There are many entries in the *Meditationes* in which he compares his law of the equilibrium of a weight on an inclined plane to Varignon's law. Here he complicates things by introducing a mean direction given by Leibniz's form of the theorem, which is far from being the easiest one. It is important to note that the sole text in which Jacob decomposes forces is Jac. Op. LXXXVII, *Extrait d'une Lettre de M. Bernoulli de Bâle, contenant l'examen de la solution de ses problèmes*, in which he writes: ...which by the doctrine of communication of motions may be solved horizontally FE $= p\ dx\!: x$, and vertically ED $= p\ dy\!: x$.[30]

Further, that text gives one of the examples, given rarely in general but frequently by Jacob, of the cohabitation of differential calculus and proportion theory. In fact, this text shows that he did not completely understand differential calculus at that moment, since he leaves out an important step, one which will be taken by his nephew, Daniel Bernoulli, in his St. 15, *Methodus universalis determinandae curvaturae fili a potentiis quamcunque legem inter se observantibus extensi, una cum solutione problematum quorundam novorum eo pertinentium* (Bernoulli 1728). It is clear from those elements—to which one must add the observation of the artificial way in which Jacob introduces levers in the study of all these problems—that the lever plays a role in the theory of proportions similar to that of the parallelogram law in differential calculus. As examples of levers introduced by Jacob there are those of the elastica that we have just seen, but there are also those more performant of the rigid body.

We have thus seen that Jacob connected all the problems he had posed and showed that their common denominator is the osculating radius, but we should not forget that he also posed the problem of the isoperimeters, linking them to the preceding problems. He even found those various elements starting from the isoperimeters.[31]

In his Op. XCIII, *Solutio propria Problematis Isoperimetrici,* Jacob publishes the table he had prepared in Med. CCXLVI, *Quaedam formulae aequationum differentio-differentialium reductae aequationes differentiales primi generis* (Fig. 14).

This argument gave rise to two other main texts of Jacob: his letter to his brother Johann Bernoulli, with his own solution of the problem of the isoperimetrical problem,[32] and Op. XCVI, *Analysis magni Problematis Isoperimetrici.*

This last text gives the complete mathematical developments. I just quote the first sentence to show that Jacob is conscious of the importance of what he presents:

---

[30] Jac. Bernoulli, Op. LXXXVII (*Opera*, 1744, p. 833): *laquelle par la doctrine de la communication des mouvemens se peut résoudre en horizontale FE* $= p\ dx\!: x$, *et en verticale ED* $= p\ dy\!: x$.

[31] The interested reader should read H. Goldstine's comment in *Streitschriften,* (Goldstine 1991, pp. 54–64).

[32] *Jacobi Bernoulli epistola ad fratrem suum Johannem Bernoulli, cum annexa solutione propria problematis isoperimetrici* (Goldstine 1991, pp. 471–484).

	Si	erit quantitas
1.	$dy = pdx : \sqrt{(aa - pp)}$	$\int pdy$ Maximum & $\int(dt:p)$ Minimum
2.	$dy = (a - p)dx : \sqrt{(2ap - pp)}$	$\int pdy$ Minimum
3.	$dy = apdx : \sqrt{((bb - aa)pp - 2aabp + a^4)}$	$\int(dt:p)$ Maxim.
4.	$dy = adx : \sqrt{(pp - aa)}$	$\int(dy:p)$ Maxim. & $\int pdt$ Minim.
5.	$dy = (p - a)dx : \sqrt{(2ap - aa)}$	$\int(dy:p)$ Minim.
6.	$dy = adx : \sqrt{(bb - 2bp + pp - aa)}$	$\int pdt$ Max.
7.	$dy = qdt : \sqrt{(aa + qq)}$	$\int qdy$ Maxim. & $\int ydq$ Min. *vel* Max
8.	$dy = (a - q)dt : \sqrt{(bb - 2aq + qq)}$	$\int qdy$ Min. & $\int ydq$ Max. *vel* Min.
9.	$dy = adt : \sqrt{(aa + qq)}$	$\int(dy:q)$ Max. & $\int xdq$ Min. *vel* Max.
10.	$dy = (aq - bb)dt : b\sqrt{(bb - 2aq + qq)}$	$\int(dy:q)$ Min.
11.	$dy = adt : \sqrt{(aa + bb - 2bq + qq)}$	$\int xdq$ Max. *vel* Min.

Fig. 14 Jacob's tables (Jac. Bernoulli, Op. XCIII, 1700, Opera, 1744, pp. 876–878). Public domain

> I call it a great problem, not so much because of the difficulty of solving it, (although it is so great that it can easily compete with the most difficult), but because it opens the way for us to advance the ends of science and explore new regions.[33]

This text is in my opinion the first of the two texts, together with the one about infinite series, truly written in the differential language he really knows, not in another guise in order to hide his knowledge.

[33] Bernoulli (1701), Op. XCVI (*Opera*, 1744, p. 895): *Magnum Problema appello, non tam ob solvendi difficultatem, [etsi tantam, ut certare facile possit cum difficillimis,] quam quod ad provehendes scientia fines, novasque regiones ulstrandas viam nobis aperit.*

5 Third Period: From 1728, When Daniel Bernoulli and Leonhard Euler Reread the Work of Jacob Bernoulli, and the Birth of the *Methodus inveniendi*

5.1 Daniel Bernoulli

In St. 15, *Methodus universalis determinandae curvaturae fili a potentiis quamcunque legem...*, mentioned above, Daniel Bernoulli takes up again the leading idea of Jacob's Med. CCXLV (*Filum* ACDEFGB *extremitatibus suis* A *et* B *suspensum ab infinitis potentiis* C, D, E, F, G *juxta directiones quas vis* HE, HD, IE, KF, LG *agentibus extenditur*), but in the meantime Daniel has discovered the power of the parallelogram law of forces, which he had demonstrated axiomatically two years earlier in his St. 9, *Examen principiorum mechanicae, et demonstrationes geometricae de compositione et resolutione virium.*

Daniel begins by considering a string on which four forces act in two points, and he writes the equality of the forces acting on one point of the middle of the rope. He then translates to infinitesimal elements and gives the expression in function of the osculating radius. Using his knowledge of the composition of forces, he can limit himself to two types of forces: vertical forces, and those perpendicular to the curve. All other cases will be compositions of those two. In the corollaries, he particularizes to the case of all perpendicular forces and finds the elastica, the velaria and the lintearia. He then particularizes to only parallel forces, and he finds all catenaries.

Next, he particularizes first to a tarpaulin without weight, which he says is the result of his uncle Jacob, and then to the tarpaulin under its own weight, finding the result of his father Johann.

He shows that in the case of forces perpendicular to the curve, that is to say for the lintearia, the radius of curvature is $F = \frac{a}{R}$; in the case of forces of directions different of that of the perpendicular but parallel to each other, as in the catenary, "the force is in compound reciprocal ratio of the radius of curvature and of the square of the sine of the angle that the force makes with the axis".[34]

Then he treats the elastica, reproducing the demonstration of his uncle. He particularizes to Jacob's case of a blade without weight, then to the blade under its own weight, but this he postpones until later. He will never do it, but Euler will solve that problem, as we shall see.

Finally, he treats a problem proposed by Jacob at the very beginning of this long story but forgotten, namely the extensible catenary.

---

[34] Bernoulli (1728, p. 65): *esse ubique potentiam in ratione reciproca composita ex radio osculi et quadrato sinus anguli, quem potential facit cum axe.*

5.2 Leonhard Euler, E.8

Euler's aim in E.8, *Solutio problematis de invenienda curva quam format lamina utcunque elastica in singulis punctis a potentiis quibuscunque sollicitata*[35] is to treat the blades of any elasticity whatsoever, but also all the forms taken by a perfectly flexible string under the action of any forces whatsoever. He is thus on the same line as Daniel Bernoulli, but his mathematical treatment is much more elaborate.

I shall only add one remark about a detail concerning the composition of forces. In order to let whatever a force act on an elastic blade, Euler introduces the force due to two weights, one of which is acting through a pulley and the other one acting vertically (thus one horizontally and the other vertically).

5.3 Leonhard Euler, E.27

Euler's E.27, *Problema isoperimetrici*, is an important work on the method to find the curves that enjoy a property of maximum or minimum, or a solution of the problem of isoperimeters in its broadest sense. At the end of that work, he gives an *additamentum* about the elastica. The position of this text has always surprised me, but reading the works of Jacob Bernoulli to which Euler refers one realizes that he posed the question of the isoperimeters hoping to find a general solution for the problem of the elastica, which encompassed, he realized, those of the catenary, the velaria, and the as well as the lintearia. And he had gone rather far in the solution of the problem of the isoperimeters and its particularisation to various problems. Nevertheless, Euler misses one more element in order to close that study, and as Euler himself says it, Daniel Bernoulli will provide that element.

5.4 The Letters Exchanged by Euler and Daniel Bernoulli

In his letter written on 5 May 1739, Euler tells Daniel Bernoulli that he does not find the quantity that has to be maximized in the case of the elastica. Jacob could not find it either, and that is the reason why he tried various criteria about the centre of gravity of the beam or of the enclosed surface. Euler knows that it is the "quantity of potential force", but he cannot formulate it. It is of course impossible to solve that problem using isoperimeters if one does not know the quantity to maximize. This also explains the elements of the quarrel between Jacob and Johann Bernoulli concerning the position of the centre of gravity, which makes it possible to solve the other problems via the isoperimeters but not the elastica.

---

[35] The works of Euler cited in what follows are available at the Euler Archive, http://eulerarchive.maa.org/.

In a letter of 20 October 1742 Daniel tells Euler that the potential live force of the elastica is $\int \frac{ds}{R^2}$.

5.5 *Euler, E.65*

In his introduction to E.65, *Methodus inveniendi lineas curvas maximi minimive proprietate gaudentes sive solutio problematis isoperimetrici latissimo* sensu *accepti*, Euler does justice to Daniel Bernoulli and poses the problem of the elastica in the following manner: among all curves of the same length going from A to B, where the tangents to the curve are given, to find the one that renders $\int \frac{ds}{R^2}$ minimal. This time the problem will be well posed, and all cases analyzed. He will also add the study of the curvature of a blade under its own weight, etc.

6 Conclusion

The problems treated in this article may, at first sight, seem far from the history of architecture. And I concede that architecture itself was absent from this study. But the reader must ask himself the question: what would architecture be without the theory of elasticity? I am sure that Nervi could never have built anything without it.

References

Benvenuto, Edoardo. 1991. *An Introduction to the History of Structural Mechanics,* Part I. Springer.

Bernoulli, Daniel. 1728. St. 15, Methodus universalis determinandae curvaturae fili a potentiis quamcunque legem inter se observantibus extensi, una cum solutione problematum quorundam novorum eo pertinentium. In *Commentarii Academiae scientiarum imperialis Petropolitanae* III, 1729–1731, 62–69.

Bernoulli, Jacob. 1694a. Curvatura Laminae Elasticae. Ejus Identitas cum Curvatura Lintei a pondere inclusi fluidi expansi. Radii Circulorum Osculantium in terminis simplicissimis exhibiti, una cum novis quibusdam Theorematis huc pertinentibus, etc., Op. LVIII. In *Acta Eruditorum, Junii*, 1694, 262–276; *Opera*, 1744, 576–600.

Bernoulli, Jacob. 1694b. Op. LX, *Constructio Curvae Accessus & Recessus aequabilis, ope rectificationis Curvae cujusdam Algebraicae*, Acta Eruditorum, 9/1694, 336; *Opera,* 1744, vol. II, 608–612.

Bernoulli, Jacob. 1695. Explicationes, Annotationes, et Additiones ad ea, quae in Actis sup. anni de Curva Elastica, Isochrona Paracentrica, & Velaria, hinc inde memorata, & partim controversa leguntur; ubi de Linea mediarum directionum, aliisque novis, Op. LXVI. In *Acta Editorium Decembris* 1695, 537–553; *Opera*, 1744, 639–663.

Bernoulli, Jacob. 1697. Solutio Problematum Fraternorum, una cum Propositione reciproca aliorum, Op. LXXV, Acta Editorium Maji 1697, 211–218; *Opera*, 1744, 768–778.

Bernoulli, Jacob. 1698. Extrait d'une Lettre de M. Bernoulli de Bâle, contenant l'examen de la solution de ses problèmes. *Journal des Sçavans*, 4 and 11 August 1698, 355–364; *Opera*, 1744, 829–839

Bernoulli, Jacob. 1705. Op. CII, Veritable Hypothèse de la Résistance des Solides, avec la Démonstration de la Courbure des corps qui font Ressort. In *Mémoires de mathématique et de physique de l'Académie royale des sciences* 1705, 176–186; *Opera*, 1744, 976–989.

Bernoulli, Jacob. 1700. Op. XCIII, Solutio Propria Problematis Isoperimetrici. In *Acta Eruditorum*, Juni, 1700, 261–266; *Opera*, 1744, 874–887.

Bernoulli, Jacob. 1701. Op. XCVI, *Q.D.O.M.B.V. Analysis magni Problematis Isoperimetrici,* Basileae 1701 (UB Kd III 17, 12); *Opera*, 1744, 895–920.

Bernoulli, Johann. 1696. Supplementum defectus geometrie Cartesiana circa Inventionem Locorum. Annotata quaedam in Schediasmata Leibnitianum & Tschirnbausianum in ultimo Actorum Novem. Edita. De complanatione superficierum Conoidearum & Sphaeroidearum. Problema novan Mathematicis propositum. In *Acta Editorum*, June, 264–269.

Galilei, Galileo. 1637. *Discorsi e dimostrazioni matematiche intorna à due nuove scienze.* Leiden: Elsevier. Eng. trans. *Dialogues concerning Two New Sciences.* Henry Crew and Alfonso de Salvia, trans. New York: MacMillan, 1914.

Goldstine, H. H. 1991. *Die Streitschriften von Jacob und Johann Bernoulli*, Variationsrechnung, bearbeitet und kommentiert von H. H. Goldstine mit historischen Anmerkungen von P. Radelet-de Grave. Basel: Birkhäuser.

Huygens, Christiaan. 1694. Excerpta ex Epistola C. H. Z ad G. G. L. In *Acta Eruditorum septembre 1694*, 339–341.

Leibniz, Georg Wilhelm. 1684. Demonstrationes novae de Resistentia Solidorum. In *Acta Eruditorum julii 1684*, 319–325.

Mariotte, Edmé. 1700. *Traité du mouvement des eaux et des autres corps fluides, divsisé en V* partes. Paris: Jombert.

[*Opera*] Bernoulli, Jacob, Johann Jacob Battier. 1744. *Jacobi Bernoulli, Basileensis, Opera*, 2 vols. Geneva: Sumptibus Hæredum Cramer & Fratrum Philibert.

Spiess, Otto, ed. 1955. *Der Briefwechsel von Johann I. Bernoulli*, Bd. 1. Basel: Birkhäuser.

Varignon, Pierre. 1702. De la résistance des solides en général pour tout ce qu'on peut faire d'hypothèses touchant la force ou la ténacité des fibres des corps à rompre ; et en particulier pour les hypothèses de Galilée et de Mr. Mariotte. In *Mémoires de l'Académie de Paris,* 66–93.

Chapter 8
The Theory of Vaults of Pierre Bouguer

1 Bouguer's Work on Vaults

When he writes his article "Sur les lignes courbes qui sont propres à former les voûtes en dôme" (1736) on the curved lines suitable for dome vaults, Pierre Bouguer is 36 years old and his thoughts are principally turned, under the influence of questions asked by the Academy of sciences of Paris, to application domains linked to navigation. A vast and varied domain, his researches range from astronomical measurements ("De la méthode d'observer exactement sur mer la hauteur des astres" 1729) in order to determine hours at sea, the measure of magnetic declination ("De la méthode d'observer en mer la déclinaison de la boussole" 1731) in order to get a better determination of the position at sea to the calculation of the length of the ship's mast ("De la mâture des vaisseaux" 1727), and the estimation of the lines of pursuit for one boat chasing another ("Sur de nouvelles courbes auxquelles on peut donner le nom de lignes de poursuite" 1732). In this text, a probable motivation of Bouguer becomes clear: the analysis of curves. The physical problem of one ship pursuing another appears to be only a pretext.

> Since it is easier to find the manifestations of all curved lines, the very discovery of these lines no longer attracts the same attention. But we must nevertheless continue to notice those which have curious properties, those principally which form entire families, or which being capable of an infinity of genera, are at the same time rectifiable, and which only contain spaces from which we can have quadrature...[1]

Thus writes Bouguer at the beginning of his article, even before justifying, with a physical description, the name of the curves shown in Fig. 1. He calls the curves he intends to study "lines of pursuit".

---

[1] Bougeur (1732, pp. 15–16): *Depuis qu'on trouve avec plus de facilité les symptômes de toutes les lignes courbes, la découverte même de ces lignes n'attire plus la même attention. Mais on doit cependant continuer à remarquer celles qui ont des propriétés curieuses, celles principalement qui formant des familles entières, ou qui étant susceptibles d'une infinité de genres, sont en même temps rectifiables, & qui ne renferment que des espaces dont on peut avoir la quadrature.*

P. Radelet-de Grave, *Between Mechanics and Architecture*, Mathematics and the Built Environment 8, https://doi.org/10.1007/978-3-031-73530-1_8

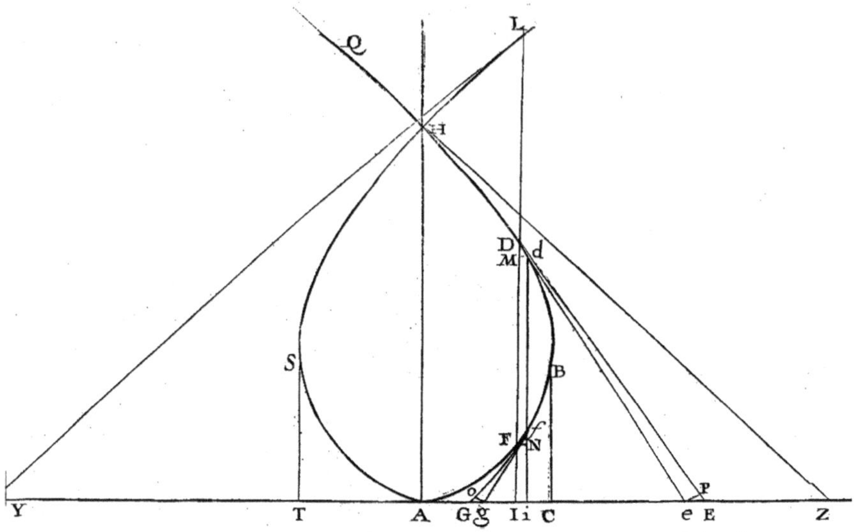

Fig. 1 Example of a line of pursuit (Bouguer 1732, Fig. 1). Public domain

In the same manner, since one knows that the most stable, infinitely small vault under its own weight is the inverse catenary, the study of that curve has spilled a lot of ink. The wish to generalize vaults to have a certain thickness may also give rise to the study of a whole family of curves. One remark in the "Essai analytique sur la mécanique des voûtes" by Charles-François-Ferdinand le Prud'homme d'Hailly, Viscount de Nieuport, written 44 years after the one by Bouguer, is that the problem of vaults is but a pretext of a mathematical analysis. It is difficult to imagine that all curves found by de Nieuport and reproduced in Fig. 2 make it possible to visualize vaults.

Nieuport purpose was to show the curves that form the vault. Explains his purpose this way. Bouguer's aim is the same: "I'll show that an infinity of curves are suitable to form that kind of vault".[2] Nevertheless, one would go too far in saying that it is a pure mathematical game. Because it is precisely there that Bouguer is situated, between that purely theoretical research, that mathematical game and a practical objective, the construction of a vault or the design of the hull of a ship (Fig. 3). As he will do in his *Traité du Navire* (1746), published after his return from Peru.

Shipbuilding furnishes an example of historical development similar to that of architecture of vaults because there practice also preceded theoretical development.

[2] Bouguer (1736, p. 149): *Je montrerai qu'une infinité de lignes courbes sont propres à former ces sortes de Voutes.*

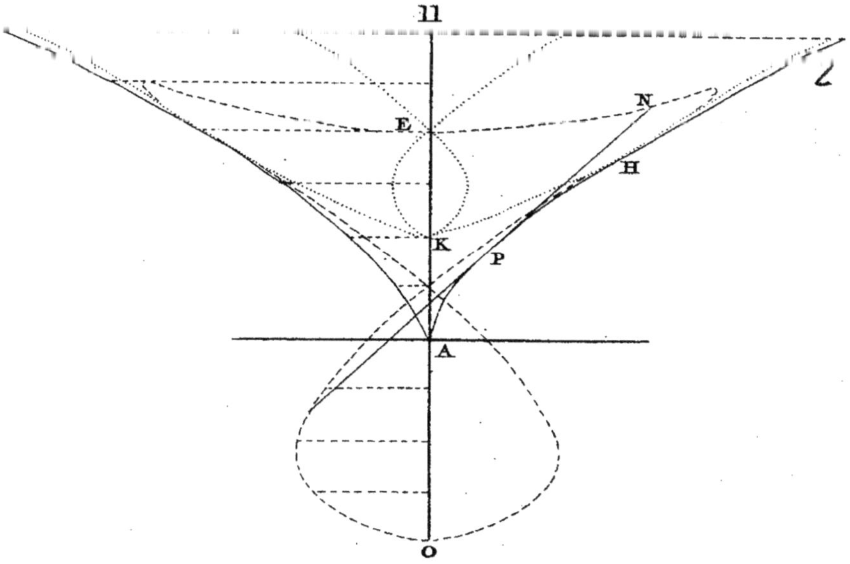

Fig. 2 Figures of vaults proposed by Nieuport (1778). Public domain

Fig. 3 A design of the hull of ship taken from the *Traité du Navire* (Bouguer 1746, Fig. 10). Public domain

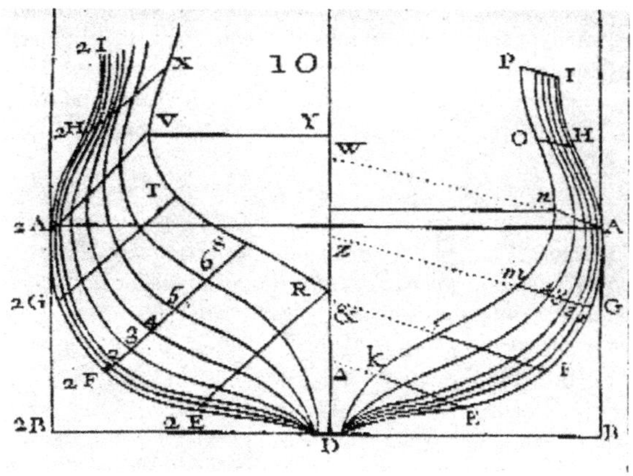

2 History of the Calculation of Domes

At the beginning of his work about vaults, Bouguer writes:

> Many people have treated with great care simple arch vaults: the recent volumes of the Memoires of the Academy contain excellent pieces on that argument, among which we must cite with distinction those of M. Couplet. Only the dome vaults remain, which nobody has examined.[3]

At that time, the scientific analyses, even of simple vault and arches is rather recent. That analysis is précised by Bélidor, five years before Bouguer, in the second book of his *Science des ingénieurs, qui traite de la mécanique des voutes*:

> we will no doubt agree that there is a method for considering subjects relating to architecture, by which we are sure not to go wrong as soon as we know how to use the knowledge acquired by study of mathematics. The principles they present to us are so fertile that there is nothing to which they cannot be applied, especially those of mechanics. It is in vain that anyone will want to persuade us that the practice left to itself can arrive at the point of perfection: experience often proves the contrary, and very shortly I am going to show an example about vaults, apropos to showing how serious it is not to follow without examination the principles which are only authorized by usage.[4]

He will come in that way to criticize the construction widely used but simplistic and oversized, which give the thickness of the *engaged pier* and which he attributes to Blondel. It was in fact published for the first time in (1643) by the Jesuit François Derand,[5] and was taken up again by Bélidor (Fig. 4), who writes:

> Divide the arch into three equal parts, leading one of the strings through the point of the transom, take outside on the same continuation, a line which is equal to it; the straight line drawn straight through the end of this line will determine the thickness of the jamb; as if, dividing the arc ACDB into three equal parts at the points CD, I lead the rope DB passing through the point of the transom at B, I only have to take outside on the same straight line, continued, the part BE equal to BD, and leading the two perpendiculars EG and BF, they will determine the thickness of the jamb BGEF, which will be proportional to the thrust of the arc ACDB.[6]

---

[3] Bouguer (1736, p. 149): *Beaucoup de gens ont traité avec beaucoup de soin des Voutes en simple Arc: les derniers Volumes des Mémoires de l'Académie contiennent d'excellents morceaux sur cette matiére, entre lesquels on doit citer avec distinction ceux de M. Couplet. Il ne reste que les Voutes en Dome que personne, que je scache, n'a examinées.*

[4] Bélidor (1830, p. 100): *on conviendra sans doute qu'il y a une méthode pour considérer les sujets qui se rapportent à l'architecture, par laquelle on est sûr de ne pas donner dans le faux, dès qu'on saura se servir heureusement des connaissances acquises par l'étude des mathématiques. Les principes qu'elles nous présentent sont d'une si grande fécondité, qu'il n'y a rien à quoi ils ne soient applicables, principalement ceux de la mécanique. C'est vainement qu'on voudra nous persuader que la pratique abandonnée à elle-même peut arriver au point de perfection: l'expérience prouve souvent le contraire, et j'en vais faire voir un exemple au sujet des voûtes, qui viendra fort à propos pour faire sentir combien il est de conséquence de ne pas suivre sans examen les principes qui ne sont autorisés que par l'usage.*

[5] Derand (1643, p. 16).

[6] Bélidor (1830, p. 106): *Partagez l'arc en trois parties égales, et menant une des cordes par le point de l'imposte, prenez en dehors sur la même continuée, une ligne qui lui soit égale; la droite menée à plomb par l'extrémité de cette ligne déterminera l'épaisseur du piédroit; comme si, divisant l'arc ACDB en trois parties égales aux points CD, je mène la corde DB passant par le point de l'imposte en B, je n'ai qu'à prendre en dehors sur la même droite, continuée, la partie BE égale à BD, et menant les deux perpendiculaires EG et BF, elles détermineront l'épaisseur du piédroit BGEF, qui sera proportionnée à la poussée de l'arc ACDB.*

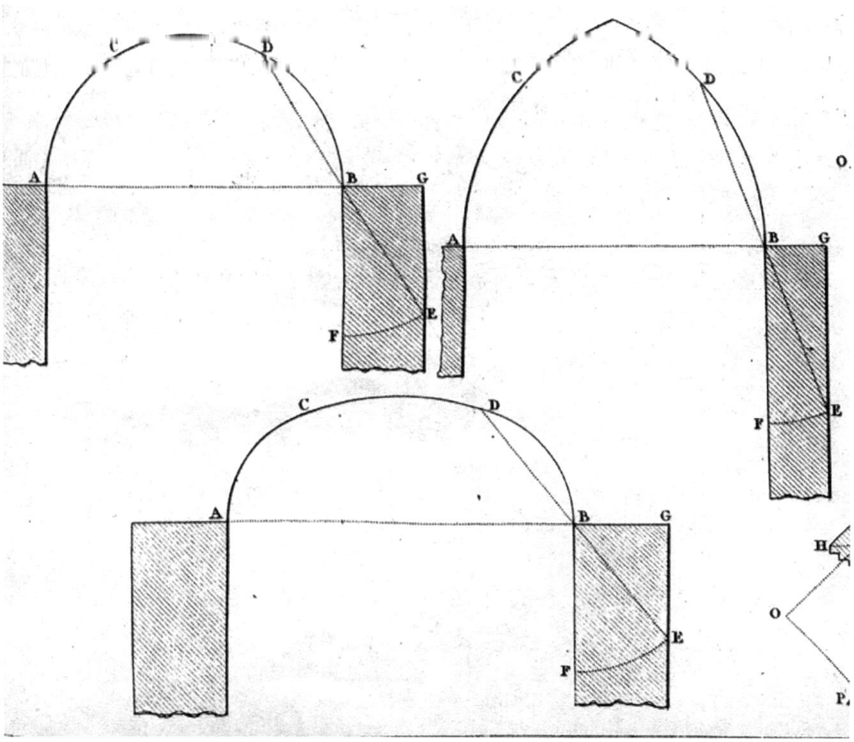

Fig. 4 Bélidor's illustration of Derand's law (Bélidor 1830, Figs. 5, 6, 9). Public domain

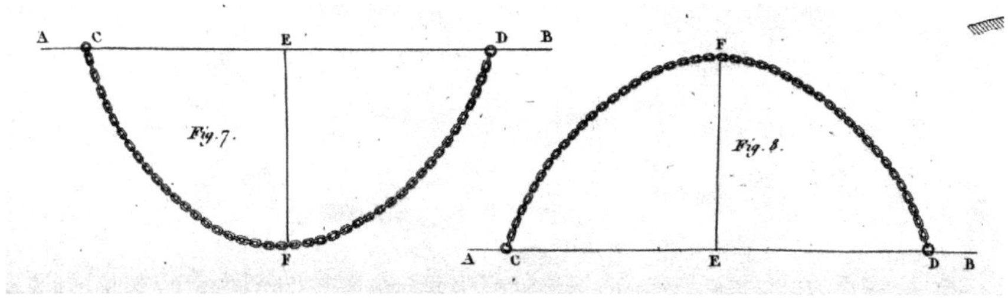

Fig. 5 Bèlidor's presentation of the catenary and the arch (Bélidor 1830, Figs. 6, 7). Public domain

Bélidor grants the greatest importance to the problem of the thickness of the pedestal, and he adds a chapter concerning "the curve that one should give to a vault, in order to, having all voussoirs equal, they would be in equilibrium".[7] He explains that it is also the curve adopted by a catenary under its own weight (Fig. 5).

[7] Bélidor (1830): *la courbe qu'il conviendrait de donner à une voûte, pour que, tous les voussoirs étant égaux en pesanteur, ils fussent en équilibre.*

Bélidor gives the result that had already been given by Hooke[8] in 1678, with the following qualification: "one knows that one does not change anything to an equilibrium of forces just changing their directions in their opposites".[9] The same result was also mentioned by Gregory in his article on the Catenary in the *Philosophical Transactions* (1697) and in 1705 it was demonstrated analytically in the case of an infinitely thin vault by Jacob Bernoulli, at the end of his *Meditationes* CCLXXXV, but this was not published until 1744 among his *Varia Posthuma* in his *Opera.* It is therefore probable that neither Bélidor nor Bouguer (who nevertheless was in contact with the Bernoulli family) knew anything about that work. Bélidor does not know any practical method for drawing that curve:

> If one wishes to construct a natural vault whose width and height are given, one must draw on a vertical surface a line CD equal to the width of the vault, drop from the middle of that line a perpendicular EF equal to the height that one wants to give to the vault, then attach the end of a chain to the point C, and carry the other end in the direction of D, so that, increasing or decreasing the length of the chain, its own weight makes it pass through the point F when it is fixed in the points C and D. After that, one may, with a pencil that one draws along the chain (without making it wobble), trace a curve, and upon it establish the figure of the false arch hanger of the vault, the section of the voussoirs, and the rest.[10]

For aesthetic reasons, Bélidor does hesitate to modify the curve obtained this way:

> if we wanted to make use of this curve, I believe that we would be obliged to bring its two ends G and H together, so that they were arranged like EA and FC, and not like EG and FH, which would not be suitable in the execution because the start of the vault would make a haunch with the jamb, which would disturb the view. It is good to take advantage of what theory can teach; but when it comes to practice, we can without qualms not follow it exactly in order to relate objects to use.[11] (Fig. 6)

---

[8] Truesdell (1960, p. 57).

[9] Bélidor (1830, p. 148): *on sait qu'on ne dérange rien dans l'équilibre des puissances en changeant seulement leur direction en son contraire.*

[10] Bélidor (1830, p. 149–150): *Si l'on veut construire une voûte naturelle dont la largeur et la hauteur soient données, il faut sur une surface verticale tracer une ligne CD égale à la largeur de la voûte, abaisser du milieu de cette ligne une perpendiculaire EF égale à la hauteur qu'on veut lui donner, ensuite attacher l'extrémité d'une chaîne au point C, et porter l'autre extrémité vers D, de manière qu'en augmentant ou diminuant la chaîne, son propre poids la fasse passer par le point F lorsqu'elle sera arrêtée aux endroits C et D. Après cela, on pourra, avec un crayon que l'on conduira tout au long de la chaîne (sans pourtant la faire vaciller), tracer une courbe, et là-dessus établir la figure du faux cintre de la voûte, la coupe des voussoirs et le reste.*

[11] Belidor (1830, p. 149): *si l'on voulait faire usage de cette courbe, je crois qu'on serait obligé de rapprocher ses deux extrémités G et H, afin qu'elles fussent disposées comme EA et FC, et non pas comme EG et FH, ce qui ne conviendrait pas dans l'exécution, à cause que la naissance de la voûte ferait un jarret avec le piédroit, ce qui choquerait la vue. Il est bon de profiter de ce que la théorie peut enseigner; mais quand il s'agit de la pratique, on peut sans scrupule ne la pas suivre exactement pour rapporter les choses à l'usage.*

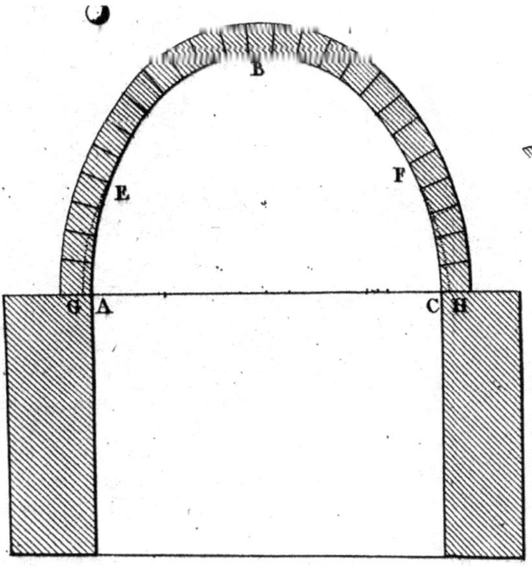

Fig. 6 Bélidor's adjustment of the curve to better suit the arch being constructed (Bélidor 1830, Fig. 5). Public domain

3 Bouguer's Work

The problem of the arch is a problem of two dimensions. Bouguer proposes to pass to three dimensions looking for the curve whose rotation will give the dome that is most stable under the action of its own weight. His work is, roughly speaking, divided into three parts. In the first, he studies the conditions of stability with help of the parallelogram law of forces. In order to do that, he simplifies the problem in the following way:

> BbA is the curve that forms the dome by its revolution around its vertical axis AD. This curved line passes through all points B, b, etc. through the middle of the thickness KL, HI, etc. of the vault, a thickness which we regard here as very small, and which effectively is, in fact, compared to the dimensions of the dome. All joints, such as KL, HI, etc. of the voussoirs are also assumed here to be perpendicular to the same curve BbA, as they ordinarily are.[12]

Those elements are to be added to the preceding remark in order to specify his idealized model of a dome: "I will always assume that the stones or voussoirs have their surfaces infinitely polished".[13]

[12] Bouguer (1736, p. 149): *BbA est la courbe qui forme le Dome par sa révolution autour de son axe, la verticale AD. Cette ligne courbe passe dans tous les points B, b, &c. par le milieu de l'épaisseur KL, HI, &c. de la Voute, épaisseur que nous regardons ici comme très petite, & qui l'est toujours en effet par rapport aux dimensions du Dome. Tous les joints, comme KL, HI, &c. des Voussoirs sont aussi supposés ici perpendiculaires à la même courbe BbA, comme ils le sont ordinairement.*

[13] Bouguer (1736, p. 149): *Je supposerai toûjours que les pierres ou les Voussoirs ont leurs surfaces infiniment polies.*

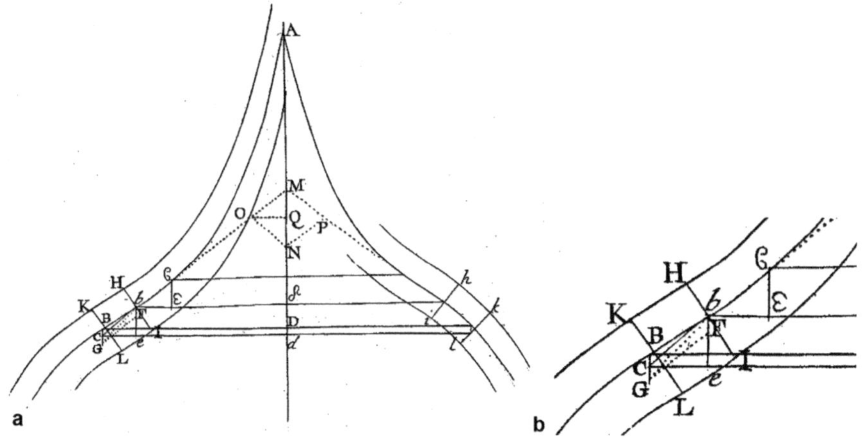

Fig. 7 **a**, left Bouguer's illustration of forces on the voussoirs; **b**, right detail (Bouguer 1736, Fig. 1). Public domain

> If we then consider a part *HAh* of the dome, it is obvious that it will push all the *HL* voussoirs which are immediately below, according to the perpendicular *bC* to the joint *HI*, or according to the extension of the short side *bb* of the curve. But as we consider lower points, the direction must change, because the gravity of each seat is successively added to the force made by the upper part. This part pushes at point *b* according to *bC*, and the force is expressed, if desired, by *bC* itself. But if we assume all the gravity of the voussoir *HL* united in point *b*, which it is permitted to do as soon as the thickness *Bb* of the voussoirs is infinitely small, we will only have to represent this gravity by the small vertical part *bF*; and if we compound it with the force *bC* that the upper part makes, we will have in the diagonal *bG* of the parallelogram *CbFG* the direction of the total force that the upper part makes increased from below by a seat, that is to say, the force that the entire *KAk* part makes. The direction of pressure is thus continually diverted; it forms a curve, which can be confused with the *AbbB* curve, but which can also be different from it, as it is here (Fig. 7, Fig. 8).[14]

He considers then three cases: the concave vault, the cone and the convex vault. It is remarkable that Bouguer is the only one, as far as I know, to consider concave vaults. If one compares the drawing he supplies to illustrate the ship hulls he studied

---

[14] Bouguer (1736, p. 149–150): *Si l'on considère après cela une partie HAh du Dome, il est évident qu'elle poussera tous les Voussoirs HL qui sont immédiatement au dessous, selon la perpendiculaire bC au joint HI, ou selon le prolongement du petit côté bb de la courbe. Mais à mesure qu'on considérera des points plus bas, la direction doit changer, parce que la pesanteur de chaque assise s'adjoûte successivement à l'effort que fait la partie supérieure. Cette partie pousse au point b selon bC, & l'effort est exprimé, si on le veut, par bC même. Mais si l'on suppose toute la pesanteur du Voussoir HL réunie dans le point b, ce qu'il est permis de faire aussi-tôt que l'épaisseur Bb des Voussoirs est infiniment petite, on n'aura qu'à représenter cette pesanteur par la petite partie verticale bF; & si on la compose avec l'effort bC que fait la partie supérieure, on aura dans la diagonale bG du parallelogramme CbFG la direction de l'effort total que fait la partie suppérieure augmentée par en bas d'une assise c'est-à-dire, l'effort que fait toute la partie KAk. La direction de la pression se trouve ainsi continuellement détournée; elle forme une courbe, qui peut se confondre avec la courbe AbbB, mais qui peut aussi en être différente, comme elle l'est ici.*

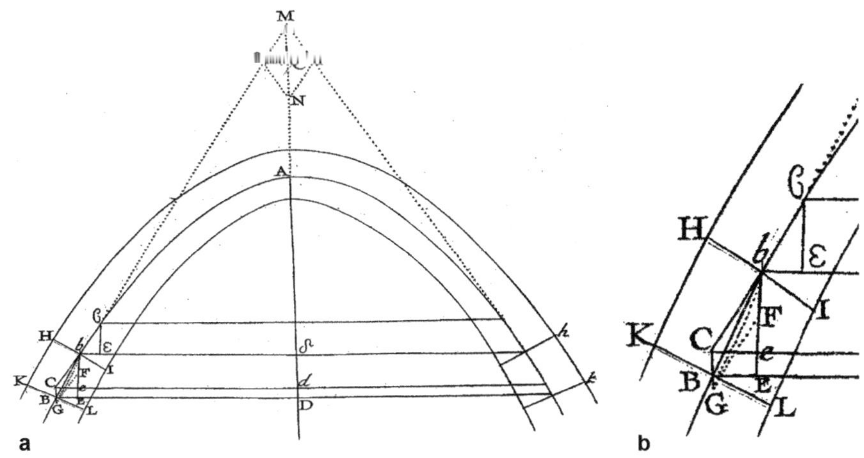

Fig. 8 **a**, left Bouguer's illustration of forces on the voussoirs; **b**, right detail (Bouguer 1736, Fig. 2). Public domain

as the optimal forms in his *Traité du Navire* one may suspect how he got that idea (Fig. 9).

Bouguer finds that there is an instability and even a rupture when the resultant applied to a voussoir of the total push of the superior part of the vault (bC) and of the weight (bF) of the considered voussoir which leads the voussoir towards the outside of the vault. There is no such risk in the concave or conic vaults, the curve always

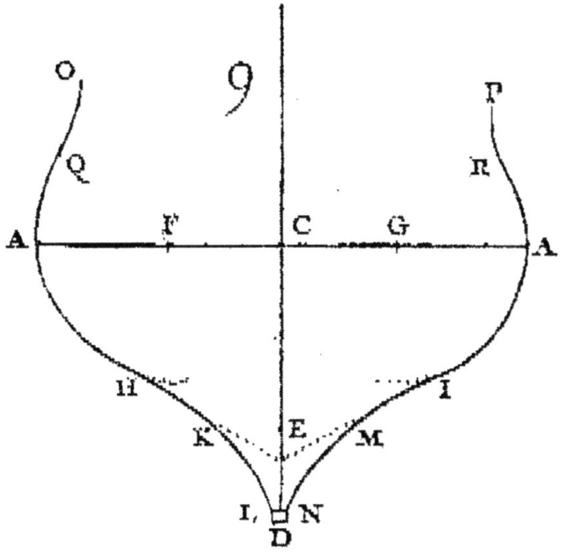

Fig. 9 Figure taken from Bouguer's *Traité du Navire* showing the idea for "concave vaults" (Bouguer 1746, Fig. 9). Public domain

passes above the direction given by the thrust and, thus *a fortiori,* the resultant of the last one composed with the weight of the voussoir.

Those vaults are always stable. In *Nuove ricerche sull'equilibrio delle volte* (1785), Mascheroni will mention that result of Bouguer, underlining its relevance. In the case of convex vaults, the curve passes under the direction given by the thrust.[15]

In this case, one can save the situation when the resultant (*bG*) of the total pressure (*bC*) and of the weight of the voussoir (*bF*) passes under the curve. Bouguer formulates that condition by comparing the vertical distance *CG,* which equals *bF* and *CB* and asks that $CG \geq CB \cdot eC$, which, expressing the magnitudes, gives him the following analytical expression:

$$\frac{\varepsilon y dx\sqrt{dy^2 + dx^2}}{\int \varepsilon y\sqrt{dy^2 + dx^2}} \geq ddx$$

where ε designates the thickness of the voussoirs.

Among the curves that satisfy that condition, he looks for, and that is the aim of the second part of his work:

> the last of all, [which] has its short sides, like *bB,* exactly located on the directions *bG.* the vertical joints do not support any part of the force, since the voussoirs are not pushed inwards. Also, the slightest external agent is capable of overthrowing this last vault; and although it supports itself, it is always ready to fall.[16]

He searches at three dimensions the limit case corresponding to the catenary at two dimensions. The illustration of that last case, given by Poleni in his *Memorie istoriche, della Gran Cupola del Tempio Vaticano* (1748) speaks for itself: his vault is composed of spheres in equilibrium, but the slightest breath makes it collapse (Fig. 10).

This limit case, the one of the most convex curves, which he qualifies as least, corresponds to the equality CG = CB, which he expresses in the following way:

$$\frac{\varepsilon y dx\sqrt{dy^2 + dx^2}}{\int \varepsilon y\sqrt{dy^2 + dx^2}} = ddx$$

The third part of Bouguer's text is dedicated to applications of the formula:

$$\frac{\varepsilon y dx\sqrt{dy^2 + dx^2}}{\int \varepsilon y\sqrt{dy^2 + dx^2}} \geq ddx$$

---

[15] For more on Mascheroni, see "On Mascheroni's "New Researches on the Equilibrium of Vaults", Chap. 14 in this present volume.

[16] Bouguer (1736, p. 152): la *dernière de toutes* [qui] *a ses petits côtés, comme bB, exactement situés sur les directions bG. les joints verticaux ne supportent dans celle-ci aucune partie de l'effort, puisque les Voussoirs ne sont point poussés en dedans. Aussi le moindre agent extérieur est-il capable de renverser cette dernière Voute; & quoiqu'elle se soûtienne, elle est toûjours prête à tomber.*

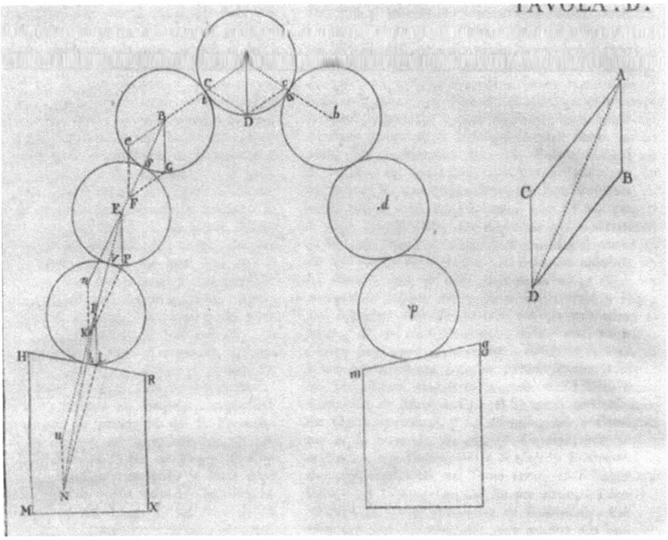

Fig. 10 Poleni's vault made of spheres (Poleni 1748, Tav. D). Public domain

He addresses two problems: Given the curve of the vault, how to determine its thickness, and the inverse problem, given the thickness, how to determine the curve.

Mascheroni disagrees with Bouguer concerning the considerations on the thickness of the vault. In his estimation, that Bouguer oversimplifies the computations of the volumes of the voussoirs and maintains that the optima curve no longer passes through the centres of gravity. More precisely, he holds that the centres of gravity are no longer where he says they are (Fig. 11).[17]

Finally, in order to better determine the last of all the curves which, by rotation, generate a stable dome, he separates the variables of his equation and tries integration by series.

While Bouguer does not say so explicitly, Benvenuto remarks that the curve he finds is effectively an extension to the dome of the analogy between catenary and arc,[18] which means that its equilibrium is broken by the smallest displacement.

In 1776, Bossut[19] takes up again the study of the dome in equilibrium under its own weight, thanks again to the parallelogram law of forces to which he joins the ray of curvature. He does not consider the action of the adjacent voussoirs and observes that the curve he finds is not the catenary.

[17] Bouguer considers a superposition of rectangular parallelepipeds. Mascheroni shows that they must be curved in order to form a vault, and in this case their centres of gravity are moved.

[18] Benvenuto (2012, p. 345).

[19] Published in Bossut (1778,1779).

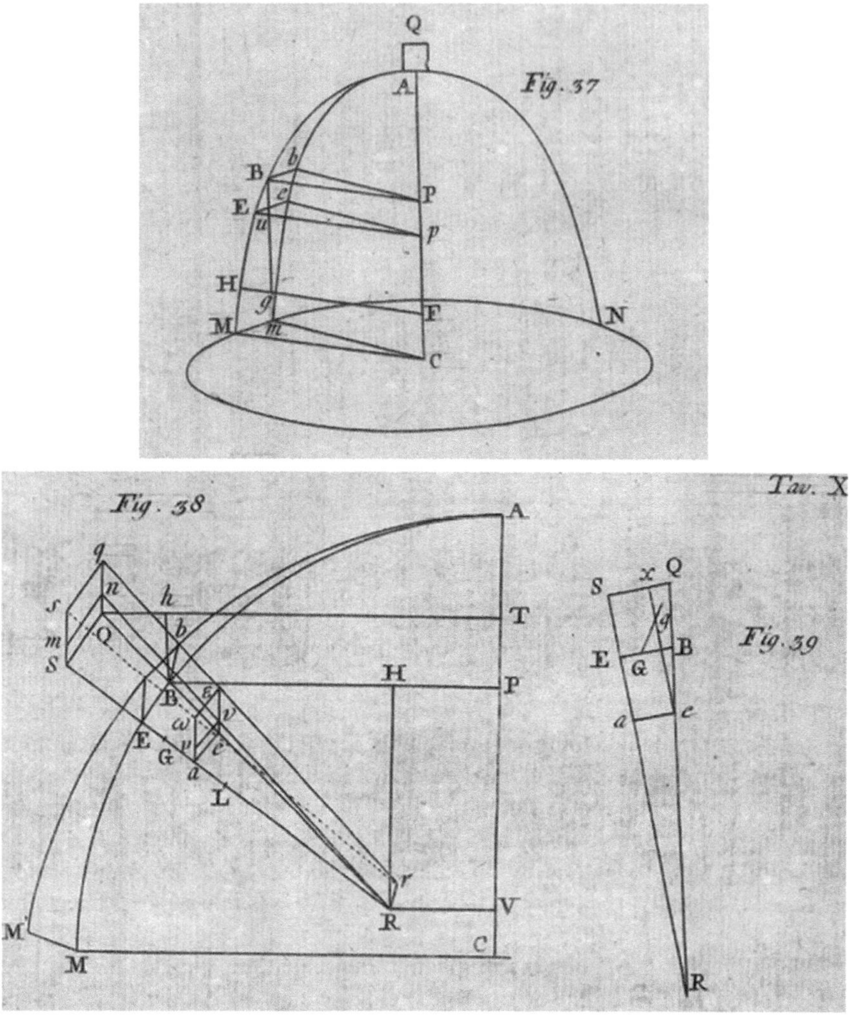

Fig. 11 Mascheroni's figures for his problems XXXIV-XXXVI addressing Bouguer's ideas (Mascheroni 1785, Tav. IX, X). Public domain

4 The Calculus of Vaults and the Principles of Mechanics

4.1 The Parallelogram Law of Forces

Like Bouguer, Bossut used the parallelogram law of forces (Fig. 12).

It appears in the same context in de la Hire's text of 1695 and is taken up again by Couplet and Bélidor. It is worthwhile emphasizing that the form given to that law in the domain of the calculus of vaults is particular. It is under that form given by de

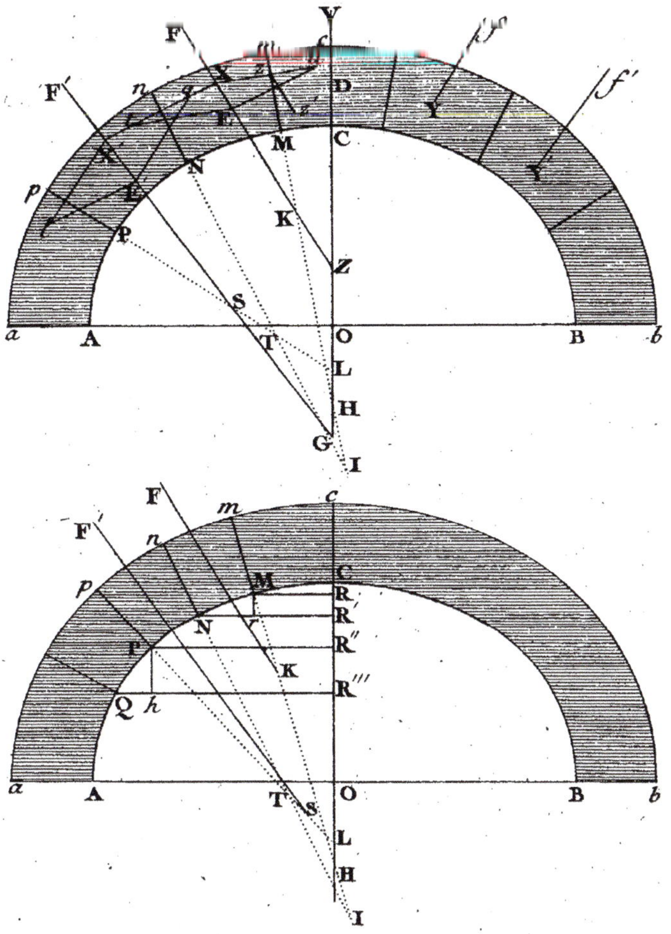

Fig. 12 Bossut's use of the parallelogram law of forces (Bossut 1779, Figs. 1, 3). Public domain

la Hire that Bélidor states it in his "Principles drawn from mechanics.[20] He shows that three forces in equilibrium are in the same relationship as the sides of a triangle that are perpendicular to the direction of three forces in equilibrium. This form is obviously equivalent to Varignon's law of sines.

The reason for that predilection is that it allows to use the direction of the joints that are always taken perpendicular to the intrados curve or to the optimal curve that one is looking for and this makes it easy to the radius of curvature of the curve one is looking for.

[20] Bélidor (1830, p. 106–107): *Principes tirés de la mécanique.*

4.2 The Voussoir and the Wedge or Inclined Plane

We find in the history of the Academy for Sciences for 1704:

> It is obvious that all the voussoirs have a wedge shape, wider at the top than at the bottom, by virtue of which they lean and support each other, and reciprocally resist the force of their weight, which would cause them to fall.[21]

This way of considering the vault also explains why those authors prefer the parallelogram law of forces that uses, as we saw, the triangle whose sides are perpendicular to the forces. This triangle corresponds to the inclined plane.

For most of the authors, the vertical support or counterfort works like a lever, but at first sight, it has nothing to do with the problem of finding the optimal curve to give to an arch or a dome with which we are concerned.

4.3 The Lever

Nevertheless, the lever also plays a role in the equation of the problem of the infinitely thin vault under its own weight carried out by Jacob Bernoulli. He uses the lever to express the condition of equilibrium at rotation of a voussoir and uses the difference of forces to express the equilibrium at translation. In his *Meditationes CCLXXXV* (1744) Jacob Bernoulli finds the catenary by requiring the equilibrium of the voussoir which can only slide vertically (translation equilibrium) on the inclined plane of the neighbouring segment or pivot around its support D (rotation equilibrium) (Fig. 13).

5 The Principle of Virtual Works and the Mechanisms of Rupture

Jacob Bernoulli, we just saw, describes two mechanisms of rupture, one by translation and the other one by rotation. In *La science des ingénieurs*, Bélidor describes another one using the counterfort (Fig. 14):

> having a semicircular vault *BDI*, of which each quarter of a circle *BD* and *DI* is devised equally in two in the *C* and *H*, experiment shows that it is always at the same places *FC* and *&H* that the vault splits when its push is greater than the resistance of the counterforts.[22]

[21] Anon. Hist. Paris (1704, p. 93): *il est évident, que tous les Voussoirs ont une figure de Coin, plus large par haut que par bas, en vertu de laquelle ils s'appuyent & se soutiennent les uns les autres, & résistent réciproquement à l'effort de leur pesanteur qui les porteroit à tomber.*

[22] Belidor (1830, p. 111): *ayant une voûte en plein cintre BDI, dont chaque quart de cercle BD et DI soit divisé en deux également au point C et H, l'expérience montre que c'est toujours aux endroits FC et fH que la voûte se désunit quand sa poussée est au dessus de la résistance des piedroits.*

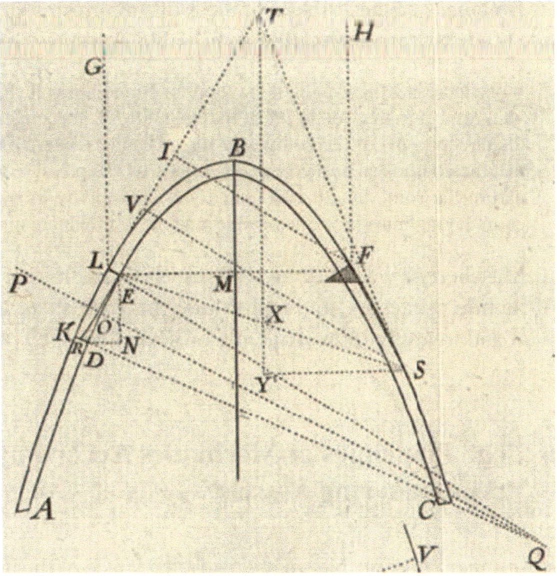

Fig. 13 Bernoulli's figure of the catenary (Bernoulli 1774, Tab. XLVIII). Public domain

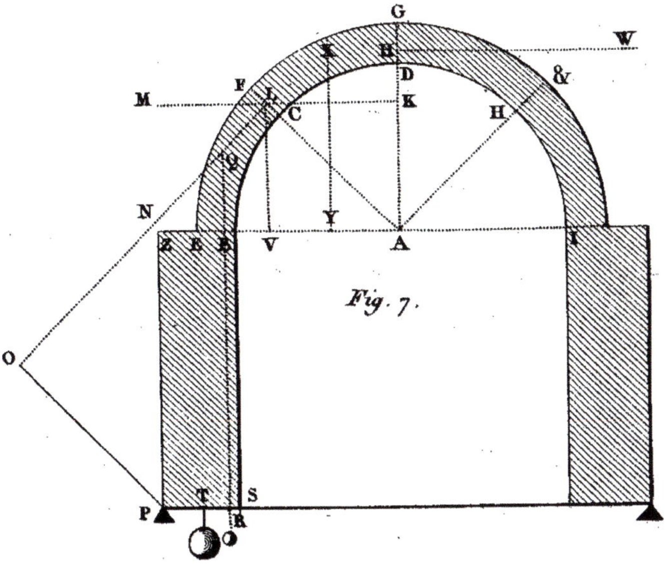

Fig. 14 Bélidor's study of rupture mechanisms (Bélidor 1830, Fig. 7). Public domain

Bouguer also describes a process of rupture similar to that of Jacob Bernoulli, but he clearly distinguishes the two possible motions:

> If one increases the weight of the superior part *HA* too much, the force *bC* becomes too great with respect to the weight *bF* of the Voussoir *HL*, the angle *CbB* becomes too acute, and so the direction *bG* of the compound force becomes external with respect to *bB*, what shows that the voussoir is pushed further outside by the pressure along *bC* than does the superior part of the vault, than it is solicited versus inside by its own weight. And in this case the entire laying must protrude outside, and the entire dome must fall.[23]

Mascheroni was the first to think about exploiting the results of the observation of rupture processes, and to consider the vault as an assembly of two bars by a ball joint and to apply it the principle of virtual works.

6 The Principles of Mechanics According to Bouguer in His "Manoeuvring Vessels"

If Bouguer was not able, in the restricted space of the article that we have commented on, give his point of view on either on the principles of mechanics or the role of practice, this is not the case in his later works. One finds in his "manoeuvring vessels" of (1757) an entire book "in which we give the knowledge of mechanics and dynamics useful or necessary to navigators, with the solution of several important naval problems".[24] The first part is a treatise on statics that can also be applied to the construction of vaults. There one finds again the elements that we just mentioned regarding the study of vaults. After having treated pulleys, Bouguer approaches the decomposition of forces or of motions with the case of the inclined plane, then he goes on to the composition of forces and of motions. He ends with the part that interests us: the study of the lever. Like many authors of that time, he sums up the force to the motion it stimulates.

With regard to the decomposition of forces, Bouguer states the theorem of de la Hire, "Another method to determine the relation that must exist between the weight or the forces that are in equilibrium, and application of that method to the inclined plane,[25] having recalled the usual form of that law, giving for two forces, the diagonal of the parallelogram they compose (Fig. 15).

---

[23] Bouguer (1736, p. 154): *Si l'on augmente trop la pesanteur de la partie supérieure HA, l'effort bC devient trop grand par rapport à la pesanteur bF du Voussoir HL, l'angle CbB devient trop petit, & alors la direction bG de l'effort composé se trouve extérieure par rapport à bB, ce qui montre que le Voussoir est plus poussé en dehors pas la pression selon bC que fait la partie supérieure de la Voute, qu'il n'est sollicité à avancer en dedans par sa propre pesanteur. Or dans ce cas l'assise entière doit saillir en dehors, & le dome doit tomber.*

[24] Bouguer (1757, p. XV): *dans lequel on donne les connoissances de Méchanique & de Dynamique utiles ou nécessaires aux navigateurs, avec la solution de plusieurs Problêmes importants de Marine.*

[25] Bouguer (1757, p. 24): *Autre méthode pour déterminer le rapport qui doit se trouver entre les poids ou les forces qui sont en équilibre, & application de cette méthode au plan incliné.*

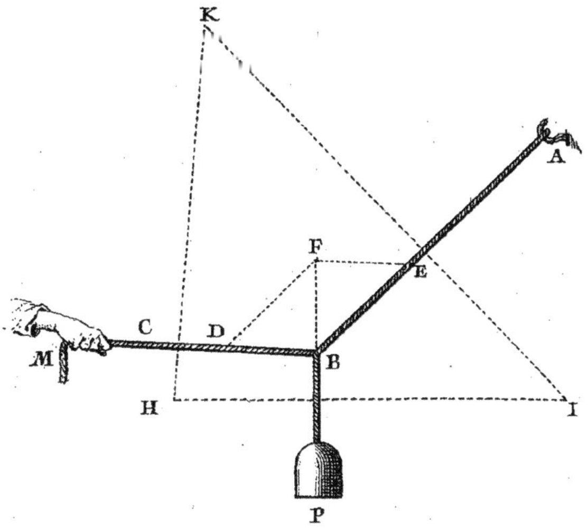

Fig. 15 Bouguer's illustration of the method for determining the ratio between the weights or forces which are in equilibrium, and the application of this method to the inclined plane (Bouguer 1757, pp. 24–25 and Fig. 18). Public domain

> But if one wishes to find that relation in a way that is even simpler, one just has to raise three perpendiculars to the three directions *BP, BA,* and *BC,* letting them pass through whatever point one wishes. Those three perpendiculars *HI, KI* and *HK* will form a triangle, and the length of each side will represent the force in the direction to which it is perpendicular. That means that *HI* expressing the greatness of the weight *P,* the two other sides *KI* and *HK* will express the magnitude of the forces that the ropes *BA* and *BC* have to sustain.[26]

But that point is not the most original. At the end of Chap. 4, we find the remark: "that at the equilibrium the moments of the forces or the product of those forces by the distance of their direction to the resting point are always exactly equal".[27] Bouguer adds to that the definition of the moment: "One must remark that the moment is a particular quantity of which one only finds rotary motion or rotation".[28] The text is certainly posterior to the one Euler dedicates to "a new problem of mechanics" (Euler 1750), and where, in the case of the motion of a body like earth around an axis of rotation, he requires the equilibrium of momenta in order to avoid the displacement of the axis. But Bouguer goes on in Chap. 5: "On the equilibrium of a great number of

---

[26] Bouguer (1757), p. 25: *Mais si on veut trouver ce rapport, d'une manière encore plus simple, on n'a qu'à élever trois perpendiculaires aux trois directions* BP, BA, & BC *en les faisant passer par quel point on voudra. Ces trois perpendiculaires* HI, KI & HK *formeront un triangle, & chacun représentera par sa longueur l'effort à la direction duquel elle sera perpendiculaire. C'est-à-dire que* HI *exprimant la grandeur du poids* P, *les deux autres côtés* KI & HK *exprimeront la grandeur des efforts qu'ont à soutenir les cordes* BA & BC.

[27] Bouguer (1757, p. 32): *Que dans l'équilibre les moments des forces ou les produits de ces forces par la distance de leurs directions au point d'appui sont toujours exactement égaux.*

[28] Bouguer (1757, p. 33): *il faut bien remarquer que le moment est une quantité particulière dont on ne trouve d'exemple que dans les méchaniques, & qui n'a lieu que lorsqu'une force tend à produire quelque mouvement de tournoyement ou de rotation.*

forces applied to a lever",[29] to explain that in that problem of statics, one cannot only write the equilibrium of forces, but, in order for the lever not to turn, it is necessary to have the equilibrium of moments. Only in *Theoria motus corporum solidorum seu rigidoru* (1765) will Euler, in a general manner, put on the same footing these two fundamental laws of statics, as Jacob Bernoulli had done, 60 years before, in the particular.

7 The Reality Test for Vault Theory

One can find multiple examples for the verification of stability examined in a theorical manner. One most often cited is that by Poleni's *Memorie istoriche, della Gran Cupola del Tempio Vaticano* (1748). Poleni uses the theory of the catenary in order to judge the stability of that cupola (Fig. 16).

On his advice, authorities agreed to put iron hoops around the cupola to hold it together. Some years later, in 1774, Bossut applied his theory in order to verify the thickness of the piers of the Sainte-Geneviève church built by Soufflot. Three more years later, Frisi applied his theory to the verification of the stability of the stability of the dome of Milan.

8 Theory and Practice in the *Traité du Navire*, the Role of Principles by Bouguer

The introduction that Bouguer gives to his *Traité du Navire* describes in an extraordinary *crescendo* the relations between theory and practice. This introduction would also find its place at the beginning of a treatise concerning vaults. He wrote:

> No subject was more in need of being illuminated with the light of Mathematics; and it is certain that none has been more deprived of it until now. Dejected at the sight of the first obstacles, seamen too soon stopped making new efforts; they took a course which could only be dictated by despair: that of giving themselves over to the most imperfect practice, denying themselves any help from theory.[30]

His discourse is inspired by military architecture, that is, in particular, by the calculations of stability.

---

[29] Bouguer (1757, p. 35): *De l'équilibre entre un grand nombre de puissances appliquées à un levier.*

[30] Bouguer (1746, p. viii): *Aucune matière ne demandoit davantage à être éclairée de la lumiere des Mathématiques; & il est certain qu'aucune n'en a été plus privée jusques à présent. Rebutés à la vûë des premiers obstacles, les Marins cesserent trop tôt de faire de nouveaux efforts; & ils prirent un parti qui ne pouvoit être dicté que par le desespoir, celui de se livrer à la Pratique la plus imparfaite, en s'interdisant tout secours de la part de la Théorie.*

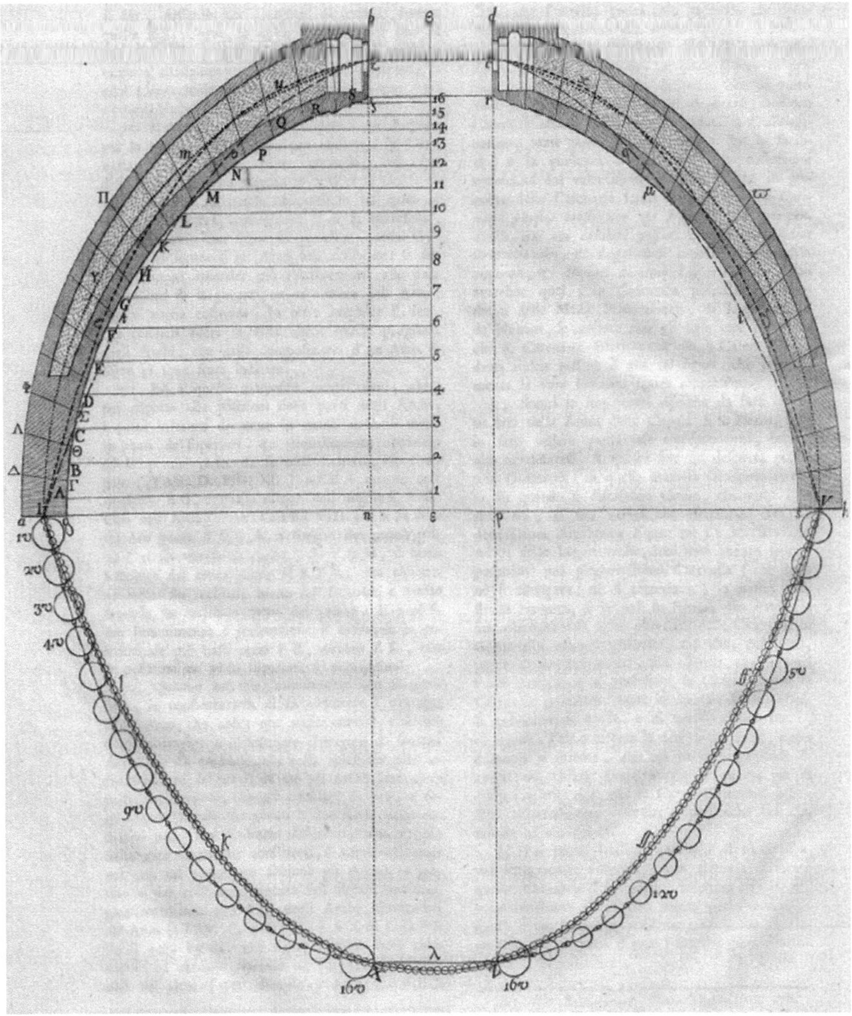

Fig. 16 Poleni's illustration of the hanging chain and inverted cupola (Poleni, 1748, Tav. E). Public domain

> We had an example before our eyes, and we did not want to pay attention to it. We did not consider that although military architecture is so very much easier, nothing has ever been carried out with regard to it, except in the direction of the geometry and mechanics.[31]

Mathematics presented the way, and Bouguer attacked the arguments of its opponents by beginning with the problem posed by approximations. Those theories use

[31] Bouguer (1746, p. viii): *On avoit un exemple sous les yeux, & on ne voulut y faire attention: On ne considera point que quoique l'Architecture militaire soit extrêmement plus facile, rien ne s'est exécuté de tout tems à son égard, que par la direction de la Géométrie & des Mécaniques.*

models, such as that of the infinitely thin vault with perfectly polished voussoirs. This must be known so as not to distort the judgment of the relationship between theory and practice. Bouguer comments in this way:

> A proposition can be true in theory and, at the same time, false in practice. In fact, there is hardly a single question of those which involve physics which is resolved with all rigor, despite the frequent applications which have been made, as if at will, in recent times by modern analysis. We always allow ourselves some softening; we sometimes skilfully dismiss particularities which would make the discussion too thorny; and after that we should not be surprised if the experiment does not respond to all we had promised ourselves. But let us not conceal any circumstance; let us try to resolve the problems in all their difficulty, and we will see a continual agreement between theory and practice; the contrary would imply contradiction.[32]

Thus, disagreements between theory and practice should not lead to rejecting the use of mathematics as many are tempted to do:

> Let us say, if we wish, that we have not yet been fortunate enough to draw from mathematics all the insight we need for the Navy; various examples prove this only too well. But let us agree at the same time that it is not possible that a science which is entirely occupied with the care of weighing, measuring, and comparing magnitudes, is not of the utmost usefulness, provided we know how to make a legitimate application in a subject where it is a question of regulating one or the other, such a large number of parts, and of establishing equilibrium between such a large number of different forces. This is what we would have soon experienced through a happy experiment, if we had worked harder to give geometry credit; by taking it out of the confines of its speculations, an effort had been made to seriously extend its uses.[33]

He then reproaches the professionals, first concerning the secrecy surrounding the knowledge of shipbuilding; this precludes an accumulation of knowledge based on experiments:

> One feel, as we do, how harmful this silence of people in the profession is: we see clearly that it prevents us from taking advantage of the factual knowledge that they must have at

[32] Bouguer (1746, p. xii–xiii): *Une proposition peut être vraye dans la Théorie & fausse en même temps dans la Pratique. A peine est-il en effet une seule question de celles qui sont mêlées de Physique, qui soit résoluë en toute rigueur, malgré les fréquentes applications qu'on a faites, comme à l'envi, dans ces derniers tems de l'Analyse moderne. On se permet toujours quelsques adoucissemens, on écarte quelquefois avec adresse les particularités qui rendroient la discussion trop épineuse; & il ne faut pas s'étonner après cela, si l'expérience ne répond pas à tout ce qu'on s'étoit promis. Mais qu'on ne dissimule aucune circonstance, qu'on tente de résoudre les Problêmes dans toute leur difficulté; & on verra un continuel accord entre la Théorie & la Pratique; le contraire impliqueroit contradiction.*

[33] Bouguer (1746, p. xiii–xiv): *Qu'on dise, si on veut, qu'on n'a pas encore été assez heureux pour tirer des Mathématiques, toutes les lumieres dont on a besoin pour la Marine; différens exemples ne le prouvent que trop; mais qu'on convienne en même temps qu'il n'est pas possible qu'une science qui est toute occupée du soin de peser, de mesurer, & de comparer les grandeurs, ne soit de la dernière utilité, pourvû qu'on en sache faire une application légitime dans un sujet où il s'agit de regler les unes sur les autres, un si grand nombre de parties, & de mettre l'équilibre entre un si grand nombre de différentes forces. C'est ce qu'on n'eût pas tardé à éprouver par une heureuse expérience, si ont eût travaillé d'avantage à mettre la Géométrie en crédi t; & qu'on se fût efforcé en la faisant sortir de l'enceinte de ses spéculations, d'en étendre sérieusement les usages.*

> least acquired through their long use. They argue willingly and heatedly about things of little consequence, while most of the construction remains buried under thick darkness; instead, if everyone communicated what experience has taught them, … we would not take long to experience the considerable fruit that would arise from this happy communication. The builders, on the contrary, … even maintain a secrecy so profound that their particular practices, although they are always only some slight modifications of the general maxims, constitute a very extraordinary heritage, which is almost never transmitted except from father to son.[34]

His second reproach regards the casual nature of practice often covers over the lack of method:

> One sees that we defer every day in very important circumstances to simple trial and error, to that which is most prone to deception: we change the shape of the upper parts of ships; we add a new deck to them or we remove it; we also completely alter the shape of their hull, and we agree to make all these changes without knowing what the immediate effect will be … while we could determine this in a manner as precise as it is infallible, by using the slightest knowledge of geometry and the simplest operations of arithmetic.[35]

Finally, Bouguer denounces the deep deficiency camouflaged by trial and error and secrecy:

> It is of very little use, or rather it becomes useless, to resort to the advice of more than one builder; we do nothing other than indulge in greater uncertainties. It always happens that they think differently from each other; and nevertheless each one also commends to his favour his own practice or the boring and long enumeration of the ships that he has already built. As it is impossible for them to reconcile, because they have no means to do so, no common principle on which they agree or from which they can start, no rule, no clue even to discern the truth, or to make it known, they are reduced to continually repeating the same assertions instead of proof.[36]

---

[34] Bouguer (1746, p. xvi): *On sent comme nous, combien ce silence des gens du métier est nuisible: on voit assez qu'il empêche de profiter des connoissances de fait qu'ils ont au moins dû acquerir par leur long usage. Ils disputent volontiers & avec chaleur sur des choses de peu de conséquence; pendant que l'essentiel de la Construction reste enseveli sous d'épaisses ténébres: au lieu que si chacun communiquoit ce que lui a appris l'expérience, … on ne tarderoit pas à éprouver le fruit considérable qui naîtroit de cette heureuse communication. Les Constructeurs au contraire, … observent même un secret si profond, que leurs pratiques particulieres, quoiqu'elles ne soient toujours que quelques legeres modifications des maximes générales, constituent comme un héritage tout extraordinaire, qui ne se transmet presque jamais que de pere en fils.*

[35] Bouguer (1746, p. v xvii): *On verra qu'on défere tous les jours dans des circonstances très-importantes au simple tatonnement, à celui qui est le plus sujet à tromper: on change la forme des Navires par leurs hauts, on leur ajoute un nouveau pont ou on le retranche, on altere aussi totalement la figure de leur carène, & on consent à faire tous ces changemens sans sçavoir quel en sera l'effet immédiat … pendant qu'on pourroit se déterminer d'une maniere aussi précise qu'infaillible, en employant les moindres connoissances de Géométrie & les plus simples opérations d'Arithmétique.*

[36] Bouguer (1746, p. xviii): *Il sert très-peu, ou plûtôt il devient inutile de recourir à l'avis de plusieurs Constructeurs; on ne fait autre chose que se livrer à de plus grandes incertitudes. Il arrive toujours qu'ils pensent différemment les uns des autres; & néanmoins chacun allegue également en sa faveur sa propre pratique ou l'ennuyeux & long dénombrement des Navires qu'il a déjà construits. Comme il leur est impossible de se concilier, parce qu'ils n'ont aucun moyen pour le faire, nul principe commun dont ils conviennent ou dont ils puissent partir, nulle regle, nul indice même pour discerner le vrai, ou pour le faire connoître, ils sont réduits à répéter continuellement les mêmes assertions, au lieu de preuves.*

They lack the theory which serves as a foundation:

> They cannot be better compared than to a large number of people, who, while busy with commerce, know neither the use of scales nor that of measurements in the exchange of their goods.[37]

To conclude:

> Practice alone and devoid of all the aid of theory cannot reveal the true rules in such a subject. The ship is such a composite whole that any change made to a single part is the beginning of an infinity of different arrangements or combinations, each of which must have a particular success. We cannot, for example, change the width of the hull without having to change all the other parts.[38]
>
> It is therefore certain that although Practice is extremely necessary and cannot be too highly recommended, it is nevertheless of too limited use when left alone in the present circumstance. Its too great limitation is a vice inseparable from its nature, and makes itself felt in an infinity of other connections. This is because it is originally sterile; it is because it only provides, when it is absolutely alone, knowledge which does not produce any other, or which always remains confined within the first limits that we have given them. ... It will not be possible to take another step forward, for lack of knowing or even suspecting that all these actions are subject to certain laws which it would be important to know. On the contrary, let us bring in the light of speculation; it is true, and it cannot be said too often, that it will always be necessary to add the experiences or knowledge of fact in which the practice consists; but theory will then use them as principles; it will draw certain deductions from them; and extending its views to all other cases, and to infinity, because it is not constrained by the same limits as practice, it will effectively take the place of an infinity of other experiences that have never been possible to execute.[39]

---

[37] Bouguer (1746, p. xviii): *On ne peut mieux les comparer qu'à un grand nombre de personnes, qui tout occupées de commerce, ne connoîtroient dans l'échange de leurs marchandises ni l'usage des balances ni celui des mesures.*

[38] Bouguer (1746, p. xx): *Que la Pratique livrée à elle seule & dénuée de tous secours de la Théorie, ne peut pas faire découvrir les vraies regles en un pareil sujet. Le Navire est un tout si composé, que chaque changement fait à une seule partie, est le commencement d'une infinité de dispositions ou de combinaisons différentes, dont chacune doit avoir un succès particulier. On ne peut, par exemple, toucher à la largeur de la carène sans se mettre dans le nécessité de changer toutes les autres parties.*

[39] Bouguer (1746, p. xxii–xxiii): *Il est donc certain que quoique la Pratique soit d'une extrême nécessité & qu'on ne puisse trop la recommander, elle est cependant d'un usage trop peu étendu, lorsqu'on la laisse à elle seule dans la circonstance présente. Sa trop grande limitation est un vice inséparable de sa nature, & qui se fait sentir dans une infinité d'autres rencontres. C'est qu'elle est originairement stérile, c'est qu'elle ne fournit lorsqu'elle est absolument seule, que des connoissances qui n'en produisent point d'autres, ou qui restent toujours renfermées dans les premières bornes qu'on leur a données.... Il ne sera pas possible d'avancer d'un pas de plus, faute de sçavoir ou même de soupçonner que toutes ces actions sont soumises à certaines loix qu'il seroit important de connoître. Qu'on fasse intervenir au contraire les lumières de la Spéculation; il est vrai, & on ne sauroit le dire trop de fois, qu'il faudra toujours y joindre les expériences ou ces connoissances de fait dans lesquelles consiste la Pratique; mais la Théorie s'en servira ensuite comme de principes, elle en tirera des inductions sûres; & étendant ses vûës à tous les autres cas, & jusqu'à l'infini, car elle n'est pas arrêtée par les mêmes bornes que la Pratique, elle tiendra effectivement lieu d'une infinité d'autres expériences qu'il jamais été possible d'exécuter.*

References

Anon. Hist. Paris. 1704. Sur la figure de l'extrados d'une voute circulaire, dont tous les voussoirs sont en équilibre entre eux. *Histoire de l'Académie de Paris pour 1704,* 93–96.

Bélidor, Bernard Forest de. 1830. *La science des ingénieurs dans la conduite des travaux de fortification et d'architecture civile,* 2nd ed. Claude-Louis Navier. Paris: Firmin Didot.

Benvenuto, E. 2012. *An introduction to the history of structural mechanics.* Part II, Vaulted structures and elastic systems: Heidelberg, Springer-Verlag.

Bernoulli, Jacob. 1744. Med. CCLXXXV et V.P. XXIX, 5/12/1704, "Problema de Curvatura fornicis, cujus partes se mutuo proprio pondere suffulciunt sine opere caementi". *Jacobi Bernoulli Basilieensis Opera,* Genève, Cramer et Philibert, 1119–1123.

Bossut, Charles. 1778. Recherches sur l'équilibre des voûtes, *Mémoires de l'Académie Royale des Sciences", année 1774,* 534–566.

Bossut, Charles. 1779. Nouvelles recherches sur l'équilibre des voûtes en dôme. *Mémoires de l'Académie de Paris, année 1776,* 587–596.

Bouguer, Pierre. 1727. *De la mâture des vaisseaux, pièce qui a remporté le prix de l'Académie royale des sciences proposé pour l'année 1727, selon la fondation faite par feu M. Rouillé de Meslay....* Paris: C. Jombert.

Bouguer, Pierre. 1729. *De la méthode d'observer exactement sur mer la hauteur des astres, piece qui a remporté le prix proposé par l'Academie Royale des Sciences pour l'année 1729.* Paris: C. Jombert.

Bouguer, Pierre, 1731. De la méthode d'observer en mer la déclinaison de la boussole. *piece qui a remporté le prix proposé par l'Academie Royale des Sciences pour l'année 1731.* Paris: C. Jombert.

Bouguer, Pierre, 1732. Sur de nouvelles courbes auxquelles on peut donner le nom de lignes de poursuite. *Mémoires de mathématique et de physique tirés des registres de l'Académie royale des sciences,* 1–14.

Bouguer, Pierre. 1736. Sur les lignes courbes qui sont propres à former les voûtes en dôme. *Mémoires de l'Académie Royale des Sciences, année 1734,* 149–164.

Bouguer, Pierre. 1746. *Traité du Navire, de sa construction et de ses mouvements.* Paris, C. Jombert.

Derand, Françoise. 1643. *L'architecture des voutes ou l'art des traits et coupe des voutes.* Paris: Sebastien Cramoisy, Imprimeur du Roy.

Euler, Leonhard. 1750. Découverte d'un nouveau principe de mécanique, E177. *Mémoires de l'Académie des Sciences de Berlin 1750,* 185–207, in *Leonardi Euleri Opera Omnia,* série. II, vol. 5, 81–108.

Euler, Leonhard. 1765. *Theoria motus corporum solidorum seu rigidoru,* E289. Greifswald: Rostock, A.F. Röse. In *Leonardi Euleri Opera Omnia,* série. II, vol. 3–4.

Gregory, David. 1697. Catenary. *Philosophical Transactions,* 637–652.

Mascheroni, Lorenzo. 1785. *Nuove ricerche sull'equilibrio delle volte.* Bergamo, Fr. Locatelli.

Nieuport, Viscount de (Charles-Ferdinand-Antoine-Florent-François Le Prudhomme d'Hailly). 1778. Essai analytique sur la méchanique des voûtes. *Mémoires de l'Académie de Bruxelles pour 1778,* 2, 43–136.

Bouguer Pierre, 1757. *Manœuvre des vaisseaux ou Traité de méchanique et de dynamique; Dans lequel on réduit à des solutions très-simples les problêmes de marine les plus difficiles, qui ont pour objet le mouvement du navire.* Paris: H. L. Guerin & L. F. Delatour.

Poleni, Giovanni. 1748. *Memorie istoriche, della Gran Cupola del Tempio Vaticano.* Padova.

Truesdell, Clifford A. 1960. *The rational mechanics of flexible or elastic bodies, 1638–1788: introduction to Leonhardi Euleri opera omnia vol. X et XI seriei secundae.* Zurich: Orell Füssli.

Chapter 9
The Entasis of Columns from Blondel to Lagrange: Mathematical Games in Search of the Greatest Stability

1 Introduction

In Antiquity, the column was slightly thickened at mid-height or at the third of the shaft, either to compensate for an optical illusion, or to recall the shape of a tree, the human body, or a muscle contracted by effort (Fig. 1). This objective will be taken up again in the Renaissance by Andrea Palladio,[1] in 1570, referring to the twelfth chapter of Book III of Vitruvius, the illustrations of which are lost. Palladio advised following the curvature outlined by a thin wooden lamina fixed at the top and bottom and bottom of the column (Fig. 2).

The ratios between the largest and smallest diameters of the column shaft are given by Vitruvius according to the height of the columns. One finds the same words of this chapter of Vitruvius repeated again in the *Opticorum libri sex* (1613) by François d'Aguilon,[2] without any other commentary.

2 Giacomo Barozzi da Vignola

In contrast, in his *Regola delli cinque ordini d'architettura*, Vignola, the architect of the church of the Gesù in Roma, initiated the evolution in 1563. He renounced the sole usage of proportions for describing the shape of the shaft of the column and avails himself of two mathematical constructions. It may be that these constructions had already been used in Antiquity since they were part of the mathematical toolkit of that epoch. Written traces of them are unfortunately lost.

---

[1] Palladio (1570, pp. 15–16).

[2] D'Aguilon (1613, p. 232).

P. Radelet-de Grave, *Between Mechanics and Architecture*, Mathematics and the Built Environment 8, https://doi.org/10.1007/978-3-031-73530-1_9

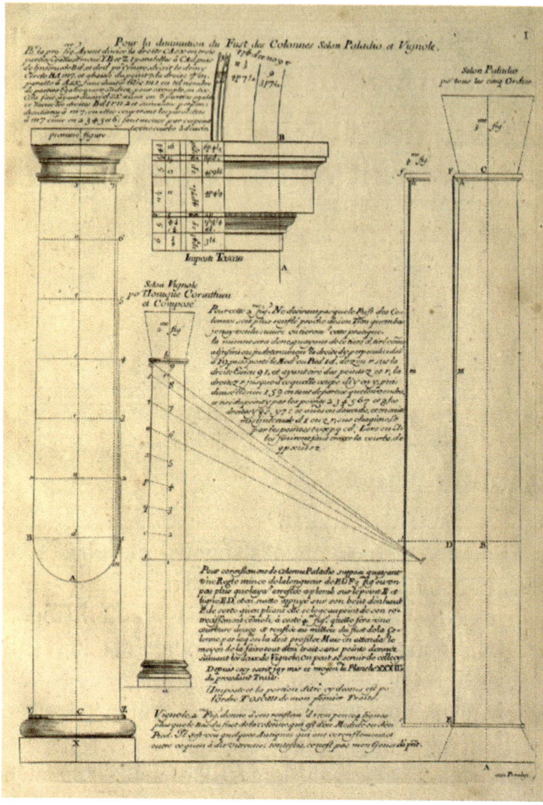

Fig. 1 Methods of entasis of the shaft of a column in Palladio and Vignola (Bosse 1664, Planche I). Public domain

In the French edition of Vignola published by Pierre Le Muet in 1632,[3] we find the following description (Fig. 3):

> The diminution of the columns is done in several ways, of which there are two that are best. For the first the column AB is either divided at a third in height or at the point E. Make a semi-circle EDC and from the point of the diminution the line AG is drawn perpendicularly. Then after that the arc GE is divided into as many equal parts as we want, such as six, and EA is also divided into that many equal parts. Where the perpendicular and horizontal lines intersect is drawn the line LHIKLM. This will give the diminution as desired, and this way can be used in the Tuscan and in the Doric.[4]

---

[3] The French edition by Le Muet is actually slightly different from the Vignola original; we might say it is an "interpreted edition" because Vignola is not so mathematical.

[4] Vignola (1632, p. 58): *La diminution des Colomnes ce faict en plusiers facons dont voicy des deux meilleures pour la premiere soit la Colomne A.B. a diviser du tiers en hault soit au point E. faict e demy cercle E.D.C et du point de la diminution. soit abessee perpendiculairement la ligne A.G.. soit puis apres l'arc G.E. divise en tant de parties esgales qu'on voudra comme en six et soit E.A. divise aussy en autant daves parties esgales et ou les lignes perpendiculaires et Orisontales se coupperont soit menee la ligne L.H.I.K.L.M. A qui donnera la diminution comme on la disire et de cette façon on s'en peult servir au Toscan et au dorique.*

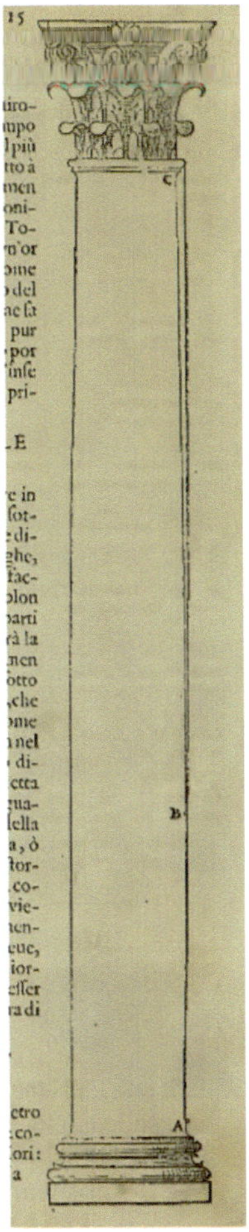

Fig. 2 Drawing accompanying the section that Palladio devoted to *gonfiezza, e diminution delle colonne* in I quattro libri dell'architettura (1570). Public domain

The column is therefore straight up to the third of the height of its shaft, and this construction gives the entasis of the shaft from the third of its height up to the neck (Fig. 4).

He continues:

> With regard to the other way, I myself discovered it, and although it is less well known it is nevertheless easy to understand from the lines. So, I say that having determined the measures of the column we must draw a line to infinity starting with C and continuing with D. Then transfer the measure CD to point A until it intersects the perpendicular at point B; that AB is continuous up to E. From there one can draw as many lines as one wants that start from the perpendicular and front to the circumference of the column on which, applying the measure

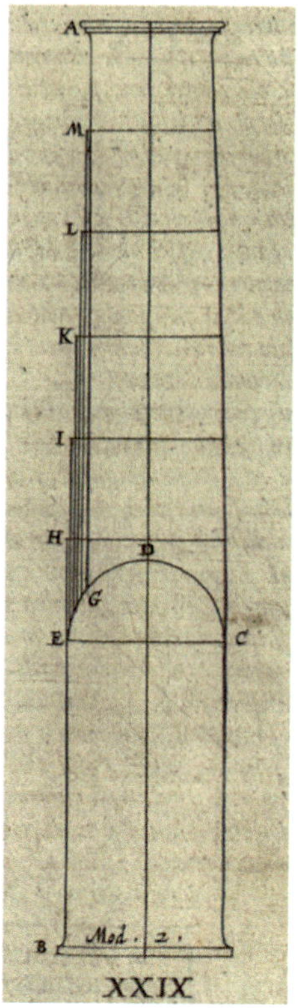

Fig. 3 Illustration of the first construction of Vignola in the edition by Le Muet (1632). Public domain

CD, one finds the swelling of the column at both the top and at the bottom. This manner can be applied to the Ionic, Corinthian and Composite.[5]

We do not want to cast doubt on Vignola's sincerity, but the construction of the first conchoid of Nicomedes, which is given here in a very confused way, has been known since Antiquity. One must first fix the radius of the largest circle of the shaft, which here is the first third, and extend the straight line; Vignola says to infinity, but he contents himself with the apex of a large triangle that occupies the centre of the figure. Once the radius is fixed he carries segments equal to that radius starting from the axis of the column to different heights and following the direction of the straight lines issuing from that point "to infinity". Since their slopes are different,

[5] Vignola (1632, p. 60): *Quand a cette autre façon je l'ay trouvée de moy mesme et quoy quette soit moins conneue elle est neantmoins facile a concevoir par les lignes. je diray donc qu'aynt resolu les mesures de la colonne on doibt tirer une ligne a linfiny en commancant par C. et continuant par D. puis reportant la mesure C.D. au point A. jusques a ce quelle coupe la perpendiculaire au point B. et que A.B. soit continuee jusques en E. de la on poura tirer tant de lignes au'on voudra qui partiront de la perpendiculaire & front a la circonference de la colonne sur lesquelles apliquant la mesure C.D. on trouvera tant en hault quen bas l'enfleure de la colonne et cette manière peult estre apliquée a l'Ionique Corinthien et Composé.*

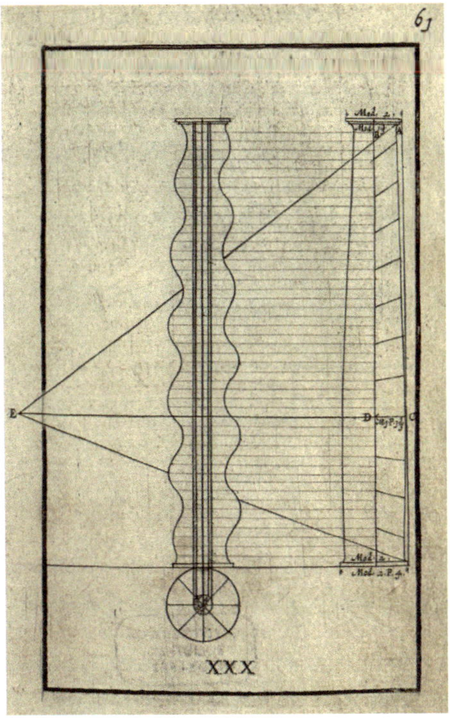

Fig. 4 Illustration of the second construction of Vignola in the edition by Le Muet (1632). The image has been retouched in order to eliminate the drawing of a twisted column that was superimposed over the construction of the swelling of the column. Public domain

the distances to the ends of the segment equal to the maximum radius at the axis of the column will be likewise all different and determined.

It was also studied by Gille Personne de Roberval (1602–1675) in his *Traite de Indivisibles* (1693), of uncertain date, published posthumously (Fig. 5). This curve has two branches but only the external curve is retained for the entasis.

3 "The Gift to All Architects"

The text published by Jacques Curabelle in 1664 is not known except from the reactions that it solicited from Abraham Bosse and François Blondel. Curabelle challenged the architects to find the solution to a paradox "regarding the perfection of the swelling or εντασιξ of the columns, touched on imperfectly by Vitruvius, and not yet solved and only imperfectly regulated, so that it can be architectonically perfect".[6]

[6] Blondel (1673, p. 2): *touchant la perfection de l'enflûre ou entasis des Colonnes, touchéè imparfaitement par Vitruve & non encore resolvë ny reglée qu'imparfaitement, quoy qu'Architectoniquement elle le puisse estre parfaitement.*

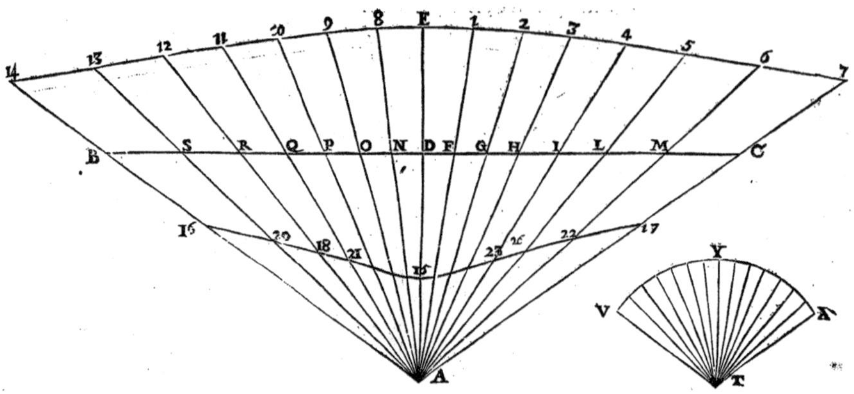

Fig. 5 The conchoid of Nicomedes by Roberval, in the *Traite des indivisibiles* (Roberval 1693, p. 211). Public domain

Blondel assembled his writings on the subject in his *Résolution des quatre principaux problèmes d'architecture*, which was published much later, in 1729, in the *Mémoires des Académie des Sciences de 1666 à 1669*. This is his first problem: "describe geometrically in several ways, in a single line, the outline of the swelling and diminution of the columns".[7] The text is composed of five discourses. In the first discourse, which is an introduction, we sense the beginning of a turning point:

> And as architecture has received what it has of the good and magnificent only from the mathematical sciences, which by the unmistakable truth of their demonstrations completely fill the capacity of our mind, and leave nothing to be desired on the subject that they explain. It is easy to understand that among them she [architecture] must still wait for what is lacking in her perfections; and the light by which one knows the difference between being able to give a reason for one's work and working while groping and blind, can only be given to one by the free hands of geometry.[8]

There will therefore no longer be any question of letting oneself be guided exclusively by aesthetics, as he specifies a little further on:

> ... these sorts of dispositions, which are only for the satisfaction of the eye, and which have no certain foundation or have been found in nature, depend so much on taste, and on the diversity of opinions, that a column may appear to some too *svelte* [slender] as the Italians call them, or loose, while others find it too squashed.[9]

---

[7] Blondel (1673, n.p.): *Decrire Géometriquement en plusieurs maniéres, & tout d'un trait, le Contour de l'enflûre & diminution des Colonnes.*

[8] Blondel (1673, n.p.): *Et comme l'Architecture n'a receû ce qu'elle a de bon & de magnifique que des sciences Mathematiques, qui par l'indubitable verité de leurs démonstrations remplissent entiérement la capacité de nostre esprit, & ne luy laissent rien à desirer sus le sujet qu'elles luy ont expliqué: Il est aisé de comprendre que d'est d'elles qu'elle doit encore attendre ce qui manque à sa perfections; & que cette lumiére, par qui l'on connoist la difference qu'il y a de pouvoir rendre raison de son Ouvrage, ou de travailler en tastonant, & à l'aveugle, ne luy peut vernir que des mains liberals de la Géometrie.*

[9] Blondel (1673, n.p.): *...parce que ces sortes de dispositions, qui ne sont que pour la satisfaction de l'œil, & qui n'ont point de fondement certain ny arresté dans la nature, dépendent tellement du*

The second discourse gives the text of a letter that Blondel sent to Abraham Bosse on this subject. He speaks of the *estrenne à tous les architectes* ("gift to all the architects"), without giving the name of the author, but in a copy of the letter that is found in the Fowler Architectural Collection of Johns Hopkins University one finds the name of the architect Jacques Curabelle, the adversary of Desargues.

Blondel remarks that the curve drawn by Vignola in his second construction is the first conchoid of Nicomedes, known in Antiquity. In effect, he recalls, this curve allowed Nicomedes to solve the famous problem of the duplication of the cube. The problem requires inserting two mean proportionals between A and $2A$, A being the side of the cube. In effect:

$$\frac{A}{x} = \frac{x}{y} = \frac{y}{2A}.$$

If one raises the first term to the cube, the result will be equal to the product of the three ratios, which makes x and y disappear:

$$\frac{A^3}{x^3} = \frac{Axy}{xy2A} = \frac{1}{2}.$$

This is equivalent to our way of putting the problem of duplication in the equation.

$$2A^3 = x^3.$$

He has thus found x such that

$$x = A\sqrt[3]{2}.$$

The problem of the duplication of the cube therefore comes down to finding two mean proportions between two given lengths. To do this, Nicomedes, following Pappus, gives the two λεντησ ΑΔ and ΔΓ and finds, by means of the geometric construction shown in Fig. 6, the segments AM and GK such that:

$$\frac{\Delta\Gamma}{\Gamma\mathrm{K}} = \frac{\Gamma\mathrm{K}}{\mathrm{MA}} = \frac{\mathrm{MA}}{\mathrm{A}\Delta}.$$

In the course of this complicated construction, details of which are beyond the scope of this article, Nicomedes must move the segment ΑΔ φορ the transfer of K in the direction ΘK. It is for this reason that he appeals to a conchoid, which is the locus of points a fixed distance away from a line as measured along a line from the focus point.[10]

goust, & de la diversité des opinions, qu'une Colonne peut paroistre aux uns trop Svelte, *ainsi que les Italiens les appellent, ou déliée, que d'autres la trouveront trop écrasée.*

[10] For more information see Van der Waerden (1963, pp. 236–237), or Pappus of Alexandria (1982, pp. 42–44).

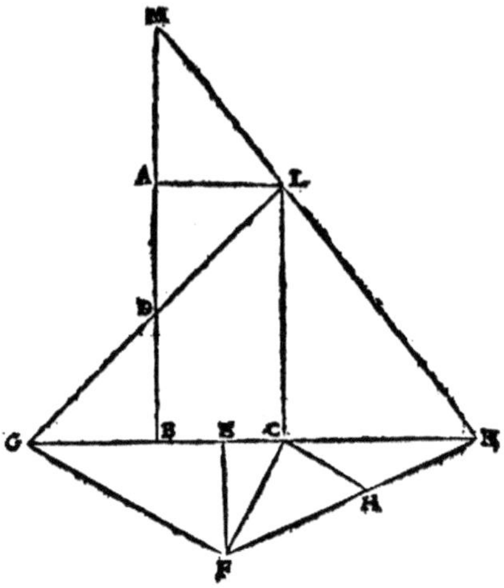

Fig. 6 Duplication of the cube by Nicomedes (Pappus 1588, p. 56v). Public domain

This curve had already been honoured by François Viéte, who used it as a new axiom in geometry in his "Supplementum geometriae" (1649),[11] in order to realise the *neusis* or intercalation of a fixed distance between two curves or straight lines.

In his *Traité des manières de dessiner les ordres de l'Architecture Antique* (1664), Bosse describes the "manner of tracing or drawing all in one line as with a compass the line which forms one side and the other of the column either from their third downwards or if they are reflected by the third or by their middle".[12] He shows an instrument that makes it possible to trace the conchoid (Fig. 7).

The third discourse of Blondel leads to Vignola's first construction. Blondel saw the composition of two movements, and he erroneously deduces that it is a parabola. He refers to Galileo, but forgets to verify that the second movement is uniformly accelerated. In contrast, in what follows he reasons correctly and shows that that curve, which he first qualifies as "a species of spiral or oval", is "a spiral or elliptical line, in which the portions of the axis are the inverse sines of the arcs, which are between them as the ordinates".[13]

The imprecision of the names of the curves is off-putting for the modern reader and may perhaps lead to seeing more errors in the text than there actually are. The discourse ends with the description of an instrument for tracing this curve, which is shown in his Planche I at the end of the treatise.

The fourth discourse is devoted to an application of conic sections to drawing the diminution of the columns. Blondel relates that in his day, the law of the parabolic

[11] Viéte (1649, p. 240).

[12] Bosse (1664, p. XXXIII): *Maniére de tracer ou de décrire tout d'un trait comme au compas la ligne qui forme de costé et d'autre le fust des colonnes soit depuis leur tiers vers le bas soit qu'elles soient reflées par le tiers ou par leur milieu.*

[13] Blondel (1673, p. 6): *Spirale ou Ovale ... une ligne Spirale ou Eliptique, dans laquelle les portions de l'Axe sont les Sinus verses des Arcs, qui sont entre eux comme les ordonnées.*

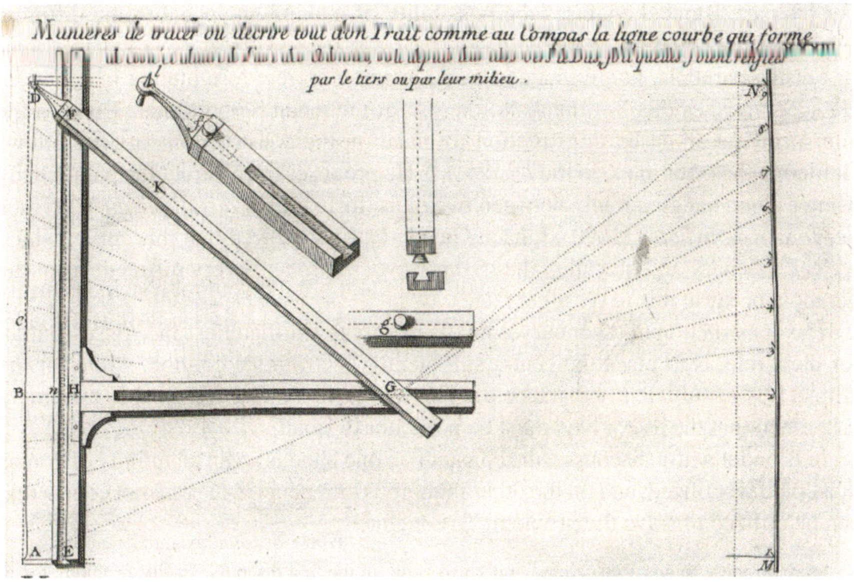

Fig. 7 Illustration of the instrument that makes it possible to draw the first conchoid of Nicomedes, from Bosse's *Traité des maniéres de dessiner les orders de l'architecture antique* (1664). Public domain

descent has not been completely accepted and that there is a rival theory that would have the descent occur along a spiral:

> … that same Galileo having demonstrated that on the hypothesis of the fall of weights in proportion to odd numbers, the line of projection is parabolic, which in the other hypothesis is spiral; and as the parabola and the spiral have, moreover, a great number of common properties, they have caused a great geometer of our time to say, that *the parabola is only a developed spiral*; there is a great deal of evidence that the effect of one in the diminution of columns could also be happily produced by the other.[14]

A note in the margin identifies the "great geometer" as Grégoire de Saint-Vincent. In 1647 this author published his great work, *Problema Austriacum plus ultra quadratura circuli* (1647), in which he dedicated a long chapter to the "Spiralis et parabola symbolizatio".[15]

In this work, in addition to profound studies of series and the logarithmic function that he recognized in the area under the hyperbola, Saint-Vincent analyses in succession the properties of the circle, the ellipse, the parabola and above all the hyperbola.

---

[14] Blondel (1673, p. 10): *…le mesme Galilée ayant démontré que sur l'hypothese de la chèûte des poids en proportion des nombres impairs, le ligne de Projection est Parabolique, qui dans l'autre hypothese estoit Spirale; & comme la Parabole & la Spirale ont d'ailleurs un si grand nombre de proprietez communes, qu'elles ont fait dire à un grand Geometre de nostre temps, che la Parabole n'estoit* qu'une Spriale dévelopée*; il y a grande apparence que l'effect de l'une en la diminution des Colonnes pourroit aussi estre heureusement produit par l'autre.*

[15] Saint-Vincent (1647, pp. 664–673).

For his part, Blondel analyses the possibility of making an appeal to the different curves for the columns.

For the parabola, he gives a construction and a correct formulation in equation. He accomplishes this last thanks to the use of the mean proportional. Then, in the aim of giving an easier construction to make on the construction site, he follows Galileo in the error made in the *Discorsi.*[16] He proposes to construct the parabola by using a chain hanging freely between two nails. In 1671, in *La statique ou la science des forces mouvantes* (1690), Ignace-Gaston Pardies showed that this curve, which we call catenary and he called the *cosinus hyperbolique*, is very different from the parabola because it is never algebraic.[17]

The construction of other curves involved only few difficulties since the essential, for the circle, is to place the centre sufficiently far from the column, while for the ellipse it is to sufficiently distance the foci and have a major axis that is extremely large such that the minor axis must be particularly small.

In Blondel's fifth discourse, the constant on one hand is again an infinity of conics that could be utilized, and on the other numerous other curves that are not conics that can be utilized to solve the problem. To conclude:

> From all these things, one can judge if I was right, in the first discourse of this problem, to wish that the Author of *the Paradox proposed to be solved for all Architects*, had made a little more clearly understood the nature of the line with which one can, as he says, describe *architectonically perfectly* the restriction and diminution of columns; then after someone has described *architectonically perfectly* an infinity in the above mentioned ways, it will always depend on the will of the Author of *the Paradox* to reserve an infinity of others, and to say that we have not yet found his.[18]

We will conclude the analysis of this text by remarking that Blondel does a good job, as he said he would do in the introduction, of replacing the aesthetic criteria with mathematical criteria but—and this is what leads us to speak of 'mathematical games'—nothing justifies those criteria. His conclusion is the proof. This is an inventory of curves having a part that is comparable to that which is dictated by aesthetic reasons.[19]

It is necessary again to remark the influence of both Grégoire de Saint-Vincent and Galileo in a field where we certainly wouldn't expect to find them.

---

[16] The error occurs during the second day: Galileo (1638, p. 146); Eng. trans. Galileo 1914, p. 149.

[17] Pardies (1690, p. 141).

[18] Blondel (1673, p. 15): *De toutes ces choses, on peut juger si je n'ay pas eû raison de souhaiter au premier Discours de ce Probleme, que l'Auteur* du Paradoxe proposé à résoudre à tous les Architects, *se fust un peu plus clairement fait entendre de la nature de la ligne, que l'on peut, comme il dit, décrire* Architectoniquement parfaitemente *pour le renflement & diminution des Colonnes; puis qu'après que quelqu'un en aura décrit* Architectoniquement parfaitemente *une infinité par les maniéres cy-dessus dites, il dépendra toûjours de la volonté de l'Auteur* du Paradoxe *de s'en réserver une infinité d'autres, & dire qu'on n'aura pas encore trouvé la sienne.*

[19] For a discussion of the aesthetic aspect, see Gros (2008).

In his *Cours d'architecture* of 1675, Blondel contents himself, as did Perrault in 1673 in his edition of Vitruvius,[20] with repeating what Vignola added to Vitruvius. In this they will be followed by Bélidor in his *Science des ingénieurs* of 1729.

4 Pieter van Musschenbroek and Leonhard Euler

Even if their objective was never to determine the shape of the shaft of a column but simply to determine the solidity of a given shape generally cylindrical or parallelepiped, we cannot omit a glance at the work of two authors who will allow us to approach the problem of entasis in a radically different way.

We have first of all a text by Pieter van Musschenbroek (1692–1761), *Physicae experimentales*, and more precisely, the part of it entitled *Cohaerentia corporum firmorum dissertationes*, which was published in 1729. In this work he makes a series of experiments on compression and draws from them an experimental law:

> Parallelepipeds of the same wood, different in length and thickness, pressed according to length, exert forces of resistance which are in the inverse ratio of the double of the lengths, in the simple ratio of the thickness of the side which is not bent, and in the double ratio of the thickness of the side which is bent.[21]

This can be written as:

$$F_R = \frac{bc^2}{a^2}.$$

So that one side is not bent, he erroneously supposes, like Jacob Bernoulli had done before him, that the neutral fibre coincided with one of the vertical surfaces of the parallelepiped.

Van Musschenbroek also identifies the phenomena of buckling, the curvature assumed by the piece of wood under the effect of weight, and rupture. He will thus set Euler on the path of exploring critical load and buckling. In the lengthy *additamentum I* in one of his greatest works, the *Methodus inveniendi* (1744), Euler finds the general equation that gives the curvature of an elastic lamina, then classifies all the types of curves that can be thus generated. He begins logically by calculating the minimal deflection, so infinitesimal of this lamina. He finds, and is astonished at, a value that is finite, not infinitesimal:

$$F = \frac{Ek^2\pi^2}{a^2}$$

---

[20] Perrault (pp. 80–84).

[21] Musschenbroek (1729, Prop. 119, p. 660): *Parallelopipeda ejusdem Ligni, quoad longitudinem & crassitiem diversa, pressa secundum longitudinem, exercent vires resistentiae, quae sunt in ratione inversa duplicata longitudinum, in ratione simplici crassitiei lateris, quod non inflectitur, & in ratione duplicata crassitiei lateris, quod inflectitur.*

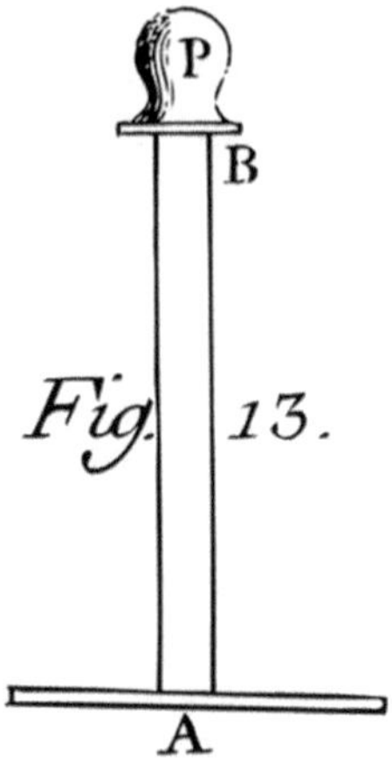

Fig. 8 Euler's illustration for his section on the strength of columns in the *Methodus inveniendi* (1744, Fig. 13). Public domain

where a is the length of the lamina and Ek^2 the absolute elasticity (our module of elasticity). After this, he devotes a brief section to the force of columns (Fig. 8).

In effect, the remark that has just been made teaches him that for a column there exists a minimal load that provokes buckling and that one can load the column with a weight inferior to that load without affecting the column.

In the same *additamentum*, Euler studies the oscillations of the elastic lamina and observes the different modes of oscillation of the lamina that has two nodes at the ends, three nodes with one at the center, then four, five, etc.

Thirteen years later, Euler took this up again and developed the brief section on the critical load in an article entirely devoted to "the strength of columns" (1759). Here he explains the astonishment that we underlined above in comparing the action of a horizontal force with that of a vertical force where the compression is acting on the column. He shows that, for a horizontal force, there is a deflection as soon as a force, as small as one wishes, is applied such that, in the case of compression, one must wait until the famous critical load is reached for there to be deflection (Fig. 9).

> We thus see that, however small the force F, which is supposed to be acting horizontally, it must always produce some inflection, since the space Dδ is proportional to the force F itself. But it is never the same when the force is acting vertically, or the column is bearing a weight so that it is in pressed down from the top.[22]

In this text, Euler first envisions the cylindrical column:

> I will restrict this research first, as I have done other times, solely to columns which are cylindrical, or which have throughout their entire length the same moment of stiffness, and then I will try to carry out these same researches to columns in which the thickness is

[22] Euler (1759, p. 256): *Nous voyons donc que, quelque petite que soit la force F, qui est supposée agir horizontalement, elle doit toujours produire quelque inflexion, puisque l'espace D est proportionnel à la force F même. Mais il n'en est pas de même lorsque la force agit verticalement, ou que la colonne a à soutenir un poids, dont elle est pressée par en haut.*

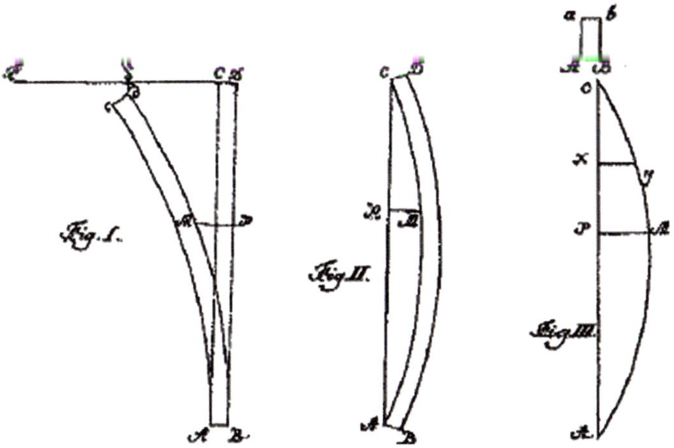

Fig. 9 Illustration from Euler's "Sur la force des colonnes" (1757, Table III). Public domain

> variable. However, one will see that this problem, being generally proposed, exceeds the limits of analysis, which obliges me to develop only some special cases...[23]

He established a general equation that cannot be solved in a general way:

$$\frac{-dx^2}{ddy} = \frac{Ekkf(x)}{Py}.$$

The special case that he proposes very often takes him back to the cylindrical column. He adds that columns tapered at the top or the bottom have no strength. He again finds the same critical load for the parabolic column.

5 Joseph-Louis de Lagrange

Fortified by the work of Euler and the progress of analysis, Lagrange takes up the problem in 1770 in a text whose title is slightly different from that of Euler: *Sur la figure des colonnes* (On the shape of columns) (1773). This slight modification tells us that Lagrange, in all reference to Euler, responds more closely to our problem of entasis:

---

[23] Euler (1759, p. 259): *Je restraindrai cette recherche d'abord aux colonnes cylindriques, ou qui ayent per toute leur longueur le même moment de roideur; & ensuite je tâcherai de pousser ces mêmes recherches aux colonnes dont l'épaisseur est variable. Or on verra que ce problème étant généralement proposé surpasse le bornes de l'analyse, ce qui m'oblige de n'en développer que quelques cas particuliers…*

> One is accustomed to giving to columns the shape of a conoid that has its greatest width towards the third of its height, and becomes slenderer as it goes towards the two extremities, from which results that which are commonly called the *swelling* and *diminution* of columns. But no one that I know has yet given a satisfactory reason for this practice.[24]

Lagrange then mentions the different texts that we have analysied and repeats following Blondel:

> It would be very easy to invent several other means to fulfill the same object. ... since it involves passing a curved and concave line down the axis through three given points. But is there not in the very nature of the thing some principle which could serve to determine the question? Among those which serve as the foundation of architecture, there is only one which has fixed and invariable rules, and is therefore susceptible of calculation: it is solidity...[25]

Finally he poses the problem:

> It is a problem of *maxima* and *minima* to determine the curve which, by its rotation around its axis, will form a column capable of supporting the greatest possible load, the height and mass of the column being given. This is, it seems to me, the true point of view from which we must consider the question of swelling and diminishing columns.[26]

Lagrange is inspired by Euler to calculate the critical load but he adds that more than the small critical load determined by Euler, there is an infinity of others acting in different ways of oscillation. In effect, the equation giving the shape of the curve adopted by the column is:

$$y = f \sin\left(\frac{m\pi x}{a}\right)$$

and the critical load

$$P = \left(\frac{m^2\pi^2 K}{a^2}\right)$$

where K is the module of elasticity and where each value of $m = 1, 2, 3, \ldots$ gives a particular shape of the curve corresponding to a mode of oscillation that is different and acting by a different weight (Fig. 10).

---

[24] Lagrange (1773, p. 125): *On a coutume de donner aux colonnes la figure d'un conoïde que ait sa plus grande largeur verse le tiers de sa hauteur, et qui aille de là en diminuant vers le deux extrémités; d'où résulte ce qu'on appelle vulgairement la* renflement *et la* diminution *des colonnes; mais personne que je sache n'a encore donné une raison satisfaisant de cette pratique;...*

[25] Lagrange (1773, p. 126): *Il sèrait très aisé d'inventer plusiers autres moyens pour remplir le même objet. ... pouisqu'il s'agit de faire passes une ligne courbe et concave verse l'axe par trois points donnés. Mais n'y aurait-il pas dans la nature même de la chose quelque principe qui pût servir à déterminer la question ? Parmi ceux qui servent de fondement à l'architecture, il n'y en a qu'un seul qui ait des règles fixes et invariables, et par conséquent susceptibles de calcul: c'est la solidité...*

[26] Lagrange (1773, p. 126–127): *...c'est un Problème de* maximis *et* minimis *de déterminer la courbe qui par sa rotation autour de son axe formera une colonne capable de supporter la plus grande charge possible, la hauteur et la masse de la colonne étant données; c'est là, ce me semble,*

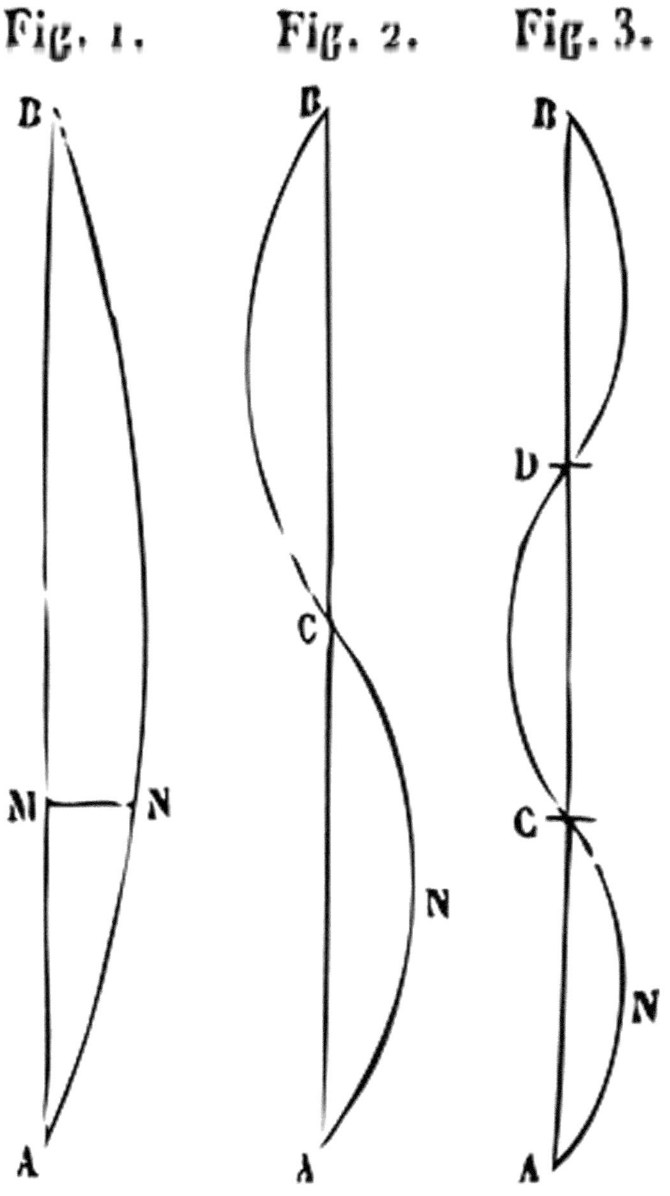

Fig. 10 Shapes of the three first ways of deformation of the column, corresponding to $m = 1, 2, 3$ given by Lagrange in "Sul la figure des colonnes (1773). Public domain

It emerges from the rest of Lagrange's paper, which is lengthy and dense with calculations, taken up again several times and not without errors, as remarked by Thomas Clausen in "Über die Form architektonischer Säulen" (1851) and more recently by Joseph Keller in "The Shape of the Strongest Column" (1960), that the cylindrical shape is preferable to the truncated cone, as it is to the different conoids formed by the rotation of a conic.

6 Conclusion

In a lengthy note added to his commentary on the article by Lagrange, Clifford Truesdell stated that the problem was late solved by Weinberger and Keller.[27] This last shows how for a given volume and length, the shape supporting the greatest critical load is triangular. This might astonish. He also shows that in the case as in those of columns with a circular or rectangular section, the entasis increases the critical load that is supported.

As for the modern approach, it eludes the solution to this genre of problem of optimisation since it begins with a shape and characteristics of materials to determine subsequently, by the method of finite elements, whether the column resists the solicitations placed on it or not.

Finally, it seems that a detailed analysis of the text of Lagrange and a precise identification of the errors that he made is still lacking, just as is a calculation of optimisation of the shape of the column. But this is hardly surprising when we read, in the article by Karl-Eugen Kurrer (2008), about the difficulties encountered in the elaboration of the theory of buckling, which is the prerequisite.

References

Bélidor, Bernard Forest de. 1729. *La science des ingénieurs dans la conduite des travaux de fortification et d'architecture civile.* Paris: Claude Jombert.

Blondel, François. 1673. *Résolution des quatre principaux problèmes d'architecture*, in *Mémoires des Académie des Sciences de 1666 à 1669.* Paris: Imprimerie Royale.

Blondel, François. 1675–1683. *Cours d'Architecture Enseigne Dans l'Académie Royale d'Architecture*, 2 vols. Paris: Pierre Aubouin & François Clousier.

Bosse, Abraham. 1664. *Traité des manières de dessiner les ordres de l'Architecture Antique en toutes leurs parties.* Paris.

Clausen, Thomas. 1851. Über die Form architektonischer Säulen. *Bull. Cl. Physico Math. Acad. St. Pétersbourg* 9 (1851): 369–380.

la véritable point de vue sous lequel on doit envisager la question de renflement *et de la* diminution *des colonnes.*

[27] See Truesdell (1960, pp. 355–356) and Keller (1960). Neither Truesdell nor Keller give a reference to the article by Weinberger.

D'Aguilon, François. 1613. *Opticorum libri sex philosophis iuxta ac mathematicis utilis*. Antwerp: Plantin Press for Jan Moretus.

Euler, Leonhard. 1744. *Methodus inveniendi lineas curvas maximi minimive proprietate gaudentes sive solutio problematis isoperimetrici latissimo sensu accepti*, E65. Lausanne-Geneva. In: *Opera omnia* series 1, vol. 24. Basel: Birkhäuser.

Euler, Leonhard. 1759. Sur la force des colonnes. *Memoires de L'Academie des Sciences et Belles-Lettres*, vol. 13, 252–282. Eng. trans. "On the Strength of Columns", trans. J. A. Van den Broek *American Journal of Physics* 15, 315 (1947). https://doi.org/10.1119/1.1990953.

Galilei, Galileo. 1638. *Discorsi e dimostrazioni matematiche intorno a due nuove scienze attenenti alla meccanica e i movimenti locali*. Leyden: Elsevier. Eng. trans. *Dialogues Concerning Two New Sciences*, trans. Henry Crew and Alfonso de Salvio. New York: MacMillan, 1914.

Gros, Pierre. 2008. Le dessin de l'entasis. In *La colonne. Nouvelle histoire de la construction*, ed. Roberto Gargiani. Lausanne: Presses polytechniques et universitaires romandes, 22–31

Keller, Joseph B. 1960. The shape of the strongest column. *Archive for Rational Mechanics and Analysis* 5: 275–285.

Kurrer, Karl-Eugen. 2008. Research d'une théorie juste du flambage pour les colonnes en acier (1910 à 1943). In *La colonne. Nouvelle histoire de la construction*, ed. Roberto Gargiani. Lausanne: Presses polytechniques et universitaires romandes, pp. 324–335.

Lagrange, Joseph-Louis. 1773. Sur la figure des colonnes. *Miscellanea Tourinesia* (Royal Society of Turin), Tomus V 123 (1770–1773); also *Oevures* Vol. 2, 125–170.

Musschenbroek, Pieter van. 1729. *Physicae experimentales, et geometricae, de magnete, tuborum capillarium vitreorumque speculorum attractione magnitudine terrae, cohaerentia corporum firmorum dissertationes ut et ephemerides meteorologicae Ultrajectinae*. Leiden: Samuelem Luchtmans.

Palladio, Andrea. 1570. *I quattro libri dell'architettura*. Venice: Domenico de'Franceschi.

Pappus of Alexandria. 1588. *Pappi Alexandrini Mathematicae collectiones. A Federico Commandino Urbinatae in latinum conuersae, & commentarijs illustratae*. Venice: Franceso dei Franceschi Sanese.

Pappus of Alexandria. 1982. *La collection mathématique*. Paris: P. ver Ecke.

Pardies, Ignace-Gaston. 1690. *La statique ou la science des forces mouvantes*. Paris: Florentin & Pierre Delaulne.

Perrault, Claude. *Les dix livres d'architecture de Vitruve, corrigez et traduits nouvellement en françois avec des notes et des figures*. Paris: Jean-Baptiste Coignard.

Roberval, Gille Personne de. 1693. *Traite des indivisibiles*. In *Divers ouvrages de Monsieur Roberval, Recueils de l'Académie royale des sciences*, vol. 6. Paris : L'Académie royale des sciences.

Saint-Vincent, Grégoire de. 1647. *Problema Austriacum plus ultra quadratura circuli*. Antwerp: Jan van Meurs.

Truesdell, Clifford A. 1960. *The rational mechanics of flexible or elastic bodies 1638–1788: Introduction to Vol. X and XI, Leonhard Euler, Opera Omnia*, series II, vol. XI, sect. II

Van der Waerden, B. L. 1963. *Science Awakening*. Leyden.

Viéte, François. 1649. Supplementum geometriae. In *Francisi Vietae, Opera mathematica*. Leyden.

Vignola, Jacopo Barozzi da. 1632. *Règles Des Cinq Ordres D'Architecture de Vignole*. Paris: Pierre Le Muet.

Chapter 10
A Study of Coulomb's Application of the Rules of Maxima and Minima to Problems of Statics

1 Preamble

The theory of stability of arches seems to come from three main domains: statics; elasticity, taken in the sense of the study of deformations; and resistance of materials, which I contrast to elasticity because it is more generally concerned with the rupture of materials.[1] Contributions to these domains are analyzed in Appendix 1, in which for each domain distinctions are drawn between mechanics and architecture, or if one prefers, between theory and practice. The three domains are complementary and cannot be rigorously separated. One finds them explicitly mentioned in the titles of three historical studies by authoritative authors:

- Stephen P. Timoshenko, *History of Strength of Materials, with a Brief Account of the History of the Theory of Elasticity and Theory of Structures* (1953);
- Clifford A. Truesdell, *The Rational Mechanics of Flexible or Elastic Bodies 1638–1788*, Introduction to *Leonardo Eulerii Opera Omnia*, Ser. II, vol. 11 (1960);
- Edoardo Benvenuto, *An Introduction to the History of Structural Mechanics.* Part I, *Statics and Resistance of Solids*; Part II, *Vaulted Structures and Elastic Systems* (1991).

Timoshenko centered his study on resistance of materials, while Truesdell, who tried to complete and to broaden Timoshenko's point of view, centered his work on elasticity. Truesdell added the study of oscillatory movements, so stability was not at the center of his concerns. Stability was returned to a place of honor by Benvenuto who, in order not to repeat things already explained by Truesdell, avoided elasticity.

The "Essai sur une application des règles de Maximis et Minimis à quelques problèmes de statique, relatifs à l'architecture" (1773)[2] by Charles-Augustin Coulomb

[1] To these three domains should be added geometrical themes, but these are beyond the scope of this chapter.

[2] In what follows, quotes from the original will be cited in the English translation published by Jacques Heyman (1972), with the original text given in a footnote. I am grateful to Santiago Huerta for help in obtaining Heyman's volume.

P. Radelet-de Grave, *Between Mechanics and Architecture*, Mathematics and the Built Environment 8, https://doi.org/10.1007/978-3-031-73530-1_10

(1736–1806) is well known and its importance is acknowledged by the three authors just mentioned, among others. Timoshenko's study is the most complete but he doesn't examine the structure of Coulomb's essay. Truesdell's text sets aside retaining walls and the stability of arches, and Benvenuto's study sets aside the whole beginning and the part concerning arches without friction and cohesion. Each of these authors considers Coulomb's text to be important but for different reasons, and therefore they on comment different parts. Let us see what each has to say about Coulomb's text.

Timoshenko, who bases his history of resistance of materials[3] in the eighteenth century on Coulomb's text, is enthusiastic:

> No other scientist of the eighteenth century contributed as much as Coulomb to the science of mechanics of elastic bodies. The principal advances are included in his paper of 1773 (Timoshenko 1953, p. 48).

Truesdell addresses this:

> This memoir of Coulomb has been praised very highly. On the basis of it Timoshenko asserts: [he then quotes Timoshenko's passage we just reproduced.] Opinion aside, this estimate is misleading, since, as we have seen, there is scarcely a hint of elasticity in the paper. In one respect Coulomb's work, since it reverts to purely statical arguments in the manner of Galileo, Leibniz, and Varignon, is a step backward. In another, however, it is of the very greatest importance, since it reintroduces Parent's concept of interior shear stress, essential to any complete theory of deformable bodies. This aspect justifies the more definite praise of St. Venant, which is restricted to § VII: it is only in our century that his paper, which in the three pages entitled 'Remarks on rupture' contains so many things on this subject, has finally been studied and understood (Truesdell 1960, p. 400–401).

Let us give the last word to Benvenuto:

> Here not only the resolution of the problem but even its formulation is a departure from the chain of ideas which, till then, had governed the science of masonry construction. There is a radical change of orientation, as well as a richer, more realistic physical model. While previous authors had tried to adapt the form or dimensions of a vault so that its stresses should conform to a given static scheme, Coulomb does the opposite. Given a vault of defined shape and dimensions, what stresses arise? (Benvenuto 1991, p. 394).

2 The Structure of the Coulomb's *Essai*

In 1729 and 1730, Pierre Couplet (1617–1743) had begun to put together the various elements that Coulomb organises here. For a very long time after, those same elements constituted the academic basis for treatises on masonry construction. This can be seen, for instance, in the fact that the first chapter or first division of the textbook

[3] The third chapter of Timoshenko's book (1953), which is dedicated to the eighteenth century, opens with some remarks on the gap, which he estimates to be around one century, between theory and practice. He then studies the works of Antoine Parent about the rupture of beams in order to approach Coulomb's work on resistance (the first part of our chapter) and on torsion. Finally, he isolates two themes of the work we are going to analyse: retaining walls and the arches, of which Timoshenko starts retracing the history before commenting on Coulomb.

published in 1939 by Gustave Magnel (1889–1955),[4] a well-known teacher at the University of Ghent, was still structured in a similar manner and in which are found the same elements as in Coulomb's *Essai.* Magnel added the study of silos and of walls destined to retain water instead of earth. We must add that in Magnel's book those chapters are part of the second division, entitled "metallic constructions", a field which only appeared in the 1830s.

3 Commentary on the Essay

Coulomb starts by stating that his aim is to examine "the effect of friction and of cohesion in some problems of statics".[5] In doing so he differs from his predecessors, who limited their study to the stability of an arch under its own weight, whose voussoirs are perfectly rigid and whose joints are perfectly smooth. Coulomb's predecessors had observed that in order for such an arch to be stable, it was necessary that the last voussoir, which rests horizontally on the piers, had an infinite weight. If this is not so, the horizontal component of the arch will cause the last voussoir to slip and the arch will collapse. Those earlier scholars overcame that difficulty by saying that the joints are never perfectly smooth so there always is some friction that allowed that weight not to be infinite.

Next, Coulomb announces that he will first study cohesion and then the simple case of masonry piers:

> Since this problem only requires quite simple considerations, which will help in the understanding of all the other sections of this Note, we will try to indicate the principles of its solution.[6]

There is therefore some principal idea that is elaborated in the case of simple piers and upon which the more elaborate cases that follows are based. Let us read his presentation of cohesion and of his method of calculating:

> Suppose that a masonry pier is cut by a plane inclined to the horizontal, in such a way that the two portions of the pier are connected at the cut by a certain cohesion, while all the rest of the material is of perfect strength, that is, united by an infinite cohesion.[7]

---

[4] For more about Magnel, see "Gustave Magnel (1889–1955): A Life in Structural Engineering", Chap. 21 in this present volume.

[5] Coulomb, 1773, p. 343 (Heyman 1972, p. 41): *L'influence du frottement & de la cohesion, dans quelques problèms du Statique.*

[6] Coulomb, 1773, p. 343 (Heyman 1972, p. 41): *Comme ce problème n'exige que des considérations assez simples, qui servent à faire entendre toutes les autres parties de cet essai, tâchons de développer les principes de sa solution.*

[7] Coulomb, 1773, p. 343 (Heyman 1972, p. 41): *Si l'on suppose un pilier de maçonnerie coupé par un plan incliné à l'horizon, en sorte que les deux parties de ce pilier soient unies dans cette section, par une cohésion donnée, tandis que tout le reste de la masse est parfaitement solide, ou lié par une adhérence infinie.*

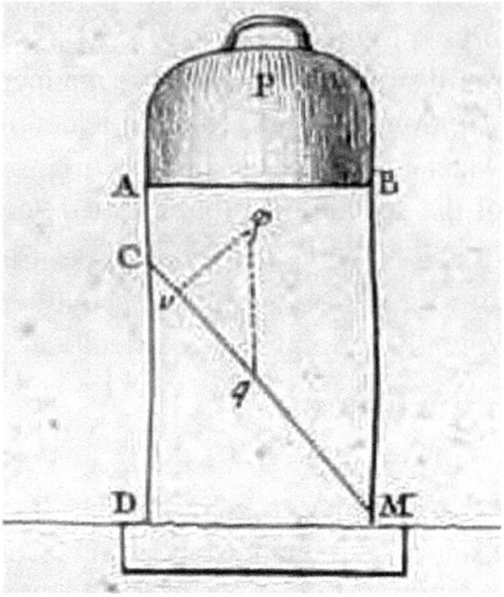

Fig. 1 Coulomb's Fig. 5 (1773: Tab. XV). Public domain

Coulomb considers a pier divided into two parts that are taken to be perfectly rigid (Fig. 1).

The pier can only break along the imposed direction: that of the section. Next, he imagines a weight applied to the pier:

> Suppose then that the pier is loaded with a weight, the weight will tend to make the upper portion of the pier slide along the inclined plane dividing it from the lower portion. Thus, in the state of limiting equilibrium, the component of the load resolved parallel to inclined plane will exactly equal the cohesion.[8]

If the pier does not break, this means that the "cohesion" along the chosen direction is sufficiently great. "Cohesion" thus opposes itself to shear along the chosen direction. That is, *it is thus a force of "cohesion"*, equilibrated by the component along that direction of the weight of the superior part of the pier and of the added weight. Unfortunately enough, the distinction between force and pressure is not clear in the text. We will see this once more with the sleight of hand done by Coulomb in his calculation.

He continues:

> If it is now noted that, for a homogeneous pier, the strength is in fact the same everywhere, then, for the pier to carry a given load there must be no section on which the load component can cause the upper portion slide.[9]

---

[8] Coulomb, 1773, p. 343 (Heyman 1972, p. 41): *qu'ensuite on charge ce pilier d'un poids: ce poids tendra à faire couler la partie supérieure du pilier sur le plan incliné, par lequel il touche la partie inférieure. Ainsi, dans le cas d'équilibre, la portion de la pesanteur, qui agit parallèlement à la section, sera exactement égale à la cohérence.*

[9] Coulomb, 1773, p. 344 (Heyman 1972, p. 41): *Si l'on remarque actuellement, dans le case de l'homogénéité, que l'adhérence de pilier est réellement égale pour toutes les parties; il faut, pour que le pilier puisse supporter un fardeau, qu'il n'y ait aucune section de ce pilier, sur laquelle l'effort décomposé de sa pression puisse faire couler la partie supérieure.*

This last statement must be true regardless of the direction chosen, and this is most important because it will allow him to make use of the rules *"de Maximis et de Minimis"*:

> Thus, to determine the greatest load that can be carried by a pier, that section must be sought, among all sections, for which the coherence is in equilibrium with a minimum load, since then any load other than that determined from this condition would be too small to fracture the pier.[10]

This is exactly what Coulomb writes, but it is clear that his idea is that any "cohesion" higher than the one that was determined in that way will not allow the fracture of the pier. The focus of Coulomb's reasoning is not the variation of the weight that loads the pier but the direction in which the cohesion is the weakest.

Coulomb evades the problems linked to friction:

> As well as the strength that derives from cohesion, I have considered also that due to friction. The same principles suffice to satisfy both cases...[11]

Next, he applies his method to retaining walls and then to arches. He recalls that until that time only arches stable under their one weight had been taken into account and that in 1697 David Gregory (1659–1708) had shown that the solution of that problem was the catenary. We know that Robert Hooke (1635–1703) and Philippe de la Hire (1640–1718) knew that even earlier. Let us look at all this in more detail.

4 Preliminary Propositions I to III

Three propositions summarise the fundamental equations of statics.[12] Coulomb works only in a plane but that restriction does not limit his study because the same problem is repeated in parallel sections. On the contrary it is rather difficult to understand if he makes a distinction between force and pressure (Fig. 2):

> I. Let the plane *abcde*, acted upon by any forces in the same plane, be in equilibrium on the line *AB*; then the resultant of all these forces will be perpendicular to the line *AB*, and will act in between the points *a* and *e*.[13]

---

[10] Coulomb, 1773, p. 344 (Heyman 1972, p. 41): *Ainsi, pour déterminer le plus grand poids que puisse supporter un pilier, il faut chercher parmi toutes ses sections celle dont la cohésion est en équilibre avec un poids qui soit un minimum: car, pour lors, toute pression, au-dessus de celle déterminée par cette condition seroit insuffisante pour rompre le pilier.*

[11] Coulomb, 1773, p. 344 (Heyman 1972, p. 42) *Outre la résistance qui provient de la cohésion, j'ai eu égard à celle dûe au frottement. Les mêmes principes suffisent pour remplir les deux conditions.*

[12] They correspond to the section concerning the conditions of stability in the textbook by Gustave Magnel (1935). I shall not enter into an historical study of those principles.

[13] Coulomb, 1773, p. 346 (Heyman 1972, p. 43): *I. Soit le plan abcde, sollicité par des forces quelconques situées dans la direction de ce plan, en équilibre sur la ligne AB; la résultante de toutes ces forces sera perpendiculaire à la ligne AB, & tombera entre les points a et e.*

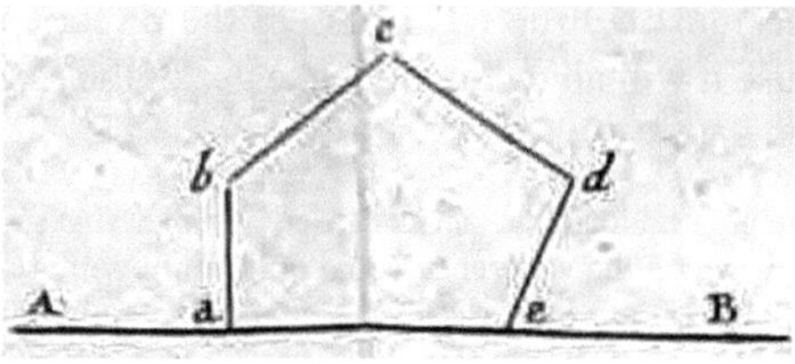

Fig. 2 Coulomb's Fig. 4 (1773: Tab. XV). Public domain

This first proposition contains two conditions of equilibrium: translational equilibrium and equilibrium of forces:

$$\sum_i \vec{F_i} = 0 \tag{1}$$

From this follows it that the resultant must be perpendicular to the surface of support, here the line AB, and the rotational equilibrium or equilibrium of moments:

$$\sum_i \vec{F_i} . \vec{d_i} = 0 \tag{2}$$

which requires the application point to be between the limits of that surface, that is, between the points *a* and *e*.

He continues:

> II. If all forces acting in the plane are resolved into two directions, one parallel and the other perpendicular to *AB*, the sum of the components parallel to *AB* will be zero, and the sum of the components perpendicular to *AB* will equal the thrust exerted on the line *AB*.[14]

As Leonhard Euler (1707–1783) did for the first time in 1750 in an article entitled "Discovery of a New Principle of Mechanics" (Euler 1752), Coulomb takes the vectorial character of forces into account, so that the projection of the relation of equilibrium projected on three orthogonal planes provides with him three equations in space. However, as Coulomb restricts his problem to the plan, he obtains only two equations.

Finally:

> III. If the thrust exerted on the line *AB* is denoted by *P*, the same plane can be supposed to be acted upon by all the applied forces and also by the the reaction to the thrust. But if all these forces, including the reaction to the thrust, are resolved in any two directions at right angles,

[14] Coulomb, 1773, p. 347 (Heyman 1972, p. 44): *II. Si toutes les forces, qui agissent dans ce plan sont découpées suivant deux directions, l'une parallèle à AB, l'autre qui lui soit perpendiculaire, la somme des forces décomposées, parallèlement à AB, sera nulle, & la somme des forces, perpendiculaire à AB, égalera la pression qu'éprouve la ligne AB.*

> it follows from equilibrium and from the fact that the two directions are perpendicular that the sum of the components in each direction is zero.[15]

Adding the reaction of the plane against the external forces, Coulomb can require that, at equilibrium, the sum of all acting forces be zero.

In what follows, Coulomb insists on the equilibrium of moments or equilibrium at rotation. The same condition of Eq. (2) is repeated in terms of moments in paragraph VII, "Remarks on fracture of bodies", where we find:

> …and further, by the principles of statics, the sum of the moments about C of all the tensile and compressive forces must equal the moment of the ϕ about the same point[16]

Historical Digression I[17]

In 1704 Jacob Bernoulli calculated the equilibrium of an infinitely thin arch stable under its own weight, taking into account both the translational equilibrium (Eq. (1) or parallelogram law of forces) and the rotational equilibrium (Eq. (2) or equilibrium of moments) (Fig. 3).

---

[15] Coulomb, 1773, p. 347 (Heyman 1972, p. 44): *III. Si la pression qu'éprouve la ligne AB est exprimée par P, le même plan pourra être supposé sollicité par toutes les forces qui lui sont appliquées, et de plus par la réaction de la pression. Mais si toutes ces forces, ainsi que la réaction de la pression, sont décomposées suivant deux directions quelconques perpendiculaires l'une à l'autre; il suit de l'équilibre et de la perpendicularité des deux directions, que la résultante suivant chaque direction, sera nulle.*

[16] Coulomb, 1773, p. 350 (Heyman 1972, p. 46): *mais par les principes de Statique l'on a encore la somme des* momentum *autour du point G* [recte C] *de toutes les forces, soit de traction, soit de pression, égale au* momentum *du points ϕ autour de même point.*

[17] The essence of what follows is given by Truesdell in his introduction to Euler's *Opera omnia* (Truesdell 1960).

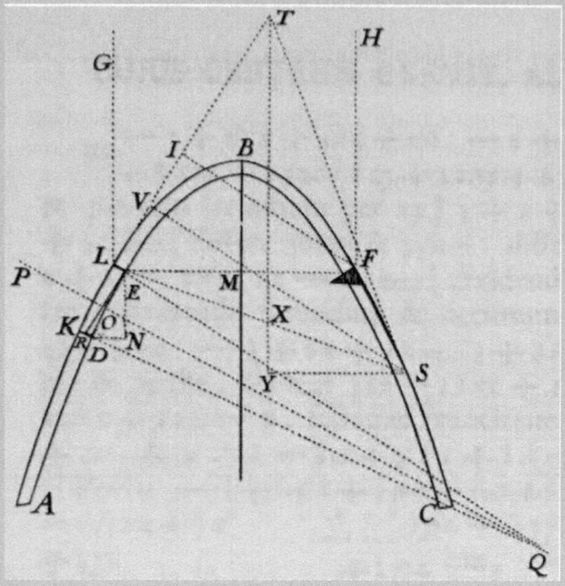

Fig. 3 Jacob Bernoulli's Fig. 46 (1704, Tab. XLVIII). Public domain

Jacob required, on one hand, the impossibility for a voussoir KDEL to fall down inside the arch (translation) and, on the other, the impossibility for the same voussoir to rotate about the point D (rotation).

This text is part of Jacob's *Meditationes* and was published with other *Varia Posthuma* in Jacob's *Opera* in 1742. It was known by Coulomb, who noticed an error in his calculation that consequently caused Jacob to find the catenary as a solution for the equilibrium to translation but prevented him from finding it to be the solution for the equilibrium to rotation.

But the generality as well as the complementarity of those principles would only be understood much later. In his famous "Discovery of a new principle of mechanics" (Euler 1752), written in 1750, Euler writes for the first time the vector relation $\overrightarrow{F} = m\overrightarrow{a}$ in terms of the components. The projection of that relation gives him three equations that make it possible to write the equilibrium to translation (Eq. 1). In what follows, he also uses the equilibrium of momenta (Eq. 2) but does not put this relation, in any case already known to Archimedes, on the same level. It is only in a later text, whose title translates as "Natural basic principles of the study of the state of equilibrium, and the motion, of bodies that are both perfectly flexible and elastic" (Euler 1771), which Euler had just published (but which Coulomb seems not to have known) that he finally puts both relations on the same footing, thus making it possible to understand where Jacob Bernoulli's solution was ahead of his time.

5 Propositions IV (on Friction), and V (on Cohesion)

> Friction and cohesion are not active forces like gravity, which always exerts its full effect, but only are passive forces; these two forces can be measured by the limits of their strength. For example, when it is stated that, for certain polished woods, friction on a horizontal plane is three pounds for a body weighing nine pounds, this implies that a force less than three pounds will not disturb its equilibrium state.[18]

This last sentence contains the germ of his method of calculation: as long as one remains below a certain limit, nothing happens. We shall come back to that point.

Forces of friction and cohesion must be studied experimentally; we could say that they depend on the constitutive laws of the materials. Coulomb dedicates the second part of each of his paragraphs respectively to experimental results of tests of friction and cohesion.

For friction, he proposes the law of Guillaume Amontons (1633–1705), who inaugurated the study of friction in 1699. Amontons' law states that friction is proportional to the pressure. He remarks that this law is but a rough approximation and that it should be modified in function of more numerous and more precise experiences. But in the case of friction, he does not give any result of measurements.

For cohesion, he explains that:

> Cohesion is measured by the resistance that solid bodies offer to the simple separation of their parts. Since each element of solids, if they are homogeneous, has this same strength the total cohesion is proportional to the number of parts to be broken, and hence to the area of rupture.[19]

Writing this, he proposes an approximate constitutive law. He adds, "I made some tests to try to find the value of this cohesion".[20]

Contrary to the case of friction, he gives numerous experimental results. As Truesdell puts it, cohesion, a tension or force opposed to a pressure, is expressed for the first time per unit of surface.

---

[18] Coulomb, 1773, p. 347 (Heyman 1972, p. 44): *Le frottement et la cohésion ne sont point des forces actives comme la gravité, qui exerce toujours son effet en entier, mais seulement des forces coërcitives; l'on estime ces deux forces par les limites de leur résistance. Lorsqu'on dit, par exemple, que dans certains bois polis, le frottement sur un plan horizontal d'un corps pesant neuf livres, est trois livres; c'est-à-dire que toute force au-dessous de trois livres ne troublera point son état de repos.*

[19] Coulomb, 1773, p. 348 (Heyman 1972, p. 44): *La cohésion se mesure par la résistance que les corps solides opposent à la désunion directe de leurs parties. Comme chaque élément des solides, lorsqu'ils sont homogènes, est doué de cette même résistance; La cohésion totale est proportionnelle au nombre des parties à désunir, & par conséquent à la surface de rupture des corps.*

[20] Coulomb, 1773, p. 348 (Heyman 1972, p. 44): *J'ai cherché à déterminer par quelques expériences, la force de cette cohésion.*

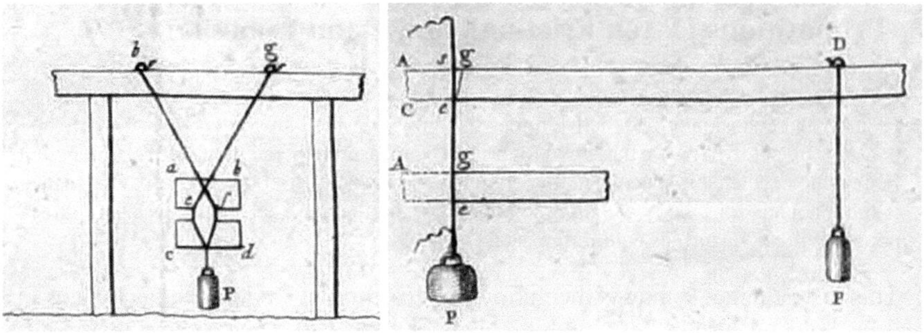

Fig. 4 Coulomb's Figs. 1, 2 and 3 (1773: Tab. XV). Public domain

One finds a confirmation of the importance that Coulomb attributes to this fact by noticing that in the different experiments through which he compares rupture by traction to transversal rupture due to shear, he considers identical surfaces of rupture. As shown in our Fig. 4, in Columb's Fig. 1 the surface *ef* is equal to the surfaces *ge* of his Figs. 2 and 3. It is also in the crucial role played by those surfaces that one finds the justification of the restriction of the study of stability to the plane and thus of the preliminary propositions.

6 Proposition VII, Remarks on the Rupture of Bodies

In proposition VII, Coulomb discusses the rupture of a cantilevered beam (Fig. 5).

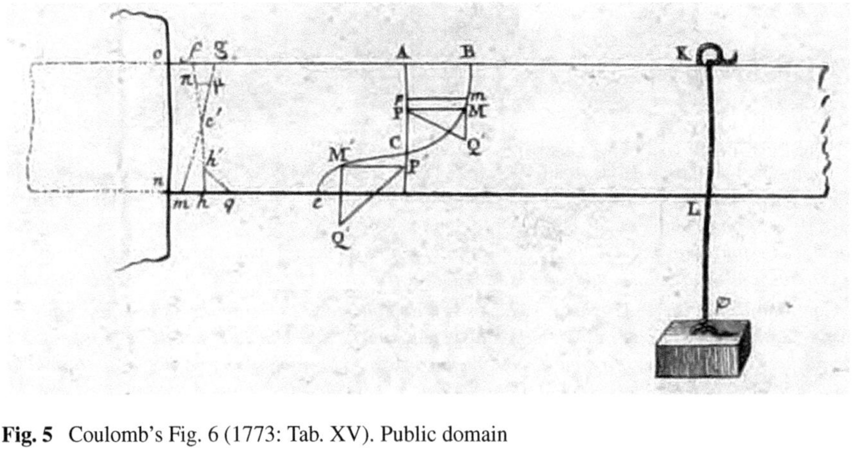

Fig. 5 Coulomb's Fig. 6 (1773: Tab. XV). Public domain

Historical Digression II

Galileo had already compared rupture by traction with transversal rupture (Fig. 6).

Fig. 6 Galileo's figure of rupture (1638, p. 114). Public domain

His study was taken over by Edme Mariotte (1620–1684) and Gottfried Wilhelm Leibniz (1646–1716). Leibniz marks a turning point, or a change of area, from the other scholars mentioned in Table 1. He shows that rupture does not occur in an instant along the whole slice but that a deformation occurs first. In order to obtain that, he introduces along the section of the beam springs that are progressively extended. As David Speiser stated it, in that way, Leibniz provides the opportunity to introduce differential calculus into that problem (Fig. 7). The path that he opened would be pursued by Jacob Bernoulli in his study of the elastica, about which he published a first text in 1694.

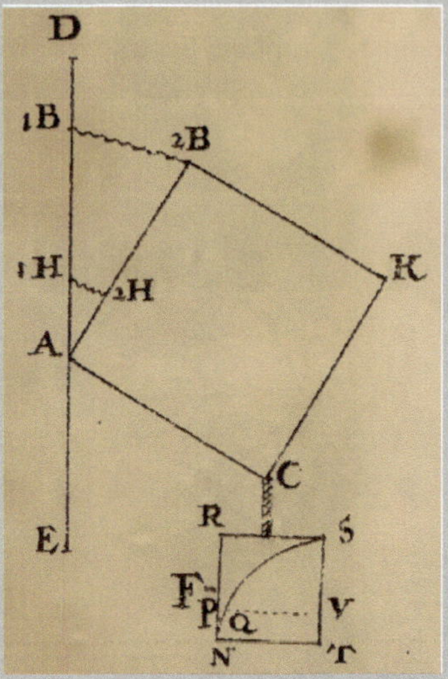

Fig. 7 Leibniz's Fig. 3 (1684: Tab. IX). Public domain

The discontinuous aspect rupture by Galileo does not allow the use of differential calculus. But this point of view allows another type of calculation, the one proposed by Coulomb in the text we are studying.

The construction of proposition VII, as we have already recalled, is remarkable. Coulomb starts by reducing the problem to the plane, saying that the same situation will occur in each longitudinal and vertical section of the beam.

> It is clear, by considering only one vertical face of the body, all the others being equal and parallel, that all the points of the line *AD* will be stressed in preventing the rupture of the body by the weight ϕ; it follows therefore that an upper part *AC* of the line acts in tension directed along *QP*, while the lower parts acts in compression along $Q'P'$. If all these forces, whether tensile or compressive, are resolved in two directions, vertical and horizontal, denoted by *QM* and *PM*, and if through all points *M* is constructed the curve *BMCe*, then this curve will give the locus of all the stresses acting perpendicular to the line *AD*. Thus the portion *ADKL* must be supposed to be acted upon by all the horizontal forces *PM*, by all the vertical forces *MQ*, and by the weight ϕ; it follows from article 3 that since there is equilibrium, the sum of the horizontal forces is zero, and that therefore the total tensile force corresponding to *ABC* must equal to the total compressive force *CeD*. Further, by the same article, the sum of the vertical forces *QM* equal the weight ϕ; and further, by the principles of statics, the sum of the moments about *C* of all the tensile and compressive must equal the moment of the weight ϕ about the same point ... We have to satisfy the three conditions above, therefore, whatever the relations between the deformation and the cohesion of the elements of a body.[21]

---

[21] Coulomb, 1773: 350 (Heyman 1972, p. 46): *Il est évident, en ne considérant qu'une face verticale de ce solide, les autres étant égales et parallèles, que tous les points de la ligne AD résistent pour empêcher le poids* ϕ *de rompre le solide; que par conséquent une partie supérieure AC de cette ligne fait effort par une traction dirigée suivant QP, tandis que la partie inférieure fait effet, par une pression dirigée suivant* $Q'P'$*. Si l'on décompose toutes les forces, soit de traction, soit de pression,*

This text could be read in the following way: Independent of the constitutive law, which is to say independent of the relation between cohesion and dilatation of the elements of a solid, there are three conditions to fulfill that follow from fundamental laws. As Coulomb works in thc plane, equilibrium of forces gives two relations—the horizontal and vertical projections—and the equilibrium of moments gives a third relation.

Let us underline an important result Coulomb obtains during his reasoning. He shows that the areas under the curves of tension and compression must be equal. This means that in his Fig. 6 (shown below in our Fig. 11) the area *ABC* equals the area *CcD*. This had been reported by Parent in 1713 as a consequence of the equilibrium of momenta.

Coulomb pursues the comparison of his theoretical results with the experiments of the preceding proposition. In order to do this, he gives himself a particular constitutive law, the one that had been given by Hooke in 1678: if a material is pulled, the elongation is proportional to the applied force.

Historical Digression III

This law of elongation, given by Hooke in 1678, was taken up by Leibniz in his 'Demonstrationes novae de resistentia solidorum' (1684). The generality of that law was questioned ten years later by Jacob Bernoulli when he tried to determine the equation of the elastic beam (Fig. 8).

---

suivant deux directions, l'une verticale & l'autre horizontale, exprimée par QM & PM; & si par tous les points M l'on fait passer une ligne BMCe, cette courbe sera le lieu géométrique de tous les efforts perpendiculaires qu'éprouve la ligne AD. Ainsi, la tranche ADKL doit être supposée sollicitée par toutes les forces horizontales PM, par toutes les forces verticales MQ, & par la pesanteur du poids par conséquent, puisqu'il y a équilibre, il faut, art.3, que la somme des puissances horizontales soit nulle; que, par conséquent, l'aire des tensions ABC égale l'aire des pressions Ced [recta CeD] *Il faut de plus, par le même article, que la somme des forces verticales QM soit égale au poids; mais par les principes de la statique l'on a encore la somme des* momentum *autour du point G* [recta C] *de toutes les forces, soit de traction soit de pression, égale au* momentum *du poids autour du même point... Nous avons donc, quel que soit le rapport entre la dilatation des éléments d'un solide & leur cohésion, les trois conditions précédentes à remplir.*

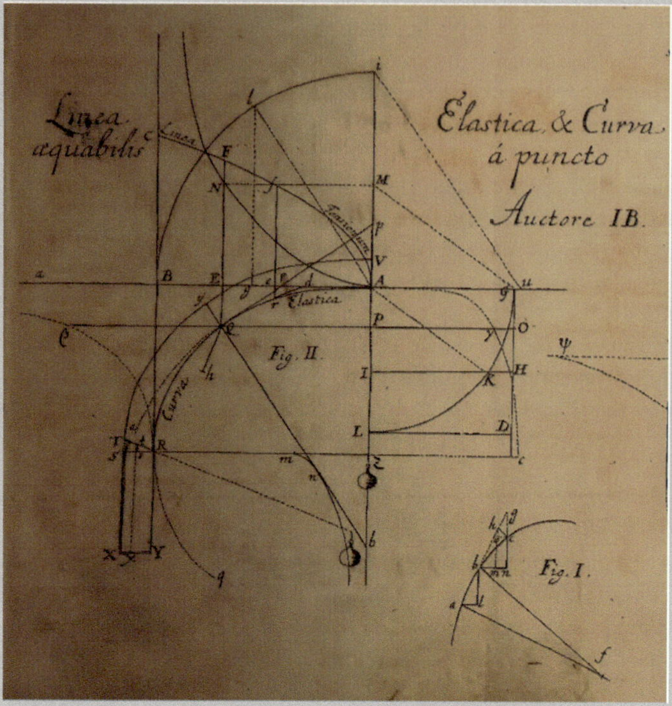

Fig. 8 Jacob Bernoulli's Fig. II (1694a, Tab. VI). Public domain

In the first part of his text, Jacob considers any law of tension, then proposes a first simplification considering, in his second part, conics (or laws in powers) and finally returns to the only case he is able to solve completely, namely Hooke's law, the linear law. At least he has underlined the fact that Hooke's law is only a special case and that materials generally used follow much more complex laws of tension.

In the case of Hooke's law, things are simplified because the areas under tension and under compression are triangular. In other words, as shown in the figure proposed by Truesdell in order to illustrate Parent's reasoning (Fig. 9), the area of the compression triangle *CBX* is equal to the area of the dilatation triangle *TAC*. If one considers also that compression and dilatation are done in the same way except for the sign, we may express the same thing by saying that the neutral fiber is at the middle of beam. But one may imagine cases in between where compression and dilatation occur linearly but differently; that is, cases in which the neutral fiber would not be in the middle of the beam.

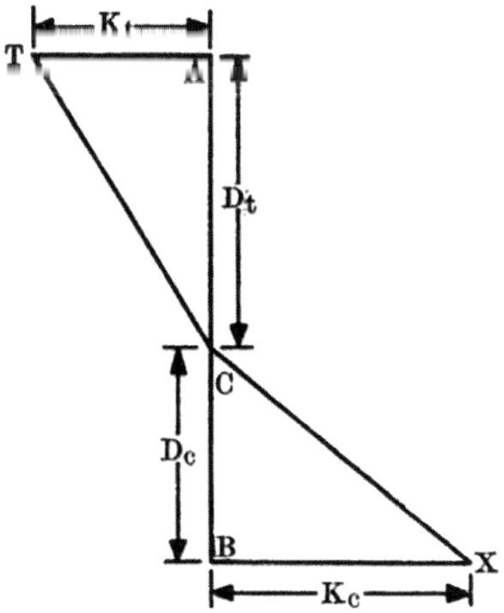

Fig. 9 Truesdell apropos of Parent. Author, after Truesdell, 1960, p. 113, Fig. 41

Historical Digression IV

Jacob Bernoulli, and Leibniz before him, had omitted compression and considered only tension. For that reason, they placed the neutral fiber at the internal surface of the beam, therefore at the intrados. Jacob was brought back to the right path by Huygens, who caused him to observe that error, giving rise to Jacob's second article on the elastica (1695) (Fig. 10).

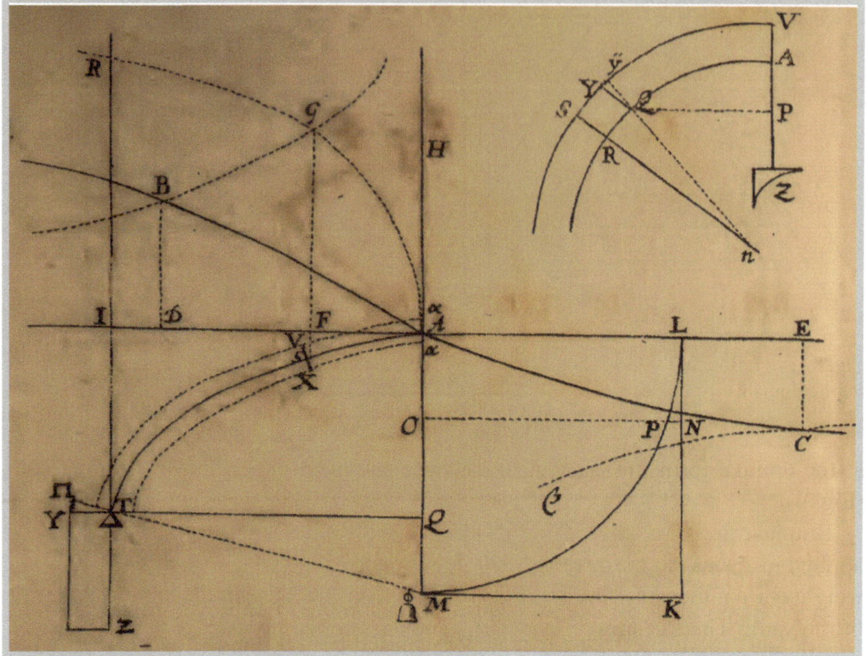

Fig. 10 Jacob Bernoulli's Fig. 2 (1695: Tab. X). Public domain

This research would occupy Jacob Bernoulli until the end of his days. One sees in the figure from this second article that the curve of tension has been completed by a curve of negative tensions or curve of compressions. Coulomb particularizes the equation of equilibrium of moments to Hooke's case and discusses the value of this model, comparing his new theoretical results with the experimental results he had given before. He deduces that no stone exists that really satisfies Hooke's equation.

During this discussion, Coulomb twice emphasises the fact that his equation of equilibrium of moments is established "at the instant of fracture" or "when about to break".[22] In contrast to Leibniz, who had made evident a continuous process leading to rupture, Coulomb insists, on the other hand, that it is discontinuous.

Historical Digression V

One finds this idea first mentioned in 1702 by Pierre Varignon (1654–1722), who compares longitudinal and transversal rupture. Regarding the weights that cause them, he states:

[22] Coulomb, 1773, p. 351 (Heyman 1972, p. 47): *dans l'instant de rupture … prêt à se rompre.*

"Let thus be whatever body … resisting successively to the action of the weights P and Q, the first one pulling perpendicularly to its length and the other one along that length as trying to pull it out the wall, each of them being the greatest that the body is able to sustain without breaking in ABC, as I suppose it to occur under the effort of only one weight a tiny bit bigger then P."

As we shall see, this idea is certainly not foreign to the method proposed by Coulomb, a method that he first states in a simple case.

7 Proposition VIII, Strength of Masonry Piers

Coulomb supposes that "the cohesion gives an equal strength whether the force acts parallel or perpendicular to the plane of fracture"[23] and, as already said, that inside the parts both above and below the sliding plane the cohesion is infinite (see Fig. 1). He neglects the weight of the pier. Let us overlook the doubtful quality of that model and the apparent miscalculation, or sleight of hand,[24] that will appear next in order to concentrate on the method of calculation that he wants to introduce.

Coulomb considers: the weight P supported by the pier; the angle DMC, which he calls x, between the direction of sliding and the horizontal base; a, the length of that base; and δ, the cohesion, which is, as Coulomb said at the very beginning of his text, the force that opposes the sliding (shear) of the upper block $ABMC$ on the lower one CMD. His calculation establishes one or more relations of equilibrium in the case of a fixed angle of sliding x. By canceling the derivative, he finds a minimal value within the acceptable values. The maximum weight that can be supported must be kept inferior to this minimum. This value is thus *valid for any value of the angle* x. By projection on the sliding line or surface[25]:

for the cohesion[26]:

$$\delta CM = \frac{\delta a}{\cos x};$$

for the component of the force:

$$rq = P\sin x;$$

[23] Coulomb, 1773, p. 352–253 (Heyman 1972, p. 48): *oppose une égale résistance, soit que la force soit dirigée parallèlement ou perpendiculairement au plan de rupture…*

[24] I say 'sleight of hand' because a similar mistake will be repeated with other words in the last example of the arches and the result is ultimately correct.

[25] It is in this ambiguity that lies the justification for the sleight of hand. Coulomb considers the same problem in all the plane sections parallel to each other. Unfortunately, he doesn't say so explicitly.

[26] Here and elsewhere the presentation of the formulae has been slightly modernized.

in case of equilibrium

$$P\sin x = \frac{\delta a}{\cos x}$$

or

$$P = \frac{da^2}{\sin x \cos x}.$$

Coulomb emphasizes:

But since the column must be able to carry the weight P without breaking, whatever the section CM, then the weight P must always be less than the quantity given by $da^2/\sin x\cos x$ whatever the value of x.[27]

Angle x was fixed because of the need to put it in the equation, but in order for the pier not to break it is necessary for the weight P to remain less than the reaction of the weaker section. Coulomb finds that minimal reaction by setting the derivative equal to zero; this therefore happens when P is determined as a minimum from the equation $P = da^2/\sin x\cos x$, which gives:

$$dP = \frac{\delta a^2\left[-dx(\cos x)^2 + dx(\sin x)^2\right]}{(\sin x \cos x)^2}.$$

This derivative becomes zero for $\sin x = \cos x$, that is, for an angle of 45°. The maximal applicable weight is $2\delta a^2$.

In Proposition IX, Coulomb adds a term that takes into account friction following Amontons' law, that is, proportional to the compressive force. Coulomb consequently modifies the equation of equilibrium:

$$P\sin x = \frac{\delta a^2}{\cos x} + \frac{P\cos x}{n}$$

where n is a constant, and $\frac{1}{n}$ is the coefficient of proportionality between compressive force and friction. The equation may be written:

$$P = \frac{\delta a^2}{\cos x\left(\sin x - \frac{\cos x}{n}\right)}.$$

The minimal reaction will thus be such that:

$$dP = dx\left(\sin x - \frac{\cos x}{n}\right) - dx\cos x\left(\cos x - \frac{\sin x}{n}\right) = 0$$

[27] Coulomb, 1773, p. 353 (Heyman 1972, p. 48): *Mais comme la colonne doit être en état de supporter le poids P sans se rompre, quelle que soit la section CM, Il faut que le poids P soit toujours plus petit que la quantité donnée par* $\frac{daa}{\sin x\cos x}$, *quelle que soit la valeur de x* (my emphasis).

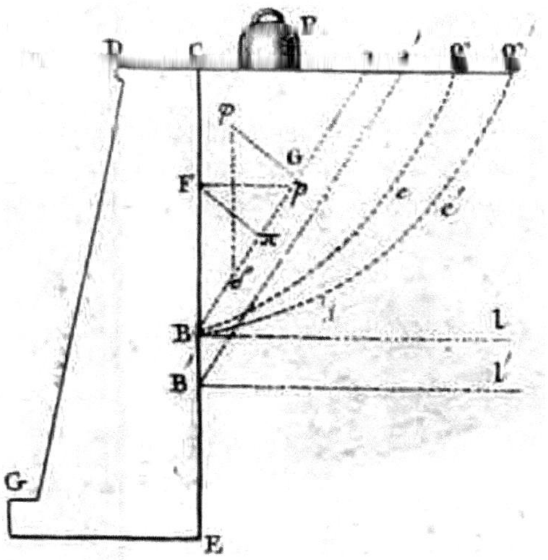

Fig. 11 Coulomb's Fig. 7 (1773: Tab. XV). Public domain

or

$$\tan x = \frac{1}{\sqrt{1 + \frac{1}{n^2}} - \frac{1}{n}}.$$

He then compares his theoretical results with the experiments made by Pieter van Musschenbroek (1692–1761) and published 44 years earlier in that scientist's *Physicae experimentales et geometricae* (1729).

8 Proposition IX, on Earth Pressures and Retaining Walls

Let us see now how Coulomb uses this calculation in the case of retaining walls:

> Let a heavy triangular right-angled prism *CBa* be supported on the line *Ba* by a force *A* acting through *F* perpendicular to the vertical line *CB*; suppose also that it is acted upon by its own weight ϕ, and restrained on the line *Ba* by its cohesion on that line and by friction.[28]

Coulomb once more privileges a direction that will then be varied in order to find the minimum of resistance (Fig. 11).

[28] Coulomb, 1773: 357 (Heyman 1972, p. 51): *Si l'on suppose qu'un triangle CBa rectangle, solide et pesant, est soutenu par la ligne Ba par une force A appliquée en F, perpendiculairement à la verticale CB; qu'en même temps il est solicité par sa pesanteur* ϕ, *et retenu sur la ligne Ba, par sa cohésion avec cette ligne, et par frottement.*

Let $CB = a$, $Ca = x$, then $\delta\left(a^2 + x^2\right)^{\frac{1}{2}}$ will give the coherence of the line *aB*; the weight ϕ of the triangle *CBa* will equal $gax/2$, where g denotes the density of the triangle.

If the forces A and ϕ are resolved in two directions, parallel and perpendicular to the line *Ba*, the triangles $\phi G\delta$, and $F\pi p$, which represent these resolved forces, will be similar to the triangle *CaB*. Thus the forces will be given by the following expressions[29]:

$$\phi G,\ \text{force perpendicular to } aB,\ \text{component of } \phi\text{:}\ \frac{\phi x}{\left(a^2 + x^2\right)^{\frac{1}{2}}}$$

$$G\delta,\ \text{force parallel to } aB,\ \text{component of } \phi\text{:}\ \frac{\phi a}{\left(a^2 + x^2\right)^{\frac{1}{2}}}$$

$$F\pi,\ \text{force perpendicular to } aB,\ \text{component of } A\text{:}\ \frac{Aa}{\left(a^2 + x^2\right)^{\frac{1}{2}}}$$

$$\pi p,\ \text{force parallel to } aB,\ \text{component of } A\text{:}\ \frac{Ax}{\left(a^2 + x^2\right)^{\frac{1}{2}}}.$$

If $\frac{1}{n}$ expresses the constant ratio of friction to pressure, we shall find the force that exerted in order to make the triangle slide on *aB* expressed in the following way:

$$\frac{\left[\phi a - Ax - \frac{\phi x - Aa}{n} - \delta\left(a^2 + x^2\right)\right]}{\left(a^2 + x^2\right)^{\frac{1}{2}}}.$$

In the case of equilibrium, this expression will be equal to zero, so that we get:

$$A = \frac{\left[\phi\left(a - \frac{x}{n}\right) - \delta\left(a^2 + x^2\right)\right]}{\left(x + \frac{a}{n}\right)}.$$

This gives a value of *A* at equilibrium, when the total force is zero. He then considers the increase of the force applied or exerted by the wall and imagines that the triangle of earth goes up instead of down. This second part of the calculation was not taken into account in the case of the pier. It is this part of the calculation that will allow Coulomb to fit the acceptable values between an upper and a lower limit:

> But if it be assumed that the force applied at *F* increases to a value such that the same triangle is on the point of moving in the direction *Ba*; then, denoting this force by A', the force along Ba will be

[29] Coulomb, 1773, p. 357 (Heyman, 1972, p. 51): *Si l'on decompose la force A et la force* φ *suivant deux directions, l'une parallèle à la ligne Ba, l'autre qui lui soit perpendiculaire, les triangles* $\phi G\delta$, $F\pi p$, *qui expriment ces forces décomposées, seront semblables au triangle CaB; l'on aura donc pour ces forces les expressions suivantes* [formula].

$$\frac{\left[A'x - \phi a - \frac{\phi x - Aa}{n} - \delta\left(a^2 + x^2\right)\right]}{\left(a^2 + x^2\right)^{\frac{1}{2}}};$$

from which, for equilibrium,

$$A' = \frac{\left[\phi\left(a + \frac{x}{n}\right) + \delta\left(a^2 + x^2\right)\right]}{\left(x - \frac{a}{n}\right)},$$

an expression which becomes infinite when $x = \frac{a}{n}$.[30]

This gives a new balance of forces with a change of signs and a value of A' when this force is zero, at equilibrium. Coulomb deduces that the force that is applied or exerted by the wall must be between the two values A and A'[31]:

> It has thus been shown that when cohesion and friction contribute to the stability of the triangle, then the limits of the force which can be applied at *F* perpendicular to *CB* without moving the triangle will lie between A and A′.[32]

However, as for the column, this first step of the calculation was done in the case of a fixed sliding surface. It is thus necessary, Coulomb tells us, in order to have a force *A* sufficient to support the entire mass that

> the surface must be sought, among all the surfaces *CBeg*, for which the thrust on the line *CB* [of the wall] is a maximum. Similarly, to find the greatest force which can act through *F* without disturbing equilibrium, another curve $Be'g'$ must be sought such that the force A' sufficient to cause the surface $CBe'g'$ to slide along $Be'g'$ is a minimum. The limits of the horizontal force that can be applied at *F* without moving the soil [*le fluide*] will be bounded by the limits *A* and A', where *A* will be a maximum and A' a minimum.[33]

---

[30] Coulomb, 1773, p. 358 (Heyman, 1972, p. 52): *Mais si l'on suppose que la force appliquée en F, vienne à augmenter, au point qu'elle soit prête à mettre le même triangle en mouvement suivant la direction Ba; pour lors, en nommant A′ cette force, l'on aura* [formula] *pour l'effort suivant Ba; d'où l'on tire, dans le cas d'équilibre,* [formula], *quantité qui seroit infinie si x égaloit a/n.*

[31] By the way, this gives a limit value when the adherence and friction are zero, $A < \frac{ga^2}{n} < A'$.

[32] Coulomb, 1773, p. 358 (Heyman 1972, p. 52): *Il est donc démontré que lorsque la cohésion & le frottement contribuent à l'état de repos du triangle, que les limites de la force que l'on peut appliquer en F, perpendiculairement à CB, sans mettre le triangle en mouvement, seront comprises entre A & A′.*

[33] Coulomb, 1773, p. 359 (Heyman 1972, p. 52): …*toutes les surfaces* CBeg, *celle dont la pression sur la ligne* CB *[du mur] est un maximum. De même si l'on vouloit déterminer la plus grande force qui puisse agir en F, sans troubler l'état de repos, il faudroit chercher une autre courbe* Be′g′, *telle que* A′ *suffisante pour faire couler la surface* CBe′g′ *suivant* Be′g′ *soit un minimum; et les limites de la force horizontale, que l'on peut appliquer en* F, *sans mettre le fluide en mouvement, seront comprises entre les limites* A & A′, *où* A *sera un* maximum *et* A′ *un* minimum.

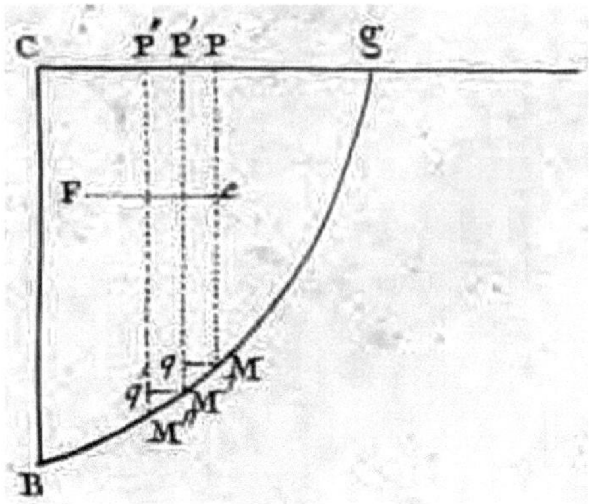

Fig. 12 Coulomb's Fig. 8 (1773: Tab. XV). Public domain

Therefore,

$$A_{max} < F < A'_{min}.$$

This statement is very general and shows clearly how his method of calculation works. It is too general for the problem of retaining walls. For this reason, in a first moment, Coulomb considers only the weaker horizontal force that the retaining wall sustaining the earth mass is able to resist without breaking equilibrium, and thus he only searches for *A*. He greatly simplifies the problem by admitting that the sliding curve *Beg* is a straight line, an assumption that is correct in the case of soil. His treatment of this problem then becomes essentially the same as for the column.

True to his habit, he compared his theoretical results with those provided by experience, in this case the walls built by the engineer Marshal Sébastien Le Prestre de Vauban (1633–1707).

9 Proposition XV, on the Surface of Greatest Thrust in Cohesive Fluids

Coulomb returns to the most general case and tries to determine the curve, not the triangle, that gives rise to the greatest pressure against the wall (Fig. 12).

How to put the problem in equation is sketched out and given in the absence of cohesion, but:

> The extension of these remarks, as well as the application of the above formulae, requires a special piece of work, and would take me away from the simplicity which I laid down for myself in this memoir.[34]

[34] Coulomb, 1773, p. 370 (Heyman 1972, p. 60): *Le développement de ces remarques, de même que l'application des formules qui précedent, demandent un travail exprès, & m'éloigneroit de la simplicité que je me suis prescrite dans ce Mémoire.*

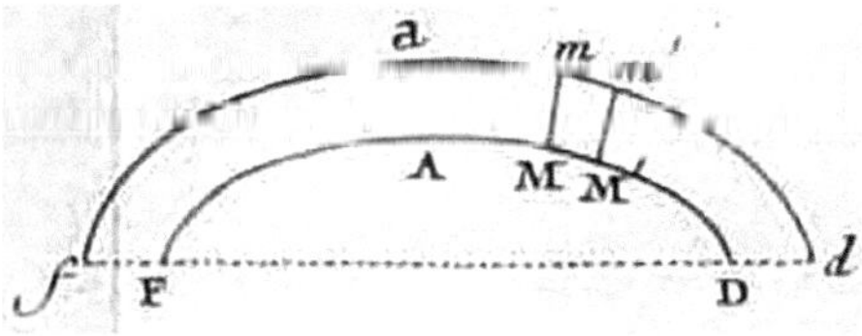

Fig. 13 Coulomb's Fig. 9 (1773, Tab. XVI). Public domain

10 Proposition XVI, on Arches

The first paragraph describes an arch in a general way by means of two curves: *FAD*, the inner curve, or intrados, and *fad*, the outer curve, or extrados (Fig. 13).

Let the lines *Mm* be perpendicular or of some other orientation with respect to the inner curve. The forces acting on an element *AaMm* of the arch or voussoir are gravity, cohesion and friction. Let us remark that Coulomb does not use this terminology, particular to the domain of arches.

11 Proposition XVII, on Arches with Joints that Have Neither Friction nor Cohesion

Historical Digression VI

This problem is the first that was considered at the time of the study of the stability of arches. The model consists in considering perfectly rigid voussoirs and perfectly smooth joints. First Hooke, then La Hire, Jacob and Johann Bernoulli, and David Gregory after them, had said, shown, affirmed, and demonstrated that the catenary, turned upside down, must give the curve of the arch most stable under its own weight. In fact, the catenary is the curve adopted by the perfectly flexible chain suspended between two points under its own weight. When reversed and when the tensions are changing into pressures, one finds the depiction of the most stable arch we have just described.

Coulomb puts the problem in equations and observes the identity of his result with that given by Euler in his "Solutio problematis de inveniendi curva quam format lamina elastica in singulis punctis a potentiis quibuscunque sollicitata" (1728), in which he generalizes the problem of the catenary to whatever applied forces and not only to gravity that is necessarily vertical. Coulomb remarks that the way he elaborates his equation, as shown here after, has nothing to do with the one of Euler.[35]

Coulomb considers all the external forces applied to the elements of an infinitely thin arch whose joints are perpendicular to the intrados (Fig. 14).

[35] Truesdell (1960) observes that it is the same as the one given by Johann Bernoulli in his Lecture to de l'Hôpital.

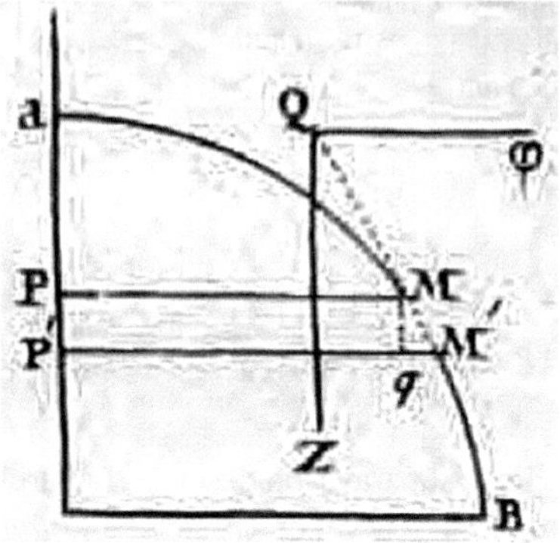

Fig. 14 Coulomb's Fig. 10 (1773: Tab. XVI). Public domain

He resolves all those forces into two components, one vertical and the other horizontal. He writes the two equations of equilibrium and divides the one by the other, member by member, obtaining:

$$\frac{dx}{dy} = \frac{F_{hor}}{F_{ver}}.$$

In the corollaries, he particularises to an infinitely thin arch on which only gravity acts and whose horizontal thrust is constant. This case had already been treated by Jacob Bernoulli in 1704.

Next, he considers an arch of finite thickness. In his first remark, he underlines the link between what he had just written and the preliminary remarks. He says that he had only sought to require that the resultant of forces applied to a voussoir be perpendicular to the joint: this corresponds to the first condition of the first preliminary proposition. He adds that at the same time he made sure that the resultant does not fall outside the base. He shows that the line formed by the resultants, if they are perpendicular to the joints, is a curve parallel to the intrados. As he had required that they start from the summit, the line remains well between the limits. Let us remark that the condition on thrusts is the corollary of that on tensions for the catenary, which must always be tangent to the curve. Imposing this supplementary condition on the problem allows it to be solved with help of only one equation. It was in that way that Jacob Bernoulli, as Coulomb reminds us,[36] could find the catenary with help of only one condition of equilibrium, the equilibrium at translation or the sum of applied force being zero. On the contrary, with the help of the second equation giving the equilibrium to rotation, Jacob did not find that same equation again because of an error in his calculation. Coulomb, for his part, just showed that one must find the same result. He had arrived at a profound understanding of that aspect of the problem and he will prove it once more.

In the second remark, he observes that if the angle in B is a right angle the joint should become infinite; it becomes the asymptote to the superior curve, as one

[36] We remarked on that subject at the beginning of this chapter.

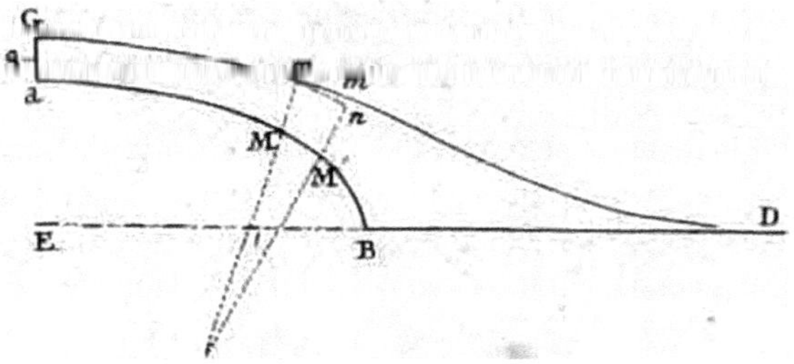

Fig. 15 Coulomb's Fig. 11 (1773; Tab. XVI). Public domain

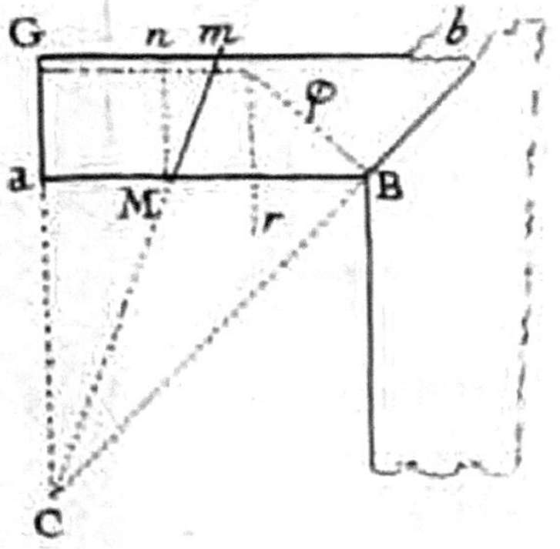

Fig. 16 Coulomb's Fig. 13 (1773: Tab. XVI). Public domain

sees in his Fig. 11 (shown here in Fig. 15), and the weight of the last voussoir, as the predecessors of Coulomb had said, should become infinite, unless—as the predecessors had already asserted and as Coulomb reiterates—the friction and the adherence play in favor of the architect.

In a last corollary, he fixes the two curves, intrados and extrados, and deduces the angle that the joints of the voussoirs must form with respect to each other. He applies this calculation to the case of the flat strip (Fig. 16).

Let us make here a digression in order to underline that the model of the arch whose voussoirs are perfectly rigid and whose joints are perfectly smooth is such that all problems related to constitutive equations are rendered null, allowing the role of fundamental equations to be underlined quite clearly. Perfectly rigid voussoirs and joints perfectly smooth are particularly simple constitutive equations. Those simplifications allow to concentrate attention on the fundamental laws of equilibrium at translation and at rotation.

We may observe the same intellectual approach in the case of the study of collisions between two marbles. In that case, authors gave themselves the following simple constitutive equations: they posited that the bodies were either soft, meaning that they remain amalgamated after the collision, or were perfectly elastic, meaning that they would recover their original configuration exactly after the collision. Once those simplifications concerning the constitutive equations implemented, the fundamental relations of conservation of quantity of motion and of energy appear much more clearly. Those fundamental laws remain true for materials obeying much more complex constitutive laws, but it was necessary to go through those simplifications in order for them to appear.

In that particular part of Coulomb's work, we must not be astonished not to find, together with friction and cohesion, the method of calculating he had introduced in his work. This method is precisely dedicated to overcome the difficulties linked to the ignorance of the exact behavior of the material, of the constitutive equations.

12 Proposition XVIII, on the Stability of Arches, Allowing for Friction and Cohesion

In this last part, Coulomb supposes, once again, a more complicated behavior, that is, the case of the joints of the arch (Fig. 17).

He appeals to his new method of calculating:

> Problem. The intrados *aB* and the extrados *Gb* are given for an arch, as are the joints *Mm* normal to the intrados; required the limits of the horizontal thrust at *f* which would support

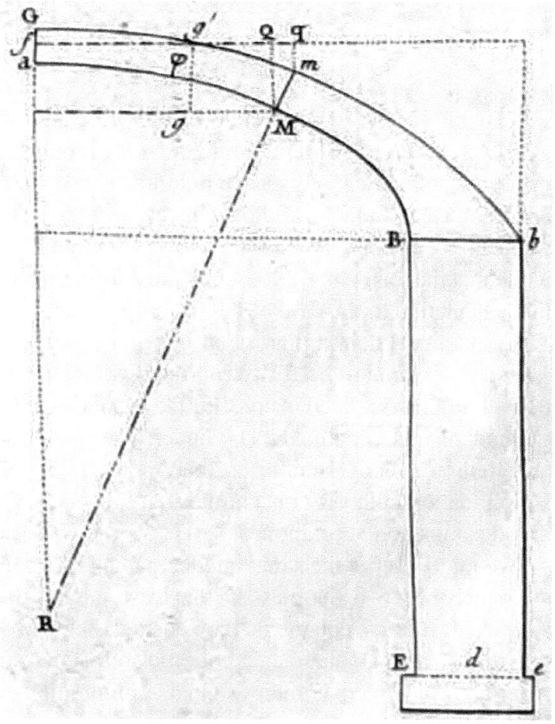

Fig. 17 Coulomb's Fig. 14 (1773: Tab. XVI). Public domain

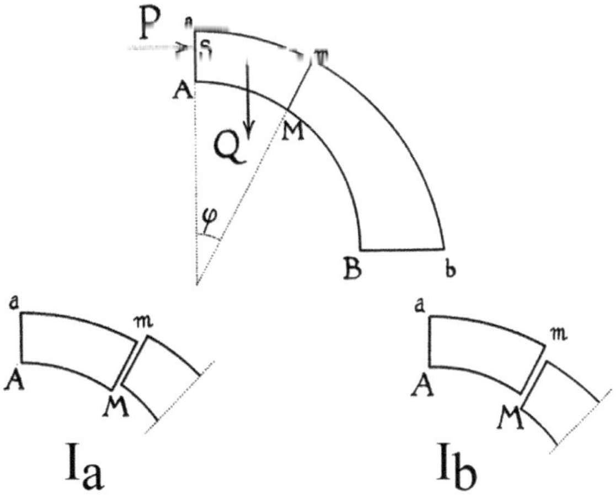

Fig. 18 Steps I_a and I_b of Coulomb's calculation. Author, after Benvenuto (1991, II, Fig. 12.14)

> the arch, when it is acted upon by its own weight and sustained by the cohesion and the friction of the joints.[37]

In order to facilitate understanding, we number the steps of the calculation.

I_a) In the first step, Coulomb proceeds in the same way as for the retaining wall. He resolves the forces and sets the sum of the horizontal components equal to zero. This gives him a value of the thrust exerted at *f* at the equilibrium (thrust due to the other part of the arch). This is the equilibrium for translation: this thrust, he says, prevents the voussoir from sliding along joint *mM* (Fig. 18).

Coulomb expresses this relation in function of the angle in *R*, center of curvature of the arch. As the position of the joint is not fixed, just as the sliding surfaces was not fixed in the case of the retaining wall, it is necessary that the pressure at the top be sufficient whatever the value of the angle at the center and thus for a maximal value A_{max}, which he finds by setting the derivative equal to zero:

$$A_{max} < \text{thrust at the summit.}$$

I_b) Coulomb pursues the same reasoning as in the case of retaining walls. Let us imagine, still, that instead of sustaining the voussoir that we are considering, the thrust becomes too great and makes this voussoir slide in the other direction, to the top. Coulomb finds then another value for *A* at the equilibrium. The permitted values must always be less than whatever the value of *A* is, and thus also less than A_{min}:

$$\text{Thrust at the summit} < A_{min}.$$

---

[37] Coulomb, 1773, p. 377 (Heyman 1972, p. 66): *Problème. Dans une voûte, la courbe intérieure aB, la courbe extérieure Gb étant données, les joints Mm, perpendiculaires aux éléments de la courbe intérieure, seront aussi donnés: l'on demande les limites de la pression horizontale en f, qui soutiendra cette voûte, en supposant qu'elle soit sollicitée par sa propre pesanteur, et retenue par la cohésion et le frottement des joints.*

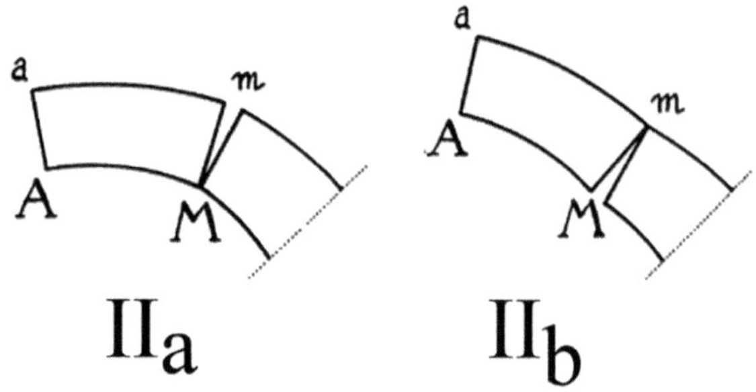

Fig. 19 Steps II_a and II_b of Coulomb's calculation. Author, after Benvenuto (1991, II, Fig. 12.14)

$\mathbf{I_{a,b}}$) In order to avoid sliding, or, if one prefers, to ensure the equilibrium of the forces present, the thrust exerted at the summit by the other part of the arch must lie between A_{max} and A_{min}:

$$A_{max} < \text{Thrust at the summit} < A_{min}.$$

This means that the force must be superior to the maximum of the forces just sufficient to sustain the voussoir when sliding down and inferior to the minimum of the forces triggering ejection of the voussoir up.

$\mathbf{II_a}$) In this manner is fulfilled the equilibrium of forces or the first condition of equilibrium of the first preliminary proposition of Coulomb or the equilibrium at translation. But there remains to establish the equilibrium at rotation, or, if one prefers, to fulfil equilibrium of moments that corresponds to the second part of the first preliminary proposition (Fig. 19).

This is what Coulomb will do next. He writes, perhaps more clearly than in the second part of his first preliminary proposition:

$$B \bullet MQ = \phi gM \ - \delta' z^2$$

This means that the momentum of the thrust at summit B multiplied by its lever arm MQ equals the weight of the voussoir ϕ times its lever arm gM. This last one is equal to the distance from its center of gravity g to the center of rotation M. The term $\delta' z^2$ indicates the cohesion that avoids the rotation of the voussoir. With this equation Coulomb finds a value at equilibrium for the thrust at summit B. If the value of B were smaller, the resultant would be applied under M, the internal point of the arch. It is thus necessary to find the maximal possible value of B, B_{max}. For the arch to be stable, it is necessary that the thrust remains superior to that value.

$$B_{max} < \text{Thrust at the summit.}$$

$\mathbf{II_b}$ But the voussoir could still rotate around the external point m of the arch or in the terms of Coulomb, the resultant could be applied above that point. In this case he obtains:

$$B \cdot m1 = \phi g q + \delta' z^2.$$

It is necessary that the resultant be applied under that point, this means that the force is kept inferior to the littlest of the B so determined.

$$\text{Thrust at the summit} < B_{min}.$$

$\mathbf{II_{a,b}}$) Finally, the equilibrium at rotation imposes that:

$$B_{max} < \text{Push at the summit} < B_{min}$$

It remains to combine the conditions $I_{a,b}$ and $II_{a,b}$:

$$\mathbf{I_{a,b}}) \quad A_{max} < \text{thrust at the summit} < A_{min}$$

and

$$\mathbf{II_{a,b}}) \quad \mathbf{B_{max}} < \text{thrust at the summit} < \mathbf{B_{min}}$$

By choosing $C_{max,max}$ equal to the greatest of A_{max} and of B_{max} and $C_{min,min}$ equal to the smaller of A_{min} and of B_{min} one should then have:

$$C_{max,max} < \text{thrust at the summit} < C_{min,min}.$$

At this point of his text, Coulomb announces that in practice one more often works by trial and error. He concludes a bit further:

> Comparing the above principles with the different approximate methods used in practice, it will be easily seen that their authors were not sufficiently aware of the two equilibrium conditions necessary for the state of stability. In the method, for example, attributed to La Hire, reported by Bélidor, and used by almost all builders, the arch is divided into three portions, and the thrust of the upper part is calculated in accordance with the first equilibrium condition; the dimensions of the piers are then found by the second equilibrium condition. Now it should be obvious that if the upper portion is divided near the keystone, and if it is supposed that the arch breaks in four portions, rather than three, the thrust of the upper portions will often be, for flat arches, much greater than that found by the method of M. de la Hire, and the size of the piers calculated by his method will often be inadequate.[38]

[38] Coulomb, 1773, p. 382 (Heyman 1972, p. 69): *En comparant les principes qui précèdent avec les différentes méthodes d'approximation usitées dans la pratique, l'on s'appercevra facilement que leurs auteurs n'ont point assez distingué les deux conditions d'équilibre nécessaires pour l'état de repos. Dans celle, par exemple, que l'on attribue à M. de la Hire, rapportée par M. Belidor, et pratiquée par presque tous les Artistes, l'on divise la voûte en trois parties, et l'on calcule la pression de la partie supérieures, en se conformant à la première condition d'équilibre, et l'on détermine ensuite les dimensions des pieds-droits, par la deuxième condition d'équilibre. Or, pour peu que l'on y fasse attention, l'on verra que si l'on divise la partie supérieure vers la clef, et que l'on suppose que cette voûte se rompe en quatre parties, au lieu de se rompre en trois, la force de pression des parties supérieures se souvent, dans les voûtes plates, beaucoup plus grande que celle*

One must add to this that while the *Traité de mécanique* published by La Hire in 1695 is most valuable, the text 'Sur la construction des voutes dans les edifices' about the *pieds droits* published in 1712 is mostly false. It is possible, reading these words of Coulomb written in 1773, to precisely measure the gap existing between theory and practice.

13 Conclusion

The method of calculation introduced by Coulomb in his *Essay de Maximis et Minimis* makes it possible to take into account constitutive equations that are more complex and closer to reality, but they still are paired with a certain ignorance of the precise behaviour of matter. In contrast to the 'Demonstrationes novae de resistentia solidorum' proposed by Leibniz in 1684, who tried to introduce a certain continuity into the phenomena of rupture, Coulomb tried to master the discontinuity by establishing limits and by inserting the admissible values between those limits. This type of calculation leads to what engineers nowadays call 'coefficients of safety,' aimed at maintaining tension in materials within acceptable limits.

Appendix 1: Three domains for the study of the stability of vaults

Full bibliographical details are given in the References at the end of this chapter.

Year	Statics		Elasticity		Resistance of materials	
	Mechanics	Architecture	Mechanics	Architecture	Mechanics	Architecture
1638					Galileo	
1643		Derand				
1673	Pardies	Blondel				
1678			Hooke			
1684					Leibniz	

(continued)

qui se détermine par la méthode de M. de la Hire, et que les dimensions des pieds-droits, difés per cette méthode, seront souvent insuffisantes.

(continued)

Year	Statics		Elasticity		Resistance of materials	
	Mechanics	Architecture	Mechanics	Architecture	Mechanics	Architecture
1686					Mariotte	
1687	Varignon					
1691	Catenary*		Catenary*			
* The problem of the catenary was posed by Jacob Bernoulli. Responses were formulated by Leibniz, Johann Bernoulli and Huygens						
1694			Jacob Bernoulli			
1695		La Hire	Jacob Bernoulli			
1697	Gregory					
1702		La Hire			Varignon	
1704	Jacob Bernoulli	Parent				Parent
1705			Jacob Bernoulli		Jacob Bernoulli	
1707						Parent
1710						Parent
1712		La Hire				
1713				Parent		
1728		Couplet	Daniel Bernoulli Euler			
1729		Belidor Couplet			Bülfinger Musschenbroek	Belidor
1730		Couplet				
1744			Euler			
1750	Euler					
1755	Aepinus					
1770			Euler			
1773		Coulomb				Coulomb
1774		Bossut		Bossut		
1775						Euler
1785		Mascheroni		Mascheroni		
1833				Metallic Constructions		

References

Aepinus, Franz Ulrich Theodore. 1755. De la figure des supports d'une voûte. *Histoire de l'Académie des Sciences et des Belles Lettres de Berlin* (1757), 386–393.

Anonymous. 1704. Sur la figure de l'extrados d'une voute circulaire, dont tous les voussoirs sont en equilibre entre eux, *Histoire de l'Académie Royale des Sciences, année 1704* (Paris, 1722), 93–96.

Anonymous. 1712. Sur la poussée des voutes. *Histoire de l'Académie Royale des Sciences, année 1712* (Paris, 1731), 74–77.

Anonymous. 1729. Sur les voutes. *Histoire de l'Académie Royale des Sciences, année 1729* (Paris, 1733), 103–112.

Anonymous. 1730. Sur les voutes. *Histoire de l'Académie Royale des Sciences, année 1730* (Paris, 1732), 107–110.

Anonymous. 1774. Sur l'équilibre des voûtes. *Histoire de l'Académie Royale des Sciences, année 1774* (Paris, 1778), 59–64.

Anonymous. 1776. Sur l'équilibre des voûtes. *Histoire de l'Académie Royale des Sciences, année 1776* (Paris, 1779), 43–45.

Benvenuto, Edoardo. 1991. *An Introduction to the History of Structural Mechanics*, 2 vols. Part I, *Statics and Resistance of Solids*; Part II, *Vaulted Structures and Elastic Systems*. Heidelberg: Springer.

Bernoulli, Daniel. 1728. St. 15, Methodus universalis determinandae curvaturae fili a potentiis quamcunque legem inter se observantibus extensi, una cum colutione problematum quorundam novorum pertinentium. *Commentarii Acad. Scientiarum Petropolitanae* III (1732), 62–69. Bülfinger, Georg Bernhard. 1729. De solidorum resistentia specimen. *Commentarii Acad. Scientiarum Petropolitanae* IV (1735), 164–181.

Bernoulli, Jacob. 1694a. Curavatura laminae elasticae, ejus identitas cum curvatura lintei a pondere inclusi fluidi expansi, radii circulorum osculantium in terminis simplicissimis exhibiti; una cum novis quibusdam theorematis huc pertinentibus &c. (Op. LVIII), *Acta Eruditorum*, 262–276; *Opera*, 576–600.

Bernoulli, Jacob. 1694b. Solutio Problematis Leibnitiani de Curva Accessus & Recessus aequabilis a puncto dato, mediante rectificatione Curvae Elasticae (Op. LIX), *Acta Eruditorum*, 276–280; *Opera*, 601–607.

Bernoulli, Jacob. 1695. Explicationes, Annotationes, et Additiones ad ea, quae in Actis sup. anni de Curva Elastica, Isochrona Paracentrica, & Velaria, hinc inde memorata, & partim controversa leguntur; ubi de Linea mediarum directionum, aliisque novis (Op. LXVI), *Acta Eruditorum*, 537–553.

Bernoulli, Jacob. 1704. VP. XXIX, Problema de Curvatura fornicis, cujus partes se mutuo proprio pondere suffulciunt sine opere caementi. *Jacob Opera*, 1119–1123 (Med. CCLXXXV).

Bernoulli, Jacob. 1705. Veritable Hypothêse de la Résistance des Solides, avec la Démonstration de la Courbure des corps qui font Ressort (Op. CII), *Mémoires de l'Académie Royale des Sciences", année 1705* (Paris, 1730), 176–186; *Opera*, 976–989.

Bernoulli, Johann. 1691–1692. Leçons à de l'Hôpital N° 36–45 (Op. CXLIX). *Opera*, III, 491–516.

Blondel, François. 1673. Résolution des quatre principaux problèmes d'architecture. *Mémoires de l'Académie des Sciences de 1666 jusqu'à 1699*, 5 (1729), 355–530.

Bossut, Charles. 1774. Recherches sur l'équilibre des voûtes. *Mémoires de l'Académie Royale des Sciences", année 1774* (Paris, 1778), 535–566.

Coulomb, Charles Augustin. 1773. Essai sur une application des règles de Maximis et Minimis à quelques problèmes de statique, relatifs à l'architecture. *Mémoires de Mathématique et de physique, présentés par divers savans*, vol. 7, année 1773 (Paris 1776), 343–382.

Couplet, Pierre. 1729. De la poussée des voûtes. *Mémoire de l'Académie royale des Sciences* (Paris, 1731), 79–117; *Histoire de l'Académie Royale des Sciences, année 1731* (Paris, 1735), 75–81.

Couplet, Pierre. 1730. Seconde partie de l'examen de la poussée des voûtes. *Mémoires de l'Académie Royale des Sciences* (Paris, 1732), 117–141.

Derand, François. 1643. *L'architecture des voutes, ou l'art des traits, et coupe des voutes*. Paris: Sebastien Cramoisy.

Euler, Leonhard. 1728. Solutio problematis de inveniendi curva quam format lamina elastica in singulis punctis a potentiis quibuscunque sollicitata (E 8), *Commentarii Acad. Scientiarum Petropolitanae* III (1732): 70–84; *Opera Omnia*, ser. II, 10, 1–16.

Euler, Leonard. 1744. *Methodus inveniendi lineas curvas maximi minimive proprietate gaudentes, Additamentum I de curvis elasticis* (E 65). Lausanne et Genève: Marc Michel Bousquet; *Opera Omnia* ser. I, 24.

Euler, Leonhard. 1750. Découverte d'un nouveau principe de mécanique (E177). *Mémoires de l'académie des sciences de Berlin* 6 (1752): 185–217; *Opera Omnia*, ser. II, 5, 81–108.

Euler, Leonhard. 1770. E 410, Genuina principia doctrinae de statu aequilibrii et motu corporum tam perfecte flexibilium quam elasticorum. *Novi Commentarii Acad. Scientiarum Petropolitanae* XV (1771), 340–370; *Opera Omnia*, ser. II, 11, 37–61.

Euler, Leonhard. 1775. E 480, Regula facilis, pro diiudicanda firmitate pontis aliusve corporis similis, ex cognita firmitate moduli, *Novi Commentarii Acad. Scientiarum Petropolitanae* XX, 271–285; *Opera Omnia*, ser. II, 11, 220–231.

Euler, Leonhard. 1776. De Gemina methodo tam aequilibrium quam motum corporum flexibilium determinandi et utriusque egregio consensu (E 481). *Novi Commentarii Acad. Scientiarum Petropolitanae* XX, 286–303; *Opera Omnia*, ser. II. 11, 180–193.

Galilei, Galileo. 1638. *Discorsi e dimostrazioni matematiche, intorno a due nuove scienze attenti alla mecanica e i movimenti locali,* Leiden: Elsevier. Eng. trans. Stillman Drake, *Two New Sciences, Including Centers of Gravity & Force of Percussion.* Madison: University of Wisconsin Press, 1974.

Gregory, David. 1697. Catenaria, *Philosophical Transactions*, 19, no. 231 (August 1697): 637–652.

Heyman, Jacques. 1972. *Coulomb's Memoir on Statics: An Essay in the History of Civil Engineering.* Cambridge University Press.

Hooke, Robert. 1678. *Lectures de Potentia Restitutiva, or of Spring, Explaining the Power of Springing Bodies.* London: John Martyn.

La Hire, Philippe de. 1695. *Traité de mécanique, où l'on explique tout ce qui est nécessaire dans la pratique des Arts, et les propriétés des corps pesants lesquelles ont eu plus grand usage dans la Physique.* Paris: Imprimerie Royale.

La Hire, Philippe de. 1702. Remarques sur la forme de quelques arcs dont on se sert dans l'Architecture, *Histoire de l'Académie Royale des Sciences", année 1702* (Paris 1743), 94–97.

La Hire, Philippe de. 1712. Sur la construction des voutes dans les edifices. *Mémoires de l'Académie Royale des Sciences, année 1712* (Paris, 1731), 69–77.

Leibniz, Gottfried Wilhelm von. 1684. Demonstrationes novae de resistentia solidorum, *Acta Eruditorum*, 319–325; *Mathematische Schriften,* 6, 106–112.

Magnel, Gustave. 1935. *Cours de stabilité des constructions.* Ghent: Van Rysselberghe & Rombaut.

Mariotte, Edme. 1686. *Traité du mouvement des eaux et des autres corps fluides.* Paris: E. Michallet.

Mascheroni, Lorenzo. 1776. *Nuove ricerche sull'equilibrio delle volte.* Bergamo: Francesco Locatelli.

Musschenbroek, Peter van. 1729. *Physicae experimentales et geometricae ... dissertationes ut et ephemerides meteoroligicae ultrajectinae.* Lyon: Samuel Luchtmans.

Pardies, Ignace Gaston. 1673. *La statique ou la science des forces mouvantes.* Paris: Sebastien Mabre-Cramoisy.

Parent, Antoine. 1704. Nouvelle statique avec frotemens et sans frotemens ou Regles pour calculer les frottemens des machines dans l'état d'équilibre, *Histoire de l'Académie Royale des Sciences, année 1704* (Paris, 1722), 173–197.

Parent, Antoine. 1707. Des resistances des poûtres par rapport à leur longueurs ou portées, & à leur dimensions & situations; & des poûtres de plus grande résistance, indépendamment de tout système physique, *Mémoires de l'Académie Royale des Sciences", année 1708* (Paris, 1730), 17–31.

Parent, Antoine. 1710. Des points de rupture des figures, de la manière de les rappeller à leurs tangentes: d'en déduire celles qui sont par-tout d'une résistance égale: avec la méthode pour trouver tant de ces sortes de figures que l'on veut: et de faire en sorte que toute sorte de figure soit par tout d'une égale résistance, ou ait un ou plusieurs points de rupture. *Mémoires de l'Académie Royale des Sciences", année 1710* (Paris, 1732), 177–194.

Timoshenko, Stephen P. 1953. *History of Strength of Materials, with a Brief Account of the History of the Theory of Elasticity and Theory of Structure.* New York: McGraw-Hill.

Truesdell, Clifford A. 1960. *The rational mechanics of flexible or elastic bodies 1638–1788. Introduction to Leonardo Eulerii Opera Omnia,* Ser. II, 11. Zurich: Orell Füssli.

Varignon, Pierre. 1687. *Projet d'une nouvelle mécanique.* Paris: Edme Martin, Jean Boudot, Estienne Martin.

Varignon, Pierre. 1702. De la résistance des solides en général pour tout ce qu'on peut faire d'hypothêses touchant la force ou la tenacité des fibres des corps à rompre; et en particulier pour les hypothêses de Galilée et de Mariotte, *Mémoires de l'Académie Royale des Sciences, année 1702* (Paris, 1743), 66–97.

Chapter 11
Jacques-Germain Soufflot and the Relationship Between Rules and Taste

1 Introduction

When, in 1770, Patte attacked the architect Jacques-Germain Soufflot by publishing a pamphlet intended to "mathematically prove" that the columns already erected to support the dome of Sainte-Geneviève, Soufflot did not respond in his own name; he entrusted the pen to Jean-Baptiste Rondelet (1743–1829), who responded with a *Réfutation* (1770) and who would continue his work after 1780. On the other hand, in 1739, Soufflot began reflecting on the relationship between the "rules" and the "taste", a reflection that he will continue practically until his death. This was the subject of four of his little-known communications at the Academy of Architecture in Lyon.

- *Mémoire sur les proportions de l'architecture* (Soufflot 1739);
- *Un parallèle des églises gothiques avec les églises modernes* (Soufflot 1741);
- *Dans l'art de bâtir le goût est-il préférable à la science des règles ou celle-ci au goût ?* (Soufflot 1744).
- *De l'identité du goût et des règles dans l'exercice de l'architecture* (Soufflot 1775).

The first three communications precede the quarrel with Patte, while the last takes place in the very midst of it. Soufflot's reflections proposed in these four texts will be the subject of this article.

A maxim attributed to Plato and known in several forms, but which is difficult to actually find, clearly describes Soufflot's main concern. One of these expressions, "Beauty is the splendor of the truth", reflects best the idea I want to emphasize here and applies particularly well to the case of architecture. Indeed, it is the question of expressing the notion that the beauty of a structure comes, for example, from the conformity of its lines to the laws of nature, or that its lines allow an immediate reading of the natural laws which underlie its stability. Soufflot's titles mentioned above show

P. Radelet-de Grave, *Between Mechanics and Architecture*, Mathematics and the Built Environment 8, https://doi.org/10.1007/978-3-031-73530-1_11

his progression towards this concept, which he expresses through reflection on the laws of nature and of taste, and the search for aesthetic beauty.[1]

Soufflot first analyzes the rules of proportion decreed by the Greeks, then the structure of Gothic churches, where the distribution of forces is apparent and which he, for that reason, finds beautiful. He then explicitly questions the relationship between rules and taste, concluding in his last text that they are the same. Although Soufflot uses the term "identity", I find that the term "reflection" better expresses his idea.

In what follows, we will analyse in detail the progression of Soufflot's reasoning toward this idea of reflection between laws and beauty.

2 9 December 1739, Mémoire sur les proportions de l'architecture

In the beginning of his first article, Soufflot contrasts two architects, François Blondel (1618–1687) and "Monsieur Perrault, director of the Royal Academy of Sciences who published the translation of Vitruvius", which is more than enough to us to conclude that he is speaking about Claude Perrault (1613–1688), brother of Charles (1628–1703), the author of four volumes of *Parallèle des Anciens et des Modernes.*[2] Soufflot somehow transcends the quarrel between the ancients and the moderns by opposing two moderns. But what a difference between the two authors!

Perrault's translation of Vitruvius's *Ten Books on architecture* (1673) is accompanied by illustrations and commentaries by Perrault himself, and Soufflot compares it to Blondel's *Cours d'architecture.* Both texts were written before 1695, the year when Philippe de la Hire (1640–1718) first introduced principles of mechanics in an architecture book. Previously, architecture was largely based on the "rules" which had been established by Vitruvius and which were based on the laws of proportions or analogies, that is, the equality of ratios, which already governed mechanics at the time of Vitruvius, by means of the law of the lever and principles of equilibrium. Proportions and analogies were the only mathematics at hand before the invention of the differential and integral calculus that Leibniz and Newton were discovering at the time Blondel and Perrault were writing. It is precisely the mathematization or mechanization of architecture that is at stake in their dispute.

Soufflot writes about François Blondel:

---

[1] It should be noted that Soufflot's last two texts, which bear most directly on the relationship between rules and taste, have not, to my knowledge, been commented on, ans they have been in most aesthetic texts, such as Gandt (2004). Rules are seen as a product of society and not as a reflection of the laws of nature. This point of view, certainly defensible, is not that of Soufflot, which I analyze here. The reader interested in a more traditional look on the subject, will consult with profit, chapter X of the very beautiful book by Rykwert (1980).

[2] *Parallèle des Anciens et des Modernes*, not to be confused with *Parallèle de l'architecture de l'antique et des modernes* due to Freart de Chambray and Errard (1689).

> It is in various chapters at the end of that excellent work that he has put forward and proved as far as he was able that proportions had in architecture, as in other arts, natural principles and that for that reason, we felt ourselves forced, even without knowing them, to admire and love the buildings in which one encounters them and to conceive a kind of disgust for those where we do not find them, even if the latter prevail on the others either by the cleanliness of the execution or by the delicacy of the ornaments.[3]

One effectively finds at the end of François Blondel's lectures, an entire book, his fifth, dedicated to proportions in architecture. The table of contents of that book (Fig. 1) gives us an idea of the breadth of the argument.

Chapter XVI clearly sums up Blondel's opinion. Its title is evocative: "Inductive reasonings to prove that proportions are the cause of beauty in Architecture; And that this beauty has no less its foundation in Nature than that of chords in music"[4] The term "induction" may be surprising; it indicates a regret on the part of Blondel, who would have liked to provide a mathematical demonstration of the following assertion. He says this explicitly at the beginning of this chapter. Blondel then lists the laws of mechanics, the lever, optics, the laws of reflection and refraction, and of music, in all of which proportions are "the cause of the consonances whose mixture makes harmonious chords, and to make of it a stable and constant principle of one of the most beautiful parts of Mathematics, that is to say Music".[5] All of these laws are mathematical and form the basis of the science of which they are the object.

He then continues his inductive reasoning:

> Now if, by a long series of observations, we come to discover and know something which is always found in the measurements of pleasant buildings, never found in those of others: Could we not rightly assume that this thing would be the cause of the beauty which generates pleasure in Architecture? …
>
> It is therefore with this in mind that I have researched, with great attention and work, the particular measures of the buildings which pass for the most beautiful among those who know them; And as in my examination of it, I found that there were certain proportions between the sizes of the mass and those of its parts, and between those of the same parts; And as I have recognized that these proportions, which are common to all Pleasant Buildings, and which are generally found in all parts, are nevertheless rarely found (not to say not at all) in unpleasant buildings.[6]

---

[3] Soufflot (1739) (Petzet 1961, p. 131): *C'est dans plusieurs chapitres à la fin de cet excellent ouvrage qu'il a avancé et prouvé autant qu'il a pû que les proportions dans l'architecture avoient comme dans les autres arts des principes naturels, que par cette raison nous nous sentions forcés même sans les connoître à admirer et à aimer les bâtimens dans les quels elles se rencontrent et à concevoir un certain desgoust pour celles où elles ne se trouvent point, lors même que ces derniers l'emportent sur les autres soit par la propreté de l'exécution soit par la délicatesse des ornemens.*

[4] Blondel (1675, p. 768): *Inductions pour prouver que les proportions sont la cause de la beauté dans l'Architecture; Et que cette beauté n'a pas moins son fondement dans la Nature, que celle des accords dans la musique.*

[5] Blondel (1675, p. 770): *la cause des consonnances dont le mêlange fait les accords harmonieux, et d'en faire un principe stable et constant d'une des plus belles parties des Mathématiques, c'est-à-dire la Musique.*

[6] Blondel (1675, pp. 770–771): *Or si, par une longue suite d'observations, l'on venoit à découvrir et à connaître quelque chose qui se trouvant toûjours dans les mesures des bâtimens agréables, ne se rencontrast jamais dans celles des autres: Ne pourroit-on pas présumer avec raison que cette*

LIVRE CINQUIEME.

De la proportion des parties de l'Architecture.

CHAPITRE I. *Suivant le sentiment de Vitruve.* p. 727
CHAP. II. *Sentimens des Interpretes de Vitruve sur les proportions des parties de l'Architecture.* p. 729
CHAP. III. *Sentimens de Leon-Baptiste Albert sur les proportions des parties d'Architecture.* p. 730
CHAP. IV. *Suite de la Doctrine de Leon-Baptiste Albert sur les proportions des parties d'un batiment.* p. 733
CHAP. V. *Sentimens des autres Architectes sur les proportions des parties d'un batiment.* p. 737
CHAP. VI. *Exemples des proportions dans quelques batimens de Palladio.* p. 739
CHAP. VII. *Exemples des proportions de quelques batimens modernes.* p. 742
CHAP. VIII. *Exemples des proportions de quelques batimens antiques.* p. 744
CHAP. IX. *Proportions du Temple de la Rotonde.* p. 748
CHAP. X. *Les proportions des beaux batimens, prouvent la necessité des proportions dans l'Architecture.* p. 754
CHAP. XI. *Application des proportions de la Musique à l'Architecture par M. Ouvrard.* p. 756
CHAP. XII. *Suite de la même pensée.* p. 758
CHAP. XIII. *Application de la proportion Geometrique aux parties de l'Architecture par M. l'Abbé de S. Hilarion.* p. 761
CHAP. XIV. *Examen des raisons que l'on apporte contre la necessité des proportions en Architecture, qui ne sont, comme on dit, approuvées que par accoutumance.* p. Id.
CHAP. XV. *Refutation de ces raisons.* p. 764
CHAP. XVI. *Inductions pour prouver que les proportions sont la cause de la beauté dans l'Architecture; Et que cette beauté n'a pas moins son fondement dans la Nature que celle des Accords dans la Musique.* p. 768
CHAP. XVII. *Proportions de la façade de l'Eglise de Milan.* p. 774
CHAP. XVIII *Refutation de quelques autres difficultés.* p. 779
CHAP. XIX. *Conclusion de tout ce discours par la recherche de ce qui peut estre la cause que certaines choses nous plaisent plus que d'autres dans l'Architecture.* p. 783
CHAP. XX. *Il faut un jugement exquis & une experience consommée pour faire un bon choix des proportions dans chaque partie du batiment.* p. 786

Fig. 1 The table of contents of Book V of François Blondel's course on architecture. Public domain

We could deduce from this that Blondel believes that there is a fundamental mathematical law of Architecture, which relates to masses and whose application makes the constructed object aesthetic. Unfortunately, he does not know this law, and it is worth noting that he does not make the law of the lever play this role, because he is aware of its insufficiency as well as its notoriety.

Let us resume the reading of Soufflot who then analyzes a small commentary by Claude Perrault to the text of Vitruvius. This text predates that of Blondel, who responded to Perrault and his supporters:

> It is also in the notes to his translation of Vitruvius's Book IV, Chap. 1, that Mr. Perault announced that the proportions in architecture were only based on habituation and had no necessary and convincing positive beauty which surpassed the beauty of other proportions, like the beauty of a diamond surpasses that of a pebble, but having been encountered in works which also had other convincing beauties, such as those of the material and the accuracy of the execution, they made the beauty of these proportions approved and loved although it had nothing positive other than this reason to love things by company and by habituation, by encounter in almost all the things that please, although we do not believe it for lack of reflection on it.[7]

There is therefore no reason nor natural law that is the foundation of architecture and explains its beauty. We only find architectural constructions beautiful because we are used to seeing them as beautiful. Soufflot paraphrases Perrault's idea somewhat, commenting:

> There is to the text *cum non esset symmetriarum ratio nata.*[8] This expression of Vitruvius's seems to favor the opinion of most of the architects who believe that the proportions of the members of architecture are something natural, such as are the proportions of magnitudes, for example, of the stars, with respect to each other, or parts of the human body. For me, I have translated according to the thought that I have that these proportions were established by the consent of the architects, who, as Vitruvius himself testifies, imitated the works of one another, and who followed the proportions that the first had chosen, not as having a positive, necessary and convincing beauty, which surpassed the beauty of other proportions,

---

chose là seroit la cause de la beauté qui engendre le plaisir dans l'Architecture? C'est donc dans cette veüe que j'ay recherché avec beaucoup d'attention et de travail les mesures particulières des batimens qui passent pour les plus beaux parmi ceux qui s'y connoissent; Et comme dans l'examen que j'en ay fait, j'ay trouvé qu'il y avoit certaines proportions entre les grandeurs de la masse et celles de ses parties, et entre celles des mêmes parties; Et comme j'ay reconnu que ces proportions, qui sont communes à tous les Édifices agréables, et qui se rencontrent généralement dans toutes les parties, ne se trouvent néanmoins que rarement, (pour ne pas dire point du tout,) dans les bâtimens désagréables.

[7] Perrault (1673, p. 100, n. 1): *C'est aussi dans les notes sur sa traduction de Vitruve L. 4 c. 1er que Mr Perault a anoncé que les proportions dans l'architecture n'estoient fondées que sur l'accoutumance et n'avoient point de beauté positive nécessaire et convinquante qui surpassat la beauté des autres proportions comme la beauté d'un diamant surpasse celle d'un caillou mais que s'estant rencontrée dans des ouvrages qui avoient d'ailleurs d'autres beautés convinquantes telles que sont celles de la matière et de la justesse de l'exécution elles ont fait aprouver et aimer la beauté de ces proportions bien qu'elle n'eut rien de positif que cette raison d'aimer les choses par compagnie et par accoutumance par rencontre presque dans toutes les choses qui plaisent bien qu'on ne le croye pas faute d'y avoir fait reflection.*

[8] Vitruvius, book IV, Chap. 1: "the determination of the exact proportions of the order had not begun". The original text has: *cum **etiamnum** non esset symmetriarum ratio nata.*

as the beauty of a diamond surpasses that of a pebble; but only because these proportions were found in works which, moreover, having other positive and convincing beauties, such as those of the material and the correctness of the execution, have made people approve of the love of beauty of these proportions, although it was nothing positive. This reason for loving things by companionship and by accustomedness is found in almost all the things that please, although we do not believe it, for lack of thought.[9]

But one clearly finds there the idea of "habituation", and Soufflot concludes firmly in favor of Blondel: "For me, I have always deferred to the sentiments of M. Blondel more than to those of M. Perault, at least for the general proportions and even for some of the particular ones.[10]

Referring to an inventor of the most beautiful sculptures of men or women, Soufflot remarks:

The skilful people who made them have in truth never strayed from certain general proportions to which nature has always been restricted, but in imitation of parts more or less unconstrained and placed appropriately have caused them to produce beauties that are as varied as they are perfect.[11]

Soufflot thus highlights the general proportions found in nature and in these sculptures, as well as the variety left to each sculptor in the choice of proportions that govern the details. He then develops this argument on the basis of elements relating to sculpture, drawn from the quarrel between Blondel and Perrault. He focuses for a moment on the overly abundant decoration of Saint-Charles, which displeased by its excess, then he relates his experiments in his experience in measuring three Roman churches: S. Andrea della Valle, S. Ignazio, and San Carlo. After having addressed, in a style very similar to that of the Blondel and Perrault quarrel, the different relationships between length, width and height of these churches, Soufflot's conclusion agrees with Blondel's opinion.

---

[9] Soufflot (1739) (Petzet 1961, p. 132): *Il y a au texte cum non esset symmetriarum ratio nata. Cette expression de Vitruve semble favoriser l'opinion de la plus grande partie des Architectes, qui croyent que les proportions des membres de l'Architecture sont quelques chose de naturel, telles que sont les proportions des grandeurs, par exemple, des Astres, à l'égard les uns des autres, ou des parties du corps humain. Pour moy j'ay traduit suivant la pensée que j'ay que ces proportions on esté établies par un consentement des Architectes, qui, ainsi que Vitruve témoigne luy-même, ont imité les ouvrages les uns des autres, et qui ont suivy les proportions que les premiers avoient choisies, non point comme ayant une beauté positive, necessaire et convaincante, et qui surpassait la beauté des autres proportions, comme la beauté d'un diamant surpasse celle d'un caillou; mais seulement parce que ces proportions se trouvoient en des ouvrages, qui ayant d'ailleurs d'autres beautez positives et convaincantes, telles que sont celles de la matière et de la justesse de l'exécution, ont fait aprouver et aimer la beauté de ces proportions, bien qu'elle n'eust rien de positif. Cette raison d'aimer les choses par compagnie et par accoûtumence se rencontre presque dans toutes les choses qui plaisent, bien qu'on ne le croye pas, faute d'y avoir fait réflexion.*

[10] Soufflot (1739) (Petzet 1961, p. 132): *Pour moy j'ay toujours déféré aux sentimens de M. Blondel plus qu'à ceux de M. Perault, au moins pour les proportions générales et même pour quelques unes des particulières..*

[11] Soufflot (1739) (Petzet 1961, p. 132): *Les habiles gens qui les ont faites ne se sont à la verité jamais éloignées de certaines proportions générales dans les quels la nature s'est toujours restreinte mais à son imitation des parties plus ou moins déliées et placées à propos leurs ont fait produire des beautés aussi variées que parfaites.*

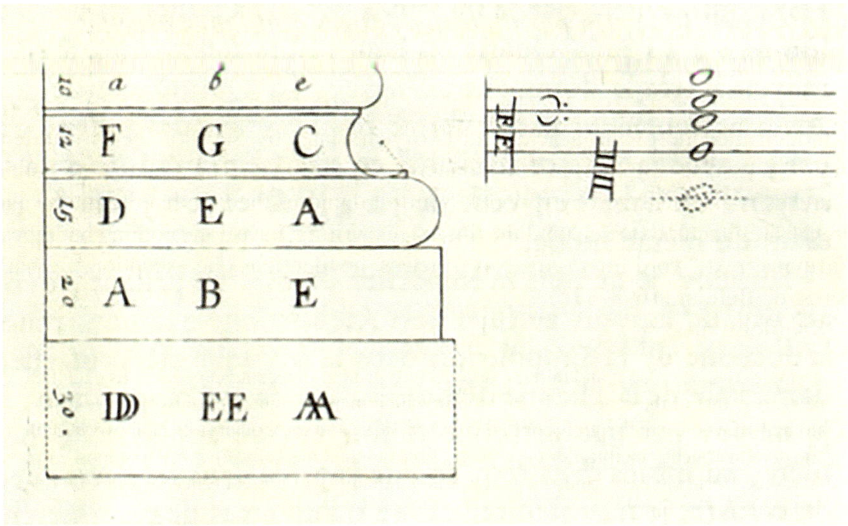

Fig. 2 Figure illustrating the similarity of the role of proportions in music and in architecture (Blondel 1683, V, vii, p. 759). Public domain

> I could also quote many others which would also be different of the ones which I praised and that one sees at least as often since they are in the most frequented neighborhoods of Rome from what I conclude boldly that there are for churches proportions that are less founded on familiarization than on something natural and as there are perfect chords in music to which we cannot refuse our agreement, there are also in architecture proportions that necessarily give pleasure to those who see them in the edifices in which they are.[12]

Blondel believed, as did René Ouvrard, author of *Architecture harmonique* (1679), that proportions play in a role in architecture similar to the one they play in music (Fig. 2).

The text seems close to the antique descriptions strictly reduced to proportions. But two differences must be underlined. The first is that proportions have something natural, as does music. It is that idea of natural proportions that is opposed to the usual concept of proportions that Perrault defends because it becomes associated with the idea of natural and mathematical law. Moreover, an essential distinction has been introduced between the general shape, proportion between length, width, and height of the entire building, between the elements of the structure and the proportions of the decorative (or non-structural objects that are left to the appreciation and to the imagination of the architect.

---

[12] Soufflot (1739) (Petzet 1961, p. 135): *J'en pourois citer beaucoup d'autres qui se trouveroient également différentes de celles dont j'ay fait l'éloge et que l'on voit pour le moins aussi souvent puisqu'elles sont dans les quartiers de Rome les plus fréquentés d'où je conclus hardiment qu'il y a pour les églises des proportions qui sont moins fondées sur l'acoutumance que sur quelque chose de naturel et que comme il y a dans la musique des acords parfaits auxquels nous ne pouvons pas refuser notre approbation il y a aussi dans l'architecture des proportions qui causent nécessairement du plaisir à ceux qui les voyent [dans] les édifices dans les quels elles se trouvent.*

3 12 April 1741, *Un parallèle des églises gothiques avec les églises moderne*

Although the title of the article read two years later by Soufflot seems different from the first, it is not so, and its concerns are the same. The example of gothic churches is also analyzed by Blondel. Soufflot deplores the fact that the Italian and French architectural books, until the time of his writing, have done nothing but repeat Vitruvius' text "word for word". Nevertheless, he mentions one exception, Bernard Forest de Bélidor (1698–1761).

Soufflot writes:

> A single treatise on civil and military architecture printed in recent years offers us excellent news. It is that of the scholar Mr. Bélidor, in which there is a quantity of experience on the strength of wood, which rightly contradicts what Bullet and some others told us above; certain and demonstrated principles for the construction and thickness of walls, which support earth or vaults; fresh and easy sands for practice and a quantity of correct reflections that no one had made before him.[13]

But what distinguishes Bélidor's *La Science des ingénieurs* (1729)? To understand this, it suffices to read the first elements of the description he gives in his preface:

> This volume is divided into six books. In the first, we teach how to apply the principles of mechanics to the construction of masonry coverings. In the second, we consider how the thrust of the vaults takes place in order to draw general rules and certain ones to determine the thickness of their counterfort.[14]

The aim of Belidor is to give the foundations for a mechanics applied to architecture. Already in 1729, at the beginning of the second book, the principle of the composition of forces following the parallelogram law. This principle is thus known by Soufflot.

Principles extracted from the *Mechanics* (Fig. 3):

> 2—It is demonstrated in *Mechanics* that three powers P, Q, R, that push or pull around a point A, along the directions AP, AQ, AR, will be in equilibrium between them, if after having drawn the parallelogram ABCD, the power P, is expressed by the AB, the power Q, by the side AD, and the power R, by the diagonal CA: or, which amounts to the same, if each power is expressed by a side of a triangle ABC, because instead of AD, one could take BC, that is equal to it; supposing thus that we be warned of this truth, here is a fundamental proposition that we can draw out of it.

---

[13] Soufflot (1741) (Petzet 1961, pp. 135–136): *Un seul traité d'architecture civile et militaire imprimé depuis quelques années nous offre d'excellentes nouveautés. C'est celuy du scavant M. Belidor dans lequel se trouvent quantité d'expérience sur la force des bois, qui démentent avec raison, ce que nous ont dit la dessus Bullet et quelques autres; des principes certains et démontrés pour la construction et l'épaisseur des murs, qui soutienent des terres ou des voûtes; des sables seuves et faciles pour la pratique et quantité de justes réflexions que personne n'avoit fait avant lui.*

[14] Bélidor (1729, preface, n.p.): *Ce volume est divisé en six livres. Dans le premier, on enseigne la manière d'appliquer les principes de la mécanique à la construction des revêtements de maçonnerie. Dans le second on considère de quelle manière se fait la poussée des voûtes afin d'en tirer des règles générales et certaines pour déterminer l'épaisseur de leur pied-droit* [...].

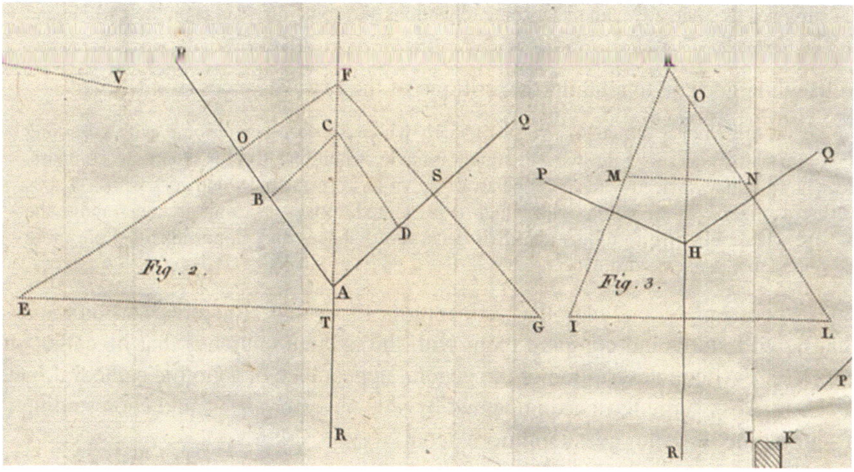

Fig. 3 Diagram illustrating the parallelogram law (Bélidor 1729, Livre II). Public domain

> Having three powers P, Q, R, which pull or push all three together around point A, I say that they will be in equilibrium, if the force with which one each acts is expressed by one of the sides of the triangle EFG, which would intersect in right angles the line of direction of each power.[15]

Bélidor's book is not the first treatise on architecture that had been published. De la Hire, let us recall, had already published his *Traité de mécanique* (1695). But it is in Bélidor's text that the architect Soufflot will find the principles necessary to develop his architecture and go beyond the simplistic stability of the triangle or the pyramid. It therefore took 34 years for de La Hire's theory to arrive on the construction site.

Soufflot mentions the unanimous criticism of his colleagues for Gothic constructions:

> Mr François Blondel is almost the only one to have said something positive about it [Gothic architecture]; he even admits in one of his chapters on proportions to have found some in the facade of the cathedral of Milan which were similar to those of antique architecture, and this is the reason he gave for the pleasure he felt as he examined it. Following his example, could we not without fault take the measurements of these monuments, the boldness of which

[15] Bélidor (1729, book II, p. 6–7): 2—*Il est démontré dans la Mécanique que trois puissances P, Q, R (Fig. 3), qui tirent ou poussent au tour d'un point A, selon des directions AP, AQ, AR, seront en équilibres entr'elles, si après avoir fait le paralellogramme ABCD, la puissance P, est exprimée par le côté AB, la puissance Q, par le côté AD, et la puissance R, par la diagonale CA: ou, ce qui revient au même, si chaque puissance est exprimée par un des côtés d'un triangle ABC, parce qu'à la place de AD, l'on pourra prendre BC, qui lui est égal; supposant donc qu'on soit prevenu de cette vérité, voici une proposition fondamentale qu'on en peut tirer.*
Ayant trois puissances P, Q, R, qui tirent ou poussent toutes trois ensembles au tour du point A, je dis qu'elles seront en équilibres, si la force avec laquelle chacune agit est exprimée par un des côtés du triangle EFG, qui couperoit en angles droits la ligne de direction de chaque puissance.

surprises us so greatly, and make some reflections and some reasonings on their whole and their parts?[16]

He then goes on to a humorous critique of the extravagances of his time:

I know that a house turned to the taste of a lady who passes for having some, with a dressing room whose plan shows the disregard that was made, in building it, of the square and compass, a new chimney placed with so-called art in one of its corners, and woodwork whose shapes are as bizarre as the colors with which it is painted, today places an architect above the one who could give to rooms and their decorations shapes and proportions that are more reasonable than extraordinary.[17]

He goes on to denounce that fad, returns to Gothic structures and announces that he will study and compare them with the ancient churches that he calls "our churches". After a description of the general appearance of a Gothic cathedral, with its Latin cross, large nave accompanied by side naves or aisles, and choir ending in a semi-dome, Soufflot goes into more detail.

The large nave is usually decorated with pillars which serve to support its vault, and those of the aisles in which one passes through large arcades which are between the pillars and are repeated by others which introduce into second aisles or into chapels. These pillars are more or less large but always placed diagonally so that they present one of their angles to the large nave. They are composed of a number of cylindrical columns; some are large and bear the impact of the vaults and arcades; The others are small and placed between the large ones which they often accompany up to the top.[18]

After a careful description of the openings that allow light into these churches, taking the examples of Saint-Jean and Saint-Nizier in Lyon, he concludes with a description of the exterior:

The sides of the church are lined with large buttresses which, although strong enough to resist the thrust of the vaults, nevertheless appear very light, because of the way in which they are constructed. There are almost always exterior galleries which cross these abutments,

---

[16] Soufflot (1741) (Petzet 1961, p. 136): *M. François Blondel est presque le seul qui en ait dit quelque chose en bien; il convient même dans un de ses chapitres sur les proportions d'en avoir trouvé dans la façade de l'église cathédrale de Milan qui étoient fort relatives à celles de l'architecture antique, et c'est la raison qu'il donne du plaisir qu'il ressentit en l'examinant. À son exemple, ne pouroit-on pas sans crime prendre les mesures de ces monumens dont la hardiesse nous surprend si fort, et faire sur leur tout et leurs parties quelques réflexions et quelques raisonnemens.*

[17] Soufflot (1741) (Petzet 1961, p. 136): *Je scay qu'une maison tournée au goût de quelque dame qui passe pour en avoir, un cabinet de toilette dont le plan marque le mépris que l'on a fait, en le construisant, de l'équerre et du compas, une cheminée nouvelle placée avec un art prétendu dans l'un de ses coins, un boisage enfin dont les formes soyent aussi bizarres que les couleurs dont on le peint, mettent aujourd'hui un architecte au-dessus de celuy qui scauroit donner à des chambres et à leurs decorations des formes et des proportions plus raisonables qu'extraordinaires.*

[18] Soufflot (1741) (Petzet 1961, p. 137): *La grande nef est ordinairement décorée de piliers qui servent à soutenir sa voûte, et celles des bas cotés dans lesquels on passe par de grandes arcades qui sont entre les piliers et sont répétées par d'autres qui introduisent dans des seconds bas côtés ou dans des chapelles. Ces piliers sont plus ou moins gros mais toujours posés diagonalement de manière qu'ils présentent un de leurs angles à la grande nef. Ils sont composés de quantité de colonnes cilindriques; les unes sont grosses et portent les retombées des voûtes et des arcades; Les autres petites et placées entre les grosses qu'elles accompagnent souvent jusqu'en haut.*

> at the height of the windows of the large nave, and others above its walls. By the first ones we go to the roofs of the aisles, by the second ones to those of the large nave.[19]

In those few lines is described, with few details, and as if were something known and understood by everybody, the transmission of thrusts in flying and close buttresses. There is the most visible application of the parallelogram law of forces that fundamentally modifies the conception of stability. But Soufflot marveled at it, as if it were something too daring.

Next Soufflot compares those Gothic churches to the great churches with large naves of his own time. His description, which includes many examples, is very similar to that of the Gothic churches he gave above, and for good reason, because he concludes:

> What can we give in this arrangement, which does not perfectly resemble those of Gothic churches? It is therefore from these churches, built long before the renaissance of ancient architecture, that we drew the idea of the distribution of ours; In this we have approved the work of these people whom we treat as crude and barbaric; we even imitated them in every way. We must therefore regard them as our masters in this respect, and, despite the contempt we have for them, we cannot refuse them this advantage, however great it may be; they are creators for this article, and we cannot accuse them of having imitated ancient temples.[20]

Soufflot continues his comparison and arrives at the facades:

> Our facades adorned with pilasters and columns in no way imitate those of the Goths, and theirs are without a doubt much superior, we like them, garnish our exteriors with abutments which support the thrust of the vaults, and are often very massive.[21]

And yet what a difference, says Soufflot, arriving at the essential paragraph of his article:

> As for the construction, if we examine that of the Gothic churches, we will see that it is more ingenious, more daring, and even more difficult than that of ours. These tribunes created in their large bays which cross the main pillars, sometimes in two places, reduce them, so to speak, to nothing. However, it is on these pillars alone that the entire weight of the vaults, however large they may be, bears and that of the framework of the roofs, which is often very

---

[19] Soufflot (1741) (Petzet 1961, p. 138): *Les côtés de l'église sont garnis de grandes buttés qui, quoique suffisament fortes, pour résister à la poussée des voûtes paroissent cependant fort légères, par la façon dont elles sont construites. Il y a presque toujours des galeries extérieures qui traversent ces buttés, à la hauteur des fenêtres de la grande nef, et d'autres au-dessus de ses murs. Par les lères on va sur les toits des bas côtés, par les 2des sur ceux de la grande nef.*

[20] Soufflot (1741) (Petzet 1961, p. 139): *Que peut-on donner dans cette disposition, qui ne ressemble parfaitement à celles des églises gothiques. C'est donc de ces églises, bâties longtemps avant la renaissance de l'architecture antique, que nous avons tiré l'idée de la distribution des nôtres; nous avons approuvé en cela l'ouvrage de ces peuples que nous traitons de grossiers et de barbares; nous l'avons même imité en tous points. Nous devons donc les regarder comme nos maîtres à cet égard, et, malgré le mépris que nous en faisons, nous ne saurions leur refuser cet avantage quelque grand qu'il soit; ils sont créateurs pour cet article, et on ne peut pas les taxer d'avoir imité les temples antiques.*

[21] Soufflot (1741) (Petzet 1961, pp. 139–140): *Nos façades ornées de pilastres et de colonnes n'imitent en rien celles des gots, et leurs sont sans contredit bien supérieures, nous garnissons, comme eux, nos dehors de buttés qui soutiennent la poussée des voûtes, et sont souvent très massives.*

> considerable. Their composition tends to make them visually light and sometimes they are indeed as strong as ours. Their position also contributes a lot, and we only see one of their four faces when entering the church and moving forward in the large nave. Their pointed arches have, like their vaults, more strength than our semi-circular arches, but are not, by far, so graceful; as for their lunettes and their ribs, there are some whose execution could not have been easy and whose cut is skilful. **They have always known how to make the main thrust fall in the place of the greatest resistance**, and we observe this rule, even in the arrangement of the rubble stones which fill the spaces between two of these ribs. The execution of our vaults is neither so delicate nor so difficult: we take care, like them, to place the weight on the strongest parts; we decorate them in a way that is larger and more noble, but sometimes a little heavy.[22]

Soufflot describes clearly the play which is made on the distribution of the loads. The sentence which I emphasized in bold shows that he understood the use that taste had to make of the law of composition of the forces or of an analogy (hydraulic for example), leading to the same results.

The following is devoted to what I will call an "old-fashioned" comparison of the proportions of different churches of his time to Gothic churches.

9 September 1744, *Mémoire pour servir de solution à cette question: savoir si dans l'art de l'architecture le goût est préférable à la science des règles ou la science des règles au goût ?*

As Alphonse de Boissieu (1807–1886), editor of this text by Soufflot, tells us in the introduction, with this text we reach the heart of our subject:

> He [Soufflot] seems to us, in his style, as in his works, imbued with this great principle which, although it may be ignored today, is nonetheless a truth in architecture as in all the arts: the beautiful is but the projection of the useful.[23]

This maxim is still known in other forms. The same idea will be used by another architect, Arthur Vierendeel[24] in the early years of the twentieth century and probably

---

[22] Soufflot (1741) (Petzet 1961, p. 140): *Quand à la construction, pour peu qu'on examine celle des églises gothiques, on verra qu'elle est plus ingénieuse, plus hardie, et même plus difficile que celle des nôtres. Ces tribunes pratiquées dans leurs gros danuir [sic] qui traversent les piliers principaux, quelquefois en 2 endroits, les réduisent pour ainsi dire à rien. C'est cependant sur ces seuls piliers que porte tout le poids des voûtes, quelque grandes qu'elles soient et celui de la charpente des couverts qui souvent est bien considérable. Leur composition tend à les rendre légers à la vue et quelquefois ils sont en effet aussi forts que les nôtres. Leur position y contribue aussi beaucoup, et on ne voit, en entrant dans l'église et en y avancant dans la grande nef, qu'une de leurs 4 faces. Leurs arcs en tiers point ont, ainsi que leurs voûtes, plus de force que nos arcs en plein cintre, mais ne sont pas, à beaucoup près, si gracieux; quand à leurs lunettes et leurs nervures, il y en a dont l'exécution ne devait pas être facile et dont la coupe est scavante. Ils ont toujours sçu faire tomber la principale poussée dans l'endroit de la plus forte résistance, et on observe cette règle, même dans l'arrangement des moelons qui remplissent les entre deux de ces nervures. L'exécution de nos voûtes n'est ny si délicate, ny si difficile: nous avons soin comme eux, d'en ranger le poids sur les parties les plus fortes; nous les décorons d'une façon plus grande et plus noble, mais quelquefois un peu lourde.*

[23] Soufflot (1744) (*Nouvelles archives*, 1832, p. 107): *Il [Soufflot] nous paraît, dans son style, comme dans ses œuvres, pénétré de ce grand principe qui, pour être méconnu aujourd'hui, n'en est pas moins une vérité en architecture comme dans tous les arts: le beau n'est que la saillie de l'utile.*

[24] For more about Vierendeel, see "Arthur Vierendeel: The Man with Iron Nerves Who Invented the Truss Without Diagonals," Chap. 19 in this present volume.

by others. In the case of Soufflot, it is a question of saying that the beautiful structures show the application of the relatively new law of composition of forces according to the law of the parallelogram. Let us also note the use of the term "science" applied to the rules which thus raises their status.

In this article, Soufflot embarks on a very schematic history of architecture that he traces back to its origins, with the aim of determining what taste is and what the rules are. In this story, naive to say the least, it even goes back to the dawn of time:

> From the necessity which they had to protect themselves from the ravages of time, industry was born; she taught them to form barracks. They first made them with pieces of wood leaned onto each other and covered with foliage or thatch like our ice boxes.[25]

So, from the start we find this pyramidal or triangular stability that our children still apply to make castles of cards. During the development seen by Soufflot, these first barracks left their mark on the roofs. It is amusing to note that his story follows the thread of the construction of a house first and then a temple, where the chosen proportions are the result of aesthetic research:

> But that was not all; it was a matter of reducing all these parts and establishing among the parts, and between the parts and the whole, a relationship which offered something graceful to the eyes.
>
> Genius emerged, and from the simple productions of industry drew what it could provide that which was most appropriate for the needs of men and that was greatest and noblest in their eyes. It was through Genius that these subjects rose above these industrious masons, whose monuments are still remembered. We live with architects; we saw the birth of taste. Taste joined industry; temples were erected, and Dorus, prince of Achaia, sovereign of the Peloponnese, had, according to Vitruvius, erected to the goddess Juno, in the famous city of Argos, the first temple of the order which, from his name, he called Doric. In imitation of it, they are built in different cities; rules were prescribed for this order, they were rectified, and they were followed in execution; they please and still please today, since we rarely move away from them, and we achieve little when we move away from them. ... I said that we saw the birth of taste, that we prescribed the rules; taste formed them, they formed taste: in my opinion it is the same thing in architecture.[26]

---

[25] Soufflot (1744) (*Nouvelles archives*, 1832, p. 108): *De la nécessité dans laquelle ils se trouvèrent de se mettre à l'abri des injures du temps naquit l'industrie; elle leur apprit à se former des barraques. Ils les tirent d'abord avec des pièces de bois inclinées les unes sur les autres et couvertes de feuillages ou de chaume comme nos glacières.*

[26] Soufflot (1744) (*Nouvelles archives*, 1832, p. 110): *Ce n'était pas tout; il s'agissait de réduire toutes ces parties et d'établir entre elles et d'elles au tout un rapport qui offrît aux yeux quelque chose de gracieux.*

Le génie se fit jour, et des simples productions de l'industrie tira ce qu'il pouvait procurer de plus commode aux besoins des hommes et ce qu'il pouvait offrir de plus grand et de plus noble à leurs yeux. Ce fut par lui que s'élevèrent au-dessus de ces maçons industrieux ces sujets dont les monuments sont encore dans la mémoire. On vit des architectes; on vit naître le goût; il se joignit à l'industrie; on éleva des temples, et Dorus, prince d'Achaïe, souverain du Péloponèse, fit, au rapport de Vitruve, élever à la déesse Junon, dans la fameuse ville d'Argos, le premier temple de l'ordre que, de son nom, il appela dorique. À son imitation, on en bâtit dans différentes villes; on prescrivit des règles pour cet ordre, on les rectifia, on les suivit dans l'exécution; elles plurent et plaisent encore aujourd'hui, puisqu'on s'en éloigne rarement, et qu'on réussit peu lorsqu'on s'en éloigne. [...] J'ai dit qu'on vit naître le goût, qu'on prescrivit les règles; le goût les forma, elles formèrent le goût: c'est selon moi une même chose dans l'architecture.

This is how rules and taste fit together and are interconnected from the very beginning of Soufflot's story. But a new element appears: genius. The author develops his idea by returning once again to the primitive triangular barracks. Putting himself in the place of the original architect, Soufflot wonders about the method followed when this architect decides for reasons of sustainability to build stone houses:

> Who will lead me into this new construction? Is it taste? Are these the rules? But, someone will ask me, how would you be guided by the rules? There are none. This is what I do not agree with. Before it was discovered and demonstrated that in every triangle the three angles are equal to two right angles, this truth nevertheless existed; in the same way, before we found that a Doric column eight diameters high, including its base and its capital, was of a beautiful proportion, this rule which guides us today did not fail to exist, though it was unknown.[27]

Soufflot here modifies the status of the rule, eliminating at the same time any possibility of a role of habituation. It becomes mathematical in essence and has always existed, even before being discovered by humans, it is a law of nature, even if Soufflot does not use these terms. It continues the reflection of the architect who begins by analyzing different situations:

> I examine these trees which support this rustic roof of which I spoke. Taken at random, they are of different sizes; one seems too thin to me, the other too strong in proportion to the burden they have to carry; the third holds a medium which pleases me more, I feel determined to approve of it, and a force whose cause I do not know commits me to the choice of this proportion.[28]

It is apparently only a simple observation, but Soufflot implies that it is actually based on previous experiments of stability. Having selected a tree trunk, he reasons goes further:

> I measure its diameter, and comparing it to its height, I find it to be about the seventh part of it, just as the feet are the seventh part of the human body. I add the bottom plinth, brick, or top shelf, and together approaching eight diameters, I conclude that this proportion is correct; I build accordingly; my work pleases and becomes a rule for those who come after me.[29]

---

[27] Soufflot (1744) (*Nouvelles archives*, 1832, p. 110): *Qui est-ce qui me conduira dans cette nouvelle construction ? Est-ce le goût ? Sont-ce les règles ? Mais, me dira-t-on, comment seriez-vous conduit par les règles ? il n'y en a point: c'est ce dont je ne conviens pas. Avant que l'on eût découvert et démontré qu'en tout triangle les trois angles sont égaux à deux droits, cette vérité n'existait pas moins; de même, avant que l'on eût trouvé qu'une colonne dorique de huit diamètres de hauteur, en comprenant sa base et son chapiteau, était d'une belle proportion, cette règle qui nous conduit aujourd'hui ne laissait pas d'exister, il s'agissait de la connaître.*

[28] Soufflot (1744) (*Nouvelles archives*, 1832, p. 112): *J'examine ces arbres qui portent ce rustique toit dont j'ai parlé. Pris au hasard, ils sont de différentes grosseurs; l'un me paraît trop mince, l'autre trop fort à proportion du fardeau qu'ils ont à porter; le troisième tient un milieu qui me plaît davantage, je me sens déterminé à l'approuver, et une force dont je ne connais pas la cause m'engage au choix de cette proportion.*

[29] Soufflot (1744) (*Nouvelles archives*, 1832, p. 112): *Je mesure son diamètre, et le comparant à sa hauteur, je trouve qu'il en est environ la septième partie, comme les pieds sont la septième partie du corps humain. J'y ajoute le socle de dessous, la brique ou le plateau de dessus, et le tout ensemble approchant de huit diamètres, je conclus que cette proportion est bonne; je bâtis en conséquence; mon ouvrage plaît et devient une règle pour ceux qui viennent après moi.*

Then the aesthetics of the columns will evolve to resemble the body of a man, the Doric, a woman, the Ionic, and a young girl, the Corinthian, their hair explaining the differences between the capitals:

> Eyes and a taste that I had for beauty, perhaps a result of the attention with which I had examined the size of men, women and girls: the rules that I did not know and whose knowledge had to be the fruit of my reflections, as my reflections had been the source of my taste; my taste and these rules therefore become the same thing, and I cannot have one without the other; hence the point of preference of this taste for proportions, over the rules for these same proportions. The rules are taste, taste is the rules.[30]

Unlike Alphonse de Boissieu, who sharply criticizes this pseudo-story, we will take a step back from Soufflot's historical discourse to see that, independent of this fanciful story, he describes the relationship between taste, the architect's personal choice, the rules that we would today call laws of nature, or principles of mechanics, and the genius that discovers these. The principles of mechanics are therefore essential and impose themselves on the architect.

Soufflot concludes with a sharp criticism of the eccentricity sought after by certain architects and denounces the lack of durability of these works. Not being based on the principles of mechanics or architecture—we should say stability—these works have no value.

The years intervening between this article and the next one will see the construction of the greater part of the Sainte-Geneviève church, including the laying of its first stone, on 6 September 1764, by Louis XV. The installation takes place when the very sophisticated foundations developed by Soufflot have already been completed but the works above ground are not yet very advanced. As can be seen in the painting by Pierre-Antoine Demachy (1723–1807), this laying of the foundation stone took place in front of a gigantic fresco on the facade of the completed church (Fig. 4).

4 20 November 1775 and 18 January 1778, *De l'identité du goût et des règles dans l'exercice de l'architecture*

Soufflot takes this subject up again 31 years later, showing that he has always kept it in mind. He was then 62 years old and in the process of building the majestic dome which was to crown the Church of Sainte-Geneviève. Pierre Patte published his devastating pamphlet five years earlier (Patte 1770). Soufflot's text is important because he gave the talk during the inauguration of the new rooms of the Royal

[30] Soufflot (1744) (*Nouvelles archives*, 1832, pp. 112–113): *yeux et un goût que j'avais pour le beau, suite peut-être de l'attention avec laquelle j'avais examiné la taille des hommes, des femmes et des filles: les règles que je ne connaissais pas et dont la connaissance devait être le fruit de mes réflexions, comme mes réflexions avaient été la source de mon goût; mon goût et ces règles deviennent donc une même chose, et je ne puis avoir l'un sans l'autre; de là point de préférence de ce goût pour les proportions, sur les règles pour ces mêmes proportions. Les règles sont le goût, le goût c'est les règles.*

Fig. 4 Laying of the first stone of Sainte-Genevieve, after Gabriel de Saint-Aubin, etched by Pierre Chenu (French, Paris 1730–late eighteenth century). Collection of the Metropolitan Museum of Art, Creative Commons CC0 1.0 Universal

Academy of Architecture, and would give it again three years later at the Académie de Lyon.

At the end of a pseudo-historical account, differing from the previous one only in elements which do not concern our subject, he synthesizes in his article of 1744 the perfect formula which he intends to develop again: "Taste formed by reason produced rules which in turn formed the taste and rectified it".[31] The sentence has been completed by the appearance of reason, and taste now derives from the rules or laws of nature. In this way, Soufflot further opposes Perrault and his supporters. To structure his argument, he lists the elements that reason must take into account in order to develop the rules and taste:

> Utility, which gives the disposition relative to the needs; solidity, which gives safety; the appropriateness, which is the relationship of the buildings with the uses and with the people;

[31] Soufflot (1775) (Petzet 1961, p. 143): *Le goût formé par la raison produisit des règles qui à leur tour formèrent le goût et le rectifièrent.*

> symmetry or the correspondence of the parts between them and with the whole, which constitute the whole and the unity.[32]

His first remark regarding utility is that it varies from place to place and across eras and concludes: "The thing made being the expression of utility, it is at the same time [the expression] of the rules".[33]

Regarding solidity he concludes:

> True architecture must therefore reject these too bold constructions, … which would make the spectator tremble for the inhabitant and the inhabitant for himself: the conservation [of the inhabitant] is the reason for solidity and that one is the expression of the rules.[34]

Note that he immediately qualifies this remark:

> We must not conclude from this that the architect who has genius and experience cannot be bold in his constructions, but there must be nothing worrying about his boldness; nor does it follow that he must, out of an ill-founded fear, or through an imitation that is not reasoned, uselessly multiply the points of support. … we know how much Hermogenes was praised for having removed in the Temple of Diana which he built in Magnesia a row of dipteral columns so that there was, as Vitruvius says, more space for circulation, and less work and expense; we know that he was also praised for having made the intercolumniation of the center of the front three diameters instead of two and a quarter for the others.[35]

In this way he exhibits the two extreme poles of what he calls solidity. The refusal of reckless boldness, or which is not subordinated to the laws of nature or statics, does not prevent a careful search for lightness.

He does not stop at the essential suitability of a building's architecture to its purpose and continues with symmetry:

> Without symmetry, which is the ratio of the corresponding masses and the proportion of the parts to each other and to the whole, there is no whole; there is no unity in an edifice; taste and rules did not govern its construction. Taste does not exist without judgment; the imprint

---

[32] Soufflot (1775) (Petzet 1961, p. 143): *L'utilité qui donne la disposition relative aux besoins, la solidité qui donne la sûreté, la convenance qui est le rapport des édifices avec les usages et avec les personnes, la symmétrie ou la correspondance des parties entr'elles et avec le tout qui constituent l'ensemble et l'unité.*

[33] Soufflot (1775) (Petzet 1961, p. 144): *La chose faite étant l'expression de l'utilité, elle est en même temps celles des règles.*

[34] Soufflot (1775) (Petzet 1961, p. 144): *La véritable architecture doit donc rejetter ces constructions trop hardies, […] qui feroient trembler le spectateur pour l'habitant et l'habitant pour lui même: sa conservation [de l'habitant] est la raison de la solidité et celle-cy est l'expression des règles.*

[35] Soufflot (1775) (Petzet 1961, pp. 144–145): *Il ne faut pas en conclure que l'architecte qui a du génie et de l'expérience ne puisse pas être hardi dans ses constructions, mais il faut que sa hardiesse n'ait rien d'inquiétant; il ne s'en suit pas non plus qu'il doive, par une crainte mal fondée, ou par une imitation qui ne seroit pas raisonnée, multiplier inutilement les points d'appui. […] on sait combien Hermogènes fut loué pour avoir supprimé dans le Temple de Diane qu'il bâtit à Magnésie un rang de colonnes du diptère pour qu'il y eut ainsi que Vitruve le dit, plus d'espace pour la circulation, et moins d'ouvrage et de dépense; on sait qu'il fut également loué pour avoir fait l'entre-colonement du milieu du frontispice de trois diamètres au lieu de deux et un quart pour les autres.*

of judgment in the whole and in parts, constitutes and makes the work of the genius pleasing, and proves the execution of the rules.[36]

What Soufflot calls symmetry is in fact the legacy of antiquity and of its theory of proportions, which has often been discussed. We could still think of it as geometry.

Rules are not that confused heap of precepts given by authors whose books often contain more words than things; these are the primitive rules that existed before we knew them, as it existed, before we demonstrated, that in any triangle the three angles are equal to two right angles; it is about these rules and all the corollaries that the succession of times has drawn from them, the sum of which has constituted good taste.[37]

Soufflot did not take his commentary on the mathematical character of rules and taste any further. He no longer mentioned Rondelet in the articles defending the stability of the dome of the Pantheon, both geometric and natural law of the vectorial composition of forces,[38] a tool that was nevertheless indispensable and well known at the time, in the evaluation of the stability of this dome. This law is also indispensable for establishing the rules and, as he has just proved, for developing good taste. But was such a mathematical discourse possible when addressing Pierre Patte, an unconditional follower of pyramidal or triangular stability? Likewise, it can be envisioned by addressing the assembly that Soufflot describes to us in his final paragraphs:

Those who indulge in these novelties are often preferred to those who want to resist the torrent that they see ready to overturn the arts; soon the rules are abandoned and taste disappears. We have experienced it in this century. But if the bizarre seduced us, it was soon despised; we owe its being forgotten to the knowledge that the Marquis de Marigny had acquired in his travels, to the love that he demonstrated for good architecture, and for the kindnesses with which he honored those who were more concerned with their glory than with the base interest which would have made them forget it; we owe it to the care that the company before which I have the honor to speak has taken to preserve good principles and constantly instill them in its students.

What should we not expect for the conservation of these principles, and for the progress of architecture, taste, and knowledge from the worthy leader of the arts who honors us today with his presence? He was kind enough to install us himself in these vast rooms built by the King's orders for our assemblies.[39]

---

[36] Soufflot (1775) (Petzet 1961, p. 145): *Sans la symmétrie qui est le rapport des masses correspondantes et la proportion des parties entr'elles et avec le tout, il n'y a point d'ensemble, il n'y a point d'unité dans un édifice; le goût et les règles n'ont pas présidé à sa construction; Le goût n'existe pas sans le jugement; l'empreinte du jugement dans l'ensemble et dans les parties, constitue et rend agréable l'ouvrage du génie et prouve l'exécution des règles.*

[37] Soufflot (1775) (Petzet 1961, p. 145–146): *Les règles ne sont pas cet amas confus de préceptes donnés par des auteurs dont les livres contiennent souvent plus de mots que de choses; il s'agit des règles primitives qui existoient avant qu'on les connut, comme il existait, avant qu'on le démontrat, qu'en tout triangle les trois angles sont égaux à deux droits; il s'agit de ces règles et de tous les corollaires que la succession des tems en a fait tirer, dont la somme a constitué le bon goût.*

[38] See "The Use of a Particular Form of the Parallelogram Law of Forces for the Building of Vaults (1650–1750)", Chap. 4 in this present volume.

[39] Soufflot (1775) (Petzet 1961, p. 146): *Ceux qui se livrent à ces nouveautés sont souvent préferés à ceux qui veulent résister au torrent qu'ils voyent prêt à bouleverser les arts; bientôt les règles sont*

All the more so, since Soufflot still has requests to make for its machine designed to test the resistance of materials.

> We dare to hope that it will subsequently put us in a position to deposit models of all the objects which can expand the light and stimulate the imagination. There we will see machines capable of demonstrating the strength of metals, the resistance of wood and different materials under the weights with which they are loaded. Experiences will reassure us, in practice, against ideas that are as adapted as they are ill-founded, which cause these metals and materials to be given dimensions much above those which are necessary, and which cause, through the fault of those who indulge in abusive uses and frivolous fears, harmful overloads and expenses that could be applied more usefully.[40]

This last sentence is a direct allusion to the oversizing of the pillars required by Patte to ensure the stability of the dome of Sainte-Geneviève:

> It is through a reasoned knowledge of practice combined with a profound theory that the architect, whose first quality must be probity, will be able without scruple to take charge of the administration of the sums that individuals will entrust to his knowledge and to his care.[41]

It is this profound theory—which, it must be admitted, Soufflot does not disclose to us—which made the construction of this immense dome possible. But there remains a secret to achieving this result, the work of the architect:

> ... but with all this knowledge, if the artist does not feel this secret influence, this fire, by which ideas develop, combine and spread, he will never reach his goal and as Vitruvius said: *Genius without labor, labor without genius, never made the artist perfect.*[42]

abandonnées et le goût s'évanoui. Nous l'avons éprouvé dans ce siècle, mais si le bizarre nous a séduit, il a été bientôt mépriséj on en doit l'oubli aux connoissances que M. le marquis de Marigny avoit acquises dans ses voyages, à l'amour qu'il a manifesté pour la bonne architecture et aux bontés dont il a honoré ceux qui se sont plus occupé de leur gloire que du vil intérêt qui la leur auroit fait oublier; on la doit aux soins que la compagnie devant laquelle j'ai l'honneur de parler a pris pour conserver les bons principes et les inculquer constament à ses élèves.
Que ne devons nous pas attendre pour la conservation de ces principes, et pour les progrès de l'architecture, du goût, et des connoissances du digne chef des arts qui nous honore aujourd'hui de sa présence; il a bien voulu nous installer lui même dans ces vastes salles construits par les ordres du Roy pour nos assemblées.

[40] Soufflot (1775) (Petzet 1961, p. 146): *Nous osons espérer qu'il nous mettra par la suite en état d'y déposer des modèles de tous les objets qui peuvent étendre les lumières et échauffer l'imagination. On y verra des machines propres à démontrer la force des métaux, la résistance des bois et des différents matériaux sous les poids dont on les charge. Les expériences nous rassureront, dans la pratique, contre les idées, aussi adaptées que mal fondées, qui font donner à ces métaux et à ces matériaux des dimensions beaucoup au-dessus de celles qui sont nécessaires, et qui occasionnent par la faute de ceux qui se livrent à des usages abusifs et à des craintes frivoles, des surcharges nuisibles et des dépenses qui pouroient être appliquées plus utilement.*

[41] Soufflot (1775) (Petzet 1961, pp. 146–147): *C'est par une connoissance raisonnée de la pratique jointe à une profonde théorie que l'architecte, dont la première qualité doit être la probité, poura sans scrupule, se charger de l'administration des sommes que les particuliers confieront à ses lumières et à ses soins.*

[42] Soufflot (1775) (Petzet 1961, p. 147): *mais avec toutes ces connoissances si l'artiste ne sent pas cette influence secrette, ce feu, par lequel les idées se développent, se combinent et s'étendent, il n'arrivera jamais à son but et comme Vitruve l'a dit: Nec ingenium sine labore, nec labor sine ingenio ... Perfectum artificem umquam fecerunt.*

5 Conclusion

One unanswered question about these texts remains. While Soufflot certainly knew the law of the parallelogram, why did he never mention it? Might he have feared it would disturb his readers to include some elaborate mathematics? This answer is justified in the case of Soufflot texts but is hardly admissible in Rondelet, when he defends Soufflot from the technical attacks of Pierre Patte. I have shown (Radelet-de Grave, 2014) that it was only in a letter addressed to the Count of Angiviller, and written very shortly before Soufflot's death, that Rondelet explicitly used the law of the composition of forces (Rondelet n.d.). The 1805 publication by Charles François Viel de Meaux (1745–1819) of a small work entitled *De l'impuissance des mathématiques pour assurer la solidité des batimens*, allows us to understand that the entire world of architects was still not ready to accept this mechanization of architecture.

References

Blondel, François. 1675–1683. *Cours d'architecture, enseigné dans l'Académie royale d'Architecture*. Paris: P. Auboin et F. Clouzier.

Bélidor, Forest de. 1729. *La Science des ingénieurs dans la conduite des travaux de fortification et d'architecture*. Paris: Jombert.

Freart de Chambray, Roland and Charles Errard. 1689. *Parallèle de l'architecture de l'antique et des modernes*. Paris: Jollain l'aîné.

de Gandt, François. 2004. Le goût et les règles: Les usages de la critique en Occident. *Extrême-Orient, Extrême-Occident* 26: 163–172.

La Hire, Phillippe de. 1695. *Traité de mécanique, où l'on explique tout ce qui est nécessaire dans la pratique des arts, et les propriétés des corps pesants lesquelles ont eu plus grand usage dans la Physique*. Paris: Imprimerie royale.

Ouvrard, René. 1679. *Architecture harmonique, ou Application de la doctrine des proportions de la musique à l'architecture,* Paris, R.-J.-B. de La Caille, 1679.

Patte, Pierre. 1770. Mémoire sur la construction de la coupole … avec Marginalia of Jean Baptiste Rondelet. Bibliothèque historique de la ville de Paris, Ms BHVP 3459. In *Jean Rondelet: The Architect as Technician*, eds. Robin Middleton, Marie-Noelle Baudouin-Matuszek. New Haven: Yale University Press, 2007, 299–301.

Perrault, Charles. 1688–1697. *Parallèle des anciens et des modernes,* 4 vols. Paris, Jean Baptiste Coignard, 1688–1697.

Perrault, Claude. 1673. *Les dix livres d'architecture de Vitruve: Corrigez et traduits nouvellement en François, avec des notes et des figures*. Paris: Coignard.

Petzet, Michael. 1961. *Soufflot Sainte-Geneviève und der französische Kirchenbau des 18. Jahrhunderts*. Berlin: de Gruyter.

Radelet-de Grave, Patricia. 2014. The Panthéon's Stability Already Questioned by Pierre Patte in 1770. In: *Masonry Structures: Between Mechanics and Architecture*. Danila Aita, Orietta Pedemonte, Kim Williams, eds., 125–187. Birkhäuser, Cham. https://doi.org/10.1007/978-3-319-13003-3_7.

Rondelet, Jean-Baptiste. 1770. *Réfutation d'un mémoire sur la construction de la coupole projettée pour couronner la nouvelle* église *de Sainte-Geneviève, où il est question de prouver que le piliers* déjà exécutés et destinés à porter cette Coupole, ont les dimensions nécessaires pour espérer d'y élever un semblable ouvrage avec solidité et à lui donner à ce sujet la solution de plusieurs problèmes utiles à la construction de édifices [...], 1770, *Marginalia* écrites sur le *Mémoire* de

Rondelet, Jean. n.d. [Lettre adressée au comte d'Angiviller, (ca. 1780)]. In *Jean Rondelet: The Architect as Technician*, eds. Robin Middleton, Marie-Noelle Baudouin-Matuszek. New Haven: Yale University Press, 2007, 307.

Rykwert, Joseph. 1980. *The First Moderns: Architects of the Eighteenth Century*. Cambridge Ma: MIT Press.

Soufflot, Jacques-Germain. 1739. *Mémoire sur les proportions de l'architecture*, Bibliothèque de l'Académie des Sciences, Belles-Lettres et Arts de Lyon, 9 December 1739, ms. 194, fols. 132r-138v. In Michael Petzet, *Soufflot Sainte-Geneviève und der französische Kirchenbau des 18. Jahrhunderts*, Berlin: de Gruyter, 1961, 131–135.

Soufflot, Jacques-Germain. 1741. *Un parallèle des églises gothiques avec les églises modernes*, Bibliothèque de l'Académie des Sciences, Belles-Lettres et Arts de Lyon, 12 April 1741, ms. 194, fols. 144r-151r. In Michael Petzet, *Soufflot Sainte-Geneviève und der französische Kirchenbau des 18. Jahrhunderts*, Berlin: de Gruyter, 1961, 135–142.

Soufflot, Jacques-Germain. 1744. *Dans l'art de bâtir le goût est-il préférable à la science des règles ou celle-ci au goût ?* Bibliothèque de l'Académie des Sciences, Belles-Lettres et Arts de Lyon, 9 septembre 1744, ms. 194, fo 126r-131r. In *Nouvelles archives statistiques historiques et littéraires du département du Rhône*, vol. 1, 1832, 107–116.

Soufflot, Jacques-Germain. 1775. *De l'identité du goût et des règles dans l'exercice de l'architecture.* L'Académie d'Architecture de Paris, 20 November 1775, and Bibliothèque de l'Académie des Sciences, Belles-Lettres et Arts de Lyon, 18 July 1778, ms. 190, fols. 53r-56v. In Michael Petzet, *Soufflot Sainte-Geneviève und der französische Kirchenbau des 18. Jahrhunderts*, Berlin: de Gruyter, 1961, 142–147.

Viel de Saint Maux, Charles-François. 1805. *De l'impuissance des mathématiques pour assurer la solidité des batimens, et recherches sur la construction des ponts*. Paris: chez l'auteur.

Chapter 12
The Catenary as Inspiration

1 The Catenary Chosen for Aesthetical Reasons in Egyptian Antiquity

At first sight the form of the catenary looks rather common, and it is indeed, as it is the form taken, for instance, by most necklaces (when the distribution of their weight is uniform) as they appear when worn in the normal fashion. Indeed, the catenary form appears as an aesthetic inspiration at least as far back as Egyptian Antiquity. One finds, at the entrance of the Egyptian Museum of Cairo, a cosmetic palette called the Narmer Palette (3000 BC) (Fig. 1), whose border is a catenary, as discovered by Tullio Viola (1987). To verify that it is a catenary, just suspend a fine chain from each end, put the picture of the palette vertically so that the rounded side is down and adjust the lowest point of the chain to that of the palette. The rest of the figures should coincide. The Egyptians would likely have done this exactly the same way.

A good number of those palettes, used for grinding cosmetics, have survived. They were used for grinding cosmetics and Petrie (1953) includes images and descriptions of each of them. Some have the form of animals, while others have geometrical figures. For some of them it is difficult to determine if they are parabolas or catenaries, but eight among them are indeed catenaries, and that we may conclude that the catenary was chosen for aesthetic reasons, because it was the form of the necklace. At the time, nobody knew anything mathematical about the form of the catenary, and they only could draw it by following the form of a hanging chain. The first geometrical curves that were studied mathematically, some 3000 years later, were algebraic curves: the straight line; the circle as studied by Euclid and Archimedes; and the conics studied by Apollonius. There is one exception: the Archimedean spiral, which is a transcendental, non-algebraic, curve. During the Middle Ages, the time of Romanesque and Gothic cathedrals, the eleventh through fourteenth centuries, when so many domes were erected, they kept studying those same curves. From the mathematical point of view, things would change only at the end of the seventeenth, as we will see.

P. Radelet-de Grave, *Between Mechanics and Architecture*, Mathematics and the Built Environment 8, https://doi.org/10.1007/978-3-031-73530-1_12

Fig. 1 Front and back sides of the Narmer cosmetic palette. Public domain

2 Vaults and Domes from Antiquity to the End of the Seventeenth Century

The corbelled vault is the clearest example to show that at the beginning, the construction technique creates the form of the vaults. The Mycenaean bridge makes that particularly clear (Fig. 2). There is certainly no aesthetic research; the only the aim is that of making a bridge with the stones available. There is certainly some catenary that could be drawn within the contours of the vault.

For barrel vaults and Nubian vaults, the circle is generally selected, but technique allows the use of other forms as well, such as ellipses or, much later, multicentred vaults. The form is thus selected through a compromise between technical and aesthetic considerations. In the case of the barrel vault, the centring must be able to sustain the work but the circle may also be chosen for aesthetic reasons, or guided by religious or cosmological ideas. Most of the well-known large arches, such as those of the Pont du Gard (first century) are circles, while large domes are based on half-spheres, like the Pantheon in Rome (second century) or the Hagia Sophia (sixth century). How they were centred is and has long been the object of many discussions.

The domes of Santa Maria del Fiore in Florence, and St. Peter's Basilica in Rome are particular cases that raise questions. As Poleni (1748) shows, the external curve of the dome of St. Peter's is a visible catenary. The same is true for Santa Maria del

Fig. 2 The Mycenaean bridge, Argolis, Peloponnesos, Greece. Public domain

Fiore. Perhaps they are evidence that Hooke's affirmation of 1676 was a statement of the *Zeitgeist*.

3 Mechanics Research on the Catenary from Simon Stevin to Robert Hooke

During the period from the mid-sixteenth to the mid-seventeenth centuries, scientists, long before architects, began to use the parallelogram law of forces in statics, and to study the properties of the hanging chain, which will later come to be called the catenary. The story starts with Simon Stevin, who first published *De Beghinselen der Weeghconst* (Stevin 1586), published an addendum to that study (1608), in which he studies the equilibrium of a beam or of a mass suspended to one or two ropes, discovering the parallelogram law (Fig. 3, left and centre). Then, with help of that law, he studies the form of a rope on which weights are hung (Fig. 3, right).

In 1638, in some fragments added to his *Discorsi*, Galileo also studies the form of a hanging chain, but making an analogy between the links of the chain and a falling body, he says:

> The curvature of the path of the shot fired horizontally appears to result from two forces, one (that of the weapon) drives it horizontally and the other (its own weight) draws it vertically downward. So in stretching the rope you have the force which pulls it horizontally and its own weight which acts downwards.[1]

[1] Galilei (1638, p. 284); Eng. trans. p. 290: *La curvità della linea del proietto orizontale par che derivi dalle due forze, delle quali una (che è quella del proiciente) lo caccia orizontalmente, e l'altra*

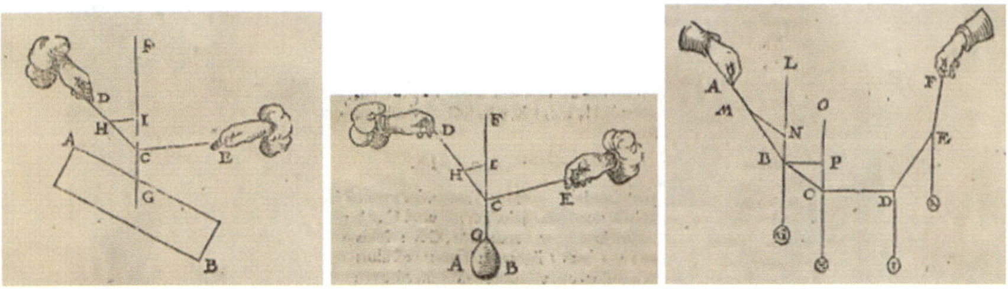

Fig. 3 Stevin's drawings showing the discovery of the parallelogram law and the study of the catenary in his *Byjwough der Weeghconst* (Stevin 1608, pp. 182–183; 1634, p. 505). Public domain

Galileo concludes that the form of the catenary is a parabola, just as the trajectory of a falling body. Galileo is right when he says that two forces act on each piece of the chain, as two forces do on the falling body. One of these forces, in both cases, is the weight of the falling body or of the piece of chain. But his error lies in the direction of the second force. For the falling body, it is inertia, which is always horizontal, but for the chain, it is the tension, which is always tangent to the chain and not horizontal as he thinks. That is the reason for his mistake; another reason is that the parabola and the catenary are not very different.

The very young Christiaan Huygens will correct Galileo's error eight years later. After having read Stevin, he demonstrates axiomatically that the form of the chain is not a parabola. In an unpublished manuscript entitled *De catena pendente*, Huygens starts by proving a general property:

> First Theorem: If a weight were suspended to two ropes, those ropes prolonged will cross each other in the vertical stemming from the centre of gravity of the weight.[2]

With the help of that property, he generalizes the problem to a series of identical weights attached to a weightless rope and determines in that way the position of a series of points on the rope. He then shows that:

> Prop. 3: The line that passes through these points differs slightly from the parabolic line; but the difference hardly matters.[3]

Huygens's manuscript was not published, but in November of the same year, he sent three letters to Mersenne.[4] The first two letters contain a more elaborate version of his ideas. There we find a beautiful drawing showing the difference between the catenary, which he has drawn point by point, and a parabola (Fig. 4). There we

---

(che è la propria gravità) lo tira in giù a piombo. Ma nel tender la corda vi sono le forze di coloro che orizontalmente la tirano, e vi è ancora il peso dell'istessa corda, che naturalmente inclina al basso.

[2] Huygens (1908, p. 37): *Si pondus suspendaur ex duobus funibus, hi producti secabunt sese in pendula ponderis gravitatis diametro.*

[3] Huygens (1908, p. 41): *Linea quae per haec puncta ducitur alde parum distare videtur a linea parabolica, quam tamen non refert.*

[4] Huygens (1888, letters 20–22, pp. 35–44).

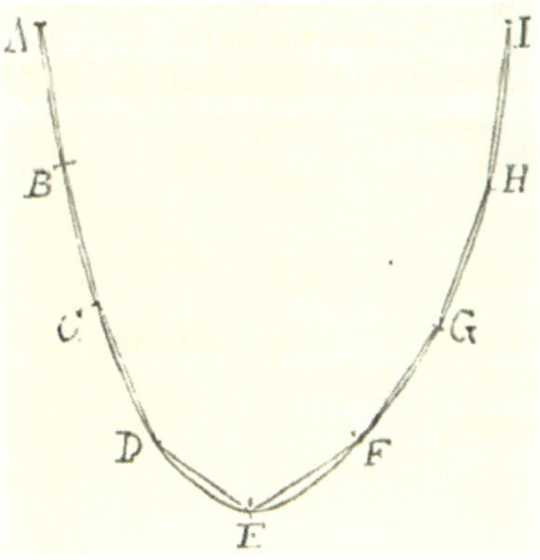

Fig. 4 Huygens's drawing showing the difference between the catenary and the parabola (Huygens 1888, p. 38). Public domain

clearly see how difficult it is to say with certainty that a catenary and not a parabola is contained within the thickness of an arch or a vault.

Huygens also shows in the same letter that if one does not consider the weights equally distant along the length of the rope, but at distances that are horizontally equidistant, as shown in Huygens's drawing or in a modern hanging bridge, then the line will be a parabola.

Twenty-three years later, in 1673, Ignace-Gaston Pardies uses the same principle and finds the same results, without having read Huygens. He adds an essential property of the catenary: the fact that the tension is tangent to the curve and that it is the same all along the catenary. This property is most important from our point of view. It explains that the most stable vault is a catenary because in that curve there are no tensions tending to push parts of the vault either outwards or inwards. We will find applications of that property, later, in works of Newton's pupils, as well in an article of Jacob Bernoulli.

Let us remark that Pardies's expression "a rope … on which hang an infinity of weights"[5] shows us how infinitesimal calculus infiltrates the study of curves in general and of the catenary in particular.

In his third letter to Mersenne, Huygens treats the same problem in a completely different manner. There he uses a principle given by Evangelista Torricelli in 1644, which says that in a static equilibrium, the centre of gravity always occupies the lowest possible place. In his first solution, Huygens had used the equilibrium of forces; here he uses an extremum principle.

At that point occurs an important and well-known event. In 1676, at the end of his book on helioscopes, Robert Hooke publishes a strange anagram:

> 2. The true Mathematical and Mechanical form of all manner of Arches for building, with true butment necessary to each of them. A Problem which no Architectonick Writer hath ever yet attempted, much less performed.
>
> Abcccddeeeeeefggiiiiiiiiillmmmmnnnnnooprrsssttttttuuuuuuuuux.[6]

[5] Pardies (1673, p. 132): *un filet … sur lequel sussent appuyées une infinità de lignes …*

[6] Hooke (1676, p. 31).

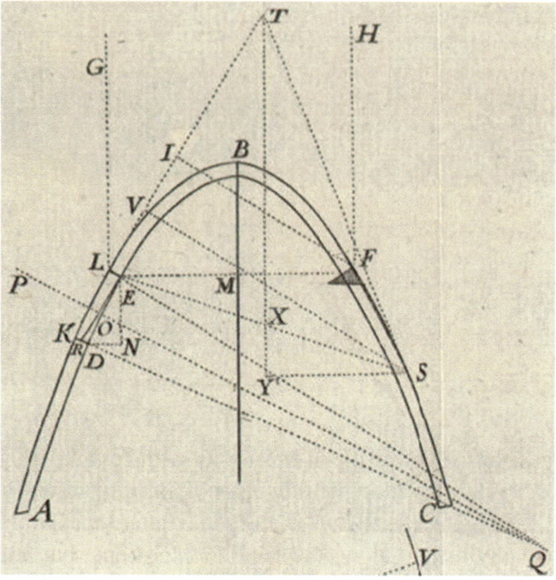

Fig. 5 Jacob Bernoulli's drawing to prove that the catenary is the strongest arch (1744: Tab. XLVIII). Public domain

Hooke writes a first version of the solution of the anagram in his journal on 26 September 1675 and Waller published the solution in his edition of Hooke's posthumous works: *Ut pendit continuum flexile, sic stabit contiguum rigidum inversum* (Waller 1705, p. xxi), which may be translated "as hangs the flexible line, so but inverted will stand the rigid arch". But the influence on architecture will not be immediate, as we will see.

Jacob Bernoulli, in one of his *Varia posthuman* (Bernoulli 1744) also uses the property of pressure or compression but in a different way (Fig. 5). He uses what engineers call the equilibrium to translation and to rotation, saying that if the vault is stable, a voussoir cannot either rotate or be translated. For the translation, he correctly finds the equation of the catenary, but he makes an error with the rotation.

Daniel Bernoulli and Leonhard Euler will continue along the line opened up by Jacob Bernoulli, and Euler will develop an important part of the theory of elasticity. They were both particularly well trained in infinitesimal calculus, but that was not the case of most of the architects of the time.

4 Back to Architectural Theory, as Elaborated in Important Treatises and Some Articles

The mathematical and mechanical discoveries that we described above gave rise to a great number of books concerning the theory of architecture. The first impulse was given by the well-known *Traité de mécanique, où l'on explique tout ce qui est nécessaire dans la pratique des arts* by Philippe de la Hire, published in 1695 (Radelet-de Grave 2013). The particularity of his treatise is to introduce, as a foundation of the theory of architecture, an equivalent of the parallelogram law of forces that is particularly well adapted to the study of the stability of vaults. Instead of considering the sides of the parallelogram as the forces to be compounded and its diagonal as the resultant, he takes a triangle whose sides are perpendicular to the three forces in equilibrium (Fig. 6).

The introduction of the law of equilibrium of forces as the foundation of stability will enable architects to study the equilibrium of vaults mathematically. Many will do so in order to confirm Hooke's unproven affirmation. Progressively, as the century progresses, mechanics and infinitesimal calculus enter the domain of architecture. In 1729, in his famous *La science des ingénieurs*, Bernard Forest de Bélidor introduces forces and La Hire's triangle law, and he shows that the inverted catenary forms the most stable vault. In the same year, Claude Antoine Couplet introduces the parallelogram law as the fundament of stability and he uses it to demonstrate the stability of certain vaults. But all the vaults he considers are circular, or as he puts it, the lines of their joints all meet in one point. As a consequence, he does not mention the catenary, but he uses differential calculus.

In an article published in 1753, *Resolutiones Problematum spectantium ad Architectura civilem*, Georg Wolfgang Krafft solves a problem by means of minima and maxima, and draws a catenary but does not mention its name (Fig. 7).

Two years later, he starts a new memoir, *De curvis funiculariis et catenaries,* which has a much more scientific character, saying that:

> As the principle of composition of forces is more than often used in the whole theoretical mechanics: it can also, in the same way, be applied to various curves, called funiculars, and I

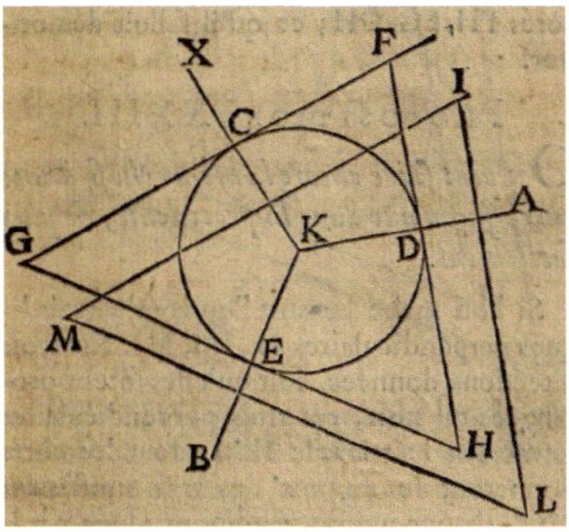

Fig. 6 La Hire's triangle law of forces (La Hire 1695, p. 73). Public domain

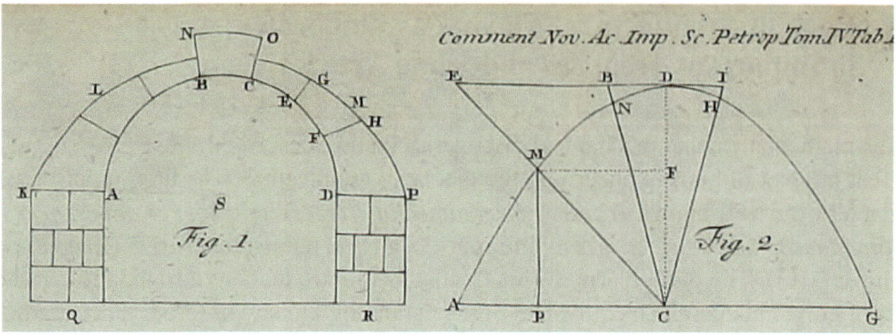

Fig. 7 Krafft draws the catenary but does not mention its name (Krafft 1753, Tab. IV). Public domain

> add that those curves derive of the laws of pure mechanics; and that is the aim of the present writing.[7]

Following Jacob Bernoulli, Krafft uses the radius of curvature, which will be essential to study elastic curves, a generalization of the catenary. Like Couplet, he uses infinitesimal calculus and he compares the equations he finds with those found by Jacob Bernoulli, Johann Bernoulli, Daniel Bernoulli, and Euler. Architecture completely disappears, leaving only the catenary and other elastic curves. But, as he already had published about architecture, the text may have had an influence on later treatises of architecture.

Approximately twenty years later, the Abbé Charles Bossut, who was a scientist, published articles in 1774 and 1776 on the equilibrium of vaults which were often quoted by architects. He uses the parallelogram law and the radius of curvature together with infinitesimal calculus to find the equation of the catenary. His 1774 text concerns the two-dimensional vaults as in all the previous works, but in his second text of 1776, he enlarges the problem to three dimensions and studies the stability of domes and cupolas.

Émiland-Marie Gauthey proclaims, in 1771, that only de la Hire's rule, meaning the parallelogram law, is applicable:

> Of all the rules and formulas that have been given on this matter, that of M. de la Hire is the only one that has prevailed among scholars, and is the most universally adopted.[8]

We will see with Rondelet that such an affirmation was not usually made public. It is amazing to see how the terms "parallelogram law" frightened, and were by-passed by, most authors. Later, in his Chap. 4, "De l'épaisseur qu'on doit donner aux culées

[7] Krafft (1755, p. 145): *Uti principii compositionis et resolutionis potentiarum ingens usus est in tota Mechanica theoretica: ita idem utilissime etiam ad curvas varias, funicularias dictas, applicari potest, adeo ut ex legibus pure mechanicis illae deriventur; quod praesenti scripto exponere constitui.*

[8] Gauthey (1771, p. 6): *De toutes les regles & formules que l'on a données sur cette matière, celle de M. de la Hire est la seule qui ait prévalu chez les Savants, & est le plus universellement adoptee.*

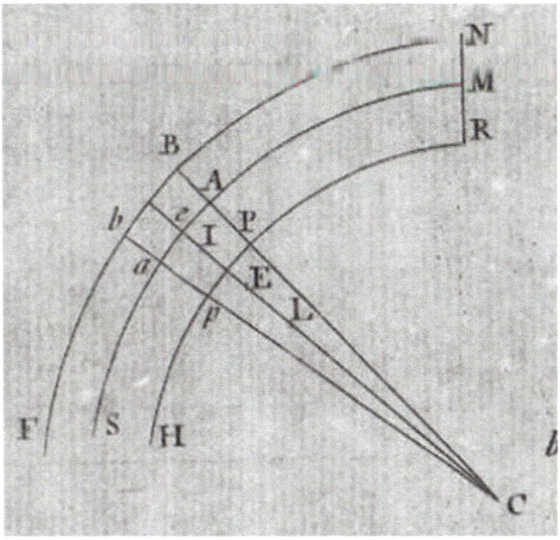

Fig. 8 Mascheroni's illustration of the catenary that forms the line of thrust passing through all the voussoirs of an arch (Mascheroni 1785, Tab. V). Public domain

des ponts" of his *Traité de la construction des ponts*, Gauthey dares to say that the catenary is the most stable vault:

> When the thickness of the vault is very small, this shape must be an inverted chain, since the vault can then be considered as an assembly of heavy spheres, infinitely small in contact with each other, the conditions of equilibrium are the same as those of a heavy and inextensible rope.[9]

Antonio Maria Lorgna and Charles Bossut, both scientists, use infinitesimal calculus, radius of curvature, etc., in their works on architecture and so do their successors, also scientists, Paolo Frisi in 1777 and Leonardo Salimbeni in 1787.

The work of Lorenzo Mascheroni, *Nuove ricerche sull'equilibrio delle volte*, published in 1785, deserves a particular remark because the author realizes what the discovery of the neutral axis of Huygens and Jacob Bernoulli implies:

> The most important point for adapting the catenary to practice is to find the thickness of the arch suitable for the equation of the catenary itself, which you want to use. To do this first of all it is certain that the catenary *MAS* (Fig. 21), if it is to be used in arches with the advantage that it has, must pass through the centre of gravity of all the elements *BPpb* of the solid arch NBPR.[10]

That means that the thrust line must be a catenary (Fig. 8).

[9] Gauthey (1832, p. 230): *Lorsque l'épaisseur de la voûte est très petite, cette forme doit être une chaînette reversée, puisque la voute pouvant alors être considérée comme un assemblage de sphères pesantes, infiniment petites en contact les unes avec les autres, les conditions de l'équilibre sont les mêmes que celles d'une corde pesante et inextensible.*

[10] Mascheroni (1785, p. 45): *Il punto più importante per adattare la catenatia alla pratica, si è trovare la grossezza dell'arco conveniente all'equazione della catenaria medesima, ch si vuole impiegare. Per ciò fare in prima luogo egli è certo, che la catenaria* MAS *(Fig. 21) se si vuole impiegare negli archi col vantaggio che le è proprio, deve passare per il centro di gravità di tutti gli elementi* BPpb *dell'arco solido* NBPR.

5 More on Architecture

Certainly, Hooke's anagram did not have a direct influence on the building of new vaults or domes, but there is at least one great architect who immediately knew its solution. This was Hooke's close friend who collaborated with him on the development of the plans of the dome of St Paul's Cathedral, completed in 1710: Christopher Wren.

The "Warrant design" of 1675 of St Paul's, as well as sketches from Wren's hand, are generally presented as the actual state of the three superposed cupolas forming the dome and, in particular, of the intermediary cupola intended to reinforce its stability. The catenary should appear as the cross-section of that intermediary cupola. But it does not.

Wren continued to transform the plans of his dome after the beginning of the construction and JacquesHeyman (1998) discusses a later cross section of the dome in which the middle vault looks completely different and appears precisely to support the lantern (Fig. 9). We see that the cross section is not a catenary but a cubic parabola $y = ax^3$, a fact that was communicated by Wren himself to the Royal Society. The reason for this is that at the time Wren didn't have the mathematical knowledge to draw the catenary. Leibniz and the Bernoulli had developed that knowledge but did not fully explain it in their publications in 1691, Huygens didn't even explain his way of tracing the catenary point by point in that publication, and neither Wren nor Hooke knew at that time about Newton's work on fluxions.

Wren's reasoning has nothing to do with aesthetics. He knows perfectly well that his curve is not exactly the right one, but he doesn't worry because it is close enough to play its role and sustain the lantern. Moreover, he hides the curve between the two other domes. For him the curve he uses is only a tool to ensure stability, and its form is deduced from the laws of stability alone. On the other hand, we observe that Hooke and Wren not only knew that the form of the vault should be the inverse of the catenary, but they also understood the reason why it must be so: because tensions in the catenary transform into pressure in the vault, giving rise to a similar equilibrium. They also knew that for the catenary, the centre of gravity was the lowest possible point on the curve, but did they know that it should be the highest for the vault? In other words, were they able to solve the problem as an extremum problem?

What about the technical point of view? An article published by HenryPitot (1726) shows how sophisticated centring was at the time (Fig. 10).

If we add to this the fact that the use of centring is almost as old as the use of vaults, and that many of those vaults had been standing for centuries, if not for millennia, it is clear that many architects prefer to go ahead that way instead of drawing the form of their vault by following the form of a catenary on a paper and turning it around to get their vault, as suggested or shown by Bélidor in 1729 (Fig. 11):

> If we want to build a natural vault whose width and height are given, we must on a vertical surface draw a line CD equal to the width of the vault, drop from the middle of this line a perpendicular EF equal to the height that we want to give it, then attach the end of a chain to point C, and carry the other end towards D, so that by increasing or decreasing the chain, its own weight makes it pass through the point F when it is fixed at places C and D. After

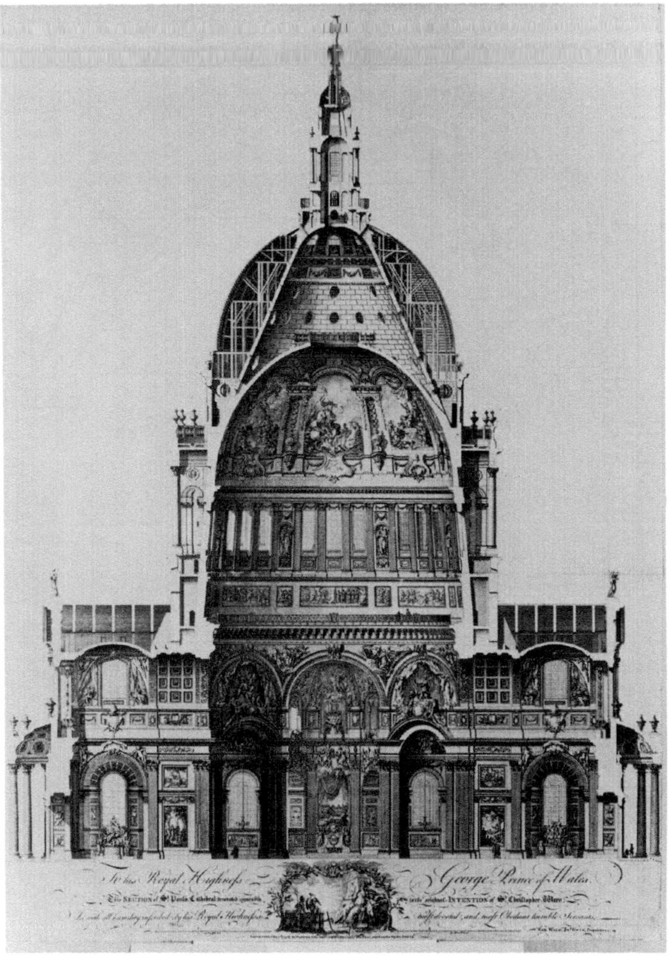

Fig. 9 Cross section of the dome posterior to the "Warrant design" of 1675, in an engraving by Samuel Wale and John Gwynn made for the Prince of Wales in 1755. Public domain

that, we can, with a pencil that we draw all along the chain (without making it wobble), trace a curve, and on that establish the figure of the false arch of the vault, the section of the voussoirs and the rest.[11]

[11] Bélidor (1729, p. 149): *Si l'on veut construire une voûte naturelle dont la largeur et la hauteur soient données, il faut sur une surface vertical tracer une ligne CD égale à la largeur de la voûte, abaisser du milieu de cette ligne une perpendiculaire EF égale à la hauteur qu'on veut lui donner, ensuite attacher l'extrémité d'une chaîne au point C, et porter l'autre extrémité vers D, de manière qu'en augmentant au diminuant la chaîne, son propre poids la fasse passer par le point F lorsqu'elle sera arrêtée aux endroits C et D. Après cela, on pourra, avec un crayon que l'on conduira tout du long de la chaîne (sans pourtant la faire vaciller), tracer une courbe, et là-dessus établir la figure du faux cintre de la voûte, la coupe des voussoirs et le reste.*

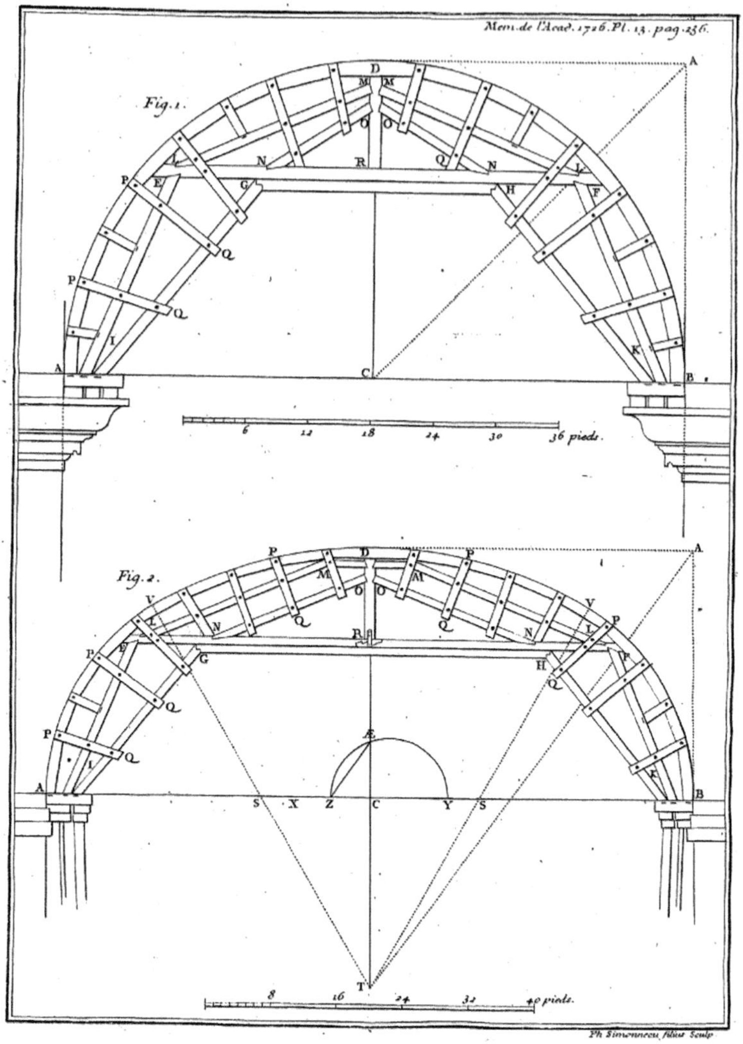

Fig. 10 Examples of centring by Henry Pitot (Pitot 1726, Tab. 13). Public domain

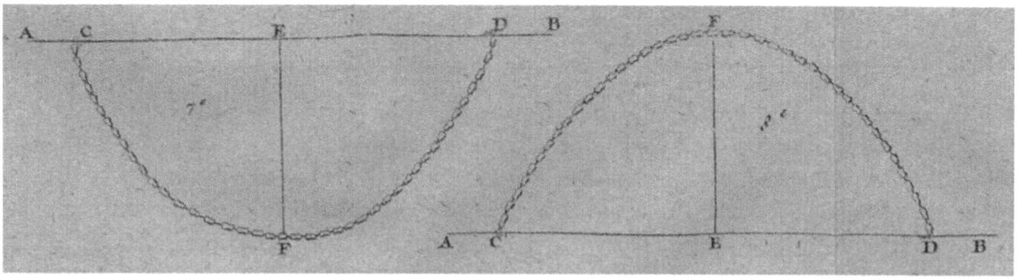

Fig. 11 Bélidor's representation of the hanging chain and its inversion to form a vault (Bélidor 1729, Bk. II, Pl. VI). Public domain

Despite the fact that in 1748, Poleni will even use that method to demonstrate the stability of the dome of St Peter's in Rome, most domes remained spherical, and in section most the vaults were still circular, elliptical or multi-centred.

Pitot's drawings, as well as other drawings of centrings, together with the fact that Pitot mentions works of Varignon, the great defender of the parallelogram law, make us suppose that it is through the study of centring that the parallelogram law will finally conquer the domain of architecture. However, the example of Sainte Geneviève, the Paris Pantheon, shows how long the path to it was (Radelet-de Grave 2015). The construction works had begun in 1755 and as soon as the piers intended to support the dome were erected, it aroused criticism from Pierre Patte, who objected that such piers would be unable to support any dome. During almost the entire time of that long dispute exclusively concerning stability, Jean-Baptiste Rondelet, who defended the ideas of architect Jacques-Germain Soufflot, never mentions the use of the parallelogram law of forces. He finally shows it in an undated private letter written around 1780 to Charles Claude de Flahaut de la Billarderie, comte d'Angiviller, *Directeur et ordonnateur general des bâtiments du roi,* who ordered the work. There was a real fear of mentioning the use of a law that seemed so artificial and was so difficult to explain to the workers on the construction site. It is then no wonder that they don't mention the catenary, which was even impossible to draw without a good grasp of the notions of mathematics, while in contrast all conic sections could be drawn mechanically with ruler and compass. We have only to think of the success of Mascheroni's *Geometria del compasso* (1797), published five years after the construction of Sainte Geneviève was completed.

6 Conclusion

After this long story, necessary to understand the various sources of inspiration in the particular case of the catenary, let us sum up that evolution. We started with an essentially aesthetic inspiration. Based on the shared appreciation of the curve of a hanging necklace, someone decided to copy this form to make the palettes used to prepare cosmetics. Today, we know that the same geometric curve ensures stability in the thin vault. This is the reason why we observe that such a curve can be included in the ancient vaults that are still standing, even if the people who built that arch didn't know anything about catenaries. However, it took generations and generations of architects to understand the composition of forces and its role in the equilibrium of a catenary and in the stability of vaults and domes. There was a final difficulty: it was possible to reproduce a curve known only from observing hanging chains, let's say through experiment, but how could they be drawn in a way that was accurate geometrically? To do that, infinitesimal calculus was necessary. After this slow evolution, and finally knowing all the characteristics of the catenary, architects such as Antoni Gaudi, Eero Saarinen, and others decided for aesthetic reasons to underline stability by showing the catenary that ensures it. In doing so they followed a maxim attributed to Plato: "Beauty is the splendour of truth".

Acknowledgements A previous version of this chapter appeared in vol. 3 of *Proceedings of the Fifth International Congress on Construction History* (Chicago, 3-7 June 2015), Donald Friedman, Brian Bowen, Thomas Leslie, John Ochsendorf, eds. We are grateful to the ICCH for permission to use the material, which has been revised for this present publication.

References

Bélidor, Bernard Forest de. 1729. *La Science des ingénieurs dans la conduite des travaux de fortification et d'architecture civile.* Paris: Jombert.

Bernoulli, Jacob. 1690. Analysis Problematis, de inventione lineae descensus a corpore gravi percurrendae uniformiter. *Acta Eruditorum*, May 1690, 217–219.

Bernoulli, Johann. 1742. Lectiones mathematicae de calculo integralium, in usum illustr. march. Hospitalii conscriptae. In: *Opera omnia, tam antea sparsim edita quam hactenus inedita (cura G. Cramer)....*, Vol. 3. Lausanne and Geneva: Bousquet.

Bernoulli, Jacob. 1744. Problema de Curvatura fornicis, cujus partes se mutuo proprio pondere suffulciunt sine opere caementi (5/12/1704). *Varia Postuma Jacob Bernoulli, Opera* II (XXIX): 1119–1123.

Bossut, Charles (Abbé). 1774. Recherches sur l'équilibre des voûtes. *Mémoires de l'Académie de Paris,* année 1774 (1778), 534–566.

Bossut, Charles (Abbé). 1776. Nouvelles recherches sur l'équilibre des voûtes en dôme, *Histoire de l'Académie royale des sciences ... avec les mémoires de mathématique & de physique*, année 1776, Paris 1779, 587–596.

Galilei, Galileo. 1638. *Discorsi e dimostrazioni matematiche, intorno a due nuove scienze attenti alla mecanica e i movimenti locali.* Leiden: Elsevier. Eng. trans. Stillman Drake. 1974. *Two new sciences, including centers of gravity & force of percussion.* Madison: University of Wisconsin Press.

Gauthey, Émiland-Marie. 1771. *Mémoire sur l'application des principes de la méchanique à la construction des voûtes et des domes…* Paris: Jombert.

Gauthey, Émiland-Marie. 1832. *Traité de la construction des ponts.* Paris: Chez Firmin Didot Fréres.

Heyman, Jacques. 1998. Hooke's Cubico-Parabolical Conoid. *Notes and Records of the Royal Society of London* 52 (1): 39–50.

Hooke, Robert. 1676. *A description of helioscopes, and some other instruments.* London: John Martyn.

Huygens, Christiaan. 1888. *Oeuvres complètes de Christiaan Huygens. Correspondance [de Christiaan Huygens], 1666–1669 / publ. par la Société hollandaise des sciences*, vol. 1. The Hague: M. Nijhoff.

Huygens, Christiaan. 1908. *Oeuvres complètes de Christiaan Huygens. / publ. par la Société hollandaise des sciences, Travaux mathematiques 1645–1651,* vol. 11. The Hague: M. Nijhoff.

Krafft, Georg Wolfgang. 1753. Resolutiones Problematum spectantium ad Architectura civilem. *Novi Commentarii academiae scientiarum Petropolitanae* 4 (1752–1753): 199–234.

Krafft, Georg Wolfgang. 1755. De curvis funiculariis et catenariis, vel illis, quae corporibus flexibilibus inducitur, cum potentiis quibusvis solicitantur. *Novi Commentarii academiae scientiarum Petropolitanae* 5 (1754–1755): 145–163.

Mascheroni, Lorenzo. 1785. *Nuove ricerche su l'equilibrio delle volte.* Bergamo: Francesco Locatelli.

Mascheroni, Lorenzo. 1797. *La geometria del compasso.* Pavia: Eredi di Pietro Galleazzi.

Pardies, Ignace-Gaston. 1673. *La statique ou la science des forces mouvantes.* Paris: Marbre-Cramoisy.

Petrie, W. M. Flinders. 1953. *Ceremonial slate palettes. Corpus of proto-dynastic pottery.* London: British School of Egyptian Archaeology.

Pitot, Henri. 1726. Examen de la force qu'il faut donner aux Cintres dont on se sert dans la construction des grandes voûtes, des Arches et des Ponts. *Mémoires de l'Académie Royale des Sciences de Paris*, 1726 (1728): 216–236.

Poleni, Giovanni. 1748. *Memorie istoriche della gran cupola del tempio vaticano, e de'danni di essa, e de'ristoramenti loro, divise in libri cinque*. Padova: Stamperia del seminario.

Radelet-de Grave, Patricia. 2013. La mécanique de Philippe de la Hire. In *Philippe de la Hire, 1640-1718, entre architecture et sciences*, ed. A. Becchi, H. Rousteau-Chambon, J. Sa-karovitch, 159–175. Paris: Picard.

Radelet-de Grave, Patricia. 2015. The Panthéon's Stability Already Questioned by Pierre Patte in 1770. In *Masonry structures: Between mechanics and architecture*, ed. Danila Aita, Orietta Pedemonte, Kim Williams, 127–185. Birkhäuser, Cham. https://doi.org/10.1007/978-3-319-13003-3_7.

Stevin, Simon. 1586. *De Beghinselen der Weeghconst*. Leyden: Franciscus Raphelengius.

Stevin, Simon. 1608. De Byvough Der Weeghconst. In: *The principal works of Simon Stevin*, ed. E.J. Dijksterhuis et al., vol. 1, 528–549. Amsterdam: E.V. Swets & Zeitlinger, 1955.

Stevin, Simon. 1634. *Les œuvres mathématiques de Simon Stevin augmentées par Albert Girard*. Leiden: Elsevier.

Stirling, James. 1717. Methodus disponendi quotcunque Sphaeras in Fornicem. In *Lineae tertii ordinis Neutoninae*. Oxford: Whistler.

Viola, Tullio, and Maria Teresa Navale. 1897. *Il profilo della tavolozza di Narner*. Memorie dell'Accademia delle scienze di Torino, vol. 10. Torino.

Waller, Richard. 1705. The life of Robert Hooke. In *The posthumous works of Dr Robert Hooke*. London: Smith and Walford.

Chapter 13
The Study of the Stability of Vaults by the Viscount of Nieuport

1 Introduction

The eventful life of Charles-François the Prud'homme d'Hailly (1747–1827), Viscount of Nieuport, Commander of the Order of Malta, a scholar of Belgian origin born in Paris, ruined by the revolution of 1789 and member of the Royal Academy of Sciences, Letters and Fine Arts of Belgium, leads from Paris to Malta where he is in the service of the Grand Master of the Order of the Knights of Saint-Jean de Jerusalem. Later, he would be a member of the Netherlands Institute as well as the academies of Zeeland and Stockholm. It is precisely this diversity that led us to look at his work, which contributes to exchanges between disciplines. His publications keep traces of his exchanges with d'Alembert and his biographers relate his relationships with Condorcet and Bossut. Nieuport will thus largely contribute to interdisciplinary exchanges as in intercultural exchanges (Fig. 1).

Like Adolphe Quetelet, I will begin my article with a quote from Jean d'Alembert's *Réflexions sur les éloges d'Académiciens*:

> The praise of a man of letters must be the account of his labours; but it may be useful to make known what he was, and to paint the man at the same time as the writer.[1]

Quetelet continues:

> And in fact, to recall only the works by which he served science, is to show him in an incomplete way, to strip him of his individuality and to satisfy the reader's curiosity imperfectly … These details, moreover, are not without interest for the general history of sciences and letters. If we want to judge the progress that a writer has made, we must necessarily take into account the circumstances in which he found himself and the obstacles he had to overcome.[2]

[1] D'Alembert (1821–1822, t. II, p. 151): *L'éloge d'un homme de lettre doit être le récit de ses travaux; mais, il peut être utile de faire connaître ce qu'il a été, et de peindre l'homme en même temps que l'écrivain.* Quoted in Quetelet (1866, p. 99).

[2] Quetelet (1866, pp. 99–109): *Et en effet, ne rappeler que les travaux par lesquels il a servi la science, c'est le montrer d'une manière incomplète, le dépouiller de son individualité et ne satisfaire*

P. Radelet-de Grave, *Between Mechanics and Architecture*, Mathematics and the Built Environment 8, https://doi.org/10.1007/978-3-031-73530-1_13

Fig. 1 Charles-François le Prud'homme d'Hailly, Viscount of Nieuport. Bust by Charles Geefs, Académie royale de Belgique. *Photo* Pascal Meurisse, reproduced by kind permission

It is this type of exercise that I will engage in here by showing the non-application by Charles-François de Nieuport of the parallelogram law of forces at a time when the historian can consider it to be a part of the common baggage.[3]

2 The Equilibrium of Segments in General

Nieuport starts from a set of three interrelated voussoirs held by two inclined planes (Fig. 2):

> 1. Let M, P, T be three voussoirs supported by the two fixed inclined planes LK, DE; their relative weights, or their forces, are expressed by the perpendiculars MN, PR, PO, TS, to the beds or joints FH, HJ. If these segments remain in this situation, they will necessarily be in equilibrium, that is, the force PO of the segment P on that M, will annul the force MN of the latter; & that likewise the force PR will be annulled by TS, which is opposite to it. Because if these two forces TS, MN were not annulled, the segment P would be pushed in the direction of their resultant, and consequently would not be able to remain in its first situation.

qu'imparfaitement la curiosité du lecteur ... Ces détails, d'ailleurs, ne sont pas sans intérêt pour l'histoire générale des sciences et des lettres. Si l'on veut juger des progrès qu'un écrivain a fait faire, il faut nécessairement tenir compte des circonstances dans lesquelles il s'est trouvé et des obstacles qu'il avait à vaincre.

[3] Elsewhere I have shown that this was not the case of Pierre Patte, opponent of Rondelet, regarding the stability of Parisian Pantheon by Soufflot. See Radelet-de-Grave (2015).

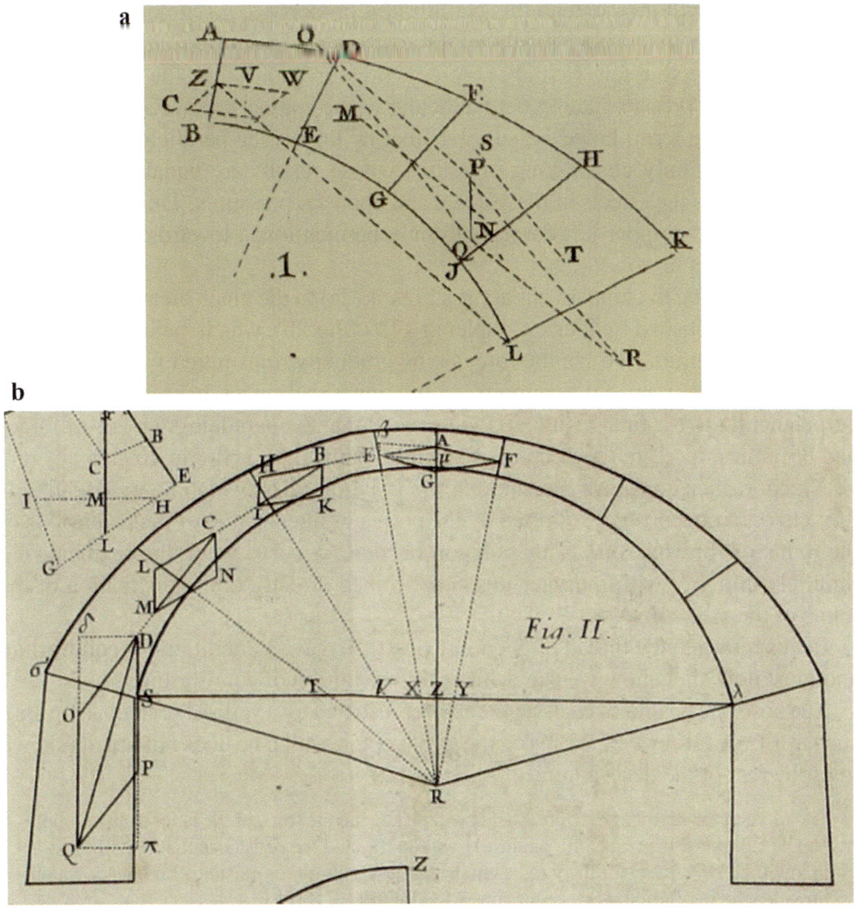

Fig. 2 **a, above** Nieuport's Fig. 1 (1778, Pl. 1); **b, below** the equivalent figure by Couplet (1729, Pl. IV). Public domain

2. The reverse of this proposition is no less true; that is, if three voussoirs are such that the action of the middle one is annulled by that of the two contiguous ones, there will be equilibrium, and the three voussoirs will remain immobile.[4]

---

[4] Nieuport (1778, p. 44): *1. Soient M, P, T (Fig. 1, gauche), trois voussoirs soutenus par les deux plans inclinés inébranlables LK, DE; que leurs poids relatifs, ou leurs efforts soient exprimés par les perpendiculaires MN, PR, PO, TS, aux lits ou joints FH, HJ; si ces voussoirs restent dans cette situation, ils seront nécessairement en équilibre, c'est-à-dire, que l'effort PO du voussoir P sur celui M, détruira l'effort MN de ce dernier; & que de même l'effort PR sera détruit par TS qui lui est opposé; car si ces deux efforts TS, MN n'étoient pas détruits, le voussoir P seroit poussé dans la direction de leur résultante, & conséquemment ne pourroit persévérer dans sa première situation. 2. L'inverse de cette proposition n'est pas moins vraie; c'est-à-dire, que, si trois voussoirs sont tels que l'action de celui du milieu soit détruite par celle des deux contigus, il y aura équilibre, & les trois voussoirs resteront immobiles.*

The author will then extend this equilibrium property to *n* segments. To do this, he simply needs to consider blocks of three segments in equilibrium as united and forming a single voussoir. He only considers pressures, or rather, the force resulting from a uniform pressure acting perpendicular to the contact surface.

The resulting term, however, suggests that he knows the parallelogram law, but he avoids it by only considering pressures whose results are equal and opposite. Figure 1a shows the effects of these equal and opposite pressures. The parallelogram that we see in the upper left corner will only be mentioned towards the end of the article and plays no role.

It is interesting to compare Nieuport's Figs. 1 (2a) to the much clearer one, at least to our eyes, published by Pierre Couplet in 1729 (Fig.2b), which makes constant use of the parallelogram. We see the force being gradually transmitted from segment to segment to finally rest on the supports, whose upper section has been calculated to be perpendicular to the final resultant. Let us recall that the parallelogram law of forces had been introduced in the domain of stability vaults by La Hire in 1695.[5]

Nieuport then considers the curve or the polygon SRN, which he will qualify as the "curve of concurrence" formed by the points of intersection of the extensions of the joints of the voussoirs. This polygon becomes a curve when the segments are infinitely thin. We will find later this curve linked directly to the inclination of the joints of the voussoirs.

Let us return to the thread of Nieuport's text. He gave a condition for equilibrium and must now show how we can achieve this condition of equilibrium.

The absolute weight of each segment is represented by a vertical segment (Fig. 3a), such as EG or HK or PQ. We avoid the term vector, which he does not use, as shown to us the rest of the description:

> The force of the segment E against that H will be EJ; and the force of the latter against E will be HF. Likewise, the force of P against H will be PL, and the resistance that H will oppose to it will be HM. Now, for there to be equilibrium (§2) all forces must be eliminated; which provides the two simultaneous equations EJ = HF; PL = HM.[6]

For a group of three segments, he therefore obtains two non-vector equilibrium equations. Let us emphasize: these are segment lengths. He will transform these equilibrium conditions among pressures by introducing a horizontal line which he calls AD, two points which do not do not appear on his drawing but which are the ends of the horizontal line FBVC.

He then notices: "Let AD be a horizontal line, passing through any point A. We will always have triangles JGE, FHK, LPQ similar to those NAB, RBC, SCD".[7]

---

[5] See "The Use of a Particular Form of the Parallelogram Law of Forces for the Building of Vaults (1650–1750)", Chap. 4 in this present volume.

[6] Nieuport (1778, p. 44): *L'effort du voussoir E contre celui H sera EJ; & l'effort de ce dernier contre E sera HF. De même que l'effort de P contre H sera PL, & la résistance que lui opposera H sera HM. Or, pour qu'il y ait équilibre, (§2) il faut que tous les efforts s'anéantissent; ce qui fournit les deux équations simultanées EJ = HF; PL = HM.*

[7] Nieuport (1778, p. 44): *Soit AD une ligne horizontale, passant par un point quelconque A. On aura toujours les triangles JGE, FHK, LPQ semblables à ceux NAB, RBC, SCD.*

a

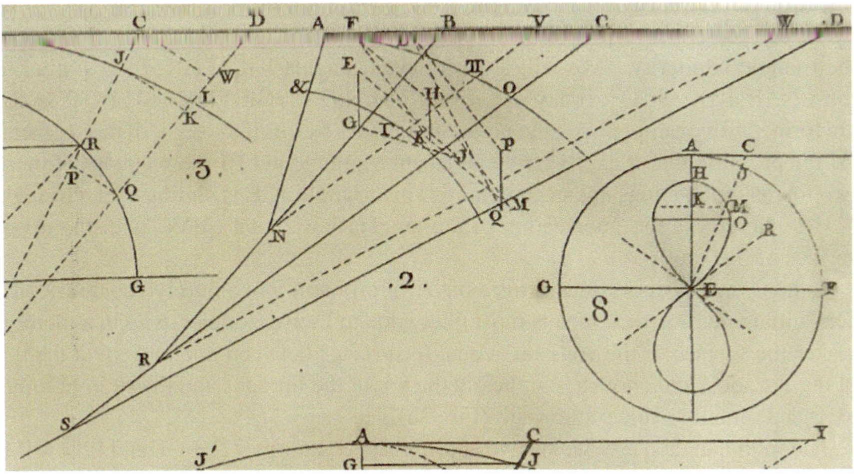

b

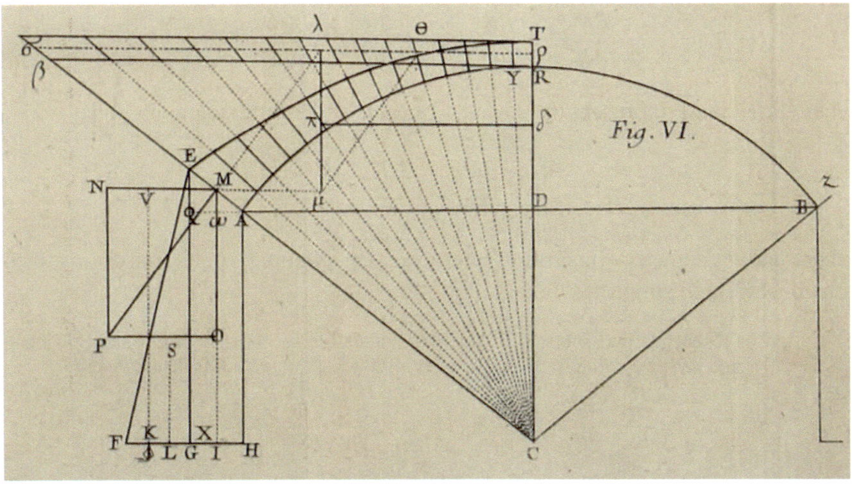

Fig. 3 **a, above** Nieuport's Fig. 2 (1778, Pl. 1); **b, below** the comparable figure by Pierre Couplet (1729, Fig. VI), by which Nieuport was probably inspired

This could pass for simple triangle similarity, but the next sentence observes:

> Thus EG, HK, PQ can be each represented by the surface of the respective segment, of which they express the absolute weight, since we suppose these voussoirs are composed of a perfectly homogeneous matter.[8]

---

[8] Nieuport (1778, pp. 44–45): *Ainsi EG, HK, PQ pouvant être représentées, chacune par la surface du voussoir respectif, dont elles expriment le poids absolu, puisqu'on suppose ces voussoirs composés d'une matière parfaitement homogène.*

He therefore represents the weight of the different segments, a physical quantity, by the surface of the respective voussoirs, a geometric magnitude. Let us then reread the triangle similarity: "We will always have triangles JGE, FHK, LPQ similar to those NAB, RBC, SCD". Today we see that the triangles JGE, FHK and LPQ (Fig. 3a) are formed, for example, with the notations of JGE, by GE the weight of the segment, GJ the pressure that it exerts on the adjacent segment and EJ the resultant of these two forces, but he does not say anything. This triangle JGE is similar to the triangle NAB, where A is the intersection of the tangent horizontal FBVC with the upper surface of the vault.

In this way Nieuport transforms a physical problem into a purely geometric one. The initial and reworked idea is most likely due to Pierre Couplet, who in a memoir makes the surface of the voussoirs equal to surfaces between one tangent at the top of the extrados, and another, parallel, at the top of the intrados and the straight joints extended on either side of the vault (Fig. 3b).

Nieuport can then rewrite his two equilibrium equations, EJ = HF and PL = HM, using similarity relations. He obtains:

$$\frac{BN}{AB}E = \frac{BR}{BC}H; \text{ and } \frac{BR}{BC}H = \frac{CS}{CD}P,$$

where E, H and P represent the surfaces of each of the three segments. That is:

$$\frac{E}{H} = \frac{AN}{BN} : \frac{BC}{BR} \text{ and } \frac{HP}{P} = \frac{BC}{CR} : \frac{CD}{CS},$$

where only geometric quantities intervene. De Nieuport then transforms these relationships and simplifies them:

> Currently, if through the points N, R, we draw, parallel to the extensions CR and DS, the lines NV and RW; we will have BN: BR = BV: BC; CS:CR = CD: CW. Thus:[9]
>
> $$\frac{E}{H} = \frac{AB}{BV} : \frac{BC}{BC} = \frac{AB}{BV}, andlikewise, \frac{H}{P} = \frac{BC}{CW} : \frac{CD}{CD} = \frac{BC}{CW}.$$

He can now deal with the problem of vaults in a purely geometric manner. In order for two neighbouring segments to be in equilibrium, their weights—that is, their surfaces – must be in the ratio of AB to BV. He points out two particular cases:

> There are two cases where this relationship is reduced to that of $\frac{AB}{AC}$; the first when the voussoirs, either finite, or infinitely small, all converge at a point; because then BN = BR, therefore BV = BC. The second, when the segments do not meet at a point, are infinitely

[9] Nieuport (1778, p. 45): *Présentement, si par les points N, R, on mène parallèlement aux prolongemens CR, DS, les lignes NV, RW; on aura BN: BR = BV: BC; CS: CR = CD: CW. Donc*

$$\frac{E}{H} = \frac{AB}{BV} : \frac{BC}{BC} = \frac{AB}{BV}; \& \, de\, même \frac{H}{P} = \frac{BC}{CW} : \frac{CD}{CD} = \frac{BC}{CW}.$$

small; since in that case SRN becomes a curve and RN its element; therefore BN + NR = BN, & BC = BV.[10] (Fig. 4a, b)

3 Vaults, When the Voussoirs Meet at a Point

Here the curve of concurrence is reduced to one point. His first problem will be: "Given the intrados of a vault, find the extrados curve, such that this vault is in equilibrium".[11]

Thanks to his relations, which have become purely geometric, he will solve the practical problems faced by the architect. He begins to translate the practical problem into his geometric language, thanks to his geometric laws of equilibrium:

> Example 1. Let HRG be a quarter of a circle, and the curvature of the intrados of a vault. Let E be the centre, F *the point of concurrence* of the joints of the voussoirs. We must ensure that the surface of each RQLJ segment (§. 3.4.) is proportional to the corresponding segment CD of the horizontal AD; that is to say, we must find the AJL curve such that the space between it and the intrados QR has this property.[12] (Fig. 4a)

The practical problem of constructing the arch intended, for example, to cross a watercourse has become geometric. He circumvented the parallelogram law by replacing the proportionality of the areas of the segments, and therefore of their weight, to lengths taken on the tangent to the vertex: "Let us assume the segments are infinitely small, which is absolutely indifferent in the present case".[13]

In fact, the joints between the segments are supposed to be meet at a point. This remains true whether the segments are finite or infinitely small. Since going into the details of this long proof adds nothing to my discussion, I will go immediately to the second example:

> Let us now suppose that the intrados curve, instead of being a circle, is an ellipse HBP, whose centre is at E; and that this point is also that of the concurrence of the joints.[14] (Fig. 5a).

The solution is of the same nature as that found for the circle.

---

[10] Nieuport (1778, p. 45): *Il y a deux cas où ce rapport se réduit à celui de [eq]; le premier lorsque les voussoirs, soit finis, soit infiniment petits, concourent tous en un point; car alors BN = BR, donc BV = BC. Le seconde, quand les voussoirs ne concourant pas en un point, sont infiniment petits; puisqu'alors SRN devient une courbe; RN son élément; donc BN + NR = BN, & BC = BV.*

[11] Nieuport (1778, p. 46): *étant donnée l'intrados d'une Voûte, trouver la courbe d'extrados, telle que cette Voûte soit en équilibre.*

[12] Nieuport (1778, pp. 46–47): *Exemple 1er. Soit HRG un quart de cercle, & la courbure de l'intrados d'une voûte. Soit E le centre, F le point de concours (Fig. 3, gauche) des joints des voussoirs. Il faut faire en sorte que la surface de chaque voussoir RQLJ soit (§. 3.4.) proportionnelle au segment correspondant CD de l'horizontale AD; c'est-à-dire, il faut trouver la courbe AJL telle que l'espace compris entr'elle & l'intrados QR, ait cette propriété.*

[13] Nieuport (1778, p. 47): Supposons les voussoirs infiniment petits, ce qui est absolument indifférent dans la cas présent.

[14] Nieuport (1778, p. 55): *supposons présentement que la courbe d'intrados, au lieu d'être un cercle, soit une ellipse HBP, dont le centre soit en E; & que ce point soit aussi celui du concours des joints.*

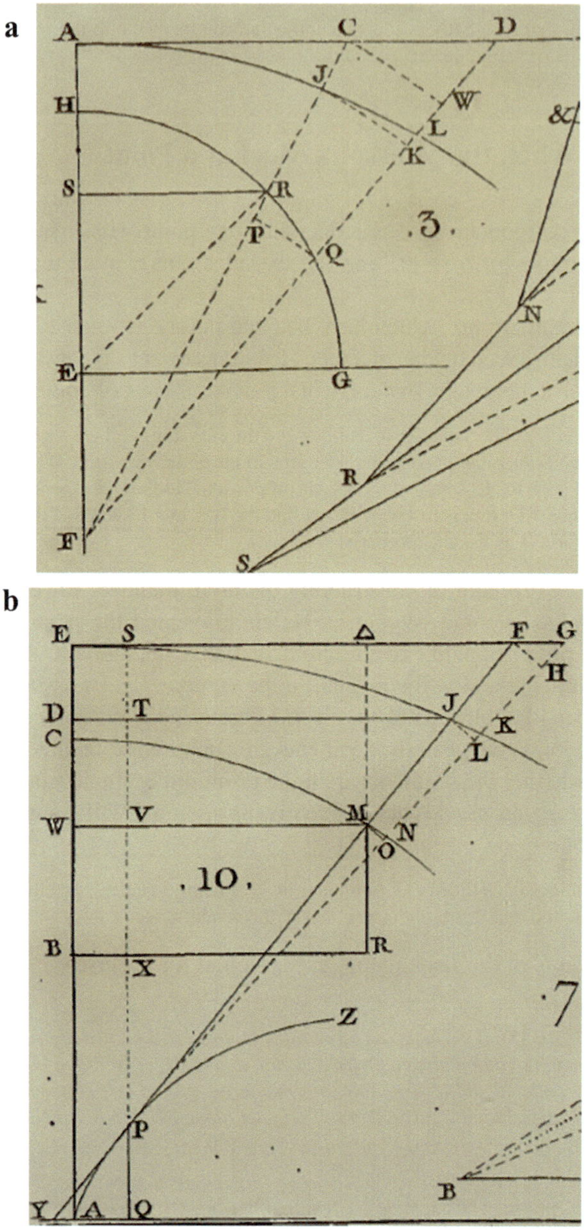

Fig. 4 **a, above** Nieuport's figure for the case where the line of concurrence is reduced to a point; **b, below** the case where the segments being infinitely small, it forms a curve (Nieuport 1778, Figs. 3, 10, Pl. 1). Public domain

a

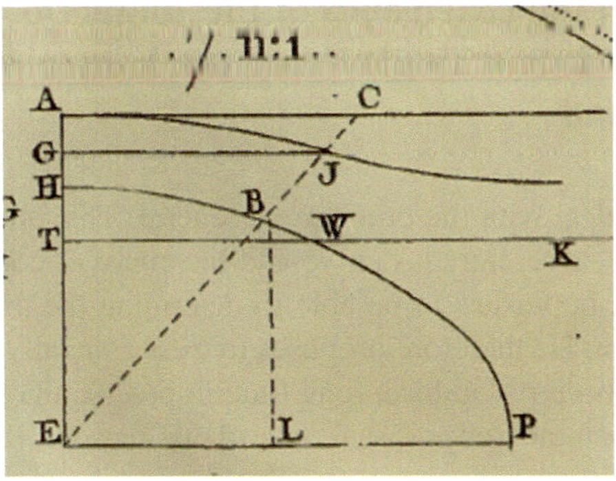

b

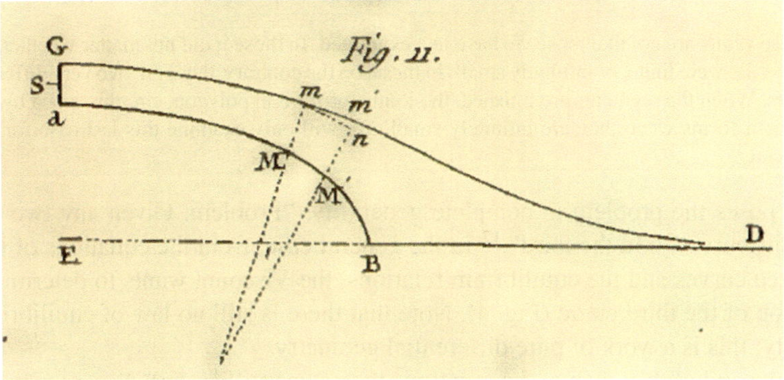

Fig. 5 Comparison of the figures when the intrados curve is an ellipse: **a, above** Nieuport (1778, Fig. 7.11:1, Pl. 1); **b, below** Coulomb (1773, Fig. 11, Pl. XVI). Public domain

The solution to this second problem was found by Charles-Augustin de Coulomb (1736–1806) in his *Essai sur une application des règles de maximis & minimis à quelques problèmes de statique à l'architecture* (Coulomb 1773) (Fig. 5b), but Nieuport does not mention this. However, it is an application of differential calculus, or more precisely the application of a variational principle.

The second problem is: "Given the extrados, determine the intrados".[15] He takes the circle as the extrados in the first example and finds a figure-eight as a solution for the intrados. This is the lemniscate of Jacob Bernoulli, an author that we will meet later.

In the second problem, he takes Nicomedes' conchoid as the extrados curve and finds the equation of the corresponding intrados curve, without naming it.

[15] Nieuport (1778, p. 56): *Etant donné l'extrados, déterminer l'intrados.*

4 Vaults, When the Extensions of the Joints Do not Converge into a Single Point, the Voussoirs Being Infinitely Small

Nieuport intends to deal with the problem in general. This problem includes three curves: the extrados curve, intrados curve and the curve of concurrence. By fixing two of these curves, he wishes to be able to determine the third by means of the equilibrium condition. He therefore proposes to treat generally and analytically the problem which had been posed for a long time in practice: Given the intrados and extrados curves, determine, for practical or aesthetic reasons, the best inclination to give to the joints:

> These vaults are not like those we have just examined. In those it did not matter whether the voussoirs were finite, or infinitely small. In these, on the contrary, these are two very different cases. When the segments are finished, the joints form a real polygon, circumscribed by the curve it forms when they are infinitely small. We will only examine this last case for the present.[16]

He poses the problem in complete generality: "Problem: Given any two of the previous curves, find the third".[17] In the general case, from the equations of two of the three curves and the equilibrium relations, the Viscount wants to determine the equation of the third curve (Fig. 4). Note that there is still no law of equilibrium or stability; this is a work of pure differential geometry.

At the end of this chapter, he mentions the catenary (Fig. 6a):

> We will perhaps be surprised not to see the catenary appearing here, to which alone in all the world is attributed the property of forming a vault which supports itself in equilibrium. But a little reflection will soon put an end to the astonishment. The real property of the catenary is that, if we have a quantity of spheres, or infinitely small circles, perfectly equal to each other, and which we place in such a way that when the lines led by the centres of two contiguous ones form this curve, they will remain in equilibrium.[18]

This description of the situation is perfect and intriguing. What did Nieuport criticize?

---

[16] Nieuport (1778, pp. 58–59): *Il n'en est pas de ces voûtes, comme de celles que nous venons d'examiner. Dans celles là peu importoit que les voussoirs fussent finis, ou infiniment petits. Dans celles-ci au contraire, ce sont deux cas bien différens. Lorsque les voussoirs sont finis, le concours des joints forme un polygone réel, circonscrit à la courbe qu'il forme quand ils sont infiniment petits. Nous examinerons seulement pour le présent ce dernier cas.*

[17] Nieuport (1778, p. 59): *Problême: Etant deux quelconques des courbes précédentes trouver la troisième.*

[18] Nieuport (1778, p. 82): *On sera peut-être surpris de ne pas voir paroître ici la chaînette, à laquelle seule tout le monde attribue la propriété de former une voûte qui se soutienne en équilibre. Mais un peu de réflexion fera bientôt cesser l'étonnement. La vraie propriété de la chaînette est que, si on a une quantité de sphères, ou de cercles infiniment petits, parfaitement égaux entr'eux, & qu'on les place de manière que les lignes menées par les centres de deux contigus, forment cette courbe, ils resteront en équilibre.*

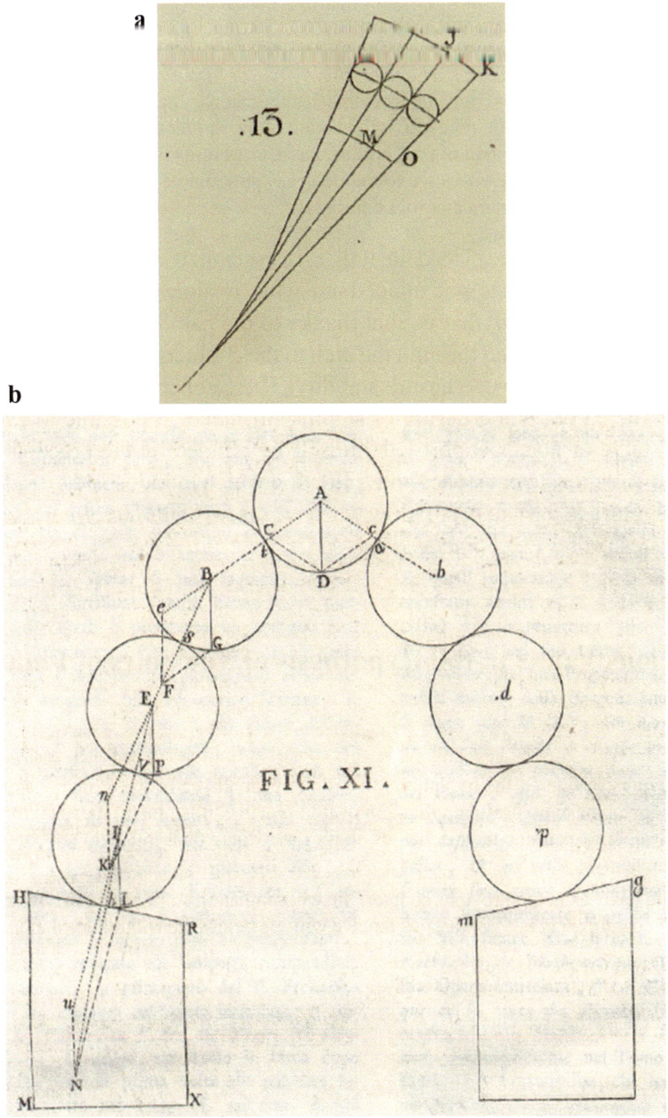

Fig. 6 Figures showing the reasoning which leads to the catenary: **a, above** in Nieuport (1778, Fig. 13, Pl. II); **b, below** in Poleni (1748, Tab. D). Public domain

Suppose now that the specific gravity of these small circles is such that the weight of each of them can be represented by the infinitely small trapezoid MOKJ, and let us form them into a vault. The equilibrium will remain as before; since neither the centres of gravity nor the weight, nor the directions of the reciprocal efforts have been changed.[19]

[19] Nieuport (1778, p. 82): *Supposons présentement que la pesanteur spécifique de ces petits cercle soit telle, que le poids de chacun d'eux puisse être représenté par le trapèze infiniment petit MOKJ, &*

Until then, we still agree with Nieuport:

> *But only the curve that passes the centres of gravity will be a catenary, and not that of the intrados or extrados.* Let us conclude from this that the catenary cannot by itself be of any use in the vaults, unless we want to research the nature of the intrados and extrados curves EJK, CMO, assuming that the curve of concurrence has always remained the AP development of the HGX catenary. But the curve we found deserves preference; since it has the advantage of making the joints perpendicular to the intrados.[20]

The sentence we have placed in italics is crucial. It allows us to see that the Viscount of Nieuport does not understand what is shown perfectly the figure of Giovanni Poleni (Fig. 6b), that is, that thanks to the parallelogram law, we can see how the force is transmitted through the arch to the counterfort, and that it is essential that it follows the catenary to ensure stability. The catenary does not give the curve to be constructed but is a criterion of stability. It makes it possible to say whether a vault is stable or not. Nieuport lets himself be blinded by the practical aspect of things, knowing how to cut stones and following which intrados and extrados curves.

We will return to the catenary below when we discuss his *Mémoire sur la propriété prétendue des voûtes en chaînette.*

5 Previous Vaults in the Hypothesis of Voussoirs of Finite Size

> After having considered the hypothesis of infinitely small segments, & demonstrated that we cannot can make no useful application of it in practice, it is time to move on to those of a finite size, & to examine the mechanics, in the case where their joints do not contribute at one point.
>
> We would have, as in the previous chapter, three combinations of intrados curves, extrados, points of concurrence; but, for reasons that we will see below, we will reduce them together; to know:
>
> 1° when the curve of concurrence is given;
>
> 2° when it is the desired curve.[21]

---

formons en une voûte. L'équilibre subsistera comme avant; puisque, ni les centres de gravité, ni les poids, ni les directions des efforts réciproques n'ont été change.

[20] Nieuport (1778, p. 82): *Mais ce sera seulement la courbe qui passe les centres de gravité, qui sera une chaînette; & non pas celle d'intrados, ou d'extrados. Concluons de ceci, que la chaînette ne sauroit être par elle même d'aucun usage dans les voûtes, à moins qu'on ne voulût rechercher la nature des courbes d'intrados & d'extrados EJK, CMO, en supposant que celle de concours restât toujours la développée AP de la chaînette HGX. Mais la courbe que nous avons trouvée, mériteroit la préférence; puisqu'elle a l'avantage de rendre les joints perpendiculaires à l'intrados.*

[21] Nieuport (1778, p. 88): *Après avoir considéré l'hypothèse des voussoirs infiniment petits, & démontré qu'on ne peut en faire aucune application utile dans la pratique, il est temps de passer à ceux d'une grandeur finie, & d'en examiner la méchanique, dans le cas où leurs joints ne concourent pas en un point.*

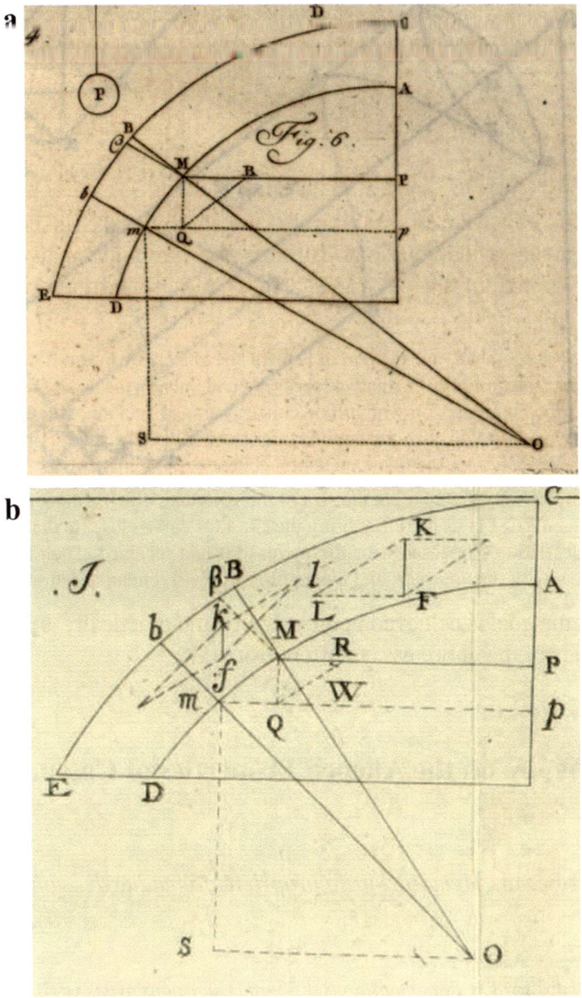

Fig. 7 Figures correcting Bernoulli's error. **a, above** Georg Wolfgang Krafft's figure (1754–55, Tab. I); **b, below** the same figure reviewed by Nieuport (1780, Pl. 1). We notice that Nieuport added two parallelograms whose meaning and usefulness are difficult to understand

As we see in this last formulation, Nieuport must renounce the generality he was looking for.

Nous aurions, comme dans le chapitre précédent, 3 combinaisons des courbes d'intrados, d'extrados, & de concours; mais, pour des raisons que l'on verra plus bas, nous les réduirons à deux; savoir,

1° lorsque la courbe de concours est donnée;
2° lorsqu'elle est la courbe cherchée.

> Let us first remember what we saw (§3) that to put finite voussoirs E, H, P etc. in equilibrium, it was necessary to ensure that
>
> E: H = AB: BV;
>
> H:P=BC:CW;
>
> And so forth.
>
> Let us develop the meaning of this proposition.[22]

He recalls its condition of equilibrium, which is closely linked to the section, not infinitely small, of the voussoirs. But this non-infinitely small character means that the curve of concurrence is a polygon. To get around the problem he uses a subterfuge:

> Suppose that the segment E, and the part of H from NB to ΓΠ, are in equilibrium, as we were taught above to do; similarly, the relative forces depend only on the size of the segments and the inclination of the planes on which they act. Any transformation, which do not disturb these two things will not disturb the equilibrium. Now if we advance ΓΠ parallel to itself, up to RC, the inclination of the plane on which H acts will not undergo any change.
>
> There will only be some [change] in the size of the voussoir, which will receive the increase between ΠΓ and RC. Let us therefore bring the extrados curve closer to that of the intrados, so that this decrease compensates for the increase it had received. Then there will be no change in the size of the voussoir; and the equilibrium will remain as before.[23]

To be sure, the goal is achieved geometrically and practically, but the elegance of the mathematics of the entire essay suffers from this.

6 The *Mémoire* on the Alleged Properties of Catenary Vaults

Nieuport describes his *Mémoire sur la propriété prétendue des voûtes en chaînettes* (1780) as:

---

[22] Nieuport1778, pp. 88–89): *Rappelons nous d'abord ce que nous avons vu (§. 3.) que pour mettre des voussoirs finis E, H, P &c. en équilibre (Fig. 2), il falloit faire en sorte que*

E: H = AB: BV;
H: P = BC: CW;
&c.

Développons le sens de cette proposition.

[23] Nieuport (1778, p. 89): *Supposons que le voussoir E (Fig. 2, gauche), & la partie de celui H, depuis NB jusqu'à ΓΠ, soient en équilibre; ce que nous avons enseigné ci-dessus à exécuter; comme les efforts relatifs ne dépendent que de la grandeur des voussoirs & de l'inclinaison des plans sur lesquels ils agissent, toute transformation, qui ne dérangera rien à ces deux choses, ne troublera pas l'équilibre. Or si on fait avancer ΓΠ parallèlement à elle-même, jusqu'en RC, l'inclinaison du plan sur lequel agit H, ne subira aucun changement.*
Il n'y en aura que dans la grandeur du voussoir qui recevra l'augmentation comprise entre ΠΓ et RC. Rapprochons donc la courbe d'extrados de celle d'intrados, de façon que cette diminution compense l'accroissement qu'il avoit reçu. Alors il n'y aura pas non plus de changement dans la grandeur du voussoir; et l'équilibre subsistera comme avant.

> ...where we demonstrate, against the fairly generally received opinion, that this curve is not that which is suitable for the intrados of a uniformly thick vault, so that all its parts are in equilibrium between them.[24]

He analyses the text by Georg Wolfgang Krafft (1701–1744), *de curvis funicu-lariis* (1754–55). In section 21 he cites Krafft in full and comments on it (Fig. 7a). He considers to be correct the text of Jacob Bernoulli (1654–1705), *Problema de Curvatura fornicis* (Bernoulli 1704), in which Jacob deduces the equation for the thin vault stable under its own weight by correctly calculating the equilibrium of a segment in translation.[25] But he also requires equilibrium in rotation, while a single equilibrium condition is sufficient in the case of a one-dimensional vault. Jacob unfortunately errs by a factor of $\frac{1}{2}$ in the calculation of this second equilibrium and therefore finds two differential equations, $\frac{ds}{s} = \frac{ddx}{dx}$ and $\frac{ds}{2s} = \frac{ddx}{dx}$, where he should have found the same equation for the two equilibrium conditions. Krafft finds the correct equation without the $\frac{1}{2}$ but he doesn't care about the possible rotation. In fact, Jacob Bernoulli was innovative in considering the two equilibria but because of this error, it was necessary to wait for Leonhard Euler (1707–1783) for this way of doing things to enter into practice.

After this long quote and his comment, the Viscount comes to his motivation:

> To better convey the truth of what I am putting forward, I am going, according to the author's [Krafft's] principle, to resolve the problem in my own way. This principle is that a vault will remain in equilibrium if each finite portion CAMB is in equilibrium with the contiguous elementary segment BMmb, which is true.[26] (Fig. 7b)

This will allow him to use his principle of equilibrium and find the correct equation of Jacob Bernoulli and Georg Wolfgang Krafft. Nieuport reports that he read Jacob Bernoulli:

> As the author [Krafft] cites Jacques Bernoulli, and even calls his solution *methodus Bernoul-liana*, it was natural to examine what this illustrious geometer says about it. It is true that he also arrives at the same result starting from the same principle; but he expressly warns that he only considers the vault in the hypothesis of infinitely small thickness (*latitudinis valdè exiguae, vel infinite parvae...* Jac. Bernoulli, Op. Volume 2, p. 1119); a hypothesis which is the only one where the difference between the intrados arch and the extrados arch becomes zero, and the voussoir is proportional to one or the other of these two arches. That is, the catenary will have the property of being the intrados of a vault in equilibrium only in the case where there cannot be a vault; since the first condition of a vault is that its constituent parts

[24] Nieuport (1780, p. 19): *où on démontre, contre l'opinion assez généralement reçue, que cette courbe n'est pas celle qui convient à l'intrados d'une voûte uniformément épaisse, pour que toutes ses parties soient en équilibre entr'elles.*

[25] For details, see "The "De curvatura fornicis" of Jacob Bernoulli: The Introduction of the Infinitely Small into the Calculus of Arches", Chap. 5 in this present volume.

[26] Nieuport (1780, p. 23): *Pour mieux faire sentir la vérité de ce que j'avance, je vais d'après le principe de l'Auteur, résoudre le problème à ma maniere. Ce principe est, qu'une voûte restera en équilibre si chaque portion finie CAMB est en équilibre avec le voussoir élémentaire contigu BMmb; ce qui est vrai.*

be wider externally than internally. Let us therefore conclude again, as I did in my Mémoire already cited, that the catenary has no property in the theory of vaults.[27]

This confirms what I said above: our Viscount is concerned about the practical aspect of the vault and its construction, as were the authors of previous generations, and does not see not the importance of the stability criterion provided by the catenary.

7 Conclusion

Do we dare to reproach Nieuport for not using the parallelogram law? I don't think so, because his reasoning is perfectly correct. He found another way of transforming this problem of statics into one of geometry, which is the role of the parallelogram law.

On the other hand, the last quote shows the limits of his method and, implicitly, the richness of the parallelogram law. This allows us to see how the forces are transmitted within the vault, and this is what gives meaning to the catenary. To do nothing say from the fact that it subjects the forces to vector calculation, which would not be the case for method of the Viscount de Nieuport.

References

D'Alembert, Jean-Baptiste Le Rond. 1821–1822. Réflexions sur les éloges académiques. In *OEuvres completes de d'Alembert*, vol. 5 vols. Paris: A. Belin, Bossange Père Et Fils, Bossange Frères.

Bernoulli, Jacob. 1704. *Problema de Curvatura fornicis, cujus partes se mutuo proprio pondere suffulciunt sine opere caementi*, VP XXIX. In *Opera*, 1119–1123.

Coulomb, Charles-Augustin. 1773. Essai sur une application des règles de *maximis* et *minimis* à quelques problèmes de statique, relatifs à l'architecture, *Mémoires de mathématiques et de physique présentés à l'Académie royale des sciences par divers savants, et lus sans ses assemblées*, vol. 7, 1773 (1776), 343–382.

Couplet, Pierre. 1729. De la poussée des voûtes. In *Histoire de l'Académie royale des sciences, avec les Mémoires de mathématiques et de physique*, 1729 (1731), 79–117.

[27] Nieuport (1780, p. 25): *Comme l'auteur [Krafft] cite Jacques Bernoulli, & appelle même sa solution methodus Bernoulliana, il étoit naturel d'examiner ce que cet illustre Géometre en dit. Il est vrai qu'il arrive aussi au même résultat en partant du même principe; mais il avertit expressément qu'il ne considère la voûte que dans l'hypothèse de l'épaisseur infiniment petite, (*latitudinis valdè exiguae, vel infinite parvae … Jac. Bernoulli Op. Tome 2, p. 1119*); hypothèse qui est la seule où, la différence de l'arc d'intrados à celui d'extrados devenant nulle, le voussoir est proportionnel à l'un ou à l'autre de ces deux arcs. C'est-à-dire que* la chaînette n'aura la propriété d'être l'intrados d'une voûte en équilibre, que dans le cas où il ne sauroit y avoir de voûte*; puisque la première condition d'une voûte est, que ses parties constituantes soient plus larges extérieurement qu'intérieurement. Concluons donc de nouveau, comme j'ai fait dans mon Mémoire déjà cité, que la chaînette n'a aucune propriété dans la théorie des voûtes.*

Krafft, Georg Wolfgang. 1754–55. De curvis funiculariis ; vel illis quae corporibus flexilibus inducuntur, cum a potentiis quibusvis sollicitantur. In *Novi commentarii Acad. scientiarum Petropolitanae*, vol. V, 1754–55 (1760), 145–163.

Nieuport, Viscount of (Charles-François the Prud'homme d'Hailly). 1778. Essai analytique sur la méchanique des voûtes, présenté le mai 1778. In *Académie Impériale et Royale des Sciences et Belles-Lettres de Bruxelles*, t. 2, 1780, 42–136.

Nieuport, Viscount of (Charles-François Le Prud'homme D'Hailly). 1780. Mémoire sur la propriété prétendue des voûtes en chaînettes, présenté le 6 novembre 1780. In *Académie Impériale et Royale des Sciences et Belles-Lettres de Bruxelles*, t. 4, 1783, 19–26.

Poleni, Giovanni. 1748. *Memorie istoriche della gran cupola del tempio vaticano, e de'danni di essa, e de'ristoramenti loro, divise in libri cinque*. Padova.

Quetelet, Adolphe. 1866. *Sciences mathématiques et physiques chez les Belges au commencement du XIXe siècle*. Brussels, 99–109.

Radelet-de Grave, Patricia. 2015. The Panthéon's stability already questioned by Pierre Patte in 1770. In *Masonry Structures: Between Mechanics and Architecture*, eds. Danila Aita, Orietta Pedemonte, Kim Williams, 127–185. Springer. ISBN: 978-3-319-13002-6, 978-3-319-13003-3 (ebook).

Chapter 14
On Mascheroni's "New Researches on the Equilibrium of Vaults"

1 Preamble

My aim in this chapter is to propose an explanation for Mascheroni's reasoning that underlies his *Nuove ricerche sull'equilibrio delle volte* (*New researches on the equilibrium of vaults*) (Mascheroni 1785). In order to do this, I would like to recall certain elements of the scientific knowledge that were at his disposal in 1785. It is not my intention to scrupulously retrace the history of those various elements but simply to underline them, as they are too numerous and to develop their story would be too long and perhaps not even that clear. Moreover, the reader will recognize elements whose history was carefully traced by either Clifford Truesdell in his introduction to vol. II.11 of Euler's *Opera* (Truesdell 1960) or by Edoardo Benvenuto in *An Introduction to the History of Structural Mechanics* (Benvenuto 1991).

The elements examined come either from texts dedicated to the stability of vaults or from texts of mechanics. The two types of literature generally do not interfere with each other, so this chapter will mainly show two similar developments, even if influences may occur.

2 Chronological Landmarks

The first author who treated the problem of vaults scientifically and thus made the first attempt at a mathematical architecture is Philippe de la Hire (1640–1780), in his *Traité de mécanique*(1695). This is significant from three points of view.

(1) From the point of view of the history of the fundamental equations of mechanics. This was eight years after Varignon's *Projet d'une nouvelle mechanique* ("Project for a new mechanics") (Varignon 1687), in which he retraced the study of simple machines or statics, not back to the principle of the lever as Hero of

P. Radelet-de Grave, *Between Mechanics and Architecture*, Mathematics and the Built Environment 8, https://doi.org/10.1007/978-3-031-73530-1_14

Alexandria had already done for some machines but to the law of the parallelogram of forces. This was fifty-five years before Euler's article "Découverte d'un nouveau principe de mécanique" ("Discovery of a new property of mechanics") (Euler 1750), in which he began to study the general motion of a body, a work that would be pursued by Poinsot (1803, 1804, 1806, 1934). Confronted in this manner with dynamics and with the different types of equilibrium, Euler intuited the necessity of taking two equilibrium relations into account, namely:

(a) Equilibrium to translation linked to the parallelogram law and to Varignon's presentation of statics;
(b) Equilibrium to rotation linked to the law of equilibrium of the lever.

At the time the work of La Hire on the law of the lever and of the moment of forces had just received a new impulse thanks to the work of Jacob Bernoulli (1655–1705).

(2) From the point of view of the history of elasticity or of elastic deformations or of the constitutive equations. This was seventeen years after the discovery of Hooke's law, first published in 1676 in the form of an anagram and solved in his *Lectures de potentia restitutiva or of spring. Explaining the Power of springing bodies* (1678, p. 31):

> The true Theory of Elasticity or Springiness, and a particular Explication thereof in several Subjects in which it is to be found: and the way of computing the velocities of Bodies moved by them. ceiiinosssttuu.

In a publication two year later, he wrote:

> About two years since I printed this theory in an Anagram at the end of my Book of the Descriptions of Helioscopes, viz. ceiiinosssttuu, id est, Ut tensio sic vis... (Hooke 1678, p. 1).

Hooke used a second anagram in his text of 1676, this time a law for building arches:

> The true Mathematical and Mechanichal form of all manner of Arches for Building, with the true butment necessary to each of them. A Problem which no Architechtonick Writer hath ever yet attempted, much less performed,

abcccddeeeeefggiiiiiiiiillmmmmnnnnnooprrssstttttuuuuuuuux.[1]

(3) From the point of view of history of architecture. This was more than 250 years after the construction of the dome of Santa Maria del Fiore in Florence by Filippo Brunelleschi (1419–1436), 150 years after the construction of the dome of St. Peter's in Rome by Michelangelo (1475–1564), and shortly after the time of the construction of St. Paul's Cathedral in London by Christopher Wren (1632–1723).

---

[1] The solution of the anagram is: *Ut pendet continuum flexile, sic stabit contiguum rigidum inversum* (As hangs the flexible line, so but inverted will stand the rigid arch).

3 The Catenary, the Elastica and Other Problems Posed by Jacob and Johann Bernoulli

This story goes a bit farther back in history than the text of La Hire. For our purpose we shall keep the following elements in mind.

Galileo had erroneously affirmed, in one of his manuscripts concerning the *Discorsi*, that the curve adopted by a rope fixed in two points under its own weight was a parabola:

The curvature of the path of the shot fired horizontally appears to.

> Just as the parabola of the projectile is described by 2 motions, horizontal and perpendicular, so the chain results from 2 forces: horizontal, by whoever pulls it from the end, and perpendicularly down, by its own weight.[2]

However, as Gottfried Wilhelm Leibniz (1646–1716) and Johann Bernoulli (1667–1748) pointed out,[3] Joachim Jungius (1587–1657) proved experimentally that Galileo was wrong.

That point is the argument of several chapters of the book by Ignace-Gaston Pardies (1636–1673), *Statique ou science des forces mouvantes* (Statics, or the science of moving forces) (Pardies 1673). Pardies presents the tension and shows that in the catenary the center of gravity of a segment of the curve always lies on a vertical passing through the intersection point of the tangents at the fixed points of the catenary. He corrects Galileo, showing that the catenary is not a parabola. This curve is the one adopted by a chain loaded with equal masses at equal horizontal distances, as for the hanging bridge (this would be clearly underlined by Euler in his *Solutio problematis de invenienda curva* of 1728). Pardies then goes on to consider curves formed by extensible ropes undergoing central forces.

The problem of the catenary would be proposed by Jacob Bernoulli to Leibniz as an application of his new calculus in 1690. Jacob posed the problem in the following way: "Find the curve matching the form of a string hanging loosely to two fixed points. I suppose also that the string is a line easily flexible in all its parts".[4] Leibniz proposed the problem, in the *Acta Eruditorum,* to the whole scientific community. Three answers were given by Christiaan Huygens (1629–1695), Johann Bernoulli, Jacob's younger brother, and Leibniz himself.

The answers of both Leibniz and Johann make use of the fact that tensions are directed along the tangents to the curve, a property that corresponds to the perfect flexibility.

---

[2] Galileo (1638b ca., VIII, p. 370): *Si come la parabola del proietto è descritta da 2 moti, orizontale e perpendicolare, cosi la catenella risulta da 2 force: orizontale, da chi la tira nell'estremità, e perpendicolare* deorsum, *da proprio peso.*

[3] Leibniz, in "De linea in quam…" (1691), and Joh. Bernoulli in Lecture XXXVI, "De Curvis Funiculariis vel Catenariis" (1691–92).

[4] Jac. Bernoulli (1690, p. 219): *Invenire, quam curvam referat funis laxus et inter duo puncta fixa libere suspensus. Sumo autem, funem esse lineam in omnibus suis partibus facillime flexilem.*

In an article entitled "Solutio Problematis Funicularii" ("Solution to the problem of the funicular" (Bernoulli 1691a, b, c), Johann answers with a geometrical construction that does not appear to use the new Leibnizian calculus, but we know from his lectures given to the Marquis de l'Hôpital that he already had such a solution at that time. He next studied the inhomogeneous or non-uniformly rope. Finally, he mentions the inverse problem: given the curve one wishes to obtain, find the repartition of the loads, which corresponds to an integration.

Jacob makes a higher bid in an article that gives "another example of the differential calculus for the measure of the logarithmic spiral, of the marine's loxodrome, and for the measure of the spherical triangles: with an annex to the problem of the funicular, etc." (Bernoulli 1691a). He studies the rope of non-uniform mass and that of uniform mass but extensible. Finally, he poses the problem of the elastica and of the velaria (sail).

Here, proposed to the scientific community, are the complete set of problems we are interested in: the *catenaria* (catenary), the *velaria* (sail), and the *elastica.* The *lintearia* (a linen cloth or tarpaulin) had already been mentioned but never was publicly posed as a problem. Johann published a geometrical "solution to the problem of the curvature adopted by a sail inflated by the wind" (Bernoulli 1692a, b, c). In the meantime, Jacob sought, in Med. CLXIV of his *Meditationes* (his unpublished scientific journal), "an estimation of the liquid pressure in a vessel of any shape" (Bernoulli 1689a). Then, in Med. CLXV (Bernoulli 1689b) he gave a solution for Johann's velaria using differential calculus. The solution in that Meditation relied on the radius of curvature, but he had just published some geometrical results on the curvature of sails (Bernoulli 1692a, b, c) in which the radius of curvature does not appear. Jacob only published his geometrical results as a list of affirmations, exactly as Johann did in his Op. IV (Bernoulli 1691a, b, c) or in Op. VII (Bernoulli 1692a, b, c) for the catenary. Nevertheless, in the same text, for his brother, Jacob also gives the characterization of the velaria in terms of the Leibnizian calculus in the following way: "If one considers little equal elements of curves, the cubes of the first differences of the ordinates are proportional to the second differences of the x coordinates".[5] That is: $(dx)^3 = ad^2y$.

At the end of this text, one finds the following remark: "as a sail swollen by the wind is to the funicular, so is the curve of a tarpaulin full of liquid to the elastica".[6]

It seems that Jacob makes a distinction here based on the direction of the force or of what he calls 'pressure'. This distinction is very promising and further on we shall see the fruits of it. Unfortunately, it is a bit tainted, in Jacob's work, by the fact that he was not sure enough of his idealization of the sail, as shown in his Med. CLXXXVIII, *De linea directionis, juxta quam velum impellitur, de impulsus vi & de veli robore* (Bernoulli 1692b).The problem caused by the idealization of the velaria, which will be recognized by Jacob, is the following: either one considers that the

[5] Jac. Bernoulli (1692a, b, c, p. 202): *Sumtis aequalibus curva portiunculis, Cubi ex primis differentiis ordinatarum sunt proportianales secundis differentiis abscissarum.*

[6] Jac. Bernoulli (1692a, b, c, p. 207): *Ut enim linteum vento tumidum* Funiculariae, *sic idem ab incombentis liquoris pondere expansum* flexi Elateris *curvaturam induit.*

wind pushes the sail but does not stay in it, that is, the wind passes through the velaria or flows laterally out of it, or one considers the velaria as an inflated balloon.

Let us summarise this[7]:

- The catenary is a perfectly flexible rope under the action of gravity.
- The velaria is a sail pushed by the wind.
- The lintearia is the curve formed by a tarpaulin filled with a liquid.
- The elastica is the curve formed by an embedded beam under the action of a weight placed I think that Jacob already looks at things in this way, grouping all those problems, even if he does not explain it clearly. I see the confirmation of this fact in a series of *Meditationes*, many of which are considered by him to be false. In this series, Jacob studies the particular geometrical properties due to the fact that the forces or pressure exerted by water are perpendicular to the lintearia. In Med CCXXVIII (Jac. Bernoulli 1694b), which deals with "how to determine geometrically an infinity of points, however close together, of the elastic curve", Jacob writes in the margin that it is false. Nevertheless, he starts from an interesting point of view: the lintearia is identical to the elastica but has an additional property: its mean impulse passes through the point of concourse of the tangents at its extremities. But all impulses are perpendicular to the curve, and this is not the case for the elastica. Jacob therefore looks for a supplementary hypothesis. This remark is similar to the one Pardies makes about the catenary: tensions are tangent to the curve if it is perfectly flexible.

I shall return to such remarks, but let us retain for the moment that Jacob classifies the different problems with respect to the direction of the applied forces, as shown by the sentence quoted above from the end of his 1692 work on the curvature of sails, together with the fact that he tries to insert the elastica into that classification.

4 Jacob Bernoulli's Treatment of the Elastica

The problem of the elastica occupied Jacob Bernoulli from 1695 until the end of his life. As late as 12 March 1705 he was still writing a letter published in the *Histoire de l'Académie de Paris* on that argument (he died on 16 August of the same year). His first text on that argument had been published eleven year earlier, with the title "Curvatura Laminae Elasticae. Ejus Identitas cum Curvatura Lintei a pondere inclusi fluidi expansi" (Jac. Bernoulli 1694a) (Fig. 1).

There he published his *theorema aureum* for the first time. He writes:

> This problem of the elastica is more difficult than the one of the funicular and not without reason. I do not mention them all in order to retain only the fact that, in the case of the catenary, we have two keys at hand that lead to two different equations, the one expressing the nature of the curve by its relation to its coordinates and the other one by its relation to

---

[7] For more about these curves, see "The Bernoullis and the Elastica from Leibniz's Nova Methodus to Euler's Methodus Inveniendi", Chap. 7 in this present volume.

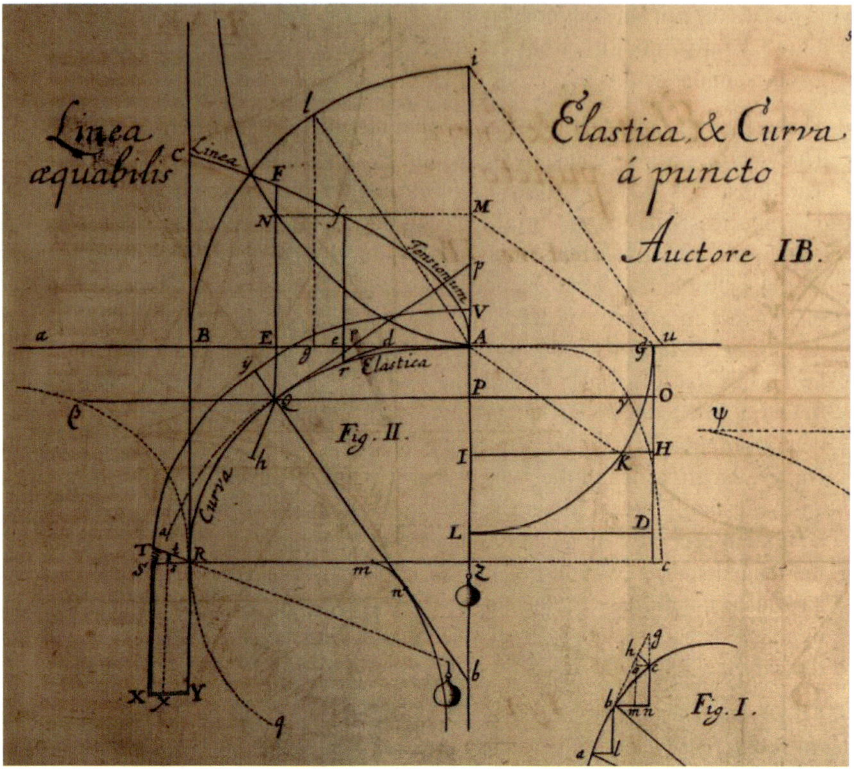

Fig. 1 The elastica of Jacob Bernoulli (Bernoulli 1694a)

its evolving string. But to determine the nature of the elastic curve we need one more key. One could think to overcome that difficulty as in the preceding problems but when going further, the victory is elusive; when that other key, which gives the relation between the evolving string or radius of the osculating circle (not in a general manner as we did in the AE of January 1691), but in simple terms purely differentials. We knew all that when we were making speculations about the string, a good deal had been given by my brother; because it was of an immense utility in the resolution of the velaria, … at last, I cannot keep from the public my aureum Theorema ….[8]

[8] Jac. Bernoulli (1694a, b, p. 263): *problema hocce plus difficultatis habere Funiculario; nec sine ratione. Ut enim alia taceam, notandum, duas praesto esse claves in* Catenarii *investigatione, quae ad duas differentes aequationes viam sternunt, quarum una naturam Curvae per relationem ipsiusmet ad coordinatas ejus, altera per relationem fili evolventis ad easdem exprimit; cum ad Curvae Elasticae naturam indagandam posteriore tantum clave aditus pateat: hinc enim manifesto sequitur, fieri bene posse, ut quispiam prioris Problematis difficultates fuperet, nec tamen protinus alterius quoque victor evadat; destitutus nimirum clavium altera, quae dictam relationem fili evolventis seu radii circuli osculatoris (non quomodocunque, ut factum in* Act. Lips. Mens. Jan. 1691, p. 22. *sed) in terminis simplicissimis et pure differentialibus exhibeat. Haec nobis jam eo tempore, quo speculationi funis vacabamus, innotuerat, eamque mox nonnullis inter peregrinandum communicavit Frater, interea dum immensa inventi utilitas in resolvenda Velaria, … tandem, ut non possim publico aureum Theorema diutius invidere.*

The term *aureum theorema* selected by Jacob in order to qualify his theorem on the radius of curvature shows how proud he was, but shows even more the importance he attached to it, as we shall see, with good reason. In fact, he had found it and used it before baptizing it that way. Of course, this theorem is only gilded under the pen of his brother Johann.

The formula giving the radius of curvature in a differential form $\frac{-dx\,ds}{ddy}$ or $\frac{dy\,ds}{ddx}$ appeared at the same time as differential calculus in Med. CLXIV (Bernoulli 1689a). Jacob had found the formula of the radius of curvature during his researches on spirals. Nevertheless, he associated that theorem with the *elastica* since it was in his first text on that curve that he published that formula for the first time. This theorem is particularly important for various reasons. It is one of the first theorems, if not the very first, expressed in a differential form. This point would not be that important if it did not allow him to translate into differential calculus all theorems established in the ancient geometry, also qualified as 'vulgar geometry', as he will say later. In fact, most of the theorems implied the evolute of the curve, which means that they implied the radius of curvature. Thus, giving the expression of that radius of curvature, reproduced above, that depends only on the curve itself, Jacob is able to translate in that new language all theorems implying the evolute.

The two keys he is speaking of are, in my opinion, on the one hand the equilibrium of the lever and, on the other, the law of tension. In this first text we see principally the law of tension. In the first part, *Generaliter*, he gives himself an unspecified law of tension. In the second part, *Specialius*, he gives himself a 'paraboloid' curve (this has to be taken in the sense of a conic) as the curve of tensions. In the third part, *Specialissime*, he takes the linear law or Hooke's law, which had already been used by Leibniz and by himself until "he arrived at the general construction of the problem", that is, until he discovered his *theorema aureum.*

Sadly enough, neither this text nor those by Johann mentioned earlier, Op. IV (Bernoulli 1691a, b, c) and Op. VII (Bernoulli 1692a, b, c), gives the development of the calculus. Both Jacob and Johann simply content themselves with stating the geometrical properties.

Let us notice that isoperimeters already appear in that text. Within the geometrical properties, one finds:

> e) Among all isoperimeters that lean on Af the elastica is the one whose center of gravity is the farthest from Af while the catenary is the one whose center of gravity is the farthest from the curve itself.

Following remarks made by Huygens, Jacob got back to work and published his second famous work on the elastica in December 1695, with the long title "Explicationes, Annotationes, et Additiones ad ea, quae in Actis sup. anni de Curva Elastica, Isochrona Paracentrica, & Velaria, hinc inde memorata, & partim controversa leguntur; ubi de Linea mediarum directionum, aliisque novis" ("Explanations, Annotations, and Additions to what was said in the *Acta* of last year about the elastic curve, the paracentric isochrone, and the velaria, here recalled, and explaining part of the quarrel; and giving something new about the middle line of direction, and other new") (Bernoulli 1695), in which he explains how to get the equation of the problem based on the principle of the lever. An angled lever supported in Q, one of its arms QP supports the applied weight and the other, namely the thickness of the beam, supports the elastic restoring forces. Further, he tacitly assumes that a plane section of the beam remains plane and thus that the elongation is inversely proportional to the radius of curvature.

5 Mathematics Appears in Architecture

The first mathematical text related to architecture, due to La Hire, dates from the same year as the second text of Jacob Bernoulli concerning the elastica, 1695. La Hire idealizes the problem in the following manner:

> One determines here the load that must be given to each stone, or voussoir as workers say, with which one forms arches or vaults, in order to get them to remain in equilibrium, even if their beds, or the surfaces on which they touch each other, be infinitely polished so that they could slide one against the other without any impediment.[9]

His idealization will permit him to draw back that problem to another one he treated before namely that of the catenary:

> A rope being presumed without weight, one determines the weights that must be applied in all its parts, so that if pulled by all those weights together, it assumes whatever shape is desired.[10]

In order to demonstrate those two propositions, La Hire uses a particular form of the parallelogram law of forces.[11] He traces the proportional ratio existing between the three forces in equilibrium back to a relation of proportion between the sides of a triangle whose sides are perpendicular to the forces in equilibrium:

> Proposition XXI: Find three forces AXB that, pulling a point K along three given directions CK, DK, EK, are in equilibrium among themselves.[12](Fig. 2)

> Proposition XXII: The same things as in the preceding being posed, one may find in another manner the ratio of the forces.[13](Fig. 3)

[9] La Hire (1695, p. 465): *On détermine icy la charge qu'on doit donner à chaque pierre ou voussoir, comme parlent les ouvriers, dont on forme des arcs ou des voutes, afin qu'elles puissent demeurer en équilibre, quand même leurs lits, ou superficies par lesquelles elles se touchent seroient infiniment polies et qu'elles pourroient glisser l'une contre l'autre sans aucun empéchement.*

[10] La Hire (1695, p. 455): *Une corde étant supposée sans pesanteur, on détermine les poids qu'il faut appliquer dans toutes ses parties, afin qu'étant tirées par ces poids tous ensemble, elle prenne une figure qu'on voudra.*

[11] See also "The Use of a Particular Form of the Parallelogram Law of Forces for the Building of Vaults (1650–1750)", Chap. 4 in this present volume.

[12] La Hire (1695, p. 70): *Il faut trouver trois puissances AXB, qui tirant un point K par trois directions données CK, DK, EK, soient en équilibre entr'elles.*

[13] La Hire (1695, p. 72): *Les mêmes choses que dans la précédente étant posées, on peut trouver d'une autre manière le rapport des puissances.*

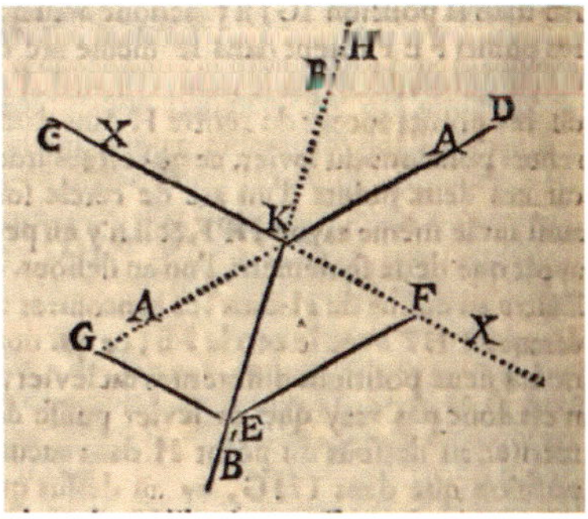

Fig. 2 La Hire's figure of the parallelogram law of forces (La Hire 1695, p. 70). Public domain

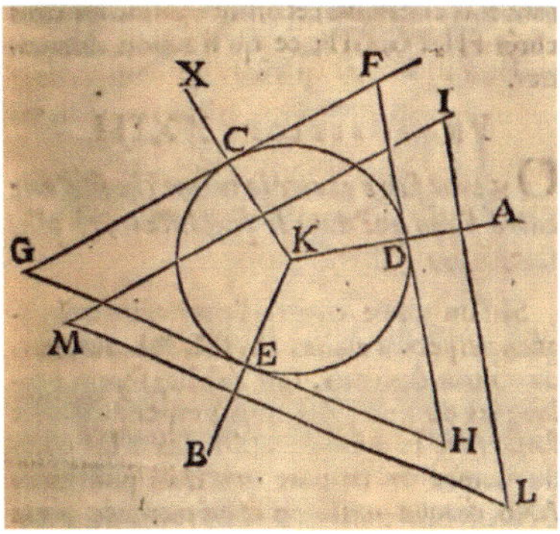

Fig. 3 La Hire's triangle law of forces (1695, p. 73). Public domain

> One supposes here that the directions of the weights are parallel to each other, thus if force X that pulls the point B along AB remains in equilibrium with weight M suspended in B, and with a force Q suspended in B, and with a force Q that pulls along the tangent BD, by the twenty-third proposition the triangle CA, whose sides are perpendicular to the directions of the forces and of the weight, will give the relation of those three forces taking the weight as a force. Thus, force X will be represented by CA, force Q by CF, and force of weight M by AF.[14]

This will allow La Hire to connect the weights attached to the rope or the weights of the voussoirs to the segments intercepted on a line tangent to the top.

The similarity between the two problems evidenced by La Hire will be more clearly stated by David Gregory (1659–1708) two years later in his article on the catenary published in the *Philosophical Transactions*:

---

[14] La Hire (1695, pp. 456–457): *On suppose icy que les directions des poids sont parallèles entr'elles, c'est-pour-quoy si la puissance X qui tire le point B selon AB demeure en équilibre avec le poids M suspendu en B, & avec une puissance Q qui tire selon la touchante BD, par la vingt-troisième proposition le triangle CAF dont les côtés sont perpendiculaires aux directions des puissances & du poids, donnera le rapport de ces trois puissances en prenant le poids pour une puissance. Ainsi la puissance X sera représentée par CA, la puissance Q par CF, & la puissance ou le poids M par AF.*

> The catenary placed in a vertical plane but inverted, makes it possible to find the figure of the thinnest arch or vault: that means spheres disposed without rigidity and without friction along the inverted catenary constitute an arc of which no part is propelled either outside or inside by the others; but whose smallest parts sustain each other by virtue of its shape.[15]

He further explains that the velaria will also assume the same shape.

6 Isoperimeters

The same year, 1697, Jacob Bernoulli went a step further with the generalization of problems in an article providing a "solution to my brother's problem" (Bernoulli 1697c), at the end of which he poses the isoperimetric problem. In fact, as we will see with Euler, all questions asked by Jacob up to that time may be put together in this way. Jacob prepares his question in his Med. CCXXXIX, "Among all isoperimetric figures or figures having the same surface find those who have some maximum of minimum. New method" (Bernoulli 1697a). In this text is found the first differential treatment of the isoperimetric problem with some examples:

(1) Among isoperimeters find the one whose center of gravity is the lowest: he finds the catenary.
(2) Among isoperimeters find the one that encloses the greatest surface: the answer is the circle.
(3) Among isoperimeters find the one whose center of gravity of the enclosed surface is the lowest: he observes that it is not the elastica, as he thought, and finds the catenary again.
(4) Among the curves enclosing a same surface, find the one having its the lowest center of gravity: he arrives at an impossibility.
(5) Among isoperimeters find those satisfying conditions of third or fourth proportionality: he again finds the catenary.

As we said in the introduction, the argument of the elastica itself immediately confronted the researcher with constitutive equations and we have effectively seen that both Jacob Bernoulli and Leibniz realized that various materials react differently and thus that the law of tension that supposed to be linear must generally be adapted to the various materials, as Jacob wanted to do it in his Op. LVIII by introducing a law of tension. My thesis is that he continues generalizing the problem via the isoperimeters in hopes that he would determine the law of tension given by nature, measuring the curvature. In order to justify that thesis allows me to take some liberty with the chronology.

[15] Gregory (1697, p. 640): *Catena in plano verticali, sed situ inverso, figuram servat nec decidit, adeoque arcum seu fornicem facit tenuissimum: Hoc est sperae minimae rigidae et lubricae in inversa Curva Catenaria dispositae, arcum constituunt cujus nulla pars ab aliis extrorsum vel introrsum propellitur; sed manentibus infimis punctis immotis, virtute suae figurae sustinetur.*

7 The New Generation with Euler and Daniel Bernoulli

In 1744, Euler published an important volume entitled *Methodus inveniendi lineas curvas maximi minimive proprietate gaudentes, sive solutio problematis isoperimetrici lattissimo* sensu *accepti* ("On the method of finding the curves that have a property of maximum or minimum or solution of the problem of isoperimeters in its broadest sense") (Euler 1744), at the end of which he gives an *additamentum* on the elastica. Euler went far into the solution of the problem of isoperimeters and of its particularization to various problems. Nevertheless, Euler lacked one element in order to close that study. As he recounted, it would be Daniel Bernoulli who furnished that key element.

In a letter dated 5 May 1739 Euler wrote to Daniel Bernoulli that he had not found the quantity to extremize (that is, to maximize or minimize) in the case of the elastica and that Jacob had not found it either, and that this is the reason why he tried various criteria regarding the center of gravity of the beam or on the surface included. Euler knows that it is the "quantity of the potential force" that has to be extremized, but he does not know how to express it. In a letter of 20 October 1742 Daniel told him that the potential live force of the elastica is **Error!**.

This correspondence would lead Euler to write the *Methodus inveniendi* (Euler 1744). Reading the works of Jacob Bernoulli to which Euler refers we realize that he posed the problem of isoperimeters in hopes of finding a solution to the general problem of the elastica that would include those of the catenary and of the velaria as well as the lintearia. At the end he includes an *additamentum* I, entitled "On elastic curves". First, he gives credit to Daniel Bernoulli for finding the key element that allowed him to conclude his study. Then he poses the problem of the elastica in the following manner: among all curves of same length going from A to B, where the tangents to the curve are given, find the curve that renders $\int \frac{ds}{R^2}$ minimal.[16] As we shall see, this text is quoted by Mascheroni.[17]

Euler studies the curves formed by an elastic lamina at rest, static. This part is subdivided in two parts. He first gives the form of the curves obtained by applying two forces in two points of an elastic lamina. Next, he studies the form of such a lamina under the action of its own weight. In the first part, he finds all forms the elastica can take and in the second he again finds the catenary.

This work closes and organizes, for the first time, the results of a research opened by Jacob and Johann Bernoulli.

[16] Euler (1744, p. 247): *ut, inter omnes curvas ojusdem longitudinis, qua non solum per puncta A & B transeant, sed etiam in his punctis a rectis positione datis tangantur, definiatur ea in qua sit valor hujus expressionis* $\int \frac{ds}{RR}$ *minimus.*

[17] See below, in the discussion of Mascheroni's Problem XXXIII.

8 Jacob Bernoulli and the Problem of the Arch

At the end of his life, on 5 December 1704, Jacob Bernoulli attacked the problem of the arch. It is the argument of the last but one of his *Meditationes*, Med. CCXLV (Bernoulli 1697d) and it would be published in Jacob's *Opera* by Gabriel Cramer as a *Varia Posthuma*: a string ACDEFGB suspended at its ends A and B with an infinity of forces C, D, E, F, G acting on it along whatever direction HE, HD, IE, KF, LG. He asks what curvature the string will take, its mean direction LP, and the force that acts along LP.This text deserves, in more than one way, our attention. Not least of its merits is that it is the first to be founded on differential calculus in the domain of architecture.

Jacob Bernoulli first specifies, at the very beginning but without justification, that the arch must be thin and that its weight must be proportional to its length. He will consider that one of the voussoirs may carry out two different motions: either it may slide if the joints are perfectly smooth, or it may rotate, inside the arch, about its lower point.

He then writes two equations, one describing the equilibrium at translation and the other the equilibrium at rotation. In the first case, he easily obtains the catenary as a solution. In the second case, unfortunately, he makes a miscalculation. Without this error, he would have found the catenary in this second case as well. With the introduction of differential calculus, we notice a change in the way of thinking: from that of La Hire, who showed the mechanical similitude of the problems, to Jacob Bernoulli, who showed that the equation describing both problems is the same.

The comparison of Bernoulli's text with that of Couplet, "De la poussée des voûtes" (Couplet 1729) published much later, but nevertheless instructive, allows us to measure how far ahead Jacob was of his time, at least regarding differential calculus. In fact, with Couplet, we come back to purely mechanical considerations closer to the style of La Hire. Couplet founds his work exclusively on the theorem of the triangle whose sides are perpendicular to the acting forces that had been enunciated by La Hire. But he demonstrates it carefully (Fig. 4).

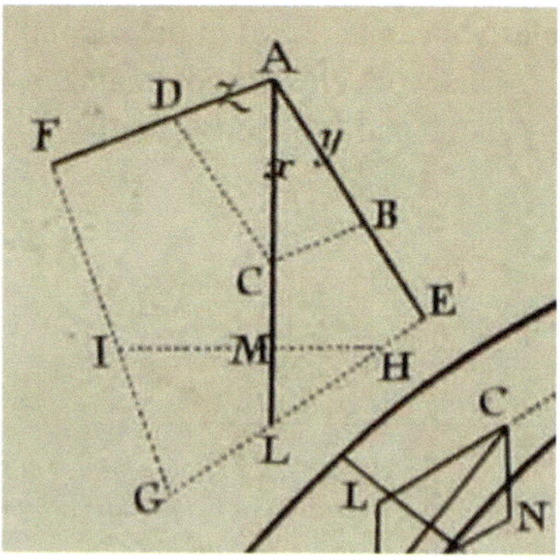

Fig. 4 Couplet's triangle of forces (Couplet 1729, Pl. 4). Public domain

Next, Couplet demonstrates a theorem which he credits to La Hire but which he improves with respect to the original one.

> When the joints of the vault converge to a same point R, as one says that they must in a vault whose intrados is circular, if one traces whatever horizontal line Sλ, the weights of the voussoirs A, B, C, D, will be expressed by the parts XY, VX, TV, ST, of that horizontal line enclosed between the extended joints of those voussoirs. Their thrust will be expressed by RX, RV, RT, RS that are the extensions of their inferior joints intercepted and included between the point of intersection R and the horizontal line Sλ, if all voussoirs are in equilibrium between each other (Fig. 5).[18]

Couplet then posed the idealized problem identical to the one posed by La Hire:

> Problem IV: Determine the uniform curvature of a vault, so that it will maintain its equilibrium, and of which we consider the voussoirs as polished, that is to say, without a connection.[19](Fig. 6)

He solved this problem in the following way:

> Solution. Take a cord hung with equal weights that will be attached all along at equal distances; those little weights will represent the effective voussoirs that we are looking for. ... The aim is thus to prove that all parts of an arch built that way will be in equilibrium; which must happen, since the loose cord naturally gives that curve, and the centrifugal forces of that loose cord, which I consider to be an inverted arch, are precisely the same as the centripetal forces of that same curve re-inverted, of which the vault in question is formed.[20]

We thus find in this text a correct mechanical justification. Unfortunately, the mathematics do not follow in order to determine the curve, the catenary. Couplet says:

> The loose cord can be considered as a polygon BCDEFGHA of an infinity of sides all equal to each other, and consequently of equal weight, or rather as a cord to which are attached at equal distances a great number of equal little weights M, N, O, P, Q, R.
>
> This posed, if from whatever point S one draws the perpendiculars SI, SK, SL, etc. on the sides of the curve, those perpendiculars will cut a horizontal line CH in equal parts, because

---

[18] Couplet (1729, p. 84): *Lorsque tous les joints de la voûte sont dirigés vers un même point* R, *comme l'on dit qu'ils doivent être dans la voûte, dont l'intrados est circulaire, si l'on tire une ligne horizontale quelconque* Sλ *les pesanteurs des voussoirs* A, B, C, D, *seront exprimées par les parties* XY, VX, TV, ST, *de cette ligne horizontale renfermées entre les joints prolongés de ces voussoirs. Leurs poussées seront exprimées par* RX, RV, RT, RS *qui sont les prolongements de leurs joints inférieurs interceptés & compris entre le point de concours* R *& la ligne horizontale* Sλ, *si tous ces voussoirs font équilibre entre eux.*

[19] Couplet (1729, p. 95): *Problème IV, Déterminer la courbure uniforme d'une voûte, telle qu'elle se maintienne en équilibre, et dont nous considéreons les voussoirs comme polis, c'est-à-dire, sans liaison.*

[20] Couplet (1729, p. 95): *Solution, Prenés une corde garnie de poids égaux qui y seront attachés tout du long à égales distances, ces petits poids représenteront les voussoirs effectifs de la voûte que nous demandons. ... Il s'agit donc de démontrer que toutes les parties de la voûte ainsi construite, seront en équilibre; ce qui doit arriver, puisque la corde lâche donne naturellement cette courbe, & que les forces centrifuges de cette corde lache que je regarde comme une voute renversée, sont précisément les mêmes que les forces centripètes de cette même courbe redressée, dont la voûte en question sera formée.*

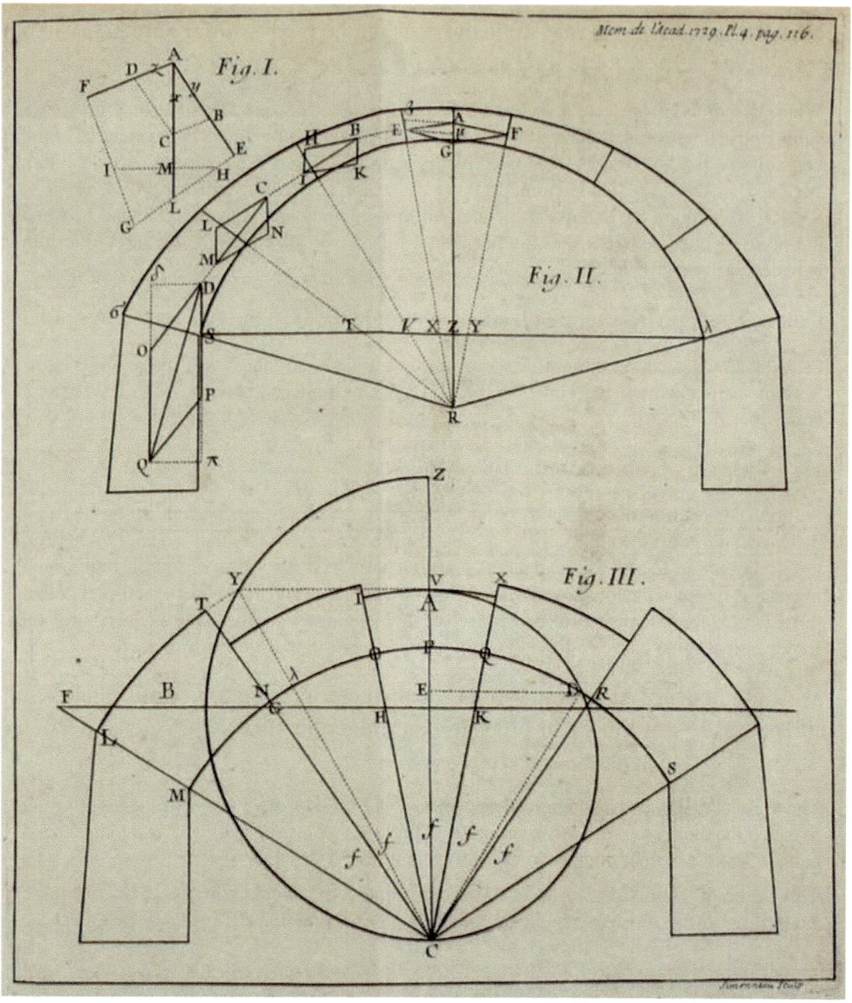

Fig. 5 Couplet's Figs. II and III (Couplet 1729, Pl. 4). Public domain

the parts of this horizontal line will represent the weights M, N, O, P, Q, R, which are all equal, as we will see.[21]

[21] Couplet (1729, pp. 95–96): *La corde lâche peut-être considérée comme un polygone BCDEFGHA d'une infinité de côtés tous égaux entre eux, & par conséquent également pesants, ou plutôt comme une corde à laquelle sont attachés à distances égales un grand nombre de petits poids égaux M, N, O, P, Q, R.*
Cela posé, si d'un point quelconque S. l'on tire des perpendiculaires SI, SK, SL, &c. sur les côtés de la courbe ces perpendiculaires couperont une ligne horizontale CH en des parties égales, parce que les parties de cette ligne horizontale représenteront les poids M, N, O, P, Q, R, qui sont tous égaux, comme nous allons le voir.

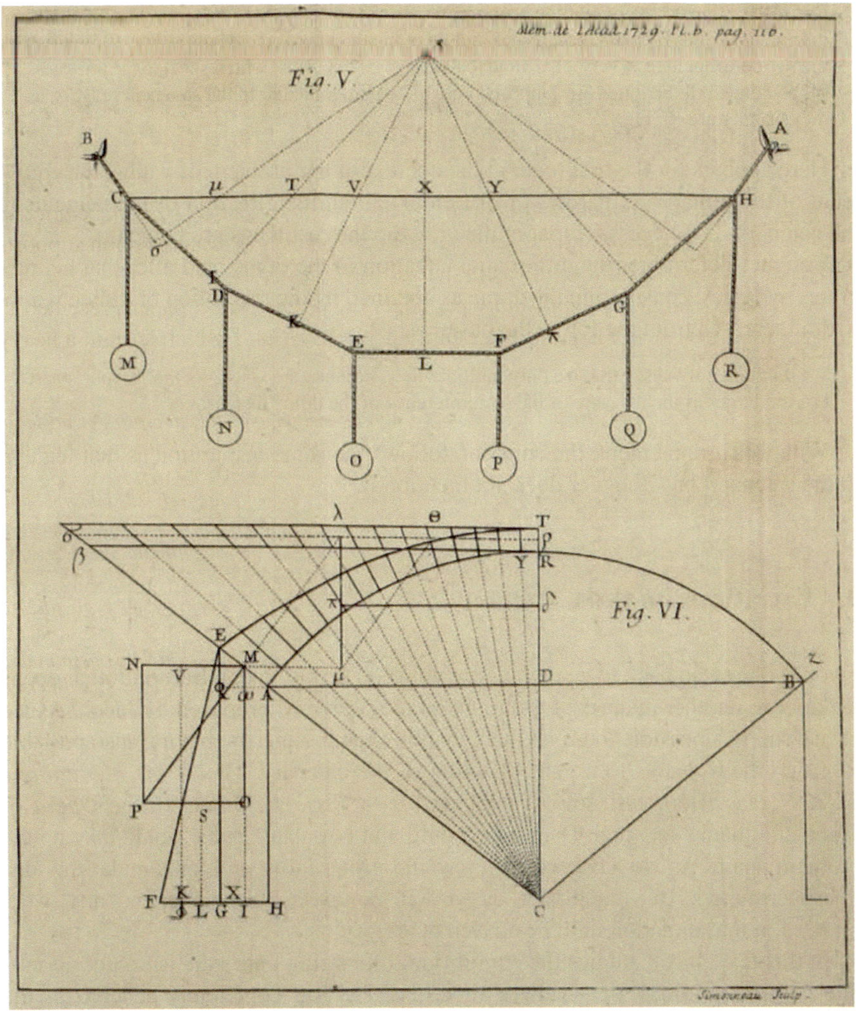

Fig. 6 Couplet's Figs. V and VI, in which we can see that Couplet also uses the fact that the inversed catenary also gives the form of the vault (Couplet 1729, Pl. 6). Public domain

Obviously supposing that there exists such a point S, he imposes that his curve is a circle. However, this note of Couplet contains some beautiful theorems and even this part contains a positive aspect, namely that the mechanical justification is in terms of forces of the coincidence between the arch stable under its own weight, and the loosely hanging cord thus also under its own weight.

In 1734, Pierre Bouguer (1698–1758) would be the first to face the problem of the dome:

> There remain only the domed vaults that nobody, as far as I know, has examined. ... I shall always suppose the stones or the voussoirs to be infinitely polished ... BbA is the curve that

> forms the dome by its revolution around its axis, the vertical AD. This curved line passes through all points B, b etc. through the middle of the thickness KL, HI of the vault, thickness that we consider here as very small, and that always is very small with respect to the thickness of the dome. All the joints are supposed here to be perpendicular to the same curve BbA, as they usually are.[22] (Fig. 7)

Here appears for the first time, in the context of the problem of vaults, the equivalent of the remarks of Pardies or of Leibniz about the direction of the tangents to the catenary. It is necessary that in the arch too the resultants are tangent.

Bouguer determines the differential equation of the dome, and after that he integrates by series, considering the dome as obtained by the revolution of a slice whose width increases from the top to the base:

> We don't care if we say that we only multiply by y, because the voussoirs that supports part HA are larger as the measure of the circumference of the dome increases.[23]

With this arrangement, the equation follows the same inspiration as that leading to the catenary, but Bouguer does not mention it.[24]

9 Variations on the Catenary

I would like to go back to an earlier, unpublished text by Jacob Bernoulli that opens the door to another manner of grouping all the problems proposed by Jacob. As we have seen, it is possible to group them thanks to isoperimeters, but it is also possible to group them thanks to a generalization of the catenary. This is the *Meditatione* CCXLV (Jac. Bernoulli 1697d) published as a *Varia Posthuma* in the *Opera* of Jacob and thanks to which Daniel Bernoulli and Leonhard Euler would take up the problem again. As we have seen, they would group those problems under the idea of isoperimeters. But much later, as we will now see, Euler, independently, went forward in that direction until he arrived at victory.

In this text, Jacob studies the equilibrium of a string under the action of an infinite number of forces of whatever directions. He thus completely generalizes the problem of the catenary. For this study, he does not use the law of the lever, but the parallelogram law of forces, which he applies under the form of the law of sines

[22] Bouguer (1734, p. 149): *Il ne reste que les Voutes en Dome que personne, que je sçache, n'a examinées. … Je supposerai toûjours que les pierres ou les Voussoirs ont leurs surfaces infiniment polies… BbA est la courbe que forme le dome par sa révolution autour de son axe, la verticale AD. Cette ligne courbe passe dans tous les points B, b &c. par le milieu de l'épaisseur KL, HI de la voute, épaisseur que nous regardons ici comme très petite, & qui l'est toujours en effet par rapport aux dimensions du dome. Tous les joints sont aussi supposés ici perpendiculaires à la même courbe BbA, comme ils le sont ordinairement.*

[23] Bouguer (1734, p. 153): *Nous n'avons que faire de dire que nous ne multiplions par y, que parce que les voussoirs qui supportent la partie HA sont plus larges à mesure que la circonférence du dome se trouve plus grande.*

[24] This aspect of the text has been very carefully studied by Edoardo Benvenuto (1991, II, p. 344).

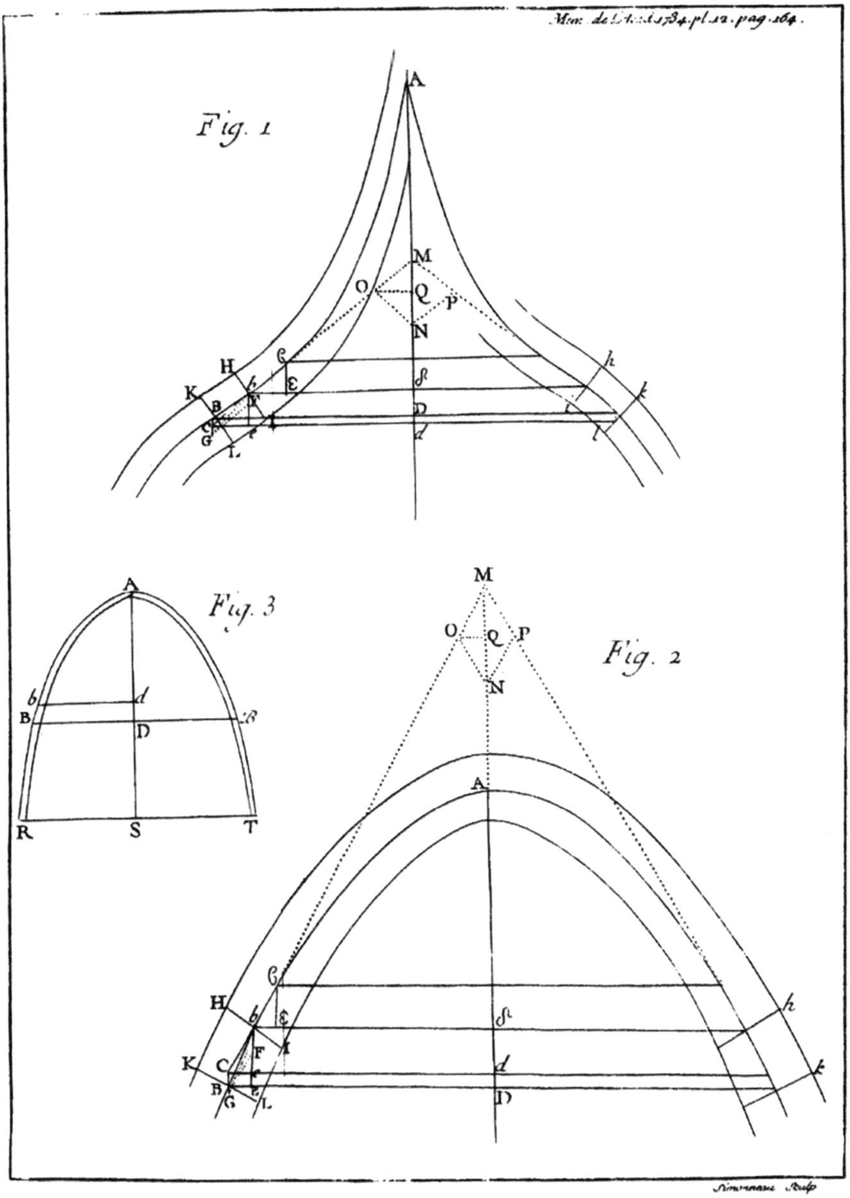

Fig. 7 Bouguer's Fig. 1 (Bouguer 1734, Figs. 1–2). Public domain

given by Varignon. He then studies some particular cases and finds the lintearia when the pressures are perpendicular to the cord, together with the elastica, the circle, the velaria, etc.

Some thirty years later, his nephew, Daniel Bernoulli returns to Jacob's guiding idea expressed in Med. CCXLV (Jac. Bernoulli 1697d) and develops that same problem in an article on "Methodus universalis determinandae curvaturae fili, a potentiis quamcunque legem inter se observantibus extensi" ("Universal method for determining the curvature of a string stretched by powers following any law among themselves" (Bernoulli 1728). However, in the meantime Daniel had discovered the power of the parallelogram law of forces, which he had demonstrated axiomatically two years before in his "Examen principiorum mechanicae" ("Examination of the principles of mechanics") (Bernoulli 1726).

He starts thus his "Methodus universalis" by considering a cord upon which four forces are acting at two points and writes the equality of the forces acting on an intermediate point of the cord. He then passes to infinitely small elements and gives the expression of the form of the cord in function of the osculating radius. Because of his knowledge of the parallelogram law, he can constrain himself to two types of forces, the one vertical and the other perpendicular to the curve; all other cases will be compositions of these two. In the corollaries, he particularizes first to various cases with only forces perpendicular to the cord. In this manner, he finds the elastica, the velaria and the lintearia. Then he considers all parallel forces and finds all the catenaries. Thus, he also finds the results of his uncle Jacob and of his father Johann.

Daniel also shows that in the case of forces perpendicular to the curve, or the cord, namely in the case of the lintearia, the radius of curvature is $F = \frac{a}{R}$; in the case of forces of direction other than the perpendicular but parallel to each other, as in the case of the catenary, "the force is composed in the inverse ratio of the osculating radius and the square of the sine of the angle that the force makes with the axis".[25]

Next, he treats the elastica, taking the demonstration of his uncle and particularizing it to the case of the lamina without weight, also of his uncle, and then to the lamina under its own weight, whose solution he postpones. He will never find it, but it will be found by Euler, as we will see.

Finally, Daniel treats a problem first proposed by Jacob at the very beginning: the problem of the stretched catenary. This is obviously not an isoperimetric problem and that is the reason why it was forgotten, showing once more that the principal idea behind those studies was to analyze of a set of isoperimetric curves under the action of various applied forces.

In the same volume of the *Commentarii*, immediately following Daniel's article, one finds Euler's paper on the same argument, the "Solutio problematis de invenienda curva" that we mentioned earlier (Euler 1728). His aim was not only to treat elastic laminae of any elasticity whatsoever, but also to find the possible forms taken by a string perfectly stretchable under the action of any forces whatsoever. He is thus in the

[25] Dan. Bernoulli (1728, p. 65): *esse ubique potentiam in ratione reciproca composita ex radio osculi et quadrato funis anguli, quem potentia facit cum axe.*

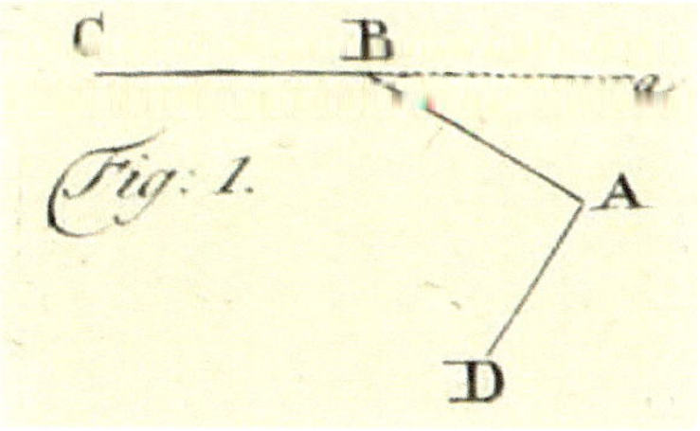

Fig. 8 Euler's Fig. 1 from "Solutio problematis de invenienda curva" (Euler 1728). Public domain

same line as Daniel Bernoulli, but the clarity of his presentation and its completeness contrasts with that of Daniel. Nevertheless, Daniel gave him many ideas.

In Euler's text we first find the Hypothesis:

> If two [rigid] rods aB and BC [on the line CBa] are joined at B in an elastic way and are to be rotated by a force AD into the position ABC, in order that the angle ABa is taken up. There will be a moment of the force AD about B considered as the product of the elastic force at B and the angle ABa. It is known that this factor is equal to the product of the force AD by AB [note that AD is at right angles to AB]. This hypothesis is generally assumed; however, the truth of this is satisfactory only if the angle ABa is exceedingly small and that can probably be shown by experiment.[26](Fig. 8)

And further on, he writes the General Problem:

> Let BMA be any elastic laminar plate and with displacements from forces applied at any fixed points whatever, moreover with that end in B fixed, and at A two weights E and F are have been hung on pulling on A, of which E is along the vertical and F the horizontal. It is required to determine the curve AMB that the nearby lamina has been turned into by being bent.[27] (Fig. 9)

Euler treats in full generality the problem of a lamina curved by two forces, one vertical and the other horizontal. The restoring force in whatever point M is v/r. He next particularizes:

I. Perfectly flexible laminae

   A. under the action of purely vertical forces:
      a. in each line element: he finds the common catenary again;
      b. in each element dx (horizontal): he obtains the parabola (the suspended bridge).

---

[26] Euler (1728, p. 71): *Hypothesis. Si duae virgae aB, BC in B elatere iunctae a potentia AD in situm ABC torqueantur, ut ang. AEa comprehendat. Erit momentum potentiae AD in B ut vis elastica in B et angulus ABa coniunctim. Hoc scilicet factum aequipollet facto ex potentia AD in AB. Assumitur vulgo haec hypothesis; eius tamen veritas si angulus ABa est vehementer parvus satis probabiliter potest physice demonstrari.* Translation by Ian Bruce, http://www.17centurymaths.com/contents/euler/e004tr.pdf.

[27] Euler (1728, p. 73): *Problema Generale. Sit lamina BMA utcunque elastica et in singulis punctis a potentiis quibusvis sollicitata, fixa autem ea sit in B, atque in A duo pondera appensa habens E et F, quorum E secundum verticalem AE, et alterum F secundum horizontalem AC trahit. Oportet determinare curvam AMB iuxta quam lamina hoc modo sollicitata flectitur,* pp. 71–73. Translation by Ian Bruce, http://www.17centurymaths.com/contents/euler/e004tr.pdf.

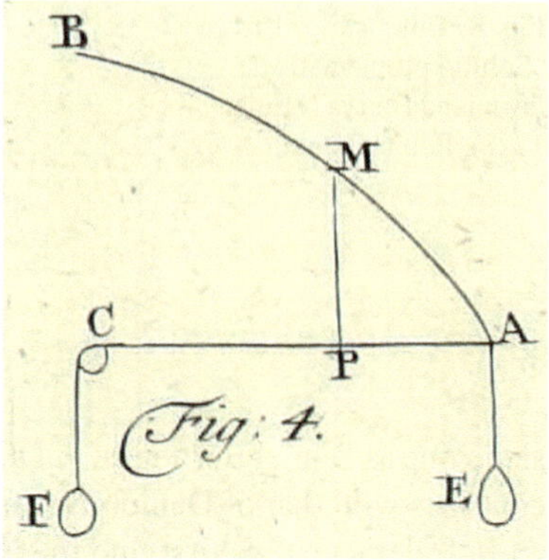

Fig. 9 Euler's Fig. 4 from "Solutio problematis de invenienda curva" (Euler 1728). Public domain

B. under the action of forces perpendicular to the curve, he gives some properties and refers to Jacob Hermann.
C. under the action of forces perpendicular to the curve and to vertical forces:
   a. he finds the lintearia full of water up to a certain height;
   b. the velaria of uniform weight;
   c. the velaria under a wind of whatever direction (catenary).
D. under the action of perpendicular and tangential forces: he does not give any name.

II. Elastic cases

A. the lamina without weight and without elasticity;
B. the lamina without weight and with uniform elastic force: he finds the elastica;
C. the lamina with weight and with uniform elastic force;
D. the lamina under its own weight.

But while the plan was there in Euler's mind, solutions are not all given. Euler would need forty-two years to complete in his program; he will finally achieve his aim in a paper entitled "Genuina principia doctrinae de statu aequilibrii et motu corporum tam perfecte flexibilium quam elasticorum" ("Natural basic principles of the study of the state of equilibrium, and the motion, of bodies that are both perfectly flexible and elastic") (Euler 1770), in which he puts a second endpoint to the elaboration of the line of research opened by Jacob Bernoulli. This consists in bringing together into a single theory all the problems of the catenary, velaria, lintearia, etc. as well as all elasticae. This work abandons the line of isoperimetric problems started by Jacob and followed by Euler in his *Methodus inveniendi* (Euler 1744) to go back to the terms of the reasoning of his earlier "Solutio problematis de invenienda curva" (Euler 1730) and of Daniel Bernoulli's "Methodus universalis" (Bernoulli 1728).

Let us recall that in the meantime Euler had published his "Découverte d'un nouveau principe de mécanique" (Euler 1750), in which he gave the three relations

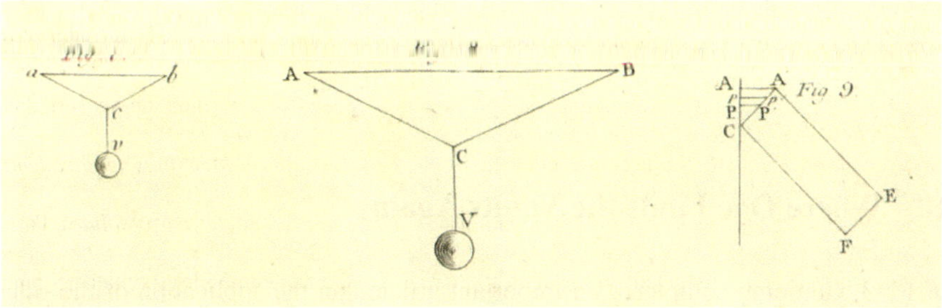

Fig. 10 Euler's Figs. 7–9 from the "Regula facilis" (Euler 1775, Tab. II). Public domain

given by projections of Newton's law$F = ma$, and where he also underlines the importance of the equilibrium of momentum. In his "Regula facilis, pro dijudicandi firmitate pontis aliusve corporis similis, ex cognita firmitate moduli" (Euler 1775), he would put both equations, namely equilibrium at translation and equilibrium at rotation, on the same level. That will allow him to determine the most general case.

The "Regula facilis" is divided in two parts. One is dedicated to equilibrium and we will analyze it; the other is dedicated to motion, and that we shall leave aside.

Euler explicitly enounces a restriction that was always done tacitly: the forces must be such that the curve remains in the plane. He starts by explicitly stating the notion of tension, in a general manner, and distinguishes in this way the famous cases that had been demonstrated (Fig. 10).

When the bodies are perfectly flexible, the tension is along the tangent.

When the body is elastic, there exists an oblique tension and there are thus two components for the tension, a tangential one and a normal one.

He bases his discussion on two factors:

- the equality of the forces in a point, which gives him two relations in the plane, one for the tangential components and one for the normal components;
- on the equilibrium of moments, which gives him a third equation.

It then becomes evident why the problem of the catenary could be easily solved without using the equilibrium of moments.

Euler then subdivides the study of elastic bodies into cases that are uniformly elastic, those which are non-uniformly elastic, and those that have an initial curvature. Those divisions are equivalent to equations of the type of Hooke's law linking the restoring force normal to the curve to the angle of deviation. Those are constitutive equations.

In the second part, he finds the particular cases that we have already largely demonstrated. He refers back finally to his classification of the elasticae that he had outlined in done in his *Methodus inveniendi* (Euler 1744) starting from the variational principal given by Daniel Bernoulli. The theoretical foundation of that text was different but the discussion concerning the curves is still valid.

In the "Regula facilis" (Euler 1775), he adds nothing new but compares the solution of the elastica given by Jacob Bernoulli starting from the law of the lever with the one he gave in his "Genuina principio" of 1770.

10 Where One Finds the Vaults Again

In 1773, Coulomb completed an important article "on the application of the rules of Maxima and Minima to problems of statics relative to architecture" (Coulomb 1773),[28] which would only be published in 1776. It was necessary, in order to write that work to have understood and assimilated the principal works of Euler that we have discussed.

Coulomb's Proposition XVII begins.

> **XVII Arches whose joints have neither friction nor cohesion**
>
> Let aB be the elevation of an arch an infinitesimal thickness, of which the joints are perpendicular to the curve aB; required the shape of the arch when acted upon by any forces.[29](Fig. 11)

We observe how precise the statement of idealization is, as well as the generalization to any forces. By decomposition and projection of the forces in action, he obtains the equation

$$\frac{dx}{dy} = \frac{\pi}{\phi}.$$

Of this he writes:

> This formula is exactly the same as that found by Euler (in the third volume of the Academy of Petersburg) for the shape of a chain acted upon by given forces. This must be so, since, on inverting the curve, and substituting tension for compression, the above theory applies equally to either case, and gives precisely the same equation. However, Euler's method has only the result in common with the method given here.[30]

The result he mentions is the one we have just spoken about.

---

[28] For a detailed study of Coulomb's "Essai", see "A Study of Coulomb's Application of the Rules of Maxima and Minima to Problems of Statics", Chap. 10 in this present volume.

[29] Coulomb (1773, p. 371), p. *Voutes dont les joints n'ont ni frottement ni cohésion. Soit aB le profil d'une voûte, d'une épaisseur infiniment petite, dont les joints soient perpendiculaires à la courbe aB; l'on demande la figure de cette voute, sollicitée par des puissances quelconques.* p. 371. English translation Jacques Heyman (1972, p. 61).

[30] Coulomb (1773, p. 372): *Cette formule se trouve exactement la même que celle qui a été déterminée par M. Euler (dans le troisième volume de l'académie de Pétersbourg) pour la figure d'une chaîne, sollicitée par des puissances quelconques. Ce qui doit effectivement arriver; car, en renversant la courbe, & substituant la tension à la pression, la théorie précédente s'applique également à l'un ou l'autre cas, & donne précisément la même expression. Au reste, la méthode de M. Euler n'a rien de commun avec celle-ci, que le résultat.* Eng. trans. Jacques Heyman (1972, p. 62).

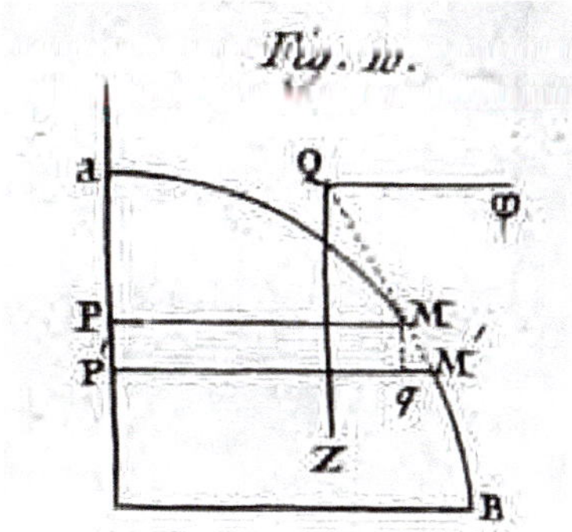

Fig. 11 Coulomb's Fig. 10 in "Essai sur une application des règles de Maximis et Minimis à quelques problèmes de statique", accompanying Proposition XVII (1773, Tab. XVI). Public domain

Next, using the radius of curvature, he deduces the equation of any arch under the action of gravity and mentions Jacob Bernoulli's study on vaults to praise him and to remark that both conditions of equilibrium would have given him the same result if he hadn't made the error in his calculation. (Let us add that long before Coulomb, in 1748, Giovanni Poleni (1683–1761) proposed his experiment in order to verify the stability of the dome of St. Peter's in Rome.)

In 1774, the abbot Charles Bossut (1730–1814) began his "Recherches sur l'équilibre des voûtes") ("Researches on the equilibrium of arches in domes") (Bossut 1774). He begins the article by reviewing the history of this study. He underlines the value of Couplet's work in the case of circular vaults. Next, he adapts the theory of La Hire, taking up his theorem but replacing the radius of the circle by the radius of curvature. In this way, he corrects the work of Couplet.

In Bossut's work, "*V, F, f, F*′, *f*, etc." (1774, p. 537) are the forces applied to each voussoir. Those forces act in any direction; Bossut does not take the symmetry into account.

Then he considers X and X', two consecutive voussoirs, respectively subjected to forces F and F' and whose joints mM, nN, pP are perpendicular to the intrados ACB, "both for the grace of the arch and for the solidity of the construction"[31] (Fig. 12).

He resolves each of the forces into two components perpendicular with respect to the two joints of the voussoir to which it is applied and imposes as a condition of equilibrium that the opposing components of two consecutive voussoirs be equal. This permits him firstly, to give the ratio of the forces in function of the angles between the forces and its components:

$$\frac{F}{F'} = \frac{\sin XtE \times \sin X'E'q}{\sin XEt \times \sin X'qE'} \quad \text{(A)}$$

Secondly, he relates these angles back to the angles between the extension of the joints between the voussoirs:

[31] Bossut (1774, p. 537): *tant pour la grâce de la voûte que pour la solidité de la construction.*

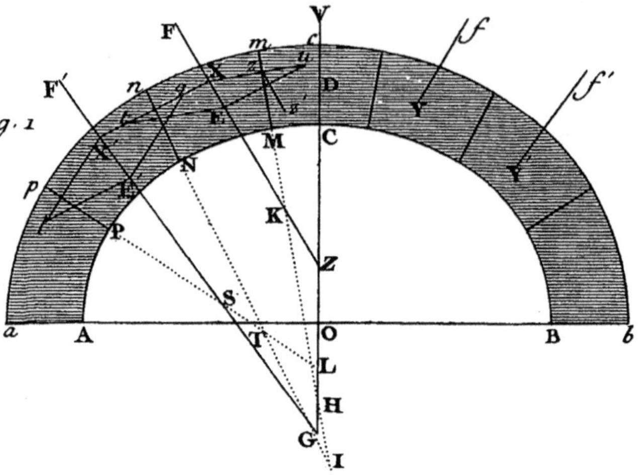

Fig. 12 Bossut's Fig. 1 from "Recherches sur l'équilibre des voûtes" (Bossut 1774, Pl. 1). Public domain

$$\frac{F}{F'} = \frac{\sin NIM \times (\cos CGF' \times \cos CLP + \sin CGF' \times \sin CLP)}{\sin PTN \times (\cos CZF \times \cos CHM + \sin CZF \times \sin CHM)}. \quad \text{(B)}$$

After that he will rewrite that equation supposing an infinite number of voussoirs, thus considering infinitely small arcs between those joints, corresponding to angles between two consecutive osculating radii. This permits him to write equation (B) in the following form:

$$\frac{F}{F'} = \left(\frac{dy'' \cos u' + dx'' \sin u'}{dy \cos u + dx \sin u} \right)$$

He must now take into account the fact that $F' = F + dF$ and that $R' = R + dR$. He poses $F = fds$ because he considers F as the resultant of an infinity of equal and parallel forces applied to ds. In this way he introduces the idea of pressure.

Thus he gives the equation of the problem in its full generality:

$$\begin{aligned} &\phi \cos u(2Rddy + dRdy) + \phi \sin u(2Rddx + dRdx) \\ &+Rdy \cdot d(\phi \cos u) + Rdx \cdot d(\phi \sin u) \quad \text{(C)} \end{aligned}$$

He next particularizes the problem in an interesting manner, in the style of Euler:

(1) To the gravitational force that is always constant, vertical and thus parallel to itself. In this case, he finds the ordinary catenary again.

(2) In the case of a vault loaded with water, that is, in the case of a force acting perpendicularly to the curve: "If a vault is loaded with water or even with liquid earth until a certain point; then the forces f would act or may be supposed to

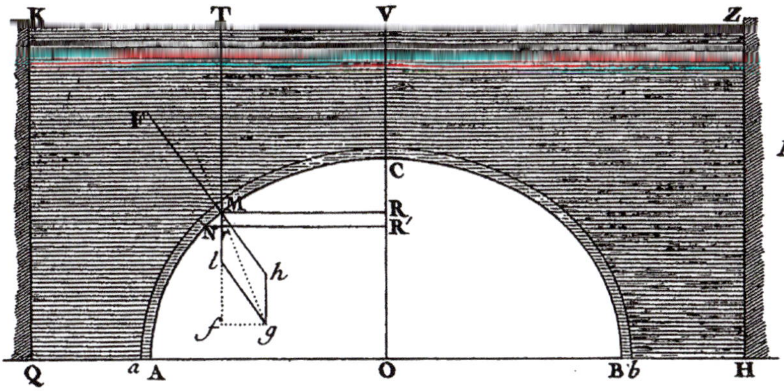

Fig. 13 Bossut's Fig. 4 from "Recherches sur l'équilibre des voûtes" (Bossut 1774, Pl. I). Public domain

act perpendicularly to the curve *ACB*".[32] In this case, he finds the equation of an elastica but does not say it explicitly. (Fig. 13)

(3) In the case of mixed forces, he means composed of the two preceding kinds.

In 1776 Bossut would use the same reasoning in order to apply it to domes and nine years later, Lorenzo Mascheroni (1750–1800) published his *Nuove ricerche sull'equilibrio delle volte*, the treatise that is the subject of our analysis.

11 Mascheroni's "New Researches on the Equilibrium of Vaults"

Lorenzo Mascheroni (Bergamo 1750-Paris 1800) was an Italian mathematician, academician and man of letters (Fig. 14).[33] His *Nuove ricerche sull'equilibrio delle volte* ("New researches on the equilibrium of vaults") was first published in 1785, with a second edition in 1829 (Fig. 15).[34] I propose to concentrate on Chaps. I and II of Mascheroni's work and on the choice of the catenary as the curve of equilibrium; on the second part of Chap. V and the study of the loaded arch; on Chap XII, as well

[32] Bossut (1774, p. 543): *Si une voûte est destinée à porter de l'eau ou même des terres liquides jusqu'à un certain point; alors les forces f agiroient ou pourroient être censées agir perpendiculairement à la courbe* ACB.

[33] For a near-contemporary account of Mascheroni's life, see Mangili (1812).

[34] The text of the second edition is identical with that published in 1785, with the addition of a eulogy in praise of Mascheroni by Ferdinando Landi added before the text and a new chapter appended to Chap. XII entitled "Sulle Curve che servono a delineare le ore ineguali degli Antichi nelle superficie piane" (On the curves that serve to delineate the unequal hours of the Ancients on plane surfaces), an independent article that doesn't have any relation to the problem of arches.

Fig. 14 Lorenzo Mascheroni in an engraving by Luigi Rados, from the 1829 edition of *Nuove ricerche*. Public domain

as, in parallel, to the corresponding problems concerning domes. First, however, it will be helpful to give an overview of all twelve chapters of the volume.

A first subdivision of the first eleven chapters is clear: the first five chapters treat the problems of arches and are thus situated in the plane and two-dimensional. The next six chapters deal with domes and are therefore three-dimensional. Now let us make a quick overview of the book and identify within in it the passages that will retain our interest.

Chap. I, On the equilibrium of lines. We will analyze this in detail.
Chap. II, On the equilibrium of arches. We will analyze this in detail.
Chap. III, On the thickness of arches. In this chapter, Mascheroni introduces the thickness of arches and enounces the principle that the catenary also ensures the stability of thick arches as long as it passes through the centers of gravity of each of the voussoirs. He is the first to be that precise concerning the position of the catenary. His predecessors said that the catenary must be kept inside the arch. As he is the first to propose this way of reasoning, he will analyze the influence of a change of position of the catenary inside the arch on the risks of collapse already inventoried in the problems X, XI and XII in Chap. II (1785, p. 25–29). This allows him to judge the relevance of the results obtained by his predecessors.

NUOVE RICERCHE

SULL' EQUILIBRIO

DELLE VOLTE

DELL' ABATE

LORENZO MASCHERONI

Professore di Filosofia nel Collegio Mariano Acc. Ecc., e corrispondente dell' Accademia di Padova.

Quid velit, & possit rerum concordia discors.
Horat. ad Iccium.

BERGAMO

PER FRANCESCO LOCATELLI

1785.

Fig. 15 Title page of Mascheroni's *Nuove ricerche*

Despite the fact that the announcement of this principle is one of the great results of this work, it is not necessary for us to analyze this chapter in greater detail. **Chap. IV**, On planes made up of voussoirs that have the strength of arches. He speaks here of arches whose voussoirs form a horizontal line. Such an arch is different from an arch loaded until the upper surface is horizontal (Fig. 16).

This chapter completes the previous one. Mascheroni analyses the risk of collapse of such an arch, a practical and classic problem that had been treated by all his predecessors.

Chap. V, On the equilibrium of ramped[35] and loaded arches. The first part of this chapter takes up the reasoning of the preceding chapters in the particular case of arches that take up the loads of neighboring arches, as one can see in his Fig. 34 (Fig. 17).

[35] *Archi rampanti*: arches that rest on supports that are not on the same level.

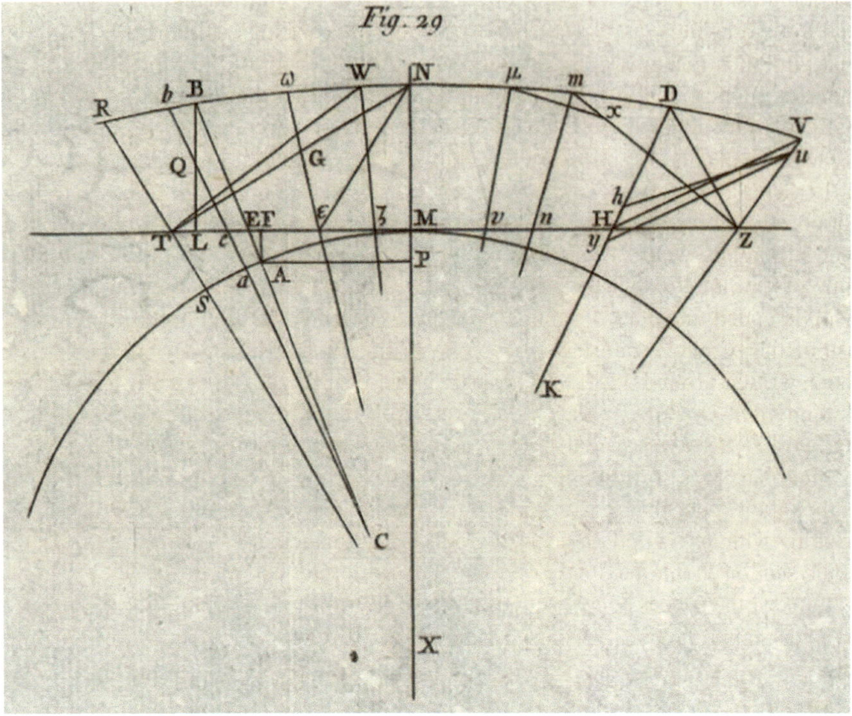

Fig. 16 Mascheroni's Fig. 29 (1785, Tav. VII). Public domain

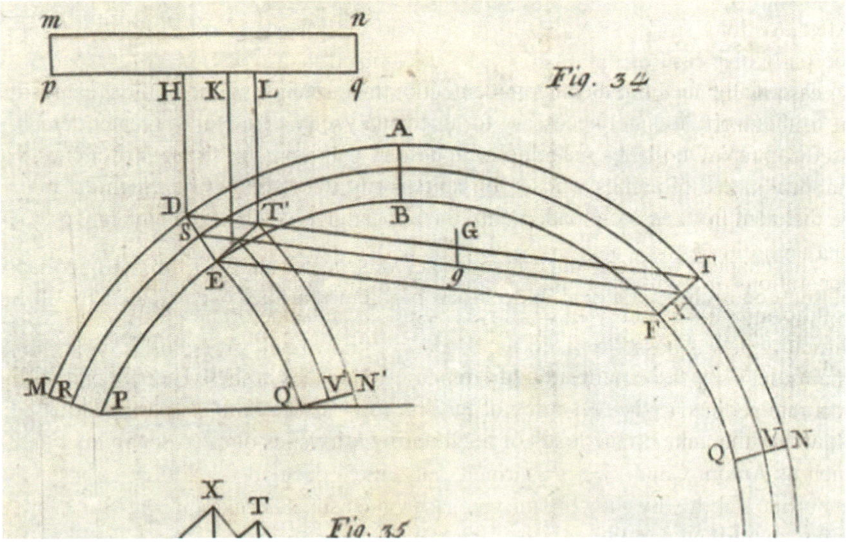

Fig. 17 Mascheroni's Fig. 34 (1785, Tav. IX). Public domain

In the second part, he studies the shape to be given to arches intended to support solid or liquid materials. He will pursue this part of his work in his Appendix (pp. 133–140), and Chap. XII will particularly retain our attention.

As already said, starting with Chap. VI, Mascheroni goes to a three-dimensional space in order to treat domes, applying to them exactly the same reasoning that he applied to arches.

Chap. VI, On domes. He looks for the curve of equilibrium of an infinitely thin dome (cf. Chap. II) and then searches for the thickness to be given to domes in order to favor their stability and to determine the position of the curve of equilibrium inside the masonry that helps foster the stability of the building (cf. Chap. III).
Chap. VII, On loaded domes.[36] As he did with arches (cf. Chap. V), here domes are loaded with either solid or fluid material. Below we will examine in detail some problems of Chaps. VI and VII.
Chap. VIII, On circular planes composed of voussoirs that have the strength of domes. This chapter parallels the treatment of arches in Chap. IV. He treats floors of towers, for example.

I quote, for the sake of completion, the title of the remaining chapters, but they are outside the study I wanted to treat here:

Chap. IX, On domes with polygonal or oval bases.
Chap. X, On annular and spiral domes.
Chap. XI, On composite arches and domes.
Chap. XII, On the curves of equilibrium in the case of a converging gravity.

For our analysis, we begin with Chap. I and to some problems upon which he founds the elements of interest we have identified.

Chap. I, On the equilibrium of lines

After having said a few words about his conception of equilibrium, Mascheroni defines relative gravity:

> By relative gravity we understand the gravity that pushes a body in the direction of inclined plane inclined to the horizon, without taking the mass of the body in consideration; if this gravity is considered together with the mass on which it acts, this product will be called relative weight.[37]

He introduces, in this way, the vectorial character of gravitational force (since the relative gravity is the component of vector g taken along the inclined plane) and of the weight (that is, the product of the mass times that relative gravity). Once this definition is given, he presents two theorems (both theorems refer to the same figure, our Fig. 18):

---

[36] This chapter is mistakenly numbered VIII in the 1785 edition.

[37] Mascheroni (1785, p. 4): *Per gravità relativa intendiamo la gravità che spinge un corpo sulla direzione di un piano inclinato all'orizzonte, non attesa la sua massa; se questa gravità si consideri insieme colla massa, nella quale agisce, questo prodotto si chiamerà peso relativo.*

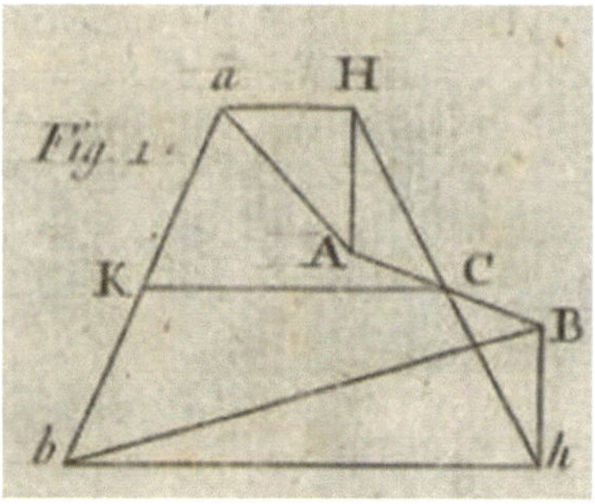

Fig. 18 Mascheroni's Fig. 1 (1785, Tav. I). Public domain

Theorem I:

If the weight B goes to b upon a plane inclined with respect to the horizon through the infinitesimal space Bb, and that in the same time by intervention of a machine, the weight A moves in direction of a on another inclined plane Aa, and if one draws the horizontals aH, bh, and the perpendiculars AH, and bh, the force of the weight A will be to the force of the weight B in composite ratio of the weight A to the weight B, and of the perpendicular displacement AH to the perpendicular displacement Bh.[38]

Mascheroni shows, in this first theorem, that the work furnished by the weight *B* which goes down from *Bh* on the inclined plane *Bb*, is recovered by the weight *A*, which ascends to a height *AH* on the inclined plane *Aa*. He does not mention the modern notion of work, but he imagines that the weights are linked by a machine.

Theorem II:

When the opposed forces in A and B are equal, the common center of gravity C does not descend.[39]

Let us understand: if the works done by the weights *A* and *B* are equal then their common center of gravity neither descends nor ascends.

Theorem III:

If a straight line makes any infinitely small motion, all points of that line carry out an equal motion in the direction of that line.[40]

Given those elements, he is able to start resolving the problems describing the equilibrium of two articulated straight bars.

---

[38] Mascheroni, 1785, p. 5: *Se il peso* B *(Fig. 1) si porti in* b *sopra un piano inclinato all'orizzonte per lo spazio infinitesimo* Bb, *e nello stesso tempo per costruzione di macchina il peso A si porti in a sopra un altro piano inclinato Aa, e se si tirino le orizzontali* aH, bh, *e le perpendicolari* AH, Bh; *sarà la forza del peso A alla forza del peso B in ragion composta del peso A al peso B, e del viaggio perpendicolare AH al viaggio perpendicolare Bh.*

[39] Mascheroni (1785, p. 5): *Quando le forze contrarie in* A e B *(Fig. 1) sono eguali, il loro centro comune di gravità C non discende.*

[40] Mascheroni (1785, p. 6): *Se una linea retta fa un moto infinitamente piccolo qualunque; tutti i punti di essa linea fanno un viaggio eguale sulle direzione della medesime linea.* This third theorem was canceled in the second edition. p. 6.

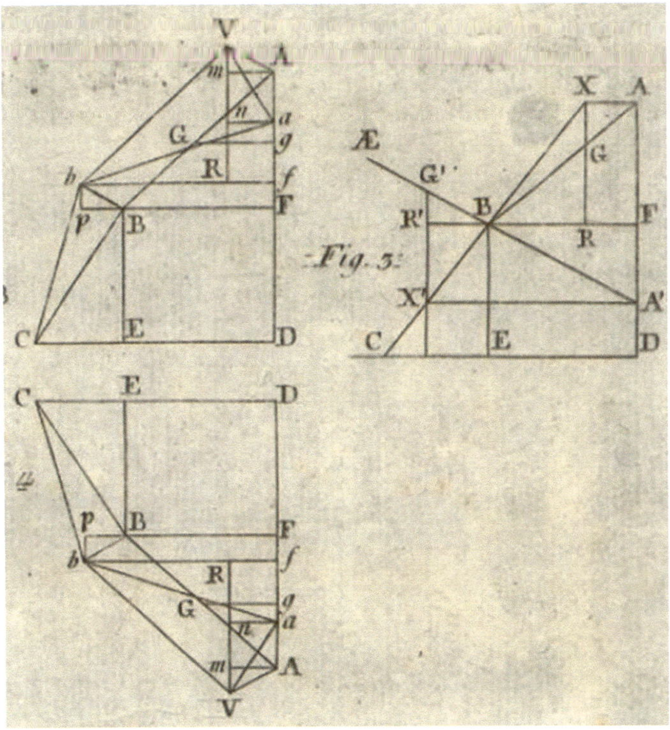

Fig. 19 Mascheroni's Figs. 3 and 4 (1785, Tav. I). Public domain

Problem I:

Let us suppose a weight A placed in A, that must go run along the line AD perpendicular to the horizon during the time that weight B placed at the extremity of an inflexible bar AB uses to turn around the centre C on a circle of radius BC; find the ratio between the two forces.[41] (Fig. 19)

Mascheroni considers a system of two bars linked by a ball joint. The weight placed at end A of one of the bars slides along a wall during the time the weight placed at the other extremity B swivels, together with the other bar around the center C.

He will once more compare the works, thus the *f dx*, or better the *p dh*, and he finds:

A.Aa : B.pb = A**Error**!: B.

[41] Mascheroni (1785, p. 7): Problema I: *Si supponga in* A *(Figs. 3. and 4.) un peso* A, *che debba scorrere per la linea* AD *perpendicolare all'orizzonte nel tempo, che il peso* B *posto all'estremità della verga inflessibile* AB *gira intorno al centro* C *col raggio* BC; *trovare la ragione delle due forze.*

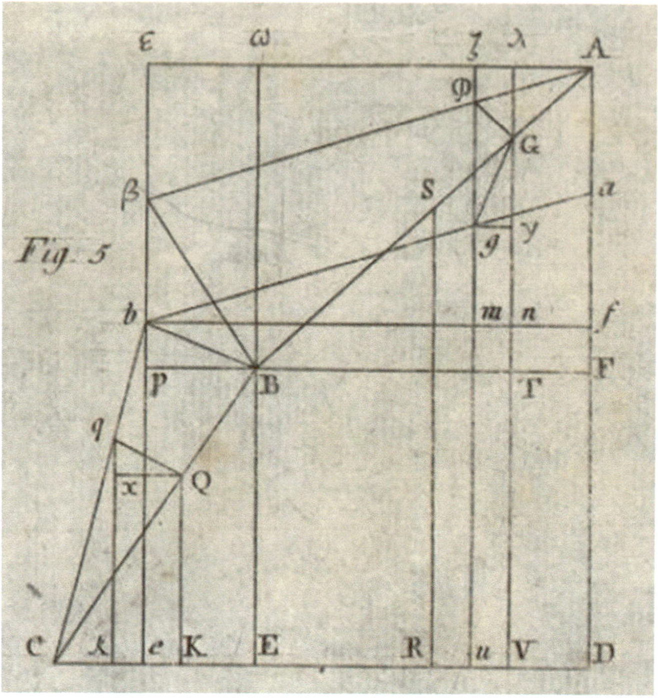

Fig. 20 Mascheroni's Fig. 5 (1785, Tav. I). Public domain

But it is the equilibrium, which he treats in a scholium, that interests him. Equilibrium occurs when this ratio equals 1. He finds the same result by the way of maxima and minima starting from the fact that the center of gravity does not descend:

A**Error**! = B.

It should be observed (see his Figs. 3 and 4 in our Fig. 19) that he simultaneously treats the cases of both figures.

Problem II:

Supposing that everything is as in the preceding problem, except that the weights are placed in G and Q, find the ratio of the two forces.[42] (Fig. 20)

The following problem is identical except that the weights have been displaced on the rods. In that new situation the same principle gives:

G.Gy : Q.qx = G**Error**!:Q**Error**!.

[42] Mascheroni, 1785, p. 10: *Supposto tutto come nel Problema precedente, se non che i pesi sieno in* G *e* Q *(Fig. 5), trovare la ragione delle due forze.*

Mascheroni remarks that the same demonstration is still valid if one turns the image upside down, as done in his figures 3 and 4 (see our Fig. 19).

In what follows in Chap. I, Mascheroni studies a system of three articulated rectilinear bars or of a threaded linear bar. The only difference that exists between these two problems is that in the second the length of the threads of the two bars are not constant. Mascheroni uses the fact that their sum is still constant. This problem was declared to be a complicated one by d'Alembert in tome VIII of his *Opuscules* (1780). Mascheroni considers than the case in which the threaded bar is curved.

As one can see in his figures 6 and 7 (Fig. 21) he shows that the demonstration is still valid if one turns it upside down.

For the sake of completeness, let us mention the well-known last theorem.

Theorem IV:

Given three bars in equilibrium, if one extends the lines AD and BC until they met in X the vertical issue of X will pass through the point G [center of gravity].[43]

Chap. II, On the equilibrium of arches

The chapter opens with the risks of collapse. I'll briefly mention three problems that analyze the three risks of collapse of an arch:

- Problem X (1785, p. 25). The culmination point of the arch descends or ascends and the arch breaks in points H and P around which it rotates (Fig. 22).
- Problem XI (1785, p. 26). The whole central block HAP descends or ascends and the feet turn around the points C and T (Fig. 23). This kind of rupture had been studied by Bossut (1774) and La Hire (1702), who related the problem to a keystone inserted into the arch. This was also generally studied by Parent (Fig. 24).
- Problem XII (1785, p. 28) Sliding of a block upon another, without rotation (Fig. 25).

Problem XIII, central in his book, is based on the theorems and on the two problems of Chap. I that we have already discussed. Mascheroni wants to determine a curve in which each point is subject to the least possible stress:

Problem XIII:

Find a curve that passes through A and C (Figs. 16 and 17) in which all the points B are such that arch CB is in equilibrium with arch BA in accordance with the conditions of Problem II.[44] (Fig. 26)

In order to solve this problem, Mascheroni relates each arc of the arch to a rectilinear bar having a center of gravity.

He thus relates Problem XIII to Problem II, which we have just analyzed. He must simply insert the coordinates of both centers of gravity in order to obtain the equation of the catenary or the funicular, which will thus be what he calls the curve

---

[43] Mascheroni (1785, p. 14): *Poste in equilibrio le tre verge, se si continuano le due direzioni* AD, BC *sino a che s'incontrino nel punto* X, *la verticale condotta per* X *passera per il punto* G.

[44] Mascheroni (1785, p. 29): *Trovare una curva, che passi per* A, *e* C *(Fig. 16. e 17.) nella quale tutti i punti* B *sen tali, che l'arco* CB *sia in equilibrio coll'arco* BA *secondo le condizioni del Problema II.*

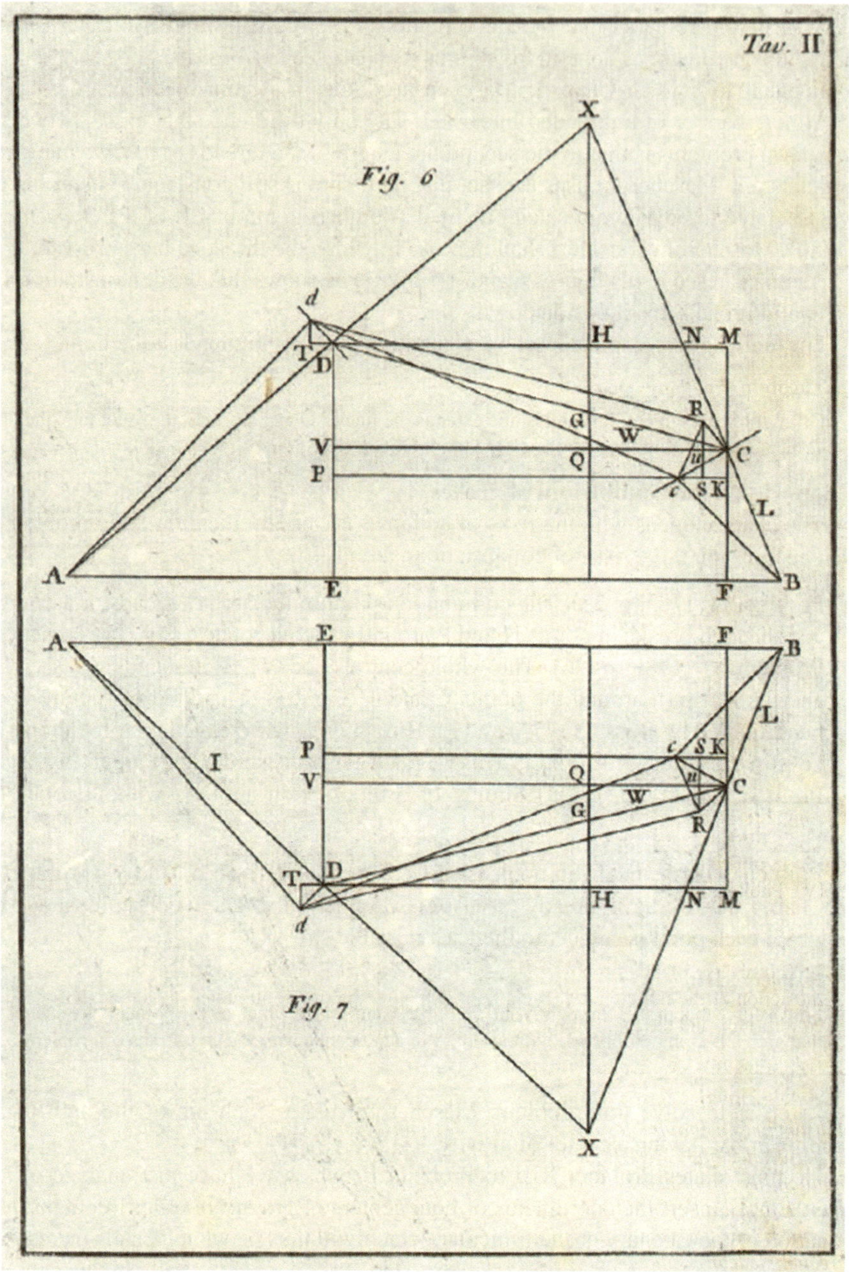

Fig. 21 Mascheroni's Figs. 6 and 7 (1785, Tav. II). Public domain

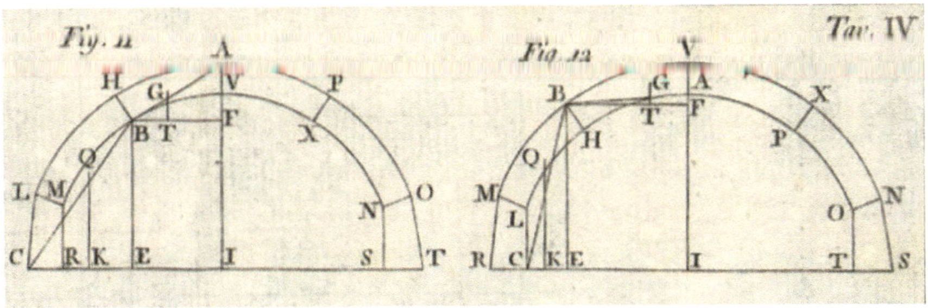

Fig. 22 Mascheroni's Figs. 11 and 12 (1785, Tav. IV). Public domain

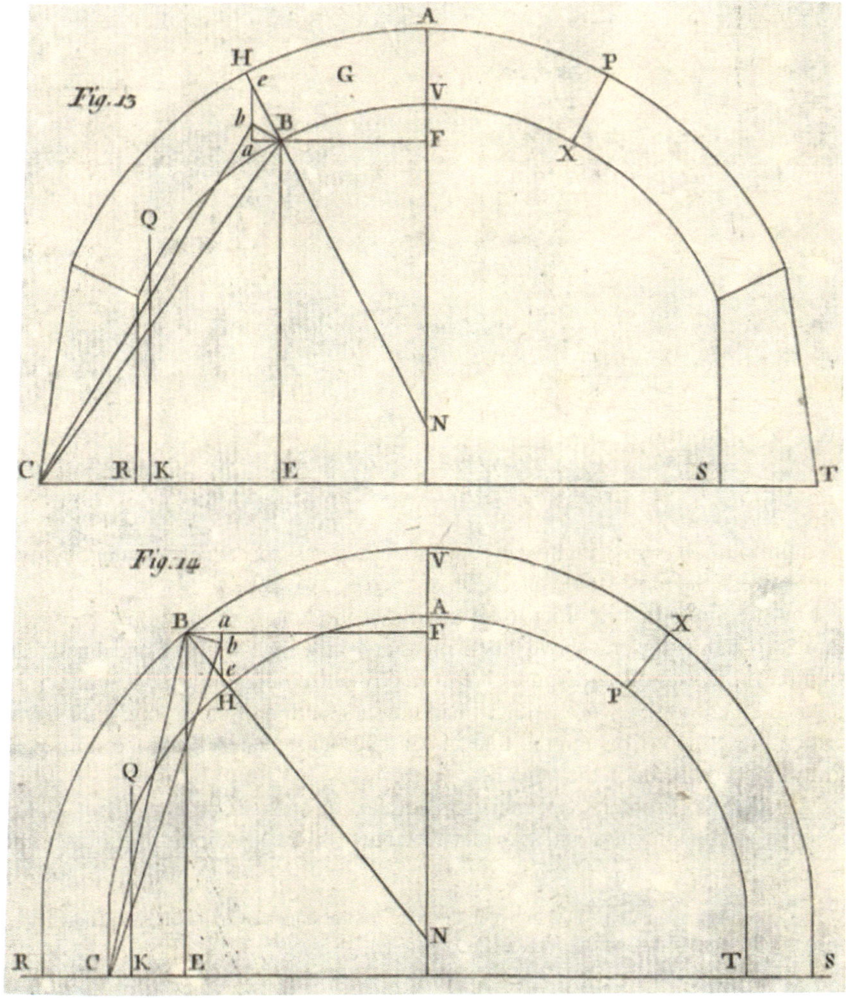

Fig. 23 Mascheroni's Figs. 13 and 14 (1785, Tav. IV). Public domain

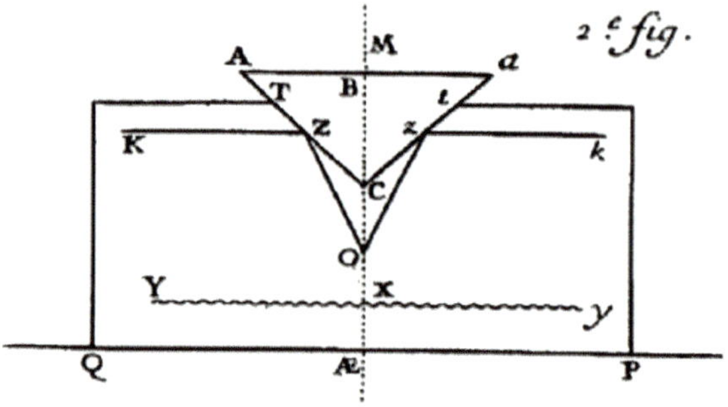

Fig. 24 Parent's 2nd figure, second memoire (1704, Pl. 6). Public domain

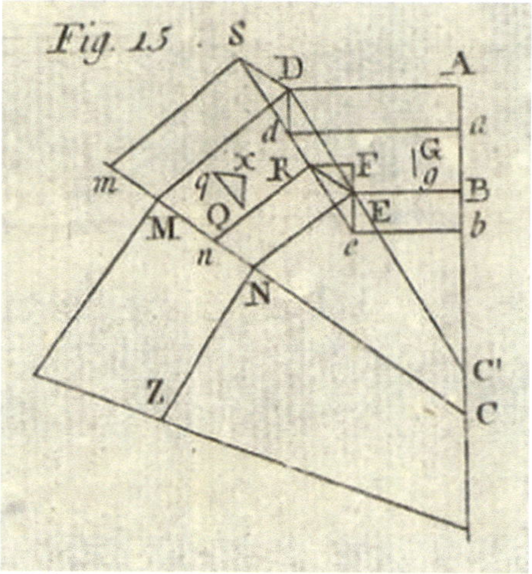

Fig. 25 Mascheroni's Fig. 15 (1785, Tav. V). Public domain

of equilibrium. It is the weighty curve without thickness, because its center of gravity is the lowest, the most stable one.

I must confess that in order to be faithful to the text, here in Fig. 26 we see the figures 16 and 17 the one above the other, as are shown the figures 3 and 4 and 6 and 7. In the 1785 edition, the figures are correctly placed one on top of the other, but in the 1829 edition they are wrong. I had the satisfaction of observing, after having written the previous paragraph, that in Mascheroni's manuscripts, the figures are always placed one upon the other.

Mascheroni's Problem XIV considers, for the first and only time in this chapter, an arch of a certain thickness and asks that the edges of the voussoirs foster equilibrium:

Problem XIV:

Find the curve of an arch ADM in which the edges of the stones DC′ and MC perpendicular to the curve serve for equilibrium.[45] (Fig. 27)

---

[45] Mascheroni (1785, p. 30): *Trovare la curva d' un arco* ADM *(Fig. 18.) nella quali i tagli delle pietre* DC′, MC *perpendicolari alla curva servano per l'equilibrio.*

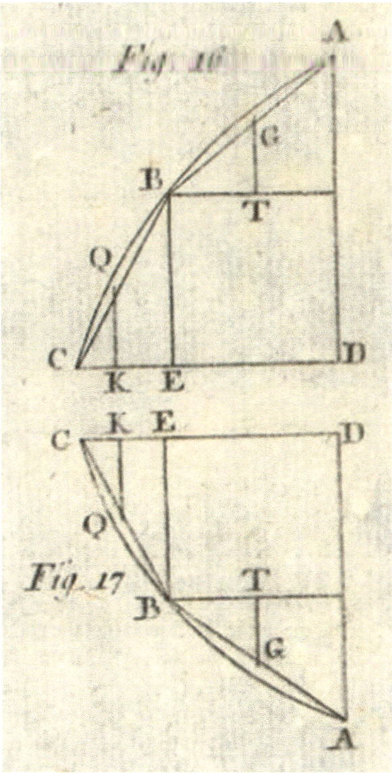

Fig. 26 Mascheroni's Figs. 16 and 17 (1785, Tav. V). Public domain

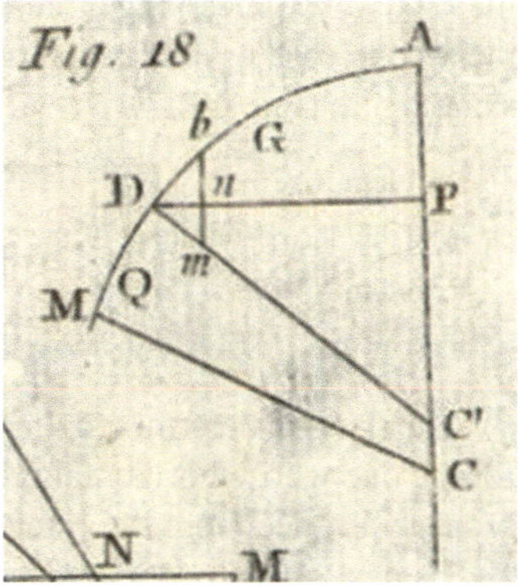

Fig. 27 Mascheroni's Fig. 18 (178, Tav. V). Public domain

He thus wants to avoid the risk described in problem XII. Applying the condition which he had just found, he again finds the equation of a catenary again. However, let us remark that he imposes the condition that the joints are perpendicular to the curve he is looking for, so that we find our tangential tensions or stresses again.

Problem XV identifies a property of the catenary, of which Mascheroni will later claim that he is the first to give:

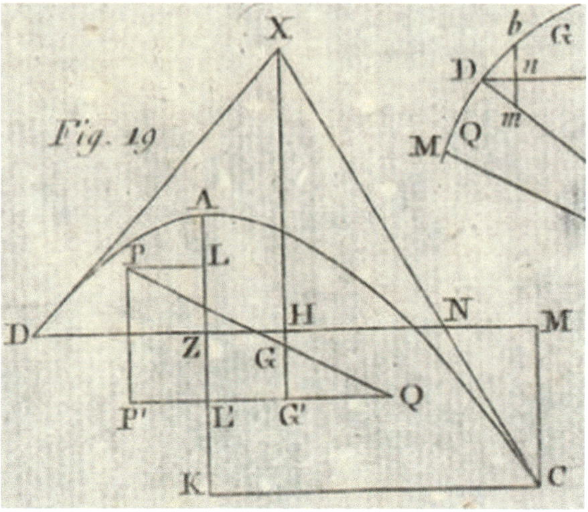

Fig. 28 Mascheroni's Fig, 19 (1785, Tav. V). Public domain

Problem XV:

Find the curve DAC such that the center of gravity G of any arc DAC is on the perpendicular XG parallel to the axis AK passing through the intersection X of the two tangents CX and DX.[46](Fig. 28)

However, this had already been found by Pardies in his *Statique ou science des forces mouvantes*in 1673 (Fig. 29).

Mascheroni next treats a series of linked problems, the first to horizontal forces, the others to forces tangential to the curve. In both cases he finds the catenary; in the second case its axis is horizontal. Problems XVI, XVII and XVIII all refer to the same figure (Fig. 30):

Problem XVI:

Find the horizontal thrust of the curve BNA presumed rigid.[47]

Problem XVII:

Find a curve ANB whose horizontal thrust is equal at each point B.[48]

After having determined, in general, the horizontal thrust of an arch, he is looking for the one whose horizontal pressure is equal in each point. The curve that has that property is once more the catenary.

His next question concerns isoperimetric curves:

---

[46] Mascheroni (1785, p. 31): *Trovare una curva* DAC *(Fig. 19.) tale, che il centro G di gravità di qualunque arco* DAC *sia nella perpendicolare* XG *parallela all' asse* AK, *la quale passa per il concorso X delle due tangenti* CX, DX.

[47] Mascheroni (1785, p. 33): *Trovare la spinta orizzontale della curva* BNA *(Fig. 9.) supposta rigida.*

[48] Mascheroni (1785, p. 33): *Trovare una curva* ANB *(Fig. 9.) d'una spinta orizzontale eguale in ogni punto* B.

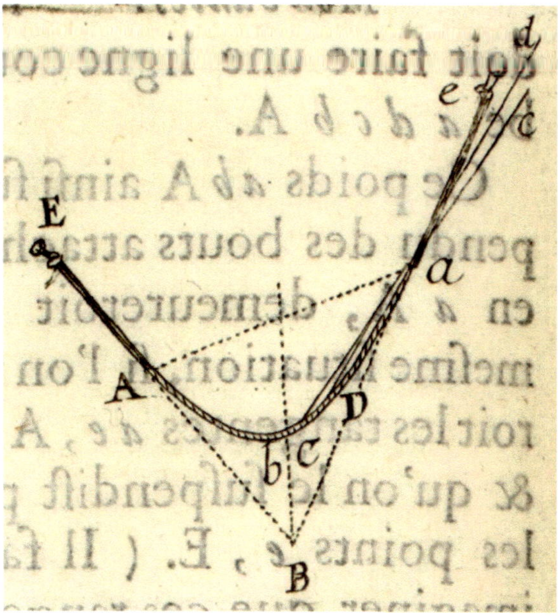

Fig. 29 The figure of Pardies (1673, p. 126)

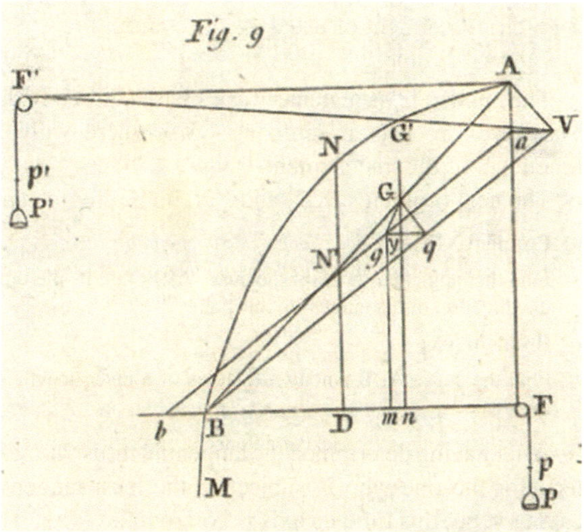

Fig. 30 Mascheroni's Fig. 9 (1785, Tav. III). Public domain

Problem XVIII:

Find a curve ANB among all the curves of a given length ANB which has the least horizontal thrust.[49]

---

[49] Mascheroni (1785, p. 34): *Trovate una curva* ANB (Fig. 9.) *tra tutte le curve della data lunghezza* ANB, *che abbia la minima spinta orizzontale.*

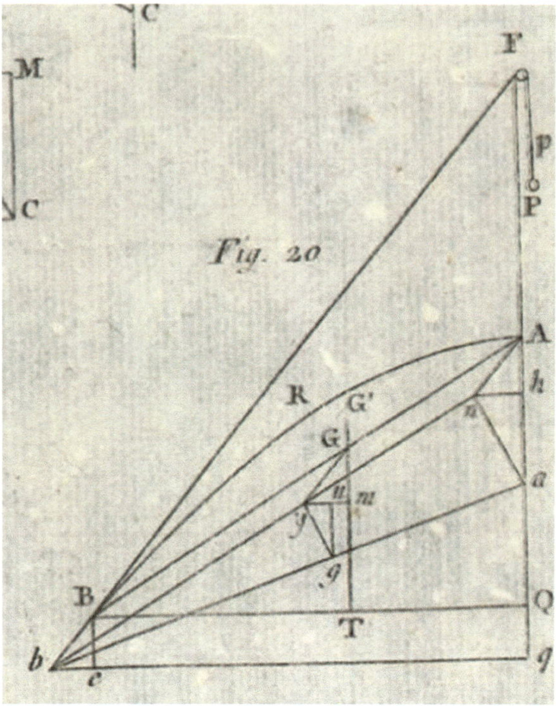

Fig. 31 Mascheroni's Fig. 20 (1785, Tav. V). Public domain

The answer is the catenary, as Euler had already demonstrated, and whose result Mascheroni repeats, minimizing $\int Sdx$ where S give the weight of the rope and is function s of the line element.

The next problems, XIX and XX, both refer to same figure (Fig. 31).

Problem XIX:

Find the tangential thrust of the curve ARB, that is, the thrust that is exercised on B along the direction of the tangent to this point.[50]

Problem XX:

Find the curve ARB among all curves of a given length ARB whose tangential thrust is minimal.[51]

After having determined the tangential thrust, he searches, out of all the isoperimeters, for the one who is subject to the least tangential thrust. He again finds the catenary, but this time its axis is horizontal.

Chapter V

In the second part of this chapter, we find two interesting problems.

[50] Mascheroni (1785, p. 34): *Trovare la spinta tangenziale della curva* ARB *(Fig. 20), cioè la spinta che esercita in* B *secondo la direzione della tangente in quel punto.*

[51] Mascheroni (1785, p. 36): *Trovare una curva ARB (Fig. 20) tra tutte le curve della data lunghezza ARB d'una minima spinta tangenziale.*

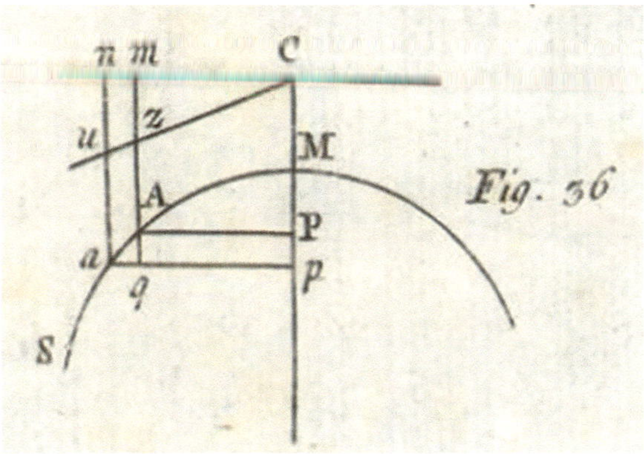

Fig. 32 Mascheroni's Fig. 36 (1785, Tav. IX). Public domain

Problem XXXII:

Find a curve of arch MAS such that, presuming the weight of the arch to be zero with respect to its load, the load itself is limited at the top by the horizontal line Cm.[52](Fig. 32)

As a solution he finds once more the homogeneous catenary.

Problem XXXIII:

Find the curve of the arch MAS that must sustain a fluid up to the height of the horizontal line Cm.[53]

In the case of an arch loaded with a fluid, Mascheroni finds the equation of the elastica. This result had been established, we have already seen, by Bossut in 1774. Mascheroni adds that in this manner one solves also the problem of the lintearia, the tarpaulin full of fluid that was posed by Jacob Bernoulli and also solved by Euler in a text already mentioned (Euler 1728).

Mascheroni underlines the fact that Euler solved that problem saying that the center of gravity of the figure bounded by the curve, the horizontal and two verticals linking them must be the lowest possible. Similarly, he said onc can solve his own problem by imposing the highest possible center of gravity.

Chaps. VI, VII and VIII, On Domes

We will limit ourselves to highlighting three problems in the chapters that deal with domes.

Problem XXXIV:

Find the equation of the curve ABM of a surface that passes through the centres of gravity of a cupola with circular base.[54](Fig. 33)

---

[52] Mascheroni (1785, p. 80): *Trovare una curva di arco* MAS *(Fig. 36.) tale, che supponendo nullo il suo peso a confronto del suo carico, lo stesso carico sia terminato superiormente dalla retta orizzontale* Cm.

[53] Mascheroni (1785, p. 81): *Trovare la curva dell'arco* MAS *che deve sostenere un fluido sino all'altezza dell'orizzontale Cm.*

[54] Mascheroni (1785, p. 85): *Trovare l'equazione della curvatura* ABM *(Fig. 37.) d'una superficie che passa per i centri di gravità degli elementi d'una cupola a base circolare.*

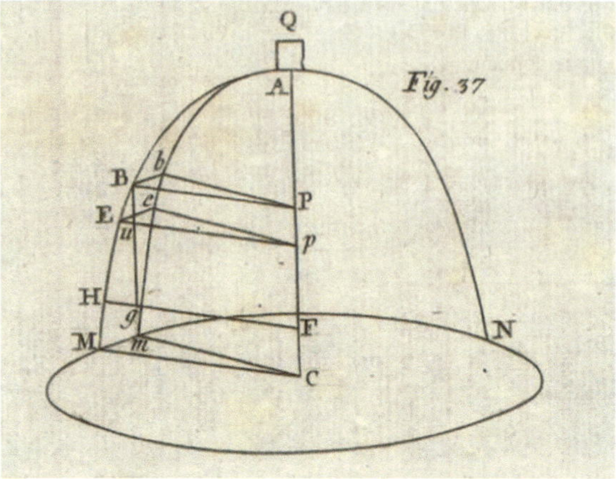

Fig. 33 Mascheroni's Fig. 37 (1785, Tav. IX). Public domain

In the case of this dome without thickness, he again finds the result obtained by Bossut (1776). Mascheroni underlines the fact that this result also gives the form of a bowl obtained by allowing a sail fixed to a circle to hang freely under its own weight.

The following two problems both refer to his figure 36 (see our Fig. 32).

Problem XLI:

Find the equation of the dome MAS loaded with a fluid, presuming the weight of the dome to be zero with respect to the load.[55]

Problem LVI:

To find the curve of the arch MAS that sustains an elastic fluid up to the horizontal line Cm.[56]

Problem XXXIII generalizes the type of fluid that composes the load.

Chap. XII, On the curves of equilibrium in case of a convergent gravity

Finally, in Chapter XII, Mascheroni comes back to Problem XXXIII and generalizes it in a remarkable way.

Problem LX:

Find the equation of the weight at equilibrium of the arch MNN′ perpendicular at each point to the rays of gravity NC and N′C′.[57](Fig. 34)

He then considers the problems of arches and domes loaded with a fluid, elastic, inelastic, homogenous or heterogenous, as a corollary of that problem and, along the way, the problem of the lintearia and of the elastica of Jacob Bernoulli.

---

[55] Mascheroni (1785, p. 104): *Trovare l'equazione per la cupola* MAS (*Fig. 36.) carica di fluido, supposto nulla il peso della cupola a confronto del carico.*

[56] Mascheroni (1785, p. 137): *Trovare la curva dell'arco MAS (Fig. 36.) che deve sostenere un fluido elastico sino all'altezza dell orizzontale* Cm.

[57] Mascheroni (1785, p. 141): *Trovare l'equazione del peso dell'arco* MNN′ *(Fig. 58.) perpendicolare in ogni luogo ai raggi di gravità NC, N′C′ per l'equilibrio.*

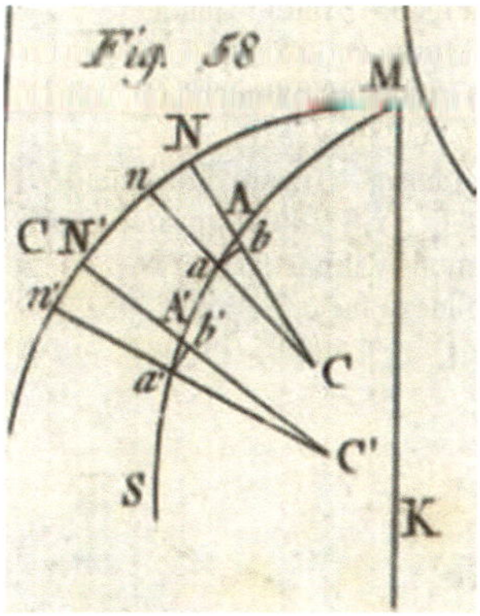

Fig. 34 Mascheroni's Fig. 58 (1785, Tav. IX) Public domain

In that way, he again finds the solutions to problems XXXIII and LVI for the arches but also of problem XLI for domes.

12 Conclusion, Part I: A Summary of Results

There was a series of results well known to Mascheroni:

- Jacob Bernoulli, in his Op. XLII (Bernoulli 1691a), had stated the connection that exists between the problems of the catenary and of the velaria, whereas his brother Johann (Bernoulli 1692a, b, c) affirmed (without formulating them explicitly because of their quarrel) the identity of their equations.
- On the other hand, Jacob had affirmed the identity of the lintearia and the elastica in his famous Op. XLVIII (Bernoulli 1692a) on the velaria, and in his Op. LVIII (Bernoulli 1694a) on the elastica, he adds the difference between the catenary and the lintearia in terms of the quantity to get extreme.

Mascheroni knew that if one reverses the catenary, one obtains an arch stable under its own weight. Let us now continue the reasoning and reverse the situations of the preceding table, or better, let us consider the solutions where one maximizes instead of minimizes a certain quantity, or as Mascheroni says, with regard to the arch loaded with liquid: given the length of the arch, the center of gravity of the fluid will remain the highest possible.

Let us recall problems I and II of the first part (see Mascheroni's Figs. 3 and 4 in our Fig. 19), where we showed that their solution was the foundation of all that follows in this treatise. Moreover, Mascheroni underlines the fact that his demonstration is valid in the case of his Fig. 3 as well as in the case of his Fig. 4. Likewise, also in the first part, he studies the equilibrium of three rods (see his Figs. 6 and 7 in our Fig. 21), and in Chap. II we have seen, that he treats simultaneously the catenary and the arch (see his Figs. 16 and 17 in our Fig. 26).

Fig. 35 Antoni Gaudí, Museo de la Sagrada Familia, polyfunicular model. Image: Par Canaan—Travail personnel, CC BY-SA 4.0, https://commons.wikimedia.org/w/index.php?curid=6964395

Finally, let us add an image of the model used by Antoni Gaudi to determine the form of the Sagrada Familia (Fig. 35).

13 Conclusion, Part II: Constitutive Equations

By tackling the problem of the elastica, one understands, as Truesdell has always pointed out, that it is appropriate to distinguish the universal laws of physics from constitutive laws that vary from one material to the other. Even without going into problems of plastic deformations, we have seen that Jacob is inevitably driven to this type of distinction.

Some of these notions are among the most complex and least intuitive of classical physics. Let us limit ourselves to mentioning the principal one: to know tension. Benvenuto said that this was the only one that can never be measured directly.

To transcribe in a mathematical language the constitutive laws that represent the physical properties of the various materials is really complex. The palliative that consists in giving a simple model or an ideal system is never easy to find.

Making an ideal system consists in formulating hypotheses on the constitutive equations, as we immediately see when Leibniz and Jacob Bernoulli admitted Hooke's law in order to determine the elastica. One could also make the same remark with the laws of collision discovered in the same period by Wallis, Wren and Huygens. Wallis considered the balls to be perfectly soft, while Wren and Huygens considered them to be perfectly elastic.

To the end of his life Jacob would pursue this study tirelessly, and was interested in the manner in which the elastica turns back to its initial position, a huge problem depending on each material, as his Med. CCLI (Jac. Bernoulli 1697b) shows:

Question: If the forces that strain the elastica are instantly removed, does it return to its straight state, at the same time in all of its parts, or faster in some and slower in others?

Benvenuto (1991) explained that those equations were entirely dependent on experiment, until Truesdell entered the question and tried to organize and axiomatize those elements of the theory too. In this sense, it is important to remark that Jacob Bernoulli furnished three important elements:

1. In Op. XLV (Jac. Bernoulli 1691b) he gave the first description of the perfectly rigid body, considering once more a lever fixed in the body.
2. In Op. LVIII (Jac. Bernoulli 1694a) he gave, the description of the perfectly elastic beam that returns to his previous state in function of the displacement it had undergone.
3. Let us add, the perfect flexibility where the tensions are along the tangents.

From this point of view, the text of Mascheroni is a model because he succeeds in focusing on the fundamental equations thanks to a perfect idealization: the voussoirs are perfectly rigid, perfectly smooth, and the applied forces are either vertical or perpendicular to the curve.

Abbreviations

TP *terminus postquem*, Date after which the text has been written.
TA *terminus antequem*, Date before which the text has been written

References[58]

Bernoulli, Daniel. 1726. Examen principiorum mechanicae, et demonstrationes geometricae de compositione et resolutione virium (St. 9), *Commentarii Academiae scientiarum imperialis Petropolitanae*, 1728, 126–142.

Bernoulli, Daniel. 1728. Methodus universalis determinandae curvaturae fili, a potentiis quamcunque legem inter se observantibus extensi (St. 15), *Commentarii Academiae scientiarum imperialis Petropolitanae*, vol. III, 62–69.

Bernoulli, Jacob. 1697a. Inter omnes Figuras Isoperimetras vel aequalium arearum invenire illam, quae habeat maximum vel minimum aliquod. Methodus Nova (Med. CCXXXIX) (T.P. 2/1697–T.A. 5/1697).

Bernoulli, Jacob. 1697b Quaestio Num elasterum, sublata subito vi tendente, eodem tempore in omnibus suis partibus in rectitudinem se restituat; an vero in aliis partibus citius, in aliis tardius? (Med. CCLI = VP IX) (T.P. 5/1697–T.A. 5/1698).

Bernoulli, Jacob. 1697c. Solutio problematum fraternorum, una cum Propositione reciproca aliorum (Op. LXXV), *Acta Eruditorum* (Mai 1697), 211–218.

[58] The dates cited are those of the composition, with the date of publication, if later, given after the source.

Bernoulli, Jacob. 1697d. Filum *ACDEFGB* extremitatibus suis *A* & *B* suspensum ab infinitis potentiis *C*, *D*, *E*, *F*, *G* juxta directiones quas vis *HE*, *HD*, *IE*, *KF*, *LG* agentibus extenditur. Quaeritur filum curvatura, ejus directio media LP, & vis qua 2dum *LP* impellitur (Med. CCXLV = VP XI) (T.P. 5/1697 - T.A. 5/1698).

Bernoulli, Jacob. 1689a. De aestimandis liquorum pressionibus in folle figurae cujuscunque (Med. CLXIV) (T.P. 1689–T.A. 9/3/1689).

Bernoulli, Jacob. 1689b. De figura veli vento inflati (Med. CLXV) (T.P. 1689–T.A. 9/3/1692).

Bernoulli, Jacob. 1690. Analysis Problematis, de inventione lineae descensus a corpore gravi percurrendae uniformiter (Op. XXXIX), *Acta Eruditorum*, May, 217–219.

Bernoulli, Jacob. 1691a. Specimen alterum Calculi differentialis in dimetienda Spirali Logarithmica, Loxodromiis Nautarum, & Areis Triangulorum Sphaericorum: una cum Additamento quodam ad Problema Funicularium, aliisque (Op. XLII), *Acta Eruditorum*, Juin, 282–290.

Bernoulli, Jacob. 1691b. Demonstratio centri oscillationis ex natura Vectis... (Op. XLV), *Acta Eruditorum*, Julii, 317–321.

Bernoulli, Jacob. 1692a Curvatura Veli (Op. XLVIII), *Acta Eruditorum*, mai, 202–207; *Opera*, 481–490.

Bernoulli, Jacob. 1692b. De linea directionis, juxta quam velum impellitur, de impulsus vi & de veli robore (Med. CLXXXVIII) (T.P. 1/1692 - T.A. 6/1692).

Bernoulli, Jacob. 1694a. Curvatura Laminae Elasticae. Ejus Identitas cum Curvatura Lintei a pondere inclusi fluidi expansi (Op. LVIII), *Acta Eruditorum*, Junii 1694, 262–276; *Opera*, 576–600.

Bernoulli, Jacob. 1694b. Curvae Elasticae puncta infinita et quantumvis proxima possunt geometrice inveniri (Med. CCXXVIII) (T.P. 9/1694–T.A. 12/1695).

Bernoulli, Jacob. 1695. Explicationes, Annotationes, et Additiones ad ea, quae in Actis sup. anni de Curva Elastica, Isochrona Paracentrica, & Velaria, hinc inde memorata, & partim controversa leguntur; ubi de Linea mediarum directionum, aliisque novis (Op. LXVI), *Acta Eruditorum*, Decembris, 537–553; *Opera*, 639–663.

Bernoulli, Johann. 1691. Solutio Problematis Funicularii (Op. IV), *Acta Eruditorum*, 274–276.

Bernoulli, Johann. 1691–2. Johannis Bernoulli Lectiones Mathematicae, de methodo integralium, aliisque, conscriptuae in usum Ill. Marchionis Hospitalii, Cum Auctor Parisiis ageret annis 1691–1692 (Op. CXLIX), *Opera Omnia*, vol. III, Lectio XXXVI – Lectio XXXVII, 491–497.

Bernoulli, Johann. 1692. Solution du Probléme de la Courbure que fait une voile enflée par le vent (Op. VII), *Journal des Sçavans*, 28 avril, 189–192.

Benvenuto, Edoardo. 1991. *An Introduction to the History of Structural Mechanics*, 2 vols. Part I, *Statics and Resistance of Solids*; Part II, *Vaulted Structures and Elastic Systems*. Heidelberg: Springer.

Bouguer, Pierre. 1734. Sur les lignes courbes qui sont propres à former les voûtes en dome, *Mémoires de l'Academie royale de Paris, année 1734* (1736), 149–164.

Bossut, Charles (Abbé). 1774. Recherches sur l'équilibre des voûtes, *Mémoires de l'Académie de Paris* (1777), 534–556.

Bossut, Charles (Abbé). 1776. Nouvelles recherches sur l'équilibre des voûtes en dôme, *Mémoires de l'Académie de Paris* (1778), 587–596.

Coulomb, Charles Augustin. 1773. Essai sur une application des règles de Maximis et Minimis à quelques problèmes de statique, relatifs à l'architecture, *Mémoires de Mathématique et de physique, présentés par divers savans*, vol. 7, année 1773 (1776), 343–382.

Couplet, Claude Antoine. 1729. De la poussée des voûtes, *Histoire de l'Académie Royale des sciences, avec les mémoires de mathématique et de physique* (1731), 79–117.

Euler, Leonhard. 1728. Solutio problematis de invenienda curva quam format lamina utcunque elastica in singulis punctis a potentiis quibuscunque sollicitata (E8), *Commentarii Academiae scientiarum imperialis Petropolitanae*, 1732, 70–84.

Euler, Leonhard. 1744. *Methodus inveniendi lineas curvas maximi minimive proprietate gaudentes, sive solutio problematis isoperimetrici lattissimo sensu accepti* (E65). Lausanne & Geneva: Marcum-Michaelem Bousquet.

Euler, Leonhard. 1750. Découverte d'un nouveau principe de mécanique (E177), *Mémoires de l'Académie de Berlin*, vol. 6 (1752), 185–217.

Euler, Leonhard. 1770. Genuina principia doctrinae de statu aequilibrii et motu corporum tam perfecte flexibilium quam elasticorum (E410), *Novi Commentarii Academiae scientiarum imperialis Petropolitanae*, Vol. XV (1771), 381–413.

Euler, Leonhard. 1775. Regula facilis, pro dijudicandi firmitate pontis aliusve corporis similis, ex cognita firmitate moduli (E 480), *Novi Commentarii Academiae scientiarum imperialis Petropolitanae*, Vol. XX (1776), 271–285.

Galilei, Galileo. 1638a. *Discorsi e dimostrazioni matematiche, intorno a due nuove scienze attenti alla mecanica e i movimenti locali,* Leiden: Elsevier. (Eng. trans. Stillman Drake, *Two New Sciences, Including Centers of Gravity & Force of Percussion.* Madison: University of Wisconsin Press, 1974.)

Galilei, Galileo. 1638b ca. Frammenti attenenti ai Discorsi. *Le opere di Galileo Galilei,* Edizione nationale, 20 vols., ed. Antonio Favaro. Florence: G. Barbera, 1898.

Gregory, David. 1697. Catenaria. *Philosophical Transactions*, vol. 19 (1695–1697), 637–652.

Heyman, Jacques. 1972. *Coulomb's Memoir on Statics: An Essay in the History of Civil Engineering.* Cambridge University Press.

Hooke, Robert. 1676. *A description of helioscopes, and some other instruments*, London: John Martyn.

Hooke, Robert. 1678. *Potentia restitutiva or spring, Explaining the Power of Springing Bodies.* London: John Martyn, 1678.

La Hire, Philippe de. 1695. *Traité de mécanique, où l'on explique tout ce qui est necessaire dans la pratique des Arts.* Paris: Imprimerie Royale.

La Hire, Philippe de. 1702. Remarques sur la forme de quelques arcs dont on se sert dans l'Architecture, *Mémoires de l'Académie de Paris*, 94–97.

Leibniz, Gottfried Wilhelm von. 1691. De linea in quam flexile se pondere proprio curvat, ejusque usu insigni ad inveniendas quotcunque medas proportionales & Logarithmos. *Acta Eruditorum* 1691 (June), pp. 277–281.

Mangili, Giuseppe. 1812. *Elogio di Lorenzo Mascheroni.* Milan: Stamperia Reale.

Mascheroni, Lorenzo. 1785. *Nuove ricerche sull'equilibrio delle volte.* Bergamo: Locatelli; 2nd ed., Milan: 1829.

Pardies, Ignace-Gaston. 1673. *Statique ou science des forces mouvantes.* Paris: Mabre- Cramoisy.

Parent, Antoine. 1704. Trouver la force avec laquelle il faut pousser un coin, pour séparer un corps ou directement, ou sur un point fixe, ou sur deux, *Mémoires de l'Académie de Paris*, 186–190.

Poinsot, Louis. 1803. *Éléments de statique.* Paris: Chez Calixte-Volland.

Poinsot, Louis. 1804. Mémoire sur la composition des momens et des aires dans la Mécanique, *Journal de l'Ecole Polytechnique*, Vol. VI, 148–181.

Poinsot, Louis. 1806. Mémoire sur la théorie générale de l'équilibre et du mouvement des systèmes, *Journal de l'Ecole Polytechnique*, Vol. VI, p. 182–241.

Poinsot, Louis. 1834. *Théorie nouvelle de la rotation des corps, Extrait d'un Mémoire lu à l'Académie des sciences de l'Institut*. Paris: Bachelier.

Truesdell, Clifford A. 1960. *The rational mechanics of flexible or elastic bodies 1638–1788. Introduction to Leonardo Eulerii Opera Omnia*, Ser. II, vol. 11. Zurich: Orell Füssli.

Varignon, Pierre. 1687. *Projet d'une nouvelle mécanique.* Paris: Veuve Ed. Martin, Jean Boudot et E. Martin.

Chapter 15
The Supports Problem: A Dead-End Street in the History of Stability of Structures

1 1695: La Hire's Problem

The birth and development of the study of elastic systems, which will progressively integrate structural engineering, has a rather long history which begins in 1695 with a problem proposed by Philippe de La Hire, left unsolved for many years. La Hire had asked the following question in Proposition XXIX of his *Traité de Mécanique*:

> Let three points BCD which are not on a straight line be given on a plane together with the three ropes BH, CH, DH which are linked to these three points by one of their extremities and which are linked together by their other extremity H, so that they form the sides of a pyramid of which the triangle BCD is the base. At point H where the three ropes meet together, let another rope HP be attached, at which the weight P is suspended; and the three ropes BH, CH, DH must be of a length such that the line HP, which is the direction of the weight, being extended in direction of the base of the pyramid, meets in I above the triangle BCD, the plane of that triangle: One must determine the relation between the three forces Y, X and Z which must sustain the weight P, along the directions BH, CH, DH.[1] (Fig. 1)

La Hire's conclusion is: "If one of those forces is therefore given as the weight *P*, it is obvious that the three others will be given too".[2]

But remarks immediately follow that make evident some difficulties:

---

[1] La Hire (1695, pp. 86–90): *Trois points BCD qui ne sont pas en ligne droite étant donnés sur un plan avec trois les cordes BH, CH, DH qui sont attachées à ces trois points par l'une de leurs extrémités, et qui sont jointes ensemble par l'autre extrémité H, en sorte qu'elles fassent les côtés d'une pyramide dont le triangle BCD est la base. Au point H où se joignent les trois cordes, soit attaché une autre corde HP d'où le poids P est suspendu; et il faut que les trois cordes BH, CH, DH soient de telles longueur, que la ligne HP, qui est la direction du poids, étant prolongée vers la base de la pyramide, rencontre en I au-dedans du triangle BCD, le plan de ce triangle: Il faut déterminer le rapport des trois puissances Y, X et Z, qui doivent soutenir le poids P, selon les directions BH, CH, DH.*

[2] La Hire (1695, p. 89): *Si l'une de ces puissances est donc donnée comme le poids P, il est évident que les trois autres YXZ le sont aussi.*

P. Radelet-de Grave, *Between Mechanics and Architecture*, Mathematics and the Built Environment 8, https://doi.org/10.1007/978-3-031-73530-1_15

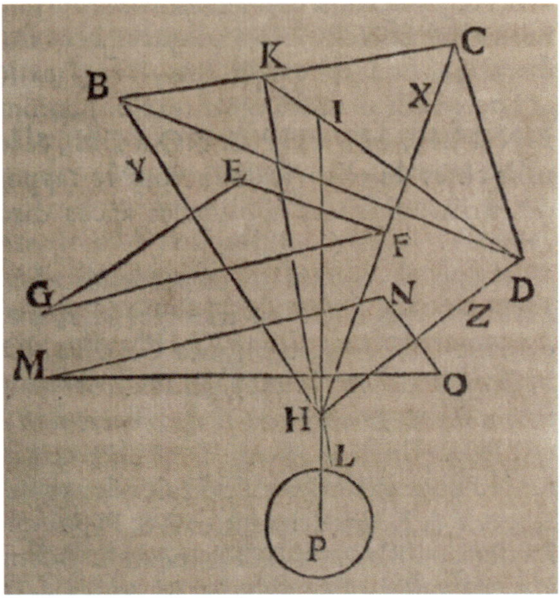

Fig. 1 1 Figure accompanying La Hire's Proposition XXIX (1695, p. 87). Public domain

> Remarks: It is obvious that if the point I would fall on one of the sides of the triangle ABC, as in K, the rope that would be stretched from the angle D of the triangle BCD, opposed to the side BC, on which falls the point I, would not help to sustain the weight P, and that only both others BH, DH would sustain it.[3]

In making this remark, La Hire solves the so-called "supports problem" (Benvenuto 1984, 1991) in a particular but incorrect way, by claiming that the third support does not carry anything. Most of those who came after him will say that the repartition of the load is indeterminate.

The problem remained buried in La Hire's *Traité* for 55 years before giving rise to something more than 25 articles.

2 1714: Johann Bernoulli and "Virtual Velocities"

In the meantime, but certainly without any connection to La Hire's text, Johann Bernoulli between August 1714 and February 1715 wrote letters to Bernard Renau d'Eliçagaray and Pierre Varignon in which appears the idea of *virtual velocities* (Radelet-de Grave 2004), which only much later will make it possible to solve the supports problem.

---

[3] La Hire (1695, p. 89): *Remarques: Il est évident que si le point I tomboit sur l'un des côtés du triangle ABC, comme en K, la corde qui seroit tenduë de l'angle D du triangle BCD, opposé au côté BC, sur lequel tombe le point I, ne serviroit de rien pour soutenir le poids P, et qu'il n'y auroit que les deux autres BH, DH qui le soutiendraient.*

In his letter to Renau, Bernoulli writes: "The energies of the powers or the forces are as the products of those powers or of those forces by the virtual velocities" [4]

And in his letter to Pierre Varignon, he writes: "In each equilibrium, there is an equality of energy of the absolute forces, that means between the product of the absolute forces by the virtual velocities".[5]

But it is in one of the formulations found in the letter to Renau that the concept of virtual velocities will become useful in architecture:

> By virtual velocities, I mean only the disposition to move that forces have in perfect equilibrium, when they are not currently moving; So in your figure 1 which is here the second, if this weight B inseparable from the line MB, is in equilibrium with the weights N and O, its virtual velocity is the small line BP, and the virtual velocities of N and O are CP and RP; and then the product of weight B by BP, which makes the energy of weight B, is equal to the two products of weight N by CP, and of weight O by RP, which are their energies. This is why to avoid equivocation, instead of saying that the powers or the forces are like the products of masses by velocities, you would have done better to express yourself by saying that the energies of the powers or the forces are like the products of these powers or of these forces by the virtual velocities" (Fig. 2).[6]

But all the results of those discussions between Johann, Renau and Varignon would only be published after Varignons's death in his *Nouvelle méchanique* (1725); even there, it was buried at the end of the book.

3 1772: The Supports Problem

La Hire's remark did not stimulate any reaction for 77 years. The first reaction was given by Vincenzo Riccati in "about the principles of mechanics" (1772, pp. 26–29). Unlike La Hire, who uses the parallelogram law of forces, or more precisely his

---

[4] Johann Bernoulli to Bernard Renau d'Eliçagaray, 12 August 1714, p. 8: *Les énergies des puissances ou des forces sont comme les produits de ces puissances ou de ces forces par les vitesses virtuelles.* The original of the letter is conserved at the Universitätsbibliothek Basel, and can be found on e-manuscripta: http://dx.doi.org/10.7891/e-manuscripta-24241.

[5] Johann Bernoulli to Pierre Varignon, 26 February 1715, p. 5: *Dans chaque équilibre il y a une égalité d'énergies des forces absolues, c'est-à-dire entre les produits des forces absolues par les vitesses virtuelles.* The original of the letter is conserved at the Universitätsbibliothek Basel, and can be found on e-manuscripta: http://dx.doi.org/10.7891/e-manuscripta-43010. Note that the letter to Varignon is dated 26 February 1715 and not 26 January 1717 as written in Varignon's *Nouvelle mécanique* (1725), a posthumous work.

[6] Johann Bernoulli to Bernard Renau d'Eliçagaray, 12 August 1714, p. 8: *J'entends par vitesse virtuelle la seule disposition à se mouvoir que les forces ont dans un parfait equilibre, où elles ne se meuvent pas actuellement; Ainsi dans votre Fig. 1 qui est ici la seconde, si ce poids B inseparable de la ligne MB, est en equilibre avec les poids N et O, sa vitesse virtuelle est la petite ligne BP, et les vitesses virtuelles de N et O sont CP et RP; et alors le produit du poids B par BP, ce qui fait l'energie du poids B, est egal aux deux produits du poids N par CP, et du poids O par RP lesquels sont leur energies; C'est pourquoy eviter l'equivoque, au lieu de dire que les puissances ou les forces sont comme les produits des masses par les vitesses, vous auriez peut-être mieux fait de vous exprimer ainsi, les energies des puissances ou des forces sont comme les produits de ces puissances ou de ces forces par les vitesses virtuelles.*

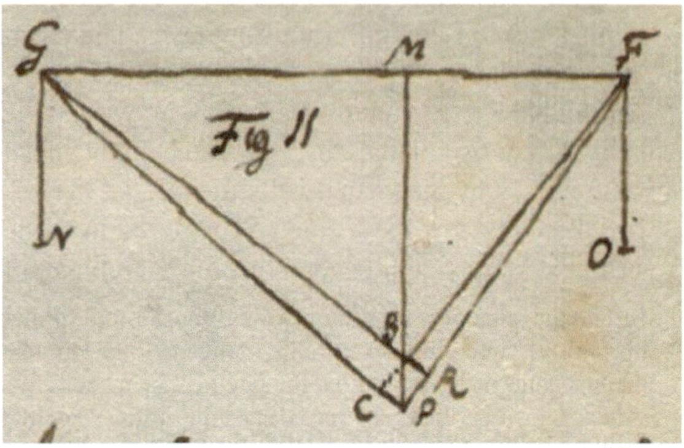

Fig. 2 Johann Bernoulli's Fig. II in the letter to Renau, 12 August 1714, p. 28. Public domain

triangle of forces,[7] Vincenzo Riccati uses a principle of virtual velocities to try to solve the problem. But he did not get any concrete solution.

The search for a solution for the supports problem would last for more than 55 years, from 1772 to 1827. But it appears that nothing really changed between 1695 and 1827, as we read in an article published in the *Annales de Mathématiques pures et appliquées* for 1826–1827 entitled "Note sur un paradoxe de statique" (Note on a paradox of statics) (1826–1827), by Joseph Diez Gergonne:

> One knows that if a heavy body is suspended freely at one or two vertical cords, or even by three vertical ropes not situated in the same plane, the tensions of those cords are completely determinate and even easy to calculate; and that it is still the same with the pressures experienced by one or two points of a horizontal plane, upon which a heavy body rests, or even for three such points, if they are not on a straight line.
>
> But it is generally admitted in statics that, if a heavy body is suspended by more than three vertical cords, or even by three cords situated in a same plane, the tensions of those cords remain indeterminate, and that the same occurs with the pressures exercised on a fixed horizontal plane, in the points upon which rests a heavy body, when those points are more than three, or, when they are only three, but on a straight line.[8]

[7] See "The Use of a Particular Form of the Parallelogram Law of Forces for the Building of Vaults (1650–1750)" Chap. 4 in this present volume.

[8] Gergonne, 1826–1827, pp. 75–79: *On sait que, lorsqu'un corps pesant est suspendu librement par un ou deux cordons verticaux, ou même par trois cordons verticaux non compris dans un même plan, les tensions de ces cordons sont complètement déterminées, et même faciles à calculer; et qu'il en est encore de même des pressions éprouvées par un ou deux points d'un plan fixe horizontal, par lesquels y pose un corps pesant, et même pour trois pareils points, lorsqu'ils ne sont pas en ligne droite. Mais il est généralement admis en statique que, si un corps pesant est suspendu par plus de trois cordons verticaux, ou même par trois cordons situés dans un même plan, les tensions de ces cordons demeurent indéterminées, et qu'il en est de même aussi des pressions exercées sur un plan fixe horizontal, dans les points par lesquels y pose un corps pesant, lorsque ces points sont au nombre de plus de trois, ou, lorsqu'étant au nombre de trois seulement, ils se trouvent appartenir à une même ligne droite.*

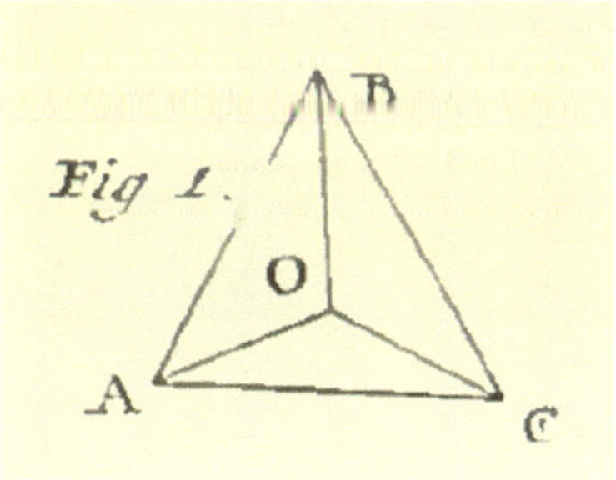

Fig. 3 Euler's Fig. 1, from "De pressione ponderis in planum cui incumbit" (Euler 1774, Tab. II). Public domain

4 1773: Leonhard Euler's Influence

At an unknown date, Leonhard Euler took up that problem in a text entitled "Von dem Drucke eines mit einem Gewichte beschwerten Tisches auf eine Fläche" (On the pressure of a table charged with a weight upon a surface); Jacob II Bernoulli (1759–1789), grandson of Johann I, published that draft in Euler's *Opera Omnia* (Euler 1794). In it Euler only treats both simple problems of three and four feet, without any generalisation. These are really difficult to follow and don't seem to have been understood by many people, but before the time of that publication Euler had already completed his solution in a text written in 1773 entitled "De pressione ponderis in planum cui incumbit" (On the pressure exerted by a weight on the plane on which it rests) (Euler 1774).

Euler starts with what he calls the easiest problem: a weight on a plane sustained by three feet A, B, C. He also traces the normal to the plane passing through the center of gravity of the weight, as shown in his Fig. 1. Point O is the projection of the center of gravity on the horizontal plane. Euler joins O to each point A, B, C and says that:

> 2. the total pressure is to the pressure in A, or in B, or in C, as the whole area of the triangle ABC to the area of the triangle BOC or AOC or AOB; from this one understands that the pressures in the various feet cannot be equal unless the point O is in the center of gravity of the triangle ABC (Fig. 3).[9]

Euler thus solves the problem of three support points and goes on with four such points, A, B, C, D, belonging to the same plane. But this time he considers that the support points may move outside their plane and skip in or out the plane by an infinitely small length Aα, Bβ, Cγ, Dδ. But with one more condition αβδγ, must all remain in a same plane. He then gives the following general principle:

[9] Euler (1774, p. 290): *tota pressio se habebit ad pressionem in puncto A, vel B, vel C, quemadmodum area totius trianguli ABC ad aream trianguli, sive BOC, sive AOC, sive AOB; ex quo intelligitur pressiones singulorum pedum inter se aequales non fore, nisi punctum O in ipsum centrum gravitatis trianguli ABC incidat.*

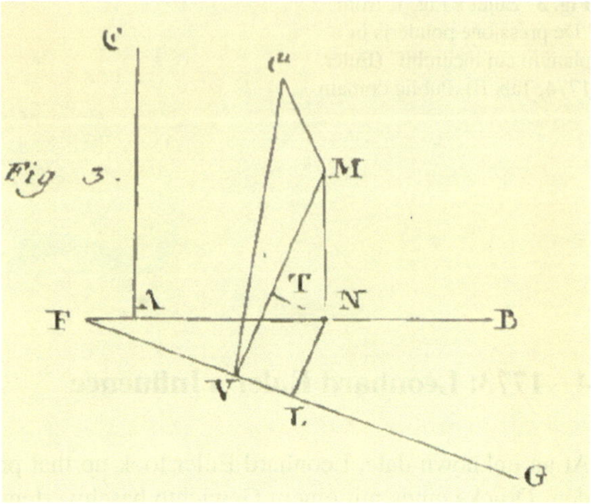

Fig. 4 Euler's Fig. 3, from "De pressione ponderis in planum cui incumbit" (Euler 1774, Tab. II). Public domain

> M being the extremity of the foot of the considered object and Mμ the pressure in M, all μ's must be in the same plane. … The line FG is the intersection of the plane in which the weight must rest and of the plane of all μ's (Fig. 4).[10]

In Problem 1:13 (1774, p. 296) Euler comes back to the case of three support points and solves it with a new method. He has called attention to this problem and many scientists will start thinking about it. What is important is the idea of leaving some motions possible for the plane containing the four support points.

A few years later Jean le Rond d'Alembert, in an article entitled "Sur quelques questions de mécanique" (On some questions of mechanics) (1780) also treated this kind of problem (Fig. 5).

D'Alembert treats the problem of three support points A, B, D, but this time they are aligned. C is the center of gravity. His question is, which part of the body does each support point sustain? D'Alembert immediately sees that the problem is indeterminate. Because he just has two equations: $x =$ weight of point A, $y =$ weight of point B and $z =$ weight of point C. Thus, one equation is:

$$x + y + z = P.$$

Now the weights are equidistant thus: $AC = a$, $BC = b$, $CD = b + c$. Thus, he has

$$xa = yb + zb + zc.$$

[10] Euler (1774, p. 292): *sit punctum M sive extremitas cuiuspiam pedis, sive elementum quodpiam basis pro quo pressio quaeritur. Concipiatur ibi perpendiculariter erecta linea M · μ ipsi pressioni proportionalis, atque necesse est, omnia ista puncta μ in quopian plano terminari … quem in sinem sit recta FG intersectio, qua planum cui pondus incumbit, a plano illo per omnia puncta μ traseunte intersecatur.*

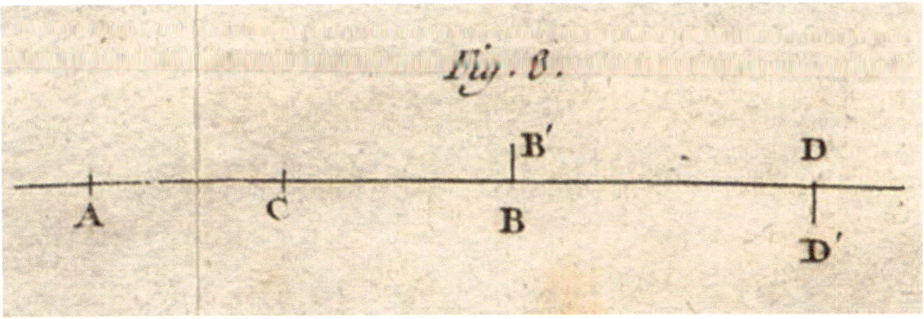

Fig. 5 D'Alembert's Fig. 8, from "Sur quelques questions de mécanique" (D'Alembert 1780, Pl. 1). Public domain

One therefore has three unknowns and only two equations. The problem is not solvable.

Let us give a summary of the solutions of the problem of support points. Two special cases are not solvable: (a) Three aligned supports points; (b) Four support points.

In 1788 Abbé Charles Bossut, in an Italian and completed version of his *Trattato elementare di Meccanica* (1788) cites both Euler (1774) and d'Alembert (1780).[11] Bossut starts considering two inclined planes in front of each other and a sphere leaning upon both of them. It is thus in equilibrium. Then he considers a body on two and three support points. Those two problems are determinate. He does the calculation that Euler had not done. He next considers the indeterminate case of the three support points on the same line, for which he refers the reader to d'Alembert (1780). Finally, he considers the case of more than three support points, just as Euler did. Bossut relates what Euler does, considering the floor under the four support points formed with clay and supposes that the pressure exerted by the support points are proportional to the quantity those support points penetrate into the clay.[12]

Paolo Delanges, in his "Memoria sulle pressioni esercitate da un corpo sostenuto da tre o più appoggi collocati nello stesso piano" (Memoir on the pressures exerted by a body supported by three or more supports located in the same plane) (1790), starts, as do many authors, with the problem of three unaligned supports points, thus forming a triangle ABC, as shown in Fig. 6. O is the projection of the center of gravity in the plane of the three unaligned supports points. QO is parallel to AB and OP to AC. G is the weight or the total pressure of the body RST.

Euler's theory says that the pressure on point A, let us call it P_A, the pressure on point B, P_B and the pressure in point C P_C is such that

---

[11] The earlier French version of Bossut's work, *Traité élémentaire de mécanique statique* (1772) precedes Euler (1774) and d'Alembert (1780) and therefore this addition does not appear in it.

[12] Bossut (1788, p. 212): *Se il corpo s'appoggia con più di tre punti, il problema, considerato sempre secondo le leggi ordinarie della Meccanica, è indeterminato, qualunque sia la disposizione rispettiva degli appoggi. Eulero, per renderlo determinato in generale, riguarda il piano su cui posano gli appoggi, come formato d'argilla, e suppone che le pressioni degli appoggi sono proporzionali alle quantità onde i piedi del corpo si conficcano nell'argilla.*

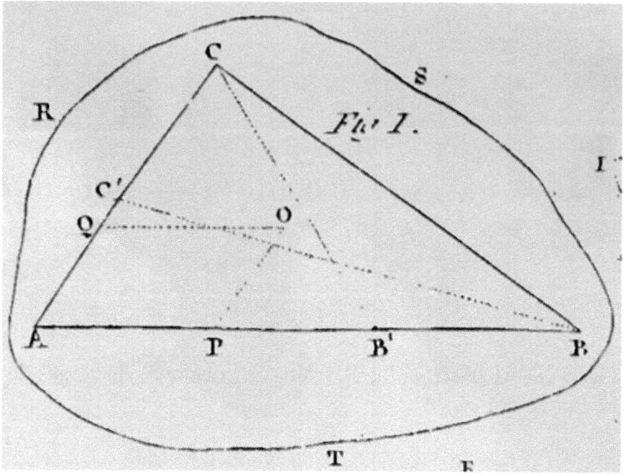

Fig. 6 Delanges's Fig. 1, from "Memoria sulle pressioni esercitate da un corpo sostenuto da tre o più appoggi collocati nello stesso piano" (Delanges 1790, p. 129). Public domain

$$PA = G(1 - AP/AB - AQ/AC) \text{ and}$$
$$PB = G \cdot AP/AB \text{ and } PC = G \cdot AQ/AC.$$

But Delanges says that we could maintain the equilibrium by changing the place of the center of gravity O on the line QO and placing C in C′ and B in B′, so the solution is not unambiguous.[13]

Several years later Delanges published "Nuove considerazione intorno alla pressione di un corpo sostenuto da tre e più appoggi in un piano orizzontale" (New considerations about the pressure of a body sustained by three or more support points on a horizontal plane) (1799).

This time he estimated that the problems is solved but he had fixed the support points. He had thus given more information in his question. This was not the case in his first article (Fig. 7).

In 1811, he published a new and final "Analisi e soluzione sperimentale del problema delle pressioni" (Analysis and experimental solution to the problem of the pressures).

---

[13] Delanges (1790, pp. 108–109): *Ma siccome immaginandosi, come è lecito, che il centro 0 di gravità cadesse in qualsivoglia altro punto della parallela Q0 anche prolungata, e l'appoggio B fosse trasportato ovunque nella AB, rimane costante la pressione in C, variando solo quelle degli appoggi AB, e che rimarrebbe similmente costante la pressione in B cangiandosi quelle in AC, se il centro O situato fosse ovunque nella parallela PO e il punto C nella retta CA; cosi è chiaro, che la teoria euleriana non dà del problema che una soluzione in parte determinata ed in parte no, mentre se fosse determinata interamente, il cangiamento di luogo del punto 0, o d'uno degli appoggi dovrebbe alterare il carico non di due, ma di tutti e tre insieme. Ciò che poi rende incerta la suddetta teoria si è, che sostituito all'appoggio B il B′, ed al C il C′ in modo che il centro O cadesse fuori delle rette CB′, BC, e che per conseguenza il corpo RST tendesse a capovolgersi, com'è manifesto, o sulla retta CB, ovvero sulla BC; nulla ostante resterebbero determinate le pressioni de' punti CAB′, o C′AB come se dovesse sostenersi in equilibrio.*

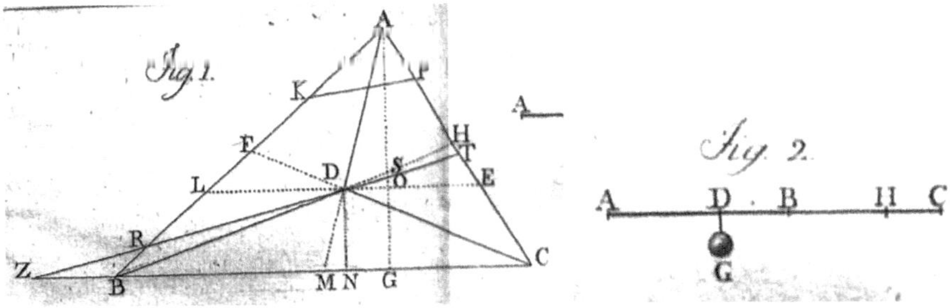

Fig. 7 Delanges's Figs. 1 and 2, from "Nuove considerazione intorno alla pressione di un corpo sostenuto da tre e più appoggi in un piano orizzontale" (Delanges 1799, Tav. IV). Public domain

As Edoardo Benvenuto (1991, p. 449) puts it: "[Delanges'] abilities are not displayed at their best in his treatment of the supports problem; in fact, his contributions are conspicuously—even ingenuously—wrong". And I would add that this provides a measure of the difficulty of a problem that looked quite easy at first sight.

5 1788: The Non-influence of Lagrange's *Mécanique Analytique*

In 1788 Lagrange's major work, *Mécanique analytique* (Analytical mechanics), came out. It contains two parts in one volume: a first part on "mechanics or statics", and a second on "mechanics or dynamics". The first to follow Lagrange in applying the new "principle of least action" to the supports problem was Paolo Paoli, in his "Memoria sopra alcuni problemi meccanici" (Memoir on some problems of mechanics) (1792).

After having recalled the work of Euler, he writes:

> A famous problem of mechanics is which is sought the pressure suffered by the various supports that sustain a body on a horizontal plane. Leonhard Euler proposed a first solution of that problem in tome XVIII of the Novi commentarii Academiae Scientiarum Imperialis Petropolitanae. This great Geometer observed that the principles of statics were not sufficient in order to solve it in all cases; he resorted to an assumption that seemed exact at first sight, but when examined attentively is seen in fact to be insufficient... Among those who treated that subject, one must, in particular mention d'Alembert...[14]

But Paoli immediately adds ironically, that contrary to the others, it was Delanges who had given a solution to the problem.

[14] Paoli (1792, pp. 534–535): *Un celebre Problema di Meccanica è quello, in cui si cerca la pressione che soffrono i vari appoggi, dai quali un corpo è sostenuto sopra un piano orizontale. Il Sig. Leonardo Euler ha preso a risolvere questo Problema nel Tomo XVIII of the nuovi Commentari di Pietroburgo. Questo gran Geometra osservando, che i principi della Statica non erano sufficienti per determinarlo in tutti i casi, ebbe ricorso ad una ipotesi, che pare a prima vista esatta, ma poi esaminata attentamente si trova affatto insufficiente. ... Tra quelli, che hanno trattato di questa materia, si deve in particolar modo rammentare il Sig. d'Alembert.*

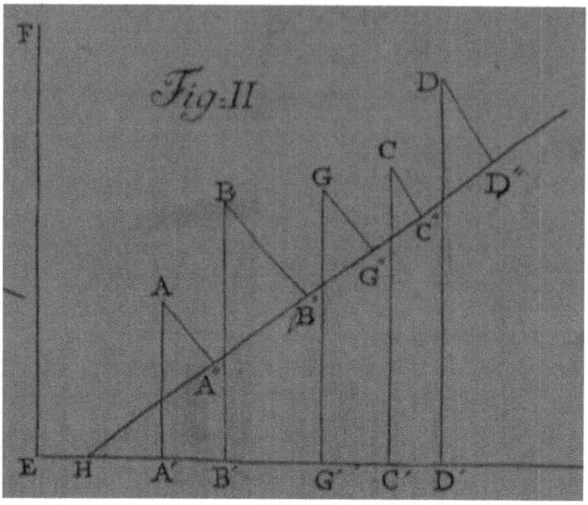

Fig. 8 Paoli's Fig. II, from "Memoria sopra alcuni problemi meccanici" (Paoli 1792, facing p. 559). Public domain

In the first part of Paoli's text, entitled "Della pressione di un corpo sopra varj punti di un paino orizontale" (On the pressure of a body upon various points of an horizontal plane) (Paoli 1792, p. 537ff), he aims at a general solution as follows (Fig. 8).

He first considers that all the weight P is concentrated in the center of gravity G:

> If we consider the weight of the body P unified at the center of gravity, which is permitted, all of force P will act on the plane in the point G perpendicularly to the plane. And this will be the force that produces the pressure at the points A, B, C, D, etc.[15]

If one wants to prevent that motion:

> First of all it will be necessary to prevent the progressive motion of the plan, which can only be vertical; thus because of the principle of virtual velocities, if one gives a little impulse to the plan, such that δz would be the little space that would be traveled vertically by the point G, one gets $P\delta x - (P' + P'' + P''' + P''''etc.)\delta z = 0$, because actually we only consider vertical motions, and because δz is the little space traveled by every point of the plan. We thus get the equation
>
> $$P' + P'' + P''' + P'''' + etc. = P \tag{1}$$
>
> That indicates, that the sum of all pressures must be equal to the weight of the body. But besides the progressive motion of the plan, the motion of rotation must also be avoided. Let us take whatever a straight line EA′ in this plan, and from each point G, A, B, C, D, etc. Let us draw from those points the perpendiculars GG′, AA′, BB′, CC′, DD′, etc., which we will call respectively y, y′, y″, y‴, y⁗, etc. Now let us suppose the axis EA′ motionless and let us give a little impulse to the plan, and let δu be the little arche described by the point G around that axis; we'll obtain $\frac{y'\delta u}{y}, \frac{y''\delta u}{y}, \frac{y'''\delta u}{y}, \frac{y''''\delta u}{y}$, etc. The little arches described respectively by the points A, B, C, D, etc.: and because of the principle of the virtual velocities, we'll obtain

[15] Paoli (1792, p. 537): *e le concepiamo il peso del corpo P riunito nel centro di gravità, lo che è permesso, la forza P agirà sul piano nel punto G perpendicolarmente ad esso, e sarà quella forza, che produce le pressioni nei punti A, B, C, D, ecc.*

$$P\delta u - \frac{P'y' + P''y'' + P'''y''' + P''''y'''' + etc.)}{y}\delta u = 0,$$

that is

$$P'y' + P''y'' + P'''y''' + P''''y'''' + etc. = Py \tag{2}$$

Which equation we could also deduce from the theory of momenta.

Similarly, if we take another axis EF perpendicular to the first one, and let us call the lines EG, EA′, EB′, EC′, ED′, etc. respectively x, x′, x″, x‴, x⁗ etc., we'll obtain[16]

$$P'x' + P''x'' + P'''x''' + Px'''' + etc. = Px \tag{3}$$

He concludes.

> III. As therefore the principles of mechanics give us only three equations in order to determine the pressures exercised by various supports of a body, it follows that it will only be possible to solve that problem when there are three supports but will remain indeterminate in case that number is greater. And in the case of three support points the pressures will have to be deduced from those equations.[17]

---

[16] Paoli (1792, p. 538): *Primieramente adunque converrà che sia impedito al piano il moto progredivo, il quale non può essere che verticale; e quindi in vigore del principio delle velocità virtuali se si dà al piano un piccolo impulso in modo, che* δz *sia lo spazietto percorso verticalmente dal punto G, si avrà* $P\ \delta x - (P' + P'' + P''' + P'''' + ec.)\ \delta z = 0$, *perchè qui non consideriamo per ora che il solo moto verticale, e perciò* δz *è lo spazietto percorso da tutti i punti del piano. Abbiamo dunque l'equazione.*

$$P' + P'' + P''' + P''''ec. = P \tag{1}$$

La quale indica, che la somma di tutte le pressioni dev' essere eguale al peso del corpo. Oltre il moto progressivo del piano deve anch' essere impedito il moto di rotazione. Prendiamo una retta qualunque EA′ in questo piano, e da tutti i punti G, A, B, C, D, ec. tiriamo su di essa le perpendicolari GG′, AA′, BB′, CC′, DD′, ec, che chiameremo rispettivamente y, y', y'', y''', y'''', *ec., Adesso supponendo immobile l'asse EA' diamo un piccolo impulso al piano, e sia* δu *l'archetto dal punto G descritto intorno a quest'asse; saranno* $\frac{y'u}{y}, \frac{y''u}{y}, \frac{y'''u}{y}, \frac{y''''u}{y}$, *ec. gli archetti rispettivamente descritti dai punti A, B, C, D, ec.: ed in forza del principio delle velocità virtuali avremo* $Pu - \frac{(P'y'+P''y''+P'''y'''+P''''y''''+etc.)}{y}u = 0$ *cioè.*

$$P'y' + P''y'' + P'''y''' + P''''y'''' + etc. = Py \tag{2}$$

la qual' equazione si poteva anche dedurre dalla Teoria de' momenti. Similmente, se prendiamo un altro asse EF perpendicolare al primo, e chiamiamo le rette EG, EA', EB′, EC′, ED′, ec. rispettivamente x, x', x'', x''', x'''' *ec., avremo.*

$$P'x' + P''x'' + P'''x''' + P''''x'''' + etc. = Px. \tag{3}$$

[17] Paoli (1792, p. 539): III. *Siccome pertanto i principi della Meccanica non ci danno per determinare le pressioni esercitate da varj appoggi di un corpo che tre equazioni, ne segue che il problema li potrà risolvere quando tre sono gli appoggi, ma resterà indeterminato, quando il numero degli*

$$P' + P'' + P''' = P$$
$$P'y' + P''y'' + P'''y''' = Py$$
$$P'x' + P''x'' + Px''' = Px.$$

Thus, the same difficulty arises with the use of Lagrange's principle of virtual velocities. The problem cannot be solved in case of more than three support points. It is still a dead-end street. Lagrange's mechanics will play an important role after the birth of the theory of elasticity.

In the case of a body resting more than one inclined plane, Paoli shows that until six support points one gets enough equations to solve the problem but that it is impossible to solve the problem for more than six support points.

In the face of this impossibility, many authors, such as Delanges (1811), Lorgna (1794) and Paoli (1802), would engage in rather sterile verbal jousting. They would all forget about the new point of view introduced by Lagrange.

In his article "Dell'azione di un corpo retto da un piano immobile esercitata ne' punti di appoggio che lo sostentano" (The action of a body supported by an irremovable plane exerted in the points of support which support it) (1794), Anton Maria Lorgna observes that "when a body is supported by a plane, the mechanics have not yet determined how much each point of this plane supports respectively".[18] After having mentioned the failure of several authors who we have already mentioned, he continues with a quote from D'Alembert, who claims "that there is still something lacking in the principles of Mechanics, and that there are cases where the laws known until now appear insufficient".[19] Signaling thus that the evil is deeper than his contemporaries thought. But his text only points out the failures of his predecessors.

The first to really understand Lagrange's *Analytical Mechanics* was Fourier in 1798. In a "Mémoire sur la statique" (Memoire on statics) (1798), he writes:

> Galileo and Descartes had then some knowledge of that truth. Johann Bernoulli, who is properly speaking the inventor, announced it at the beginning of the century, without publishing any demonstration (see the second volume of Varignon's Mechanics). This discovery was communicated to Varignon at the time he was composing his new mechanics: he was able to prove the theorem in various particular cases; but that enumeration, necessarily incomplete, didn't drive him to the general demonstration. Mechanics was then rapidly perfected: this science does not have today anymore difficulties proper and considered in its broadest connections, it reduces to a question of calculation. This result, one of the most beautiful

appoggi è maggiore. E nel caso di tre appoggi i valori delle pressioni P', P'', P''' si dovranno dedurre dall'equazioni.
$P' + P'' + P''' = P;$
$P'y' + P''y'' + P'''y''' = Py;$
$P'x' + P''x'' + P'''x''' = Px.$

[18] Lorgna (1794, p. 178): *Quando un corpo è retto da un piano, la Meccanica non è giunta ancora a determinare, quanto sostenga respettivamente ciascun punto del medesimo piano.*

[19] Lorgna (1794, p. 179), quoting D'Alembert 1780, p. 45: *On voit ... qu'il manque encore quelque chose aux principes de Méchanique, & qu'il y a des cas où les loix connues jusqu'ici, paroissent insuffisantes.*

> that has been obtained in exact sciences, is the purpose of Analytical Mechanics. The author of this great work saw in Bernoulli's longtime sterile proposition a primordial truth whose most fruitful consequences follow without exception and without effort.[20]

But despite Lagrange's text and Fourier's comments, Gianfrancesco Malfatti proposes a very long essay entitled "Tentativo sul problema delle pressioni, che soffrono gli appoggi collocati agli angoli di una figura, derivate da un peso posto dentro la sua aja" (Attempt at the problem of the pressures sustained by the supports locate at the angles of a figure, derived from a weight placed inside its plane" (1799). Gregorio Fontana (1799) proposes a new version of the problem of the door: to evaluate the pressure of a door on its hinges. Having difficulties with that problem, he asks the help of Mascheroni. But even with his help, he does not find very helpful solutions. As we have already said, the same problem was then taken up again by Paoli in "Sul problema degli appoggi" (On the supports problem) (1802), who received an answer from Malfatti in "Brevi riflessioni alla critica del tentativo sul problema delle pressioni" (Brief reflections on the critique of the attempt at the problem of pressures) (1803).

In his *Eléments de statique* (Elements of Statics), Louis Poinsot treated the "equilibrium of a body leaning against an unshakeable plane" (1803, p. 141). He then considers in turn one point of support, then two, then any number whatsoever. In this last case, he gives a supplementary condition:

> …that all forces applied to the system must reduce to one, perpendicular to the fixed plane, and whose value is not positive. In the second place, I say that its direction must meet the plane inside the polygon formed by the support points A, B, C, D, etc.[21]

The last part of the "second place" introduces what is called today the "support polygon".

In a corollary Poinsot considers "the pressures exerted by the resultant of the system of forces on the various points of support":

> But there are only three equations to solve this problem and one condition saying that the unknowns must all be positive, from which it follows that the various pressures exerted on

---

[20] Fourier (1798, p. 478): *Galilée et Descartes eurent depuis quelque connaissance de cette vérité. Jean Bernoulli, qui en est, à proprement parler, l'inventeur, l'annonça au commencement du siècle sans en publier de démonstration (voir le IIe Volume de la Mécanique de Varignon). Cette découverte fut communiquée à Varignon dans le temps qu'il composait sa nouvelle Mécanique: il parvint à prouver le théorème dans plusieurs cas particuliers; mais cette énumération, nécessairement incomplète, ne l'a point conduit à la démonstration générale. Depuis, la Mécanique a été rapidement perfectionnée: cette science n'a plus aujourd'hui de difficultés qui lui soient propres, et, considérée dans ses rapports les plus entendus, elle se réduit à une question de calcul. Ce résultat, l'un des plus beaux que l'on ait obtenus dans les sciences exactes, est l'objet de la Mécanique analytique. L'auteur de ce grand Ouvrage a vu, dans la proposition longtemps stérile de Bernoulli, une vérité primordiale, dont les conséquences les plus fécondes découlent sans exception et sans effort.*

[21] Poinsot (1803, pp. 145–146): *que toutes les forces appliquées au système doivent se réduire à une seule, perpendiculaire au plan fixe, et dont la valeur ne soit pas positive. In second lieu, je dis que sa direction doit rencontrer ce plan dans l'intérieur du polygone formé par les points d'appui A, B, C, D,* etc.

the plane remain unknown when there are more than three support points, and even if they are only three but they fall in a straight line.[22]

Up to this point, we have a good summary of what had been said in various ways by the predecessors, but Poinsot adds what is in fact the solution of the paradox:

All bodies are more or less flexible and elastic, and when they are pressed against each other by different points situated in the same plane, the total pressure is distributed in a particular manner by virtue of these physical properties, and of the three conditions given above (§125), to which the individual pressures must always satisfy. But it would be necessary to take these properties into account in order to find as many equations as there are points of contact, and thereby know the various pressures, and it is a very delicate question of physics, which we will not seek to discuss here.[23]

In 1809, Giuseppe Venturoli, puts together his lectures on mechanics in a first volume, and his lectures on hydraulics in a second[24] and publishes a second edition of his *Elementi di meccanica e d'idraulica* (Elements of Mechanics and Hydraulics). Chapter XVI of the first volume (1809, pp. 46–48) is entitled "Pressures on supports of a rigid system in equilibrium" and solves the problem of hinges:

104. Proposition I. Let be a rigid system in equilibrium, and sustained by a fulcrum or a fixed pin. Keeping the denominations used until now, the pressures p, q, r sustained by the pin in the directions of the x, y, z are determined by the equations

$$p = \sum \cdot P;\, q = \sum \cdot Q;\, r = \sum \cdot R$$

In fact, the pressures that the fulcrum sustains are equal and opposite in direction to the forces that should be applied in order to maintain the equilibrium, if the fulcrum were removed. Now for the three first equations of equilibrium (§100) it is obvious that one should apply in the direction opposite to x, y, z, the forces $\sum \cdot P$; $\sum \cdot Q$; $\sum \cdot R$.

105. Corollary. The fulcrum sustains the effort of the powers P, Q, R just as if those were directly applied to it.

106. Proposition II. Let the system be in equilibrium around the fixed axis OZ (Fig. 13) fixed in two points, or pins M, N (Fig. 9).

One calls p, q, r the pressures that support, in the direction of the x, y, z, the pin M that corresponds to the abscissa OM = ζ, and let p′, q′, r′ be the pressures sustained by the other

[22] Poinsot (1803, pp. 147–148): *Des pressions exercées par la résultante des forces du système sur les différents points d'appui.*
Mais comme on n'a que ces trois équations, avec une seule condition que les inconnues z, z', z'', z''', etc. *doivent être toutes positives, il s'ensuit que les diverses pressions exercées sur le plan demeurent indéterminées, lorsqu'il y a plus de trois appuis, et même lorsqu'il n'y en a que trois, s'ils tombent en ligne droite.*

[23] Poinsot (1803, pp. 152–153): *Tous les corps sont plus ou moins flexibles et élastiques; et lorsqu'ils sont pressés les uns contre les autres par différents points situés dans un même plan, la pression totale se distribue d'une manière particulière en vertu de ces propriétés physiques, et des trois conditions données ci-dessus (125), auxquelles les pressions individuelles doivent toujours satisfaire: mais il faudrait tenir compte de ces propriétés, pour trouver en tout autant d'équations qu'il y a de points de contact, et connaître par-là les diverses pressions; et c'est une question très-délicate de physique, que nous ne chercherons pas à discuter ici.*

[24] I was unable to find and consult this first edition published in 1806.

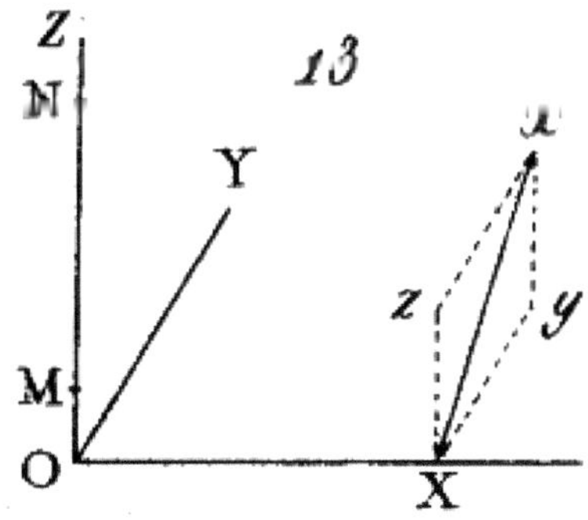

Fig. 9 Venturoli's Fig. 13 (1809, Tav. II). Public domain

pin N that corresponds to the abscissa ON $= \zeta'$. One obtains the values of the pressures from the following five equations:

$$
\begin{aligned}
p + p' &= \sum \cdot P \\
q + q' &= \sum \cdot Q \\
r + r' &= \sum \cdot R \\
p\zeta + p'\zeta' &= \sum \cdot Pz - \sum \cdot Rx \\
q\zeta + q'\zeta' &= \sum \cdot Qz - \sum \cdot Ry
\end{aligned}
$$

In fact, let us suppose that in order to maintain equilibrium, releasing the points M and N and thus rendering the system free, one would have to apply in the direction opposite to the coordinates the forces p, q, r in the point M, and the forces p′, q′, r′ at point N. Searching the values of those forces with the general equations of equilibrium (§100) we find precisely the equations we just have written here above.

107. Coroll. I. The third equation lets us know that the entire axis is pushed in the direction of the z with the force = Σ·R. For the other four one can easily determine the pressures to which each support is subject in the direction of the x and the y.

108. Coroll. II. From there we may easily solve the problem[25] in which are sought the efforts that exercises the weight of a door upon the two hinges that hold it up; or more generally the pressures exercised upon the two hinges M, N of the axis by a force parallel to the axis itself.

Let the force = –R ; the distance of its direction from the axis = x ; the interval between both hinges MN = $\zeta' - \zeta$. One finds

$$r + r' = -R;\ p' = -p = \frac{Rx'}{\zeta} - \zeta;\ q' = q = 0.$$

Thus, the door weighs with all its weight R along the vertical line of the hinges; and in addition to that the superior hinge is led outside and the inferior is pushed inside, both with force = $Rx/\zeta' - \zeta$.[26]

[25] At this point Venturoli refers the reader to Fontana (1799, p. 135).

[26] Venturoli (1809, pp. 46–48): *Cap. XVI. Pressioni sugli appoggi d'un sistema rigido equilibrato.*

6 1811: Finally, with Its Second Edition, the Influence of Lagrange's "Analytical Mechanics" Extends

In 1817 Venturoli published the third edition of his *Elementi di meccanica e d'idraulica*, in which we find exactly the same chapter on "Pressures upon supports of a rigid system in equilibrium" as in the edition of 1809.

In 1819 the *Annals of Philosophy* published a summary of a paper presented at the Royal Society by C. Bonnycastle, entitled "On pressures which sustain a heavy

---

104. Proposizione I. Sia un sistema rigido equilibrato, e sostenuto da un fulcro o pernio fisso. Tenendo ferme le denominazioni sino ad ora usate, le pressioni p, q, r *che il pernio sostiene nel senso delle* x, y, z *sono determinate dall' equazionip* $= \sum \cdot P$; $q = \sum \cdot Q$; $r = \sum \cdot R$

Difatti le pressioni che sostiene il fulcro sono eguali e contrarie a quelle forze che vi si dovrebbono applicare per mantener l'equilibrio, nel caso che il fulcro si togliesse. Ora per le tre prime equazioni dell'equilibrio (100) è manifesto che vi si dovrebbero applicare in senso contrario alle x, y, z, *le forze* $\sum \cdot \mathrm{P}$; $\sum \cdot \mathrm{Q}$; $\sum \cdot R$

105. Corollario. Il fulcro sostiene lo sforzo delle potenze P, Q, R *come se queste gli fossero immediatamente applicate.*

106. Proposizione II. Sia il sistema equilibrato attorno l'asse immobile OZ (Fig. 13) fisso in due punti, o perni M, N. *Si chiamino* p, q, r *le pressioni che sostiene nel senso delle* x, y, z *il pernio* M *cui corrisponde l'ascissa* OM $= \zeta$ *e siano similmente* p′, q′, r′ *le pressioni sostenute dall'altro pernio N cui corrisponde l'ascissa ON* $= \zeta'$. *I valori delle pressioni si hanno dalle cinque equazioni che seguono.*

$$p + p' = \sum \cdot P$$
$$q + q' = \sum \cdot Q$$
$$r + r' = \sum \cdot R$$
$$p\zeta + p'\zeta' = \sum \cdot Pz - \sum \cdot Rx$$
$$q\zeta + q'\zeta' = \sum \cdot Qz - \sum \cdot Ry$$

Difatti supponghiamo che per mantener l'equilibrio, sciogliendo i punti M, N, *e rendendo così libero il Sistema, si dovessero applicare in senso contrario alle coordinate le forze* p, q, r *nel punto* M, *e le forze* p′,q′,r′ *nel punto* N. *Cercando i valori di queste forze per le generali equazioni dell'equilibrio (100) incontriamo appunto le equazioni qui sopra scritte.*

107. Coroll. I. La terza equazione fa conoscere che tutto l'asse è sospinto nel senso delle z con forza $= \sum \cdot R$ *Per le altre quattro poi agevolmente si determinano le pressioni che soffre ciaschedun degli appoggi nel senso delle x e delle y.*

108. Coroll. II. Di qui può sciogliersi agevolmente il problema [here the reference to Fontana] *nel quale si cercano gli sforzi che esercita il peso d'una porta sui due cardini che la reggono; o più generalmente le pressioni esercitate su' due cardini M, N dell'asse da una forza parallela all'asse medesimo.*

Sia la forza $= -R$; *la distanza della sua direzione dall'asse* $= x$; *l'intervallo fra i due cardini MN* $= \zeta'' - \zeta'$ *Si troverà.*

$$r + r' = -R; p' = -p = Rx/\zeta' - \zeta; q' = q = 0$$

Così la porta grava di tutto il suo peso R la linea verticale de' cardini; ed oltre a ciò il cardine superiore è tratto in fuori, e l'inferiore è sospinto in dentro, entrambi con forza $= Rx/\zeta' - \zeta$.

Body in equilibrium when the points of support are more than three". The summary of the paper is as follows:

> March 11. A paper, by C. Bonnycastle, Esq. was read, entitled, "On the Pressures which sustain a heavy Body in Equilibrium when the Points of Support are more than three." The author, after some general remarks, observed, that this is a problem which has hitherto never been satisfactorily investigated, though its assistance is necessary in estimating the strength of bridges and materials in general, and in determining the deflection and curvature of elastic plates. The difficulties attending the investigation of this problem were referred by Mr. B. to the too great generality of the method of investigation hitherto employed, and which, for the most part, has consisted not in the direct solution of the equation, but in comparing it with another admitting of an easier solution. The author, after some further remarks, observed, that there is no method purely mathematical by which the difficulty can be surmounted; and that when abstractedly considered it appears impossible to determine the pressures which a heavy body exerts when supported on four or more fulcra. By considering, however, the circumstances under which pressure is usually generated, we shall be enabled, the author continued, to discover the law of its distribution, and this law must always govern its proportional quantities and intensities. Mr. B. then proceeded to examine the subject in this point of view, and to explain the general law of the distribution of pressure on determinate fulcra in different instances. The paper concluded with an investigation of the case, when the number of points of support is infinite; or in other words, consists of a line or plane surface (Bonnycastle 1819, p. 307).

In 1822 and 1823, Giuseppe Venturoli's books were translated into English by Daniel Cresswell. After those translations there would be further editions in Italian which contained an addition concerning "the principle of virtual velocities and its uses in mechanics". Let us now turn to that addition as translated by Cresswell concerning virtual velocities, a direct influence of Lagrange:

> Explanation and demonstration of the principle of virtual velocities
>
> 786. To the point A (Fig. 69.) let there be applied a force S, which acts in the direction of the straight line AB. Imagine the point A to be carried to another point a, very near to A. If from a there be drawn ab perpendicular to AB, Ab will be the small space described by the point A, in the direction AB, which is the direction of the force S. By the Virtual Velocity, then, of a force, is meant the indefinitely small space, which, in a given indefinitely small motion, is described by the point of application of the force, in the direction of the force itself (Venturoli 1823, p. 169).

From this first extract of Venturoli's text, we can see how close that definition is to Bernoulli's expression in his letters to Varignon and to Renau d'Elissagaray. Of course, that influence is not direct but passed through Johann's son Daniel together with Euler, his pupil and Daniel's lifelong friend, and finally through Lagrange.

Venturoli goes on defining the momentum of the force:

> 787. The product S.Ab, that is, the product of the force multiplied by its virtual velocity, is here called the Momentum of the force S.
>
> These terms are adopted for the sake of abridgement. But it is necessary to guard against confounding this meaning of the word momentum, with other different quantities, which, in other parts of the work (54.103) have been designated by the same name (Venturoli 1823, pp. 169–170).

Important for us is also the fact that Venturoli dedicates his Book IV to the "Equilibrium of Structures" (1823, pp. 65–108), in which Chap. VI is on "Equilibrium of Arches and Domes" (1823, pp. 82–84) and Chap. VII is on "Arches of finite Thickness" (1823, pp. 85–89). However, he does not apply the theory of virtual velocities to those constructions. In this text Venturoli refers to the works concerning architecture by Philippe de la Hire (1702, 1712), Claude Antoine Couplet (1729), Bernard Forest de Bélidor (1729), Pierre Bouguer (1734), Augustin Coulomb (1776), Charles Bossut (1777) and Lorenzo Mascheroni (1785).

In his "Note sur les questions de statiques dans lesquelles on considère un corps pesant supporté par un nombre de points d'appuis surpassant 3" (1825) Claude-Louis Navier mentions Euler and Fourier and affirms that following the principles of statics in the case of a number of support points higher than three, the pressure on each support is indeterminate. But he adds:

> If one supposes on the contrary the bodies as flexible and elastic, as they really are, the indeterminacy ceases entirely, and there is always, for each particular question, a number of equations equal to the one of unknowns.[27]

He gives some examples. Let us consider the easiest one:

> To indicate some applications, one supposes in the first place an elastic beam M, M′, loaded with two weights Π, Π′, placed in the middle of each half AM and AM′. One easily determines the shape assumed by the beam, and one finds the stresses exerted at the points.

These are respectively:

$$\mathrm{A} \quad \frac{32\Pi + 22\Pi'}{32}$$

$$\mathrm{M} \quad \frac{13\Pi - 3\Pi'}{32}$$

$$\mathrm{M'} \quad \frac{13\Pi' - 3\Pi}{32}$$

So that the middle point A supports 2/3 of the total weight (Fig. 10).[28]

Therefore, the way to solutions is now paved even if it is still not generally known or accepted.

A first application to architecture will be done by Gaspard-Michele Pagani in "Mémoire sur l'équilibre des colonnes" (Memoire on the equilibrium of columns)

[27] Navier, 1825, p. 35: *Si l'on suppose au contraire les corps flexibles et élastiques, comme ils le sont véritablement, l'indétermination cesse entièrement, et il y a toujours, pour chaque question particulière, un nombre d'équations égale à celui des inconnues.*

[28] Navier (1825, pp. 35–36): *Pour indiquer quelques applications très-simples, on supposera en premier lieu une verge élastique M M′, sans pesanteur, supportée au milieu A et aux extrémités M, M′, et chargée de deux poids Π, Π′ placés aux milieux des deux moitiés A M et A M′. On détermine facilement la figure affectée par cette verge, et l'on trouve que les efforts exercés aux points ... en sorte que le point milieu A soutient un peu plus des 2/3 du poids total.*

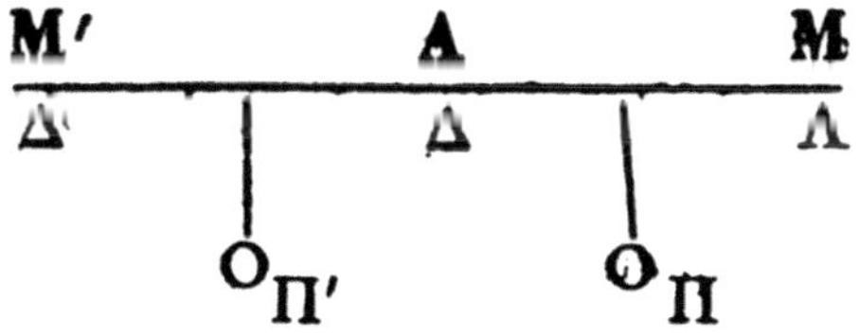

Fig. 10 Navier's figure for his first problem (1825, p. 35). Public domain

(1839). Pagani starts explaining the problem as we already told it many times and he refers to a note with title "Note sur l'équilibre d'un système dont une partie est supposée inflexible et dont l'autre partie est flexible et extensible" (Note on the equilibrium of a system part of which is supposed inflexible and the other flexible and extensible) that he had published some years before (1834).

Written in 1852, but published only in 1853, a final article on the supports problem was that by Giuseppe Fagnoli entitled "Riflessioni intorno la teorica delle pressioni che un corpo os sistema di forma invariabile esercita contro appoggi rigidi ed irremovibili dai quali è sostenuto in equilibrio" (Reflections concerning the theory of pressures that a body of a system of invariable form exerts upon rigid and fixed supports which sustain them in equilibrium) (1853). It appeared just a bit later than the first *Leçons sur la théorie mathématique de l'élasticité des corps solides* (Lectures on the mathematical theory of elasticity of solid bodies) (1852) by Gabriel Lamé (1852), who presents it as something new:

> Lectures on elasticity of solid bodies are an essential part of the lectures on mathematical physics that I teach at the Faculty of Sciences of Paris. Several very competent persons who followed those lectures advised me to publish them, and think that that work would not be without use.
>
> Following an advice, certainly given with an extreme benevolence, I do not consult my strengths: I give into my convictions about the importance and the opportuneness of the subject in question.
>
> Mathematical physics, properly speaking, is an completely modern subject, which belongs exclusively to the geometers of our century. Actually, this science contains only three chapters, variously extended, that are treated rationally, meaning, that is, that they rest only upon indisputable principles or laws. Those chapters are the theory of static electricity on the surface of conductors; the analytical theory of heat; and finally, the mathematical of elasticity of solid bodies. The last one is the most difficult, the less complete; it is also the most useful, at a time when one wants to appreciate the importance of a mathematical theory by the results it can immediately furnish to the industrial practice.[29]

[29] Lamé (1852, pp. V, VI): *Des Leçons sur l'élasticité des corps solides font partie essentielle du Cours de Physique mathématique que je professe à la Faculté des Sciences de Paris. Plusieurs personnes très-compétentes, qui ont assisté à ces Leçons, me conseillent de les publier, et pensent que cet ouvrage ne sera pas sans utilité.*
En suivant un conseil, dicté sans doute par une extrême bienveillance, je ne consulte pas mes forces: je cède à mes convictions sur l'importance et l'opportunité du sujet dont il s'agit.
La Physique mathématique, proprement dite, est une création toute moderne, qui appartient exclusivement aux Géomètres de notre siècle. Aujourd'hui, cette science ne comprend en réalité que trois chapitres, diversement étendus, qui soient traités rationnellement; c'est-à-dire qui ne s'appuient que sur des principes ou sur des lois incontestables. Ces chapitres sont la théorie de l'électricité statique à la surface des corps conducteurs; la théorie analytique de la chaleur; enfin la théorie

We may also say that the elasticity of solid bodies is the outcome of the questions asked about the supports problem. The dead-end street now opens onto a very rich avenue but it came at a cost and many will have to work hard to unearth the treasures of the theory of elasticity. Lamé explains that origin of elasticity and the path that had been made since:

> In the theory of the equilibrium and motion of the solid bodies, one considers those bodies as having a perfect rigidity; one supposes that the distances between the points-forces remain invariable, as intense as those forces are. This abstraction suffices for the problems we had in mind, and simplifies their solutions without disturbing the rigour, except for some very particular cases. But this hypothesis ignores the law according to which is transmitted, from one part of the solid body to another, the reciprocal influence which destroys the action of one force by that of the others; it is nevertheless an important phenomenon that has its limits, which are really necessary to know, since when the forces that are in equilibrium thanks to the solid acquire a sufficient degree of intensity, the body, after having more or less changed its form, finally breaks. It is the necessity of studying that phenomenon and avoiding, in the constructions, the ruptures and the permanent deformations, that gave rise to the mathematical theory of the elasticity of the solids, a theory that was extended by geometers to the research on the laws which are followed by the small motions or vibrations of elastic mediums.[30]

Let us add some bibliographical notes. Barsotti (1842) makes a summary of all articles published on the problem of the action of a door on its hinges, mentioning by the way almost all those who participated to the solution of that problem. Finally, Bertelli (1850) gives of the same evolution a more experimental point of view.

References

Barsotti, G. 1842. Degli sforzi che fanno i punti di sostegno d'una porta per reggerla in equilibrio, qualunque ne sia il numero e la posizione. *Annali Di Fisica, Chimica e Matematica* 8: 146–154.

Bélidor, Bernard Forest de. 1729. *La science des Ingénieurs dans la conduite des travaux de fortification et d'architecture civile.* La Haye: Pierre Gosse.

---

mathématique de l'élasticité des corps solides. Le dernier est le plus difficile, le moins complet; il est aussi le plus utile, *à une époque où l'on veut apprécier l'importance d'une théorie Mathématique par les résultats qu'elle peut fournir immédiatement à la pratique industrielle.*

30 Lamé (1852, pp. 4–5): *Dans la théorie de l'équilibre et du mouvement des corps solides, on considère ces corps comme ayant une rigidité parfaite; on suppose que les distances des points forces restent invariables, quelque intenses que soient ces forces. Cette abstraction suffit pour les problèmes qu'on a en vue, et simplifie leurs solutions sans troubler leur rigueur, excepté dans quelques cas très-particuliers. Mais cette hypothèse laisse ignorer la loi suivant laquelle se transmet, d'une partie à l'autre du corps solide, l'influence réciproque qui fait détruire l'action d'une force par celle des autres; c'est cependant un phénomène important et qui a ses limites, bien nécessaires à connaître, puisque, quand les forces qui se font équilibre par l'intermédiaire solide acquièrent un degré suffisant d'intensité, le corps, après avoir plus ou moins changé de forme, finit par se briser. C'est la nécessité d'étudier ce phénomène et d'éviter, dans les constructions, les ruptures et les déformations permanentes, qui a donné naissance à la théorie mathématique de l'élasticité des solides; théorie que les géomètres ont étendue à la recherche des lois que suivent les petits mouvements, ou les vibrations des milieux élastiques.*

Benvenuto, Edoardo. 1984. A brief outline of the scientific debate which preceded the works of Castigliano. *Meccanica* 19: 19–32.

Benvenuto, Edoardo. 1991. The eighteenth-century debate on the supports problem. In: *An introduction to the history of structural mechanics,* vol. 2, 441–460. Springer.

Bertelli, Francesco. 1850. Ricerche sperimentali circa la pressione dei corpi solidi ne' casi in cui la misura di essa, secondo le analoghe teorie meccaniche si manifesta indeterminata e intorno alla relazione fra le pressioni e la elasticità de' corpi medesimi. Memoria Postuma. *Memorie Dell'accademia Delle Scienze Di Bologna* 1: 433–461.

Bonnycastle, C. 1819. On pressures which sustain a heavy Body in equilibrium when the points of support are more than three. *Annals of Philosophy* 13: 307.

Bossut, Charles (Abbé). 1772 *Traité élémentaire de mécanique statique.* Paris: Claude Antoine Jombert.

Bossut, Charles (Abbé). 1777. Recherches sur l'équilibre des voûtes. *Histoire de l'Académie royale des sciences, avec les mémoires de mathématique et de physique* (1778), 534–566.

Bossut Charles. (Abbé). 1788. *Trattato elementare di Meccanica,* vol. 1. Pavia: Stamperia del Monastero di S. Salvatore.

Bouguer, Pierre. 1734. Sur les lignes courbes qui sont propres à former les voutes en dome. *Histoire de l'Académie des sciences de Paris avec les Mémoires de Mathématique e de Physique* (1734), 149–166.

Coulomb, Augustin 1776. Essai sur une application des règles de *Maximis* et *Minimis* à quelques problèmes de statique, relatifs à l'architecture. *Memoires de Mathematique et de Physique présentés à l'Académie des sciences par divers Savans,* année 1773, 343–382.

Couplet, Claude Antoine. 1731. De la poussée des voûtes. *Mémoires de l'Académie royale des sciences,* année 1729, 79–117.

D'Alembert, Jean-Baptiste Le Rond, 1780. Sur quelques questions de mécanique. *Opuscules Mathématiques, ou Mémoires sur différens sujets de géométrie, de méchanique, d'optique, d'astronomie, etc.* VIII, 36–45. Paris: Claude Antoine Jombert.

Delanges, Paolo, 1790. Memoria sulle pressioni esercitate da un corpo sostenuto da tre o più appoggi collocati nello stesso piano. *Memorie di matematica e fisica della Società italiana* 5, 107–129

Delanges, Paolo. 1799. Nuove considerazione intorno alla pressione di un corpo sostenuto da tre e più appoggi in un piano orizzontale. *Memorie di matematica e fisica della Società italiana* 8, 60–76.

Delanges, Paolo. 1811. Analisi e soluzione sperimentale del problema delle pressioni. *Memorie Di Matematica e Fisica Della Società Italiana* 15: 114–153.

Euler, Leonhard. 1774. De pressione ponderis in planum cui incumbit (E456). *Novi Commentarii Academiae Scienciarum Imperialis Petropolitanae* XVIII, pp. 32–34, 289–329; *Euler Opera Omnia* series secunda, vol. 9, pp. 407–412. Euler, *Opera Omnia,* series 2, 9, 1–34.

Euler, Leonhard. 1794. Von dem Drucke eines mit einem Gewichte beschwerten Tisches auf eine Fläche (E 682). *Opera Omnia* series secunda, vol. 9, 407–412.

Fagnoli, Giuseppe. 1853. Riflessioni intorno la teorica delle pressioni che un corpo o sistema di forma invariabile esercita contro appoggi rigidi ed irremovibili dai quali è sostenuto in equilibrio. *Memorie della Academia delle Scienze dell'Instituto di Bologna* 4, 109–138.

Fontana, Gregorio. 1799. Sopra la pressione delle porte contro il loro arpioni. *Memorie di matematica e fisica della Società italiana* VIII, part I, 135–139.

Fourier, Joseph. 1798. Mémoire sur la statique, contenant la démonstration du principe des vitesses virtuelles et la théorie des moments. *Journal de l'Ecole polytechnique,* Ve cahier, p. 20; *Œuvres de Fourier, publiées par les soins de Gaston Darboux* (Paris, 1890), vol. 2, 477–521.

Gergonne, Joseph Diez. 1827. Note sur un paradoxe de statique. *Annales De Mathematiques Pures Et Appliquées* 17: 75–79.

Lagrange, Joseph Louis, 1788. *Mécanique analytique.* Paris: Desaint.

La Hire, Philippe de. 1695. *Traité de Mécanique, Ou L'On Explique Tout Ce Qui Est Necessaire Dans La Pratique Des Arts.* Paris: L'Imprimerie Royale.

La Hire Philippe de. 1702. Remarques sur la forme de quelques arcs dont on se sert dans l'Architecture. *Mémoires de l'académie royale des sciences*, 1702, 94–97.
La Hire Philippe de. 1712. Sur la construction des voutes dans les édifices. *Mémoires de l'académie royale des sciences*, 1712, 69–77.
Lamé, Gabriel. 1852. *Leçons sur la théorie mathématique de l'élasticité des corps solides*. Paris: Bachelier.
Lorgna, Anton Maria. 1794. Dell'azione di un corpo retto da un piano immobile esercitata ne' punti di appoggio che lo sostentano. *Memorie di matematica e fisica della Società italiana* 7, 178–192.
Malfatti, Gianfrancesco. 1799. Tentativo sul problema delle pressioni, che soffrono gli appoggi collocati agli angoli di una figura, derivate da un peso posto dentro la sua aja. *Memorie di mathematica e fisica della Società italiana* VIII, part II, 319–415.
Malfatti, Gianfrancesco. 1803. Brevi riflessioni alla critica del tentativo sul problema delle pressioni, fatta dal Sig. Paoli nel tomo IX di questa Società. *Memorie di matematica e fisica della Società italiana* X, part II, 245–248.
Mascheroni, Lorenzo. 1785. *Nuove ricerche sull'equilibrio delle volte*. Bergamo: Locatelli.
Navier, Claude-Louis. 1825. Note sur les questions de statiques dans lesquelles on considère un corps pesant supporté par un nombre de points d'appuis surpassant 3. *Nouveau Bulletin de la Société philomatique de Paris*, année 1825, 35–37.
Pagani, Gaspard-Michel. 1834. Note sur l'équilibre d'un système dont une partie est supposée inflexible et dont l'autre partie est flexible et extensible. *Nouveaux mémoires de l'Académie de Bruxelles*, vol. 8, 1–14.
Pagani, Gaspard-Michele. 1839. Mémoire sur l'équilibre des colonnes. *Memorie della Regia Accademica della Reale Accademia delle Scienze di Torino*, 2nd series, vol. 1, 355–371.
Paoli, Paolo. 1792. Memoria sopra alcuni problemi meccanici. *Memorie di matematica e fisica della Società italiana* 6, 534–559.
Paoli, Paolo. 1802. Sul problema degli appoggi, *Memorie di matematica e fisica della Società italiana* IX, 92–98.
Poinsot, Louis. 1803. *Elémens de statique*. Paris: Calixte-Volland.
Radelet-de Grave, P. 2004. Déplacements, vitesses et travaux virtuels. In : *variar para encontrar / varier pour mieux trouver / The Lore of Variation: Finding Pathways to Scientific Knowledge*, C. Alvarez, J. Lacki, J.R. Martinez, P. Radelet-de Grave, eds., 177–204. Universidad Nacional Autónoma de México.
Riccati, Vincenzo. 1772. *De' principi della meccanica. Lettere di Vincenzo Riccati al P. Virgilio Cavina*. Venice: Stamperia Coleti.
Varignon, Pierre. 1725. *Nouvelle mécanique, dont le projet fut donné en 1687*. Paris: Claude Antoine Jombert.
Venturoli, Giuseppe. 1809. *Elementi di meccanica e d'idraulica*, 2nd ed. Bologna: Tipografia de' Fratelli Masi.
Venturoli, Giuseppe. 1822. *Elements of the Theory of Mechanics, translated from the Italian by Daniel Cresswell, to which is added a selection of problems in mechanics*. Cambridge: J. Smith.
Venturoli, Giuseppe. 1823. *Elements of Practical Mechanics, to which is added a treatise upon the principle of virtual velocities and its uses in mechanics, translated from the Italian by Daniel Cresswell*. Cambridge: J. Smith.
Venturoli, Giuseppe. 1846–1847. *Elementi di meccanica e d'idraulica*, 2 vols., 7th ed. Milan: presso il Libraio Carlo Branca.

Chapter 16
The Discovery of General Methods for Elastic Systems, 1852–1875

1 Prelude: Energy and Forms of Energies

In 1824, Sadi Carnot published his important work, *Réflexions sur la puissance motrice du feu et sur les machines propres à développer cette puissance.* The book opens in this way:

> No one is unaware that heat can be the cause of movement, that it even has great driving power: the steam engines, so widespread today, are clear proof of this for all to see. It is to heat that the great movements which strike our gaze on the earth must be attributed; it is to it that the agitations of the atmosphere are due, the rise of clouds, the fall of rain and other meteors, the currents of water which crisscross the surface of the globe and which man has managed to use for its use a small part; finally, earthquakes and volcanic eruptions also recognize heat as the cause.
>
> It is from this immense reservoir that we can draw the moving force necessary for our needs; Nature, by providing us with fuel from all sides, has given us the ability to create heat and the driving power that results from it at all times and in all places. Developing this power, appropriating it for our use, such is the purpose of fire machines.[1]

Even if Sadi Carnot does not use these words, he is clearly speaking of energy and of various forms of energy.

---

[1] Carnot, 1824, pp. 1–2: *Personne n'ignore que la chaleur peut être la cause du mouvement, qu'elle possède même une grande puissance motrice: les machines à vapeur, aujourd'hui si répandues, en sont une preuve parlante à tous les yeux. C'est à la chaleur que doivent être attribués les grands mouvemens qui frappent nos regards sur la terre; c'est à elle que sont dues les agitations de l'atmosphère, l'ascension des nuages, la chute des pluies et des autres météores, les courans d'eau qui sillonnent la surface du globe et dont l'homme est parvenu à employer pour son usage une faible partie; enfin les tremblements de terre, les éruptions volcaniques, reconnaissent aussi pour cause la chaleur.*
C'est dans cet immense réservoir que nous pouvons puiser la force mouvante nécessaire à nos besoins; la nature, en nous offrant de toutes parts le combustible, nous a donné la faculté de faire naître en tous temps et en tous lieux la chaleur et la puissance motrice qui en est la suite. Développer cette puissance, l'approprier à notre usage, tel est l'objet des machines à feu.

P. Radelet-de Grave, *Between Mechanics and Architecture*, Mathematics and the Built Environment 8, https://doi.org/10.1007/978-3-031-73530-1_16

Ten years later, Emile Clapeyron will complete the work of Sadi Carnot, permitting Gabriel Lamé in 1852 to completely change the way we think of constructions. In 1834, Emile Clapeyron published his *Mémoire sur la puissance motrice de la chaleur*, in which of course refers to Carnot:

> But before entering into the matter, it is useful to return to the fundamental axiom which serves as the basis of Mr. Carnot's research, and which will also be my starting point.
>
> It has long been noted that we can use heat to develop motive force, and conversely, that with motive force we can develop heat. ...
>
> Conversely, it is always possible to use the passage of heat from a hot body to a cold body for the production of a mechanical force.[2]

Both men clearly describe what we today call today a transfer of energy between different forms of energy.

2 1852–1875 the Discovery of General Methods for Elastic Systems

For the reasons we mentioned in the preface to this chapter, a new branch of the tree of knowledge grew the concept of forms of energy. In 1852, in his *Leçons_sur_la_ théorie_mathématique de l'élasticité des corps solides*, Gabriel Lamé explains the influence of the new science of Elasticity on the way of building. Let us have a look at some of the sections of Lesson 7 more precisely in relation with construction, in order to underline the important notions that were acquired:

> § 31[The work of an elastic force]. When a force pulls or presses a solid body, of which at least three points are fixed, the product of this force times the projection, in its direction, of the total displacement that it has been caused to undergo at its point of application, represents twice the work done, from the instant when the displacement and the force were equal to zero, to the instant when the displacement and the force reached their final values.
>
> §. 32 [M. Clapeyron's theorem]. Mr. Clapeyron found an expression for the same work, in which all the elastic forces developed in the interior of the solid body intervene. The equality of these two expressions constitutes a theorem, or rather a principle, analogous to that of living forces, and which appears to have equal importance for applications.[3]

---

[2] Clapeyron, 1834, p. 155: *Mais avant d'entrer en matière, il est utile de revenir sur l'axiome fondamental qui sert de base aux recherches de M. Carnot, et qui sera aussi mon point de départ. On a remarqué depuis longtemps que l'on peut employer la chaleur à développer de la force motrice, et réciproquement, qu'avec de la force motrice on peut développer de la chaleur. ...*
Réciproquement, il est toujours possible d'utiliser le passage du calorique d'un corps chaud à un corps froid pour la production d'une force mécanique.

[3] Lamé, 1852, pp. 79–80: *§. 31. [Le travail d'une force élastique] Lorsqu'une force tire ou presse un corps solide, dont au moins trois points sont fixes, le produit de cette force par la projection, sur sa direction, du déplacement total qu'elle a fait subir à son point d'application, représente le double du travail effectué, depuis l'instant où le déplacement et la force étaient nuls, jusqu'à celui où le déplacement et la force ont atteint leurs valeurs finales.*

Lamé gives than the mathematical expressions for the work of a traction (§ 33); the work of a compression (§ 34), and for the power of a spring (§ 35) all developed from Clapeyron's theorem. The he arrives to constructions:

> § 36. [Applications to constructions] When establishing any construction, whatever its type, this principle will provide a relationship which can be used to find the most suitable proportions of the different parts of this construction. Generally, we arrange the different pieces, in framework or iron, in such a way that they experience forces directed in the direction of their length; that is to say that there is, at each interior point, only one main elastic force, the same throughout the entire extent of the same piece.
>
> § 37. [Case of a triangular assembly]

Let us take the simple case of a triangular assembly ABC, composed of two pieces of frame equally inclined $\overline{AB}, \overline{AC}$, and of a horizontal part $\overline{BC}$ which connects the first two. Those three pieces are of the same nature, E is their coefficient of elasticity. A vertical force Π is exerted at vertex A, which causes it bend vertically of [a distance] a; each inclined piece, $\overline{AB}$ or $\overline{AC}$ is pressed longitudinally by a force F, such that $\Pi = 2Fcos\alpha$, α being the semi-angle in A; from which $F = \frac{\Pi}{2coscos}$; ...[4]

These cases are rather elementary, but keep in mind that the same theory leads to constructions as sophisticated as spires in civil engineering and other things.

Lamé concludes Lesson 7 with § 38 (Comparisons and generalisations), and goes on to Lesson 8 (Elastic lines and surfaces; Equilibrium of an elastic string; Vibrating string) and Lesson 9 (Equilibrium of an elastic surface; Equilibrium of a planar membrane; Vibrating membrane). The samples chosen by Lamé are not extraordinary, but bear in mind that the same theory is still in use today.

Now that elasticity has been developed, the old unsolved problems of supports are taken up again and even give rise to many new problems, such as the one posed by Clapeyron regarding the calculation of an elastic beam resting freely on unequally spaced supports (Clapeyron 1857). He speaks of the first suspension bridge over the Menai strait from Bangor (Wales) to the Isle of Anglesey (Fig. 1). Its construction had been completed in 1850. This impressive work has been described in book by Molinos and Pronnier (1857), the civil engineers who built the bridge. In their

---

§. 32. [Théorème du M. Clapeyron] M. Clapeyron a trouvé une expression du même travail, dans laquelle interviennent toutes les forces élastiques développées dans l'intérieur du corps solide. L'égalité de ces deux expressions constitue un théorème, ou plutôt un principe, analogue à celui des forces vives, et qui paraît avoir une importance égale pour les applications.

[4] Lamé, 1852, pp. 86–88: § 36. *[Applications aux constructions] Lors de l'établissement de toute construction, de quelque genre qu'elle soit, ce principe fournira une relation dont on pourra faire usage, pour trouver les proportions les plus convenables des différentes parties de cette construction. Généralement on dispose les différentes pièces, en charpente ou en fer, de telle sorte qu'elles éprouvent des efforts dirigés dans le sens de leur longueur; c'est-à-dire qu'il n'y ait, en chaque point intérieur, qu'une force élastique principale, la même dans toute l'étendue d'une même pièce.*

§ 37. Prenons pour exemple le cas simple d'un assemblage triangulaire ABC, composé de deux pièces de charpente également inclinées AB, AC, et d'une pièce horizontale BC qui relie les deux premières ces trois pièces sont de même nature, E est leur coefficient d'élasticité. Un effort vertical Π s'exerce au sommet A qu'il fait fléchir verticalement de a; chaque pièce inclinée, AB ou AC est pressée longitudinalement par une force F, telle que $\Pi = 2F\cos\alpha$, α étant le demi-angle en A; d'où $F = 2cos$; la pièce horizontale BC est tirée longitudinalement par une force $Fsin = sin\ 2cos$; ...

introduction the summarise the influence of the first railways on the development of construction theories:

> Preface. The execution of the railways had a remarkable influence on the art of construction. … All branches of industry have felt the effects but the most remarkable result, the one which will in the future constitute the hallmark of the constructions of our time, is the definitive and large-scale introduction of metal in works of art. The need to build bridges with spans unknown until this time, to cover considerable spaces, free of support points, to offer the operation of stations in harmony with the movement that they concentrate, soon led to recognition of the impotence of the means of construction which had until then offered sufficient resources. These resources had also remained those of ancient civilization, because it should be noted that since the most ancient times the field of art had been little enriched in this respect...
>
> The Asnières bridge is the first important bridge built on rational ideas that we have owned. … These constructions were then applied on a very large scale and completely popularized in the execution of the Southern Railway.
>
> The aim of the work that we are publishing is to provide on the theory and construction of metal bridges, elements and indications that are still not very widespread, and which, in view of the daily increasing importance of this type of construction, will appear interesting and provide some services. ... We have reproduced the method of the learned professor Mr. Bélanger,[5] who has taught since 1842 at the School of Bridges and Roads, and more later at the Central School, who was the first to treat this question in a general way. We then indicate the ingenious method of Mr. Clapeyron, which presents a notable abbreviation in the calculations.
>
> Construction is treated in a special way in the second part…
>
> In the third part of the work, several numerical applications of general formulas to works of great interest, and precise details on the execution of several of them…
>
> Finally, we have tried to shed some light on the choice of the best system of bridges, seeking to make it clear that this delicate question depends on a host of conditions whose relative values vary for each particular case, and each of which require new appreciations. …
>
> We have brought together in an atlas the types of all the bridges which, to date, have received an application in the arts.[6]

---

[5] This is a reference to Jean-Baptiste Bélanger (1790–1874), author of *Notes sur la mécanique appliquée aux principes de la stabilité des constructions* (1848).

[6] Molinos and Pronnier, 1857, n.p.: *Préface. L'exécution des chemins de fer a eu sur l'art des constructions une influence remarquable. … Toutes les branches de l'industrie; en ont ressenti les effets mais le résultat le plus remarquable, celui qui constituera dans l'avenir le cachet des constructions de notre époque, c'est l'introduction définitive et sur une grande échelle du métal dans les travaux d'art. La nécessité de construire des ponts avec des portées inconnues jusqu'alors, de couvrir des espaces considérables, libres de points d'appui, pour offrir à l'exploitation des gares en harmonie avec le mouvement qu'elles concentrent, fit bientôt reconnaître l'impuissance des moyens de construction qui avaient offert jusque-là des ressources suffisantes. Ces ressources étaient d'ailleurs restées celles de la civilisation ancienne, car il est à remarquer que depuis les temps les plus reculés le domaine de l'art s'était peu enrichi sous ce rapport...*
Le pont d'Asnières est le premier pont important, construit sur des idées rationnelles, que nous ayons possédé. … Ces constructions ont ensuite été appliquées sur une très-grande échelle et tout à fait vulgarisées dans l'exécution des chemins du Midi.
Le but de l'ouvrage que nous publions est de donner sur la théorie et la construction des ponts métalliques des éléments et des indications encore peu répandus, et qui, en présence de l'importance chaque jour croissante de ce genre de construction, pourront paraître intéressants et rendre quelques

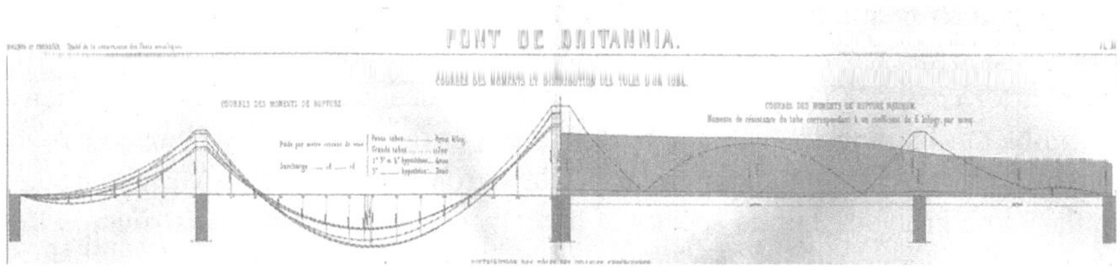

Fig. 1 The Menai suspension bridge, diagram of momentum (Molinos and Pronnier 1857, Atlas, Pl. XI). Public domain

In his article published in 1857 mentioned above, Clapeyron describes the curve of moment which we see in the figure of the Menai suspension bridge (Fig. 1). He explains that the general form of the bridge is a rectilinear beam resting on four unequally-spaced pillars.

In his article, he restricts the problem:

> In this first Memoir, which I summarise here, I examine the case of a rectilinear beam placed on two supports at its ends. Its section is constant, and it supports a uniformly distributed load; also given is the moment of the forces acting at the two ends at the right of the supports. We deduce the equation of the elastic curve that affects the axis of the beam, the mechanical conditions to which all its points are subjected, and the portion of the total weight supported by each support.
>
> The solution to the general problem is thus reduced to the determination of the moments of the forces tending to produce the rupture of the beam vertically on each of the supports on which it rests. This is achieved by expressing that the two elastic curves corresponding to two contiguous spans are tangent to each other on the intermediate support, and that the moments there are equal.[7]

---

services. ... nous avons reproduit la méthode de M. Bélanger, enseignée depuis 1842 à l'Ecole des ponts et chaussées, et plus complétement ensuite à l'Ecole centrale, par ce savant professeur, qui, le premier, a traité cette question d'une manière générale. Nous indiquons ensuite la méthode ingénieuse de M. Clapeyron, qui présente une abréviation notable dans les calculs.
La construction est traitée d'une manière spéciale dans la deuxième partie ...
Dans la troisième partie de l'ouvrage, plusieurs applications numériques des formules générales à des travaux d'un haut intérêt, des détails circonstanciés sur l'exécution de plusieurs d'entre eux, ...
Enfin, nous avons essayé de jeter quelque lumière sur le choix du meilleur système de ponts, en cherchant à établir que cette question délicate dépend d'une foule de conditions dont les valeurs relatives varient pour chaque cas particulier et exigent pour chacun d'eux des appréciations nouvelles. ...

[7] Clapeyron, 1857, p. 1077: *Dans ce premier Mémoire, dont voici le résumé, j'examine le cas d'une poutre droite posée sur deux appuis à ses extrémités, sa section est constante, elle supporte une charge répartie uniformément; on se donne en outre le moment des forces agissant aux deux extrémités au droit des appuis. On en conclut l'équation de la courbe élastique qu'affecte l'axe de la poutre, les conditions mécaniques auxquelles tous ses points sont soumis, et la partie du poids total supportée par chaque appui.*
La solution du problème général se trouve ainsi ramenée à la détermination des moments des forces tendant à produire la rupture de la poutre au droit de chacun des appuis sur lesquels elle repose. On y parvient en exprimant que les deux courbes élastiques correspondant à deux travées contiguës sont tangentes l'une à l'autre sur l'appui intermédiaire, et que les moments y sont égaux.

Let us also mention a work published by Clapeyron in 1858, *Mémoire sur le travail des forces élastiques dans un corps solide élastique déformé par l'action de forces extérieures* (1858), which is concerned with an essential problem of elasticity, the problem of springs, which had been treated for the first time by Hooke in 1678, then by Lamé (1852). In the same year as Clapeyron, Luigi Federico Menabrea takes up the same problem, but he speaks of a *Nouveau principe sur la distribution des tensions dans les systèmes élastiques*:

> Here is the statement of this new principle which I will call the principle of elasticity: *When an elastic system comes into balance under the action of external forces, the work developed by the effect of tensions or compressions of the links which unite the various points of the system is a minimum.* The differential equation which expresses this minimum will be referred to as the *elasticity equation*; we will soon see its use for determining tensions.
>
> First of all, I will give a brief idea of the demonstration of this principle. Let us consider the most general case, and let n be the number of points in the system joined together by m elastic links. Each point considered in isolation will remain in balance under the action of the external forces applied to it, and the tensions of the links which lead to it. The number of balance equations for the n points will be 3n; if p is that of the equations which must exist between the external forces, independently of the tensions, for there to be equilibrium, the number of equations which actually contain the tensions will be reduced to 3n – p. Thus, when ni will be > 3n – p, the previous equations will not be sufficient to determine all the tensions.
>
> It will be the same when the system contains a certain number of fixed points. This indeterminacy means that there is an infinity of tension values which, combined with the given external forces, are capable of keeping the system in balance. The values of the effective tensions depend on the respective elasticity of the links, and when this is determined, the same must apply to the tensions.
>
> Since, in the case we are considering, tensions can vary without equilibrium ceasing to exist, we must admit that these variations take place independently of any work by external forces; they are always accompanied by lengthening or shortening in the various corresponding links, which gives rise, in each of them, to a development of work. The variations in length of the links must be assumed to be very small so that the respective positions of the various points of the system are not significantly altered. But, since during this small internal movement the balance continues to exist and the work of the external forces is zero, it follows that the total elementary work of the tensions thus developed is also *zero*.[8]

---

[8] Menabrea, 1857, pp. 1056–: *Voici l'énoncé de ce nouveau principe que j'appellerai principe d'élasticité.* Lorsqu'un système élastique se met en équilibre sous l'action de forces extérieures, le travail développé par l'effet des tensions ou des compressions des liens qui unissent les divers points du système est un minimum. *L'équation différentielle qui exprime ce minimum sera désignée sous le nom d'*équation d'élasticité*; on en verra bientôt l'usage pour la détermination des tensions.*
Je donnerai avant tout une idée succincte de la démonstration de ce principe. Considérons le cas le plus général, et soit n le nombre des points du système réunis entre eux par m liens élastiques. Chaque point considéré isolément restera en équilibre sous l'action des forces extérieures qui lui sont appliquées, et des tensions des liens qui viennent y aboutir. Le nombre des équations d'équilibre pour les n points sera 3n; si p est celui des équations qui doivent subsister entre les forces extérieures, indépendamment des tensions, pour qu'il y ait équilibre, le nombre des équations qui contiennent effectivement les tensions se réduira à 3n – p. Ainsi, lorsque ni sera > 3n – p, les équations précédentes ne suffiront pas pour déterminer toutes les tensions.
Il en sera de même quand le système contiendra un certain nombre de points fixes. Cette indétermination signifie qu'il y a une infinité de valeurs des tensions qui, combinées avec les forces extérieures

Menabrea explains a central point of the theory of elasticity. One needs to understand that elasticity gives rise to energy, even inside a body that appears to be at rest.

In 1859, Alessandro Dorna, in his *Memoria delle pressioni sopportate dai punti d'appoggio di un sistema equilibrato ed in stato prossimo al moto*, explains that:

> Among the forces to be determined, having also dealt with the pressures which are supported by the support points, I arrived at the general formulas which produce their values; and, subsequently, a careful examination of these formulas allows me to see some simple theorems on the distribution of pressures, which are completely new, as far as I know, in the science of equilibrium.[9]

Dorna explains that Menabrea had encouraged him to present those results. H continues by explain the order in which he will proceed:

> 1. Having cited some examples to demonstrate that, assuming the support points are completely fixed, the pressures would be indeterminate, and having said how this can be demonstrated with calculation, I note which cautions should be observed so that the pressures would be indeterminate; and I suppose: 1) that each support point can move in all directions, as if it were attached to a fixed point, adjacent to the same, by means of an infinitely short elastic bar; 2) that producing (when the equilibrium between the external and internal forces of the system has been established) any of the movements that the principle of virtual moments supposes, the support points, in their quality of being fixed, will move infinitely less than all the other points of the system.[10]

---

données, sont aptes à tenir le système en équilibre. Les valeurs des tensions effectives dépendent de l'élasticité respective des liens, et lorsque celle-ci est déterminée, il doit en être de même des tensions.
Puisque, dans le cas que nous considérons, les tensions peuvent varier sans que l'équilibre cesse d'exister, on devra admettre que ces variations s'effectuent indépendamment de tout travail des forces extérieures; elles sont toujours accompagnées d'allongements ou d'accourcissements dans les divers liens correspondants, ce qui donne lieu, dans chacun d'eux, à un développement de travail. Les variations de longueur des liens doivent être supposées très-petites pour que les positions respectives des divers points du système ne soient pas sensiblement altérées. Mais, puisque pendant ce petit mouvement intérieur l'équilibre continue à exister et que le travail des forces extérieures est nul, il s'ensuit que le travail total élémentaire des tensions ainsi développées est également nul.

[9] Dorna, 1859, p. 282: *Fra le forze da determinarsi, essendomi anche occupato delle pressioni, che sono sopportate dai punti d'appoggio, pervenni alle formole generali, che ne somministrano i valori; e, posteriormente, un'attenta disamina di queste formole mi lascio scorgere alcuni semplici teoremi sulla distribuzione delle pressioni, i quali sono affatto nuovi, che io sappia, nella scienza dell'equilibrio.*

[10] Dorna, 1859, p. 282: *1.–Citati alcuni esempi atti a dimostrare che, supponendo i punti d'appoggio affatto immobili, le pressioni sarebbero indeterminate, e detto come ciò si possa dimostrare col calcolo, noto quale sia l'avvertenza da usarsi affinché le pressioni sarebbero indeterminate; e suppongo 1° che ogni punto d'appoggio possa muoversi in tutti i sensi, come se fosse attaccato ad un punto fisso, al medesimo contiguo, mediante una spranghetta elastica infinitamente breve; 2° che producendo (quando si sarà stabilito l'equilibrio fra le forze esteriori ed interiori del sistema) uno qualunque degli spostamenti, che il principio dei momenti virtuali suppone, abbiano i punti d'appoggio, nella loro qualità di fissi, a muoversi infinitamente meno di tutti gli altri punti del sistema.*

Dorna solves and explains all the difficulties linked to the problem of supports that we presented in the previous chapter on the dead-end street in the history of constructions.[11]

> Considerations on the determination of pressures, and general procedure for finding them.
>
> When the points of support of a system are considered as fixed, the pressures which they bear are generally found to be indeterminate. For example, the pressures that the hinges of a door support along the straight line that joins them cannot be given by calculation, because just one hinge is enough to support all the effort exerted along that straight line, and it makes no difference which one. Likewise, when there are three fixed points in a straight line, and a force is exerted on this perpendicular, the three pressures are indeterminate, because one can, in an infinite number of ways, imagine the force broken down into three others parallel to the same and directed against those points. The same thing happens whenever you have four or more supports, although not in a straight line. All this can be seen with simple reasoning. To deduce the same consequences with the calculation, it is necessary to replace the fixed points with forces equal and opposite to the pressures that they withstand; and, having thus made the system free, apply to it the six equilibrium equations of rigid and free systems.[12]

James Clerk Maxwell, in "On the calculation of the equilibrium and stiffness of frames" (1864) also gives a similar explanation, as does Claude-Louis Navier same year in the edition of *Résumé des Leçons données à l'école des ponts et chaussées sur l'application de la mécanique à l'établissement des constructions et des machines* edited by Saint–Venant (1864).

In 1865, James H. Cotterill published four articles in which he applies the principle of least action to several problems. In the first of these, "On an extension of the dynamical principle of least action" (1865a), he introduces his text in the following way:

> When a material body, or system of bodies, is exposed to the action of force, the points of application of that force move to a greater or lesser degree, and the motion of these points of application, by reason of the physical connexion of the points of the system, causes a general relative displacement of every one of them, thereby calling into play forces which increase with the extent of the displacement, and at length become sufficiently great to balance the applied forces; the displacement then ceases, and a state of equilibrium is attained.

---

[11] See "The Supports Problem: A Dead-End Street in the History of Stability", Chap. 15 in this present volume.

[12] Dorna, 1859, p. 285: *Considerazioni sulla determinazione delle pressioni, e procedimento generale per trovarle.*
Quando i punti d'appoggio di un sistema sono considerati come immobili, si trova che le pressioni, che i medesimi sopportano, sono generalmente indeterminate. Per esempio le pressioni, che i cardini di una porta sostengono secondo la retta che li unisce, non possono essere date dal calcolo, perché basta un cardine solo a sostenere tutto lo sforzo che si esercita secondo quella retta, ed è indifferente quale. Medesimamente, quando si hanno tre punti fissi in linea retta, ed e esercitato uno sforzo a questa perpendicolare, riescono indeterminate le tre pressioni, perché si può, in una infinità di modi, immaginare scomposta la forza in tre altre parallele alla medesima e dirette contro quei punti. La stessa cosa ha luogo ogni qualvolta si hanno quattro o più sostegni, sebbene non in linea retta. Tutto ciò si vede col semplice ragionamento. Per dedurre le stesse conseguenze col calcolo, bisogna sostituire ai punti fissi delle forze uguali e contrarie alle pressioni, che i medesimi sopportano; e, reso cosi libero il sistema, applicarvi le sei equazioni d'equilibrio dei sistemi rigidi e liberi.

> Thus, in every case of the balance of forces by the resistance of matter, a certain amount of energy is expended by the applied forces, and a certain amount of work done in partially overcoming the resisting forces; and it is a well-known case of the general law of the conservation of energy, that the energy so expended is equal to the work so done.
>
> But further, if any portion of the system be considered apart from the rest, the forces in equilibrium, and the energy expended by them considered as applied to the detached portion must be equal to the work done in that detached portion; and these are general conditions to which the forces generated at every point of the system are subject. Now Mr. Moseley has shown that if any number of pressures are in equilibrium, some of which are resistances, then each of these resistances is a minimum, subject to the conditions imposed by the equilibrium of the whole – a principle which he has called the principle of the Least Resistance; let us assume this principle, and let us further suppose for the present, that it is generalized so as to include the case of the resisting forces generated as above described; then each of those resisting forces is a minimum, subject to the general conditions stated above: and further, the relative displacements which the cause of the forces must also be the least possible, and the work done the least possible. Thus, in the assumption mentioned, to which I shall return in the sequel, it appears that the work is a minimum, subject to the law of conservation of energy and the statical conditions of equilibrium; and this principle, analogous to the dynamical principle of Least Action, it is the object of this article to consider and apply.[13]

Cotterill then demonstrates, with help of his principle of least action, the results of Clapeyron, Lamé and Moseley. The same year, in a second article (1865b), he solves with the same principle the equilibrium of arched ribs of uniform section. He writes:

> The object of this article is to show how to find approximately the thrust, shear, and bending moment on any section of an arched rib acted on by given forces under given circumstances.[14]

In his third article, "Further Application of the Principle of Least Action" (1865c), Cotterill explains his project:

> I shall now consider some cases in which it is required to find the law of variation, chiefly with a view to verify the statement made in my former article, that the results obtained are identical with those obtained by other methods.[15]

In his fourth and final article for that year, "On Elliptic Ribs" he calculates the stress on any section of an elliptical rib with the hope that the results obtained "may perhaps be of service in the theory of arches".[16]

Three years later, in 1868, Louis Frederic Menabrea published a general principle for determining pressures and tensions in an elastic system. In his introduction, Menabrea presents the problem, and shows how to manage "fixed points" in elasticity:

> Let us consider a system of material points joined together by connections and to which external forces are applied which create mutual equilibrium through the system in question; the ordinary rules of statics indicate the relationships which must take place between these forces for equilibrium to really exist. But if we want to investigate how the forces of the external forces are distributed among the connections, what are the pressures exerted on the

[13] Cotterill, 1865a, p. 299.

[14] Cotterill, 1865b, p. 380.

[15] Cotterill, 1865c, p. 480.

[16] Cotterill, 1865d, p. 21.

fixed points which can exist in the system, the problem, except in a small number of cases is generally indeterminate, as long as we consider the connections of the system as rigid, or, in other words, as being of invariable length. Indeed, the forces supported by the connections which lead to a given point must balance the external forces applied to this point. We can, when the number of these connections exceeds a restricted limit, conceive an infinity of various distributions of the forces, all of which are suitable for maintaining balance between external forces.

But this indetermination ceases from the moment we consider the elasticity of the body or, more precisely, the elasticity of the connections which unite the various points of the system together.[17]

This provides a good summary of what we read before about fixed points and the earliest expressions of the *principle of elasticity*. Let us now read the one proposed by Menabrea:

When any elastic system comes into equilibrium under the action of external forces, the total work developed in the extension and compression of the connections, as a result of the relative displacements of the points of the system, or, in other words, the work developed by inner forces, is a MINIMUM.[18]

In §1, Menabrea gives preliminary considerations, definitions, and the "equation of elasticity"[19]:

$$\sum \varepsilon\lambda d = \sum \frac{1}{\varepsilon} T\delta T = 0$$

where: λ is the variation of l, the distance between two points; T is the tension of the link between two given points corresponding to a variation λ; ε is a coefficient that depends on the nature of the connection and s called the "coefficient of resistance".

Menabrea then verifies the principle in various particular cases, and solves five problems before giving a demonstration of the principle in the case of a free system.

---

[17] Menabrea, 1868, p. 5: *Considérons un système de points matériels réunis entre eux par des liens et auxquels on applique des forces extérieures qui se font mutuellement équilibre par l'intermédiaire du système dont il s'agit; les règles ordinaires de la statique indiquent les rapports qui doivent avoir lieu entre ces forces pour que l'équilibre existe réellement. Mais si l'on veut rechercher comment les efforts des forces extérieures se répartissent entre les liens, quelles sont les pressions exercées sur les points fixes qui peuvent exister dans le système, le problème, sauf dans un petit nombre de cas est en général indéterminé, tant que l'on considère les liens du système comme rigides, ou, en d'autres termes, comme étant de longueur invariable. En effet, les efforts supportés par les liens qui aboutissent à un point donné devant faire équilibre aux forces extérieures appliquées à ce point, on peut, lorsque le nombre de ces liens dépasse une limite restreinte, concevoir une infinité de répartitions diverses des efforts qui, toutes, sont propres à maintenir l'équilibre entre les forces extérieures.*

Mais cette indétermination cesse du moment que l'on a égard à l'élasticité du corps ou, plus précisément, à l'élasticité des liens qui réunissent entre eux les divers points du système.

[18] Menabrea, 1868, p. 7: *Lorsqu'un système élastique quelconque se met en équilibre sous l'action de forces extérieures, le travail total développé dans l'extension et la compression des liens, par suite des déplacements relatifs des points du système, ou, en d'autres termes, le travail développé par les forces intérieures, est un* MINIMUM.

[19] Menabrea, 1868, p. 12.

In his last section, he extends the principle to the case of systems containing fixed points or rigid parts. Finally, in his book *Sul Principio di Elasticità. Dilucidazioni* (1870), Menabrea explains in a more accessible manner and for a larger readership the ideas he had developed in the previous articles.

References

Bélanger, Jean-Baptiste. 1845. *Notes sur la mécanique appliquée aux principes de la stabilité des constructions et à la théorie dynamique des machines*. Paris: École royale des Ponts et Chaussées.

Carnot, Sadi. 1824. *Réflexions sur la puissance motrice du feu et sur les machines propres à développer cette puissance*. Paris: Bachelier.

Clapeyron, Emile. 1834. Mémoire sur la puissance motrice de la chaleur. *Journal de l'Ecole Polytechnique*, XXIIIe cahier, tome XIV, 153–191.

Clapeyron, Émile. 1857. Calcul d'une poutre élastique reposant librement sur des appuis inégalement espacés. *Comptes Rendus Hebdomadaires Des Séances De L'académie Des Sciences* 45: 1076–1080.

Clapeyron, Émile. 1858. Mémoire sur le travail des forces élastiques dans un corps solide élastique déformé par l'action de forces extérieures. *Comptes Rendus Hebdomadaires Des Séances De L'académie Des Sciences* 46: 208–212.

Cotterill, James H. 1865a. On an extension of the dynamical principle of least action. *The London, Edinburgh and Dublin Philosophical Magazine and Journal of Science*, Series 4, 29: 299–305.

Cotterill, James H. 1865b. On the equilibrium of arched ribs of uniform section. *The London, Edinburgh and Dublin Philosophical Magazine and Journal of Science*, series 4, 29: 380–389.

Cotterill, James H. 1865c. Further application of the principle of least action. *The London, Edinburgh and Dublin Philosophical Magazine and Journal of Science*, Series 4, 29: 480–486.

Cotterill, James H. 1865d. On elliptic ribs. *The London, Edinburgh and Dublin Philosophical Magazine and Journal of Science*, Series 4, 30, 21–33.

Dorna, Alessandro. 1859. Memoria delle pressioni sopportate dai punti d'appoggio di un sistema equilibrato ed in stato prossimo al moto. *Memoria della Reale Accademia della Scienze di Torino* 18: 281–318.

Lamé, Gabriel. 1852. *Leçons_sur_la_théorie_mathématique de l'élasticité des corps solides*. Paris: Bachelier.

Maxwell, James Clerk. 1864. On the calculation of the equilibrium and stiffness of frames. *The Philosophical Magazine: A Journal of Theoretical Experimental and Applied Physics* 27 (182): 294–299.

Menabrea, Luigi Federico. 1858. Nouveau principe sur la distribution des tensions dans les systèmes élastiques. *Comptes Rendus Hebdomadaires Des Séances De L'académie Des Sciences* 46: 1056–1060.

Menabrea, Luigi Federico. 1868. *Etude de statistique physique. Principe général pour déterminer les pressions et les tensions dans un système élastique*. Torino: Imprimerie royale.

Menabrea, Luigi Federico. 1870. *Sul Principio di Elasticità. Dilucidazioni*. Torino: Stamperia reale.

de Molinos, L., and C. Pronnier. 1857. *Traité théorique et pratique de la construction des ponts métalliques*. Paris: Morel et Cie.

Navier, Claude-Louis, and Adhémar-Jean-Claude Barré de Saint-Venant. 1864. *Résumé des Leçons données à l'école des ponts et chaussées sur l'application de la mécanique à l'établissement des constructions et des machines ... avec des notes et des appendices par Barré de Saint Venant*. Paris: Dunod.

Chapter 17
The Industrial Revolution and Its Consequences. Part I: Iron Frameworks and Cullmann's "Graphostatics"

1 Introduction

In order to be able to judge the situation before the introduction of the vectorial calculus, I shall first recall a text from W. J. Macquorn Rankine. It dates from 1858, thus not before the very beginnings of vectorial analysis, which go back to the works of Hermann Grassmann and William Rowan Hamilton, but at a time when those works were only known to pure mathematicians. In 1844 Grassmann published his linear expansion theory (Grassmann 1844), at the moment when Hamilton published a series of four works on the composition of forces and quaternions (Hamilton 1844a, b, c, d). These all provided a purely mathematical point of view, but inspired a theory that will be crucial for construction theory.

In fact, some twenty years later the first edition of Culmann's "graphostatics" appears. This may be considered as a vectorial geometry intended for engineers. During the years following that first publication, Culmann follows the general motion that tends to vectorial analysis and the second edition of his work is much more analytical.

I shall analyse Culmann's work principally through five of his texts, the first edition of his graphical statics (1865), one particular article concentrating on the law of composition of forces (1870), the second edition of the graphical statics (1875) together with its translation into French (1880). But I shall try not to forget the followers of Culmann in Italy: Antonio Favaro and his lectures on graphical statics (Favaro 1877) and Luigi Cremona's elements of graphic calculus (Cremona 1874).

P. Radelet-de Grave, *Between Mechanics and Architecture*, Mathematics and the Built Environment 8, https://doi.org/10.1007/978-3-031-73530-1_17

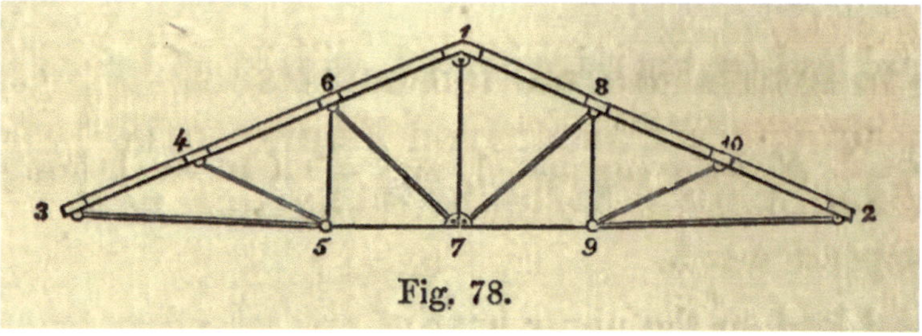

Fig. 1 A Polonceau framework in Rankine (1858, p. 147). Public domain

2 W. J. M. Rankine and August Ritter

In order to get an idea of the way men of practice solved problems that will be solved later with help of graphostatics, let us have a look works by Rankine and August Ritter

In Chap. II on stability in his *Manual of Applied Mechanics* (1858) Rankine, who will become one of the heroes of the elastic revolution,[1] computes in 1858, the framework of an iron roof of the type Polonceau. (We shall also find it later in Castigliano, 1873). Rankine explains:

> Example II. Fig. 78 [our Fig. 1] represents another form of truss common in roofs. Let W be the weight of the roof, as before, distributed over the rafters 1 2, 1 3. 2 3 is the great tie-rod; 1 7, 6 5, 8 9, suspension-rods; 7 6, 7 8, 5 4, 9 10, struts.

(1.) Primary Truss 1 2 3. The load at 1, as before, is to be taken as $= \frac{1}{2}$ W.

(2.) Secondary Trusses 7 6 3, 7 8 2. The load at 6 is to be held to consist of one-half of the load between 6 and 1 and one-half of the load between 6 and 3; that is, one-half of the load between 1 and 3, or $\frac{1}{4}$ W. The trusses are triangular, each consisting of two struts and a tie, and the stresses are to be found as in Article 149.
The suspension-rod 1 7 supports two-thirds of the load on 7 6 3, and two-thirds of the load on 7 8 2; that is,

$$\frac{2}{3} \cdot \frac{2}{4} \cdot \mathrm{W} = \frac{1}{3} \mathrm{W};$$

and this, together with $\frac{1}{6}$ W; which rests *directly* on 1, makes up the load of $\frac{1}{2}$ W; already mentioned.

(3.) Smaller Secondary Trusses 3 4 5, 9 10 2. Each of the points 4 and 10 sustains a load of $\frac{1}{6}$ W, from which the stresses on the bars of those smaller trusses can be determined. One-half of the load on 4, that is $\frac{1}{12}$ W, hangs by the suspension-rod 6 5; and this together with $\frac{1}{6}$ W, which rests directly on 6, makes up the load of $\frac{1}{4}$ W on that point, formerly mentioned. The same remarks apply to the suspension-rod 8 9.

(4.) *Resultant Stresses*. The pull between 5 and 9 is the sum of those due to the primary and larger secondary trusses; that between 5 and 3, and between 9 and 2, is the sum of the pulls due to the primary, larger secondary, and smaller secondary trusses.

[1] See "The Industrial Revolution and its Consequences: Part II. Elasticity Penetrates Construction", Chap. 18 in this present volume.

> The thrust on 1 6 is due to the primary truss alone; that on 6 4 to the primary and larger secondary truss; that on 4 3 to the primary, larger secondary, and smaller secondary trusses; and similarly for the divisions of the other rafter.[2]

Rankine follows in the calculus a traditional method explained by Camille Polonceau (1840). In section II of chapter II, Rankine treats the equilibrium of chains, cords, ribs and linear arches.

Two years later, August Ritter publishes first an article on the calculation of the construction of roofs and bridges (Ritter 1861), and then, a year later, a book on the same argument. In 1871, he'll finally gives a final last version of his book. In that last version, he will also give a calculation of the elastica that we shall analyse in the next section 21.

But let us start with the article of 1861. He opens his article with the following remark:

> In larger buildings of this type [roofs and bridges], everything depends on the correct and most economical distribution of the material, not only because the cost of the material is comparatively high compared to the labour costs, but especially because the material used forms a dead load, the smallness of which often affects the possibility of the entire undertaking. A good construction should be sufficient to ensure that the resistance of the material is fully exploited everywhere and that no superfluous material is used.[3]

Ritter immediately remarks that this is always the case for bars that are pulled or pushed along their length. On the contrary, this is never the case for bars whose resistance is desired. This desire led to an important use of H or I beams.

We may see that at the time the importance of the role of pure mechanics is supplanted by practical rules issued from practice and mechanics. A consequence of which is the emergence of a new protagonist in the figures, namely the forces or loads *P, Q, R, S, T* as we can see in his figures 1 and 2 (Fig. 2). In the same figure, the tensions are called *X, Y, Z*.

Next, Ritter introduces the "method of static momenta", which only uses the law of the lever. It has the advantage, he remarks, that it uses only the lever arms of the forces that can be measured on the figures.

He then gives some examples before applying the method to the calculus of iron bridges, as for example the calculus of a paraboloidal bridge girder illustrated in his figures 12 and 13 (Fig. 3).

---

[2] Rankine (1858, pp. 147–148).

[3] Ritter (1861, p. 410): *Bei grösseren Bauwerken dieser Art kommt alles auf richtige und möglich sparsamste Vertheilung des Materials an, nicht allein, weil die Kosten des Materials vergleichsweise gross sind gegen die Arbeitskosten, vielmehr besonders deshalb: weil das verwandte Material eine todte Last bildet, von deren kleinheit oft die Möglichkeit des ganzen Unternehms abhängt. Eine gute Construction soll der Förderung genügen: dass überall die Widerstandsfähigkeit des Materials vollkommen ausgenutzt und nirgend überflüssiges Material verwendet sei.*

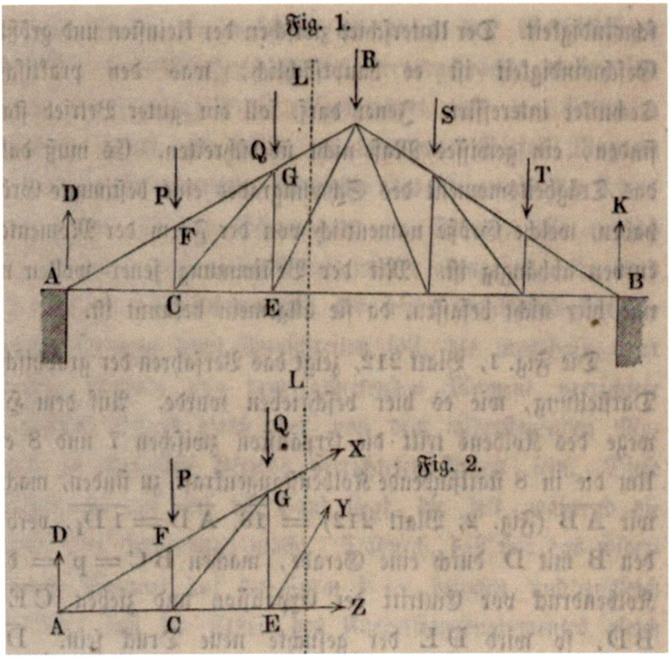

Fig. 2 Ritter's figures 1 and 2 of his text of (Ritter 1861). Public domain

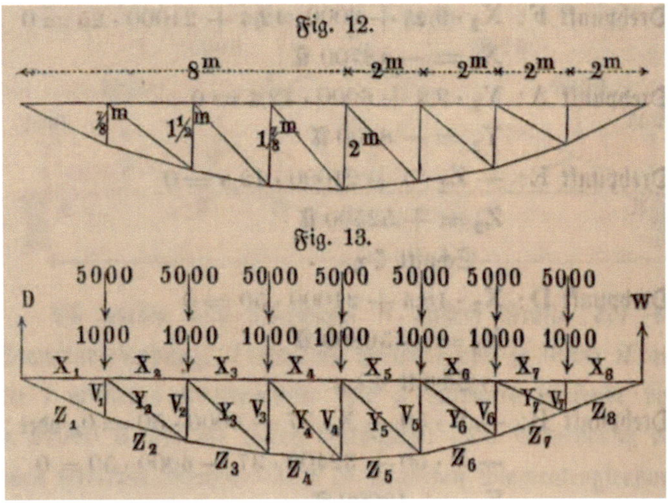

Fig. 3 Ritter's figures 12 and 13 of his text (Ritter 1861). Public domain

3 Karl Culmann's "Graphostatics"

Before entering Culmann's work, we have on the one hand to define graphical statics, or graphostatics, and on the other hand to situate that new branch with respect to parallel researches done in mathematics and in mechanics as well as in the sciences of construction.

Graphostatics is, as its name indicates, a graphic and thus geometric technique that makes it possible to judge the stability of a construction, which is in general a problem of equilibrium of forces. We have seen its prehistory in the preceding chapters.[4] We have also noted the scarce use of the parallelogram law.[5] However, the analytical expressions of that law had only been expressed by Euler in his E. 177 (Euler 1752).

From a purely mathematical point of view other dates are important. The works of Culmann, for example, were produced between 1865 and 1880, that is, means 62 years after the Lazare Carnot's geometry of position (Carnot 1803) and 38 years after Möbius's barycentric calculus (Möbius 1827a, b), some 20 years after the first edition of Grassmann's extension theory (1858) and the contemporaneous works of Hamilton on quaternions (1844a, b, c, d). Those works were the first attempts to formulate a vector analysis. This is the research contexts, very different from one another, in which we should place the research of Culmann, who is also looking for a vectorial analysis, but in the practical context of construction.

I have made a minute comparison of the texts by Culmann, in particular of the four introductions written by Culmann for the various editions of his graphostatics. I shall analyse within those texts and in the article of 1870, the evolution of his treatment of the composition of forces and of the polygon of forces. In parallel, I shall mention some articles concerning the science of construction.

4 The Intentions, Aims, and Critiques in Culmann's Preface of 1865

In his first preface, Culmann describes his aim and its situation between mechanics, geometry, and analyses. But after having mentioned some authors that we will find farther on, it is with a reference to Poncelet that Culmann opens his preface:

> ... when he [Poncelet] tried to devise geometric solutions for the various tasks that presented themselves in the field of engineering.

---

[4] See "A Study of Coulomb's Application of the Rules of Maxima and Minima to Problems of Statics", Chap. 10 in this present volume.

[5] See "The Bernoullis and the Elastica: From Leibniz's *Nova methodus* to Euler's *Methodus inveniendi*", Chap. 7 in this present volume.

> In fact, the engineer will also give preference to geometric solutions as soon as an accuracy of three digits or 1/000, which can be completely achieved, is sufficient: because he always has the drawing tools at hand, drawing is his language and skill. Due to his topographical-geodetic education, he is more accustomed than the architect and the mechanic to assessing the accuracy of sections and to clearly defining lines and points.[6]

The introduction to his first edition opens with an affirmation: "The drawing is the engineer's language." But he does an important reproach to Poncelet[7] and explains, in contrast, what his own project consists of:

> Poncelet's solutions were always just translations of previously developed analytical expressions. That this is a detour and that a geometric construction is much less easy to memorise if, when using it, one has to have an analytically developed formula in mind, the derivation of which is not present, than if the line structure given by the task itself forms the basis, from which the solution develops simply geometrically. Poncelet himself probably felt it deeply, and he studied geometry eagerly, as if he were anticipating what benefits it could provide. However, he only used it in exceptional cases to solve problems in the field of construction; he always preferred analytical solutions, which he later translated.[8]

At the time, Culmann considers analytical methods as a detour. He found that the engineer, who started from lines, had to develop his solution geometrically. He will thus try "to provide graphical statics to the technicians, as an attempt to present a geometric treatment of usual exercises from the science of engineers thanks to that totally new geometry".[9]

What does Culmann mean by a "new geometry"? The answer is given at the beginning of his introduction. He first mentions the work of Barthélémy Édouard Cousinery (1790–1851), which that he does not appreciate much but had to rank

---

[6] Culmann (p. V): *… als er [Poncelet] sich bemühte, geometrische Lösungen für die verschiedenen im Ingenieurfach sich darbietenden Aufgaben zu Ersinnen.*
In der That wird auch der Ingenieur den geometrischen Lösungen den Vorzug geben, sobald eine Genauigkeit von drei Stellen oder 1/000, die vollständig erreicht werden kann, genügt: denn die Reisswerkzeuge hat er immer bei der Hand, das Zeichnen ist seine Sprache und Vermöge seiner topographisch-geodätischen Bildung ist er mehr als der Architekt und der Mechaniker gewohnt, die Genauigkeit von Schnitten zu beurtheilen und Linien und Punkte scharf zu bestimmen.

[7] Culmann considers two texts of Poncelet in engineering science, *les Solutions graphiques des principales questions sur la stabilité des voûtes* and the *Mémoire sur la stabilité des revêtements et de leurs fondations,* both published in, "Mémorial de l'officier du Génie", XII, 1835, pp. 151–215 et vol. XIII, 1840, pp. 7–270.

[8] Culmann (pp. V–VI): *Die Lösungen Poncelet's waren immer nur Uebersetzungen vorher entwickelter analytischer Ausdrücke. Dass dies ein Umweg sei und dass eine geometrische Construction viel weniger leicht sich einprägt, wenn mann bei Anwendung derselben eine analytisch entwickelte Formel, deren Herleitung nicht gegenwärtig ist, im Kopfe haben muss, als wenn das durch die Aufgabe gegebene Liniengebilde selbst die Grundlage bildet, aus der sich die Lösung einfach geometrisch entwickelt: hat wohl Poncelet selbst innig gefühlt, und eifrig studierte er Geometrie, gleichsam ahnend, welchen Nutzen sie gewähren könnte. Allein nur ausnahmsweise bediente er sich ihrer zur Lösung eer Aufgaben aus dem Gebiete des Baufaches, immer zog er analytische Lösungen, die er nachher übersetzte, vor.*

[9] Culmann (1866, p. VI): *die graphische Statik den Technikern als einen Versuch vorzulegen, die einer geometrischen Behandlung zugänglichen Aufgaben aus dem Gebiete des Ingenieurfachs mit Hülfe der neueren Geometrie zu lösen.*

among the elders. He certainly prefers the geometry of position of Karl Georg Christian von Staudt (1798–1867) published in 1847, and the more recent one of 1866 of Theodor Reye (1838–1919), as well as the *Treatise on Conic Sections* by George Salmon (1813–1904) in its translation by Wilhelm Friedler. One should also add the works of Michel Chasles (1793–1880), the Treatise on superior geometry (Chasles 1852), and the later Treatise on conic sections (Chasles 1866).

Culmann therefore means the geometry of transformations, the changes of reference system, and the passage to imaginaries (the cyclic points). He continues by noting that some well-meaning colleagues suggested that he give the principal definitions of that geometry in the introduction. But he believes this to be impossible because:

> The substance of all those definitions and expressions constitutes precisely the geometry of position; an introduction that would treat that would not be better than an introduction treating differential and integral calculus.[10]

The bright new geometry came out of Carnots's position geometry, which came out of Leibniz's *Analysis situs*:

> Thus, in our graphical statics we presuppose the elements of geometry, which is the "geometry of position" of Dr. G. K. Ch. V. Staudt (1847), because this work is, as far as we know, the first and also the only one in which the elements of the geometry of position can be found beautifully and systematically put together with a strictly geometric derivation. The expressions we use correspond to the definitions given there: linear complexes, sheaves of rays, bundles of rays, harmonious, involutory structures, etc.[11]

For certain analytical developments, he refers to the German translation of Salmon, and launches into a story of his intent to follow his course and having the prerequisites indispensable to understand his lectures. He then turns back to graphostatics and gives the plan of his lectures:

> It [graphical statics] arose from the fact that when some tasks from the field of construction were treated graphically, it soon became apparent that certain general geometric concepts and constructions, which recur in all tasks and which have a certain connection with one another, have graphical solutions that have to be prepared in advance, in order not to have to prepare them for each particular part of the science of construction.[12]

---

[10] Culmann (p. VII): *Den Inbegriff aller dieser Definitionen und Aussagen bildet gerade die Geometrie der Lage; eine Einleitung aber, die diese behandelt, wäre nicht besser als eine jener, welche Differential- und Integralrechnung behandeln. Welcher Anfänger z. B. wird in der Einleitung.*

[11] Culmann (p. VII): *Wir setzen also in unserer graphischen Statik die Elemente der Geometrie voraus, welche die "Geometrie der Lage von Dr. G. K. Ch. V. Staudt" enthält, denn dieses Werk ist, so viel uns bekannt ist, das erste und auch das einzige, in welchem bei streng geometrischer Ableitung die Elemente der Geometrie der Lage schön systematisch zusammengestellt zu finden sind. Den in ihm gegebenen Definitionen entsprechen die von uns gebrauchten Ausdrücke: gerades Gebilde, Strahlenbüschel, Strahlenbündel, harmonische, involutorische Gebilde u. s. w.*

[12] Culmann (p. X): *Sie [graphische Statik] entstand daraus, dass bei der graphischen Behandlung einiger Aufgaben aus dem Gebiete des Baufachs es sich bald zeigte, dass gewisse allgemeine geometrische Begriffe und Constructionen, die bei allen Aufgaben wierderkehren und die untereinander in einem gewissen Zusammenhange stehen, jenen graphischen Lösungen vorausgeschickt werden müssen, wenn man sie nicht bei jedem einzelnen Abschnitt der Baukunde vorausschicken wollte.*

Culmann insists on the repetitive character of certain graphical reasonings that he will put together in his first part: "This general part forms the content of the first chapters of the graphic calculus and of the graphostatics. This part was already published in 1864."[13]

This first part had already been published in 1864 because it could be useful in domains other than only that of construction. Culmann notes with clairvoyance that, as things progress, that first part will have to be extended to new types of problems. He is completely correct, as that first part will contribute to the elaboration of vectorial analysis. The whole evolution of Culmann's thinking, over those three editions, will be to pass from a useful instrument to a fundamental mathematical theory. He explains:

> We would have completely separated this first part from the those that followed had it become voluminous enough; it now only contains the general part needed to deal with our construction tasks, but it will and must expand as graphic methods become more widely used. But then it will go beyond the treatment of the specialist, and it must be developed by the geometer and the mechanic into an entity of its own, one which relates to modern geometry as analytical mechanics relates to higher analysis.[14]

In short, Culmann begins with a tool in order to solve the repetitive practical problems of engineers, but he envisions an organized whole that would have to geometry a relation comparable to that of analytical mechanics to analysis. His aim from since the very beginning is to develop a vectorial analysis. In this first edition he gives an account of the works going in the same direction, without forgetting the special interests of his readership, who with help of that tool, should be able to communicate easily with men on the construction site.

5 A First Transition: August Ritter and Otto Mohr

The first edition of August Ritter's *Elementare Theorie und Berechnung eiserner Dach- und Brücken- Constructionen* (1862), as well as the first part of its second edition published in 1870, are not fundamentally different from his first article we commented on above (Ritter 1861). As we read in his preface, the difference lies in the introduction of elasticity in the second edition of that book.

---

[13] Culmann (p. XI): *Dieser allgemeine Theil bildet den Inhalt der beiden ersten Abschnitte des graphischen Rechnens und der graphischen Statik. Als erste Lieferung wurden diese schon im Jahre 1864.*

[14] Culmann (p. XI): *Wir würden diese erste Lieferung ganz von dem folgenden getrennt haben, wenn sie voluminös genug geworden wäre, sie enthält jetzt nur den allgemeinen Theil, dessen wir zur Behandlung unserer Bauaufgaben bedurften, allein er wird und muss sich ausdehnen, sowie die graphischen Methoden weitern Eingang finden; dann aber wird sie der Behandlung durch den speciellen Fachmann entschlüpfen, und sie muss durch den Geometer und den Mechaniker zu einem eigenen Ganzen ausgebildet werden, das sich zur neuern Geometrie verhält wie die analytische Mechanik zur höhern Analysis.*

Between 1874 and 1875 Otto Mohr published a series of three articles on arched frameworks. The first, *Beitrag zur th der Bogenfachwerksträger* (1874a), concerns the calculation of the force necessary to support laterally the pressure of a certain arched framework. To evaluate horizontal pressure H due to a certain weight P when the form and the section of the framework he gives $H = z \cdot P$ (Fig. 4).

Here Mohr uses of course the parallelogram law of forces and the principle of virtual velocities. In his next article *Beitrag zur Theorie des Fachwerks* (Mohr 1874b) he calculates:

- the change in shape of a truss that is fixed together;
- the estimation of the change of length of the redundant structural parts of a truss that is fixed together;
- determination of the stresses in the structural parts of a loaded compound truss;
- determination of the stresses of a compound framework which are caused by temperature changes and by inaccuracies in the manufacture of the lengths of the structural parts.

And a year later, in *Beitrag zur Theorie des Fachwerks* (Mohr 1875) he proceeds with the work he had started in the previous article. More precisely, he applies what he has calculated in the preceding article to arch trusses:

> The term arched truss can be used to summarize all those trusses that have to accommodate a horizontal support reaction in more than one support point. The number of fixed support points is usually two, and when calculating such trusses it is usually assumed that the distance AB between these two points is completely unchangeable. (Fig. 5).

He introduces a *Biegungspolygon* (bending polygon) in order to evaluate the influence of a little modification of the length of a piece of beam framework:

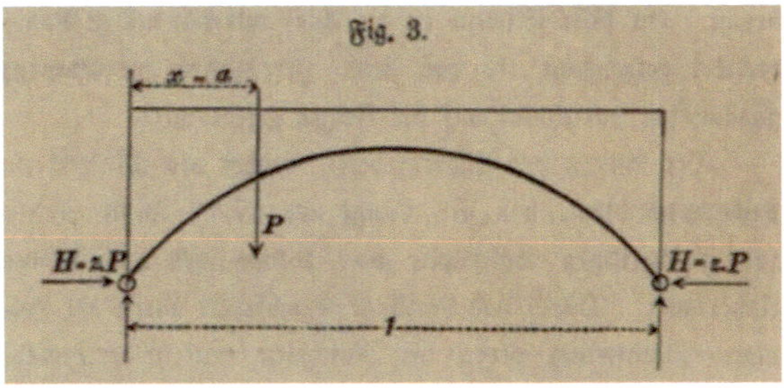

Fig. 4 Mohr's calculation of pressure H due to a certain weight P (Mohr 1874a, p. 224). Public domain

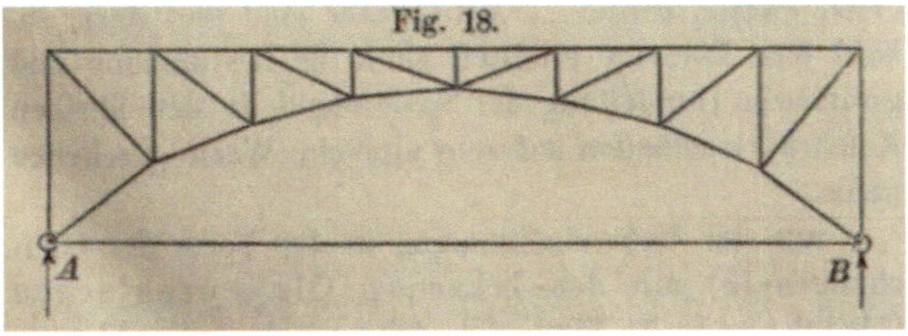

Fig. 5 Mohr's arch truss (Mohr 1875, p. 17). Public domain

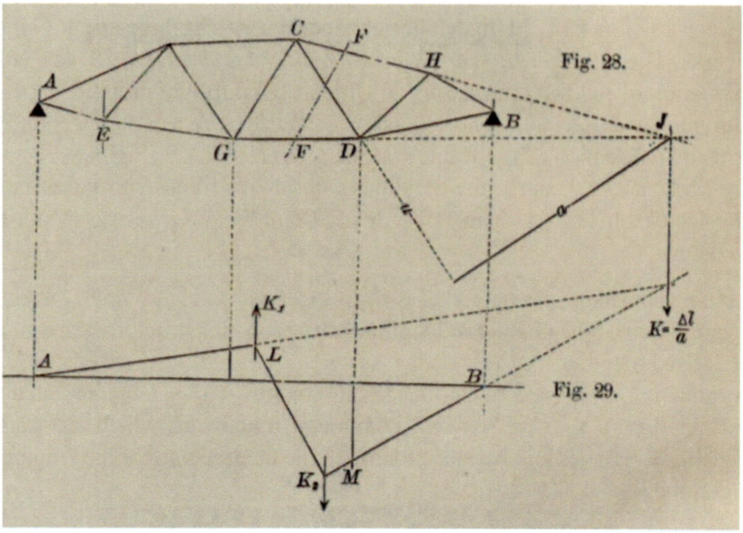

Fig. 6 Mohr's introduction of a bending polygon (Mohr 1875, p. 24). Public domain

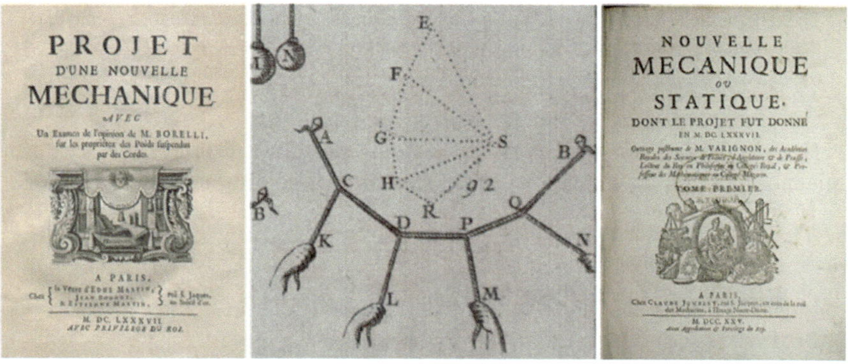

Fig. 7 **Left** The title page of Varignon's 1687 first edition of the Projet d'une nouvelle mechanique; **centre** The polygon of forces in the posthumous 1725 edition; **right**, the title page of the 1725 edition. Public domain

> The bending polygon is to be constructed, which arises when only a single construction element CD of a simple beam framework changes its length by the small measure Δl.[15] (Fig. 6)

In my opinion, this seems difficult to realize without elasticity theory.

---

[15] Mohr (1875, p. 24): *Es ist das Biegungspolygon zu konstruiren, welches entsteht, wenn nur ein einziger Konstructionstheir CD (Fig. 28) eines einfachen Balkenfachwerks seine Länge um das kleine Mass* Δl *verändert.*

6 The Preface of Culmann's Edition of 1875

Culmann's aim in his introduction of 1875 is to make give an account of the progress he had made during the eleven years between the first edition and the new one. He first observes:

> No doubt, it is in Italy that it [graphical statics] found the most fertile ground. There Cremona introduced it at the Milan Polytechnic school of, and furthermore, from a higher point of view; he considers it not only as a practical aid in order to save some calculations in some precise cases, but as the completion of the education in statical geometry for young engineers. He did not rebel against the preliminary learning indispensable for that education, but presented positional geometry himself in order to prepare his students to follow this with graphostatics.[16]

He recognizes Cremona as both more skilful and bolder in developing graphical statics, while also noting that others, including Favaro, followed the same path.

Culmann laments the fact that he was not understood in Germany and thus did not receive the necessary institutional means. He starts with a fierce criticism of the reactions in southern Germany, where only the practical aspect and the saving of certain calculations were considered, and the nature of it (*das Wesen derselben*) was only reluctantly considered:

> To understand the bodies in which forces are in equilibrium and the line structures that represent the directions and magnitudes of these forces as geometrically related structures, and to apply to them the geometry of position in which the relationships between such related structures are worked out in the most perfect way, at first sight people have definitely reacted negatively to this view.[17]

We observe that in the time between the two editions, Culmann's thought has matured, in the sense that he had explicitly understood where the heart of his calculus lies: in treating the forces as represented in magnitude and direction by lines; in short, today we would say by vectors.

Culmann continues by denouncing the German professors' desire to get around mathematics, which is nevertheless indispensable to a good engineer. It is clearly difficult for them to consider the object that represents the force, but also many other physical quantities, such as velocity, as mathematical objects, vectors, and to grasp the power of that representation. He supports his remark by a comparison with

---

[16] Culmann (1875, p. V): *Unstreitig fand sie in Italien den günstigsten Boden. Dort hat sie Cremona am Mailänder Polytechnikum eingeführt, und zwar in einer hohen Auffassung; er betrachtete sie nicht blos als praktisches Hülfsmittel um in gewissen Fällen einige Rechnungen zu ersparen, sondern als den Abschluss der geometrisch statischen Bildung junger Ingenieure. Er hat sich nicht aufgelehnt gegen die Vorkenntnisse, die sie zum Studium voraussetzen muss, sondern hat selbst Geometrie der Lage vorgetragen um seine Schüler vorzubereiten, und darauf die graphische Statik folgen lassen zu können.*

[17] Culmann (1875, p. VI): *Die Körper, an welchen Kräfte im Gleichgewicht sind, und die Liniengebilde, welche die Richtungen, und Grössen dieser Kräfte darstellen, als geometrische zusammengehörige Gebilde aufzufassen und auf diese die Geometrie der Lage anzuwenden, in der ja die Beziehungen so verwandter Gebilde auf das vollkommenste ausgearbeitet, fix und fertig vorliegen, gegen diese Auffassung hat man sich durchaus ablehnend verhalten.*

mechanics, which he had tried to teach without differential and integral calculus. He concludes by noting that anyone who wants to study mechanics in depth must know mathematics. He formulates the following wish:

> Let us hope that the geometry of position will also become established in schools and that they will eventually rise to the higher level, despite the resistance of the overly practical professors.[18]

He notes that in France it has been decided to offer for the first time a high-level mathematic-scientific education at the École Polytechnique and in the *grandes écoles* that follow it. Nevertheless, he regrets that, as he already noted in 1865, the constructions of Poncelet used in France are nothing but translations of elaborate analytical expressions, and that already in 1837 Poncelet had applied the fundamental tool of graphical statics to the funicular, or force, polygon. After that, Culmann decreed, France had stagnated until the release of Maurice Levy's *La statique graphique et ses applications aux constructions* (Levy 1874):

> In this book, Cremona's method, which is very useful for constant loading, is further explained, but the most unfavourable types of loading in the constructions are nowhere determined or taken into account, and very soon we move on to analysis in which is developed Cremona's method for constant loads. But irregular types of loads are never mentioned or considered, and one rapidly passes to Analysis.[19]

Culmann affirms, ironically, that in Germany the name graphical statics was modified to *Graphostatics*, but apart from that they did not progress a lot. He goes on to denounce a double infernal triangle: the one of architects, engineers and mathematicians associated to the one of mathematics, technics and arts. He notes series of difficulties and incompatibilities. First, "Those technicians are until now missing the indispensable training in mathematics, more precisely in geometry.[20] Then, "It is impossible to educate simultaneously the same students in art and in mathematics, one of the two disciplines must dominate; at an academy of architecture, one forms artists.[21] Finally:

> Berlin currently has the first mathematical faculty in the world, but unfortunately it is located apart from the university and trains senior teachers for the secondary schools, which are technical institutions, who understand nothing about technology. And finally, the technician

---

[18] Culmann (1875, p. VIII): *Hoffen wir, es werde sich auch die Geometrie der Lage in den Schulen einbürgern und diese trotz der widerstrebenden allzu praktischen Herrn Professoren schliesslich die höhere Stufe erklimmen.*

[19] Culmann (1875, p. VIII): *In diesem Buche ist die für constante Belastung sehr zweckmässige Methode Cremona's weiter ausgeführt, die ungünstigsten Belastungsarten der Constructionen sind aber nirgends ermittelt und berücksichtigt, und sehr bald wird zur Analysis übergegangen..*

[20] Culmann (1875, p. IX): *Es fehlte bis jetzt den Technikern dort an der nothwendigen mathematischen, namentlich geometrischen Bildung.*

[21] Culmann (1875, p. IX): *Unmöglich ist es, dieselben Schüler gleichzeitig zu Künstlern und zu Mathematikern auszubilden, die eine oder die andere Disciplin muss vorherrschen; an der Bauacademie werden Künstler erzogen.*

understands too little mathematics and is not spiritually penetrated enough by it because he cannot come into contact with those who strive for the highest goals.[22]

From those three observations, he concludes with bitterness:

A Cremona who stood high on the mathematical ladder and also achieved something in the practical area could never have emerged in Berlin. A university-educated mathematician will never draw a force diagram. Though graphics are in principle excluded from the university, Staudt did not hesitate to add figures to his geometry. ...

It is thus indispensable to educate professors who must train technicians in technical institutes, and therefore for those institutes to obtain the best professors in the direction of mathematical sciences and the most intelligent that the land can offer.[23]

He concludes by showing what Cremona owes to Maxwell, and above all, what he does not owe to him concerning the reciprocal diagrams.

7 Preface to the Edition of 1880

While the texts of the second edition and of the translation are congruent, their prefaces are not identical; the order of the ideas has been modified. Culmann's new preface is signed and dated November 1879, while the preface of the second edition is dated April 1875. In this last preface, the position of graphical statics between geometry and analysis becomes clearer.

After having given to Poncelet the principal role among his predecessors at the schools of Metz and Paris in the first edition of 1866, in the edition of 1875 among his successors in graphostatics he gives the principal role to Cremona of the universities of Milan and Rome. In 1880 he returns to Poncelet, to whom he associates Monge and Michon, as the French edition requires. He recalls the justification concerning precision that he attributes here to Poncelet, but whom he did not mention in the first German edition:

Poncelet was the first to recognize that these methods, while being much more expeditious than analytical methods, nevertheless offered a more than sufficient approximation in practice

[22] Culmann (1875, p. IX): *Berlin besitzt zur Zeit wohl die erste mathematische Facultät der Welt, allein leider sitzt sie abseits an der Universität, und bildet für die Realgymnasien, die technische Anstalten sind, Oberlehrer, die von der Technik nichts verstehen. Und schliesslich versteht der Techniker zu wenig Mathematik, wird zu wenig geistig von ihr durchdrungen, weil er mit Denjenigen nicht in Berührung kommen kann, die nach den höchsten Zielen streben.*

[23] Culmann (1875, p. X): *Nie hätte ein Cremona in Berlin entstehen können, der hoch oben auf der mathematischen Stufenleiter stehend, auf dem praktischen Gebiete auch etwas leistet; nie wird ein an einer Universität gebildeter Mathematiker einen Kräfteplan zeichnen; ist doch principiell das Graphische von der Universität ausgeschlossen, widerstrebte es nicht Staudt, Figuren zu seiner Geometrie beizugeben. ... Unumgänglich nothwendig ist es daher diejenigen Lehrer, welche Techniker bilden sollen, an technischen Anstalten zu erziehen, und diese Anstalten müssen desshalb, wenn die technischen Studien gedeihen sollen, mit den besten und tüchtigsten Lehrkräften in mathematisch naturwissenschaftlicher Richtung, die das Land überhaupt bieten kann, ausgestattet werden.*

> since, whatever one does, it will never be possible to obtain in a project reported on paper, greater accuracy than that given by a graphic plan.[24]

His praise is tempered by a reproach to Poncelet for his not using Varignon's funicular polygon:

> Nevertheless, Poncelet has not used, in order to determine the resultants, the funicular polygon, which use offers such precious resources for the graphical statics. It was reserved to his follower at the school of Metz, M. Michon, to be the first to apply it to the determination of the centres of gravity of the voussoirs, in his *Théorie des voûtes*.[25]

Varignon did not fully describe the funicular polygon in 1687 in his *Projet d'une nouvelle mécanique ou statique*, as Culmann's note might suggest, but in the posthumous work with the title, correctly cited by Culmann, *Projet d'une nouvelle mécanique ou statique, dont le projet fut donné en 1687* (Varignon 1725).[26] (Fig. 7).

Culmann continues:

> Positional geometry, to which Poncelet made so much progress, was not, however, at that time sufficiently advanced for it to be possible to completely replace ordinary geometry (*Geometrie des Maasses*[27]) in the development and the demonstration of drawings. So Poncelet resorted, as often as possible, to ordinary geometry, and when elementary methods were no longer sufficient for his demonstrations, he limited himself to translating the algebraic formulas into drawings.[28]

In this paragraph, Culmann describes the position of geometry, applied to sketches by Poncelet and which will lead to graphical statics, with respect to ordinary geometry. The German has *Geometrie des Maasse*s, meaning geometry of measurement, with respect to geometry of position *Geometrie der Lage* and with respect to algebra, distinct from analysis by the fact that it does not use differential and integral calculus. Culmann precises the idea of geometry of position mentioning the text of von Staudt

---

[24] Culmann (1880, p. IX): *Poncelet avait reconnu le premier, que ces méthodes, tout en étant beaucoup plus expéditives que les méthodes analytiques, offraient cependant une approximation plus que suffisante dans la pratique puisque, quoi que l'on fasse, il ne sera jamais possible d'obtenir dans un projet rapporté sur le papier, une exactitude supérieure à celle donnée par une épure graphique.*

[25] Culmann (1880, pp. IX–X): *Poncelet n'a cependant pas fait usage, pour déterminer les résultantes, du polygone funiculaire, dont l'emploi offre des ressources si précieuses à la statique graphique, et il était réservé à son successeur à l'école de Metz, M. Michon, d'en faire le premier l'application à la détermination des centres de gravité des voussoirs, dans sa Théorie des voûtes.*

[26] The premises of the polygon of forces are found in Varignon (1725, vol. I, pp. 181–195).

[27] Culmann alludes to *Grundzüge Einer Wissenschaftlichen Darstellung Der Geometrie Des Maasses* [mesure] of Oskar Schlömilch (1823–1901). This work contains two parts. There were many editions of it, the first dating from 1849 for the first part and 1854 for the second. A new edition had just appeared when Culmann was writing his last introduction.

[28] Culmann (1880, p. X): *La géométrie de position, à laquelle Poncelet a fait faire tant de progrès, n'était cependant pas à cette époque suffisamment avancée pour qu'il fut possible de la substituer complètement à la géométrie ordinaire (Géométrie des Maasses) dans le développement et la démonstration des épures. Aussi Poncelet recourait-il, aussi souvent que possible, à la géométrie ordinaire, et lorsque les méthodes élémentaires ne lui suffisaient plus pour ses démonstrations, il se bornait à traduire en épures les formules algébriques.*

(1847), professor of mathematics at Erlangen, but he omits to mention the *Analysis situs* that we owe to Leibniz.

Culmann next traces the genesis of graphostatics:

> When we were called, in 1855, during the creation of the Polytechnic School of Zurich, to teach the construction course (including earthworks, the construction of bridges, roads and railways), we were obliged to introduce Poncelet's graphic methods into our teaching to make up for the shortcomings of applied mechanics courses. This course then only included analytical methods in Zurich; ...
>
> This introduction of the theories of Graphical Statics into construction courses did not fail to present certain disadvantages, by unduly delaying the progress of the studies. We obtained, in 1860, the creation of a winter course (two lessons per week), compulsory for engineers, in which we dealt with those problems of statics applied to construction which lent themselves to graphic solutions, and of which this information did not find space, due to lack of time, in the technical mechanics course (then taught by Mr. Zeuner).[29]

It is that course, at first impression of a secondary importance, that will evolve and become graphostatics. The difficulties met by the students and the explanations he must give them leads him to the very practical conclusion that follows:

> In those circumstances, we always found that it was easier to recall theorems of the geometry of position, whose demonstrations could be made could be done with help the lines of the sketch [*épure*] itself, than to resort to analytical calculus, whose development necessitated the use of a separate sheet of paper.
>
> This is how we were led, so to speak irresistibly, to replace algebra as far as possible by positional geometry. During the first years, the students' knowledge in this subject left, it is true, a little to be desired.[30]

This will lead to the opening of a course of geometry of position by M. Fiedler, who taught descriptive geometry, and indirectly to the publication of the *Statique graphique*.

---

[29] Culmann (1880, pp. X–XI): *Quand nous fûmes appelés, en 1855, lors de la création de l'École polytechnique de Zurich, à professer le cours de construction (comprenant les terrassements, la construction des ponts, des routes et des chemins de fer), nous fûmes obligé d'introduire dans notre enseignement les méthodes graphiques de Poncelet pour suppléer aux lacunes des cours de mécanique appliquée. Ce cours ne comprenait alors à Zurich que les méthodes analytiques; ...*
Cette introduction des théories de la Statique graphique dans les cours de construction, ne laissait pas que de présenter certains inconvénients, en retardant outre mesure la marche des études; nous obtînmes, en 1860, la création d'un cours d'hiver (à deux leçons par semaine) obligatoire pour les ingénieurs, dans lequel nous traitions ceux des problèmes de statique appliqués à la construction, qui étaient susceptibles de solutions graphiques, et dont renseignement ne trouvait pas place, faute de temps, dans le cours de mécanique technique (alors professé par M. Zeuner).

[30] Culmann (1880, p. XI): *Dans ces circonstances nous avons toujours trouvé qu'il était bien plus facile de rappeler des théorèmes de géométrie de position, dont la démonstration pouvait se faire à l'aide des lignes mêmes de l'épure, que de recourir à des calculs analytiques dont les développements exigeaient l'emploi d'une feuille de papier séparé.*
C'est ainsi que nous fûmes amenés, pour ainsi dire irrésistiblement, à remplacer autant que possible l'algèbre par la géométrie de position. Pendant les premières années, les connaissances des élèves, dans cette matière, laissaient, il est vrai, un peu à désirer.

Unfortunately, the reactions were not the ones Culmann expected. Instead of developing a branch of knowledge for its own sake, his readers considered it as a set of tricks intended to bypass boring calculations:

> Our sketches had more success than our methods. Our publication was followed by a great number of *elementary statics,* in which while reproducing our simplest sketches (most of the time without changing anything), the authors tried to give analytical demonstrations of them.[31]

Culmann then describes the real hopes he had placed in his work:

> We believe that the truth is not there; that we will never manage to trace the lines of a drawing and to simultaneously execute the algebraic operations which the explanation of this drawing involves, nor to fully understand the meaning of each line and to represent the static relationships, if we limit ourselves to translating a formula whose developments are no longer present in memory.[32]

He hopes to be able to detach the work of the sketch from the considerations of calculation and thus to make from the elaboration of the sketch itself a calculation.

He ends by giving credit to Cremona, once more recognising his influence as well as the fact that he deemed it useful to teach what will become the graphical statics as well as the geometry of position. He had hoped that mathematicians would formulate his graphostatics analytically, as Salmon and Fiedler had formulated analytical geometry:

> But since it was not the case, we tried in the second edition, to link as closely as possible the analytical solutions to the purely geometrical solutions. The new analytical methods have the great merit of leading directly to the goal and, furthermore, of agreeing with geometric methods. In most cases, we were able to deduce the formulas of the geometrical developments which precede them. This method of proceeding has the advantage of giving theorems a form that, in many cases, immediately follows from the geometrical constructions, and, in addition, of leaving the choice, each time that we give the two solutions, between the graphical construction and the calculation. In practice it is sometimes one of the methods, sometimes the other, which leads more quickly to the goal.[33]

---

[31] Culmann (1880, p. XI): *Nos épures obtinrent plus de succès que nos méthodes. Notre publication fut suivie d'un grand nombre de Statiques élémentaires, dans lesquelles, tout en reproduisant nos épures les plus simples (le plus souvent sans y rien changer), les auteurs s'efforçaient d'en donner des démonstrations analytiques.*

[32] Culmann (1880, pp. XI–XII): *Nous estimons que la vérité n'est pas là; qu'on ne parviendra jamais à tracer les lignes d'une épure et à exécuter simultanément les opérations algébriques que comporte l'explication de cette épure, ni à se bien pénétrer de la signification de chaque ligne et à se représenter les relations statiques, si l'on se borne à traduire une formule dont les développements ne sont plus présents à la mémoire.*

[33] Culmann (1880, p. XIII): *Mais comme il n'en a rien été, nous avons essayé dans la deuxième édition, de joindre aussi brièvement que possible les solutions analytiques aux solutions purement géométriques. Les méthodes analytiques nouvelles ont le grand mérite de conduire directement au but et, en outre, de concorder avec les méthodes géométriques. Dans la plupart des cas nous avons pu déduire les formules des développements géométriques qui les précèdent. Ce mode de procéder a l'avantage de donner aux théorèmes une forme, qui, dans bien des cas, découle immédiatement des constructions géométriques, et, en outre, de laisser le choix, toutes les fois que nous donnons les deux solutions, entre la construction graphique et le calcul; dans la pratique c'est tantôt l'une des méthodes, tantôt l'autre qui conduit le plus rapidement au but.*

Analytics was considered as an obstacle to the understanding of the students, in the first introduction became yet its indispensable complement: "Thanks to the method we followed, we showed those who seek to explain a drawing analytically how analysis must be applied to bring out the identity of formulas and drawings".[34]

8 The Plans of the 1866 and 1875 Editions and Their Differences

The aims of Culmann's two editions are very different. Table 1 gives an idea of the differences in the organisation of the contents.

9 The *Graphische Statik* of 1866

We see for the first time, in the first edition of 1866, the notation $\overrightarrow{AB}$ with an arrow, which disappears in the second edition.

Since Daniel Bernoulli and Jacob Hermann in his *Phoronomia* (1716), many authors considered a force as represented by a length and direction. In 1837, Giusto Bellavitis wrote:

> In geometry when one indicates a straight line with the letters that designates their end, one considers only its length without taking the direction into account. From now on when we shall designate a line, we shall instead consider its direction as well as its length.[35]

The notation introduced by Culmann in his first edition will only reappear much later, after much hesitation. The introduction of such a notation together with an explanation of the utility of indicating the direction as well as the length of the vector or of the segment is already a sign of the will of constructing a calculus thanks to a symbolic language.

The chapters that follow treat a series of particularly important cases as the polygon of forces and the funicular polygon, which we shall find again in the following edition of 1875.

---

[34] Culmann (1880, p. XIII): *Grace à la méthode que nous avons suivie, nous avons montré à ceux qui cherchent à expliquer une épure analytiquement comment il faut appliquer l'analyse pour faire ressortir l'identité des formules et des épures.*

[35] Bellavitis (1937, p. 246): *Nella Geometria quando s'indica una retta colle lettere che ne segnano gli estremi si suol riguardare soltanto la sua lunghezza, senza far caso della direzione, d'ora in poi quando noi nommineremo une retta intenderemo invece di considerare tanto la sua lunghezza, quanto la sua direzione.*

Table 1 Covers and contents of the 1866 and 1875 editions of *Graphische Statik*

	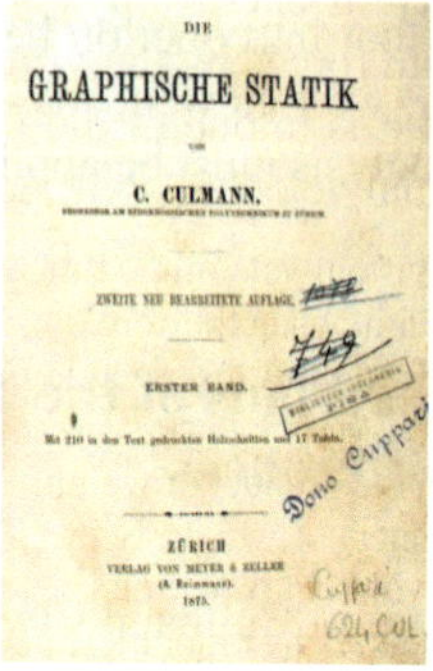
Edition of 1866 **First part. Graphic calculation** Chap. 1. Operations with lines Chap. 2. Transformations of surfaces Chap. 3. Transformations of bodies **Second part. Graphostatics** Chap. 1. Composition of forces acting on one point or in one plane Chap. 2. The projective relationship between the force polygon and the funicular polygon Chap. 3. Moment of forces and infinitely distant forces in a plane Chap. 4. Forces in a plane Chap. 5. Parallel forces Chap. 6. Centre of gravity Chap. 7. Moment of inertia Chap. 8. Construction of the central ellipse, the central ellipsoid and the core of certain bodies **Third part. The beam** …	Edition of 1875 **First part. Graphic calculation** Chap. 1. Operations on lines Chap. 2. Logarithms and slide rule Chap. 3. Transformation des surfaces Chap. 4. Transformation des volumes **Second part. Composition of forces** Chap. 1. Composition of forces acting on a point or in one plane Chap. 2. Moment of forces and forces at infinite in one plane Chap. 3. Forces in space Chap. 4. Projective relations between the force polygon and the funicular polygon **Third Part. Moment of parallel forces** Chap. 1. Parallel Forces Chap. 2. Centre of gravity Chap. 3. Moments of inertia Chap. 4. Construction of the central ellipse and of the nucleus of plane figures Chap. 5. Moments of inertia, central ellipsoid and nucleus of some bodies **Fourth part. Elements of the theory of elasticity** …

10 Culmann's *Ueber das parallelogramm und über die Zusammensentzung der Kräfte*

Culmann's introduction to his article on the parallelogram and the composition of forces (1870) explains the reasons for the enthusiasm for the demonstrations of the composition of forces:

> If one seeks to derive the theorems of graphic statics analytically, it is very easy to do so if the methods through which such great progress has recently been achieved in analytical geometry are applied, and that consists of combining the equations of lines and points in a single symbol and calculating with this. By introducing this symbolism, the composition of the forces acting on a point in space, the composition of the forces in the beam of rays is reduced to an empirical formula $\sum \alpha A$,[36] and the composition of the forces in the plane is reduced to a similar empirical formula $\sum \alpha A$ f, where α is the normal form of the equation of infinity distant point in Plucker's coordinates, a the normal equation of the direction of the force in the plane, in ordinary coordinates, and A denotes the magnitude of the force. Finally, the composition of these two molecular formulas results in the composition of the forces in space.[37]

In the French translation of *Graphische Statik* (Culmann 1880), the translators added a note concerning Plücker's coordinates:

> B.— On the systems of coordinates.
>
> As the elementary treatises on analytical geometry give few details on the coordinate systems which are used in different parts of M. Culmann's work, we believe it is useful to give some indications on these systems.
>
> In Cartesian coordinates, we consider geometric shapes as composed of points, and we determine the position of the points by means of their distances to two fixed axes in plane geometry and to three fixed planes, in space geometry.
>
> In Pluckerian coordinates, the elements that we consider are the line in the plane and the plane in space.
>
> Let us confine ourselves, for the moment, to the geometry of the plane. Let

$$Ax + By + Cz = 0$$

---

[36] Culmann means to say $\sum_i \alpha_i A_i$.

[37] Culmann (1870, p. 1): *Sucht man die Sätze der graphischen Statik auf analytischem Wege abzuleiten, so gelingt es sehr leicht, wenn jene Methoden, durch welche in der letzten Zeit in der analytischen Geometrie so grosse Fortschritte erzielt worden sind, angewendet werden; und di darin bestehen: Die Gleichungen von Linien und Punkten in einem einzigen Symbol zusammenzufassen und mit diesem zu rechnen. Durch Einführung dieser Symbolik reducirt sich di Zusammensetzung der Krafte, die im Raum auf einen Punkt wirken, di Zusammensetzung der Kräfte im Strahlenbündel auf eine Summenformel $\sum \alpha A$, und die Zusammensetzung der Kräfte in der Ebene auf eine ähnliche Summenformel $\sum \alpha A$, wenn α di Normalform der Gleichung des unendlich fernen Punktes in Plucker'schen Coordinaten,* a *di Normalgleichung der Richtunslinie der Kraft in der Ebene, in gewöhnlichen Coordinaten, und* A *die Grösse der Kraft bezeichnen. Endlich ergiebt sich aus der Zusammensetzung dieser beiden Summenformeln di Zusammensetzung der Kräfte im Raum.*

be the equation of a line in ordinary coordinates. If the ratios $\frac{A}{C}$ and $\frac{B}{C}$ were given, the line would be completely determined; we can, consequently, consider these relationships as the coordinates of the line. These are the Pluckerian coordinates; these coordinates are, as is easily recognised, the inverses, their signs changed, of the segments intercepted by the right on the coordinate axes. By calling them ξ and η, the above equation can be written:

$$\xi x + \eta y + 1 = 0$$

It is susceptible to a double interpretation, in which we find the principle of duality which plays such an important role in positional geometry. If, in fact, we consider x and y as the variables, the equation represents in Cartesian coordinates the line whose Pluckerian coordinates are ξ and η; if we consider ξ and η, as the variables, the equation represents, in Pluckerian coordinates, the point whose Cartesian coordinates are x and y. In general, any equation in ξ and η represents a geometric location, to which are tangent all the lines whose Pluckerian coordinates satisfy this equation. It is for this reason that these coordinates are also called *tangential*.[38]

After that Culmann explains the direction toward which he will orient his research, but in the article that follows, he does not develop this point. One must await the second edition in order to see the development of that idea, but his aim is clearly stated:

Calculating in that way statics may be close to analytical geometry as graphical statics is close to position geometry, in this way one puts in evidence the identity of analytical geometry

---

[38] Culmann (1880, pp. 597–598): *B.—Sur les systèmes de coordonnées.*
Comme les traités élémentaires de géométrie analytique donnent peu de détails sur les systèmes de coordonnées dont il est fait usage dans différentes parties de l'ouvrage de M. Culmann, nous croyons utile de donner quelques indications sur ces systèmes.
Dans les coordonnées cartésiennes, *on considère les formes géométriques comme composées de points, et on détermine la position des points au moyen de leurs distances à deux axes fixes dans la géométrie plane et à trois plans fixes, dans la géométrie de l'espace.*
Dans les coordonnées pluckériennes, *les éléments que l'on considère sont la droite dans le plan et le plan dans l'espace.*
Bornons-nous, pour le moment, à la géométrie du plan. Soit.

$$Ax + By + Cz = 0$$

l'équation d'une droite en coordonnées ordinaires. Si l'on donnait les rapports $\frac{A}{C}$ et $\frac{B}{C}$, la droite serait complètement déterminée; on peut, par suite, considérer ces rapports comme les coordonnées de la droite. Ce sont les coordonnées pluckériennes; ces coordonnées sont, comme il est facile de le reconnaître, les inverses, changés de signe, des segments interceptés par la droite sur les axes coordonnés. En les appelant ξ et η, l'équation ci-dessus peut s'écrire.

$$\xi x + \eta y + 1 = 0$$

Elle est susceptible d'une double interprétation, dans laquelle on retrouve le principe de dualité qui joue un rôle si important dans la géométrie de position. Si, en effet, on considère x et y comme les variables, l'équation représente en coordonnées cartésiennes la droite dont les coordonnées pluckériennes sont ξ et η; si l'on considère ξ et η, comme les variables, l'équation représente, en coordonnées pluckériennes, le point dont les coordonnées cartésiennes sont x et y. En général, toute équation en ξ et η, représente un lieu géométrique, auquel sont tangentes toutes les droites dont les coordonnées pluckériennes satisfont à cette équation. C'est pour ce motif que ces coordonnées sont aussi appelées tangentielles.

> and geometry of position, and introducing that last one, the difference between graphical statics and analytical statics disappears, both become part of geometry.
>
> If one looks for the deep reason of that scientific kinship, she lies in the fact that the parallelogram law of forces is a geometrical law that can also be demonstrated geometrically.[39]

11 The Work of Alberto Castigliano

In the third part of the thesis that Castigliano defended at the Regia scuola d'applicazione per ingegneri (Castigliano 1873), he gives an application of his theorem of minimal work. First of all, he gives some formulas to express the molecular work of a solid subject to forces of tension, compression, shear and flexion. After that, he applies his new theorem to a beam sustained by more than two supports. Along the way, he solves one of the first problems of supports.

Castigliano explains the equilibrium of this structure in the following way.

> In order to find the conditions of equilibrium of this frame, it is necessary to write that the molecular work, which is done in the deformation of it, is a minimum, and as it is symmetrical with respect to the vertical *BH,* it will be sufficient to consider the part of it which is on the right of this vertical. This molecular work is expressed in function of the tensions in the bars *EI, BE, EA, EF*, which I shall call t, t_1, t_2, t_3, and which, having to be in equilibrium around the point *E*, are linked among themselves by two equations.[40] (Fig. 8)

He generalises this method and shows all calculations in full detail. He then simplifies his calculations of metal trusses, proving along the way that it gives exactly the same results as using the theorem of least work:

> But it is easy to see that by proceeding in this way, precisely the same operations are performed as for finding bending moments relative to the points *E, F, G, B*, if the beam *BA* were fixed in D, supported at *G, F, E, A* on fixed supports and loaded with a weight uniformly distributed along its length and perpendicular to its axis. We can thus immediately apply Clapeyron's equation between the bending moments on three successive supports, which we saw in no. 13 can be obtained using the theorem of the least work. Having found the four unknown

---

[39] Culmann (1880, p. 599): *En calculant ainsi la statique peut être rapprochée de la géométrie analytique comme la statique graphique de la géométrie de position, de cette manière on met en évidence l'identité entre la géométrie analytique et la géométrie de position, et par introduction de cette dernière, la différence entre la statique graphique et la statique analytique disparaît, toutes deux deviennent partie de la géométrie.*
Si l'on recherche la raison profonde de cette parenté scientifique, elle réside dans le fait que le parallélogramme des forces est une loi géométrique qui peut également se démontrer géométriquement.

[40] Castigliano (1873, p. 38): *Per trovare le condizioni d'equilibrio di quest'incavallatura, bisogna esprimere che il lavoro molecolare, il quale si fa nella deformazione di essa, è un minimo, e siccome essa è simmetrica rispetto alla verticale BH, basterà considerarne quella parte, che è alla destra di questa verticale. Questo lavoro molecolare si esprime in funzione delle tensioni delle aste EI, BE, EA, EF, le quali io chiamerò* t, t_1, t_2, t_3, *e che dovendo farsi equilibrio intorno al punto E sono legate fra loro da due equazioni.*

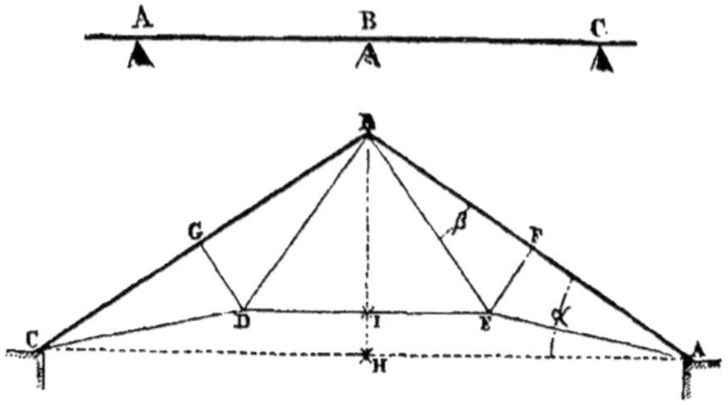

Fig. 8 **Above** A beam resting on three aligned supports; **below** Simple nesting of the Polonceau (1840) system. (Castigliano 1873). Public domain

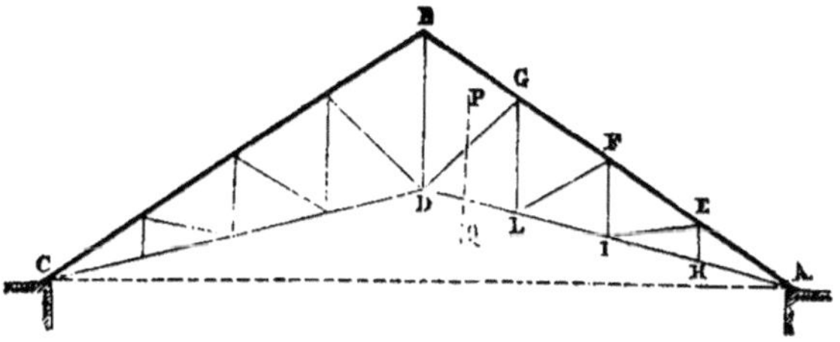

Fig. 9 Castigliano's calculation of metal trusses (Castigliano 1873). Public domain

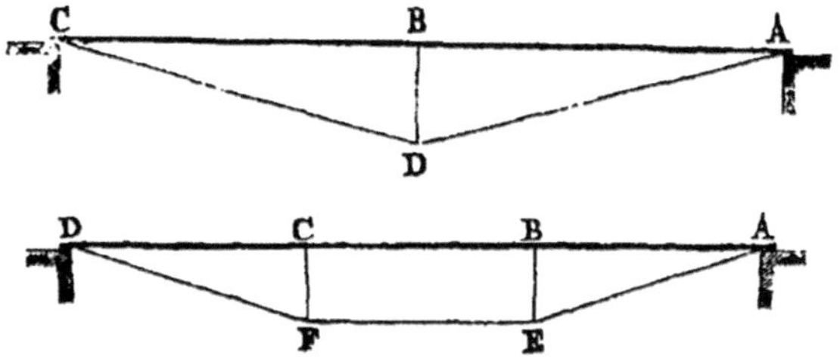

Fig. 10 **above** First reinforced beam; **below** Second reinforced beam (Castigliano 1873). Public domain

moments, we can immediately find, as we already said, the tensions within the eleven rods.[41] (Fig. 9).

As an application to the preceding theory, Castigliano applies it to two kinds of simple reinforced beams and shows that his results concord with the classically obtained ones (Fig. 10).

---

[41] Castigliano (1873, p. 48): *16. Metodo generale pel calcolo approssimato delle incavallature metalliche. Ma è facile vedere che procedendo in questo modo, si fanno precisamente le stesse operazioni come per trovare i momenti inflettenti relativi ai punti E, F, G, B, se la trave BA fosse incastrata in D, appoggiata in G, F, E, A sopra appoggi fissi e caricata d'un peso uniformemente distribuito sulla sua lunghezza e perpendicolare al suo asse. Potremo dunque applicare direttamente l'equazione di Clapeyron fra i momenti inflettenti su tre appoggi successivi, la quale abbiam veduto nel N° 13 come si ottenga col teorema del minimo lavoro. Trovati i quattro momenti incogniti, si possono trovar direttamente, come già abbiam detto, le tensioni delle undici verghe.*

12 The French Translation of Culmann's Edition of 1875

The first section of the first chapter of the French translation by Glasser, Jacquier, and Valat of Culmann's 1875 edition, *Traité de statique graphique* (Culmann 1880) is completely new, and has no equivalent in 1866 edition. It is partly inspired by a book by Albert Mousson, *Physik auf Grundlage der Erfahrung*, and Culmann quotes the first paragraph aimed at defining forces in general starting from observation:

> 36. Forces in general
>
> We only know the existence and nature of forces by observation of their effects, and we must thus find their particularities in those effects. A force is completely defined when one gives:
>
> 1° Its application point, that is, the material element upon which it acts directly;
>
> 2° Its direction, or the line following which its point of application would move, if the force would act without being disturbed;
>
> 3° Its magnitude or intensity in relation to a well determined unit of force.[42]

This already allows the reader to make the link with the calculation with lines that occupies the first part of Culmann's text. After having recalled the third law of Newton, which does not interest Culmann, as he is only interested in statics, but that will allow us to show another example of Culmann's desire to generalise, he remarks in a footnote:

> In this quote, we have left aside everything that relates to the different measurement systems, and replaced the terms second, gram, meter, with unit of time, weight, length.[43]

Mousson explains the composition of forces by bringing it back to those of movements.

> To determine the resultant of several independent forces, acting on the same molecule, the resultant, which is nothing other than the force which, acting alone, would produce the same effect. We represent the forces by lines of lengths proportional to their intensities, and we treat them as lines representing movements. The resulting movement that we obtain by the parallelogram or the polygon represents the resultant of the forces in direction and magnitude.[44]

---

[42] Mousson (1875, pp. 26–27), quoted in Culmann (1880, p. 145): 36. *Des forces en général. Nous ne connaissons l'existence et la nature des forces que par l'observation de leurs effets; c'est par conséquent dans leurs effets que nous devons rechercher leurs caractères propres. Une force est définie complètement lorsqu'on donne:*
1° Son point d'application, c'est-à-dire l'élément matériel sur lequel elle agit immédiatement;
2° Sa direction, où la ligne suivant laquelle se déplacerait son point d'application, si la force agissait seule et sans être contrariée;
3° Sa grandeur ou son intensité rapportée à une unité de force bien déterminée.

[43] Culmann (1880, p. 146): *Dans cette citation, nous avons laissé de côté tout ce qui se rapportait aux différents systèmes de mesures, et remplacé les termes seconde, gramme, mètre, par unité de temps, de poids, de longueur.*

[44] Mousson (1858, p. 27), quoted in Culmann (1880, pp. 146–147): *Pour déterminer la résultante de plusieurs forces indépendantes, agissant sur une même molécule, résultante qui n'est autre chose que la force qui, agissant seule, produirait le même effet, on représente les forces par des lignes de longueurs proportionnelles à leurs intensités, et on les traite comme des lignes représentant*

At the end of the paragraph, Culmann leaves off quoting Mousson in order to state the theorem that brings back the composition of the forces, not to the composition of motions as Mousson did, following a general thinking:

> If we represent the direction and the intensity (or magnitude) of the forces by the direction and the length of lines, the composition of the forces is nothing other than the addition of the lines, as it is set forth in no. 1, p. 4, an addition of which the parallelogram of forces is only a special case.[45]

It is a particular case, would we say today, of the vectorial law of composition. On p. 4 is found what is meant by the graphical addition of lines:

> The graphical addition simply consists of carrying, from the origin, the first line in size and direction, then from the end of this, the 2nd line also in size and direction, then the 3rd and so on for all the others. We will consider as giving, in magnitude and direction, the result of this construction or addition, the line which, starting from the origin, closes the polygon thus formed.[46]

This result acquired, Culmann reproduces both the geometrical and the analytical demonstrations one from his article of 1870. Those two demonstrations are founded on the same axioms, which that he explains in the following way:

> 37. The Parallelogram of Forces
>
> As statics does not concern itself with the movement which could be generated by the forces, but only considers them from the point of view of their equilibrium, we understand that we have endeavoured, in all the treatises for the teaching pure statics, to borrow nothing from dynamics and to find demonstrations which do not emanate from the idea of movement. Starting with the balance of three forces, we sought a direct demonstration of the parallelogram of forces. This demonstration, without the help of dynamics, does not seem possible, because all the demonstrations that we have attempted suppose the following three propositions:
>
> If three forces of intensities A, B, C, whose azimuths are α, β, γ, are in equilibrium:
>
> 1. They all pass through the same point; they are located in the same plane, and the direction of any one of them, γ, for example, $\frac{A}{B}$ is a function of the A/B ratio of the other two. Furthermore, this direction does not change when the two forces are increased in the same ratio, for example doubled.
> 2. When the directions of A and B, that is to say α, β, are assumed to be constant, there is, for each value of the ratio $\frac{A}{B}$, one and the same always real direction of the third force C; in other words, the ratio $\frac{A}{B}$ and the direction γ of C are projective.

des mouvements. Le mouvement résultant que l'on obtient par le parallélogramme ou le polygone représente la résultante des forces en direction et en grandeur.

45 Culmann (1880, p. 147): *Si l'on représente la direction et l'intensité (ou grandeur) des forces par la direction et la longueur de lignes, la composition des forces n'est autre chose que l'addition des lignes, telle qu'elle est exposée au n°1, p. 4, addition dont le parallélogramme des forces n'est qu'un cas particulier.*

46 Culmann (1880, p. 4): *L'addition graphique consiste simplement à porter, à partir de l'origine, la première droite en grandeur et en direction, puis à partir de l'extrémité de celle-ci, la 2e droite également en grandeur et en direction, puis la 3e et ainsi de suite pour toutes les autres. On considérera comme donnant, en grandeur et en direction, le résultat de cette construction ou de l'addition, la droite qui, partant de l'origine, ferme le polygone ainsi formé.*

3. When one of the three forces becomes equal to 0, the other two are equal and in opposite directions, and when two forces are equal to each other, the third divides the angle formed by their directions into two equal parts. Thus

$$\text{for A} = 0,\ \gamma = \beta - 180°.$$
$$\text{for B} = 0,\ \gamma = \alpha - 180°.$$
$$\text{if A} = \text{B},\ \gamma\gamma = \frac{1}{2}(\alpha + \beta) - 180°.$$

We sometimes try to demonstrate the first proposition by saying that we obviously change nothing by changing the unit by means of which the forces A and B are expressed, and that consequently γ must be a function of the ratio $\frac{A}{B}$. In all demonstrations we admit propositions no. 3 as axioms that are self-evident. As for proposition no. 2, it emerges from the analytical form given to most demonstrations.[47]

He then puts in parallel a geometrical demonstration and an analytical demonstration. If we admit those propositions to be right, we arrive at the triangle of forces (half parallelogram) that represents the equilibrium of three forces. This is the form of the law that was first used by architects.[48]

Culmann demonstrates that

[47] Culmann (1880, pp. 147–148): *37. LE PARALLÉLOGRAMME DES FORCES. Comme la statique ne s'occupe pas du mouvement qui pourrait être engendré par les forces, mais qu'elle les considère seulement au point de vue de leur équilibre, on comprend qu'on se soit efforcé, dans tous les traités pour l'enseignement de la statique pure, de ne rien emprunter à la dynamique et de trouver des démonstrations qui n'émanent pas de l'idée de mouvement. En commençant par l'équilibre de trois forces, on a cherché une démonstration directe du parallélogramme des forces. Cette démonstration, sans l'aide de la dynamique, ne parait pas possible, car toutes les démonstrations que l'on a tentées supposent les trois propositions suivantes:*

Si trois forces d'intensités A, B, C, dont les azimuths sont α, β, γ, sont en équilibre: 1° Elles passent toutes trois par un même point; elles sont situées dans un même plan, et la direction de l'une quelconque d'entre elles, γ, par exemple, est une fonction du rapport $\frac{A}{B}$ des deux autres. En outre cette direction ne change pas quand les deux forces sont augmentées dans un même rapport, par exemple doublées. 2° Quand les directions de A et B, c'est-à-dire α, β, sont supposées constantes, il y a, pour chaque valeur du rapport $\frac{A}{B}$, une seule et même direction toujours réelle de la troisième force C; en d'autres termes, le rapport $\frac{A}{B}$ et la direction γ de C sont projectifs. 3° Quand une des trois forces devient égale à 0, les deux autres sont égales et de sens contraire, et quand deux forces sont égales entre elles, la troisième divise en deux parties égales l'angle formé par leurs directions. Ainsi pour A = 0, pour B = 0, si A = B, [See formulas in the text.] On essaye quelquefois de démontrer la première proposition en disant que l'on ne change évidemment rien en changeant l'unité au moyen de laquelle les forces A et B sont exprimées, et que par conséquent γ doit être une fonction du rapport $\frac{A}{B}$. Dans toutes les démonstrations on admet les propositions n°3 comme des axiomes se comprenant par eux-mêmes. Quant à la proposition n°2, elle ressort de la forme analytique donnée à la plupart des démonstrations.

[48] See "The Use of a Particular Form of the Parallelogram Law of Forces for the Building of Vaults (1650–1750)", Chap. 4 in this present volume.

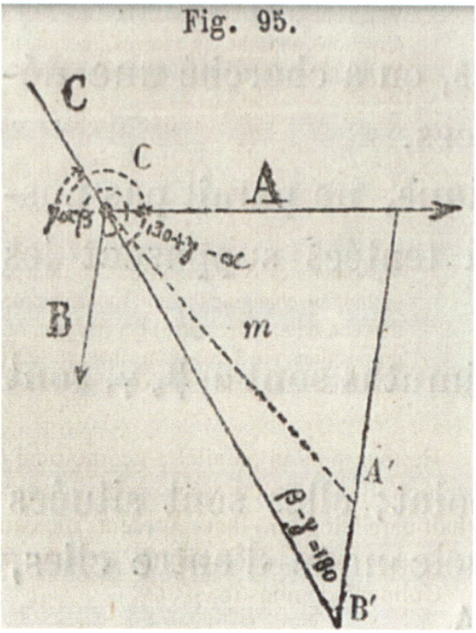

Fig. 11 The triangle of forces in equilibrium in Culmann (1880, p. 148). Public domain

> Three forces acting on the same point, and which are in equilibrium, behave in magnitude and in direction as the sides of a triangle which are respectively parallel to the forces that are proportional to the greatness of those forces.[49] (Fig. 11)

This geometrical demonstration is immediately followed by an analytical demonstration.

We observe that he gives the law of composition in the form of the triangle that corresponds to the law of sines in a form particular to Varignon in his *Projet d'une nouvelle mécanique*. Varignon's Lemma IV gives the traditional parallelogram law:

> Given point A as the centre of direction of the body EF, and that this body is pushed at the same time and uniformly by two forces applied in E and in F, following the lines EC and FB, which pass through point A, with forces which are between them as AC to AB: let us complete the parallelogram BC and look for a moment at this body as if it had no weight. Whatever angle BAC these lines make between them, this body thus pushed will follow the diagonal AD.[50] (Fig. 12)

---

[49] Culmann (1880, p. 149): *Trois forces agissant sur un même point, et qui sont en équilibre, se comportent en grandeur et en direction comme les côtés d'un triangle qui sont respectivement parallèles aux forces et sont proportionnels aux grandeurs de ces forces.*

[50] Varignon (1687, pp. 7–8): *Lemme IV. Présentement que le point A soit le centre de direction du corps EF, & que ce corps soit poussé en même-tems/uniformément par deux puissances appliquées en E & en F, suivant les lignes EC & FB, qui passent par le point A, avec des forces qui soient entr'elles comme AC & AB: que l'on acheve le parallelogramme BC, & que l'on regarde pour un moment ce corps comme l'il n'avoit aucune pesanteur. Quelque angle BAC que ces lignes fassent entr'elles, ce corps ainsi poussé suivra la diagonale AD.*

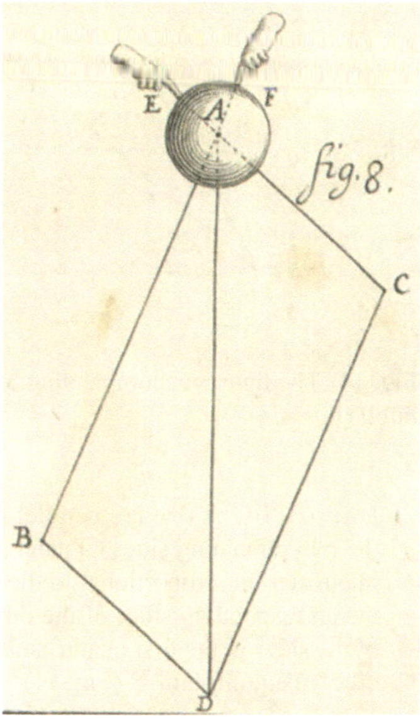

Fig. 12 The figure accompanying Varignon's Lemma IV (Varignon 1687, Pl. 1). Public domain

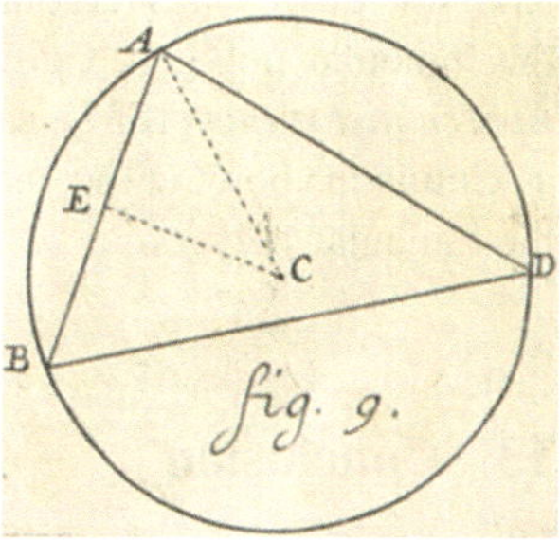

Fig. 13 The figure accompanying Varignon's Lemma V (Varignon 1687, Pl. 1). Public domain

His Lemma V makes the link with the triangle: "The three sides of any right triangle are to themselves as the sines of the angles to which they are opposite".[51] (Fig. 13).

In this manner, Varignon shown the equivalence of both theorems of the parallelogram of forces or of the triangle of forces. But it is with Möbius, in his *Lehrbuch der Statik*[52] (1837), that one better understands the link between that classical way of seeing the parallelogram law of forces and graphostatics.

§29. Equilibrium of four forces in one plane.

If of two plane quadrilaterals ABCD and abcd, the diagonals of one correspond to the diagonals of the opposite name of the other, i.e., AC with bd and BD with ac, and three sides

---

[51] Varignon (1687, p. 9): *Les trios cotez d'un triangle réctiligne, quel qu'il soit, sont entr'eax, comme les sinus des angles aus quels ils sont opposez.*

[52] *Lehrbuch der Statik,* Leipzig, Göschen.

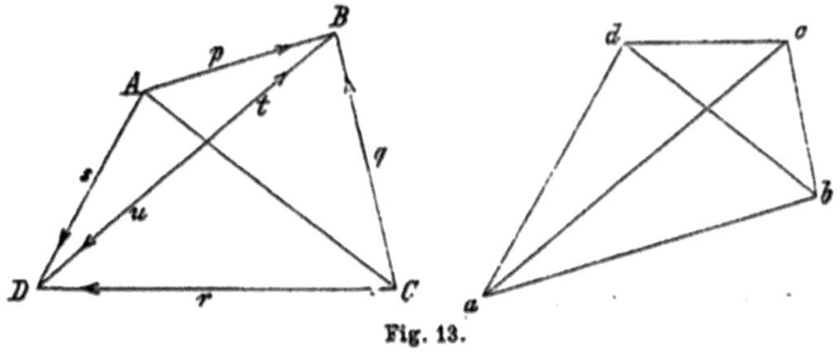

Fig. 14 The figures accompanying Möbius's equilibrium of four forces in a plane (Möbius 1837). Public domain

> DA, AB, BC of one are parallel to the sides of the same name da, ab, bc of the other, so the two remaining sides CD and cd are also parallel to each other; and four forces, whose intensities are proportional to the sides of one quadrilateral, and whose directions fall into the corresponding sides of the other quadrilateral and thereby coincide with the directions of the sides in the first quadrilateral determined by the succession of the corners, keep each other in equilibrium.[53] (Fig. 14)

Culmann shows the importance of the Pluckerian coordinates that explain analytically the duality, as Gergonne (1826) would say, between the polygon of forces and the funicular polygon. Culmann speaks of reciprocity and Michel Chasles, in his *Aperçu historique* (1837), saw there the real principle of nature. Duality does exist, as Culmann shows at the end of each paragraph, between the polygon of forces and the funicular polygon.

13 Conclusion

The fact that Culmann reprises almost literally his entire first edition with some additions and few modifications in its arrangement proves that something that seemed to be merely technical, a way to facilitate the engineer's life, in fact revealed a deeper theory that analysis permitted him to bring to light.

The differences between editions reside in the introduction of a few justifications, because those techniques are generalisable to more difficult problem as such as frameworks, beams variously charged etc. Culmann has not treated them case by

[53] Möbius (1837, I, pp. 43): *§29. Gleichgewicht zwischen vier Kräften einer Ebene. Wenn von zwei ebene Vierecken ABCD und abcd die Diagonalen des einen mit den ungleichnamigem Diagonalen des anderen, d. i. AC mit bd und BD mit ac, und drei Seiten DA, AB, BC des einen mit den gleichnamigen Seiten da, ab, bc des anderen parallel sind, so sind auch die zwei übrigen Seiten CD und cd einander parallel; und vier Kräfte, deren Intensitäten den Seiten des einen Vierecks proportional sind, und deren Richtungen in die entsprechenden Seiten des andern Vierecks fallen und dabei mit den im ersten Vierecke durch die Aufeinanderfolge der Ecken bestimmten Richtungen der Seiten übereinstimmen, halten einander das Gleichgewicht.*

case. He has given a tool to the engineer that can be adapted to the various problems of statics that he encounters if there are no deformations.

Culmann is at the crossroads of two worlds: that of the mathematicians, who don't hesitate to use analysis but do not care much about practice, and that of men on the construction site, who feel the forces but have difficulty organising their experiments. The first are held back in the elaboration of a vectorial analyses by the fact that they do not perceive or feel the object "force" and still do not know its abstract correspondent: the vector. The others are held back, in contrast, by a too concrete view of forces.

However, if we are looking at the history in its entirety, vectorial analysis and graphical statics were both born in the mechanics of the pseudo-Aristoteles, with the composition of the velocity vectors according to the parallelogram law. But it will take more than 2000 years before men will get used to the idea of representing physical quantities as displacements, velocities and forces by oriented lines of a definite length, giving rise, almost simultaneously, to vectorial analysis and graphostatics.

References

Bellavitis, Giusto. 1835. Sposizione del Metodo delle equipollenze. *Annali delle science del regno Lombardo-Veneto* VII, 243–261.

Carnot, Lazare. 1803. *Géomètrie de position*. Paris: Duprat.

Castigliano, Alberto. 1873. *Dissertazione intorno ai sistemi elastici, presentata alla Commission Esaminatrice della R. Scuola d'applicazione degli Ingenieri in Torino per ottenere la Laurea di Ingenere Civile*. Torino: Vincenzo Bona.

Chasles, Michel. 1837. *Aperçu historique sur les origines et le développement des méthodes en géométrie*. Brussels: Hayez.

Chasles, Michel. 1852. *Traité de géométrie supérieure*. Paris: Gauthier-Villars.

Chasles, Michel. 1866. *Traité des sections coniques, faisant suite au Traité de géométrie supérieure*. Paris: Gauthier-Villars.

Cremona, Luigi. 1874. *Elementi di calcolo grafico*. Torino: Stamperia reale di G. B. Paravia.

Culmann, Karl. 1866. *Die graphische Statik*. Zürich, Meyer und Zeller.

Culmann, Karl. 1870. Ueber das Parallelogram and über die Zusammensetzung der Kräfte. *Vierteljahrsschrift der Naturforschenden Gesellschaft in Zürich* 15: 1–24.

Culmann, C. [Karl]. 1875. *Die graphische Statik*, 2nd ed. Zürich, Meyer und Zeller.

Culmann, C. [Karl]. 1880. *Traité de statique graphique / par C. Culmann... ; traduit sur la deuxième édition allemande*. Trans. G. Glasser, J. Jacquier, A.Valat. Paris: Gunod.

Euler, Leonhard. 1752. E. 177, Decouverte d'un nouveau principe de Mecanique. *Mémoires de l'Académie des sciences de Berlin* 6, 185–217.

Favaro, Antonio. 1877. *Lezioni di statica grafica*. Padova: Sacchetto.

Gergonne, Joseph Diaz. 1826. Considérations philosophiques sur les élémens de la science de l'étendue. *Annales* 16: 209–232.

Grassmann, Hermann. 1844. *Die Wissenschaft der extensiven Grösse oder die Ausdehnungslehre eine neue Mathematische Disciplin dargestellt und durch Anwendungen erläutert*. Leipzig: Wigand.

Hamilton, William Rowan. 1844a. On the Composition of Forces. *Proceedings of the Royal Irish Academy* 2: 166–168.

Hamilton, William Rowan. 1844b. On a new Species of Imaginary Quantities connected with a theory of Quaternions. *Proceedings of the Royal Irish Academy* 2: 424–434.

Hamilton, William Rowan. 1844c. On quaternions, or on a new system of imaginaries in algebra. *The London, Edinburgh and Dublin Philosophical Magazine and Journal of Science* (3rd Series), vol. XXV, 10–13.

Hamilton, William Rowan. 1844d. On a theory of quaternions. *British Association Report*, 1844, Part II, 2.

Hermann, Jacob. 1716. *Phoronomia, sive de viribus et motibus corporum solidorum et fluidorum.* Amsterdam, Wetstenios.

Levy, Maurice. 1874. *La statique graphique et ses applications aux constructions.* Paris: Gauthier-Villars.

Möbius, Augustus Ferdinand. 1827a. *Der barycentrische Calcül ein neues Hülfsmittel zur analytischen Behandlung der Geometrie.* Leipzig: Johann Ambrosius Barth.

Möbius, Augustus Ferdinand. 1827b. *Lehrbuch der Statik.* Leipzig: Göschen.

Mohr, Otto. 1874a. Beitrag zur Theorie der Bogenfachwerksträger. *Zeitschrift Des Archit- und Ingenieur-Vereins in Hannover* 20: 223–238.

Mohr, Otto. 1874b. Beitrag zur Theorie des Fachwerks (1). *Zeitschrift Des Archit- und Ingenieur-Vereins in Hannover* 20: 509–526.

Mohr, Otto. 1875. Beitrag zur Theorie des Fachwerks (2). *Zeitschrift Des Archit- und Ingenieur-Vereins in Hannover* 21: 17–38.

Mousson, Albert. 1858. *Physik auf Grundlage der Erfahrung, Band 1.* Zürich: Schulthess.

Polonceau, Camille. 1840. Notice sur nouveau système de charpente en bois et fer. *Revue générale de l'architecture et des travaux publics*, 27–32

Rankine, W. J. Macquorn. 1858. *A Manual of Applied Mechanics.* London and Glasgow: William Griffin and Co.

Ritter, August. 1861. Über die Berechnung eiserner Dach- und Brücken-Constructionen. *Zeitschrift des Architekten und Ingenieur-Vereins zu Hannover*, Bd. VII heft IV, 410–426.

Ritter, August. 1862. *Elementare Theorie und Berechnung eiserner Dach- und Brücken- Constructionen.* Hannover, Rümpler (2nd ed. 1870).

Salmon, George. 1869. A Treatise on Conic Sections. An Account of Some of the Most Important Modern Algebraic and Geometric Methods. London: Longmans, Green. German trans. Wilhelm Friedler, *Analytische Geometrie der Kegelschnitte mit besonderer Berücksichtigung der neueren Methoden*, 1873.

Varignon, Pierre. 1687. *Projet d'une nouvelle mécanique.* Paris: Martin et Boudot.

Varignon, Pierre. 1725. *Projet d'une nouvelle mécanique ou statique, dont le projet fut donné en 1687.* Paris: Jombert.

Von Staudt, Karl Georg Christian. 1847. *Die Geometrie der Lage.* Nuremberg: Bauer und Raspe.

Chapter 18
The Industrial Revolution and Its Consequences. Part II: Elasticity Penetrates Construction

1 W. J. Macquorn Rankine

In his introduction to his *Laws of the elasticity of solid bodies* (1850), W. J. Macquorn Rankine gives an overview of the science of elasticity. His point of view is larger than construction history. But as he introduces new concepts that will deeply modify the art of building, as we just saw in the prelude, we shall have an attentive look at his work.

In his introduction, Rankine introduces the new science:

1. The science of the elasticity of solid bodies, considered with reference to its most important application, the determination of the strength of structures, consists of three parts:

> First. The investigation of what may be specially termed the Laws of Elasticity; that is to say, the mutual relations which must exist between the elasticities of different kinds possessed by a given solid, and between the different values of those elasticities in different directions.
>
> Secondly. The integration of the equations of equilibrium and motion of the particles of an elastic solid. The results of this process enable us to determine the relative displacements of the particles from their natural positions in a solid body of a given material and figure, subjected to a given combination of forces.
>
> Thirdly. The application of the results derived from the first two branches of the theory to our experimental knowledge of the pressures and relative displacements to which the particles of known materials may safely be subjected in practice. This enables us to compute the strength of actual structures.[1]

It should be recalled, as Rankine knows, that elasticity gives rise to energy even inside a body that seems at rest. That energy was attributed to very small internal motions. This is really a revolution and Rankine decides to precise the vocabulary of that new situation:

> I shall therefore use it, throughout this paper, in the restricted sense of *relative displacement of particles*, whether consisting in dilatation, condensation, or distortion; while under the

[1] Rankine, 1851, p. 47.

P. Radelet-de Grave, *Between Mechanics and Architecture*, Mathematics and the Built Environment 8, https://doi.org/10.1007/978-3-031-73530-1_18

term pressure I shall include every kind of force which acts between elastic bodies, or the parts of an elastic body, as the cause or effect of a state of strain, whether that force is tensile, compressive, or distorting.[2]

He goes further as follows:

The nature and magnitude of a simple and uniform strain are defined by three things.

First. The direction of the lines along which the particles of the body are displaced from their natural position.

Secondly. The direction along which the rate of variation of the displacement from point to point is a maximum. This direction is normal to a series of planes of equal displacement, and may be called the *strain-normal.*

Thirdly. The amount of that rate of variation, being the differential coefficient of the displacement with respect to distance along the strain-normal.[3]

This definition may sound a bit cumbersome to us, but the reader must remember that we are around 1850 and vector calculus is still in limbo. Nevertheless, as described by Rankine, stress and pressure have the properties of vectors:

The component strains into which all others can be resolved with respect to a given set of axes are thus reduces to six, three longitudinal and three transverse.

7. A pressure, like a strain, is defined by three things:

1st. the direction of the pressure.

2nd. The position of the surface at which the pressure is exerted

3rd. The amount of pressure as expressed in units of force per unit of area of the surface of action

A pressure on a plane, in whatsoever direction it may act, may be resolved into three rectangular components, one normal to the plane and two tangential. The normal pressure may be compressive or tensile: when compressive, it is considered as positive; when tensile, negative.[4]

Again, because of the lack of vector calculus, the rest of the text is rather verbose and difficult to read.

Five years later, in 1855 Rankine gives an important talk at the royal Society of London with title "On Axes of Elasticity and Crystalline Forms (1855), in which he defines fundamental notions of Elasticity starting with an important link to mathematics:

The term "*axes of elasticity,*" therefore, may now be taken in a more extended sense, to signify *all directions with respect to which certain kinds of elastic forces are symmetrical*; or speaking algebraically, *directions for which certain functions of the coefficients of elasticity are null or infinite.*

The theory of axes and coefficients of elasticity is specially connected with that branch of the calculus of forms which relates to linear transformations, and which has recently been so greatly advanced by the researches of Mr. Sylvester, Mr. Cayley, and Mr. Boole. In such applications of that calculus as occur in this paper, the nomenclature of Mr. Sylvester is

[2] Rankine, 1851, p. 49.

[3] Rankine, 1851, p. 49.

[4] Rankine, 1851, p. 51.

followed; and by the adoption of the "*Umbral Notation*" of that author immense advantages are gained in conciseness and simplicity.

In this paper, the word "*Strain*" will be used to denote the change of volume and figure constituting the deviation of a molecule of a solid from that condition which it preserves when free from the action of external forces; and the word "*Stress*" will be used to denote the force, or combination of forces, which such a molecule exerts in tending to recover its free condition, and which, for a state of equilibrium, is equal and opposite to the combination of external forces applied to it.

In framing a nomenclature for quantities connected with the theory of elasticity, $\theta\lambda\iota'\psi\iota\varsigma$ is adopted to denote strain, and $\tau\alpha\sigma\iota\varsigma$ to denote stress.[5]

Rankine develops in this text the foundations of elasticity, but the new vocabulary he introduces is rather cumbersome and will not be integrated in the common language. Two years later, in 1858, he publishes an impressive *Manual of Applied Mechanics* where he dedicates an important part II to the theory of structures. It is divided in three chapters. At the end Chap. I, "Definition and General principles", he explains:

From the foregoing considerations, it is evident that the theory of structures may be divided into two divisions, relating, the first to stability, or the property of resisting displacement of the pieces, and the second to STRENGTH and STIFFNESS, or the power of each piece to resist fracture and disfigurement.[6]

In Chap. II, "Stability", Rankine analyses some frames. With Chap. III, "Strength and stiffness", we enter a new world with new concepts as stiffness, resilience and shearing. Some of these have already been defined and used in the context of pure physics. It seems that here, with Rankine, they enter the domain of architecture. He tries to be rather general and treats liquids as well, but I'll restrict myself to elements used in architecture, thus principally to solids.

Section 1. Summary of the theory of elasticity as applied to strength and stiffness.

236. The Theory of Elasticity relates to the laws which connect the stresses, or pressures and tensions, which act at the surface and in the interior of a body, with the alterations of dimensions and figure which the body and its parts simultaneously undergo. That theory, therefore, is the foundation of the principles of the strength and stiffness of materials of construction

237. Elasticity is the property which bodies possess of occupying, and tending to occupy, portions of space of determinate volume and figure, at given pressures and temperatures, and which, in a homogeneous body, manifests itself equally in every part of appreciable magnitude …

241 Rigidity or stiffness. A solid body, besides resisting change of volume like a liquid, possesses also rigidity, or the property of resisting change of figure. …

242. Strain and Fracture. In popular language the words *strain* and *stress* are applied indifferently to denote either the system of forces at the surface of a solid body whereby its volume and figure are altered, or the alteration of volume and figure of the body and its parts thereby produced. For the sake of clearness in scientific language, certain authors have recently endeavoured to appropriate the word *strain* to the alterations, of what nature soever,

[5] Rankine, 1855, pp. 119–120.

[6] Rankine, 1858, p. 130.

in the volume and figure of a solid body and of its parts, produced by forces applied to it, and the word stress as formerly defined. This nomenclature will be used in the present treatise. *Fracture* of a solid occurs when a strain is carried so far as to cause actual division of the solid into parts…

243. Perfect and Imperfect Elasticity. Plasticity. A body is said to be perfectly elastic, which, if strained at a constant temperature by the application of a stress, recovers its original volume, or volume and figure, when such stress is withdrawn. Deviations from this property constitute imperfect elasticity …The elasticity of every solid is sensibly perfect when the strain does not exceed a certain limit…

244. The **Ultimate Strength** of a solid is the stress required to produce fracture in some specified way. … The **Proof Strength** is the stress required to produce the greatest strain of a specific kind consistent with safety; that is, with the retention of the strength of the material unimpaired. A stress exceeding the proof strength of the material, although it may not produce instant fracture, produces fracture eventually by long-continued application and frequent repetition … **Resilience** or **Spring** is the quantity of mechanical work required to produce the proof strain … Each solid has as many different kinds of stiffness, toughness, strength, and resilience as there are different ways of straining it...[7]

Rankine gives also the resistance of different materials to direct compression and crushing, to bending and to twisting and benching, cross-bending and to shearing.

In 1861 August Ritter had published an article about on the construction of roofs and bridges, and a year later a first edition of his book, *Elementare Theorie und Berechnung eiserner Dach- und Brücken- Constructionen* on the same argument. In the second edition of this book published in 1870, Ritter adds an important §51 (pp. 324–338) concerning the elastic line (Fig. 1).

Ritter presents the elastic line as what we call the neutral fibre of a beam but he doesn't mention a neutral fiber. He considers the line element MN as an arc of a circle having O the meeting point of both normal CD and EF and ON = ρ. Calling τ the moment of inertia of the cross-sectional area he finds, after a calculus made with rather strange notations, this result, expressed with our usual notations:

$$E\frac{d^2y}{dx^2} = K(l - x)$$

$$E\frac{dy}{dx} = K\left(lx - \frac{x^2}{2}\right)$$

$$Ey = K\left(\frac{lx^2}{2} - \frac{x^3}{6}\right).$$

For $x = l$ we get $y = s$ where s is the lowering of the extremity B:

$$s = \frac{Kl^3}{3E}$$

For Ritter s is the important quantity.

[7] Rankine, 1858, pp. 270–273.

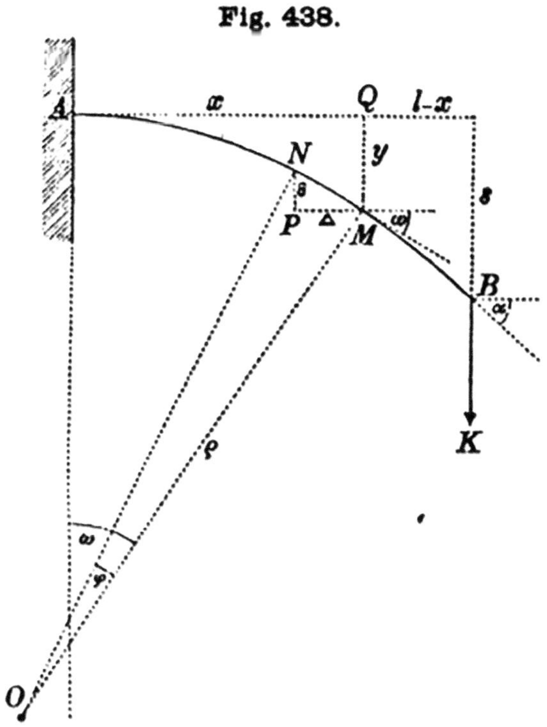

Fig. 1 The elastic curve of Ritter (1870, p. 325). Public domain

We have seen that in 1850, Rankine had difficulties to express the fact that the pressure was a vector, saying that a pressure, like a strain, is defined by three things. In a text published in 1870, "On Diagrams of Forces in Frameworks", he writes:

> As Mr. Clerk Maxwell, in a paper lately published in the Transactions of this Society, has done me the honor to refer to me as having been the first to show how to combine in one diagram a system of line representing the directions and magnitudes of all the forces acting in a given frame, I which to put on record, in the Proceedings of the Society, the time and manner of my first publication of the method in question. It was in the year 1856, in a lithographed synopsis of lectures which I delivered in the University of Glasgow, entitled Mechanical Laws, Formulae, and Tables.[8]

In doing this Maxwell contributed much to the invention of the vector.

2. 1873–1934 Structural Engineering: The Importance of the Virtual, Betti and Castigliano

We can immediately see the difference with the previous period. Right at the beginning of his article written in 1872 and published the next year, on the "theory of elasticity", Betti explains only that the deformations are infinitely small with respect to other motions.

[8] Rankine, 1870, p. 171.

1. Deformation of a body

Solid bodies are not perfectly rigid as they are considered in Rational Mechanics, but all of them, under the action of certain forces, without breaking their connection, can change within certain limits of shape and extension, and therefore the distances of their points can vary infinitely close, that is, the magnitudes of their linear elements. Let us limit ourselves to considering these deformations in the case in which the ratios between the variations of the linear elements and the elements themselves are quantities so small that the powers of higher order can be neglected compared to those of lower order.[9]

In fact, Betti recalls here the principle of elasticity,[10] but he adds:

2. Potential of the elastic forces

A deformation produced by a solid and elastic body, if it does not exceed certain limits, gives rise to internal forces that tends to bring back the body to its primitive state, those forces are called elastic forces. So during the deformation and the return of the body to its primitive state, the temperature is maintained constant, the mechanical work done by the external forces to produce the deformation is equal to the mechanical work done by the external forces when producing the deformation is equal to the mechanical work done by the elastic forces to draw back the body to its primitive state. In fact, we have in this case a closed and reversible cycle of transformations, and through the second principle of thermodynamics, if Q is the heat absorbed or emitted by the body during the deformation, Q' the one emitted or absorbed when turning back to its primitive state and T the constant temperature at which the transformations occur, we'll have: $QT\text{-}Q'T{=}0$. Thus, $Q = Q'$.

Thus we have neither production nor consumption of heat, and thus the labour consumed must be equal to the labour produced.[11]

We also find:

3. Equations of equilibrium for solid, elastic and homogeneous bodies.

To determine the relations that must exist between the forces that act on a solid, elastic and homogeneous body, and the deformations of its elements, in order to obtain a equilibrium,

[9] Betti, 1872, p. 5: *1. Deformazione di un corpo. I corpi solidi non sono perfettamente rigidi come si considerano nella Meccanica rationale, ma tutti sotto l'azione di certe forze, senza rompere la loro connessione, possono mutare entro certi limiti di forma e di estensione, e quindi possono variare le distanze dei loro punti infinitamente vicini, cioè le grandezze dei loro elementi lineari. Limitiamoci a considerare queste deformazioni nel caso in cui i rapporti tra le variazioni degli elementi lineari e gli elementi stessi siano quantità, talmente piccole che si possano trascurare le potenze di ordine superiore di fronte a quelle di ordine inferiore.*

[10] For more on the theory of elasticity, see "The Discovery of General Methods for Elastic Systems, 1852–1875", Chap. 16 in this present volume.

[11] Betti, 1872, pp. 14–15: *Potenziale delle forze elastiche. Una deformazione prodotta da un corpo solido elastico, se non oltrepassa certi limiti, dà origine a forze interne che tendono a riportare il corpo nello stato primitivo le quali si dicono forze elastiche. So durante la deformazione e il ritorno del corpo allo stato primitivo, la temperatura si mantiene costante, il lavoro meccanico fatto dalle forze esterne nel produrre la deformazione è uguale al lavoro meccanico fatto dalle forze elastiche nel ricondurre il corpo allo stato primitivo. Infatti abbiamo in questo caso un ciclo chiuso invertibile di trasformazioni, o per il secondo principio della Termodinamica, se Q è il calore assorbito o emesso dal corpo durante la deformazione, Q' quello omesso o assorbito nel ritornare allo stato primitivo e T la temperatura costante alla quale avvengono le trasformazioni, avremo:* $QT\text{-}Q'T = 0$ *onde:* $Q = Q'$.
Dunque non si ha nè produzione nè consumo di calore, e quindi il lavoro consumato dev'essere uguale al lavoro prodotto.

we will use the following principle of Lagrange: in order a system, whose virtual motions are invertible, to be in equilibrium it is necessary and sufficient that the mechanical work done by the force in whatever virtual motion, be equal to zero.[12]

It is interesting from our point of view to see and to understand the role played by the Lagrangian mechanics. The first edition of the *Mécanique analytique* was published in 1788 and had already been republished twice, in 1811 and 1815, but it had no influence on the understanding of the problem of the supports.[13] Betti shows now how appropriate but difficult this mechanics is for elasticity:

If in a solid and homogeneous body, two systems of displacements are in equilibrium with two systems of forces applied to the surfaces, the sum of the products of the forces' components of the first system with the components of the displacements in the same points of the second, is equal to the sum of the products of the forces' components of the second system with the components of the displacements in the same points of the first.[14]

The next year, in 1873, two young men, both 26 years old, present their dissertations at the Regia scuola d'applicazione per ingegneri to obtain the degree of civil engineer. Their arguments were similar. Valentino Cerruti's argument was "articulated elastic systems"; Alberto Castigliano's title was "About elastic systems". Cerruti was awarded the highest score, Castigliano the second, perhaps because his work was new and maybe not fully understood. Effectively, he gave a new theorem of the minimum work which gives him the same results than the virtual displacement.

The thesis contains an introduction and three parts. The introduction starts with a remark that sums up the previous developments. We shall see that it was to have important consequences for the history of constructions.

Introduction

1. The determination of the efforts, that support the various parts of systems of bodies used in practice, is a major question for engineers. If the bodies in nature were absolutely rigid, there would only be few cases in which it would be possible to determine those efforts: but as in nature all bodies are elastic, it is possible to determine completely the efforts of all parts of whatever a system; as the modification of the form of some of those parts draws necessarily with it the modification of all others, and from those changes of form depends precisely the dimensions of the efforts.[15]

---

[12] Betti, 1872, p. 69: *Equazioni di equilibrio dei corpi solidi elastici omogenei. Per determinare le relazioni che debbono esistere tra le forze che agiscono sopra un corpo solido elastico omogeneo, e le deformazioni degli elementi dello stesso, affinche si abbia equilibrio, ci varremo del seguente principio di* Lagrange*: affinche un sistema, i cui moti virtuali sono invertibili, sia in equilibrio è necessario e sufficiente che il lavoro meccanico fatto dalle forze in un moto virtuale qualunque, sia uguale a zero.*

[13] For more on this, see "The Supports Problem: A Dead-End Street in the History of the Stability of Structures", Chap. 15 in this present volume.

[14] Betti, 1872, p. 89: *Se in un corpo solido elastico omogeneo, due sistemi di spostamenti fanno equilibrio a due sistemi di forze applicate alle superficie, la somma dei prodotti delle componenti delle forze del primo sistema per le componenti degli spostamenti negli stessi punti del secondo, è uguale alla somma dei prodotti delle componenti delle forze del secondo sistema per le componenti degli spostamenti nei medesimi punti del primo.*

[15] Castigliano, 1873, p. 5: *1. La determinazione degli sforzi, che sopportano le diverse parti dei sistemi di corpi adoperati nella pratica, è una delle questioni più importanti per gli ingegneri. Se*

Next Castigliano gives a summary of what his predecessors did. He mentions Navier and Saint–Venant, Vène, Cournot, Mossotti, and gives the principal of elasticity or of least work of Federico Luigi Menabrea. Finally, he presents his work in the following manner:

> This little work is divided in three parts: in the first I explained the way of solving all problems related to elastic systems taking as unknown the variations of the coordinates of the vertexes: in the second I furnished a demonstration of the fact that the theorem of the minimal work will always lead to the same results as those to which drove the preceding method; in the third I did some applications of the same theorem.[16]
>
> Theorem of the minimal work
>
> 5. Let us consider a system composed of elastic verges attached together with joins and let us retain the preceding denominations: I say that we can determine the tensions Tpq in order them to render minimal the expression $\sum \frac{T_{pq}^2}{\epsilon_{pq}}$, supposing that between those tensions we must have the equations (1), in which we consider as constants all the external forces X_p, Y_p, Z_p, and all angles α_{pq}, β_{pq}, γpq, the values of the tensions, that we obtain like that, coincide perfectly with those obtained by the method of displacements.[17]

He then gives the equations (1):

> Let us call $V_1, V_2, V3, \ldots Vp, \ldots$ the various summit of the system, X_p, Y_p, Z_p, the components of the force applied at the summit *Vp*, α_{pq}, β_{pq}, γ_{pq}, the angles that the rod does with the axis T_{pq} the tension of that rod. The equations of equilibrium at the summit *V*, will be[18]:

$$X_p + \sum T_{pq} \cos\cos\alpha_{pq} = 0; \quad Y_p + \sum T_{pq} \cos\cos\beta_{pq} = 0;$$
$$Z_p + \sum T_{pq} \cos\cos\gamma_{pq} = 0. \qquad (1)$$

---

i corpi in natura fossero assolutamente rigidi, non vi sarebbero che pochi casi in cui tali sforzi si potrebbero determinare: ma in natura i corpi essendo tutti elastici, riesce possibile determinare completamente gli sforzi di tutte le parti di un sistema qualunque; poichè la modificazione della forma di alcune di queste parti trae seco necessariamente quella di tutte le altre, e da questi cambiamenti di forma dipende appunto la grandezza degli sforzi.

16 Castigliano, 1873, p. 7: *Quest'opuscol o è diviso in tre parti: nella prima ho esposto il modo di risolvere tutti i problemi relativi ai sistemi elastici prendendo per incognite le variazioni delle coordinate dei vertici: nella seconda ho procurato di dimostrare che il teorema del minimo lavoro condurrà sempre agli stessi risultati a cui conduce il metodo precedente: nella terza ho fatto alcune applicazioni del teorema medesimo.*

17 Castigliano, 1873, p. 14: *Theorema del minimo lavoro: 5. Consideriamo un sistema formato di verghe elastiche congiunte a snodo e teniamo le denominazioni precedenti: io dico che se determino le tensioni* T_{pq} *in modo che rendano minima la funzione* $\sum \frac{T_{pq}^2}{\epsilon_{pq}}$, *supponendo che fra quelle tensioni debbano aver luogo le equazioni [1], nelle quali però si considerino come costanti tutte le forze esterne* $\mathrm{X_p, Y_p, Z_p}$, *e tutti gli angoli* $\alpha_{pq}, \beta_{pq}, \gamma_{pq}$, *i valori delle tensioni, che cosi ottengo, coincidono perfettamenti con quelli ottenuti col metodo degli spostamenti.*

18 Castigliano, 1873, p. 9: *Chiamiamo* V1, V2, V3, ... Vp, ... *i diversi vertici del sistema, Xp, Yp, Zp le componenti della forza applicata al vertice* Vp, pq, pq, pq, *gli angoli che la verga V, V, fa cogli assi,* T_{pq} *la tensione di questa verga. Le equazioni di equilibrio pel vertice V, saranno:* Eq. (1).

He transforms this in λ:

$$X_p + \sum \in_{pq} \lambda_{pq} \cos\cos\alpha_{pq} = 0;$$
$$Y_p + \sum \in_{pq} \lambda_{pq} \cos\cos\beta_{pq} = 0;$$
$$Z_p + \sum \in_{pq} \lambda_{pq} \cos\cos\gamma_{pq} = 0, \quad (2)$$

where λ is the lengthening or shortening and $\in = \frac{E\omega}{l}$.

Castigliano closes his thesis by explaining some particular cases of frames, solving by the way, the problem of support.

In his article, *Intorno all'equilibrio dei sistemi elastici* (1875) published in the important journal, the *Atti della Reale Accademia di Torino,* Castigliano first recalls the work of Vène of 1827, that of Cournot of 1828, that of Pagani of 1834, and those of Menabrea published in 1857, 1858 and 1867. As he explains in his introduction, this article is a revised version of the thesis we analysed in detail. He explains the new publication:

> Since the time I published my thesis, I always thought about it when it was possible; and even if I was often distracted from it for months because of my various occupations of my work and others, nevertheless I think I found some new and more simples demonstrations or even more rigorous than the ones I had given before: furthermore, in order to render more evident to the engineers the advantage that can be drawn from the theorem of the least work, I applied it to the to the examination of the stability of the ribs of the canopy at the station of Bra.[19]

Approximately one year later, Castigliano publishes a more complete version containing the theorem of the least work, *Nuova teoria intorno all'equilibrio dei sistemi elastici*:

> If one applies to any elastic body forces such that the body would be in equilibrium if it were rigid, this will deform and find a new equilibrium condition. As in such a deformation the points of application of the external forces are moving, they produce some work.
>
> But the deformations of the body give birth to internal forces or elastic ones, which also produce some work, because their points of application are moving. It is commonly said that this work is equal to the one of the external forces.
>
> In order that this assertion to be accepted without difficulty, it is necessary to explicate the sense of the words. Now, when we say that forces of a certain intensity are applied to a body, we mean to say that infinitely small forces are applied to the body, which increase by infinitely small degrees until they reach their final values. In this way elastic forces also increase by infinitely small degrees, and in each instant they form a system equivalent to the one of the external forces.
>
> Now, it is well known that these systems of forces are doing the same work.

---

[19] Castigliano, 1875, p. 383: *Da quando ho stampata la mia dissertazione, io son sempre venuto meditandola, quando mi è stato possibile; e benché ne sia stato distolto per mesi intieri dalle molteplici occupazioni del mio impiego e da altre estranee, pure mi pare di aver trovato alcune nuove dimostrazioni più semplici o più rigorose di quelle, che io aveva dato dapprima: inoltre per rendere più evidente agli Ingegneri il partito, ch'essi possono trarre dal teorema del minimo lavoro, ne ho fatto l'applicazione all'esame della stabilità delle centine della tettoia nello scalo di Bra.*

We shall call it the *work of the elastic forces* or *work of deformation of the body*.

…

This said, the two new theorems are:

1° If for whatever elastic system the work of deformation expressed in function of the external forces, one derives with respect to one of those forces, the derivative, that one obtains, expresses the displacement of the application point of the force projected upon its direction.

2° If one derives the same expressions of the work of deformation one derives with respect to the moment of a couple, the derivative, that one obtains, expresses the rotation of the line that joints the points of application of both forces of the couple.

Those theorems, which importance is obvious, are only true if the deformations are tiny, so that the powers of the displacements and of the rotations higher than the first are negligible with respect to that one.

They can be joined into a single theorem, which I'll call the theorem of the derivatives of the work of deformations, or more briefly, the theorem of the derivatives of the work.[20]

It is worth mentioning that Menabrea contested the priority of this theorem in a letter presented to the Accademia dei Lincei dated 27 March 1875:

He [Castigliano], if I have understood correctly, claims priority for a special method tending towards the same object as *the principle of the minimum work of elastic reactions* which I have called since 1858 *the principle of elasticity*, and which he would develop in one of his dissertations ("On elastic systems") presented to the examining commission of the Royal School of Engineers in Turin in 1873; while a similar method would be found exposed in the

[20] Castigliano, 1876, pp. 127–129: *Se ad un corpo elastico qualunque si applicano delle forze tali, che si farebbero equilibrio quando il corpo fosse rigido, questo si deforma e si stabilisce in una nuova condizione d'equilibrio. Siccome in tale deformazione i punti d'applicazione delle forze esterne si spostano, queste fanno un lavoro.*
Ma la deformazione del corpo fa nascere delle forze interne od elastiche, le quali fanno anch'esse un certo lavoro, perché i loro punti d'applicazione si spostano. Dicesi comunemente che questo lavoro è uguale a quello delle forze esterne.
Affinché quest'asserzione sia senza difficoltà accettata, è necessario precisar bene il senso delle parole. Ora, quando noi diciamo che ad un corpo si applicano delle forze d'una certa intensità, vogliam dire che si applicano al corpo delle forze infinitamente piccole, le quali si fanno crescere per gradi infinitamente piccoli sino a raggiungere i loro valori finali. In questo modo anche le forze elastiche crescono per gradi infinitamente piccolo, e in ciascun istante esse formano un sistema equivalente a quello delle forze esterne.
Ora, è noto che tali sistemi di forze fanno lo stesso lavoro.
Noi lo chiameremo lavoro delle forze elastiche *o* lavoro di deformazione del corpo…
Ciò posto, i due nuovi teoremi sono i seguenti:
1° Se per un sistema elastico qualunque il lavoro di deformazione espresso in funzione delle forze esterne si differenzia rispetto ad una di queste forze, la derivata, che si ottiene, esprime lo spostamento del punto d'applicazione della forza proiettati sulla sua direzione.
2° Se la medesima espressione del lavoro di deformazione si differenzia rispetto al momento di una coppia, la derivata, che si ottiene, esprime la rotazione della linea, che congiunge i punti d'applicazione delle due forze della coppia.
Questi teoremi, la cui importanza è evidente, sono veri soltanto se le deformazioni sono piccolissime, per modo che le potenze degli spostamenti e delle rotazioni superiori alla prima siano trascurabili rispetto a questa.
Essi possono riunirsi in un solo, ch'io chiamerò teorema delle derivate del lavoro di deformazione o più brevemente theorema delle derivate del lavoro.

> note "On the determination of tensions and pressures in elastic systems", read by me after this Accademia de' Lincei in the session of 3 January, without my having made sufficient mention of his dissertation, having limited myself to citing it in a footnote.[21]

In fact, in 1858, Menabrea had introduced a *Nouveau principe sur la distribution des tensions dans les systèmes élastiques*, which he presents in the following way:

> Here is the statement of this new principle that I shall call principle of elasticity. When an elastic system goes to equilibrium under the action of external forces, the work developed by the effect of the tensions or compressions of the links that unite the various points of the system is a minimum. The differential equation that expresses that minimum will be given the name equation of elasticity; we shall soon see the way to use it in order to determine the tensions.[22]

He continues by explaining that in the most general case, the system has n points linked by m elastic connections, there are thus $3n$ equations of equilibrium. Given p as the number of equations linking the external forces, there are thus $3n - p$ equations containing the tensions. If $m > 3n - p$ or if the system contains a certain number of fixed points there is no unique solution. Then:

> Since, in the case we are considering, the tensions can vary without the balance ceasing to exist, we will have to admit that these variations take place independently of any work of external forces;
>
> To express this consequence, let T be the tension of any link, δl the elementary variation of the length of this line; the work developed as a result of the corresponding voltage variation will be $\mathrm{T}\delta l$, and consequently, for the entire system, we will have
>
> (1) $\sum \mathrm{T}\delta l = 0$
>
> Let l be the extension or shortening that the link initially experienced under the action of the tension T, we have, independent of the sign,
>
> (2) $\mathrm{T} = \varepsilon l$
>
> where ε is a coefficient that I will call relative elasticity coefficient, and which is a function of the modulus of elasticity, the section, and the length of the connection.[23]

---

[21] Menabrea, 1875, p. LXII: *Egli [Castigliano], se ho bene inteso, fa richiamo per la priorità di un metodo speciale tendente allo stesso oggetto che il principio del minimo lavoro delle reazioni elastiche da me chiamato fin dal 1858 principio di elasticità, e che egli avrebbe svolto in una sua dissertazione (Intorno ai sistemi elastici) presentata alla commissione esaminatrice della R. Scuola degli ingegneri in Torino nel 1873; mentre un consimile metodo si troverebbe esposto nella nota Sulla determinazione della tensioni e delle pressioni nei sistemi elastici, da me letta posteriormente a questa Accademia de' Lincei nella seduta del 3 gennaio p.p., senza che io vi abbia fatto abbastanza ampia menzione della sua dissertazione, essendomi limitato a citarla in una nota a piè di pagina.*

[22] Menabrea, 1858, p. 1056: *Voici l'énoncé de ce nouveau principe que j'appellerai principe d'élasticité. Lorsqu'un système élastique se met en équilibre sous l'action de forces extérieures, le travail développé par l'effet des tensions ou des compressions des liens qui unissent les divers points du système est un minimum. L'équation différentielle qui exprime ce minimum sera désignée sous le nom d'équation d'élasticité; on en verra bientôt l'usage pour la détermination des tensions.*

[23] Menabrea, 1858, pp. 1057–1058: *Puisque, dans le cas que nous considérons, les tensions peuvent varier sans que l'équilibre cesse d'exister, on devra admettre que ces variations s'effectuent indépendamment de tout travail des forces extérieures; Pour exprimer cette conséquence, soient T la tension d'un lien quelconque, δl la variation élémentaire de la longueur de ce line; le travail développé*

Along the way, Menabrea proves the relations obtained by Dorna in his *Memoria sulle pressioni sopportate dai punti d'appoggio di un sistema equilibrato ed in istato prossimo al moto* (1859).

Castigliano will come back to the problem in two publications of 1879 and 1882. In 1879, he puts all his results together in his *Theorie de l'équilibre des systèmes élastiques, et ses applications.* He recalls his three theorems: (1) about the derivatives of the work, first part; (2) about the derivatives of the work, second part; and (3) the least work theorem. He attributes a particular case of the first to the English astronomer Charles Green and he gives its generalization in this text. The second is the reciprocal of the first and Castigliano gave it in his thesis of 1873. The same theorem was extended in his article *Nuova Teoria* (1876). The third one can be seen as a corollary of the second and it was perceived, says Castigliano by different researchers such as Cournot, Pagani, Mossotti, Dorna and Menabrea:

> Theorem of the derivatives of the work of deformation
>
> First part – If the work of deformation of a system is expressed with respect to the relative displacements of the external forces applied at its summits, a formula is obtained whose derivatives with respect to those displacements give the value of the corresponding forces.
>
> Second part – If, on the contrary, the work of deformation of an articulated system is expressed in function of the external forces applied to the summits, a formula is obtained whose derivatives, with respect to the displacements, give the value of the corresponding forces.[24]

After having given precisions concerning the conditions for the validity of the theorem he gives the Theorem of least work:

> If one looks for the minimum of the function $\frac{1}{2}\sum \frac{T_{pq}^2}{\varepsilon_{pq}}$, which expresses the work of deformation of an articulated system taking into account the $3n - 6$ equations between tensions of all bars of the system, one obtains for these tensions the values that occur in the system after the deformation.[25]

---

par suite de la variation de tension correspondante sera $T\delta l$*, et par conséquent, pour l'ensemble du systèm, on aura*
(1) $\sum T\delta l = 0$
Soit l l'extension ou l'accourcissement qu'a primitivement éprouvé le lien sous l'action de la tension T, on a, indépendamment du signe,
(2) $T = \varepsilon l$
ou ε est un coefficient que j'appellerai coefficient d'élasticité relatif, et qui est fonction du module d'élasticité, de la section et de la longueur du lien.

[24] Castigliano, 1879, p. 26: *Théorème des dérivées du travail de déformation. Première Partie - Si l'on exprime le travail de déformation d'un système articulé, en fonction des déplacements relatifs des forces extérieures appliquées à ses sommets, on obtient une formule, dont les dérivées, par rapport à ces déplacements, donnent la valeur des forces correspondantes.*
Seconde Partie – Si l'on exprime, au contraire, le travail de déformation d'un système articulé en fonction des forces extérieures, on obtient une formule, dont les dérivées, par rapport à ces forces, donnent les déplacements relatifs de leurs points d'application.

[25] Castigliano, 1879, pp. 30–31: *Théorème du moindre travail. Si l'on cherche le minimum de la fonction* $\frac{1}{2}\sum \frac{T_{pq}^2}{\varepsilon_{pq}}$*, qui exprime le travail de déformation d'un système articulé, en tenant compte des* $3n-6$ *équations entre les tensions de toutes les barres du système, on obtient pour ces tensions les valeurs qui ont lieu dans le système après la déformation.*

Fig. 2 Castigliano's truss (1879, Pl. 2). Public domain

In his second chapter Castigliano proves that the same theorems are valid for recessed systems. The third chapter gives the general equations for the elastic equilibrium of solids. Here Castigliano is unable to give one general equation, so that he particularizes the problem to a rather important number of cases and this is the reason of the fourth chapter, "Applications approximatives":

> 1. Experiment result. In the problems that we solved in the previous chapter, we saw that we can determine the elastic forces which take place in a prism subjected to traction, bending or torsion, while it is only acted upon by forces applied on its bases and distributed according to a given law.
>
> However, it never happens, in practice, that we know the law according to which the forces applied to the bases of a prism are distributed, so that, strictly speaking, the solutions that we reported in the previous chapter would be of no use if knowledge of this law was absolutely necessary for the determination, at least approximately, of the elastic forces taking place inside. Fortunately, we can do without it, by virtue of the following result that experience has demonstrated, and which is of the greatest importance.[26]

The fifth chapter gives the theory of trusses: "We know that lattice beams are called beams composed of two parts having a section a T, connected to each other by inclined bars as seen in Figure 2 of Plate II".[27] (Fig. 2).

Such a truss is a non-deformable system; its only deformations are those due to the elasticity of its materials. Castigliano only furnishes approximate solutions for particular cases.

The seventh chapter expounds the theory of beams whose axis is rectilinear. Castigliano takes up again the original problem of Galileo and of Jacob Bernoulli. But he treats it, of course, in the context of elastic systems with all the context that it implies.

---

[26] Castigliano, 1879, p. 142: *1. Résultat d'expérience. Dans les problèmes que nous avons résolus au chapitre précédent, ous avons vu qu'on peut déterminer les forces élastiques qui ont lieu dans un prisme soumis à traction, à flexion ou à torsion, alors qu'il n'est sollicité que par des forces appliquées sur ses bases et distribuées suivant une loi donnée.*
Or, il n'arrive jamais, dans la pratique, qu'on connaisse la loi suivant laquelle les forces appliquées sur les bases d'un prisme y sont distribuées, en sorte que, rigoureusement parlant, les solutions que nous avons rapportées au chapitre précédent seraient sans utilité, si la connaissance de cette loi était absolument nécessaire pour la détermination, au moins approximative, des forces élastiques ayant lieu à l'intérieur. Heureusement qu'on peut s'en passer, en vertu du résultat suivant que l'expérience a démontré, et qui est de la plus haute importance.

[27] Castigliano, 1879, p. 175: *On sait qu'on appelle poutres à treillis, des poutres composées de deux parties ayant une section a T, reliées l'une à l'autre par des barres inclinées comme on le voit dans la figure 2 de la planche II.*

In Chap. VIII, “Theory of beams with rectilinear axis, loaded vertically”, Castigliano reprises, in the new context of a theory of elasticity, the problem of the loaded column.[28]

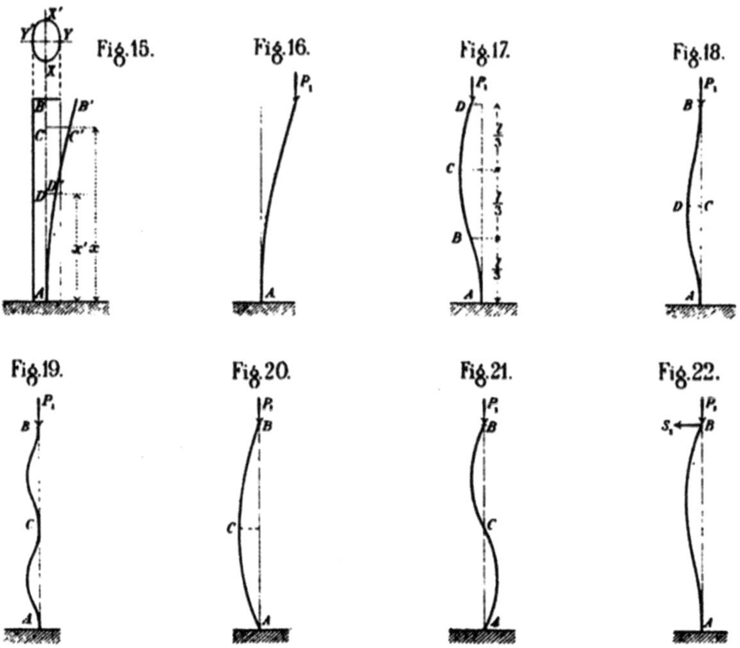

The second part of Castigliano’s book is dedicated to numerical applications of the developed theory.

To be complete, let us mention that in 1882, Castigliano published a last note on the argument, *Intorno ad una proprietà di sistemi elastici.*

References

Betti, Enrico. 1872. Teoria della elasticità. *Nuovo cimento*, serie 2, t. VII–VIII, 5–12, 69–97.

Castigliano, Alessandro. 1873. *Intorno ai sistemi elastici.* Torino: Vincenzo Bona.

Castigliano, Alessandro. 1875. Intorno all’equilibrio dei sistemi elastici. *Atti Della Reale Accademia Delle Scienze Di Torino* 10 (1874–1875): 380–423.

Castigliano, Alessandro. 1876. Nuova teoria intorno all’equilibrio dei sistemi elastici. *Atti della Reale Accademia delle scienze di Torino* XI (1875–1876), 127–286.

Castigliano, Alessandro. 1879. *Theorie de l’équilibre des systèmes élastiques, et ses applications.* Torino: Frédéric Negro.

Castigliano, Alessandro. 1882. *Intorno ad una proprietà di sistemi elastici.* Torino: Ermanno Loescher.

[28] For more on the loaded column, see “The Entasis of Columns from Blondel to Lagrange: Mathematical Games in Search of the Greatest Stability”, Chap. 9 in this present volume.

Cerruti, Valentino. 1873. *Sistemi elastici articolati*. Torino: Vincenzo Bona.

Dorna, Alessandro. 1859. Memoria sulle pressioni sopportate dai punti d'appoggio di un sistema equilibrato ed in istato prossimo al moto. *Memorie della Reale Accademia delle Scienze di Torino* (2) 18: 281–318.

Menabrea, Luigi Federico. 1858. Nouveau principe sur la distribution des tensions dans les systèmes élastiques. *Comptes rendus hebdomadaires des séances de l'Académie des sciences* 46: 1056–1060.

Menabrea, Luigi Federico. 1875. Lettera del 27 marzo 1875. *Atti della Accademia reale dei Lincei*, II, serie II, (1874–75), LXII–LXVI.

Rankine, W. J. Macquorn. 1851. The laws of the elasticity of solid bodies. *Cambridge and Dublin Mathematical Journal* 6: 47–80, 175–181, 185–186.

Rankine, W. J. Macquorn. 1855. On Axes of Elasticity and Crystalline Forms. June 21, William. J. M. Rankine, to the Royal Society. In *Miscellaneous Scientific Papers by W. J. Macquorn Rankine*, 119–120. London: Charles Griffin and Co.

Rankine, W. J. Macquorn. 1858. *A Manual of Applied Mechanics*. London: Griffin.

Rankine, W. J. Macquorn. 1870. On diagrams of forces in frameworks. *Proceedings of the Royal Society of Edinburgh* 7: 171–172.

Ritter, August. 1870. *Elementare Theorie und Berechnung eiserner Dach- und Brücken- Constructionen*, 2nd ed. Hannover: Carl Rümpler.

Chapter 19
Arthur Vierendeel: The Man with Iron Nerves Who Invented the Truss Without Diagonals

1 Introduction

The man stands very straight in the middle of the bridge. A slight smile brushes his face as, twenty yards from him, the coachman sets in motion the heavy convoy. We are in 1902, in Avelgem, and it has been six years since Arthur Vierendeel developed "his" truss without diagonals, which is used here for the first time on a bridge in Belgium. He tested it until it broke at the Tervueren World's Fair; the results obtained conformed perfectly with his calculations: the elastic limit was reached precisely under the expected load. Two of his bridges have been in the United States for two years. He is not at all worried when he sees the approaching convoy and prepares to greet the coachman who, worried by the rumours, demanded that the inventor be on the bridge as he passed. It is true, he thinks, that in the Warren, Pratt and Howe trusses the arrangement of the diagonals allows play in the compressions and extensions in the bars, but are these arrangements very different from those that Palladio had already proposed for wooden bridges? However, the other beams are made of iron; they follow other laws and therefore a different aesthetic. Why refuse the economy of the diagonals they allow?

Vierendeel's smile broadens. He sees again, at the Royal Circus, 26 years ago, the regiment of grenadiers take their place in a single movement, but with a heavy heart, to make sure that the bold iron structure proposed by the young, 25-year-old engineer was capable of resisting. Back then he had full confidence in his calculations, and even today no doubt creeps into his mind. He continues to reflect, bringing together the ideas that will come to underpin the course in construction stability that he has been giving in Louvain for nearly ten years. He thus follows the recommendations of the latest bill on academic degrees, which attributed the life and interest in lectures to the personal researches of the master. Scientific vocations arise, not from the oratory effects of a well-prepared lecture, but from the simple presentation, often improvised, of a recent discovery, of a new idea by a researcher still vibrating of the joy of his invention.

P. Radelet-de Grave, *Between Mechanics and Architecture*, Mathematics and the Built Environment 8, https://doi.org/10.1007/978-3-031-73530-1_19

Vierendeel will therefore have to make the students feel that it is calculation which ensures stability and provides the architect with the key to aesthetic freedom: "aesthetics first, then calculation, and this secondary role which is assigned to it is still a very fine one".[1] Aesthetics must allow the gaze to understand stability. This is why the professor criticises the Eiffel Tower:

> The Eiffel Tower is a tower made of thin lattices, this is the big aesthetic mistake that was committed by its builders; ...For this tower (tower as for all metal constructions in general) it was necessary to opt for large and robust ribs rising proudly from the bottom to the ridge and connected from distance to distance by powerful arcatures. ... The mind most foreign to things of construction would have felt the unshakeable stability of the framework by virtue of the constructive intuition which is in each of us.[2]

We must therefore compose between calculation and aesthetics depending on the material used:

> If the mathematical formula leads to an arrangement which is not beautiful, we must complete it according to our eye, and this by virtue of the principle that the rational which is not beautiful is not completely rational; the beautiful alone is completely rational.[3]

Such reasoning can only set Arthur Vierendeel in opposition to the school of scientific builders for which all that is rational is beautiful.

The convoy safely reached the other side of the bridge:

> In reality, the rational is not always beautiful, it is sometimes, but on the other hand the Beautiful is always rational, a formula which is close to that of Plato: Beauty is the splendour of truth".[4]

---

[1] Vierendeel (p. 778): *Donc, l'esthétique d'abord, le calcul ensuite et ce rôle secondaire qui lui est assigné est encore très-beau.*

[2] Vierendeel (p. 786): *La Tour Eiffel est une tour en minces treillis, c'est la grosse faute esthétique qui a été commise par ses constructeurs; ...Il fallait pour cette tour (tour comme pour toutes le constructions métalliques en général) le parti-pris des grandes et robustes nervures montant fièrement de fond jusqu'au faîte et relies de distance en distance par de puissantes arcatures. ... L'esprit le plus étranger aux choses de la construction eût senti l'inébranlable stabilité du canevas en vertu de cette intuition constructive qui est en chacun de nous.*

[3] Vierendeel (p. 799): *Si la formule mathématique conduit à une disposition qui n'est pas belle il faut la compléter au gré de notre œil et ce en vertu de principe que le rationnel qui n'est pas beau n'est pas le rationnel complet; le beau seul est le rationnel complet.*

[4] Vierendeel (p. 790): *En réalité le rationnel n'est pas toujours beau, il l'est quelquefois, mai par contre le Beau est toujours rationnel, formule qui se rapproche de celle de Platon: Le Beau est la splendeur du Vrai.*

Fig. 1 Arthur Vierendeel (1852–1940). Public domain

2 Vierendeel's Character and His First Works

2.1 Early Life and Education

The birth of Jules Arthur Meunier was registered in Louvain on 10 April 1852. He was given the name Vierendeel when he was five years old, when his mother married a locksmith who bore that name (Fig. 1).[5] He did his Humanities in Grammont and in 1874, earned the diploma of "ingénieur des arts, des manufactures, du génie civil et des mines" at the University of Louvain in Belgium. He was then engaged at the metal engineering company Nicaise et Decluve in La Louvière, where he stayed from 1876 to 1885.

2.2 The Royal Circus of Brussels (1876)

One year after his arrival at Nicaise et Decluve, at the age of 25, he won the competition for the building of the Cirque royal de Bruxelles, one of the first great metallic frameworks erected in Belgium.

The impression of lightness given by the building was so strong that the owners asked him to stop the works, but he went on. A committee was then appointed by the government to inquire about the stability but didn't find any error in his

[5] The biographical details are taken from an unpublished lecture given by Lederer (1991), on 15 March 1991 at Louvain-la-Neuve during a colloquium organised on the occasion of the 100th anniversary of Vierendeel's appointment as Professor at the Université catholique de Louvain; see also Lederer (1969, 1979).

Fig. 2 A drawing of a greenhouse in of Laeken in Brussels by Vierendeel himself (Vierendeel 1901, Pl. 56). Public domain

computations. Nevertheless, their opinion was that the building would not resist the first representation. Vierendeel still went on without changing anything to his plans. A week before the inauguration, the minister gave the order to a regiment of grenadiers to test the resistance of the Circus by taking a seat all together. The Circus resisted and would still be there if it had not been decided to pull it down because it no longer corresponded to the taste of the time. We do not have any reproduction or pictures of the Royal Circus, but the drawing of the greenhouses of Laeken in Brussels by Vierendeel himself recalls his conception of the lightness of an edifice (Fig. 2). He had the tenacity of a Flemish, the people said about the architect. This tenacity and a total confidence into his computations would be characteristic of Vierendeel all his life.

In 1885, he became director of the new technical administration for the province of western Flanders of the Ministry of public works. In 1889, when Louis Cousin retired, he asked Vierendeel to take over his courses on resistance of materials, on stability of constructions and on history of architecture as well. In the context of his work for the Ministry, he established the system of roads for the then new automobiles and studied the installation of a network for the distribution of electricity in his province. During his career as professor, he recommended that professors should be active engineers and contributed to the division of the faculty of applied science from that of science.

The next buildings Vierendeel built were the Eden Theater in 1878, the frame of the Kortrijk station (designed by Fouquet) in 1879 and the boilerwork hall of the metalworking firm Ateliers de la Louvière, in 1881.

3 The First Trusses and the Invention of Vierendeel

In his *Esquisse d'une histoire de la technique* (1921) Vierendeel gives a short history of the first iron bridges. We'll select the main points that are relevant to our story.

The Britannia Bridge (Fig. 3) built in 1846–1850 by Stephenson, Fairbairn, and Hodgkinson over the 280 m of the strait of Menai was a big success. It was formed of two parts, each spanning a length of 140 m.

However, with its plain walls it looked like an open-air tunnel. Trusses were already known at that time but it was not until the success of the Britania Bridge that they were applied to bridges. A first type of truss was designed by the Belgian architect Alfred Neuville and applied in 1846 to a bridge over the Lys at Grammene (Fig. 4). In 1912 that bridge would be replaced by a Vierendeel bridge.

Fig. 3 Britania Bridge. Public domain

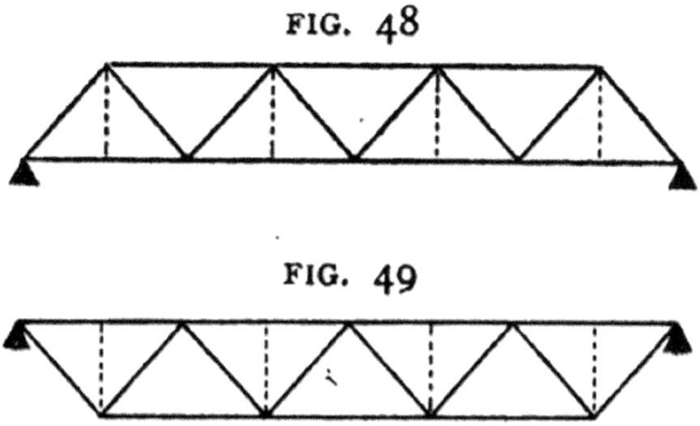

Fig. 4 The Neuville truss (Vierendeel, *Cours de stabilité des constructions*, tome 2, p. 69). Public domain

Fig. 5 The Monié Truss (Vierendeel, *Cours de stabilité des constructions*, tome 2, p. 70). Public domain

FIG. 50

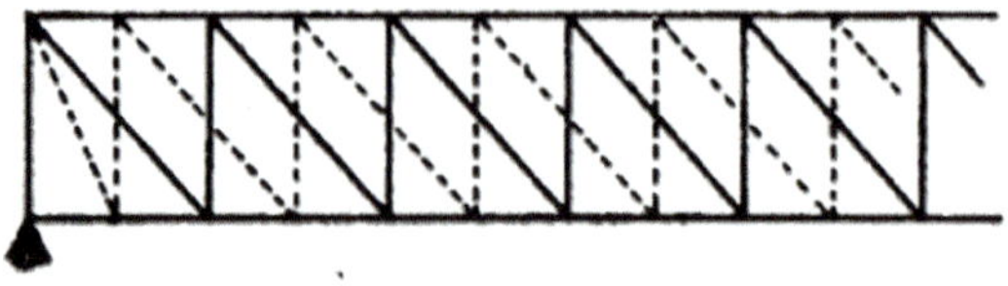

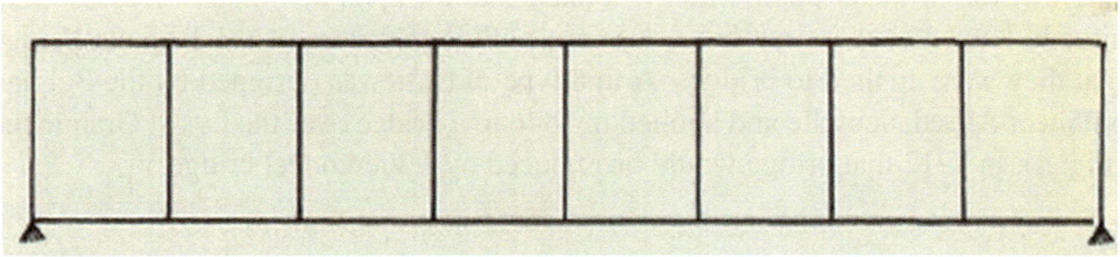

Fig. 6 The scheme of the original Vierendeel girder (Vierendeel 1936, p. 573). Public domain

Vierendeel wrote about Neuville's trusses:

> The assembly joints of the inclined bars on the horizontal bars were forged, they were quite expensive to execute but perfectly solid as evidenced by the long life of the Neuville beams.[6]

The British architect James Warren used the same truss in 1851 at Newark and in 1853 at Crumlin. But in Warren's trusses the pieces were tied together non-rigidly (jointed assembly). Other architects added vertical beams. In 1858 another Belgian engineer, Monié, proposed a new type of truss where the beams form one, two or even three series of "N"s (Fig. 5).

All those trusses have the particular characteristic of being composed of triangles. Such a piece is always rigid if its pieces form knee points or if they are rigidly fixed.

3.1 *What is a Vierendeel Truss?*

In contrast to the trusses above, a Vierendeel girder is composed of rectangles (Fig. 6).

To be rigid the parts of the rectangle must be rigidly fixed together in the corners, otherwise the truss would collapse. But let us see how Vierendeel describes his trusses in the three patents he took out.

---

[6] Vierendeel (1921, pp. 370-373): *Les nœuds d'assemblage des barres inclinées sur les barres horizontales étaient forgés, ils étaient assez coûteux d'exécution mais parfaitement solides ainsi que le prouve la longue durée des poutres Neuville.*

3.2 The Discovery of the Truss

The inventor of the beam without diagonals never told how he got that idea. Nevertheless, his biography furnishes us an interesting element concerning those circumstances. At the beginning of his career, when he was employed at the Ateliers de La Louvière, he spent some weeks adjusting the tensions in the diagonals of a 75-m bridge on the Meuse at Heumen, in the province of Gueldre in Holland. The method consisted in hitting the diagonals in order to make them resonate and judging the intensity of the frequency of oscillation and thus of the tension existing in the diagonal thanks to the pitch of the sound that they emitted. This experiment familiarized him with that complex problem and made him understand the imprecision of those measures.

It is therefore no wonder that the first application of the Vierendeel truss was done six years before the redaction of the patent. In fact, he already used the crucial idea in 1890, when he had to make the plans for the tower of the Basilica of Our Lady of Dadizele (Fig. 7) The construction of the church had begun in 1860 but was left unfinished because the foundations were not strong enough to support the tower. There he applied the concept that he would use for his trusses, six years before his first patent.

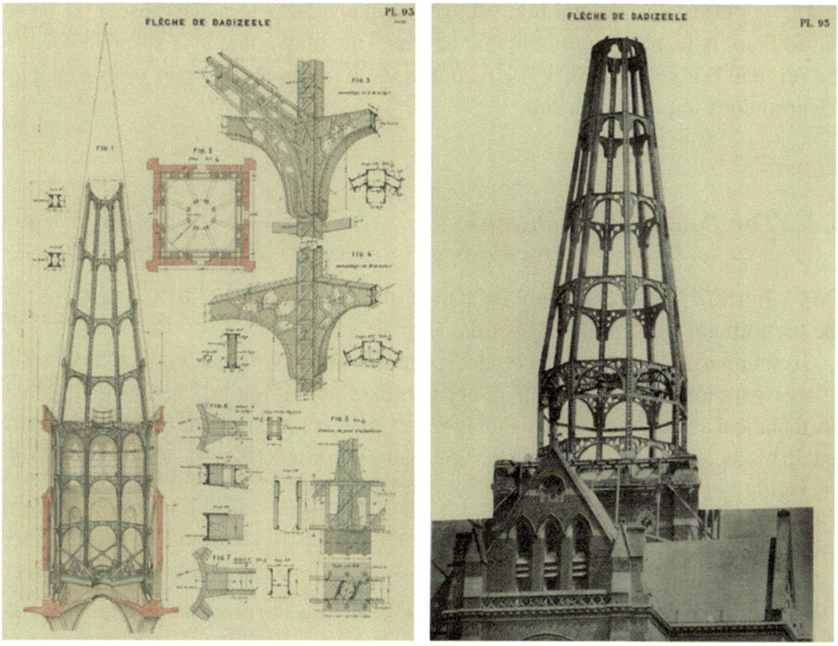

Fig. 7 Bell-tower of the Basilica of Our Lady of Dadizele (Vierendeel 1901, pl. 93, 95). Public domain

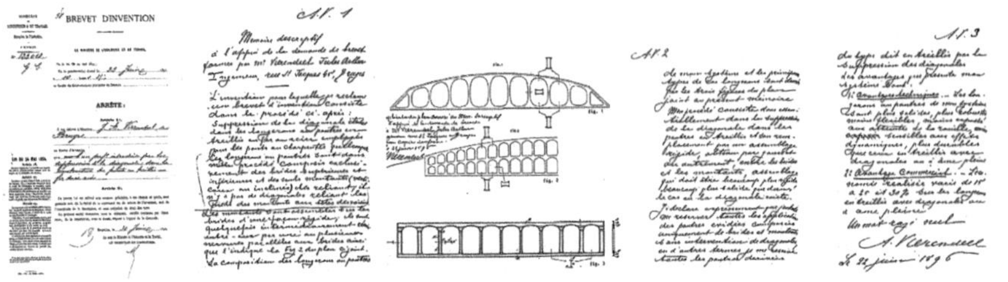

Fig. 8 The patent of invention no. 122.018 of the Vierendeel's truss. Public domain

The precise date of birth of the truss is difficult to establish. According to him, he spoke about it for the first time in public in 1895.[7] It is certain that Vierendeel entered a campaign to defend his beam in 1897, as evidenced by a conference given at the International Congress of Architects in August 1897 in Brussels:

> And now the lattice: great artistic subjugation of iron, because the lattice with its stiff, straight forms, without variety, without elasticity, is a constructive device which is nothing less than aesthetic, but fortunately we are delivered from the lattice; Here is a bridge whose iron forms are made without the intervention of any diagonal, of any lattice, and this achievement is obtained with an economy of material and without sacrificing strength, as proven by the experiments, currently underway, in Tervuren, on a bridge with a span of 32 meters.[8]

He would have to defend himself against many detractors. The discourse he set out to them is mainly of an aesthetic nature. This may come as a surprise because the original bridge is not very pretty, but as we'll see in the last part of this chapter, the aesthetics improve over time.

3.3 *The Patent of Invention*

On 22 June 22 1896 he took out the patent of invention no. 122.018, where we find the first official description of his truss (Fig. 8).

To summarize the patent, Vierendeel proposes a truss without diagonals composed of only parallel bars and of vertical uprights assembled rigidly. Then he enumerates the technical and commercial advantages. He insists on the importance of the rigidity, and in his lectures one finds many designs of these corners (Fig. 9), which must assure the rigidity of the junction between bars and uprights.

---

[7] Vierendeel (1900, p. 163).

[8] Vierendeel (1897, p. 14): *Et maintenant le treillis: grande sujétion artistique du fer, car le treillis avec ses formes raides, droites, sans variétés, sans élasticité, est un dispositif constructif qui n'est rien moins qu'esthétique; mais heureusement, du treillis nous sommes délivrés; voici un pont dont les fermes en fer sont réalisées sans intervention d'aucune diagonale, d'aucun treillis, et cette réalisation est obtenue en faisant une économie de matière et sans sacrifier de la solidité, ainsi que le prouvent les expériences, actuellement en cours, à Tervueren, sur un pont de 32 mètres de portée.*

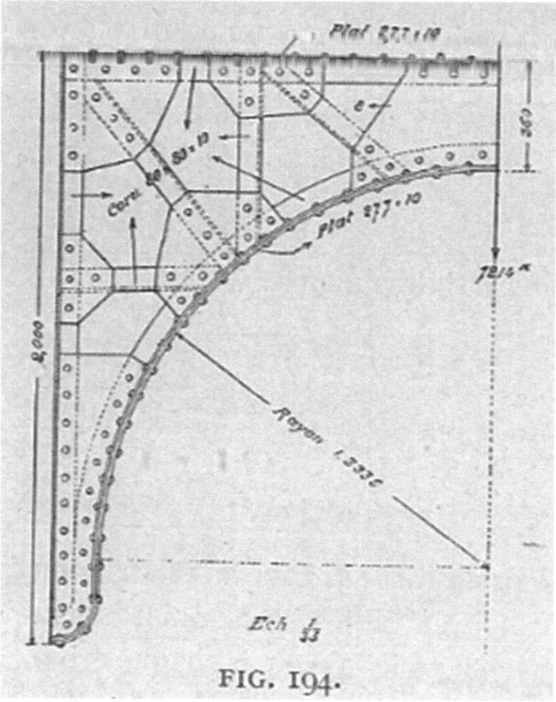

Fig. 9 Drawing by Vierendeel of a corner insuring the rigidity and thus the solidity of the link between bars and uprights (Vierendeel 1901, p. 722). Public domain

In the solidity of those corners resides the secret of the Vierendeel truss. Effectively, if one presses on the apex of a triangle, or on the upper element of a truss with a diagonal, the system resists but if one presses the upper element of a rectangle, or of a Vierendeel truss, unless the corners have been consolidated, it deforms.

3.4 The Experimental Truss of Tervuren Near Brussels (November 1897)

Slightly more than a year after having taken out his first patent, Vierendeel allowed a bridge spanning 31.5 m to be built following his new conception. He paid it with his own money. His aim was to subject the bridge to experiments, and so did he for three and a half months, until its rupture. He started at the end of August 1897 to improve the resistance to fatigue, by loading the bridge with 157 t. He left that load on it until November 10. The day after he added 47 t and on the next day he removed the entire 204 t load. The bridge returned to its initial position exactly like a spring, having conserved its elasticity and resistance. On 23 November the bridge was loaded again until rupture occurred on November 26. Every element – for example, for example, the deflection for a given load and at given places – had been calculated before and

Fig. 10 The Vierendeel bridge of Tervuren before the experiment (Lambin and Christophe 1898). Public domain

Fig. 11 The Vierendeel bridge of Tervuren after the experiment (Lambin and Christophe 1898). Public domain

compared with the experimental results. The fatigue values were slightly smaller than the estimates. The deformations occurred following the theoretical descriptions and rupture occurred where indicated by the computations. Vierendeel wrote that the computations had been published six months before the experiments were performed, so the results of the experiments proved that his theory was correct (Figs. 10 and 11).

3.5 First Improvement Patent

The next year, on 29 May 1897, he took a first improvement patent, in which the description is more precise (Figs. 12 and 13):

> Descriptive memorandum filed in support of an improvement patent application filed by Mr. Jules Arthur Vierendeel for a NEW SYSTEM OF IRON OR STEEL BEAMS OR SIDES. Main patent no. 122.018 of June 22, 1896.[9]

---

[9] Vierendeel, Brevet de perfectionnement, n. 128.573: *Mémoire descriptif déposé à l'appui d'une demande de brevet de perfectionnement formée par Monsieur Jules Arthur Vierendeel pour un NOUVEAU SYSTEME DE POUTRES OU LONGERONS EN FER OU ACIER. Brevet principal n° 122.018 du 22 juin 1896.*

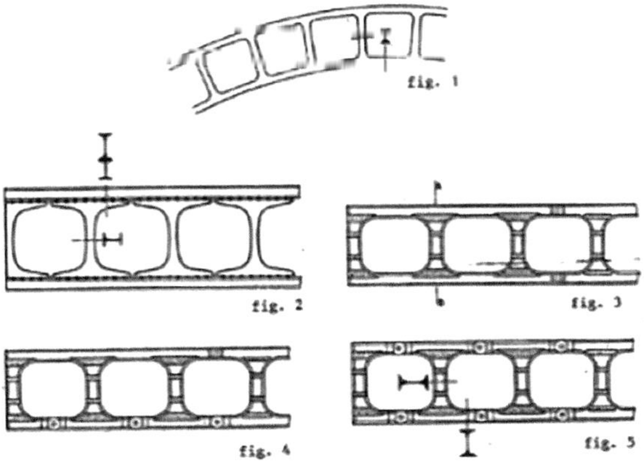

Fig. 12 Various forms of the first truss as shown in improvement patent no. 128.573. Public domain

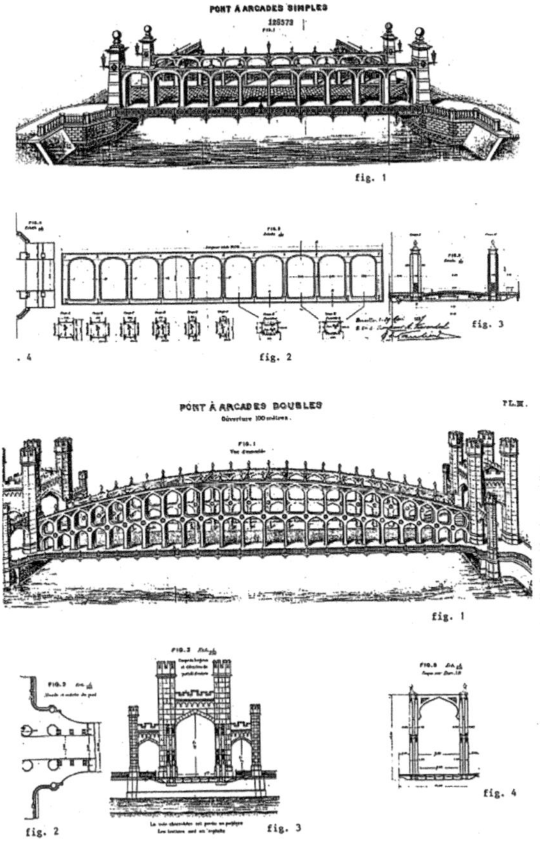

Fig. 13 Figures giving an idea of what he could do with his truss, added by Vierendeel to his first improvement patent. Public domain

There are two important aspects in the description. First, Vierendeel refuses triangulation and for that reason suppresses the diagonal:

> The process consists in the removal of the diagonal or in other words in the removal of any triangulation in the chords or lattice beams, straight or curved, of iron or steel, in the applications listed above.[10]

Second, refusing triangulation means that he is left with rectangles and for that reason the rigidity of the joints is very important:

> My process therefore consists essentially in the removal of the diagonal in the lattice beams and its replacement by a rigid assembly, obtained by gusset or otherwise, between the flanges and the uprights, an assembly which must be much more rigid, much more solid than in the case where the diagonal exists …[11]

Vierendeel sees various advantages—theoretical, technical, and commercial—in his truss:

> Theoretical advantage: The calculation of my chords with jointed flanges or completely rigid flanges can be done without resorting to any hypothesis whatsoever, which is not at all the case for our lattice beams of the rigid type used in Europe. This calculation is just as exact, just as rigorous as that of a fully jointed lattice beams of the type used in America.
>
> My completely rigid chord therefore enjoys the theoretical advantage of fully jointed chords, and it avoids the practical drawbacks of these: difficulty of execution and faulty behavior in current service. As my chord is rigorously calculable, it is therefore much more secure than rigid lattice chords as they are made in Europe. My system constitutes the truly rational constructive type for iron and steel chords.
>
> Stronger, more robust, better connected in all their parts, less flexible, less subject to vibrations, less exposed to the attacks of rust, less sensitive to the effects and as a final consequence of all these advantages, they are more durable than the longitudinal truss beams with diagonals (European type or American type). More exactly and is of a more rational construction than the chords and lattice beams with diagonals, and it follows that it can achieve the same solidity with an economy of weight, resulting in an economy of cost of up to 10 to 25 and even 30% depending on the various cases of practice.[12]

---

[10] Vierendeel, Brevet de perfectionnement, no. 128.573: *Le procédé consiste dans la suppression de la diagonale ou en d'autres termes dans la suppression de toute triangulation dans les longerons ou poutres en treillis, droites ou courbes, en fer ou en acier, dans les applications ci-dessus énumérées.*

[11] Vierendeel, Brevet de perfectionnement, no. 128.573: Mon procédé consiste donc essentiellement dans la suppression de la diagonale dans les poutres à treillis et son remplacement par un assemblage rigide, obtenu par gousset ou autrement, entre les brides et les montants, assemblage qui doit être beaucoup plus rigide, beaucoup plus solide que dans le cas où la diagonale existe …

[12] Vierendeel, Brevet de perfectionnement, no. 128.573: *Avantage théorique: Le calcul de mes longerons avec brides articulées ou brides complétement rigides peut être fait sans recourir à aucune hypothèse quelconque, ce qui ni est pas du tout le cas pour nos longerons en treillis du type rigide usité en Europe. Ce calcul est tout aussi exact, tout aussi rigoureux que celui d'une poutre en treillis complétement articulé du type usité en Amérique.*
Mon longeron complétement rigide jouit donc de l'avantage théorique des longerons complétement articulés, et il en évite les inconvénients pratiques qui sont: difficulté d'exécution et façon défectueuse de se comporter en service courant. Mon longeron étant calculable d'une façon rigoureuse présente par le fait beaucoup plus de sécurité que les longerons en treillis rigides tels qu'on les exécute en

3.6 The Second Improvement Patent for the Transition to Reinforced Concrete

Finally on 16 April 1899, he took a second improvement patent. This one enlarges the application of the patent to trusses made of reinforced concrete. It will take some years, until 1940, to fully elaborate that new technique:

> Application for an improvement patent for the invention patent issued on 22 June 1896 n° 122018 for improvement introduced by the removal of the diagonal in the construction of iron and steel lattice bridges by Mr. Arthur Vierendeel.
>
> My improvements consist in forming the beams of a frame made of rolled bars rigidly connected together, which is embedded in a cement mortar providing the external appearance of the beam.[13]

3.7 The Theoretical Advantages

From Vierendeel's point of view, the great advantage of his truss is a theoretical one: it can be rigorously computed and if some hypotheses are made because the computations are too long to be performed, the incidences of these hypothesis are well known. This is not the case with the triangular framework. He says:

> The rigid joint lattice type is not very rational from either a theoretical and practical point of view: we do not know how to calculate it with a sufficient approximation (the fundamental assumption of the calculation, the joint, is the opposite of reality) nor execute it properly (the flanges, diagonals and sometimes the uprights do not have the double tee profile), nor mount and adjust it exactly (the triangulation of the drawing is by no means carried out with efficiency in the executed parts); in short, faulty calculation and equally faulty execution, and in these faults there is little room for improvement, the type does not lend itself to it, without unhesitatingly adopting the joint of the Americans.[14]

Europe. Mon système constitue le type constructif vraiment rationnel pour les longerons en fer et en acier.
Plus solides, plus robustes, mieux reliés dans toutes leurs parties, moins flexibles, moins sujets aux vibrations, moins exposés aux atteintes de la rouille, moins sensibles aux effets et comme conséquence finale de tous ces avantages, ils sont plus durables que les longerons en treillis avec diagonales (Type Européen ou type Américain). Plus exactement, et est d'une construction plus rationnelle que les longerons et poutres en treillis avec diagonales, et il en résulte qu'il peut réaliser la même solidité avec une économie de poids, d'où résulte une économie d'argent pouvant atteindre de 10 à 25 et même 30% selon les divers cas de la pratique.

[13] Vierendeel, Brevet de perfectionnement, no. 144.061: *Demande d'un brevet de perfectionnement au brevet d'invention pris le 22 Juin 1896 N° 122018 pour perfectionnement introduit par la suppression de la diagonale dans la construction des ponts en treillis de fer et acier par Monsieur Arthur Vierendeel.*
Mes perfectionnements consistent à former les poutres d'une ossature, faite de fers laminés rigidement reliés ensemble, qui est noyée dans un mortier de ciment réalisant l'aspect extérieur de la poutre.

[14] Vierendeel 1900, p. 164: *Le type en treillis à nœuds rigides est peu rationnel au double point de vue théorique et pratique: on ne sait ni le calculer avec une approximation suffisante (l'hypothèse*

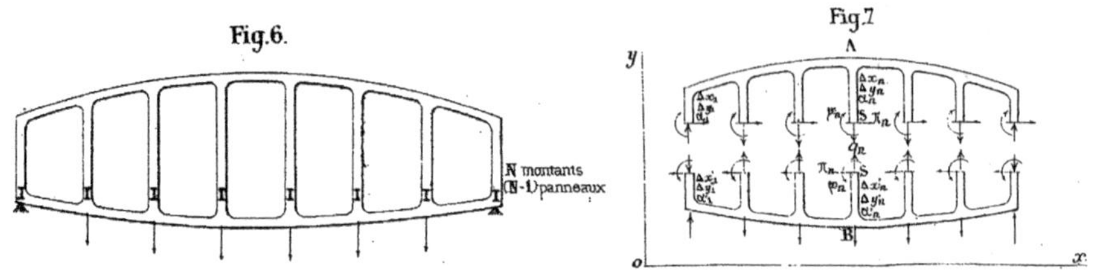

Fig. 14 Vierendeel's Figs. 6 and 7 (Vierendeel 1900, Pl. 234). Public domain

Without going into the details, let us have a look at the way he computes his trusses (Fig. 14):

> Let (Fig 6) be a beam of any shape carrying any loads; we relate it to two rectangular coordinate axes, we cut it into two sections by a section meeting all the uprights (Fig 7) which are N in number; each isolated section will be in exactly the same mechanical situation as when it was whole with the other provided that, on each section made, we apply in the form of external forces the equivalent of the internal forces which act there when solidarity exists between the two sections, we thus have on the center of gravity of section S of upright no. n:
>
> a vertical component q_n.
>
> a horizontal component π_n.
>
> a moment μ_n.
>
> Three unknowns for each upright, in total the problem therefore has 3 N unknowns.[15]
>
> The equations determining these unknowns will be provided to us by the fact that the expression of displacements (linear and angular) of a section S of an upright, expression calculated according to the dimensions of the upper section and the applied forces, is equal to the expression of displacements of the same section calculated according to the dimensions of the lower section and the forces applied to it.[16]

---

fondamentale du calcul, l'articulation, est le contre-pied de la réalité), ni l'exécuter convenablement (les brides, les diagonales et quelque fois les montants n'ont pas le profil double té), ni le monter et régler exactement (la triangulation du dessin n'est nullement réalisée avec efficacité dans les pièces exécutées); en résumé, calcul défectueux et exécution tout aussi défectueuse, et à ces défectuosités il n'y a guère d'amélioration possible, le type ne s'y prête pas, à moins de prendre franchement l'articulation des Américains.

[15] Vierendeel (1900, p. 169): *Soit (Fig 6) une poutre de forme quelconque portant des charges quelconques; nous la rapportons à deux axes coordonnées rectangulaires, nous la coupons en deux tronçons par une section rencontrant tous les montants (Fig 7) qui sont au nombre de N; chaque tronçon isolé sera exactement dans la même situation mécanique que lorsqu'il était solidaire de l'autre pourvu que, sur chaque section faite, nous appliquions sous forme de forces extérieures l'équivalent des forces intérieures qui y agissent lorsque la solidarité existe entre les deux tronçons, nous avons ainsi sur le centre de gravité de la section S du montant N°n:*
Une composante verticale q_n
Une composante horizontale π_n
Un moment μ_n
Trois inconnues pour chaque montant, au total le problème comporte donc 3 N inconnues.

[16] Vierendeel (1900, pp. 169–170): *Les équations déterminant ces inconnues nous seront fournies par le fait que l'expression des déplacements (linéaires et angulaire) d'une section S d'un montant, expression calculée en fonction des dimensions du tronçon supérieur et des forces appliquées, est*

He thus obtains three equations for each upright except the first, that is, 3(N–1) equations for a problem of 3N unknowns. Three more equations are given by the classic laws of statics: the sum of horizontal components of forces is zero and so is the sum of the vertical ones; the sum of the moments is also zero.

Let us go back to the comparison between a triangular and a rectangular framework as Vierendeel presents it (Fig. 15).

Vierendeel applies the same reasoning to both frameworks, he cuts them in the middle of the uprights and observes (Fig. 16):

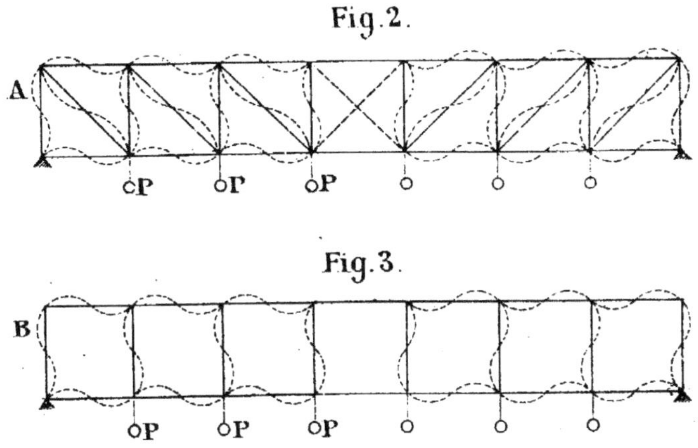

Fig. 15 Repartition of the stresses in a truss with diagonal (above) compared to the same in a Vierendeel truss (below) (Vierendeel 1900, Pl. 234). Public domain

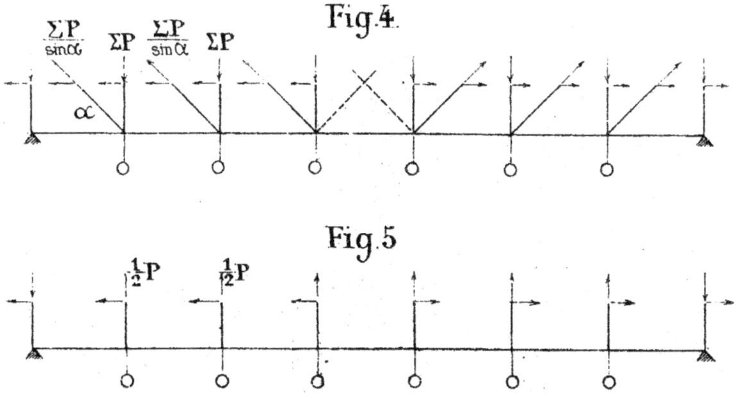

Fig. 16 Stresses in a truss with diagonal compared to those in a Vierendeel truss (Vierendeel 1900, Pl. 234). Public domain

---

égal à l'expression des déplacements de la même section calculée en fonction des dimensions du tronçon inférieur et des forces y appliquées.

1. To resist horizontal shearing there are two bars for each joint in the triangular framework and only one in the rectangular one. The two bars of the triangular one share the stress in a proportion impossible to determine.
2. Each bar of the triangular framework supports the totality of the shear stress in the rectangular one only the uprights support it.
3. All uprights of the triangular framework are submitted to buckling. This is not the case in the rectangular one.

His conclusion is, as we have already seen, that the rectangular framework can be computed accurately and that the triangular one cannot.

4 The Administration of Bridges and Roadways

In 1885, Vierendeel was appointed Chief Engineer director of the technical Administration of the province of West Flanders, precisely at the times of the beginnings of the automobile. In the context of the Administration, before the war of 1914–1918 war, he planned and organized the road network for the new cars, turning 2271 km of dirt roads into paved roads. But after the war, it was necessary to rebuild that road network in the region from Dixmude et Ypres, along the river Yser, as that region had been the scene of fierce battles where tanks had circulated.

Vierendeel also studied the installation of a first network for the distribution of electricity in his province. At the time, he also built pylons without diagonals for wireless transmission (TSF, *transmission sans fil*).

Nevertheless, the first bridge using Vierendeel's truss was not built in Belgium, but in the United States. It was the Kinzua Viaduct built in Pennsylvania between 1900–1902:

> The United States' civil engineering development of the Vierendeel began in 1901. A series of towers, forming the Kinzua Viaduct for the Erie Railroad in McKean County, Pennsylvania, used vertical tapered Vierendeels up to 285 feet in height. This solution was a radical departure from conventional tower design. The selection of rigid joints by engineer C.R. Grimm eliminated the need for a complex diagonal bracing system, while providing a cleaner appearance. Continued use of vertical Vierendeels occurred in the towers for many suspension bridges, where lateral load and vibration resistance require substantial rigidity without objectionable diagonals.[17]

Later on, the Belgian Railroad Society built forty Vierendeel bridges between 1932 and 1939, of which twenty-six such bridges were built across the Albert Canal. (Fig. 17). It was built in 1902. Today it has been replaced by a Vierendeel bridge of the bowstring type that was not designed by our hero.

About the bridge at La Louvière (Fig. 18), Vierendeel wrote:

> We ourselves realized in 1913 the longest straight beam for the time in reinforced concrete, that is, the La Louvière footbridge. Its span is 56 m, it supports a walkway 3 m wide, the

[17] Wickersheimer (1976, p. 57). See also Grimm (1901).

Fig. 17 The bridge of Avelghem over the Escaut spanning 42 m, 1902. This was the first bridge Vierendeel built (Vierendeel 1921, p. 383, Fig. 5). Public domain

Fig. 18 The bridge of La Louvière, 1913 (Vierendeel 1921, p. 383, Fig. 1). Public domain

test of which was 400 kg. per square meter; the two beams proved to be perfectly stable and rigid.[18]

The Soyer Bridge was built in reinforced concrete near Tournai in 1922 (Fig. 19). Sadly enough, it was demolished. This bridge is completely different from the other Vierendeel bridges and testifies to the spirit of research of its author. There is a similar one in Auckland, New Zealand, the Grafton Bridge, built in 1910.

The Scheepsdale Bridge, a swing bridge over the Bruges-Ostend canal (Fig. 20) was built in 1932, but sadly demolished in November 2009 despite the protests of the international community for construction history.

[18] Vierendeel (1921, p. 389): *Nous avons nous-même en 1913 réalisé la plus longue poutre droite jusqu'ici exécutée en béton armé, c'est celle de la passerelle de La Louvière (fig. I): sa portée est de 56 métres, elle supporte une passerelle de 3 métres de largeur dont l'épreuve était de 400 kilog. par métre carré; les deux poutres se sont montrées parfaitement stables et rigides.*

Fig. 19 The Soyer Bridge over the Escaut, Tournai (Postcard). Public domain

Fig. 20 Scheepsdale Bridge near Bruges. Photo Karel Roose (Creative Commons Attribution-Share Alike 3.0 Unported license)

5 The University of Louvain

In 1886, Vierendeel succeeded Victor Cousin as Professor at the University of Louvain, where he taught construction stability. During his university career, he advocated education given by active engineers. In the same optic, Vierendeel worked for the separation of the faculty of applied science from that of pure science.

In 1884, in his laboratory at the University, he subjected a steel sample to progressive traction and represented graphically the link between the traction and the elastic, permanent and plastic deformations of the sample. He determined in that way and with high precision the limits of elasticity, of compression and of rupture.

5.1 *Buckling Theory*

Another aspect of his theoretical work attracts our attention. Driven by his reflections on lattice beams, Vierendeel was interested in buckling in the various parts of the trusses designed by others and in his own. His first publication dedicated to buckling appeared in 1904. He picked the work back up on 29 August 1907, after the famous collapse in 50 s of the bridge over the Saint Laurent in Quebec, a collapse that occurred during construction and killed 86 workers. The suddenness of the catastrophe drove him to attribute it to buckling, which also explained other catastrophes (Vierendeel 1908). He then concentrated on that problem and dedicated nine articles to it.

In the lectures that he gave on that topic, he obtains by dynamical reasoning, a generalisation of Euler's formula for the critical load (Fig. 21). Our man loves polemics and begins his work by criticising Euler. That criticism that will lead to reactions from many engineers, whom he categorises as Eulerians opposed to anti-Eulerians. His critique turns on the way of posing the problem:

> The first theoretical attempt to solve the problem of the stability of the pieces loaded in compression was made around 1770 by the great geometer Euler (1707-1783); we are going explain his theory briefly. He supposes a piece AB (Fig 2) of uniform section which, first straight, had assumed a certain curvature as an effect of some kind of cause, and he looks for the minimum value of the forces P directed along its primitive axis and able to maintain the assumed curvature. We shall underline that there is in this way of posing the question a defect of reasoning called a petition of principle, to which we shall come back soon.[19]

We are facing here a kind of problem that occurred often in history. Vierendeel reproaches Euler for an argument surely used by the Eulerians, but not by Euler

[19] Vierendeel (1931, pp. 274–275): *La première tentative théorique de résoudre le problème de la stabilité des pièces chargées de bout a été faite vers 1770 par le grand géomètre Euler (1707–1783), nous allons en deux mots exposer sa théorie: il suppose une pièce AB (Fig 2) de section uniforme, qui, primitivement droite, a pris une certaine courbure sous l'effet d'une cause quelconque et il recherche la valeur minimum des forces P dirigées suivant son axe primitif et à même de maintenir l'inflexion prise; nous signalerons qu'il y a dans cette façon de poser le problème une défectuosité de raisonnement nommée pétition de principe sur laquelle nous reviendrons dans un instant.*

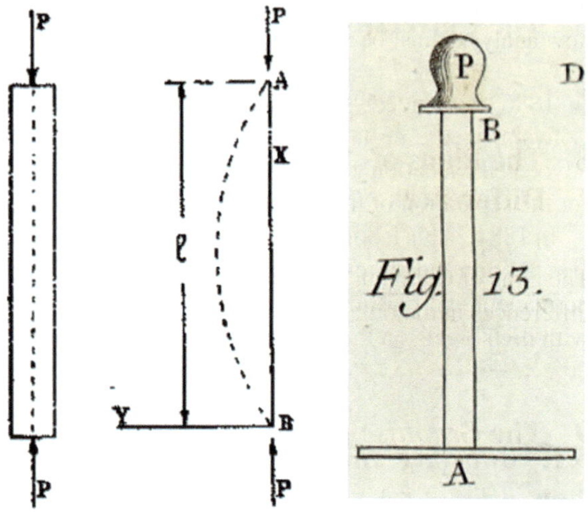

Fig. 21 A sample intended to test buckling and Vierendeel's drawing accompanying his theory (Vierendeel 1931, Fig. 2, p. 274) (left and center); Fig 13 from Euler's *Methodus inveniendi* (1744, Tab. IV) (right). Public domain

himself. Euler is looking for the load that produces an infinitesimal deflection of a column and he finds:

$$P = \pi^2 \frac{EI}{l^2}.$$

He does not speak of conserving that deflection. Euler, like Vierendeel, considers that phenomenon to be a dynamic one and does not attribute it to a load that was not well centered on the column or to an existing curvature. Vierendeel certainly did not go back to Euler's text.

Vierendeel estimates that:

1. Buckling is a dynamic bending resulting from the sudden loading of a piece loaded vertically.
2. The critical load of buckling is not influenced by an accidental eccentricity of the action of the load, or by a defect in the rectitude of the axis of the bar, or by any variation of its homogeneity; it is the expression of a very well-defined phenomenon, and distinct from the ordinary static bending.[20]

Vierendeel puts forward the existence of a particular phenomenon, again taking the problem of a kind of column but this time loaded somewhat eccentrically. He finds for the eccentricity d of the charge $d = f \cos \frac{l}{2}\sqrt{\frac{P}{EI}}$. This relation must still be valid when $d = 0$:

[20] Vierendeel (1931, p. 281) *1. Le flambage est une flexion dynamique résultant de la mise en charge brusque d'une pièce chargée de bout. 2. La charge critique de flambage n'est pas influencée par une excentricité accidentelle de l'action de la charge, ou par une défectuosité dans la rectitude de l'axe de la barre, ou par une variation quelconque dans son homogénéité; elle est l'expression d'un phénomène bien défini, distinct de la flexion statique ordinaire.*

$$f \cos \frac{l}{2}\sqrt{\frac{P}{EI}} = 0.$$

When the arc is less than $\frac{\pi}{2}$, the cos > 0 and f must be zero, there cannot be any bending but if

$$\frac{l}{2}\sqrt{\frac{P}{EI}} = \frac{\pi}{2} \text{ or } P = \frac{\pi^2 EI}{l^2}$$

then there is bending, and that is exactly Euler's formula. Vierendeel transforms it by inserting in it the modulus of elasticity and the fatigue $\frac{P}{\Omega}$, a formula that he found empirically:

$$E = A - B\frac{P}{\Omega}.$$

He also introduces $I = \Omega r^2$ and sets $\pi^2 = 10$. He obtains:

$$\frac{P}{\Omega} = \frac{10A}{10B + \left(\frac{l}{r}\right)^2}.$$

Then he will find that same formula in a dynamic way considering an initial displacement z and a section C whose center of gravity is at a distance y from the axis of the bar. He designates as M the bending moment acting on section C (Fig. 22).

The force that bends the bar is the difference between the effective load P and the portion $\frac{P}{g}\frac{d^2z}{dt^2}$ of this force employed to imply the acceleration to the mass $\frac{P}{g}$; the force

FIG. 3

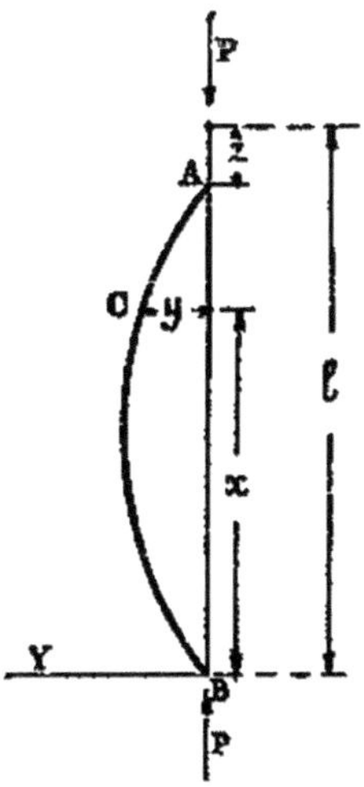

Fig. 22 Vierendeel's Fig. 3 (Vierendeel 1931). Public domain

acting in this way on the section C is thus $P - \frac{P}{g}\frac{d^2z}{dt^2}$ and its moment with respect to the center of gravity of the section C is:

$$M = \left(P - \frac{P}{g}\frac{d^2z}{dt^2}\right)y = Py\left(1 - \frac{1}{g}\frac{d^2z}{dt^2}\right).$$

He obtains this, writing that:

$$\frac{EI}{\rho} = -M,$$

where ρ is the ray of curvature. The equation of the elastica at all instants t is rigorously:

$$\frac{EI\frac{d^2y}{dx^2}}{\left[1 + \left(\frac{dy}{dx}\right)^2\right]^{\frac{3}{2}}} = -Py\left(1 - \frac{1}{g}\frac{d^2z}{dt^2}\right).$$

He simplifies, neglecting $\frac{dx}{dy}$ with respect to 1. The denominator is now 1 and the equation becomes

$$\frac{EI}{1 - \frac{1}{g}\frac{d^2z}{dt^2}}\frac{d^2y}{dx^2} = -Py.$$

This gives him

$$P = \frac{1}{1 - \frac{1}{g}\frac{d^2z}{dt^2}} \cdot \frac{\pi^2 EI}{l^2}.$$

He shows that the expression that he has found is a good approximation of this, and so is Euler's formula. His critique is thus essentially directed against the Eulerians and their static conception of buckling.

5.2 Aesthetics in Architecture

In order to vie for the *Prix du Roi* (King's Prize), which he would indeed win, Vierendeel wrote *La Construction architecturale en fonte, fer et acier* (1901) completed by a good number of particularly beautiful plates.

The author begins with an history of metal followed by what he calls an "*architectonographie*" of cast iron, iron, and steel. He then analyses some 52 achievements in these materials. Here is an example (Fig. 23):

Fig. 23 Paris, reading room of the old National Library, drawing by Vierendeel (1901, Pl. 16). Public domain

> When one presents himself rue de Richelieu, in Paris, and enters the lecture hall of the national Library, one suddenly and invincibly experiences a profound impression, a violent emotion shakes the being, one opens his eyes, raises his head, and looks in an admiring astonishment.
>
> One loses the perception of the world that surrounds him and unconsciously absorbs himself in the contemplation of the room that envelops him.
>
> One is really under a spell; the spell of a beauty never before seen. The moment is delicious. One perceives the silent harmony of lines, of surfaces and well pondered masses.
>
> A great room, slender columns, elegant arches, graceful cupolas, a clear and perfect unity, a light ideally sweet and calm, shimmering of Moorish azulejas, an ensemble noble and discretely rich, a suave composition ... One admires and never tires of admiring.[21]

Calculation is a way to set the imagination free, but one must be precise about what aesthetics means. Vierendeel does this by using a counter example:

[21] Vierendeel (1901, p. 99): *Quand on se présente rue de Richelieu, à Paris, et qu'on entre dans la salle de lecture de la Bibliothèque nationale, on éprouve brusquement et invinciblement une impression profonde, une commotion secoue l'être, on ouvre les yeux, on relève la tête et on regarde dans un étonnement admiratif.*
On perd la perception du monde qui vous entourent et on s'absorbe inconsciemment dans la contemplation de la salle qui vous enveloppe.
On est vraiment sous le charme; le charme d'une beauté non encore vue. Le moment est délicieux. On perçoit la silencieuse harmonie des lignes des surfaces et des masses bien pondérées.
Une grande salle, de sveltes colonnes, des arcs élégants, des coupoles gracieuses, une unité claire et parfaite, une lumière idéalement douce et calme, des chatoiements de Azulejas mauresques, un ensemble noble et discrètement riche, une composition suave ... On admire et on ne se lasse pas d'admirer.

> The Eiffel tower is a tower of thin lattice, and there lies the big aesthetic error that was made by its constructors: those thin lattices diminish the architectural cachet of the tower and justify in large part all the criticisms that were made by many architects and artists.
>
> It was necessary for that tower (just as for all metallic constructions in general) to have the great and sturdy ribs rising proudly from the bottom to the ridge and connected from distance to distance by powerful arcs.[22]

We might say that this is a matter of taste but Vierendeel immediately explains why there is an error of taste:

> Even the spirit most foreign to the things of construction would have understood this framework, he would have been struck by the simplicity of the canvas, he would have grasped its mechanism, he would have felt its unshakeable stability by virtue of this constructive intuition which is in everyone, from there a pleasure, an intimate satisfaction and above all this deep impression that iron is a powerful element and that it can produce powerful things.[23]

The error of taste lies in the fact that the architect does not make perceptible what he has learned thanks to the calculation of the stability proper to that particular edifice. The author then immediately explains his thought more precisely in order to avoid a too important influence of the mathematical formula.

> There is a certain school of scientific builders for which all that is rational is beautiful.
>
> The deplorable effect produced by the structural tie rods peremptorily demonstrates the fallacy of this proposal and this is fortunate, because if it were founded, goodbye to inspiration and feeling in architectural matters; all that would remain would be the cold scientific formula.[24]

---

[22] Vierendeel (1901, p. 786): *La tour Eiffel est une tour en mince treillis, c'est là la grosse faute esthétique qui a été commise par ses constructeurs; ce sont ces treillis qui diminuent le cachet architectural de la tour et justifie en grande partie toutes les critiques dont elle a été l'objet de la part de beaucoup d'architectes et d'artistes.*
Il fallait pour cette tour (tout comme pour toutes les constructions métalliques en général) le parti-pris des grandes et robustes de nervures montant fièrement de fond jusqu'au faîte et reliés de distance en distance par de puissantes arcatures.

[23] Vierendeel (1901, p. 786): *L'esprit le plus étranger aux choses de la construction eût compris cette ossature il eut été frappé de la simplicité du canevas, il en eut saisi le mécanisme, il en eut senti l'inébranlable stabilité en vertu de cette intuition constructive qui est en chacun de nous, de là un plaisir, une satisfaction intime et par dessus tout cette impression profonde que le fer est un élément puissant et qu'il peut produire de puissantes choses.*

[24] Vierendeel (1901, p. 790): *Il existe une certaine école de constructeurs scientifiques pour laquelle tout ce qui est rationnel est beau. L'effet déplorable produit par les tirants de charpente démontre péremptoirement la fausseté de cette proposition et c'est heureux, car si elle était fondée, adieu l'inspiration et le sentiment en matière architecturale, il ne resterait plus que la froide formule scientifique.*

6 The Evolution of the Vierendeel's Bridges and of His Criteria of Elegance

Vierendeel built some fourteen bridges for the society of Belgian railways, twenty-six on the Albert Canal and six on the Scheldt. He also built some in China and in Africa. Reproduced here are some pictures of bridges constructed by Vierendeel and collected by Karel Roose. Others are taken from Vierendeel's publications. We reproduce them in an almost chronological order, allowing us to see the evolution of Vierendeel towards more and more solid but also elegant bridges.

The simplicity of the definition given in the patent allows different types of bridges from completely straight transitioning to the bowstring type. Figure 24 shows roughly the evolution of Vierendeel trusses from a simple succession of rectangles to a bridge of the type bowstring, passing through an intermediate state with an upper span curved. Next, we shall show some examples of those various types of bridges, in order to show the evolution of their elegance.

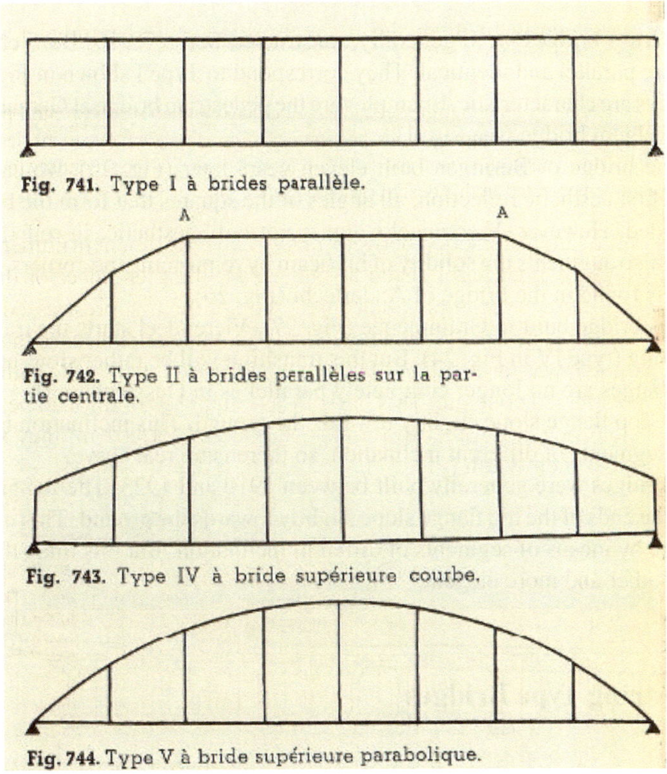

Fig. 24 Four types of Vierendeel trusses: Type 1, with parallel flanges; Type II, with parallel flanges in the central part; Type IV, with a curved upper chord; Type V, with a parabolic upper chord (Vierendeel 1936, p. 573). Public domain

Fig. 25 The bridge of Beringen built in 1909 and today no longer existing. Postcard. Public domain

Type I truss bridges were generally constructed before 1910. Both chords and verticals are parallel and identical. They correspond to Type I shown in Fig. 24. The squared bays are characteristic. Examples are the pedestrian bridge at Geraardsbergen and the Avelgem bridge (see Fig. 17).

With the bridge of Beringen built eleven years later (Fig. 25), we can see the result of a first aesthetic reflection, all angles of the squares that form the beam have been rounded. However, Vierendeel's aim is not only aesthetic; in rounding those angles, he also augments the solidity of his beam by reinforcing the corners. A similar treatment is found in the bridge of Anderlecht (Fig. 26).

With his bridge built in Grammenée (Fig. 27), Vierendeel starts the transition to the bowstring (type IV in Fig. 24). But this transition will be rather slow, as we shall see. The flanges are no longer completely parallel as in Ousselghem (Fig. 28). The ends of the top flange slope slightly towards the ground. This inclination is done by means of segments of different inclination, so there is no real curve.

These bridges were generally built between 1910 and 1923. The flanges are not parallel. The ends of the top flange slope slightly towards the ground. This inclination is still done by means of segments of different inclination. But this time, the form is much smoother and more elegant.

7 Bowstring Type Bridges

Most of these bridges were built between 1913 and 1940. In the previous sections we saw the beginnings of the evolution towards the form of those bridges. One may now distinguish four steps in that evolution.

Fig. 26 The bridge of Anderlecht, probably built before 1910, has the particularity of having the railway upon the beam, requiring a much more significant stability of that beam. Like the bridge of Beringen, it has reinforced corners. Photo: Nenea Hartia (Creative Commons Attribution-Share Alike 4.0 International license)

Fig. 27 1923–5, The bridge of Grammenée. This bridge still exists today; it was duplicated and the copy is identical to the original. Today it allows cars pass on one side and trains on the other. Photo: Karel Roose (Creative Commons Attribution ShareAlike 3.0 License)

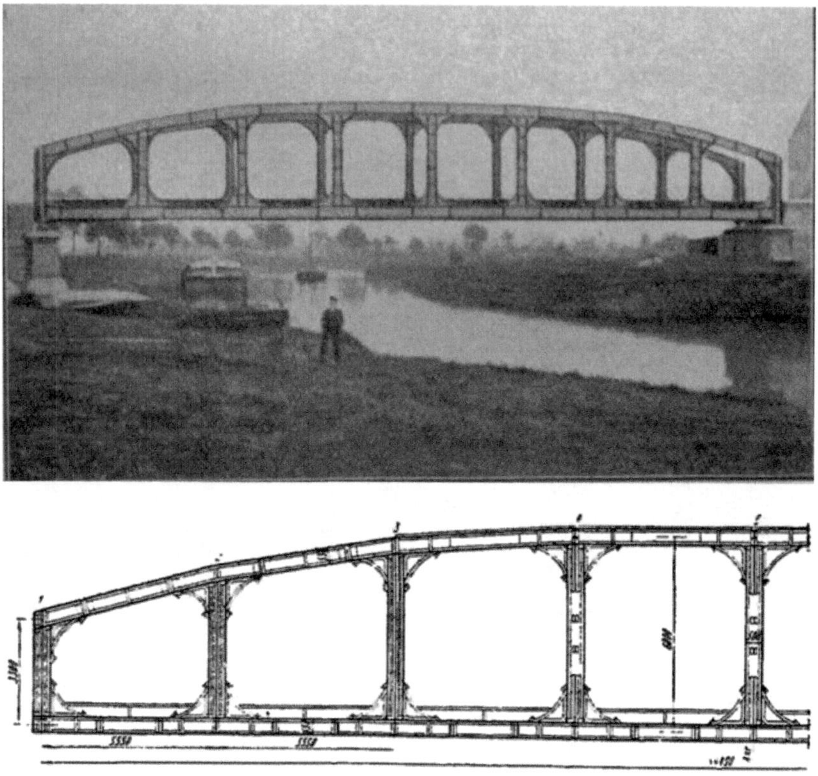

Fig. 28 The bridge of Ousselghem built in 1910 on the Leie (no longer in existence) (Vierendeel 1936, p. 574) (above); Its technical design by Vierendeel (p. 42) (below). Public domain

7.1 The Upper Flanges Are Formed of Two Parabolas Supposed to Be Parallel but Cannot Be

After having rounded the angles, Vierendeel changes the beams so that the upper span is formed of two parabolas, supposed to be parallel. However, this cannot be, because two parabolas are always similar and cannot therefore be parallel (Fig. 29, above). Nevertheless, Vierendeel, as well as many other architects, tried to give the illusion of such a parallelism. Vierendeel tries to have an upper flange in the shape of a parabola whose two upper and lower sides appear parallel.

The first idea that comes to mind to achieve this parallelism is to take the same parabola twice and shift one of the two upwards, for example.

The Val Benoît bridge (Fig. 30) Vierendeel's approach to the problem. On the right end of the bridge, we can see that the parabolas are not parallel anymore and that the crossing of the two parabolas with the horizontal beam is not very elegant. Vierendeel will turn that defect into an advantage.

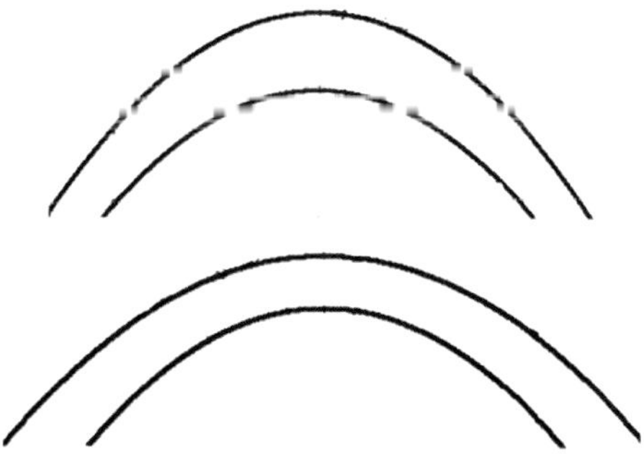

Fig. 29 Two identical parabolas but offset do not give the illusion of parallelism (above); By modifying the parameter of one of the two parabolas we can give the illusion of parallelism (below). Author

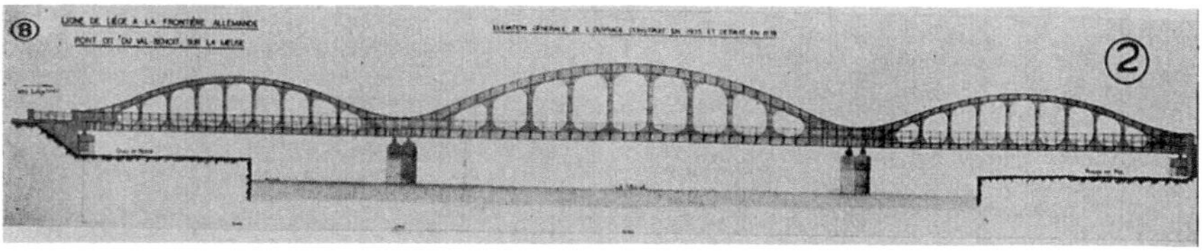

Fig. 30 The Val Benoît bridge, built in 1935, no longer existing today (Desprets 1951, p. 55). Public domain

Fig. 31 Bocholt Bridge 1934. Entirely welded, 55 m (Vierendeel 1936, Fig. 748, p. 574)

In the Bocholt Bridge (Fig. 31) the same problem arises here Vierendeel chooses to arrange it by securing the ends of the bridge into the quay wall.

7.2 *The Upper Flanges Are Formed of Two Parabolas Approximately Parallel and an Extension*

To give the illusion of a parallelism up to a certain distance from the summit, it is necessary for the upper parabola to have a slightly larger distance between foci

Fig. 32 The bridge of Gellik Lanaken was built between 1932 and 1939 and is still in use. Photo: EvilFreD (Creative Commons Attribution—Partage dans les Mêmes Conditions 3.0 (non transposée))

(Fig. 29, below). But the illusion is restricted to a certain region around the summit, which is the reason for the large feet that those bridges have. One can see clearly, in the Bridges of Angleur and Gellik (Fig. 32) how Vierendeel solves, or better, uses the difficulty. These were the first two bridges of this kind that were built.

These bridges are similar to the preceding ones, but they are completed by an extension that extends over the bank, assuring the bond with the bank. The extension is not an extension of the parabola. It is a new curve that fits more of less with the parabola but anchors the bridge to the bank. In those two cases they are decorated by one or two holes.

7.3 *The Upper Flanges Are Formed of Two Similar Parabolas*

The bridge of Hasselt (Fig. 33) was destroyed by bombs in 1940 and rebuilt in 1941. We clearly see the two similar parabolas that seem parallel but who move away from each other in order to form an extension which strengthens the support. But this time, the lower parabola is maintained and underlined. The other one is prolonged by a new curve in order to give weight to the support.

One sees the aesthetic game played by Vierendeel in his project of an iron bridge of thousand meters (Vierendeel 1896) (Fig. 34). He exploits the increasing distance

Fig. 33 The Hasselt Bridge over the Albert Canal. Postcard. Public domain

between the two parabolas to ballast its bridge by increasing the weight on the banks. Moreover, in that unique and unrealized project, he gives his bridge four resting points, two on each side.

In his design for the 1000-m bridge, Vierendeel emphasizes the lower parabola and raises the end of the upper one to make the support of the bridge heavier. He also embellishes it, as often, with a circular cut-out, such as in the bridge of Ile Monsin, in Herstal-Liège. We find the same idea on several bridges such as those in Bruges, Laeken and Eigenbilzen.

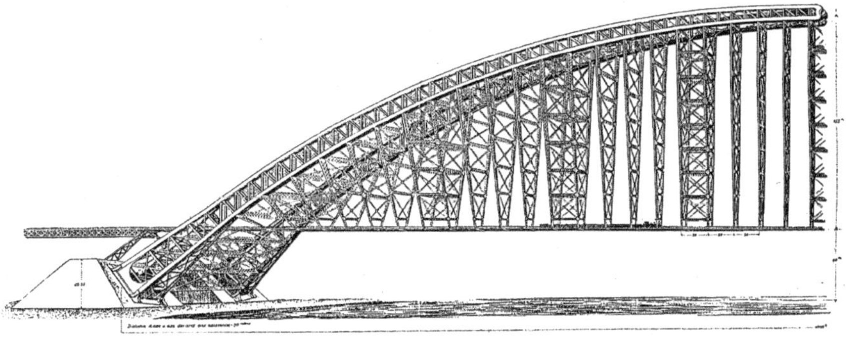

Le Pont en fer de mille mètres (VUE DE LA MOITIÉ DU PONT)
Projet de M. Vierendeel.

Fig. 34 Project of an iron bridge of 1000 m, an ambitious project that was never built (Vierendeel 1896, p. 90 and pl. 3). Public domain

Fig. 35 Ghlin bridge. Photo: Michielverbeek (Creative Commons Attribution-Share Alike 4.0 International license)

7.4 *The Upper and Lower Flanges Are Made of Two Similar Parabolas and an Intermediate Parabola is Drawn*

Vierendeel goes on playing the game of parabolas introducing in the upper flanges, between the two parabolas which mark the upper and lower edges of the flange, a third parabola "parallel" to the lower one, thus accentuating the clearance obtained by increasing the parameters of the successive parabolas, as seen in the bridges of Oelegem and Ghlin (Fig. 35).

For these two bridges, Vierendeel adds a third, central and perfect, parabola. The other two parabolas, upper and lower, are a bit distorted in order to give more weight to the supports and thus to give more stability to the bridge.

8 The Astonishing Posterity of Vierendeel's Truss

Already during Vierendeel's lifetime, his bridge was used to cross the Lualaba River in the Congo (the Bukama bridge, 1928); in later years his framework was also used to build towing boats and barges on that river. After his death, many articles, theses and books (such as those by Gustave Magnel,[25] for example) were written about Vierendeel's truss, but it seems that fewer bridges were built following his idea.

During his lifetime Vierendeel was a member of an international jury nominated by the Chinese government to study the bridges to build across the Yellow River.

[25] Magnel (1934) and Magnel (1935, vol. 1, pp. 240–242; vol. 2, pp. 193–257).

Table 1 Structures using Vierendeel trusses

Architect/Engineer	Structure, Opening date	Location
Louis Kahn/Komendant and Dubin	Salk Institute, 1965	La Jolla, California
NBBJ (Naramore, Bain, Brady, and Johanson)	Seattle-First National Bank, 1969	Seattle
Minoru Yamasaki/Fazlur Rahman Khan	World Trade Center, 1971/1973	New York
Foster & Parners/Ove Arup and Partners	Hong Kong Shanghai Bank, 1985	Hong Kong
John Portman/Weidlinger Associates, Inc	Marriot Marquis Hotel, 1985	New York
Foster and Partners/Ove Arup and Partners	Commerzbank headquarters, 1997	Frankfurt
Rafael Viñoly Architects	Kimmel Center for the Performing Arts, 2001	Philadelphia
Francisco Javier Manterola Armisen	Bridge over the river Ebro L.A.V. Zaragoza—Lérida, 2001	Osera de Ebro (Sp)
Snøhetta a.s., Norway	New Bibliotheca Alexandrina, 2001	Alexandria, Egypt
	Earthquake simulator laboratory	Berkeley
HDR, Inc. and others	Mike O'Callaghan–Pat Tillman Memorial Bridge, 2010	Colorado River between Arizona and Nevada

Long after his death, in 1989, Qian Lin Xi built nine bridges of 100 m each. Since that time things have changed, and many architects have used Vierendeel's truss in more recent constructions. Table 1 provides some examples.

Vierendeel retired from the Administration in 1927 but taught until he was 83 years old. His last year of teaching, 1935, would see the last revised version of his textbook on the stability of constructions. He died at Uccle on 8 November 1940.

9 Conclusion

I hope to have been able to underline the fact that Vierendeel made no claim that his truss was to be preferred to those using triangulation, but that the theory he gave for his truss without diagonals was more precise.

Vierendeel was an ardent defender of Plato's adage, "Beauty is the splendor of truth". This allows us to better understand the link that he establishes between the beautiful and the rational. He wrote:

> In reality the rational is not always beautiful, it is sometimes, but on the contrary the Beautiful is always rational, an expression which is similar to that of Plato: Beauty is the splendor of truth ... We must not hesitate, if the mathematical formula leads to an arrangement which is

not beautiful, we must complete it according to our eyes and this by virtue of the principle that the rational which is not beautiful is not completely rational; the beautiful alone is completely rational.[26]

That is the work or research both on stability (calculus) and aesthetics that I tried to trace observing the evolution of the bridges designed by Vierendeel, more precisely studying their upper flange. That work that aims at bringing to light and to emphasizing the lines along which the load descends, simplifying the structure, and in that way increasing both solidity and aesthetics.

Acknowledgements This chapter is composed of three unpublished lectures: Genova, 29 November 1999, "Arthur Vierendeel (1852-1940), Pour une architecture du fer"; Madrid, January 2003, International Conference on Construction History, "The Surprising Posterity of Arthur Vierendeel"; Brugg (Switzerland), 27 September 2008, "Arthur Vierendeel, l'homme aux nerfs d'acier".

References

Desprets, R. 1951. Le pont du Val-Benoit sur la Meuse, Considérations sur les dispositifs adoptés pour ses reconstructions successives. *L'ossature Métallique* 16 (2): 53–63.

Euler, Leonhard. 1744. Additamentum I de curvis elasticis. In *Methodus inveniendi lineas curvas maximi minimive proprietate gaudentes.* Lausanne and Genève.

Grimm, C. R. 1901. The Kinzua Viaduct of the Erie Railroad Company. *Transactions of A.S.C.E*, Vol. XLVI.

Lambin, Albert, and Paul Christophe. 1898. *Le Pont Vierendeel: rapport sur les essais jusqu'à la rupture effectués au parc de Tervueren, par M. Vierendeel, sur un pont métallique de 31m.50 de portée avec poutres à arcades de son système.* Brussels: Goemaere

Lederer, André. 1969. Vierendeel, Jules Arthur, chevalier. *Biographie nationale* XXXV: 730–742.

Lederer, André. 1979. Un grand constructeur, Arthur Vierendeel. *Colloque d'Histoire des sciences III, Louvain la Neuve, 17 mars 1977, Les sciences exactes et naturelles à l'Université de Louvain de 1835 à 1940.* Louvain: Editions Nauwelaerts.

Lederer, André. 1991. Arthur Vierendeel 1852–1940, sa vie, son œuvre. Lecture given on 15 March 1991.

Magnel, Gustave. 1934. *Le calcul pratique des poutres Vierendeel.* Ghent: Van Rysselberghe & Rombault.

Magnel, Gustave. 1935. *Cours de stabilité des constructions.* Ghent: Van Rysselberg & Rombaut.

Vierendeel, Arthur. 1896. *Le pont de mille mètre.* Louvain: A. Uystpruyst.

Vierendeel, Arthur. 1897. *Architecture du fer et de l'acier. Conférence donnée au Congrès international des Architectes réuni en août 1897, à Bruxelles.* Brussels and Paris: Lyon-Claesen.

Vierendeel, Arthur. 1900. Théorie Générale des Poutres Vierendeel. *Mémoires et Compte Rendu des Travaux de la Société des Ingénieurs Civils de France* 75: 163–218.

Vierendeel, Arthur. 1901. *La construction architecturale en fonte, fer et acier.* Louvain: Uystpruyst; Paris: Vve Dunod.

[26] Vierendeel (1901, p. 790): *En réalité le rationnel n'est pas toujours beau, il l'est quelquefois, mais par contre le Beau est toujours rationnel, formule qui se rapproche de celle de Platon: Le Beau est la splendeur du vrai... il ne faut cependant pas hésiter, si la formule mathématique conduit à une disposition qui n'est pas belle il faut la compléter au gré de notre œil et ce en vertu du principe que le rationnel qui n'est pas beau n'est pas le rationnel complet; le beau seul est le rationnel complet.*

Vierendeel, Arthur. 1908. Effondrement du pont de Quebec. *Annales des travaux publics de Belgique*. Brussels.

Vierendeel, Arthur. 1921. *Esquisse d'une Histoire de la Technique*, Collection Lovanium. Paris-Brussels: Vromant.

Vierendeel, Arthur. 1931. *Cours de Stabilité des Constructions*. Tome I, *Résistance des matériaux*, 5th ed. Louvain: Uystpruyst; Paris: Dunod.

Vierendeel, Arthur. 1936. Poutre Vierendeel et Poutres Triangulées. *L'ossature Métallique* 5: 572–576.

Wickersheimer, David J. 1976. The Vierendeel. *Journal of the Society of Architectural Historians* 35 (1): 54–60.

Chapter 20
Pier Luigi Nervi's Idea of the *Verità delle Strutture*

Beauty is truth, truth beauty, that is all ye know on earth, and all ye need to know
John Keats (1795–1821)

1 Introduction

In his beautiful introduction to Pier Luigi Nervi's lecture, Sergio Poretti writes:

> Historical structural rationalism, firmly anchored to the consideration of the coherence of form, structure and construction as an indispensable value, rediscovers in Nervi a master. It is the search for this coherence that is the true fundament of his architecture: which translates into that principle of economy which, while guiding the designer's ethics and measuring the correctness of the solution, at the same time founds the aesthetics of the work as emulation of the secret harmony of the universe. (Trentin and Trombetti 2010, x)

In what follows, I'll try to support and develop this idea. But before entering the subject, I had like to recall just one biographical fact. Pier Luigi Nervi attended the Civil Engineering School of Bologna from which he graduated in 1913. His teachers of mathematics were Salvatore Pincherle and Federigo Enriques, both well-known mathematicians, the first through his work on functional analysis and the second for his work on algebraic geometry. Although Nervi never, as far as I know, published his calculations, it is clear that he was perfectly familiar with the mathematics he needed and that they are the background of all quotations I'll present in the following.

The paper will be devised in two parts. In the first one, I'll show how Nervi correlates the ideas of efficiency, economy, and beauty in some of his publications. In the second part, I'll show how he applies them in some of his works. Occasionally, I'll show how he considered history.

I shall make frequent use of a wonderful book recently published: *La lezione di Pier Luigi Nervi* (Trentin and Trombetti 2010). It reproduces, in Italian and in English, the notes Roberto Einaudi took when from December 1959 to March 1960 he followed the lectures given by Nervi at the University of Rome.

P. Radelet-de Grave, *Between Mechanics and Architecture*, Mathematics and the Built Environment 8, https://doi.org/10.1007/978-3-031-73530-1_20

2 How Does Nervi Express His Ideas on Structures in His Writings?

2.1 Structure and Beauty

The first expression of that relation is given by Nervi in his Preface to the book *The Works of Pier Luigi Nervi*, published in 1957:

> The first step in any design was the search for the most suitable structural solution technically and economically; then followed patient and passionate work: the detailing and calculation of the various structural elements so as to refine the form and thus meet the static and structural requirements.
>
> In examining and deciding upon the most suitable structural scheme, the most varied solutions gradually emerged. In so doing, I was always anxious to arrive at an independent decision free from existing aesthetic theories as well as from solutions I or others had already discovered (Nervi 1957, p. vi)

Nervi also emphasizes his views on structures in the first lecture he gave on 1 December 1959 at the faculty of architecture Valle Giulia, of the Università degli studi di Roma "la Sapienza". Roberto Einaudi reports them as follows:

> Structural aesthetics are a natural aesthetics. Structure is a serious thing. The more you obey natural laws, the more beautiful the result. The more the airplane became useful the more it became beautiful. The form and structure of the ships and airplanes of today evolved from the demands of nature: the same should be true of architecture. It is ridiculous to make a complicated structure that spans 15-30 feet, but in large constructions it is the structure that must be the first design consideration. Man can joke, nature cannot. Structural aesthetics do not permit fantasy. (Trentin and Trombetti 2010, p. 64)

On 18 December he adds:

> An architecture with beauty as its only goal is an empty architecture. An architecture that is born through need (structure, plumbing, spaces, etc.) will evolve its own beauty ... It is wrong to approach beauty for beauty's sake, to be preoccupied with the end result first, instead of realizing what is behind a true, meaningful beauty. Never, never, take a form and calculate it to support the loads. (Trentin and Trombetti 2010, p. 74)

2.2 Structure and Economy

After having linked beauty and efficiency of the structure, Nervi goes on linking technology, beauty and economy of the structure, in the lecture given on 22 December: "The Machine Hall in Paris, built in the 1890s, is pure technology yet the result is of great beauty. Technology has no other goal than economy" (Trentin and Trombetti 2010, p. 78). And some weeks later:

> Concrete can span enormous distances, and it can be distorted to fantastic shapes. But this is to ignore statics which guides the resulting form of concrete, just as it guides the forms of other materials. In reinforced concrete one uses the form that is most useful, most economical ...

> The static fact is the guide to aesthetics in architecture ... Once the choice of which structural system to use has been made, the guide is statics, economy. (Trentin and Trombetti 2010, p. 88)

I'll say more on economy when looking at particular constructions.

3 How Does Nervi Express Those Ideas in His Structures?

3.1 The Flow of Forces

> Simplicity in transmitting loads is best. (Trentin and Trombetti 2010, p. 138)
>
> Frequently one can see beautiful joints where all the reactions and stresses are clearly visible (Nervi 1957, p. 142)

The Giovanni Berta Stadium project (today the Artemio Franchi Stadium) in Florence, one of the first Nervi realized, was obtained after a competition project. In a lecture given on 19 January, Nervi says about that construction: "Here the problem was primarily of statics, where I balanced the forces resulting from the cantilever roof" (Trentin and Trombetti 2010, p. 98) (Fig. 1).

Figure 2 explains how it was done combining horizontal traction exercised by the wall and a pressure given by the support acting diagonally. Nervi explains it more carefully:

Fig. 1 Pier Luigi Nervi, the Giovanni Berta Stadium during construction. Public domain

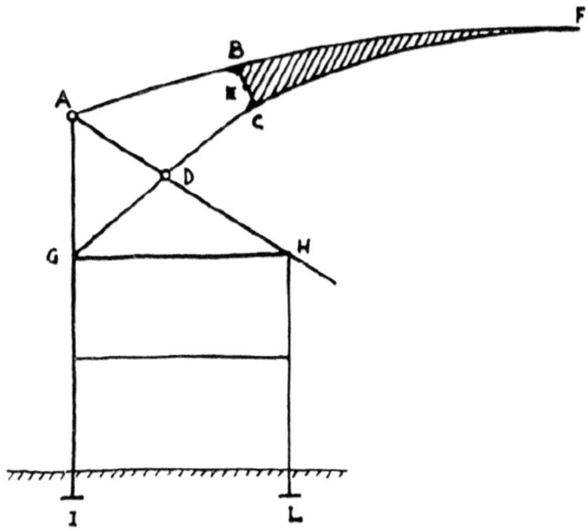

Fig. 2 Illustration accompanying the text of Nervi (Michelucci 1932, p. 112). Public domain

> The BCF cantilever is engaged to the elastic structure with two ABCD hinges, creating in E, the centre of gravity of the interlocking section, statically determined reactions. The procedures of the elastic arch with two hinges make it possible to determine the reactions in A and in D, and therefore the stresses, statically determined, in AD, AG, GD, DH, GH, and the reactions in I and L. The shape of the assembly: ABCD elastic frame and cantilever, such that the moments in C and B of the rods AB and DC are minimal and practically zero. This condition corresponds to an aesthetic sense of balance and naturalness in addition to the constructive economy. The HL abutment corresponds approximately to the center of gravity of the cover so that there are no negative reactions in I. This is also a condition of balance and economy.[1]

Figure 3 shows the amplitude and aesthetics of that free space. This space allows people to move easily to reach their place.

Some 30 years later, between 1957 and 1959, Pier Luigi Nervi and his son Antonio built the Flaminio stadium, for the Olympic games in Rome. The terraces are much larger than the previous ones and the cantilever roof is much longer. It has been recently replaced by a membrane structure in order to cover the entire stadium.

> Statics is a fountain of ideas that will never betray you. (Trentin and Trombetti 2010, p. 142)

[1] Michelucci (1932, p. 112): *La mensola piena BCF si innesta sulla struttura elastica a due cerniere ABCD originando in E, baricentro della sezione d'incastro, reazioni staticamente determinate.*
I procedimenti dell'arco elastico a due cerniere permettono di determinare le reazioni in A e in D, e quindi le sollecitazioni, staticamente determinate, in AD, AG, GD, DH, GH, e le reazioni in I e L. La forma del complesso: telaio elastico ABCD e mensola, è tale che i momenti in C e in B delle aste AB e DC sono minimi e praticamente zero.
Questa condizione corrisponde a un senso estetico di equilibrio e di naturalezza oltre alla economia costruttiva.
Il pilastro HL corrisponde approssimativamente al baricentro della copertura cosicché non si hanno in I reazioni negative. Condizione anche questa di equilibrio ed economia.

Fig. 3 Free space left under the stepped rows of seats (Michelucci 1932, p. 112) Public domain

3.2 *Pier Luigi Nervi and History*

Forces flow down through Nervi's terraces exactly as forces flow down through the buttresses of Gothic cathedrals. The cantilever modifies the problem, but, as Nervi explains it, he solved that problem in the same spirit, drawing those new forces down to the ground.

Nervi has a strong stand against history, which that surprised me first because it contradicts remarks similar or even identical to the one I just made about Gothic cathedrals. In a lecture given on 22 March, he says: "Simplicity in transmitting loads is best" (Trentin and Trombetti 2010, p. 138).

However, his terminology is changing and he is gradually replacing the word "history" with "culture". In a lecture given on 23 February, he says:

> Culture is the worst possible enemy the architect can have when he has to start a project … One must obliterate the past; the problem is to design a building for the present, not for the past. (Trentin and Trombetti 2010, p. 122)

I interpret his remarks as follows. Historical monuments that are standing for centuries or millennia are useful to the architect as they allow him to sharpen their view and familiarize themselves with stability. On the other hand, it is useless for them to know who built, why and precisely when. What he wrote in his introduction to *The Works of Pier Luigi Nervi* explains that clearly:

> It is more difficult to define a good structural organism. Let us look at the great architectural creations of the past such as the Gothic cathedrals or the great domes of the Renaissance. We are filled with unbounded admiration for their designers, whose intelligence, sensitivity, intuition and genius made them such immensely complex and magnificent buildings.
>
> It could be argued that the main characteristic of these works, which makes them still speak to us across the centuries, lay in the fact that their builders realised their design in a form which shows an intuitive understanding of the laws of equilibrium and gravity normally alien to human understanding. Seen thus all great buildings of the past were successful just in so far as they were the result of the complex mental process inherent in grasping their statical

basis-something accessible only to a powerful and exceptional brain. As a direct result of the revolution in building which started in the middle of the last century, architecture today not only has immense potentialities but is also endangered by structural theories, steel and reinforced concrete. Complex theories and mathematical formulae were discovered which enabled forces to be evaluated even in highly complicated structures. The importance of this advance can hardly be overestimated (Nervi 1957, p. vi)

In his lectures, he says:

For me there is a great pleasure to see a beautifully carved capital a handsome Gothic vaulting ... Perhaps culture is valuable in permitting the enjoyment of a poem or a melody, but what difference does it make to this enjoyment knowing who wrote or who composed it? (Trentin and Trombetti 2010, p. 124)

3.3 *Vaults*

Between 1935 and 1942, Nervi built, together with Arturo Danusso, eight hangars for airplanes in Orvieto, Orbetello, and Torre del Lago. The structure of the first hangar in Orbetello, built in 1935, was made with timber formwork and concrete poured on site. During the war, between 1939 and 1941, because of the high cost of materials, Nervi found that method much too expensive and changed it for the second hangars. The structure remained almost the same and Nervi maintained the catenary form for the transversal section of the hangar but he then used pre-cast concrete components and for the next series of hangars built in the same years, he used prefabricated concrete components, initiating a new way of building. Nervi often insists on the fact that those prefabricated concrete components were made on the site and that the forms could be reused many times.

There is a great economic advantage in a form well built that can easily be used again and again (Trentin and Trombetti 2010, p. 134).

3.4 *Domes*

Both for aesthetic and structural reasons, the parabola is better than a circle for the elevation of a vault (Trentin and Trombetti 2010, p. 148).

The aim of what follows is to question that remark. In 1943 Nervi made the design for a large vaulted hall having a clear span 990 feet for which he intended to use prefabricated reinforced concrete units. There appeared the famous corrugations in the celling, intended not only to be decorative but principally to absorb sound. He will apply that idea for the first time in 1957 in the Palazzetto dello sport, built in Rome for the Olympic games. In a lesson given on 4 March, Nervi says:

In the Palazzetto dello sport, the acoustics were terrible: a dome concentrates sound. The ribbing of the Palazzetto did absolutely nothing to reduce the bad acoustics: it gave a substantially continuous cupola. The problem was to avoid the concentration of sound. One can go to

> a discontinuous structure; a structure which is not smooth, but is undulated; not just slightly but radically. (Trentin and Trombetti 2010, p. 134)

The idea was to extend the surface and thus the absorption of sound. At the same time, corrugations deflected the sound in different directions. In the next lecture, talking this time of the Palazzo dello sport of the Eur (1959), he adds: "Sound enters the corrugations of the roof and is absorbed. Corrugations have an acoustical value as well as structural value" (Trentin and Trombetti 2010, p. 134).

We may add that such a ceiling also allows the passage of ducts for air conditioning and electrical cables. Its aesthetic effect is indisputable. In a lecture given on 5 February, we read:

> Cupola Palazzetto dello sport; ... The dome of reinforced concrete was difficult to build, and very costly. What if we could build it in pieces as in a mosaic? We could produce precast pieces that would be joined by poured concrete. But how could this be done for a dome, a three-dimensional surface? (Trentin and Trombetti 2010, p. 106)

With those two buildings – the Palazzetto dello sport in 1957 and the Palazzo dello Sport of the Eur between 1957 and 1959—Nervi develops his idea of prefabricated concrete units, executed on site with reusable forms, adding thin steel armatures into the mixture of sand, gravel and a high proportion of cement of the thin concrete units. That is what he calls *ferrocemento*. The units had to be thin and light, in order for men to be able to carry them. This allowed him to build both surbased cupolas, the first of 60 m and the second of 100 m in diameter. This procedure took the name *Sistema Nervi*. All vertical sections of both domes are catenaries and all horizontal sections are circles, both ensuring static stability.

3.5 Catenaries

> In a cupola all horizontal sections are circles. As long as these horizontal sections remain circular, there is no need to follow the curve of pressures. The dome can take any form desirable in section. If we were to follow the curve of pressures, the domes then we would have a parabola dome. But why don't we ever see one? Obviously we can violate this law. This is because when we have a circular plan section we have continuous rings that redistribute the forces equally on their whole perimeter. If the section of the dome corresponds to the curve of pressures the parallel rings don't enter into function. But if instead the dome does not follow the curve of pressures, then all points on a ring will either tend to go towards the center or towards the outside and the parallel ring will either be in compression or tension. (Trentin and Trombetti 2010, p. 140)

Nervi built many domes and made designs for some more domes; I counted 28, but they were not all realized. The vertical cross-section of 26 of them is a catenary or as he said corresponds to the line of pressure. From the two exceptions, one is surely a parabola. It is a competition project for a Sports Centre in Vienna. The form of the other one is difficult to establish with the documents I have but it is surely not a catenary.

Galileo gave a nice property of the parabola that permits to detect them easily. You have to trace a tangent and find its intersection point A with the axe of symmetry of the curve. You also trace the perpendicular to that axis issued from the point of contact of the tangent and the curve. Point B will be the intersection of that perpendicular with the axis of symmetry. If the curve is a parabola, its summit must be at the middle point of the segment joining the points A and B. I checked that way, for instance the section of the first hangar Nervi built in 1935 in Orbetello and it is surely not a parabola. To detect if the curve is a catenary, you just reverse the picture of the dome and check if it coincides with the curve of a light chain hanging between any two points of the curve, as shown in Fig. 4.

As the vertical sections of almost all his domes are catenaries, it is rather strange that Nervi never writes the word catenary, as in a lecture given on 19 January: "In a compression system the arch must follow the curve of compression, not the designer's whim. The architect must have a static comprehension" (Trentin and Trombetti 2010, p. 88). On the contrary, as we can see in the quote above, he mentions the parabola. But, as he says in a lecture, "Calculations can always change and withstand slight differences" (Trentin and Trombetti 2010, p. 82). He could assume the difference between catenary and parabola to be very small. That is true when they are not too large, but they are large in the samples we checked.

Let me take one more sample linked to the use of catenaries and showing his incredible sense of statics. Nervi made some designs for the Centre National des Industries et Techniques (1958), the first building to be built at the Défense in Paris. He explains that:

> The beams hold the springing points of the three arches together. The structural calculation is not based on the theory of shell construction, but on the principle of the cross-vaulting (Nervi 1957, p. 136)

However, this advice was not followed by his colleagues, as he explains:

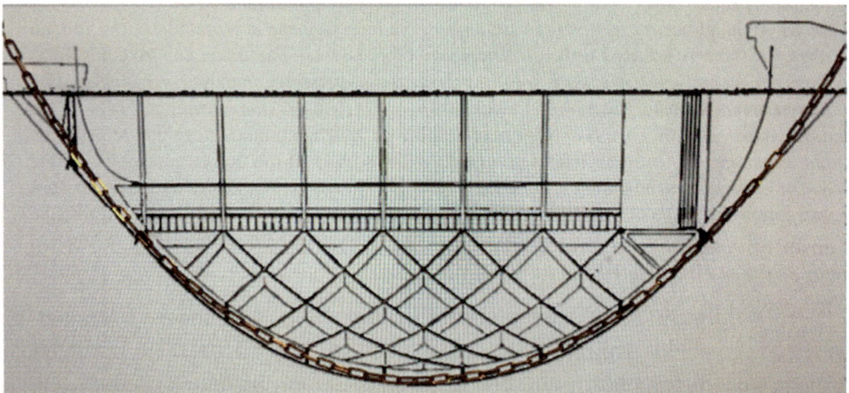

Fig. 4 The inverted section of one of Nervi's hangars and the corresponding catenary. Author

> In the 200m (660′) span triangular building in Paris, they have tried to transform what is essentially a groined vault into a thin shell. The result was that the supports became bulky and the ribbing remained weak. (Trentin and Trombetti 2010, p. 140)

In fact, his proposal seems to have been discarded and in my opinion his design looks much better than what was realized.

4 The Aula Paulo VI or Paul VI Audience Hall (1971)

Considered by many to be Nervi's masterpiece, the hall is completely hidden and lost in the dense urban fabric of the Vatican City. It is impossible to get a general view of the outside, except from the sky, and that view is not very impressive. The inside, on the contrary, is amazing and studying its structure, I realized that the Aula was the outcome of all the experiments Nervi had carried on in all the constructions I mentioned and in many others since the hangars of 1935.[2]

> The only way to learn is through experience and a gradual increase in scope. (Trentin and Trombetti 2010, p. 136)

This is exactly what Nervi did his whole life long. The Aula is a large dome; its interior is covered with decorative undulations, as we saw in the Palazzetto dello sport, in order to avoid the converging of sound, but also giving the possibility of hiding various ducts. The maximal vertical sections of the dome are all catenaries (Fig. 5). New difficulties lie in the fact that the lateral walls are converging toward the large stage as in a movie hall. The Pope sits on that stage. The effect is surely grand and impressive. There was a long discussion about the form and decoration of both lateral windows of the Aula and Einaudi tells us that Nervi was particularly proud of the result shown in Fig. 6 and the unusual ribs make me think of Nervi's remark: "In the rose window of the Sainte Chapelle in Paris, the aesthetics and the statical lines cannot be separated. This comes from a profound knowledge of structure."

5 Conclusion

I had a look at Nervi's work from an a priori rather narrow point of view, flow of forces and catenaries. But they represent a large part of the ways Nervi chose to ensure the stability of his buildings. As he said: "Humanity has created its own Master, the natural laws" (Trentin and Trombetti 2010, p. 120).

Flow of forces and catenaries are highly representative of those laws of nature in architecture and in order to obtain an esthetical result the designer ought to let those laws being visible in his works and ought to avoid useless decorations. That is Nervi's message.

---

[2] For more details, see Norcia (2015).

Fig. 5 Pier Luigi Nervi, Aula Paolo VI, View of the covering and the elliptical window (Redazione ArchiDiAP). Public domain

Fig. 6 Pier Luigi Nervi, Aula Paolo VI, the ellipsoidal window from the exterior (Redazione ArchiDiAP). Public domain

I based my research not only on Nervi's constructions but also on some of his designs. I made large use of the lectures he gave in 1959–60 in Rome where, I am convinced, he aimed principally at making that important message clear to his students. And I shall finish with a last quotation explaining the title of this chapter:

> What is beauty in architecture? It does not begin with the relation between solid and voids. It begins with a truth more fundamental: the Structure is the truth. An architecture of solids and voids without truthful structure has no meaning (Trentin and Trombetti 2010, p. 139).

Acknowledgements A previous version of this chapter appeared in vol. 2 of Building Knowledge, Constructing Histories (*Proceedings of the 6th International Congress on Construction History* (6ICCH 2018), July 9–13, 2018, Brussels, Belgium), Ine Wouters, Stephanie Voorde, Inge Bertels,

Bernard Espion, Krista Jonge, Denis Zastavni, eds. We are grateful to the editors and to Routledge/Taylor & Francis for permission to use the original material, which has been revised for this present publication.

References

Antonucci, Micaela, Annalisa Trentin, and Tomaso Trombetti, eds. 2014. *Pier Luigi Nervi. Gli stadi per il calcio*. Bologna: Bononia University Press.

Halpern, Allison B., David P. Billington, and Sigrid Adriaenssens. 2013. The ribbed floor slab systems of Pier Luigi Nervi. In *Proceedings of the International Association for Shell and Spatial Structures (IASS) Symposium 2013*, 1–7. Wroclaw, Poland.

Michelucci, Giovanni. 1932. "Lo stadio "Giovanni Berta" in Firenze, dell'ing. Pier Luigi Nervi. http://www.sba.unifi.it/upload/scienzetecnologiche/nervi/architettura_1932_03.pdf.

Nervi, Pier Luigi. 1933. Arte e tecnica del costruire. *Quadrante*, n. 2, June, 28.

Nervi, Pier Luigi. 1945. *Scienza o arte del costruire? Caracteristiche e possibilità del cemento armato*. Roma: la Bussola. Rpt. Torino: CittàStudi Edizioni, 2014.

Nervi, Pier Luigi. 1955. *Costruire correttamente. Caratteristiche e possibilità delle strutture cementizie armate*. Milano: Hoepli; Reed. Milano: Revue, 1965.

Nervi, Pier Luigi. 1957. *The works of Pier Luigi Nervi*. New York: Frederick A. Praeger Inc.

Norcia, Elisabetta. 2015. Aula delle udienze pontificie. https://archidiap.com/opera/aula-delle-udienze-pontificie/.

Trentin, Annalisa, and Tomaso Trombetti, eds. 2010. *La lezione di Pier Luigi Nervi*. Milano-Torino: Bruno Mondadori, Ricerca.

Chapter 21
Gustave Magnel (1889–1955): A Life in Structural Engineering

1 Beginnings

Born in Essen, near Anvers, on 15 September 1889, he died in Ghent on 5 July 1955 (Fig. 1). By the time Gustave Magnel earned his degree in civil engineering at the Rijksuniversiteit in Ghent in 1912, he had already written seven papers on very different topics, including metal wear due to abrasive materials, metal welding, ultra-violet rays, and some theoretical problems of mathematics.[1] These early papers testify to the wide range of his interests. During the first world war he joined the D. G. Somerville Enterprises in London. He was rapidly promoted to chief engineer, but Prof. François Keelhoff called him back to Ghent University in 1919.

The nine writings from this period focus on the themes of elasticity and the calculation of arches. We can see in one of these papers, entitled "Méthode rapide de calcul des lignes d'influence d'arcs prismatiques surbaissés à deux encastrements" (Magnel 1917), the concept that will recur from then on: the mathematics of the mathematician are not the same as the mathematics of the engineer. He wittily explains his idea in a note on the teaching reform in engineering schools: "A force may be a vector for the mathematician, but for the engineer it is something that pushes or pulls".[2]

In 1948 he explained his concept more thoroughly to the science class of the Royal Academy of Belgium, of which he has been a member since 1946:

[1] The author's earlier studies of Magnel appeared as "Gustave Magnel (1889–1955), a scientific biography", in *Essays in the history of the theory of structures. In honour of Jacques Heyman*, ed. Santiago Huerta, pp. 389–427 (Madrid: Instituto Juan de Herrera, 2005) and "Cemento precompresso e sperimentazione, il ruolo di Gustave Magnel", in *Teoria e pratica del costruire: saperi, strumenti, modelli. Esperienze didattiche e di ricerca a confront / Theory and practice of construction: knowledge, means, models. Didactic and research experiences*, Atti del Seminario Internazionale, Ravenna, 27–29 October 2005, ed. Giovanni Mochi, vol. 2, pp. 849–857 (Bologna: Edizioni Moderna).

[2] Magnel (1950a): *Een kracht kan wel een vector zijn voor een wiskundige, maar voor de ingenieur is het iets dat duwt of trekt.*

P. Radelet-de Grave, *Between Mechanics and Architecture*, Mathematics and the Built Environment 8, https://doi.org/10.1007/978-3-031-73530-1_21

Fig. 1 Gustave Paul Robert Magnel beneath a picture of the Sclayn bridge, one of his major works. SECO Archives

> Unfortunately, in those days, only theoretical hyperstatic systems created by mathematicians were available. But mathematicians rarely worry about the time required to solve a numerical problem:
>
> Treatises from this period said that to solve a *n* times hyperstatic system, it was sufficient to write down *n* times that the derivative of the deformation strain with respect to each of the hyperstatic unknown quantity is equal to zero, and solve the *n* equations of *n* unknown quantities such obtained. They sometimes added that this calculation could easily be done with matrices. Let's imagine the poor engineers in front of a problem with, let's say, 20 unknown quantities; it would have taken them months only to solve the equations, and months prior to that only to write them down (Magnel 1948a)

Magnel had a constant desire to provide professional engineers with simple mathematical tools that could ease computations. Nonetheless, he was perfectly aware of the developments of the theoretical mathematics that he used to teach during his classes. In 1923 he became first secretary and then president of the committee of the Belgian Association for Standardization (later the Belgian Institute for Standardization) in charge of establishing the rules for the calculation and building of works in reinforced concrete. Already in his first report he pointed out the question of the shrinkage in concrete that generates tension in the reinforcement: a problem which would grow more complex with the emergence of prestressing technology.

2 First Academic Experiences

Upon his return to Ghent University, Magnel became head of the laboratory of material strength and assistant in the preparatory classes for engineering schools. He thus linked, from the very first years of his career, experimental research with teaching.

It is important to notice that all his activities were always strongly connected to the professional practice of the building engineer.

Magnel was immediately convinced of the interest in reinforced concrete, and in 1922 a ministerial decision authorized him to teach an open course entitled "La pratique du calcul du béton armé" in the Faculty of Sciences, a course that had never been taught in Belgium before. That same title appears again on volume I of his famous textbook (Magnel 1923). The sections of the volume, whose first edition was 160 pages long, are dedicated to the practice of organic calculation, the Belgian requirements for reinforced concrete works (this part being the consequence of his work at the Belgian Association for Standardization), the requirements for cement suppliers, the theory of reinforced concrete, and finally, the scientific dosage of concrete. This volume would be re-published four times and would grow from 160 to 515 pages without the topics being changed. In the last edition of 1949, the various parts of the volume are always the same. Nevertheless, Magnel progressively included in his volume the new experiments and gains in knowledge in the different topics, whereas the new fields are progressively approached in the three other volumes of the complete *Pratique du Calcul du Béton Armé*. The second volume is dedicated to the analysis of continuous beams, hyperstatic plates, and the composed warped bending. The third one is about the problem of the calculation of arches, a problem that we also find in numerous papers, and finally the fourth volume is about prestressed concrete. This fourth volume came out some time after the first three.

These volumes, some of which are completed by volumes of illustrations, are meant for a student readership and they gather all his personal expertise. To place the student in front of experience, of reality, is also to place him in front of the real difficulties of calculation. Calculation is no longer only theoretical, and is thus no longer modelled. It is important that the student not lose him time in painstaking and useless calculations:

> To permit the realisation of the essential point of our program, it is important that the student lose as little time possible in the material execution of calculation, and is thus able to concentrate his attention on the technical questions that require a serious effort of thought. So that the student can concentrate all of his intelligence on various technical problems, there is a way to avoid tiresome calculations whose execution do not teach him anything, as for example, those necessary to solve, by successive approximations, the third-degree equations to which leads the study of the parts subject to compound flexion. The aim of the course in reinforced concrete is not that of learning to solve a third-degree numeric equation. This is a secondary question that can be avoided with the correct use of abaci and tables that have already been tested in well-run studies.
>
> It is thus important to put within the student's reach a book that expounds a series of rapid methods for solving frequent problems which would otherwise require long calculations that do not teach anything to those who undertake them.
>
> There exist numerous books that expound practical methods of calculating reinforced concrete. If we had believed it necessary to publish a new one, it is because in the first place it will facilitate the task from the pedagogical point of view; in the second place because numerous methods are our own and have already demonstrated their usefulness in the frequent use that has been made of them in numerous studies; and finally, because

few authors among those we know have made the effort to examine in detail the degree of approximation of their methods.[3]

All of Magnel's courses were published and each enjoyed numerous revised editions. From this point of view, he remained faithful to a principle that he himself stated in 1905: "Professors must publish their courses and recommend the book of another author. The students must be forbidden to take notes; when they take notes they cannot reflect".[4]

3 The Laboratory for Reinforced Concrete: Teaching, Research and Experiments

Magnel's courses, largely illustrated with examples from real practice, show that an engineer does not become a professional only by reading books.

As early as 1923 Magnel had thought about creating a laboratory for reinforced concrete. The fact that the desire to create a laboratory arose at the same time as the beginning of his course shows that Magnel thought of these two methods of teaching—theoretical and experimental—as complementary. He stated this clearly in his introduction to his *Pratique du Calcul du Béton Armé*:

The course in the Practice of the Calculation of Reinforced Concrete, which we will be giving at Ghent University for the first time this winter, would be sterile if it were not accompanied

[3] Magnel (1923): *Afin de nous permettre de réaliser ce point capital de notre programme, il importe que l'étudiant perde le moins de temps possible à l'exécution matérielle des calculs et puisse ainsi porter toute son attention sur les questions techniques qui demandent un sérieux effort de l'esprit.*
Afin que l'étudiant puisse concentrer toute son intelligence sur les divers problèmes techniques, il y a lieu d'éviter de lui faire faire des calculs pénibles dont l'exécution ne lui apprendrait rien, tels p. ex. ceux nécessaires pour résoudre par approximations successives l'équation du 3eme degré à laquelle conduit l'étude des pièces soumises à flexion composée. Le but d'un cours de béton armé n'est pas d'apprendre à résoudre une équation numérique du 3eme degré. C'est là une question accessoire, qu'il y a lieu d'éviter par l'emploi judicieux d'abaques et de tableaux, qui ont fait leurs preuves dans des bureaux d'études bien conduits.
Il importe donc de mettre entre les mains des étudiants un livre exposant une série de méthodes rapides pour la résolution des problèmes courants qui, autrement, nécessite de longs calculs qui n'apprennent d'ailleurs rien à celui qui les entreprend.
Il existe de nombreux livres exposant des méthodes pratiques de calcul du béton armé. Si nous avons cru devoir en publier un nouveau, c'est d'abord parce que cela nous facilitera la tâche au point de vue pédagogique, ensuite parce qu'un assez grand nombre de nos méthodes nous sont personnelles et ont déjà prouvé leur utilité dans l'emploi courant qu'en font plusieurs bureaux d'études, enfin parce que bien peu d'auteurs, parmi ceux que nous connaissons, se donnent la peine d'examiner en détail le degré d'approximation de leurs méthodes.

[4] Magnel: *De professoren moeten ofwel hun cursussen uitgeven of het boek van een andere auteur aanbevelen. Het zou bijna de studenten moeten verboden worden nota's te nemen; als men een cursus moet opnemen kan man niet nadenken.*

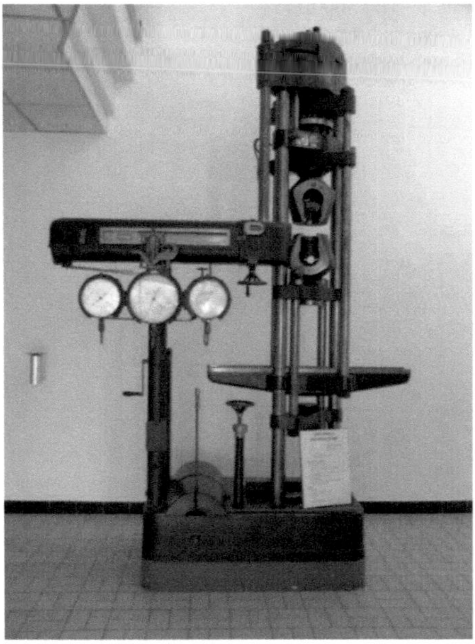

Fig. 2 The 30 ton Losenhausen universal testing machine (1926). Photo: author

> by practical exercises designed to give students the opportunity to apply to concrete cases the methods taught.[5]

Unfortunately, the Ministry of Education did not grant his requests. He was finally listened to by Edouard Anseele, the Minister for Public Works, who in 1926 allowed him to install his laboratory in the old Hotel Flandria,[6] near Ghent's Sint-Pieters train station. The numerous instruments of the laboratory included "a complete installation for normal cement acceptance tests with machines for tensile and compression tests" and, most importantly, "various measuring devices and in particular six precision devices for measuring concrete shrinkage".[7] Magnel was a forerunner in the study of the shrinkage of concrete, as well as the phenomenon of creep in both concrete and steel. In the laboratory there was also a universal testing machine (Fig. 2) and a 300-ton press (Fig. 3).[8]

---

[5] Magnel (1923): *Le cours de Pratique du Calcul du Béton Armé, que nous donnerons à l'Université de Gand pour la première fois cet hiver, serait stérile s'il n'était accompagné d'exercices pratiques destinés à donner aux élèves l'occasion d'appliquer à des cas concrets les méthodes enseignées.*

[6] This building is still extant.

[7] Magnel: *Une installation complète pour essais normaux de réception de ciment avec machines pour essais de tractions et de compression … Divers appareils de mesure et notamment six appareils de précision pour la mesure des retraits du béton.*

[8] These machines are now found in the atrium of the Magnel Laboratory in Zwijnaarde. I thank Prof. Luc.

Taerwe for his hospitality and for the noteworthy documentation that he put at my disposal: Taerwe (1999, 2004, 2005).

Fig. 3 The 300-ton Losenhausen press (1926). Photo: author

Magnel's reasoning is identical to that of a practicing engineer. He is concerned about lost time, and about what is gained. He takes up this idea once again in his *Cours de stabilité des constructions*, adding there the important idea of the order of magnitude:

> The young engineer often imagines that it is possible to make "exact" calculations. This will only be true if the data and the theories applied are correct and if we have considered all the circumstances of the problems to be solved. We do not try to do our calculations "exactly", that is, we do not lose time doing complex calculations where there are faster methods that give a maximum error of 5%. What good is it, starting from necessarily false data, to apply a rigorous theory that gives, by admitting the accuracy of the data, an error of only 1%, while the inaccuracy of the data still prevents us from knowing to about 20% what the tensions in the construction studied are? Any theory that gives a solution from the data to within 5% seems to us acceptable and even preferable to a theory that is supposed to be exact or that requires considerable time and exposes the engineer to making numerous material errors.[9]

[9] Magnel (1935): *Le jeune ingénieur s'imagine souvent qu'il est possible de faire des calculs « exacts». Ce ne serait vrai que si les données et les théories appliquées étaient exactes et si nous tenions compte de toutes les circonstances des problèmes à résoudre. Ne cherchons pas à faire nos calculs « exactement» c.à.d. ne perdons pas notre temps à faire des calculs complexes là où il existe des méthodes plus rapides donnant une erreur maximum de 5%. A quoi bon, en partant de données forcément fausses, appliquer une théorie rigoureuse donnant, en admettant l'exactitude des données, une erreur de 1% seulement, alors que l'inexactitude des données nous empêche quand même de savoir à 20% près, quelles sont les tensions dans la construction étudiée. Toute théorie donnant à partir des données une solution à 5% près nous paraît acceptable et même préférable à une théorie soit disant exacte là où cette dernière demande un temps considérable et expose l'ingénieur à commettre des erreurs matérielles nombreuses.*

Fig. 4 Interior of the laboratory of Sint Pieternieuwstraat (Magnel 1953a)

On the other hand, the laboratory had to address its own needs. Our professor, an experimental researcher, thus had to come into contact with the private companies in order to obtain contracts for quality control. It was in this way that that laboratory received the commission to control the quality of all the masonry concrete produced in Belgium. It might seem surprising that Magnel was also the president of the professional organization of producers of concrete, but he was always careful to avoid interfering in matters of business. He only ever engaged in the role of consulting engineer, and his laboratory only carried out quality control.

The spaces in the Hotel Flandria soon proved to be too small for the laboratory. Luckily, in 1934 the University of Ghent designed a large building to be built in St. Pietersnieuwstraat, the Technicum Complex, destined to host technical laboratories such as those of hydraulics and reinforced concrete (Figs. 4 and 5). Even though only a little reinforced concrete was used in the building's construction, Magnel collaborated with Jean Norbert Cloquet in its design. The laboratory moved there in 1937, and Magnel worked there until his death.

This laboratory, which later on moved to Zwijnarde, still exists and is now named after its founder. Magnel assigned other tasks to it in addition to teaching. In fact, the laboratory was able to do some tests for third persons, for a certain fee:

1° testing the received materials: cements, bricks, tiles, stones, concrete and concrete products.

2° tests aiming at determining the best qualities of some materials in order to produce concrete on big building sites

3° tests aiming at estimating the value – or non-value – of some new products such as special steels, waterproofers, etc.[10]

[10] Magnel (1939, pp. 2026–2027): *1° Essais de réception des matériaux de construction: ciments, briques, tuiles, pierres naturelles, béton et produits en béton; 2° Essais tendant à déterminer le*

Fig. 5 Buckling test in the laboratory of *Sint Pieternieuwstraat* (Magnel 1953a). Public domain

In addition, the laboratory was in charge of the control of all Belgian slag cement plants, Magnel being the president of their professional union.

4 Later Academic Career

In the meantime, Magnel's academic career proceeded regularly. He became a teacher in 1927 and then associate professor in 1932. Two years later he published a book on the practical calculation of Vierendeel beams (Magnel 1934) (Fig. 6). Here again we recognize his desire to make engineers' calculations problems easier. This time he no longer deals with reinforced concrete but with special metallic beams that do not have diagonals, according to Vierendeel's patent and are thus non-trussed beams. Unlike Magnel's previous books, this one would not be re-published. Metallic beams are substituted at least in Magnel's concerns, by prestressed concrete beams.

In 1935 Magnel published the first volume of another work in four volumes: the *Cours de stabilité des constructions* (Magnel 1935), which would be re-published several times, as were the volumes on the practice of the calculation of reinforced concrete. In 1937 he became a full professor.

---

meilleur parti à tirer de certains matériaux en vue de la fabrication du béton sur les grands chantiers; 3° Essais tendant à établir la valeur – ou la non valeur – de produits nouveaux tels qu'armatures spéciales, hydrofuges, système de hourdis creux, etc.

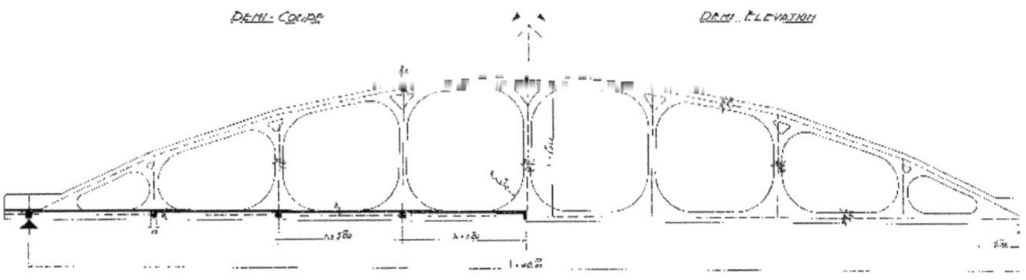

Fig. 6 Vierendeel beam from Magnel's book *Le calcul pratique des poutres Vierendeel* (Magnel 1934). Public domain

5 The *SECO*: Building Sites Security and Insurance

On 21 December 1934, Magnel, together with Eugene François, Professor at the Université Libre de Bruxelles, founded the SECURITAS bureau, predecessor of the present SECO bureau, following the example of the SECURITAS office which had been founded in Paris in 1927. The society's first charter explains its goals:

> The goal of the bureau is to perform technical control of the projects, the computations, and the execution of the works in building industry and civil engineering, in exchange of a fee. The control is aimed at reducing the risk of accidents and at making possible the widest expansion of insurance policies at reduced costs. …
>
> The control service is optional, even for the members of the bureau. No one is required to submit; the bureau does not provide design and drawing services nor does it conduct construction operations. It only gives advice and suggests cautiousness on the basis of scientific knowledge and expertise. Therefore, it does not take on pecuniary responsibilities (SECO Archives).

The presentation booklet published by SECURITAS in December 1935 (Fig. 7) points out that Belgian law was extremely strict towards architects and contractors, who therefore deserve to be provided with good insurance:

> Art. 1382: Any human act which causes damages to anyone compels those responsible for the act to repair it.
>
> Art. 1792: If a construction built with forfeit fees is ruined in total or in part due to poor construction techniques, or even to poor ground, architects and contractors are responsible for ten years.
>
> Art. 2270: After ten years, architects and contractors are relieved from the guarantee of the buildings that they achieved or directed.

This booklet also includes information about the composition of the bureau:

The "Bureau Securitas" stems from a large association comprising architects, contractors, consultant engineers, joined by a technical committee formed of professors who teach stability and construction in all our universities and at the military school.

Magnel was a consultant for the Bureau and his laboratory carried out tests on structural components, especially when innovative procedures were involved, thus making possible the insurance of the building process.

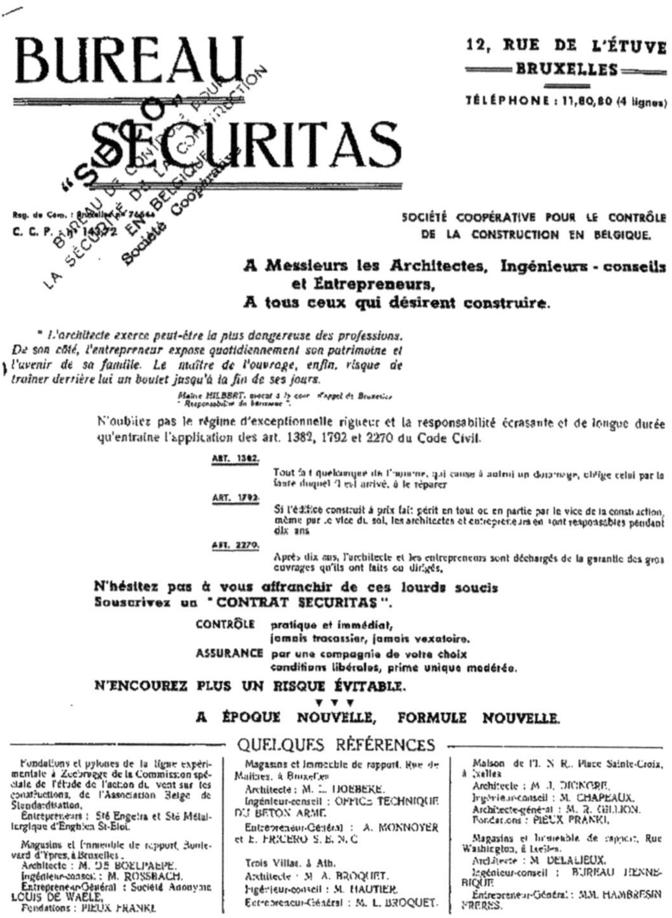

BUREAU SECURITAS

12, RUE DE L'ÉTUVE
BRUXELLES
TÉLÉPHONE : 11,80,80 (4 lignes)

Reg. de Com. : Bruxelles n° 7464
C. C. P. n° 14392

SOCIÉTÉ COOPÉRATIVE POUR LE CONTRÔLE DE LA CONSTRUCTION EN BELGIQUE.

A Messieurs les Architectes, Ingénieurs - conseils et Entrepreneurs,
A tous ceux qui désirent construire.

" L'architecte exerce peut-être la plus dangereuse des professions. De son côté, l'entrepreneur expose quotidiennement son patrimoine et l'avenir de sa famille. Le maître de l'ouvrage, enfin, risque de traîner derrière lui un boulet jusqu'à la fin de ses jours.

Maître HILBERT, avocat à la cour d'appel de Bruxelles " Responsabilité du bâtisseur ".

N'oubliez pas le régime d'exceptionnelle rigueur et la responsabilité écrasante et de longue durée qu'entraine l'application des art. 1382, 1792 et 2270 du Code Civil.

ART. 1382.
Tout fait quelconque de l'homme, qui cause à autrui un dommage, oblige celui par la faute duquel il est arrivé, à le réparer

ART. 1792.
Si l'édifice construit à prix fait périt en tout ou en partie par le vice de la construction, même par le vice du sol, les architectes et entrepreneurs en sont responsables pendant dix ans

ART. 2270.
Après dix ans, l'architecte et les entrepreneurs sont déchargés de la garantie des gros ouvrages qu'ils ont faits ou dirigés.

N'hésitez pas à vous affranchir de ces lourds soucis
Souscrivez un " CONTRAT SECURITAS ".

CONTRÔLE pratique et immédiat, jamais tracassier, jamais vexatoire.
ASSURANCE par une compagnie de votre choix conditions libérales, prime unique modérée.

N'ENCOUREZ PLUS UN RISQUE ÉVITABLE.

A ÉPOQUE NOUVELLE, FORMULE NOUVELLE.

QUELQUES RÉFÉRENCES

Fondations et pylones de la ligne expérimentale à Zeebrugge de la Commission spéciale de l'étude de l'action du vent sur les constructions, de l'Association Belge de Standardisation.
Entrepreneurs : Sté Engeira et Sté Métallurgique d'Enghien St-Eloi.

Magasins et Immeuble de rapport, Boulevard d'Ypres, à Bruxelles.
Architecte : M. DE BOELPAEPE.
Ingénieur-conseil : M. ROSSBACH.
Entrepreneur-Général : Société Anonyme LOUIS DE WAELE.
Fondations : PIEUX FRANKI.

Magasins et Immeuble de rapport, Rue de Malines, à Bruxelles.
Architecte : M. L. HOEBEKE.
Ingénieur-conseil : OFFICE TECHNIQUE DU BETON ARME.
Entrepreneur-Général : A. MONNOYER et E. FRICERO S. E. N. C.

Trois Villas, à Ath.
Architecte : M. A. BROQUET.
Ingénieur-conseil : M. HAUTIER.
Entrepreneur-Général : M. L. BROQUET.

Maison de l'I. N. R., Place Sainte-Croix, à Ixelles.
Architecte : M. J. DIONGRE.
Ingénieur-conseil : M. CHAPEAUX.
Architecte-général : M. R. GILLION.
Fondations : PIEUX FRANKI.

Magasins et Immeuble de rapport, Rue Washington, à Ixelles.
Architecte : M. DELALIEUX.
Ingénieur-conseil : BUREAU HENNEBIQUE.
Entrepreneur-Général : MM. HAMBRESIN FRERES.

Fig. 7 First presentation of *SECURITAS* office (SECO Archives)

The Cancer Institute of the Public Health System in Brussels, the Koekelberg Basilica, and the Nord-Midi railroad junction were among the first works submitted to the control of the SECURITAS Bureau. The latter two works are among Magnel's achievements.

The full assembly of the SECURITAS Bureau gathered on 12 May 1938, including 125 industrialists, engineers, professors, architects and contractors, for a total of 555 social parts among the 588 that had been distributed. They decide to change the name of the bureau: SECURITAS became SECO, control bureau for construction security in Belgium.

During the war, while the construction business was struck by a severe recession, Magnel dedicated himself intensely to the SECO bureau, which was challenged with the problems of building sites insurance in war time. In 1941 insurance companies asked for a 20% raise in insurance premiums. A long negotiation began between

Magnel, SECO, insurers and re-insurers. Magnel tried to set up a handy classification of the different risks that may occur, in order to make it easier to set the insurance fees for architects and contractors: "The SECO bureau is working on the preparation of a code thanks to which a risk factor could be assigned to the different kinds of risk that insurance companies are confronted with" (SECO Archives). The classification would be finished in 1942, and included six categories:

1st category: Houses, Condominiums, Private buildings for offices, apartments or theatres. Except for: those of more than 14 levels or higher than 80 m. or having spans longer than 25 m:

(a) less than 4 levels; less than 25 m high; spans shorter than 15 m;
(b) more than 4 levels; higher than 25 m; spans longer than 15 m.

2nd Category: Public buildings except for: towers, churches, and buildings higher than 80 m or having spans longer than 25 m:

(a) less than 25 m high, with spans shorter than 15 m;
(b) towers, churches and buildings between 25 and 80 m high with spans shorter than 25 m.

3rd Category: Permanent industrial plants: less than 60 m high (80 m for chimneys) with spans shorter than 25 m:

(a) everything, except for towers, reservoirs, silos, chimneys, and special buildings;
(b) towers, reservoirs, silos, chimneys, and special buildings.

4th Category: Large structures, except when pier walls or supporting walls or abutments' support is more than 10 m deep, and when bridges have spans longer than 50 m:

(a) Any kind of bridge;
(b) (1) Piers or supporting walls bordering public roads that are not provided with machines for maintenance; (2) Piers in ports and similar works that are meant to support maintenance machines;
(c) sewage, aqueducts and similar works.

5th Category: Transformations, with exception in height and identical span to the corresponding transformed category:

A—involving vitals of extant constructions:

(a) when the ten-year guarantee amount is based on the sole cost of the transformation works;
(b) when the ten-year guarantee amount is based on the cost of the transformation works in addition to the estimated value of the building or the parts of the building that are being transformed;

B—not involving vitals of extant constructions:

(a) when the ten years guarantee amount is based on the sole cost of the transformation works;

(b) when the ten years guarantee amount is based on the cost of the transformation works in addition to the estimated value of the building or the parts of the building that are being transformed.

6th Category: Works that are not included in the first five categories, i.e.:

1st Category: more than 14 levels or higher than 80 m or spans longer than 25 m;

2nd Category: higher than 80 m (100 m) and spans longer than 25 m;

3rd Category: temporary industrial plants, constructions higher than 60 m (80 m for chimneys), with spans longer than 25 m;

4th Category: piers, supporting walls, abutments etc. ... more than 10 m and spans more than 50 m;

5th Category: same as the corresponding category that includes the transformed construction, and all special works, in particular soldered constructions, prestressed concrete, innovative designs, etc. (SECO Archives).

This classification is actually the first step towards the present-day agreement procedure for contractors. Today's procedure differs from this previous one, since the agreement is nowadays given to the contractor himself, while here the risk is estimated for every single construction process. Furthermore, the present agreement has two clauses: a financial clause and a building-type clause. Out of the financial clause, the contractor is allowed, according to his category, to build works whose cost is not superior to a fixed amount. The building-type clause is similar, in essence, to Magnel's concept but because of the technical progress it was obviously largely completed during later years.

The foundation of the SECO bureau by Gustave Magnel and Eugène François is a great achievement whose originality must not be underestimated. A deep knowledge of the construction world and business was necessary to dare to start up such a project.

6 The Prestressing Technology

The first patent related to reinforced concrete, a skeleton of iron immersed in concrete for the construction of tubs and basins, was deposited by Joseph Monier in 1867. The first bridge in reinforced concrete was constructed in 1887, and in 1899 Arthur Vierendeel extended his patent of the truss without diagonals to the same truss covered in concrete.[11] In 1920 reinforced concrete "invaded" all sorts of constructions. Magnel had become familiar with reinforced concrete while working for Somerville during the first world war; Somerville taught this subject to young graduates of English universities, where it was neglected (Fig. 8).

In spite of his interest in reinforced concrete, in 1924 Magnel urged caution:

[11] For a detailed study of Vierendeel and his work, see "Arthur Vierendeel: The Man with Iron Nerves Who Invented the Truss Without Diagonals", Chap. 19 in this present volume.

Fig. 8 An advertisement for the D. G. Somerville contracting company featuring reinforced concrete. Public domain

The engineers of the old continent – and above all the young ones – must avoid allowing themselves be hypnotized by a theory that, which, whatever it is, is always only an approximation and often a rather crude approximation, especially when it comes to a material as complex as reinforced concrete. In this area, above all, let us not ask the theory for more accuracy than it can reasonably be expected to give and, let us only admit these results after experience has shown them to be acceptable.[12]

Already two years earlier, in 1922, he had defined the limits:

There exists a fundamental difference between constructions in concrete and those in metals. Reinforced concrete is incapable of resisting, without rupturing, great stresses, while steel is a ductile material, stretching more than 20% before breaking by tension; furthermore, concrete can be crushed by compression, which is not the case with steel.[13]

---

[12] Magnel: *Les ingénieurs du vieux continent -et surtout les jeunes- doivent éviter de se laisser hypnotiser par une théorie qui, quelle qu'elle soit n'est jamais qu'une approximation et souvent une approximation assez grossière, surtout quand il s'agit d'une matière aussi complexe que le béton armé. Dans ce domaine surtout ne demandons pas à la théorie plus d'exactitude qu'elle n'en saurait raisonnablement donner et, n'admettons ces résultats qu'après que l'expérience nous les aura montrés acceptables.*

[13] Magnel (1922): *Il existe une différence fondamentale entre les constructions en béton et, les constructions métalliques. Le béton armé est incapable de prendre, sans se fissurer, de grandes*

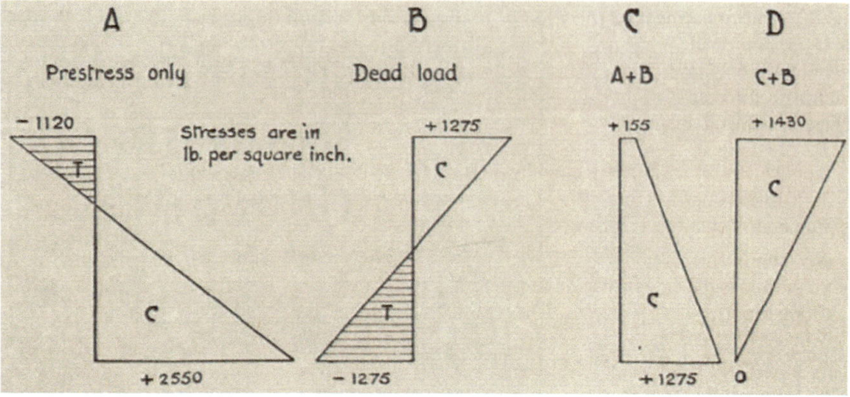

Fig. 9 Diagram of the stresses on concrete. *A*, prestress; *B*, the weight of the concrete (dead load); *C*, the sum of the weight and the prestress; *D*, the sum of the weight, the prestress, and an external load, considered in this case as equal to the weight (Magnel 1945–1946). Public domain

It was precisely these limits that led Eugène Freyssinet (1879–1962), shortly after 1928, to start off in the direction of prestressed concrete. In his book *Une révolution dans les techniques du béton* (1936), Freyssinet writes:

> I will show that these highly compact concretes, associated with steels with high elastic limit employed under tension, constitute an absolutely new means of construction, which presents itself in conditions very different from those of ordinary reinforced concrete, both from the point of view of mechanical properties as well as cost.[14]

Later, Magnel explained clearly that: "Concrete is a material whose resistance to compression is three times its resistance to tension".[15] In reinforced concrete, the concrete works in compression, and in case of tension, it is the reinforcing that works; this may cause cracking in the concrete, but the whole can resist within certain limits. The idea of prestressed concrete consists in inserting the steel reinforcing tendons with a high coefficient of elasticity into cavities in the concrete, then subjecting to very high degrees of tension. In this way the concrete is compressed and can support tensions that are less than the sum of the prestress and the limit of the tensile resistance of the concrete. Magnel summarized this in the diagram shown in Fig. 9.

Many engineers had had the idea of prestressing concrete, and Magnel did not attribute the idea either to himself or to Freyssinet. Instead, he thought that credit

déformations, tandis que l'acier est une matière ductile, s'allongeant de plus de 20% avant de se rompre par traction; en outre, le béton peut s'écraser par compression, ce qui n'est pas le cas de l'acier.

[14] Freyssinet (1936, p. 3): *Je ferai voir que ces bétons à haute compacité, associés à des aciers à haute limite élastique employés sous tension, constituent un moyen de construction absolument nouveau, qui se présente dans des conditions très différentes de celles du béton armé ordinaire, tant au point de vue des propriétés mécaniques qu'à celui des prix de revient.*

[15] Magnel (1953, p. 3): *Le béton est une matière dont la résistance à la compression est de très multiple fois la résistance à la traction.*

was due to Freyssinet for the idea of using steel of a high degree of elasticity, an idea which by itself makes the prestressing possible. Magnel had immediately intuited the advantage of this idea, and in 1940 published a research article on the problem of cracking in concrete and the use of steel with a high elastic limit steel (Magnel 1940). It must also be said that this idea responded to the criticisms that Magnel had expressed of reinforced concrete.

In 1950 Magnel wrote:

> Frequently enough, some people – not much aware of the last building techniques evolution – ask us if we are the inventors of prestressed concrete.
>
> We always answer that this new material does not have a precise father; the concept is more than 50 years old and we could easily mention several patents from the end of last century, related to prestressed concrete. (Magnel 1950b)

It was also during the second world war, in 1941, that Magnel began to study prestressed concrete, five years after the publication of Freyssinet's book introducing the notion of prior stressing or pre-stressing. Magnel actually recognized the anteriority of the discovery by Freyssinet, with whom he had a very friendly relationship. He openly and honestly acknowledged Freyssinet's merit:

> It is nowadays commonplace to say that Mr Freyssinet is the inventor of prestressed concrete, nonetheless it is not exact. The concept of prestressing is as old as reinforced concrete itself, and there are several patents related to this topic even though Mr Freyssinet did not care. The success of this French scientist lies in the demonstration that it is impossible to realize a pre-stressing which won't weaken in time without using steels of a very high elastic resistance, able to support tensions of about 80 or 100 kg/mm^2. We are not pointing out such facts in order to reduce Mr Freyssinet's awesome merits, but to restore historical truth. It is Mr Freyssinet who made prestressing possible in every day engineering art. He invented a technique and some equipment that are commonly used in France. (Magnel 1949)

In order to understand Freyssinet's merit and then appreciate Magnel's, we have to know better what prestressing is. Magnel explained the facts in a very entertaining way (Fig. 10):

> All of you already practiced prestressing when you moved a bunch of books from one shelf to another. You had to press the books with your hands, but when you wanted to move too many books in once, it did not work; you were no longer able to apply a compression strong enough to balance the tensile force due to the books weight loading the beam formed by the books themselves. There is an ultimate solution for those who do not respect books: you can drill a hole through the books and insert a metallic tie with two fine nuts; by tying the nuts the books can be compressed and the "beam" can be put on two supports. The tie action is the "pre-stress", i.e., a stress which is applied prior the upcoming of the stresses of external loads (Magnel 1948b)

The concept of reinforced concrete is therefore intended to compensate for the inadequate tensile strength by having steel bars in tension. The disadvantage is that when steel is stressed, structures hold securely but the concrete cracks. Let's listen to the teacher while he explains what to do:

> Imagine that we could put a concrete beam – eventually a not reinforced one – under a general compression equal in every direction up to, for instance, 150 kg/cm^2.

Fig. 10 Magnel explaining prestressing technique

> If we then apply an exterior load to this beam, it will not crack as long as in each of its facets this load does not create a tension superior to the value of 150 kg/cm^2 plus the value of the concrete tensile strength.
>
> This is what prestressed concrete is all about. The general compression of 150 kg/cm^2 is the *pre-stressing* (Magnel 1953b, pp. 7–8)

However, the shrinkage of concrete while drying, and creep, an irreversible plastic deformation either of the concrete or the steel, add some complexity to this simple idea, progressively weakening the prestressing. This is the object of Freyssinet's research. Let's read how he explains, in 1936, the concept of prestressing:

> A good use of high resistance materials is thus due: for steel, to the maintenance of the concrete lengthening inside handy usual limits; for concrete, to maintaining the total tension stress below a limit much inferior to the tensile strength of concrete.
>
> Theoretically, this double condition can be fulfilled by using steel, not to support the tensions and elongations that concrete cannot bear, but using it in such a way that it permanently stresses the concrete in the opposite direction than the loads stresses do, i.e., compression in the tensed parts, and reciprocally tension in the compressed parts.
>
> This result may be attained by tensing the steel before casting the concrete. In this case, reinforcement extremities are temporarily anchored and tensed by a known stress, thanks to jacks supported by some abutments. We then wait for the stiffening of the concrete, enough for the reinforcement extremities anchorage, and then the jacks are released. The reinforcement, which tends to shrink applies to the concrete a compression equal to the tension. Another procedure consists in applying a tension to some steel ties by grabbing support on already hardened concrete.
>
> In this way we create in the steel-concrete organism a double system of permanent identical stresses, but oriented in the opposite direction than the one that appears in usual reinforced concrete, due to the well-known bad consequences of shrinkage. The first result of creating a prestressing is thus the elimination of the harmful effects of shrinkage. For a long time, in many countries, practical means to achieve this goal have been searched for. But all these efforts failed and were abandoned, except for some applications to pipe fretting. (Freyssinet 1936)

7 The Magnel-Blaton Patent

We now have good ingredients to produce prestressed concrete, but still the prestressing has to be done. Edouard Anseele, in his commemoration of Magnel, had this to say about his work on prestressed concrete:

> The laboratory of Professor Magnel was the seat of a great number of experiments on reinforced concrete and prestressed concrete, to the study and dissemination of which he devoted himself during and especially after the second world war. He studied, from the theoretical and experimental points of view, the anchoring of the cables, the creep of the steel and the concrete, the risk of buckling of a prism of prestressed concrete, the influence of friction on the stress distribution in cables that are not straight, the fatigue of the beams in prestressed concrete, the moment distribution in hyperstatic prestressed concrete girders and the behavior of prestressed metal girders. He developed new calculation methods, devoted a whole series of contributions to this new construction material, and gave numerous lectures, both abroad and in Belgium. He developed what he called the Belgian prestressing system: anchors using sandwich plates, layers of wires, and pre-tensioning the wires two by two. It was and is used in many countries.[16]

Magnel and Freyssinet both invented equipment to compress concrete, thanks to strongly tensed steel wires. Freyssinet's is the earlier method but war led Magnel to develop his own system.

> Since it was impossible during the war to get Freyssinet's equipment, and we were willing to improve, if possible, the work of this great scientist, we helped to build some other kind of different device and created the sandwich wire (Magnel 1953b, pp. 21–22)

The procedures of Freyssinet and Magnel are similar. Let's study the Magnel-Blaton procedure first, and then discuss the differences between this and that of Freyssinet.

---

[16] Anseele (1956, p. 32): *Het laboratorium van Professor Magnel was de zetel van een groot aantal wetenschappelijke proefoefeningen op het gewapend beton en op het spanbeton, aan de studie en verspreiding waarvan hij zich gedurende en na de tweede wereldoorlog in het bijzonder heeft gewijd. Hij bestudeerde, theoretisch en experimenteel, de verankering der kabels, de kruip van het staal en van het beton, het knikgevaar bij voorgespannen betonprisma's, de invloed der wrijving op de spanningverdeling in kabels die niet recht zijn, de vermoeiing van spanbetonliggers, de momentenverdeling in hyperstatische spanbetonliggers en het gedrag van voorgespannen metalen liggers. Hij ontwikkelde nieuwe berekeningsmethodes, wijdde een hele reeks bijdragen aan het nieuw constructiemateriaal en gaf talloze voordrachten, zowel in het buitenland als in België. Hij ontwikkelde wat hij het Belgische voorspanningssysteem noemde: verankeringen met behulp van sandwichplaten, geklasseerde draden, voorspanning der draden twee aan twee. Het werd en wordt in vele landen toegepast.*

8 The Magnel-Blaton System[17]

This procedure, based on the use of highly elastic resistant steel, differs from other systems by its own following characteristics:

Wires arrangement:

The wires of every cable are 4-5 mm apart from one another and arranged in layers of 4. Plain grids maintain the arrangement and the parallelism of the wires all along the cable's length.

Thanks to this arrangement, positioning some special grids at the main deviation points is easy, and this will make possible the control and the reduction of friction.

Furthermore, the covering of the prestressed cables will be easy and reliable.

Laboratory experiments have shown that thanks to the perfect cover of each wire, the resulting adherence would be such that the beam could be loaded, after the release of the anchorages, up to its service load, without noticing any sliding of the wires.

These anchorages are metallic and their quality and precision are therefore easy to check:

They are shaped as a parallelepiped and made of high resistant steel, with two notches on each of the main faces.

Two wires will be placed in each notch and fixed with a small steel wedge.

Each anchorage (named 'sandwich') thus holds eight wires and by superimposing several pieces, the desired cable can be fixed.

Partial prestressing can easily be done, since the wires are anchored by pairs, some of them may be put apart temporarily.

A thick cable is easily manufactured by simply superimposing a great number of sandwich plates.

A steel plate with a central rectangular hole comes between the sandwich plates and the concrete and helps to distribute on its whole surface the compression strength applied by the cable.

Tension by pairs:

Thanks to the notch-wedge system, a jack seizes two wires and applies the desired tension and elongation.

It is connected to an oil pump provided with a manometer for control. This very light device, since it only has to apply small loads, can be easily manipulated.

Due to its light charges, it is possible to couple it with a dynamometric device which controls very precisely the tension applied to each pair of wires.

The jack is equipped with a balancing system in order to balance the load applied to both wires and ensure equal tension in each.

Removing the metallic ducts:

Expensive metallic ducts around the cables, preventing the evaporation of the humidity surplus of the injection product, have been abandoned a long time ago.

Rubber bars of a rectangular section are inserted in the lining before concrete casting.

They are removed one or two hours after casting; let's point out by the way that a 400-500 kgs force is sufficient to pull them out since the problem is not friction but rubber elasticity.

---

[17] I wish to thank Philippe Blaton, who helped me very much indeed during my research on this procedure.

Cables, manufactured in advance, beside the beam, are then introduced in the hole left by the rubber bars.

When concrete has acquired enough resistance anchorages are placed and tension is applied. Let's notice that these anchorages can be very easily positioned, between the wires layers which extend outside the beam.

Beams with exterior cables:

Because of the rectangular section of the cables, it is possible to place them outside the beams, on each side of the web.

Horizontal in the central part of the beam, the cables are deviated towards the top of the extremities thanks to hooks that cross the beam's web. The hooks have special notches to receive them.

Furthermore, it is possible to prepare the concrete in several segments, which are next aligned, joined together with cement mortar, and assembled by the prestressing.

The advantages of this procedure are:

(1) the possibility to precast these segments at the plant, for a better quality control;
(2) the lack of congestion on building site;
(3) the easy positioning of the beams: the segments can in fact be put directly in their final position, the alignment being done and the tension being applied up there.

Advantages of the Blaton-Magnel System.

(1) Tension by pairs, making possible:
   a. easy manipulation of the tensing devices;
   b. tension guarantee;
   c. use of low-cost dynamometric devices.
(2) Wire arrangements, making possible:
   a. accurate covering, effective protection leading to eventual adherence;
   b. friction reduction and friction control;
   c. possibility to place special grids at the deviation points;
(3) Removal of the metallic ducts, making possible:
   a. cost reduction;
   b. no possible humidity in the injected ducts;
   c. no slag possibly blocking wires when casting concrete.
(4) Anchorage:
   a. metallic, making control quality easy;
   b. standardized: single type for any case;
   c. use of oval wires or with tolerance without great problems;
   d. possibility of partial prestressing;
   e. possibility of very thick cables.
(5) Beam with exterior cables:

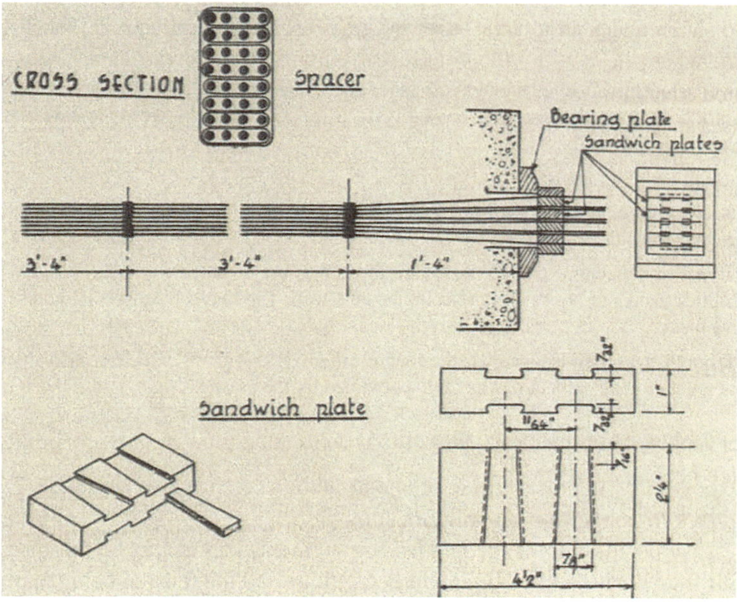

Fig. 11 General description of the elements of the Magnel-Blaton procedure (Magnel 1953a). Public domain

a. components prefabrication. Light components, prefabricated in plant;
b. exterior cables, no inserting—cables are protected by any mortar layer or any other material (asbestos, etc.);
c. cables deviation by the means of hooks, thus cable profile adapted to any necessity, continuity or fitting being very easy.[18]

The elements that comprise the Magnel-Blaton system are shown in Fig. 11, while Fig. 12 shows sandwich plates with 64 wires.

In Freyssinet's system, instead of the sandwich plates a blocking system of a conical shape is used (Fig. 13). This applies tension to all the cable wires simultaneously, and thus requires very heavy strength from the jacks (Figs. 14 and 15).

In 1942, Magnel began a campaign in favour of prestressed concrete that would later lead him as far as the United States and Canada, where he gave lectures in front of assemblies of 700 people and participated in television shows.

Prestressing would always be his favourite research topic, both theoretical and experimental, together with laboratory tests. In 1953, two years before he died, he published an interesting paper on hyperstaticity in prestressed concrete (Magnel 1953b) in which he points out the necessity to provide engineers with a mathematical

[18] This text comes from a document given to me by Philippe Blaton. I think it is very close to the patent text.

Fig. 12 The grooved anchorage of two cables, each composed of 64 wires (Magnel 1953a). Public domain

Fig. 13 Freyssinet's anchorage (Magnel 1953a). Public domain

tool that will make possible the computation of such structures, which can become very complex in case of prestressing:

> The difficulties of calculation are the following: whereas in an isostatic system the application of a pre-stress to an element does not alter exterior reactions and thus does not create exterior bending moments, in an hyperstatic system the application of prestressing, by the linear and angular deformations that it generates, alters the exterior reactions and consequently creates exterior bending moments that we named "parasitic moments". (Magnel 1953b)

Fig. 14 The jacks used in Freyssinet's system (Magnel 1953a). Public domain

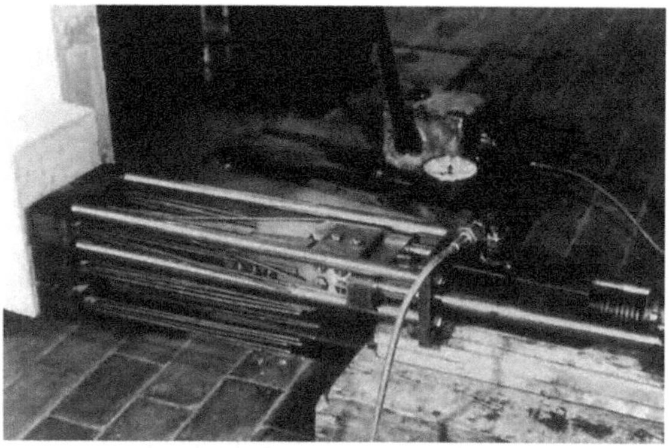

Fig. 15 The jack designed by Magnel and used to pull two wires at once (Magnel 1953b). Public domain

9 Concrete Works to Which Magnel Made a Contribution

The practice and expertise that Magnel tried to communicate to his students was surely the result of work on construction sites. Magnel actually took part in the erection of several structures in prestressed concrete, most often as a consultant, including the beginning of the building of the Koelkeberg Basilica in Brussels in 1926, and during the restoration of the Franciscan church in Saint Nicolas-Waas in 1928. Let us look at some of his most significant contributions.

9.1 The Bridge in Rue du Miroir

In 1940, Magnel studied the problem of prestressing from the theoretical point of view. In 1941, on the occasion of the construction of the Nord-Midi railway line that was to connect the stations of Bruxelles Nord and Bruxelles Midi, Magnel obtained funding from the Fonds National de recherche scientifique (the National Scientific Research Fund), the Société de chemins de fer Belge (the Belgian railway company) and the Office National de la junction (the National Junction Office), to construct the bridge in Rue du Miroir, which was built in 1943–1944.

The Blaton-Aubert firm was given the job of constructing the bridge, and the new procedure was put to work. This construction would serve not only to refine the Magnel-Blaton procedure, but also two types of experiments. With the first, the bridge, which supported six railway tracks, was subdivided into six platforms constructed according to different procedures. The first was in normal reinforced concrete; the next two used "isteg" steel with two different kinds of concrete; another used "toristeg" steel; and the last two were in prestressed concrete with two different kinds of concrete (porphyry or sand from the Rhine). The goal was to compare costs and quantities of material. Prestressed concrete was a little more expensive than the concrete with isteg steel, but much thinner (Fig. 16). The other two were more expensive and thicker. This experiment made it possible to emphasise the financial advantages and lower costs of the slab in prestressed concrete.

The second experiment was carried out in the Ghent laboratory. An entire 65 m. beam was subjected to a failure test. Magnel was not the first Belgian to carry out this kind of experiment on beams of actual scale. With the same procedure, but at his own expense, Vierendeel had set up his truss without diagonals in 1897.

The results of all these various experiments were not only published in specialist journals but were also cited in Magnel's course on the theory of prestressed concrete,

Fig. 16 Prestressing cables before concreting the prestressed bridge of the Rue du Miroir (Magnel1953a). Public domain

Fig. 17 Positioning of the first beam of the Eeklo bridge (Magnel 1953a). Public domain

the first edition of which appeared the same year as the construction of the experimental bridge and the laboratory tests. These last did not appear in the first edition, but were added in subsequent additions.

The success of this construction suggested the realization of the road bridges of Zammel built in 1944 and Eeklo, built in 1945–1946 (Fig. 17).

Many other bridges would follow. Some were made with a personal contribution by Magnel, such as the road bridge of Baudour, the Maghin bridge in Liège, the Malden bridge in Holland, and the Moerkerke bridge. Others were built without his collaboration but using the Magnel-Blaton procedure, such the bridge in Rue De Smet, and the W7 and W9 bridges near Ghent.

Meanwhile, Magnel was particularly proud of his participation on the construction of two bridges that were built around the 1950s. The first one is the Sclayn bridge over the Meuse, near Andenne, which was built by the firm Blaton-Aubert. This is the bridge that appears in the background of the portrait of Magnel himself in Fig. 1. When it was built, in 1949, this bridge was the first one in the world with continuous prestressed beams. It is 126 m long (Figs. 18, 19, and 20).

As R. H. Evans wrote:

> He [Magnel] was greatly interested in continuous structures in prestressed concrete, and published his theory on this subject in 1947, the famous Sclayn Bridge being a notable example of his theoretical work in this direction. (Evans 1956)

The Sclayn Bridge has a pre-existing central pier and two buttresses. These were not touched during the bombing of the earlier bridge during the war, and it was convenient to maintain them. The new bridge had to be continuous, that is, the two

Fig. 18 The application of prestressing to the Sclayn bridge (Magnel 1953a). Public domain

Fig. 19 The prestressing cables of the Sclayn bridge (Magnel 1953a). Public domain

bays that formed an obtuse angle over the central pier had to be prestressed together. Magnel was opposed to this idea:

> The writer was opposed to building a continuous structure of prestressed concrete until he was satisfied that sound design method had been developed and that adequate tests had been made to check the agreement between theory and practice (Magnel 1950a, b)

As shown in the theory published in the second edition of his course on prestressed concrete (Magnel 1948a, b, c), it would appear that in his opinion the tests carried out were insufficient. It must be said that the theory showed him that in this kind of hyperstatic beams there appeared moments of bending and external reactions that were difficult to evaluate theoretically, and thus they should have reasonably been

Fig. 20 The completed Sclayn Bridge (Magnel 1953a). Public domain

evaluated experimentally. In effect, the moments acting on the central pier were quite significant:

> It may be interesting to note that the secondary moment found above the central pier was 9.270 tm which should be compared with the moment at the same point due to the dead load, equal to 13.630 tm. This comparison shows the great importance of the secondary moments, which cannot be neglected in this design (Magnel 1950a, b)

We can see the same interaction between theory and experiment in numerous other projects in which Magnel participated as either consulting engineer or director of the Ghent laboratory. For example, the construction of the first prestressed concrete bridge in the United States, the Walnut Lane Bridge in Philadelphia, used the Magnel-Blaton procedure. It was the result of Magnel's campaign of 1946 in the United States in favour of prestressed concrete. This bridge was composed of thirteen beams 47 m long placed alongside each other to form a total width of 16.61 m; they were also prestressed across their width (Fig. 21).

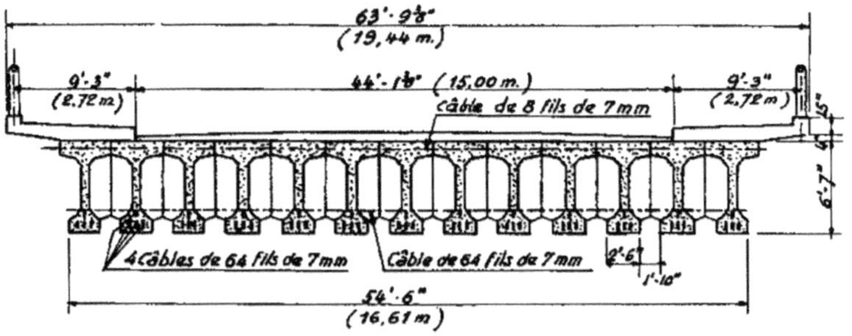

Fig. 21 A cross section of the Walnut Lane Bridge, showing the thirteen beams side by side (Magnel 1951a). Public domain

Fig. 22 Test beam of the Walnut Lane Bridge (Magnel 1953a). Public domain

Fig. 23 The completed Walnut Lane Bridge (Magnel 1953a). Public domain

In addition, during its construction, a failure test was made on one of the 47 m long beams. The test was executed in Magnel's laboratory in Ghent. This work surely is one of the highest achievements in Magnel's career (Figs. 22 and 23).

9.2 Underpinning During Restoration Works

The tower of Saint Nicolas Church in Ghent needed a complete and delicate restoration. In his preliminary studies, Magnel suggested creating two belt-beams in prestressed concrete, one of which was to be placed at the very bottom of the tower itself, that is, at the top of the four pillars of the crossing. The construction works would be carried out after his death, and they would follow Magnel's suggestion, the necessity of which he had convincingly explained:

Fig. 24 The Pont des trous in Tournai (Magnel 1953a). Public domain

> We cannot temporarily put the tower on some supports, close to the actual supports that have proved to be so efficient for centuries.
>
> We must first build, towards the bottom of the tower, a solid support in reinforced concrete, in the shape of a rigid frame, hidden inside the thickness of the masonry. It is the construction of this frame – we might say "of this rectangular belt" – that requires the use of prestressed concrete, because its execution requires – just like a real underpinning –the demolition of almost the full thickness of the peripheral walls of the tower, and this can be done only by emptying successive niches and by filling them with concrete as soon as they are dug. In this concrete we will place some thin metal pipes that, all together, will form some hollow tubes running through the whole length of the four sides of the frame and make possible the insertion of the necessary cables for prestressing (Magnel 1944).

A similar operation was done to raise the two towers of the Pont des Trous in Tournai, Belgium (Fig. 24).

9.3 Other Works

In 1947–48 Magnel collaborated on the construction of a new plant for the Union Cotonnière (Cotton Union) in Ghent (Fig. 25), a building of 30,000 square meters on a single ground level. It is interesting to notice that the large size of this plant immediately suggested standardization and work division in modules:

> In order to save money, we decided to build the construction with as much standardization as possible…
>
> The building's design was based on a 3.60 m module, which best suits the dimensions of the machines that have to be installed (Magnel 1948c)

These concepts were used again in the airplane hangars built in 1947 at Melsbroek air base.

Let us give a final example that shows the speed with which the use of prestressed concrete spread. In 1951, when the International Prestressing Meeting was organized

Fig. 25 The prestressed beams of roof of the Union Cotonnière (Magnel 1953a). Public domain

in Ghent to commemorate the 75th anniversary of the Association of Ghent Engineers, Magnel collaborated in the construction of a monumental spiral staircase to be located in the showroom of General Motors in Antwerp. As a spiral stair, it is subject to torsion, and since reactions to torsion were not well known, the tests for this elegant helical stair, whose prestressing is no longer obvious, were carried out on a full-size replica (Figs. 26 and 27).

He described the construction of the staircase in a humorous paper featuring a dialog between a Cadillac placed in the showroom and the staircase itself:

> Cadillac: We have been looking at each other for two days now; why not make friends? You look just as aristocratic as I do. By the way, who is your father?

Fig. 26 The stair subjected to testing (Magnel 1953a). Public domain

Fig. 27 The stair in construction for the General Motors showroom in Antwerp (Magnel 1953a). Public domain

Staircase: Can't you guess that? Don't you know we are brothers? Our father is the famous general – General Motors. As far as I know, we don't have the same mother. As a matter of fact, from what I hear from conversations around me, I have several mothers, and my friends call them "the Architects".

Cadillac: I am thinking of the trouble the doctor must have had in your delivery, owing to your shape being unusual.

Staircase: It was indeed a very delicate and difficult operation by a doctor named Contractor. Since he was afraid of the intricate operation required in bringing me to daylight, he consulted a specialist called the Professor, who often comes here: he decided to make an experimental delivery in his laboratory in Ghent, of a staircase born of less aristocratic parents. The professor even tested the health of the child and gave useful advice to the doctor for my birth and the strengthening of my delicate body by an operation called pre-stressing. The professor said I could not live long without being prestressed. I must say to his credit that I feel much stronger now (Magnel 1951b)

In what turned out to be his last project, Magnel had committed himself to a very important design for the Brussels Universal Exhibition in 1958: the telecommunications tower, an awesome design for a tower 507 m. high, topped with an antenna 200 m. high (Fig. 28). Unfortunately, the tower was never built.

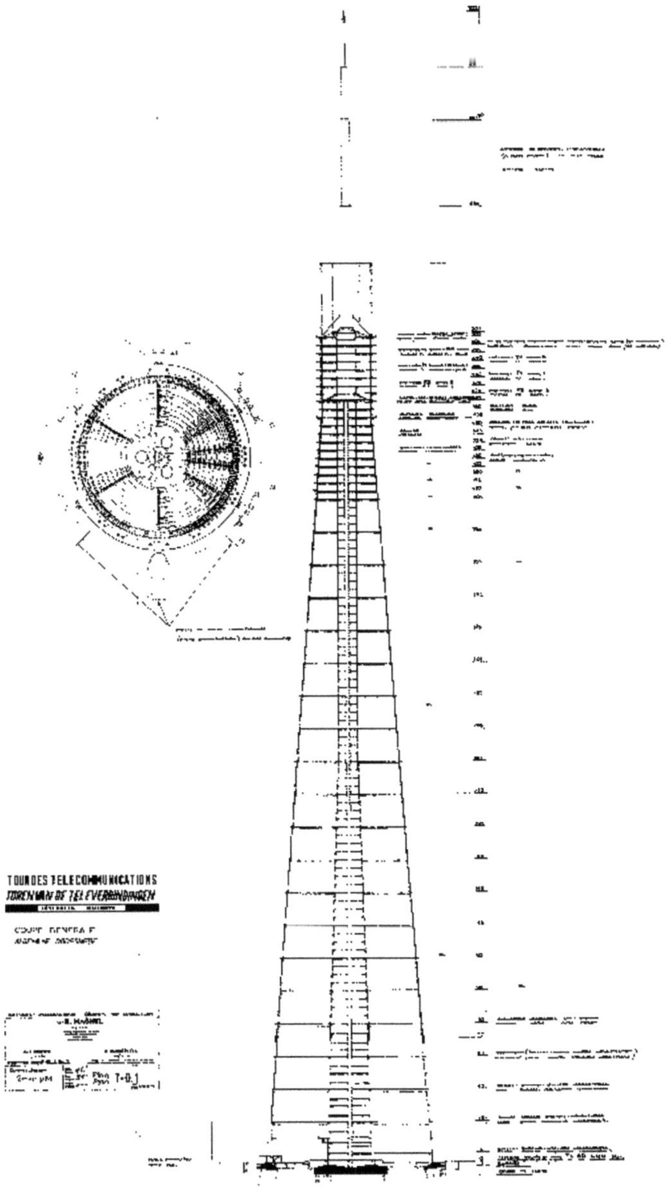

Fig. 28 Telecommunication tower for the Universal Exhibition in 1958 (Magnel Laboratory Archives)

10 Conclusion

Even if he was not the inventor of prestressed concrete, as he himself used to admit, it seems fair to state that Magnel proved, throughout Belgium, the United States, Canada and other countries, how appealing this new building material is. He did so by giving lectures but also, and mainly, through actual practice. He collaborated in the realization of a great number of structures in prestressed concrete. Furthermore, he demonstrated to contractors the reliability of this material by running accurate tests in his laboratory in Ghent.

While studying Magnel's work we can't help but be struck by both the wide range of his interests and the coherence of his thoughts. His interests covered beams and bridges construction, first in metal, then Vierendeel, and then in prestressed concrete, since he immediately understood the advantages of this material, as well as in laboratory tests to ensure the safety of the structures, and the institution, within the SECO Bureau, of an insurance system for contractors who had to provide a ten-year guarantee. In addition to all this, he also was a professor and a researcher who continue to innovate until the end of his life. He was furthermore concerned with profit, and reduced the necessary amount of construction material thanks to prestressing. He also reduced the time required for computing by providing engineers with handy, quick and efficient methods of calculation. We can thus conclude that Magnel's work concerned every facet of the engineer's profession, without exception.

Acknowledgements Special thanks to Mr. Y. Pianet, former director of the Bureau SECO, who allowed me into the SECO Archives, to Philippe Blaton for information about the Magnel-Blaton patent, and also to Mr. A. Broucke and Prof. D. Vandepitte for the abundant information they gave me and for their careful reading of this text.

References[19]

Anseele, Edouard. 1956. Speech. In *In Memoriam Gustave Magnel, 1889–1955*, 31–34. Ghent: Vanmelle.

Evans, R.H. 1956. Speech. In *In Memoriam Gustave Magnel, 1889–1955*, 51–55. Ghent: Vanmelle.

Freyssinet, Eugène. 1936. *Une révolution dans les techniques du béton*. Paris: Librairie de l'enseignement technique.

Magnel, Gustave. 1917. Méthode rapide de calcul des lignes d'influence d'arcs prismatiques surbaissés à deux encastrements. *Le génie civil* 21 (May 1917).

Magnel, Gustave. 1922. Recherches expérimentales sur le coefficient d'élasticité du béton (Prix Lemaire), *Revue universelle des mines, de la métallurgie, des travaux publics, des sciences et des arts appliqués à l'industrie* 1 (January 1922): 163ff.

[19] A complete bibliography of Magnel's writings can be found in Patricia Radelet-de Grave, "Gustave Magnel (1889–1955), a scientific biography", in *Essays in the history of the theory of structures. In honour of Jacques Heyman*, ed. Santiago Huerta, pp. 389–427 (Madrid: Instituto Juan de Herrera, 2005).

Magnel, Gustave. 1923. *La pratique du calcul du béton armé*, vol. 1. 1st edition. Ghent: Editions Fecheyr.

Magnel, Gustave. 1934. *Le calcul pratique des poutres Vierendeel* (1 vol. text + 3 vols. illustrations). Ghent: Van Rysselberghe & Rombault.

Magnel, Gustave. 1935. *Cours de stabilité des constructions*. Ghent: Van Rysselberghe & Rombaut.

Magnel, Gustave. 1939. Les laboratoires de béton de l'Université de Gand. *Béton armé: revue mensuelle technique et documentaire des constructions en béton armé* 277 (July 1939).

Magnel, Gustave. 1940. Le problème de la fissuration du béton et l'emploi des aciers à haute limite élastique. *Annales Des Travaux Publics De Belgique* 41: 589–659.

Magnel, Gustave. 1944. Le béton précontraint dans un problème de reprise en sous-œuvre. Restauration de l'Eglise St. Nicolas à Gand. *L'architecture française: architecture urbanisme decoration* 43.

Magnel, Gustave. 1945–1946. Prestressed concrete: Some new developments. *Concrete and Constructional Engineering* 40 (December 1945): 221–232, 249–254; 41 (January 1946): 10–20.

Magnel, Gustave. 1948a. L'évolution du béton armé. *Bulletin de l'Académie royale de Belgique* (classe des sciences) 12.

Magnel, Gustave. 1948b. *Stabilité des constructions*, vol. III, 2nd ed. Ghent: Fecheyr.

Magnel, Gustave. 1948c. La nouvelle usine de l'union cotonnière, à Gand. Application du béton précontraint. *La technique des travaux* 24 (7–8): 214–220.

Magnel, Gustave. 1949. L'avenir du béton précontraint. *Annales des travaux publics de Belgique*, numero jubilaire issue 1843–1948.

Magnel, Gustave. 1950a. De hervorming der ingenieursscholen. *AVOSUG* 3 (2).

Magnel, Gustave. 1950b. L'acier précomprimé – nouvelles considérations. *L'ossature métallique* 9.

Magnel, Gustave. 1951a. Essai à Philadelphie d'une poutre en béton précontraint de 47 m de portée. *La Technique Des Travaux* 27 (1–2): 49–61.

Magnel, Gustave 1951b. A concrete example. *The Link* 2 (7).

Magnel, Gustave. 1953a. *Le béton précontraint*, 3rd ed. Ghent: Rombaut-Fecheyr.

Magnel, Gustave. 1953b. L'hyperstaticité en béton précontraint. *Schweizer Archiv für angewandte Wissenschaft und Technik*, 284–290.

Taerwe, Luc. *De betekenis van Prof. Magnel als leermeester, onderzoeker en ontwerper*, unpublished text, November 1999.

Taerwe, Luc. 2004. *De gouden medaille Gustave Magnel: overzicht van de eerste negen bekroonde werken*, unpublished text, December.

Taerwe, Luc. 2005. Contributions of Gustave Magnel to the development of prestressed concrete. In *Proceedings of the Ned H. Burns Symposium on Historic Innovations in Prestressed Concrete, Special Publication, American Concrete Institute.*

Index

A

Aepinus, Ulrich Theodor, 295

B

Baldini, Ugo, 75
Bélanger, Jean-Baptiste, 428
Bélidor, Bernard Forest de, 47, 54–56, 112, 114, 116, 117, 125, 226–229, 234–237, 257, 293, 306, 307, 327, 330–332, 420
Bellavitis, Giusto, 453
Benvenuto, Edoardo, 1, 2, 109, 110, 117, 193, 212, 213, 233, 265, 266, 291, 292, 355, 370, 398, 399, 404, 411
Bernoulli, Daniel, 2, 61–63, 123, 124, 131, 132, 155, 161, 204, 215, 216, 218–220, 295, 326, 328, 365, 370, 372–375, 453
Bernoulli, Jacob, 47, 48, 70, 110, 112, 116, 117, 119, 121–123, 128, 130, 131, 135–139, 146, 147, 152, 153, 155, 157–159, 165, 180, 181, 184, 185, 193, 195–197, 199, 200, 203–206, 209–213, 215, 217–219, 228, 236, 238, 240, 257, 271, 272, 275, 277–280, 288, 295, 325, 326, 328, 329, 345, 351, 356–362, 364–366, 370, 372, 374, 376, 377, 395–399, 479
Bernoulli, Johann, 37, 39–44, 50–52, 60, 66, 110–112, 115, 116, 123, 129, 130, 155, 158, 162, 168, 169, 172, 181, 185, 187–189, 197, 200, 201, 207, 208, 216, 219, 287, 295, 328, 357, 365, 404–406, 414
Bertelli, Francesco, 422
Betti, Enrico, 471–473
Blondel, François, 5, 54, 68, 117, 226, 251–257, 260, 294, 300–308, 480
Boissieu, Alphonse de, 310, 313
Bonnycastle, C., 418, 419
Bosmans, Henri, 80, 87–89
Bosse, Abraham, 248, 251, 253–255
Bossut, Charles, 70, 71, 126, 128, 136, 140, 147–152, 154, 202, 233–235, 240, 295, 328, 329, 337, 377–379, 387, 395, 396, 409, 420
Bouguer, Pierre, 126, 127, 223–226, 228–234, 238–244, 369–371, 420
Braun, Joseph, 75
Brigode, Simon, 81
Brouwers, Louis, 88
Bülfinger, Georg Bernhard, 295

C

Carnot, Lazare, 441
Carnot, Sadi, 425, 426
Castigliano, Alberto, 438, 457, 458, 471, 473–480
Cerruti, Valentino, 473
Chasles, Michel, 443, 464
Clairaut, Jean-Baptiste, 49
Clapeyron, Émile, 426–430, 433, 457
Clausen, Thomas, 262
Cotterill, James H., 432, 433
Coulomb, Charles Augustin, 138, 143, 144, 148, 152, 265–274, 276, 277, 280–295, 345, 376, 377, 420, 441
Couplet, Claude Antoine, 65–68, 116, 124–126, 136, 139–148, 150, 152,

P. Radelet-de Grave, *Between Mechanics and Architecture*, Mathematics and the Built Environment 8, https://doi.org/10.1007/978-3-031-73530-1

154, 155, 226, 234, 295, 327, 328, 366–369, 377, 420
Cousinery, Barthélémy Édouard, 442
Cramer, Gabriel, 122, 136, 138, 366
Cremona, Luigi, 437, 447, 449, 452
Cresswell, Daniel, 419
Culmann, Karl, 58, 437, 441–444, 447–453, 455–457, 459–462, 464, 465

D
D'Aguilon, François, 77–82, 84, 247
D'Alembert, Jean-Baptiste Le Rond, 337, 387, 408, 409, 411, 414
Delanges, Paolo, 409–411, 414
Derand, François, 5, 20–22, 54, 68, 77, 117, 226, 294
Dhombres, Jean, 78
Dolmans, Pierre, 76, 81, 88
Dorna, Alessandro, 431, 432, 478
Duhem, Pierre, xi

E
Euler, Leonhard, 26, 47, 64, 65, 70, 123, 131, 153, 155, 194, 204, 209, 212, 215, 218–220, 239, 240, 257–260, 270–272, 287, 295, 326, 328, 351, 356, 357, 359, 364, 365, 370, 372–376, 378, 394, 395, 407–409, 411, 419, 420, 441, 501, 502

F
Fagnoli, Giuseppe, 421
Favaro, Antonio, 437, 447
Fontana, Gregorio, 68, 415, 417, 418
Fourier, Joseph, 414, 420
Freyssinet, Eugène, 544–547, 550–552
Frisi, Paolo, 240, 329

G
Galilei, Galileo, 17, 213, 323
Gauthey, Émiland-Marie, 328, 329
Gergonne, Joseph Diez, 406, 464
Grassmann, Hermann, 437, 441
Gregory, David, 48, 49, 111–113, 116, 158, 159, 162, 176–186, 188, 228, 269, 287, 295, 363, 364
Gros, Pierre, 256

H
Hamilton, William Rowan, 5, 437, 441
Hèrigone, Pierre, 20
Hermann, Jacob, 116, 123, 374, 453
Heyman, Jacques, 265, 267–271, 273, 276, 280–286, 291, 293, 330, 376, 531, 562
Hooke, Robert, 34, 110, 111, 113, 136, 138, 176, 177, 228, 269, 277, 278, 280, 287, 294, 323, 325–327, 330, 356, 361, 375, 398, 430
Huygens, Christiaan, 22–29, 31, 80, 88, 111, 112, 117, 129–131, 155, 158–164, 179, 188–190, 194, 198, 205, 206, 279, 295, 324, 325, 329, 330, 357, 361, 398

K
Keller, Joseph P., 262
Kochanski, Adam Adamand, 77, 85
Krafft, Georg Wolfgang, 70, 136, 138, 147, 153, 327, 328, 349, 351, 352
Kurrer, Karl-Eugen, 262

L
Lagrange, Joseph-Louis, 259–262, 411, 414, 415, 418, 419, 473, 480
La Hire, Philippe de, 44–47, 68, 110, 111, 115, 117–121, 124–127, 138–141, 148, 150, 152, 155, 235, 238, 269, 287, 293–295, 300, 307, 327, 328, 340, 355–357, 362, 363, 366, 367, 377, 387, 403–405, 420
Lamé, Gabriel, 421, 422, 426, 427, 430, 433
Leibniz, Gottfried Wilhelm von, 37, 78, 110–112, 135, 145, 155, 157–160, 162, 165–168, 172, 177, 179, 180, 189–191, 197, 200, 202, 205, 206, 209, 212, 214, 216, 266, 275–277, 279, 280, 294, 295, 300, 330, 357, 359, 361, 364, 370, 398, 441, 443, 451
Léotaud, Vincent, 77
Levy, Maurice, 448
Lorgna, Anton Maria, 329, 414

M
Magnel, Gustave, 267, 269, 514, 531–562
Malfatti, Gianfrancesco, 415
Maltese, Giulio, 117
Mariotte, Edme, 214, 275, 295

Mascheroni, Lorenzo, 113, 132, 137, 232–234, 290, 295, 329, 333, 355, 365, 379–397, 399, 415, 420
Maxwell, James Clerk, 432, 449, 471
Menabrea, Luigi Federico, 430, 431, 433–435, 474–478
Mersenne, Marin, 22–24, 29, 129, 324, 325
Möbius, Augustus Ferdinand, 464
Mohr, Otto, 444–446
Molinos, L. de, 427–429
Moseley, Henry G. J., 433
Mousson, Albert, 459, 460
Musschenbroek, Pieter van, 257, 283, 295

N
Navier, Claude-Louis, 420, 421, 432, 474
Nervi, Pier Luigi, 220, 519–528
Neuville, Alfred, 487, 488
Newton, Isaac, 37, 39, 48, 57, 62, 112, 158, 159, 177–180, 184, 186–188, 209, 300, 325, 330, 375, 459
Nieuport, Viscount of (Charles-François Le Prud'homme D'Hailly), 136, 138, 147, 149–155, 224, 225, 337–352

O
Ouvrard, René, 305

P
Pagani, Gaspard-Michele, 420, 421, 475, 478
Palladio, Andrea, 1, 2, 247–249, 483
Paoli, Paolo, 411–415
Pappus of Alexandria, 253
Pardies, Ignace-Gaston, 29–33, 113, 115, 116, 136, 138, 162, 168, 169, 186, 188, 256, 294, 325, 357, 359, 370, 392, 393
Parent, Antoine, 68, 119, 131, 266, 277–279, 295, 387, 390
Patte, Pierre, 44, 299, 313, 316–318, 333, 338
Perrault, Charles, 300
Perrault, Claude, 257, 300, 303
Pitot, Henry, 330, 332, 333
Poinsot, Louis, 356, 415, 416
Poleni, Giovanni, 54, 68–72, 112–114, 176–178, 232, 233, 240, 241, 322, 333, 347, 348, 377
Polonceau, Camille, 438, 439, 458
Poncelet, Jean Victor, 441, 442, 448–451
Pronnier, C., 427, 429

Q
Quetelet, Lambert Adolphe Jacques, 150, 337

R
Radelet-de Grave, Patricia, 44, 318, 333, 404, 562
Rankine, W. J. Macquorn, 437–439, 467–471
Reye, Theodor, 443
Riccati, Vincenzo, 405, 406
Ritter, August, 438–440, 444, 470, 471
Roberval, Gilles Personne de, 2, 11, 17–20, 116, 122, 251, 252
Rondelet, Jean-Baptiste, 299, 316, 318, 328, 333, 338
Rubens, Pieter Paul, 78, 80–83

S
Saint-Venant, Adhémar Jean Claude Barré de, 432, 474
Saint-Vincent, Grégoire de, 77–80, 82, 84–89, 255, 256
Salimbeni, Leonardo, 329
Salmon, George, 443, 452
Soufflot, Jacques-Germain, 240, 299–301, 303–318, 333, 338
Speiser, David, 275
Stevin, Simon, 2, 6–13, 25, 122, 323, 324
Stirling, James, 52–54, 68–70, 178, 187–189

T
Taylor, Brook, 186, 187
Timoshenko, Stephen P., 265, 266
Torricelli, Evangelista, 129, 325
Truchet, Jean (Père Sébastien), 104–107
Truesdell, Clifford A., 109, 110, 112, 117, 119, 123, 127, 131, 158, 177, 187, 188, 193, 228, 262, 265, 266, 271, 273, 278, 279, 287, 355, 398, 399

V
Vallery-Radot, Jean, 75–77, 84, 85
Varignon, Pierre, 2, 34–38, 46, 50, 54, 56–62, 122, 123, 202, 214, 216, 235, 266, 280, 295, 333, 355, 356, 372,

404, 405, 414, 415, 419, 446, 450, 462, 463
Vauban, Sébastien Le Prestre de, 286
Venturoli, Giuseppe, 416–420
Viel de Saint Maux, Charles-François, 318
Vierendeel, Arthur, 310, 483–499, 501–508, 510–516, 538, 539, 542, 553, 562
Viéte, François, 254
Vignola, Jacopo Barozzi da, 247, 248, 250, 251, 253, 254, 257
Viola, Tullio, 75, 321
Von Staudt, Karl Georg Christian, 443, 451

W

Waller, Richard, 326
Wallis, John, 93–105, 107, 398
Warren, James, 483, 488
Wren, Christopher, 34, 330, 356, 398

Batch number: 10726058

Printed by Printforce, the Netherlands